Handbook on Characterization of Biomass, Biowaste and Related By-products

Ange Nzihou
Editor

Handbook on Characterization of Biomass, Biowaste and Related By-products

Volume 1

Springer

Editor
Ange Nzihou
RAPSODEE CNRS, UMR-5302
Université de Toulouse, IMT Mines Albi
Albi Cedex, 09, France

ISBN 978-3-030-35019-2 ISBN 978-3-030-35020-8 (eBook)
https://doi.org/10.1007/978-3-030-35020-8

This Springer imprint is published by the registered company Springer Nature Switzerland AG
The registered company address is: Gewerbestrasse 11, 6330 Cham, Switzerland

Dedicated to
Emilie, Arthur and Laurence
Emilienne
Daniel, who would love to hear how the idea of this book blossoms

Preface

In the challenging transition toward sustainable development, one of the key pillars is the valorization of non-food biomass and biowaste into energy and valuable molecules and materials with emphasis on practices that prevent harmful emissions. The range of biomass and biowaste feedstocks available is much larger than ever before. This opens new opportunities for innovation and development of new technologies, processes, approaches and market in the context of sustainable development and the circular economy. In this context, understanding the behavior and properties of these streams is essential. It requires a combination of techniques to get a good insight of the complex materials that biomass and biowaste usually are. The handbook provides insights on standard characterization techniques, together with the advanced and emerging ones.

With this contribution to the state of the art, we planned this work to fill an important gap in the literature. We expect this reference work to be the mortar that holds the building blocks on understanding the properties and characteristics of biomass and biowaste for their better selection and use for dedicated purposes.

It is divided into 15 chapters written by 87 authors from 16 countries.

The creation of this handbook, gathering various fields of engineering sciences (chemical, environmental and mechanical), has been a complex and demanding venture. It would not have been possible without the dedication, countless effort and time from *Dr. Augustina Ephraim and Dr. Nathalie Lyczko* who helped to develop the concept of the book and addressed a number of issues raised by the contributors. *I would like to express my deepest gratitude to them both.*

Special thanks to chapter chairs, *Patricia Arlabosse, Hélène Carrere, Marco J. Castaldi, Jun Dong, Capucine Dupont, Catherine Dupré, Augustina Ephraim, Thierry Ghislain, Deepu Gopakumar, Carol Sze Ki Lin, Shengyong Lu, Hélène Métivier, Doan Pham Minh, Elsa Weiss-Hortala*, for guiding the work of the respective chapter that they handled to their outstanding completion.

I'm grateful to the *87 authors and to the reviewers* who dedicated their valuable time and expertise to create this extraordinary source of information for a wide audience.

The enthusiastic support from *Dolorès Liret and Kattialyn Gossiaux* is gratefully acknowledged.

My warm thanks to *Fritz Schmuhl* for his unswerving support and advices throughout this project. My thanks go to *the Springer Nature editorial and production teams* helping to complete this project.

The high rate of acceptance on my solicitation from the biomass and biowaste community is a strong indicator of the importance of the need that the *handbook on characterization of biomass, biowaste and related by-products* fills.

Albi, France

Ange Nzihou
Distinguished Professor of Chemical
and Environmental Engineering

Contents

Contributors

Philippe Accart RAPSODEE CNRS, UMR-5302, Université de Toulouse, IMT Mines Albi, Albi Cedex, 09, France

Andrés Anca-Couce Institut für Wärmetechnik/Institute of Thermal Engineering, TU Graz, Graz, Austria

Georgia Antonopoulou Institute of Chemical Engineering Sciences (FORTH/ICE-HT), Patras, Greece

Patricia Arlabosse RAPSODEE CNRS, UMR-5302, Université de Toulouse, IMT Mines Albi, Albi Cedex, 09, France

Muthupandian Ashokkumar Faculty of Science, School of Chemistry, The University of Melbourne, Melbourne, VIC, Australia

Hassen Benbelkacem Laboratoire DEEP—EA 7429 (Déchets Eaux Environnement Pollutions—Waste Water Environment Pollutions), Université de Lyon—INSA Lyon, Villeurbanne, France

Céline Boachon RAPSODEE CNRS, UMR-5302, Université de Toulouse, IMT Mines Albi, Albi Cedex, 09, France

Alfons Buekens State Key Laboratory of Clean Energy Utilisation, Zhejiang University, Hangzhou, China;
Department of Chemical Engineering, Vrije Universiteit Brussel, Brussels, Belgium

Hugh D. Burrows CQC, Department of Chemistry, University of Coimbra, Coimbra, Portugal

Rachel Calvet RAPSODEE CNRS, UMR-5302, Université de Toulouse, IMT Mines Albi, Albi Cedex, 09, France

Maria G. Campos CQC, Department of Chemistry, University of Coimbra, Coimbra, Portugal

Hélène Carrère Laboratoire de Biotechnologie de l'Environnement, INRA, Narbonne, France

Marco J. Castaldi Department of Chemical Engineering, The City College of New York, New York, NY, USA

Vincent Chatain Laboratoire DEEP—EA 7429 (Déchets Eaux Environnement Pollutions—Waste Water Environment Pollutions), Université de Lyon—INSA Lyon, Villeurbanne, France

Tong Chen State Key Laboratory of Clean Energy Utilisation, Zhejiang University, Hangzhou, China

Anthony Chesnaud MINES ParisTech, PSL Research University, MAT—Centre des matériaux, CNRS UMR 7633, Evry, France

Yong Chi Department of Energy Engineering, Zhejiang University, Yuquan Campus, Hangzhou, China

Simona Ciuta RRT Design and Construction, New York, USA

Jean-Michel Commandré Biomass Energy Unit, CIRAD, Montpellier Cedex 5, France

Lucy Culleton National Physical Laboratory, Teddington, UK

Sylvie Del Confetto RAPSODEE CNRS, UMR-5302, Université de Toulouse, IMT Mines Albi, Albi Cedex, 09, France

Cédric Delattre Université Clermont Auvergne, CNRS, Institut Pascal, Aubière Cedex, France

Alba Dieguez-Alonso Institute of Energy Engineering, Technische Universität Berlin, Berlin, Germany

Papa Niokhor Diouf SEREX, Rimouski College Technology Transfer Center, Amqui, Canada

Jean-Louis Dirion RAPSODEE CNRS, UMR-5302, Université de Toulouse, IMT Mines Albi, Albi Cedex, 09, France

Jun Dong RAPSODEE CNRS, UMR-5302, Université de Toulouse, IMT Mines Albi, Albi Cedex, 09, France;
Department of Energy Engineering, Zhejiang University, Yuquan Campus, Hangzhou, China

Céline Druilhe UR OPAALE, Irstea, Rennes, France

Nathalie Dumont Laboratoire DEEP—EA 7429 (Déchets Eaux Environnement Pollutions—Waste Water Environment Pollutions), Université de Lyon—INSA Lyon, Villeurbanne, France

Capucine Dupont Department of Environmental Engineering and Water Technology, IHE Delft Institute for Water Education, Delft, The Netherlands

Catherine Dupré GEPEA UMR 6144 CNRS, Saint-Nazaire Cedex, France

Xavier Duret Department of Chemical Engineering and Biotechnological Engineering, Université de Sherbrooke, Sherbrooke, Canada

Telma Encarnação CQC, Department of Chemistry, University of Coimbra, Coimbra, Portugal

Augustina Ephraim RAPSODEE CNRS, UMR-5302, Université de Toulouse, IMT Mines Albi, Albi Cedex, 09, France

Marilyne Fauchon Biodimar, LEMAR UMR 6539, Institut Européen de la Mer, Université de Bretagne Occidentale, Brest, France

Clément Gaignard Université Clermont Auvergne, CNRS, Institut Pascal, Aubière Cedex, France

Thierry Ghislain Department of Chemical Engineering and Biotechnological Engineering, Université de Sherbrooke, Sherbrooke, Canada

Maria Gonzalez Martinez Department of Environmental Engineering and Water Technology, IHE Delft Institute for Water Education, Delft, The Netherlands

Deepu Gopakumar School of Industrial Technology, Universiti Sains Malaysia, Penang, Malaysia

Laurène Haurie RAPSODEE CNRS, UMR-5302, Université de Toulouse, IMT Mines Albi, Albi Cedex, 09, France

Pinjing He College of Environmental Science and Engineering, Tongji University, Shanghai, China

Claire Hellio Biodimar, LEMAR UMR 6539, Institut Européen de la Mer, Université de Bretagne Occidentale, Brest, France

Junko Ito Algae Biomass and Energy System R&D Center, University of Tsukuba, Tsukuba, Japan

Julie Jimenez Laboratoire de Biotechnologie de l'Environnement, INRA, Narbonne, France

Guneet Kaur School of Energy and Environment, City University of Hong Kong, Kowloon, Hong Kong;
Sino-Forest Applied Research Centre for Pearl River Delta Environment, Department of Biology, Hong Kong Baptist University, Kowloon Tong, Hong Kong

Céline Laroche Université Clermont Auvergne, CNRS, Institut Pascal, Aubière Cedex, France

Eric Latrille Laboratoire de Biotechnologie de l'Environnement, INRA, Narbonne, France

Jean-Michel Lavoie Department of Chemical Engineering and Biotechnological Engineering, Université de Sherbrooke, Sherbrooke, Canada

Jack Legrand GEPEA UMR 6144 CNRS, Saint-Nazaire Cedex, France

Chong Li School of Energy and Environment, City University of Hong Kong, Kowloon, Hong Kong;
Agricultural Genomics Institute at Shenzhen, Chinese Academy of Agricultural Sciences, Shenzhen, Guangdong, People's Republic of China

Xiaotong Li School of Energy and Environment, City University of Hong Kong, Kowloon, Hong Kong

Carol Sze Ki Lin School of Energy and Environment, City University of Hong Kong, Kowloon, Hong Kong

Xiaoqing Lin State Key Laboratory of Clean Energy Utilisation, Zhejiang University, Hangzhou, China

Shengyong Lu State Key Laboratory of Clean Energy Utilisation, Zhejiang University, Hangzhou, China

Gerasimos Lyberatos Institute of Chemical Engineering Sciences (FORTH/ICE-HT), Patras, Greece;
School of Chemical Engineering, National Technical University of Athens, Athens, Greece

Nathalie Lyczko RAPSODEE CNRS, UMR-5302, Université de Toulouse, IMT Mines Albi, Albi Cedex, 09, France

Hanna J. Maria International and Inter University Centre for Nanoscience and Nanotechnology, Mahatma Gandhi University, Kottayam, Kerala, India;
School of Chemical Sciences, Mahatma Gandhi University, Kottayam, Kerala, India

Hélène Métivier Laboratoire DEEP—EA 7429 (Déchets Eaux Environnement Pollutions—Waste Water Environment Pollutions), Université de Lyon—INSA Lyon, Villeurbanne, France

Philippe Michaud Université Clermont Auvergne, CNRS, Institut Pascal, Aubière Cedex, France

Rajesh Munirathinam RAPSODEE CNRS, UMR-5302, Université de Toulouse, IMT Mines Albi, Albi Cedex, 09, France

Ioanna Ntaikou Institute of Chemical Engineering Sciences (FORTH/ICE-HT), Patras, Greece

Ange Nzihou RAPSODEE CNRS, UMR-5302, Université de Toulouse, IMT Mines Albi, Albi Cedex, 09, France

Stewart Oakley Department of Civil Engineering, California State University, Chico, CA, USA

Khai Lun Ong School of Energy and Environment, City University of Hong Kong, Kowloon, Hong Kong

Alberto A. C. C. Pais CQC, Department of Chemistry, University of Coimbra, Coimbra, Portugal

Konstantina Papadopoulou School of Chemical Engineering, National Technical University of Athens, Athens, Greece

Séverine Patry RAPSODEE CNRS, UMR-5302, Université de Toulouse, IMT Mines Albi, Albi Cedex, 09, France

Doan Pham Minh RAPSODEE CNRS, UMR-5302, Université de Toulouse, IMT Mines Albi, Albi Cedex, 09, France

Guillaume Pierre Université Clermont Auvergne, CNRS, Institut Pascal, Aubière Cedex, France

Daniel Pleissner Faculty of Sustainability, Sustainable Chemistry (Resource Efficiency), Institute for Sustainable and Environmental Chemistry, Leuphana University Lüneburg, Lüneburg, Germany

Yohan Richardson Biomass Energy and Biofuels Laboratory 2iE-CIRAD (LBEB), International Institute for Water and Environmental Engineering (2iE Foundation), Ouagadougou, Burkina Faso

Christine Rolland RAPSODEE CNRS, UMR-5302, Université de Toulouse, IMT Mines Albi, Albi Cedex, 09, France

Lina María Romero Millán RAPSODEE CNRS, UMR-5302, Université de Toulouse, IMT Mines Albi, Albi Cedex, 09, France

Louise Roques RAPSODEE CNRS, UMR-5302, Université de Toulouse, IMT Mines Albi, Albi Cedex, 09, France

Abdoul Razac Sane RAPSODEE CNRS, UMR-5302, Université de Toulouse, IMT Mines Albi, Albi Cedex, 09, France

Rababe Sani RAPSODEE CNRS, UMR-5302, Université de Toulouse, IMT Mines Albi, Albi Cedex, 09, France

Martial Sauceau RAPSODEE CNRS, UMR-5302, Université de Toulouse, IMT Mines Albi, Albi Cedex, 09, France

Benoît Serive USR 3151, CNRS Sorbonne Université, Station Biologique de Roscoff, Roscoff Cedex, France

Patrick Sharrock RAPSODEE CNRS, UMR-5302, Université de Toulouse, IMT Mines Albi, Albi Cedex, 09, France

Sabu Thomas International and Inter University Centre for Nanoscience and Nanotechnology, Mahatma Gandhi University, Kottayam, Kerala, India;
School of Chemical Sciences, Mahatma Gandhi University, Kottayam, Kerala, India

Eric Trably Laboratoire de Biotechnologie de l'Environnement, INRA, Narbonne, France

Anne Trémier UR OPAALE, Irstea, Rennes, France

Christelle Tribout UPS, INSA, LMDC, Université de Toulouse, Toulouse Cedex 4, France

Fei Wang College of Energy Engineering, Zhejiang University, Yuquan Campus, Hangzhou, China

Makoto M. Watanabe Algae Biomass and Energy System R&D Center, University of Tsukuba, Tsukuba, Japan

Elsa Weiss-Hortala RAPSODEE CNRS, UMR-5302, Université de Toulouse, IMT Mines Albi, Albi Cedex, 09, France

Claire E. White Department of Civil and Environmental Engineering and the Andlinger Center for Energy and the Environment, Princeton University, Princeton, NJ, USA

Mingxiu Zhan College of Metrology and Measurement Engineering, China Jiliang University, Hangzhou, China

Mengmei Zhang State Key Laboratory of Clean Energy Utilisation, Zhejiang University, Hangzhou, China

Yaqian Zhao School of Civil Engineering, UCD Dooge Centre for Water Resource Research, University College Dublin, Dublin 4, Ireland;
State Key Laboratory of Eco-Hydraulic Engineering in Arid Area, Xi'an University of Technology, Xi'an, People's Republic of China

Chapter 1
Biomass Categories

Augustina Ephraim, Patricia Arlabosse, Ange Nzihou, Doan Pham Minh and Patrick Sharrock

Abstract Biomass resources for the production of renewable energy, chemicals and polymeric materials are abundant. In this chapter, these resources will be categorized into woody biomass, agricultural residues and waste, municipal solid waste, sewage sludge and aquatic plants. The origins, use and typical composition (physical, chemical and biological) of the different biomass types will be presented.

1.1 Introduction

Biomass refers to organic materials that are derived from plants or animals, i.e. all materials of biological origin that are not fossilized [1, 2]. Biomass may be divided into two broad groups [3]: virgin biomass includes terrestrial biomass (e.g. trees, crops, vegetables and fruits) and aquatic biomass (e.g. algae and water plants).

Waste includes municipal waste (municipal solid waste (MSW), sewage sludge, landfill gas), agricultural waste (livestock, manure, agricultural crop residue) and industrial wastes (e.g. demolition wood, waste oil or fat).

Traditionally, biomass in the form of fuelwood, agricultural residues and animal dung has been used by society for thousands of years as a source of energy for cooking and heating. The majority of households in the developing world continue to rely on such biomass for cooking as shown by Table 1.1 [4].

"Modern" use of biomass can be divided into four major categories [3]:

- Chemicals such as methanol, fertilizer and synthetic fiber
- Energy such as heat
- Electricity
- Transportation fuel such as gasoline and diesel.

A. Ephraim (✉) · P. Arlabosse · A. Nzihou · D. Pham Minh · P. Sharrock
RAPSODEE CNRS, UMR-5302, Université de Toulouse, IMT Mines Albi, Campus Jarlard, 81013 Albi Cedex, 09, France
e-mail: aug.ephraim@gmail.com

A. Nzihou (ed.), *Handbook on Characterization of Biomass, Biowaste and Related By-products*, https://doi.org/10.1007/978-3-030-35020-8_1

Table 1.1 Estimated biofuel consumption, by region (Tg/year)

Region	Fuelwood	Crop residues	Dung	Charcoal
North America	41	0	0	0
Latin America	80	0	0	16
Africa	371	52	0	14
Europe	147	0	0	0
South Asia	344	76	75	3
East Asia	193	323	0	0
Southeast Asia	164	43	0	6
Oceania	10	0	0	0
World	1351	495	75	39

Reproduced with permission from Wiley [4]

Figures 1.1 and 1.2 present some of the thermal, chemical and biological processes involved in converting biomass into useful energy and chemical products [5, 6].

Over the last few decades, waste management has become a major issue in most developed and developing countries. According to a recent World Bank report, 1.3 billion tonne per year of municipal solid waste (MSW) is currently generated worldwide and is expected to double by the year 2025 [1]. As this high level of waste production results in significant economic and environmental costs, many countries, particularly in Europe, have set goals to become "Recycling Societies"—one that does not only avoid producing waste but also uses it as a resource [7]. To achieve this, a number of European Directives have been introduced which aim to increase levels of recycling and recovery rates as well as the production of renewable energies from waste in order to minimize the amount of landfilled waste, thus minimizing greenhouse gas emissions [8, 9]. The main areas of legislation that are considered important for this chapter are [10, 11]:

- The Renewables Directive, RED (2009/28/EC); which is designed to help states progress towards meeting the EU 2020 target of 20% energy derived from renewable sources. According to the directive, eligible feedstocks include the biodegradable fraction from industrial and municipal wastes and residues from agriculture and forestry.
- The Landfill Directive, (1999/31/EC); enforces targeted reduction of biodegradable waste in landfills.
- The Waste Framework Directive, WFD (75/442/EEC); establishes principles of the waste hierarchy—re-duce, reuse, and recycle—to encourage re-use and recycling of waste as well as minimization of waste disposal.
- The Waste Incineration Directive, WID (2000/76/EC); governs the "thermal treatment" of waste, which includes combustion (incineration), gasification and pyrolysis. WID lays out strict specifications on the operating conditions of the thermal facilities (e.g. gas temperature and emission limits).

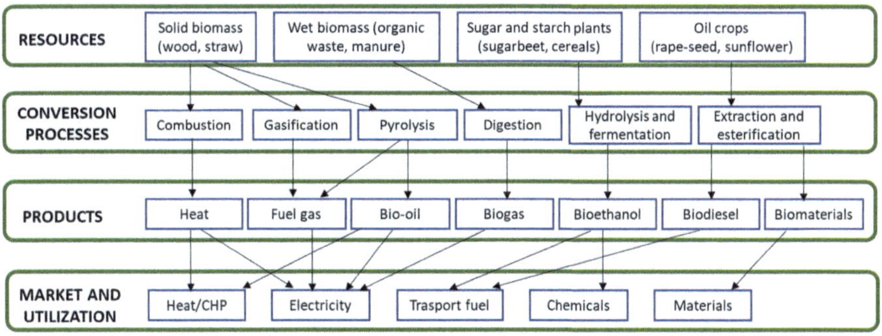

Fig. 1.1 Processes to convert biomass into useful energy and products

Fig. 1.2 Biomass conversion process to chemical and biomaterial products. Reproduced with permission from Chemical Society reviews [6]

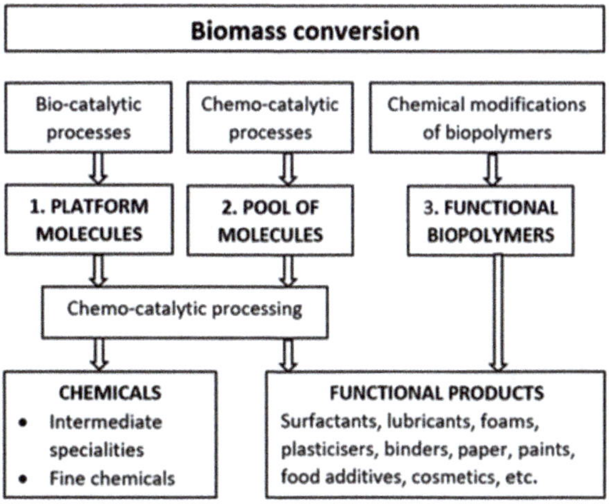

These directives have made biowaste valorization more economically attractive and has led to the development of various energy recovery technologies as shown in Table 1.2 [12].

This chapter therefore aims to present the classification of various virgin and waste biomass, with brief discussions on their origin, composition, and valorization processes.

1.2 Woody Biomass

Woody biomass is biomass from trees, bushes and shrubs [13], and can be broadly categorized as (i) forest and plantation wood, (ii) wood processing industry by-products and residues, and (iii) used wood. Figure 1.3 illustrates the various sub-classification of each woody biomass group, which will be discussed in this

Table 1.2 Energy recovery technologies

Established technologies	Type of waste	Product	Application
Anaerobic digestion and hydrolysis	Putrescibles (e.g. Food and animal waste, sewage sludge)	Biogas (methane)	Power generation, fertilizer, cooking gas
Fermentation	Cellulosic waste (e.g. Paper, agro-industrial waste, sewage sludge)	Bio-ethanol	Liquid fuel
Incineration	MSW, RDF, chemicals, clinic waste and sewage sludge	Heat, carbon dioxide, water vapour, ash	Power generation, heating
Emerging technologies			
Gasification and Pyrolysis	RDF, ASR, MSW (for gasification only)	Syngas, pyrolysis oil (bio-oil), char, ash. By-products: metals, chemicals	Transport fuel, chemicals, ammonia and fertilizers, electricity, heat

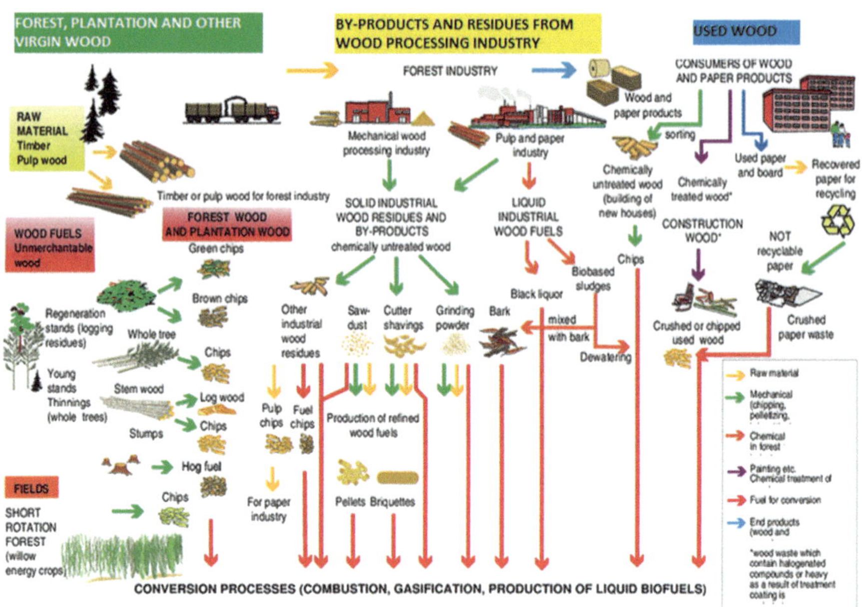

Fig. 1.3 Classification of woody biomass. Reproduced with permission from Springer Nature [13]

section. According to the World Energy Council, woody biomass provides about 90% of the primary energy annually supplied (56 EJ) by all forms of biomass worldwide [2].

1.2.1 Forest and Plantation Wood

According to a survey by Indufor in the year 2012, the world's total area of industrial fast-growing forest plantations is 54.3 million ha [14]. Figures 1.4 and 1.5 show the breakdown of the forest plantations by region. We can observe that Asia has the largest industrial forest plantations (17.7 million ha), followed by North and Latin America (12.8 million ha, each). In Africa, Oceania and Europe, there are about 5, 3.7 and 2.0 million ha of industrial forest plantations, respectively. The countries with the largest plantation area are the United States (US), China and Brazil, with each having over 5 million ha.

The wood obtained from industrial forest plantations can be grouped into softwood and hardwood. This nomenclature does not necessarily related to the wood density. On one hand, hardwoods are produced by angiosperm trees, which yield flowers and have broad leaves. On the other hand, softwoods are from gymnosperm trees that have needles and exposed seeds, but do not have leaves.

The data presented in Table 1.3 were obtained from a detailed survey of 61 countries on various forest species, summarized here as softwoods (e.g. Douglas fir, pine and spruce.) and hardwoods (e.g. willow, eucalyptus and beech). An estimated 1.4 billion m3 of raw wood were harvested from these planted forests in 2005, about 47% of which was devoted to industrial roundwood (e.g. timber, pulpwood, and chips), 39% to pulp and paper and 10% to bioenergy production [15]. Tables 1.4 and 1.5 display the typical properties of various softwood and hardwood species, respectively.

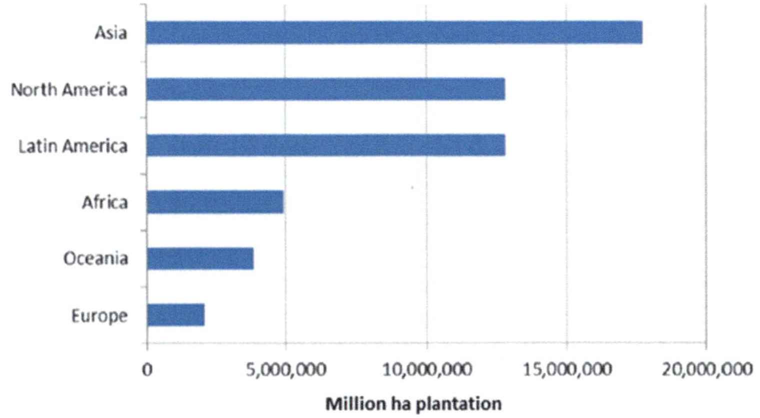

Fig. 1.4 Industrial forest plantations by region, 2012 [14]

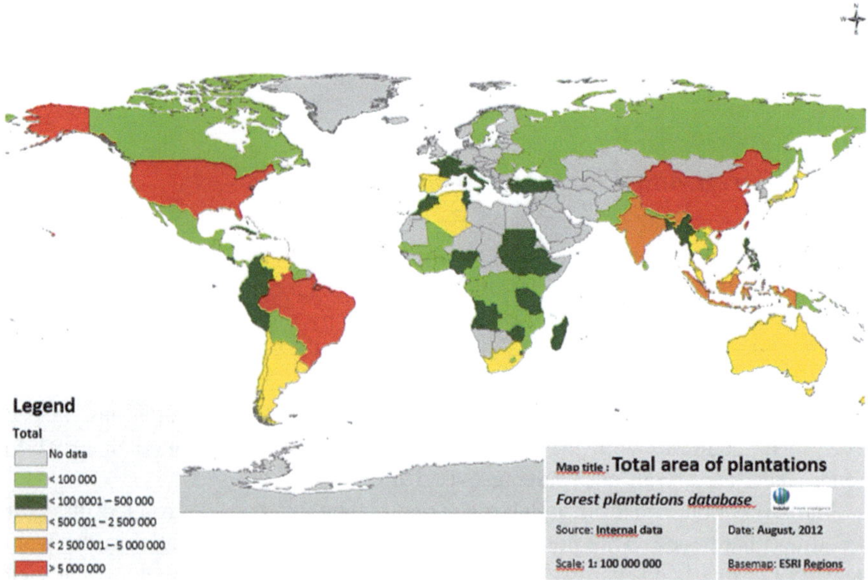

Fig. 1.5 Total area of global industrial forest plantations, 2012 [14]

Table 1.3 Planted forest area, hardwoods and softwoods (million ha)

Region	Softwoods	Hardwoods	Total
Africa	1.7	7.8	9.5
Asia	34.2	90.6	124.8
Northern, Central and Eastern Europe	62.4	12.1	74.5
Southern Europe	4.6	4.7	9.3
North and Central America	26.1	1.7	27.8
South America	5.4	5.6	11.0
Oceania	2.9	0.7	3.6
World	137.3	123.2	260.7

Reproduced with permission from Forest Products Society [15]

1.2.2 Wood Processing By-products and Residues

As shown in Fig. 1.3, the mechanical wood processing industry, as well as the pulp and paper industries, generate solid wood residues (e.g. sawdust, cutter shavings and grinding powder) and liquid wood fuels (e.g. black liquor and bio-based sludges). Refined wood fuels such as pellets and briquettes can be produced from the wood residues for energy purposes.

Table 1.4 Chemical composition of softwood species [16]

Property		Unit	Douglas Fir	Pine	Spruce
Fuel properties					
Proximate analysis	Moisture content	wt% (ar)		8.83	13.05
	Ash content	wt% (dry)	0.48	0.70	0.56
	Volatile matter (VM)	wt% (daf)	84.77	84.26	85.97
	Fixed carbon (100–VM)	wt% (daf)	15.23	15.74	14.03
Ultimate analysis	Carbon	wt% (daf)	52.04	52.01	49.26
	Hydrogen	wt% (daf)	6.30	6.25	5.88
	Nitrogen	wt% (daf)	0.09	0.14	0.13
	Sulphur	wt% (daf)	0.02	0.10	0.02
	Oxygen (calculated by difference)	wt% (daf)	41.55	41.5	44.71
Calorific values	Net calorific value (LHV)	MJ/kg (daf)	19.47	19.36	18.43
	Gross calorific value (HHV)	MJ/kg (daf)	20.95	20.68	19.79
Chemical analyses					
Halides	Chlorine (Cl)	mg/kg (daf)		602.7	66.7
Ash composition	CaO	wt% (ash)	37.08	27.56	36.43
	SiO_2	wt% (ash)	12.26	36.20	18.48
	K_2O	wt% (ash)	17.00	7.57	9.48
	SO_3	wt% (ash)	11.20	1.62	
	MgO	wt% (ash)	5.86	3.26	3.72
	Fe_2O_3	wt% (ash)	4.24	2.74	1.29
	Al_2O_3	wt% (ash)	2.83	6.86	1.50
	P_2O_5	wt% (ash)	1.86	3.39	3.16
	Na_2O	wt% (ash)	3.16	0.67	0.39
	TiO_2	wt% (ash)	0.08	0.12	
	Mn	mg/kg (ash)		7745.0	
	Cu	mg/kg (ash)		241.7	523.8
	Pb	mg/kg (ash)		234.0	25.0
	Cr	mg/kg (ash)		70.0	127.0
	Cd	mg/kg (ash)		10.0	0.9
	Hg				1.2

ar as received basis
daf dry ash-free basis

According to IEA Bionergy, in the year 2008, the estimated global production of wood pellets was 11.5 megatonnes (Mt) and the estimated amount traded was 4 Mt [17]. With an average energy density of 17.5 gigajoules (GJ) per tonne, this amounts to 200 terajoules (TJ) produced and 70 TJ traded.

Table 1.5 Chemical composition of hardwood species [16]

Property		Unit	Willow	Eucalyptus	Beech
Fuel properties					
Proximate analysis	Moisture content	wt% (ar)	12.77	11.10	11.63
	Ash content	wt% (dry)	1.96	1.57	0.67
	Volatile matter (VM)	wt% (daf)	83.54	84.90	83.14
	Fixed carbon (100–VM)	wt% (daf)	16.46	15.10	16.86
Ultimate analysis	Carbon	wt% (daf)	49.80	51.14	48.77
	Hydrogen	wt% (daf)	6.10	6.10	6.02
	Nitrogen	wt% (daf)	0.62	0.27	0.30
	Sulphur	wt% (daf)	0.05	0.04	0.03
	Oxygen (calculated by difference)	wt% (daf)	43.43	42.45	44.88
Calorific values	Net calorific value (LHV)	MJ/kg (daf)	18.49	18.93	17.85
	Gross calorific value (HHV)	MJ/kg (daf)	19.84	20.28	19.16
Chemical analyses					
Halides	Chlorine (Cl)	mg/kg (daf)	152.7	548.2	64.6
	Fluorine (F)	mg/kg (daf)	26.3		4.2
Major elements	Calcium (Ca)	mg/kg (dry)	5408.6	2215.8	2542.8
	Potassium (K)	mg/kg (dry)	2702.2	1584.2	1313.5
	Magnesium (Mg)	mg/kg (dry)	497.9	488.0	409.5
	Phosphorus (P)	mg/kg (dry)	782.4	348.0	98.0
	Silicon (Si)	mg/kg (dry)	445.2	103.0	162.5
	Sodium (Na)	mg/kg (dry)	185.5	454.0	41.9
	Aluminum (Al)	mg/kg (dry)	57.2	91.0	32.5
	Iron (Fe)	mg/kg (dry)	58.0	14.0	68.3
	Titanium (Ti)	mg/kg (dry)	3.6		2.5
Minor elements	Arsenic (As)	mg/kg (dry)	0.7	0.0	1.0
	Cadmium (Cd)	mg/kg (dry)	2.3	0.1	
	Cobalt (Co)	mg/kg (dry)	0.6		12.3
	Chromium (Cr)	mg/kg (dry)	11.1	1.4	2.3
	Copper (Cu)	mg/kg (dry)	6.3	16.0	2.0
	Manganese (Mn)	mg/kg (dry)	12.2	18.0	67.0
	Nickel (Ni)	mg/kg (dry)	23.6	1.3	2.4
	Lead (Pb)	mg/kg (dry)	96.0	0.8	0.9
	Vanadium (V)	mg/kg (dry)	0.2		0.1
	Zinc (Zn)	mg/kg (dry)	98.1	16.0	5.1
	Barium (Ba)	mg/kg (dry)	2.6		16.7
	Mercury (Hg)	mg/kg (dry)	0.1	0.1	
	Selenium (Se)	mg/kg (dry)			1.4
Other elements	Tin (Sn)	mg/kg (dry)	0.7		0.1
	Strontium (Sr)	mg/kg (dry)	14.2		5.2
	Boron (B)	mg/kg (dry)	9.0		4.4
	Antimony (Sb)	mg/kg (dry)	2.9		26.0

(continued)

Table 1.5 (continued)

Property		Unit	Willow	Eucalyptus	Beech
Ash composition	SO_3	wt% (ash)	2.35		
	Cl	wt% (ash)	0.49		
	P_2O_5	wt% (ash)	9.51	29.11	
	SiO_2	wt% (ash)	7.62	17.83	20.00
	Fe_2O_3	wt% (ash)	0.55		1.40
	Al_2O_3	wt% (ash)	1.10	7.87	7.00
	CaO	wt% (ash)	36.47	26.52	26.10
	MgO	wt% (ash)	3.52	7.25	9.20
	Na_2O	wt% (ash)	1.80	4.98	1.80
	K_2O	wt% (ash)	15.98	7.20	23.50
	TiO_2	wt% (ash)	0.05		

1.2.3 Used Wood

Used wood are recovered wood fuels that originate from socio-economic activities outside the forest and wood-processing sectors. Such wastes come from construction sites, demolition of buildings and containers [18]. Used wood may be thermally converted to energy or transformed into chips, pellets, briquettes or powder for recycling purposes.

In Europe, used woods are divided into 3 different classes based on their level of contamination; class A, B and C. Classes A and B are classified under EN 14961-1 (Solid biofuel standard) [19] and class C under EN 15359 (Solid recovered fuel standard) [20]. Table 1.6 provides the typical properties of the classes A, B and C used woods.

Class A

This is virgin wood that has only been mechanically treated. It includes chemically untreated by-products or residues from forest and wood processing industry as well as chemically untreated used wood. Class A wood such as sawmill co-products has a current market as a fuel for co-firing at coal power stations, fuel for other stand-alone biomass plants and raw material for a variety of competing markets, including animal bedding, horticultural use, and, most significantly, the panel board mills [10]. Class A wood is treated as a clean fuel and thus no Waste Incineration Directive (WID) is applied.

Class B

Class B wood is coated, lacquered or otherwise chemically treated wood. Coating does not contain halogenated compounds (for example PVC) and preservatives. Class B includes chemically treated by-products and residues from forest and wood

Table 1.6 Chemical composition of used wood [16]

Property		Unit	Class A	Class B	Class C
Fuel properties					
Proximate analysis	Moisture content	wt% (ar)	13.05	7.38	16.73
	Ash content	wt% (dry)	0.56	2.49	1.77
	Volatile matter (VM)	wt% (daf)	85.97	79.46	79.58
	Fixed carbon (100–VM)	wt% (daf)	14.03	20.54	20.42
Ultimate analysis	Carbon	wt% (daf)	49.26	50.54	53.93
	Hydrogen	wt% (daf)	5.88	5.68	5.95
	Nitrogen	wt% (daf)	0.13	1.44	0.34
	Sulphur	wt% (daf)	0.02	0.06	0.09
	Oxygen (calculated by difference)	wt% (daf)	44.71	42.28	39.69
Calorific values	Net calorific value (LHV)	MJ/kg (daf)	18.43	18.94	19.05
	Gross calorific value (HHV)	MJ/kg (daf)	19.79	20.18	20.41
Chemical analyses					
Halides	Chlorine (Cl)	mg/kg (daf)	66.7	1187.2	316.9
	Fluorine (F)	mg/kg (daf)		22.1	13.4
Major elements	Calcium (Ca)	mg/kg (dry)	6376.6	4050.0	2200.0
	Silicon (Si)	mg/kg (dry)	1790.0	2550.0	2150.0
	Potassium (K)	mg/kg (dry)	1443.7	735.0	285.0
	Magnesium (Mg)	mg/kg (dry)	506.6	450.0	215.0
	Iron (Fe)	mg/kg (dry)	211.4	510.0	2200.0
	Phosphorus (P)	mg/kg (dry)	375.6	100.0	62.5
	Aluminium (Al)	mg/kg (dry)	136.4	455.0	310.0
	Sodium (Na)	mg/kg (dry)	94.0	670.0	250.0
	Titanium (Ti)	mg/kg (dry)	12.3	275.0	2.9
Minor elements	Arsenic (As)	mg/kg (dry)	0.3	8.9	1.3
	Cadmium (Cd)	mg/kg (dry)	0.2	1.3	0.4
	Cobalt (Co)	mg/kg (dry)	0.3	2.2	0.7
	Chromium (Cr)	mg/kg (dry)	2.4	34.5	10.0
	Copper (Cu)	mg/kg (dry)	2.8	21.0	10.2
	Manganese (Mn)	mg/kg (dry)	316.0	76.5	41.5
	Nickel (Ni)	mg/kg (dry)	1.8	4.6	5.6
	Lead (Pb)	mg/kg (dry)	1.7	170.0	81.3
	Vanadium (V)	mg/kg (dry)	0.6	0.8	1.3
	Zinc (Zn)	mg/kg (dry)	41.9	315.0	256.0
	Barium (Ba)	mg/kg (dry)	73.2	345.0	75.1
	Molybdenum (Mo)	mg/kg (dry)	0.1	0.6	
	Selenium (Se)	mg/kg (dry)	0.1	0.2	

(continued)

Table 1.6 (continued)

Property		Unit	Class A	Class B	Class C
Other elements	Tin (Sn)	mg/kg (dry)	0.5	1.1	1.0
	Boron (B)	mg/kg (dry)	7.2		
	Antimony (Sb)	mg/kg (dry)	0.1		1.6
	Boron (B)	mg/kg (dry)		7.4	3.4
	Strontium (Sr)	mg/kg (dry)		17.5	
Ash composition	CO_2	wt% (ash)	23.07	23.07	1.10
	SO_3	wt% (ash)	1.24	1.24	6.58
	Cl	wt% (ash)	0.12	0.12	11.80
	P_2O_5	wt% (ash)	3.16	3.16	18.04
	SiO_2	wt% (ash)	18.48	18.48	13.31
	Fe_2O_3	wt% (ash)	1.29	1.29	4.20
	Al_2O_3	wt% (ash)	1.50	1.50	17.65
	CaO	wt% (ash)	36.43	36.43	3.63
	MgO	wt% (ash)	3.72	3.72	6.21
	Na_2O	wt% (ash)	0.39	0.39	13.44
	K_2O	wt% (ash)	9.48	9.48	
	TiO_2	wt% (ash)			0.52
	Pb	mg/kg (ash)	25.0	25.0	7.0
	Cd	mg/kg (ash)	0.9	0.9	0.0
	Cu	mg/kg (ash)	523.8	523.8	109.0
	Hg	mg/kg (ash)	1.2	1.2	
	Mn	mg/kg (ash)			0.5
	Cr	mg/kg (ash)	127.0	127.0	63.0

processing industry (e.g. furniture, kitchen wood) as well as chemically treated used wood excluding demolition wood. It does not come under WID and thus can be used as a feedstock for industrial wood processing operations, such as the manufacture of panel products including chipboards and fireboards [21]. However as it is difficult to separate out the clean waste wood from the contaminated, in countries like the UK, Class B wood is usually fed to a mass burn incinerator or directly disposed to landfill [10].

Class C

Class C consists of wood that has been coated or treated with halogenated compounds such as PVC but does not contain preservatives. An example of such wood is demolition wood whose origin is difficult to verify. Class C wood, which is classified as a solid recovered fuel according to EN 15359 [20], must be incinerated in compliance with WID.

1.3 Agricultural Residues and Waste

Agricultural waste consists of organic material, such as manure from livestock, slurry, silage effluent and crop residues. Figure 1.6 shows that agricultural and forestry waste is one of the major waste categories generated throughout Europe. Individual countries within Europe show a variation in the arising of agricultural waste due to the different extents of agriculture areas within the economy and different farming methods. Examples of agricultural waste tonnage available in Europe are: Spain, estimated at 114 million tonnes/year; France, 377 million tonne/year; UK, 87 million tonne/year [22]. In this section, we will focus our discussion on crop residues and animal dung.

1.3.1 Crop Residues

Crop residues are organic materials that are left after a crop has been harvested or processed into a usable re-source. These residues include husks, seeds, bagasse, molasses and roots, which are mainly derived from cereals, sugar crops, roots and tubers, vegetables, fruits and oil crops.

The majority of crop residues are landspread, while some are used as animal feed, compost or as fuel for bio-gas production [23]. Other, less frequent uses of crop residues are as building material and sources of extraction of organic compounds [24].

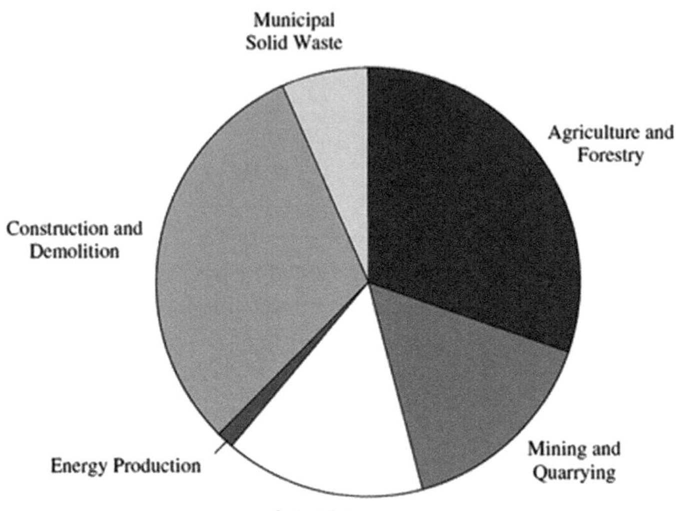

Fig. 1.6 Total waste generated by sector in the EU (15 members 2001) [22]

As shown in Table 1.7, crop residues are usually high in nutrients such as carbon (C), nitrogen (N), phosphorous (P) and potassium (K), and thus can substitute commercial fertilizers for improving crop yields and soil health. It has been reported that the integrated use of crop residues and mineral fertilizer reduces the cost and amount of fertilizer required by the crops [25–27]. Consequently, landspreading is considered to be the best practiced environmental option [28].

Regarding biogas production, crop residues may be suitable fuels if well converted. This is mainly due to their high carbon, hydrogen and volatiles content, as shown in Table 1.7. For example, 1 kg of pre-treated crop waste and water hyacinth has the potential of producing 0.037 and 0.045 m^3 of biogas, respectively [29].

1.3.2 Animal Dung

Animal dung or feces is indigestible plant material released from the intestine of an animal. It is generally used as manure for landspreading or as a fuel source. For these purposes, the most commonly used dung are those derived from cattle (cows and buffaloes), pig and poultry (chickens).

Poultry litter, consisting of a mixture of bird droppings and wood shavings has received interest due to its high generation rates and high calorific value, which therefore makes it more suitable as a fuel than as manure for landspreading [23, 30].

Table 1.7 Chemical composition of various crop residues [16]

Property		Unit	Sugarcane bagasse	Almond shell	Rice husk	Live pits
Fuel properties						
Proximate analysis	Moisture content	wt% (ar)	21.53	10.13	10.60	7.04
	Ash content	wt% (dry)	5.70	2.38	18.03	2.30
	Volatile matter (VM)	wt% (daf)	83.77	78.89	76.95	79.47
	Fixed carbon (100–VM)	wt% (daf)	16.23	21.11	23.05	20.53
Ultimate analysis	Carbon	wt% (daf)	49.57	49.90	46.14	49.89
	Hydrogen	wt% (daf)	5.97	6.15	6.37	6.32
	Nitrogen	wt% (daf)	0.40	0.71	0.90	0.92
	Sulphur	wt% (daf)	0.08	0.03	0.20	0.07
	Oxygen (calculated by difference)	wt% (daf)	43.98	43.21	46.39	42.80
Calorific values	Net calorific value (LHV)	MJ/kg (daf)	17.89	18.55	16.42	20.09
	Gross calorific value (HHV)	MJ/kg (daf)	19.19	19.91	17.79	21.22

(continued)

Table 1.7 (continued)

Property		Unit	Sugarcane bagasse	Almond shell	Rice husk	Live pits
Chemical analyses						
Halides	Chlorine (Cl)	mg/kg (daf)	1030.1	74.6	960.1	663.8
Biochemical composition	Cellulose	wt% (dry)	37.27	35.70	33.70	28.10
	Hemicellulose	wt% (dry)	35.80	28.83	22.00	37.20
	Lignin	wt% (dry)	20.13	28.60	22.83	28.25
Ash composition	SO_3	wt% (ash)	3.57	1.40	0.77	0.56
	P_2O_5	wt% (ash)	3.19	5.57	0.87	2.46
	SiO_2	wt% (ash)	47.23	8.81	89.39	30.82
	Fe_2O_3	wt% (ash)	10.01	2.22	0.40	6.58
	Al_2O_3	wt% (ash)	13.07	1.96	0.22	8.84
	CaO	wt% (ash)	4.56	14.50	1.30	14.56
	MgO	wt% (ash)	3.34	4.77	0.57	4.24
	Na_2O	wt% (ash)	0.80	1.41	0.35	27.80
	K_2O	wt% (ash)	9.97	34.36	5.04	4.40
	TiO_2	wt% (ash)	2.16	0.10	0.02	0.34

Furthermore, it has been shown that composting of poultry dung for land use has its disadvantages including the loss of nitrogen and other nutrients during composting [30]. The poultry litter has higher biogas yield potential than cattle and pig dung, after undergoing anaerobic digestion [29]. This biogas can be used in turn to generate heat for space heating within the farm or to produce electricity. Some characteristics of animal dung have a significant impact on the amount of biogas produced. These characteristics include the carbon/nitrogen (C/N) ratio, volatile matter content, and toxicity [29].

C/N ratio

A C/N ratio between 20 and 30 is considered to be optimum for anaerobic digestion [29]. If the C/N ratio is very high, the nitrogen will be quickly consumed by methanogenic bacteria (methanogens) in order to meet their protein requirements, which will lower their reaction with the left over carbon content of the fuel, and thus lower gas production. However, if the C/N is very low, the excess nitrogen will be re-leased in the form of ammonia (NH4), which may raise pH value of the digester content above 8.5, and therefore create a toxic environment for the methanogens. By observing Table 1.8, the C/N ratio of cow dung is 20 whereas for pig and chicken dung, the C/N ratios are 15 and 8 respectively. For the case where the C/N ratio is significantly low, the fuel can be mixed with those of high C/N ratio, such as crop residues, in order to bring the feedstock pH to a desirable level.

Table 1.8 Chemical composition of animal manure [16]

	Property	Unit	Cow	Pig	Chicken
Fuel properties					
Proximate analysis	Moisture content	wt% (ar)	48.64	56.10	38.10
	Ash content	wt% (dry)	33.38	25.30	24.55
	Volatile matter (VM)	wt% (daf)	83.22	79.06	80.66
	Fixed carbon (100–VM)	wt% (daf)	16.78	20.94	19.34
Ultimate analysis	Carbon	wt% (daf)	47.60	50.30	45.79
	Hydrogen	wt% (daf)	6.66	6.12	6.16
	Nitrogen	wt% (daf)	2.41	3.37	5.72
	Sulphur	wt% (daf)	0.50	0.67	0.92
	Oxygen (calculated by difference)	wt% (daf)	42.83	39.54	41.41
Calorific values	Net calorific value (LHV)	MJ/kg (daf)	19.46	20.05	18.12
	Gross calorific value (HHV)	MJ/kg (daf)	20.91	21.19	19.35
Chemical analyses					
Halides	Chlorine (Cl)	mg/kg (daf)	18,600.6	10,098.6	6251.4
	Bromine (Br)	mg/kg (daf)		7.2	12.2
	Fluorine (F)	mg/kg (daf)		37.6	13.5
Major elements	Aluminium (Al)	mg/kg (dry)		597.8	735.2
	Potassium (K)	mg/kg (dry)	11,300.0	15,290.5	30,013.2
	Sodium (Na)	mg/kg (dry)	2400.0	2771.8	4417.7
	Calcium (Ca)	mg/kg (dry)	5800.0	9841.4	64,079.1
	Silicon (Si)	mg/kg (dry)		5613.6	3817.8
	Magnesium (Mg)	mg/kg (dry)		6037.8	6960.8
	Iron (Fe)	mg/kg (dry)		2208.9	1043.5
	Phosphorus (P)	mg/kg (dry)		9247.1	21,184.2
	Titanium (Ti)	mg/kg (dry)		35.2	34.4
Minor elements	Arsenic (As)	mg/kg (dry)		2.1	
	Cadmium (Cd)	mg/kg (dry)	0.7	0.2	0.2
	Cobalt (Co)	mg/kg (dry)		1.3	1.6
	Chromium (Cr)	mg/kg (dry)	35.0	7.2	19.1
	Copper (Cu)	mg/kg (dry)	56.0	81.0	70.3
	Manganese (Mn)	mg/kg (dry)		232.2	408.6
	Nickel (Ni)	mg/kg (dry)	12.0	6.1	19.6
	Lead (Pb)	mg/kg (dry)	31.0	4.1	3.3
	Vanadium (V)	mg/kg (dry)		2.3	3.7
	Zinc (Zn)	mg/kg (dry)	253.0	324.8	351.2
	Barium (Ba)	mg/kg (dry)		21.6	22.8
	Molybdenium (Mo)	mg/kg (dry)		3.6	5.1
	Selenium (Se)	mg/kg (dry)		1.3	1.4

Volatile matter content

Volatile matter is the organic combustible part of the fuel which is liberated when heated to about 550 °C. In the absence of air, the higher the volatile matter content in a unit mass of fresh dung, the higher the gas production.

Table 1.9 Composition of office paper waste (OP), newspaper waste (NP) and cardboard waste (CB) and Whatman No. 1 filter paper (FP) as reference according to [42]

Parameter	OP	FP	NP	CB
TS (%)	95.3 ± 0.2	95.5 ± 0.1	93.2 ± 0.4	95.4 ± 0.3
VS (% TS)	98.5 ± 0.2	100	96.1 ± 0.3	87.2 ± 0.2
Ash (% TS)	1.4 ± 0.0	None	3.9 ± 0.1	12.8 ± 0.2
Lignin (% TS)	1.4 ± 0.5	None	23.4 ± 0.5	17.8 ± 0.5
Cellulose (% TS)	84.9 ± 1.3	100	68.5 ± 1.1	56.9 ± 0.8
Hemicellulose (% TS)	12.3 ± 0.6	None	13.1 ± 0.3	10.7 ± 0.3
COD (g 0_2/gDM)	1.07 ± 0.02	1.14 ± 0.03	1.21 ± 0.03	1.10 ± 0.02

TS total solids, *VS* volatile solids, *COD* chemical oxygen demand, *DM* dry matter

Toxicity

A high concentration of mineral ions, heavy metals, antibiotics and detergents that may be pre-sent in the animal dung, may inhibit the growth of microbes in the digester [29]. For example, heavy metals such as copper, chromium and nickel, in small quantities are essential for the growth of bacteria but their higher concentrations (i.e. 100, 200 and 200–500 mg/L respectively) have toxic effects. Table 1.9 displays the typical mineral and heavy metal contents of animal dung.

1.4 Municipal Waste

According to the European Commission [31], municipal waste is defined as below: "Municipal waste covers household waste and waste similar in nature and composition to household waste".

However, European Commission specifies that this definition has evolved over time by formalizing it along the 3 main dimensions for waste statistics: waste origin, waste materials and waste collectors. More details can be found elsewhere [31].

According to the EEA Report No. 2, 2013 [2], municipal waste is defined as below:

> Municipal waste is mainly produced by households, though similar wastes from sources such as commerce, offices and public institutions are included. The amount of municipal waste generated consists of waste collected by or on behalf of municipal authorities and disposed of through the waste management system.

According to the EU's Landfill Directive, municipal solid waste is defined as "waste from households, as well as other waste which, because of its nature or composition, is similar to waste from households" [32, 33].

U.S. Environmental Protection Agency considers that "Municipal Solid Waste (MSW)—more commonly known as trash or garbage—consists of everyday items we use and then throw away, such as product packaging, grass clippings, furniture, clothing, bottles, food scraps, newspapers, appliances, paint, and batteries. This comes from our homes, schools, hospitals, and businesses" [34].

From these definitions, it appears that municipal wastes can be listed as: food and kitchen waste; green waste (yard trimmings); wood and wood-based materials; paper; cardboard; glasses; plastics; metals; rubber and leather; textiles; and miscellaneous inorganic wastes. Among these wastes, food and kitchen waste; green waste (yard trimmings); wood and wood-based materials; paper; cardboard can be regrouped in the category of biowaste considering their non-fossil origin. Thus, municipal waste is generally a very complex medium which contains various components. Study on municipal waste needs specific conventions and definitions. For example, there are different methods for the analysis of municipal waste composition, which depend on the organism or country considered, as previously reviewed by Lisa and Anders [35].

The nature and the generation rate of municipal waste vary geographically as a function of continent, region, country and even department/state of each country. It depends on several factors such as income level, living standards, economic activities, urbanization rate, local regulation rules etc.

The next section will address some general statistics on municipal waste generation as well as its evolution with time. According to the World Bank [36], in 2012, the total amount of municipal solid waste generated in the world reached about 1.3 billion ton per year, amounting to a footprint of 1.2 kg per person per day. The generation of MSW by region is as follows (million ton of municipal waste per year):

- The organization for economic co-operation and development (OECD) countries [37]: 572.
- South Asia: approximately 70.
- East Asia and the Pacific Region: approximately 270.
- Eastern and Central Asia: at least 93.
- Sub-Saharan Africa: approximately 62.
- Middle East and North Africa: 63.
- Latin America and the Caribbean: 160.

Because of the rapid evolution of world population, the World Bank foresees that the total amount of MSW can reach 2.2 billion tons per year by 2025.

Figure 1.7 shows the amount of municipal waste generated in 32 European countries from 2001 to 2010 as well as the management of the waste. It is evident that the total amount of municipal waste in these countries did not significantly evolve during this decade. However, the management remarkably changed with the reduction of landfill and the increase of incineration and recycling fractions. The average amount of municipal waste generated per capita during this decade varied strongly between countries, from around 300 kg per capita in Latvia, Estonia, Poland, Slovakia to around 520 kg per capita in France, Spain, Germany or even around 700 kg per capita in Cyprus and Switzerland [32].

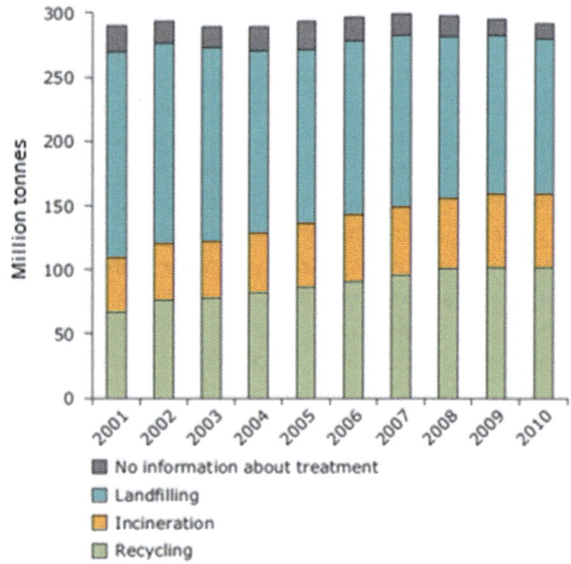

Fig. 1.7 Municipal waste amount and management in 32 European countries, 2001–2010 [32]

China is known as the country having the highest population these last decades. In parallel with its population evolution, urbanization and living standards in China have evolved quickly which has had an impact on the generation of municipal waste and its treatment in this country. Cheng and Hu [38] reported the total amount of municipal solid waste collected and treated in China during 1980–2005 period with estimates made for the 2010–2015 period (Fig. 1.8). The amount of collected MSW in this country was increased by a factor of roughly 2 for every 10 years lap. The World Bank reported also that the total municipal waste in China reached around 190 million tons per year in 2010s [36].

The evolution of MSW generation in USA from 1960 to 2014 linearly in-creased up to 2000s [39]. Then, it was practically unchanged around 250 million tons per year for 2000–2014 period. In parallel, a continuous increase was observed for the average amount of MSW generated per capita per year from 1960 to 1990s. Then, this average amount was stagnated around 750 kg per year.

In the case of France, the average amount of municipal waste generated per capita per year reached around 570 kg during the last 15 years. In 2011, 38.5 million tons of household waste and similar waste were collected and sent to treatment and valorization plants. Note that household waste represents around 80% and similar waste occupies around 20%. About the composition of household waste, a national campaign on the characterization of this waste was carried out in 2007 by ADEME [40] and the results are shown in Fig. 1.9. Its composition did not significantly evolve from 1993 to 2007. It is worth noting that putrescible wastes, paper and paperboard represent around 55% of the total household waste generated. In the case of UK, paper and cardboard wastes represent 23.6% of the household waste generated during the year for 2001–2003 period while kitchen and garden wastes represent 35.1%, which were close to the case of France [41].

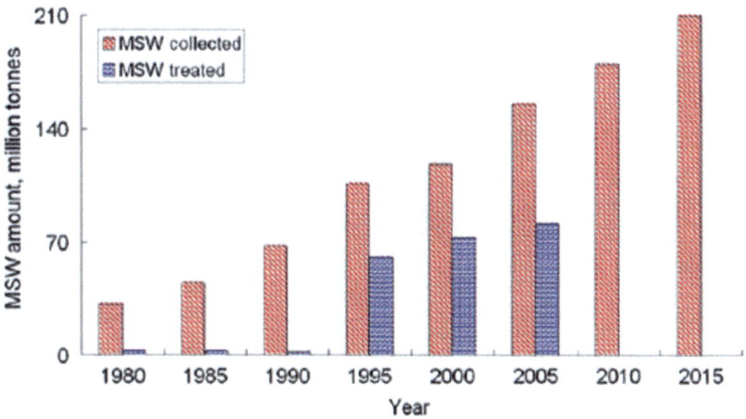

Fig. 1.8 Municipal solid waste collected and treated in China. Reproduced with permission from Elsevier [38]

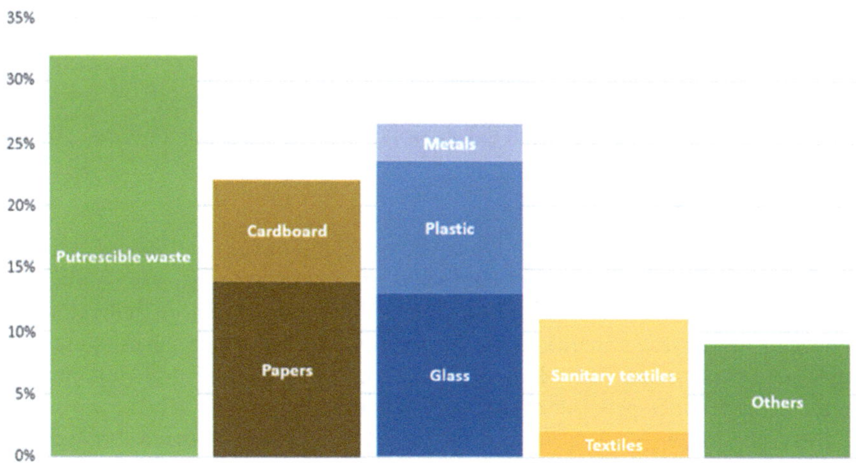

Fig. 1.9 Composition of household waste in France according to a characterization. Adapted from ADEME [40]

The evolution of paper and cardboard wastes, food wastes and garden wastes in USA is presented in Fig. 1.10 [37]. The quantity of these wastes increased up the year 2000. Then they stagnated, or even decreased in the case of paper and cardboard wastes, which probably due numerical developments. These three kinds of wastes represented around 55% of the total amount of municipal waste in USA during the last decade. This is comparable to European countries, i.e. France or UK. Taking into account the forecast of the World Bank, the total amount of paper and cardboard, food waste and garden waste may reach 1.2 billion of ton by 2025.

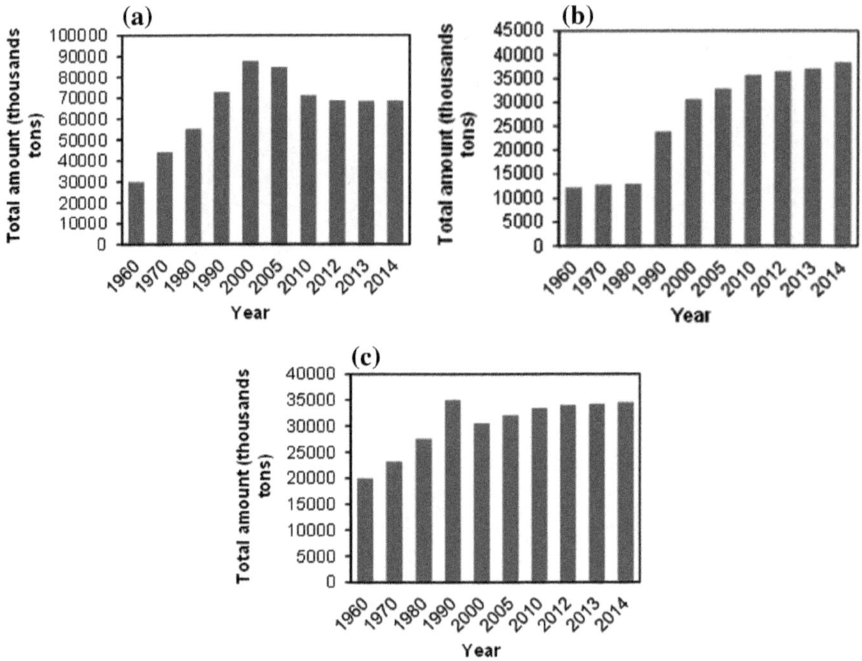

Fig. 1.10 Evolution of paper and cardboard wastes (**a**), food wastes (**b**) and garden wastes (**c**) in USA

Garden waste generally has the typical composition of woody biomass, which is presented in Sect. 1.2. Paper and cardboard are fabricated from woody biomass with various compositions of cellulose, hemicellulose and lignin. In general, lignin content in these wastes increases by the following order: office paper waste content < newspaper waste < cardboard. For example, Yuan et al. [42] reported the composition of these paper and cardboard wastes.

The composition of food waste is much more complex compared to paper and cardboard wastes. Food waste has local, seasonal and punctual properties and the composition can be very variable between continent, country, urbanization level etc. As example, a recent study in UK [43] shows that food waste generation per household per year increased by the following order (for 2008–2012 period): January to March < April to June < October to December < July to September. Other factors which influence on the food waste generation in UK are: levels of deprivation, region and nation, and population density. In another report, the Waste and Resources Action Programme (WRAP) reported the results from a wide survey of municipal waste of the Wale during 2009 [44]. The objective was to fully characterize food waste before their anaerobic digestion. The results obtained were as follows:

- Solid content: varied from around 20–34 wt% depending on the community. Also, food waste generated during winter had more solid content than during summer.
- Total carbohydrate content: varied from around 35–135 g/kg during summer and around 100–210 g/kg during winter, depending on the community exanimated. Large standard deviation was observed due to the heterogeneity of samples and the difficulty to homogenize samples.
- Lipid content: varied from around 8–130 g/kg during summer and around 25–110 g/kg during win-ter. Large standard deviation was observed.
- Protein content: varied from around 10–150 g/kg during summer and around 20–80 g/kg during winter. Large standard deviation was observed.
- CHNSO: The content of C, O, H, N and S were found to be approximately 48, 39, 7, 3 and <1 wt%, respectively.

1.5 Sewage Sludge

To protect public health and the environment, the water that is of no immediate value to the purpose for which it was used or in the pursuit of which it was produced, e.g. wastewater, has to be collected and treated in wastewater treatment plants (WWTPs) before being released to the environment. A wide variety of processes are available and provide different levels of wastewater treatment, known as preliminary, primary, secondary and advanced/tertiary treatments [45]. These treatments produce several inevitable by-products, including screenings, grit, scum and sewage sludges. Of all these constituents, sewage sludges are by far the largest in volume. Contaminants initially present in the wastewater, e.g. heavy metals, hazardous air pollutants (HAPs) and PCBs, are concentrated in the sludges.

In developed countries, one considers that the treatment of the wastewater of one so-called 'population equivalent' or PE, produces annually between 12 and 31 kg of sludge, expressed on dry basis [46, 47]. However, the amount of sludge produced depends on many factors, including the composition and volume of the incoming wastewater as well as the type of treatment processes implemented in the WWTPs, and hence is quite variable. Thus, for example, per capita annual sludge production in China was only 4.5 kg in 2013 [48]. Regional discrepancies between old (EU-15) and new European states (EU-12) members are al-so reported in the European Community [49]. Differences are due to variations in percentages of population served by centralized wastewater treatment plants, variation in influent characteristics (for instance, Chinese influents are characterized by particularly low chemical oxygen demand), treatments applied in the WWTPs as well as contribution of the industrial sectors. Accordingly, the total amount of urban sludge annually generated on a dry basis was around 6.25 million tons in China in 2013 [48], 10.9 million tons for EU-27 countries in 2005 [49] and 7.18 million of biosolids[1] in 2004 [51]. Throughout the world, stringent regulation, higher wastewater treatment coverage and implementation of advanced treatment processes will increase significantly the annual sewage sludge production.

As highlighted in Table 1.10, raw sludges contain a large amount of water, up to 98% for a secondary sludge, and high levels of organic matter. Furthermore, sludges are nutrient rich materials [52], including macronutrients (especially N, P, K, Ca, S and Mg) and micronutrients (such as Fe, B, Mn, Zn, Cu, Ni) that are essential for plant growth and plant metabolism.

Varieties of technologies are available to reduce the volume (thickening, conditioning, mechanical and/or thermal dewatering) and improve the quality (anaerobic or aerobic digestion, composting or liming). Choice and combination are closely related with the final disposal practices (Fig. 1.11). Obviously, the size of the WWTP is a key driver for the selection of an appropriate technology. For instance, due to a high capital cost, thermal drying is usually implemented in big WWTPs (PE > 100,000). However, solar dryers constitutes a good alternative to thermal dryer for small to medium size WWTPs, i.e. between 2000 and 50,000 PE.

To comply with the waste management hierarchy, full recycling of the organic matter, on condition that potential risks associated with the presence of pollutants are effectively managed, is widespread world-wide (Fig. 1.12). More or less stringent limits regarding storage of sewage sludge have been introduced in most countries, progressively restricting the amount of sewage sludge and organic wastes sent to landfills (Fig. 1.13). However the nature of the sludge, rich in nutrients but also loaded with mineral and organic contaminants, has led countries to seek different pathways for sludge disposal. In the coming decades, changes in regulations could negatively influence agricultural reuse. The most probable developments will concern possible controls on pathogen content, protection of human health and the environment from risks that can be posed by chemicals (new European REACH[2] regulation) and incentives on renewable energy. Anaerobic digestion of sewage sludge or co-digestion of sludge with food waste, organic fraction of municipal solid wastes or agricultural by-products for energy recovery is encouraged in many countries.

In developed countries, the cost related to sludge processing and management accounts for 50% of the whole operation expenses at the wastewater treatment plant, 30% of the total electricity consumption and up to 40% of the total wastewater treatment emissions [51]. Accordingly, sludge management remains one of the most complex environmental, technical, financial and regulatory challenges.

---

[1]Two different terms have been used historically: after proper treatment and processing to meet U. S. Environmental Protection Agency, sewage sludges were referred to as biosolids [50]. Consequently, biosolids do not represent the total resources. Nowadays, both terms are often used interchangeably.

[2]***R**egistration*, **E**valuation, **A**uthorisation and restriction of **CH**emicals.

Table 1.10 Composition of municipal sewage sludge [53]

Parameter	Type of sludge		
	Untreated primary sludge	Digested primary sludge	Secondary sludge
Total dry solids (% of TS)	2.0–8.0	6.0–12.0	0.8–**1**.2
Volatile solids (% of TS)	60–80	30–60	59–88
Grease and fats (% of TS)	7–35	n/a	5–12
Protein (% of TS)	20–30	15–20	32–41
Cellulose (% of TS)	8.0–15.0	8.0–15.0	7–9.7
Phosphorus (% of TS)	0.8–2.8	1.5–4.0	2.8–11.0
Nitrogen (% of TS)	1.5–4	1.6–6.0	2.4–5.0
Potassium (% of TS)	0–1	0–3.0	0.5–0.7
pH	5.0–8.0	6.5–7.5	6.5–8.0

Reprinted with permission from Elsevier
n/a data not available, *TS* total solids

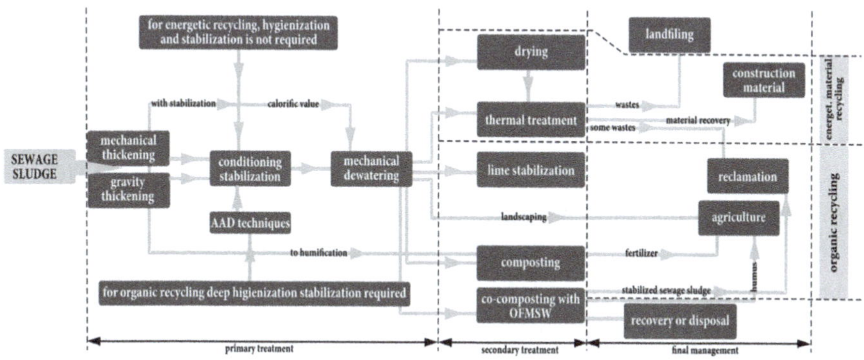

Fig. 1.11 Alternatives for the sewage sludge treatment and disposal strategies (AAD: Advanced aerobic digestion, OFMSW: Organic fraction of municipal solid waste). Reprinted with permission from Elsevier [53]

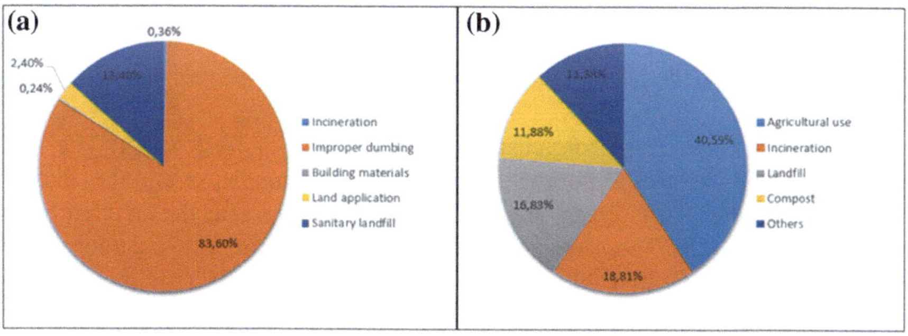

Fig. 1.12 Situation of sludge disposal (**a**) in China in 2013 [48] and **b** EU-27 in 2005 [49]

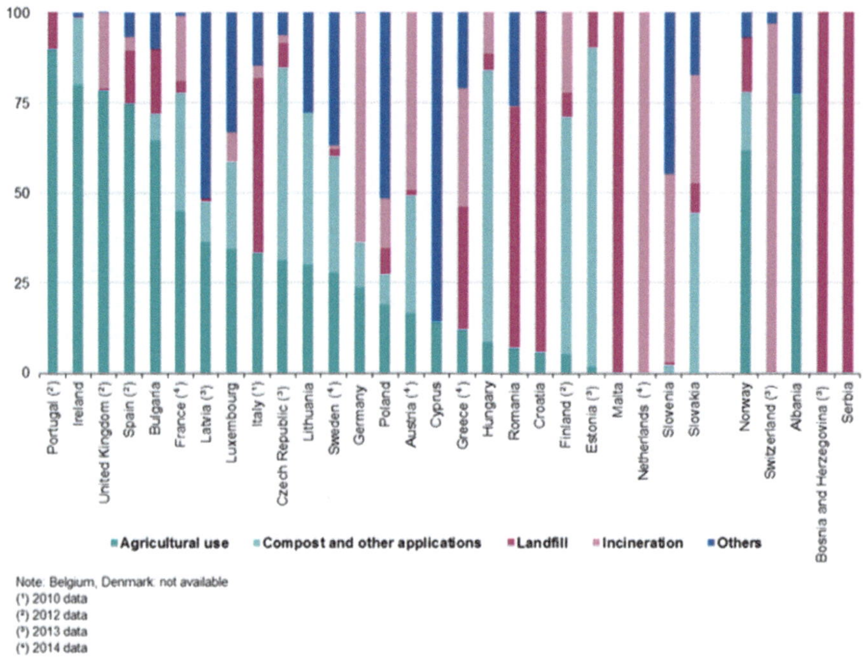

Fig. 1.13 Situation of sludge disposal in EU countries in 2015

1.6 Microalgae and Aquatic Plants

Photosynthetic autotrophs or phototrophs are organisms that can convert light energy into chemical energy and food. Algae and aquatic plants are important autotrophs that thrive in aquatic environments (shallow coastal zones, wetlands, rivers, lakes and oceans) and provide food and habitat for other organisms. A key difference between these two autotrophs is that aquatic plants have a vein-like vascular system whereas algae generally don't [54]. This section will provide a general analysis of the key characteristics and use of microalgae and aquatic plants.

1.6.1 Microalgae

Microalgae are microscopic organisms that live in salt or fresh water. The three most important classes of microalgae in terms of abundance are the diatoms (Bacillariophyceae), the green algae (Chlorophyceae), and the golden algae (Chrysophyceae) [55]. These species store energy in the form of oils, carbohydrates and proteins as shown in Table 1.11.

Table 1.11 Chemical composition of algae on a dry matter basis (%) [55]

Species of a sample	Proteins	Carbohydrates	Lipids	Nucleic acid	Oil content
Anabaena cylindrica	43–56	25–30	4–7	–	–
Botryococcus braunii	–	–	–	–	25–75
Chlamydomonas rheinhardii	48	17	21	–	–
Chlorella pyrenoidosa	57	26	2	–	–
Chlorella sp.	–	–	–	–	28–32
Chlorella vulgaris	51–58	12–17	14–22	4–5	–
Crypthecodinium cohnii	–	–	–	–	20
Cylindrotheca sp.	–	–	–	–	16–37
Dunaliella bioculata	49	4	8	–	–
Dunaliella primolecta	–	–	–	–	23
Dunalierlla salina	57	32	6	–	–
Euglena gracilis	39–61	14–18	14–20	–	–
Isochrysis sp.	–	–	–	–	25–33
Monallanthus salina	–	–	–	–	>20
Nannochloris sp.	–	–	–	–	20–35
Nannochloropsis sp.	–	–	–	–	31–68
Neochloris oleoabundans	–	–	–	–	35–54
Nitzschia sp.	–	–	–	–	45–47
Phaeodactylum tricornutum	–	–	–	–	20–30
Porphyridium cruentum	28–39	40–57	9–14	–	–
Prymnesium parvum	28–45	25–33	22–38	1–2	–
Scenedesmus dimorphus	8–18	21–52	16–40		–
Scenedesmus obliquus	50–56	10–17	12–14	3–6	–
Scenedesmus quadricauda	47	–	1.9	–	–
Schizochytrium sp.	–	–	–	–	50–77
Spirogyra sp.	6–20	33–64	11–21	–	–
Spirulina maxima	60–71	13–16	6–7	3–4.5	–
Spirulina platensis	46–63	8–14	4–9	2–5	–
Synechoccus sp.	63	15	11	5	–
Tetraselmis maculata	52	15	3	–	–
Tetraselmis sueica	–	–	–	–	15–32

Adapted with permission from Elsevier

Biomass from microalgae has many applications. The usual methods by which microalgae biomass is cultivated, harvested, processed and converted into useful products [56]. Such products include biofuels (e.g. biodiesel, biomethane and bioethanol), nutritional supplements, fertilizers, and phytoremediation media.

1.6.2 Aquatic Plants

Aquatic plants use the sun's energy and carbon dioxide in the atmosphere to produce starch and sugar via photosynthesis. These plants are known for their high productivity levels, photosynthetic efficiencies, as well as the wide range of chemicals they produce [56]. Since they grow in wetlands, aquatic plants do not compete for land that could be used for growing crops or forests [57]. Of the many aquatic plants that exist, Cattails and Duckweed have shown strong potential as resources for various environmental and economic applications, which will be discussed below.

Cattails

Cattails, also known as Typha species, have been recognized as being suitable biomass crops for wetlands due to their superior productivity (40+ Mt/ha standing crops), pest resistance, adaptability, and chemical composition [58].

Typha latifolia is one of the most widely studied cattail species. It is native throughout the United States, Eurasia, and North Africa. It has been classified as a serious weed in Hungary, a principal weed in Australia, Germany, Italy, Rhodesia, Spain, Tunisia, and a common weed in Argentina, Iran, Kenya, Portugal, and the US [59].

The roots of *Typha latifolia* contain 30% starch, 7.8% crude protein, 1% crude sugar, 0.7% glucose, 0.7% oxalic acid. Aerial portions contain 1.5–3.5% fats, 7–12% crude protein, 38–48% carbohydrates. Based on this composition, as well as the fact that cattails contain roughly 47.6% cellulose and 21.9% lignin, they can therefore be a good feedstock for ethanol production [57, 60]. Furthermore, the pollen, which is used as a medicine and foodstuff, contains 19% crude protein, 17.8% carbohydrates (glucose, fructose, arabinose, rhamnose, xylose) and 1.1% lipids. *Typha latifolia* has also been shown to have potential for use in phytoremediation of constructed wetlands [61].

Duckweed

Duckweed or Lemnaceae is another aquatic plant that has a great potential for biofuel production [62–64]. It is a small, free-floating aquatic plant with fast reproduction, and high resistance to bacteria [65].

The starch content in Duckweed varies within a wide range of 3–75% dry weight, which depends to the species of the individual strains [58]. Recently, researchers at Amity University in India and University of Jena in Germany were able to demonstrate that the absorption of heavy metals and salt (NaCl) in water by the duckweed species Lemna minor lead to an increase in its starch content to approximately 50% of dry mass. This result demonstrates the dual advantage of using Duckweed as a low-cost water purifier and feedstock for bioethanol production [66].

1.7 Conclusion

In this chapter, different categories of biomass for the production of renewable energy, chemicals and biomaterials have been presented. These include woody biomass, agricultural waste, municipal waste, sew-age sludge, algae and aquatic plants. Although woody biomass has traditionally been the major source of energy for cooking and heating in many countries, the rise of waste management directives is promoting the use of waste biomass (i.e. agricultural waste, municipal waste and sewage sludge). Furthermore, algae and aquatic plants represent a new generation of biomass resource for renewable energy and chemicals production, and thus their use offers promising economic and environmental benefits.

References

1. Loppinet-Serani, A., Aymonier, C., Cansell, F.: Current and foreseeable applications of supercritical water for energy and the environment. Chemsuschem **1**(6), 486–503 (2008)
2. World Energy Council: "Biomass," Energy Resources. https://www.worldenergy.org/data/resources/resource/biomass/. Accessed 24 Mar 2018
3. Basu, P.: Biomass Gasification and Pyrolysis: Practical Design and Theory. Academic Press, Burlington, MA (2010)
4. Fernandes, S.D., Trautmann, N.M., Streets, D.G., Roden, C.A., Bond, T.C.: Global biofuel use, 1850–2000. Global Biogeochem. Cycles **21**(2) (2007)
5. Biomass (Energy Engineering), What-When-How: In Depth Tutorials and Information. http://what-when-how.com/energy-engineering/biomass-energy-engineering/. Accessed 24 Mar 2018
6. Gallezot, P.: Conversion of biomass to selected chemical products. Chem. Soc. Rev. **41**(4), 1538–1558 (2012)
7. Europäische Kommission (ed.): Being Wise with Waste: The EU's Approach to Waste Management. Luxembourg: Publ. Off. of the European Union (2010)
8. Knoef H.: Handbook Biomass Gasification. BTG Biomass Technology Group (2012)
9. IEO, International Energy Outlook. Washington DC; Report No.: DOE/EIA-0484 (2013)
10. Evans, L., Okamura, S., Poll, J., Barker, N.: Evaluation of Opportunities for Converting Indigenous UK wastes to Fuels and Energy. AEA/ED45551/Issue 1 (2009)
11. Waste Management World: The Biowaste Directive, Recycling, 01-Jan-2010. https://waste-management-world.com/a/the-biowaste-directive. Accessed 24 Mar 2018
12. Burnley, S., Phillips, R., Coleman, T., Rampling, T.: Energy implications of the thermal recovery of biodegradable municipal waste materials in the United Kingdom. Waste Manag. **31**(9–10), 1949–1959 (2011)
13. Alakangas, E.: European standards for fuel specification and classes of solid biofuels. In: Grammelis, P. (ed.) Solid Biofuels for Energy. Green Energy and Technology. Springer, London (2011)
14. Forest Stewardship Council (FSC): Strategic Review on the Future of Forest Plantations. Helsinki, Finland, A12-06869 (2012)
15. Carle, J., Holmgren, P.: Wood from planted forests: a global outlook 2005–2030. Food Prod. J. **58**(12), 6–18 (2008)
16. Phyllis 2: Database for Biomass and Waste. Energy Research Centre of the Netherlands. Available: https://phyllis.nl/ (2018)

17. Junginger, M., Van Dam, J., Zarrilli, S., Ali Mohamed, F., Marchal, D., Faaij, A.: Opportunities and barriers for international bioenergy trade. Energy Policy **39**(4), 2028–2042 (2011)
18. FAO Forestry Paper: What Wood Fuels can do to Mitigate Climate Change. FAO Forestry Paper 162, Rome (2010)
19. EN 14961-1: Solid biofuels—Fuel Specifications and Classes - Part 1: General Requirements (2010)
20. EN 15359,: Solid Recovered Fuels—Specifications and Classes (2011)
21. WRAP and BSI: PAS 111: Specification for the Requirements and Test Methods for Processing Waste Wood. Available: http://drafts.bsigroup.com/. Accessed Jan 2011
22. European Commission: Waste Generated and Treated in Europe. In: Luxembourg: Office for Official Publications of the European Communities, Available: https://ec.europa.eu/eurostat/documents/3217494/5646057/KS-55-03-471-FR.PDF (2003)
23. Williams, P.T.: Waste Treatment and Disposal, 2nd edn. Wiley Ltd., USA (2005)
24. Smil, V.: Crop residues: agriculture's largest harvest. Bioscience **49**(4), 299–308 (1999)
25. International Atomic Energy Agency (IAEA): Management of Crop Residues for Sustainable Crop Production, Vienna, IAEA Techdoc-1354 (2003)
26. Krupnik, T.J., Six, J., Ladha, J.K., Paine, M.J., van Kessel, C.: An assessment of fertilizer nitrogen recovery efficiency by grain crops across scales. In: Mosier, A.R., Syers, K.J., Freney J.R., (eds.) Agriculture and the Nitrogen Cycle, The Scientific Committee Problems of the Environment. Island Press, Covelo, California, USA (2004)
27. Agele, S.O., Adeyemo, A.J., Famuwagun, I.B.: Agricultural wastes and mineral fertilizer on soil and plant nutrient status, growth and yield of tomato. Arch. Agron. Soil Sci. **57**(1), 91–104 (2011)
28. Great Britain, Department of the Environment: Waste Management Planning: Principles and Practice: A Guide on Best Practice for Waste Regulators. H.M.S.O., London (1995)
29. Lam, J., ter Heegde, F.: Domestic Biogas Compact Course: Technology and Mass-Dissemination Experiences from Asia. University of Oldenburg (2011)
30. Kelleher, B.P., Leahy, J.J., Henihan, A.M., O'Dwyer, T.F., Sutton, D., Leahy, M.J.: Advances in poultry litter disposal technology—a review. Bioresour. Technol. **83**(1), 27–36 (2002)
31. European Commission, "Eurostat: Guidance on Municipal Waste Data Collection (2017)
32. European Environment Agency (EEA): Managing Municipal Solid Waste—A Review of Achievements in 32 European countries. European Environment Agency (2013)
33. Branchini, L.: Chapter 2: Municipal Waste Overview. In Waste-to-Energy. Springer International Publishing, Switzerland (2015)
34. US. EPA: Municipal Solid Waste. https://archive.epa.gov/epawaste/nonhaz/municipal/web/html/. Accessed 23 Nov 2017
35. Lisa, D., Anders, L.: Methods for household waste composition studies. Waste Manag. **28**, 1100–1112 (2008)
36. Hoornweg, D., Bhada-Tata, P.: What a waste—a global review of solid waste management. In: Urban development Series Knowledge Papers. The World Bank (2012)
37. OECD.: List of OECD Member countries—Ratification of the Convention on the OECD. Available http://www.oecd.org/about/membersandpartners/list-oecd-member-countries.htm. Accessed 23 Nov 2017
38. Cheng, H., Hu, Y.: Municipal solid waste (MSW) as a renewable source of energy: current and future practices in China. Bioresour. Technol. **101**(11), 3816–3824 (2010)
39. EPA, United States Environmental Protection Agency: Advancing Sustainable Materials Management: 2014 Fact Sheet. United States Environmental Protection Agency (2016)
40. Agence de l'environnement et de la maîtrise de l'énergie (ADEME): Déchet (2014)
41. Burnley, S.J.: A review of municipal solid waste composition in the United Kingdom. Waste Manag. **27**(10), 1274–1285 (2007)
42. Yuan, X., et al.: Effect of pretreatment by a microbial consortium on methane production of waste paper and cardboard. Bioresour. Technol. **118**, 281–288 (2012)

43. Bridgwater, E., Quested, T.: Synthesis of Food Waste Compositional Data 2012. WRAP. Available: http://www.wrap.org.uk/sites/files/wrap/hhfdw-synthesis-food-waste-composition-data.pdf (2013)
44. Esteves, S., Devlin, D.: Food Waste Chemical Analysis. WRAP. Available: http://www.wrapcymru.org.uk/sites/files/wrap/Technical_report_food_waste_characterisation_Wales_2009 x2.9086.pdf (2010)
45. Tchobanoglouss, G., Burton, F.L., Stensel, H.D.: Wastewater Engineering. Treatment and Reuse, 4th edn. McGrawHill, New York (2003)
46. Turovskiy, I.S., Mathai, P.K.: Wastewater Sludge Processing. Wiley, Hoboken (USA) (2006)
47. Foladori, P.: Sludge Reduction Technologies in Wastewater Treatment Plants. IWA Publishing, London (UK) (2010)
48. Yang, G., Zhang, G., Wang, H.: Current state of sludge production, management, treatment and disposal in China. Water Res. **78**, 60–73 (2015)
49. Kelessidis, A., Stasinakis, A.S.: Comparative study of the methods used for treatment and final disposal of sewage sludge in European countries. Waste Manag. **32**(6), 1186–1195 (2012)
50. EPA: Biosolids Generation, Use and Disposal in the United States. EPA530-R-99–009 (1999)
51. Seiple, T.E., Coleman, A.M., Skaggs, R.L.: Municipal wastewater sludge as a sustainable bioresource in the United States. J. Environ. Manag. **197**, 673–680 (2017)
52. Ericksson, J.: Concentrations of 61 Trace Elements in Sewage Sludge, Farmyard Manure, Mineral Fertiliser, Precipitation and in Oil and Crops. Swedish Environmental Protection Agency, Report 5159 (2001)
53. Kacprzak, M., et al.: Sewage sludge disposal strategies for sustainable development. Environ. Res. **156**, 39–46 (2017)
54. Rideau Valley Conservation Authority: Algae and Aquatic Plant: Educational Manual. Available: http://www.southfrontenac.net/en/town-hall/resources/Algae-Manual-Concise-Final.pdf (2016)
55. Demirbas, A., Fatih Demirbas, M.: Importance of algae oil as a source of biodiesel. Energy Convers. Manag. **52**(1), 163–170 (2011)
56. Raymond, L.P.: Aquatic Biomass as a Source of Fuels and Chemicals. SERI/TP-231-1699, 7146949 (1983)
57. Acosta, M.J.: Advances in Energy Research. Nova Science Publication, New York (2012)
58. Pratt, D.C., Dubbe, D.R., Garver, E.G., Johnson, W.D.: Cattail (*Typha* spp.) Biomass Production: Stand management and sustainable yields. Final report (1988)
59. Holm, L.G., Pancho, J.V., Herberger, J.P., Plucknett, D.L.: A Geographical Atlas of World Weeds. Wiley, New York (1979)
60. Küçük, M.M., Demir, H., Genel, Y.: Supercritical fluid extraction of reed (*Thypa*). Energy Sources **27**(5), 445–450 (2005)
61. Succuro, J.S.: The Effectiveness of using *Typha latifolia* (Broadleaf Cattail) for Photoremediation of Increased levels of Lead-Contamination in Soil. Humboldt State University (2010)
62. Cui, W., Cheng, J.J.: Growing duckweed for biofuel production: a review. Plant Biol. **17**, 16–23 (2015)
63. Xu, J., Zhao, H., Stomp, A.M., Cheng, J.J.: The production of duckweed as a source of biofuels. Biofuels **3**(5), 589–601 (2012)
64. Su, H., et al.: Use of duckweed (*Landoltia punctata*) as a fermentation substrate for the production of higher alcohols as biofuels. Energy Fuels **28**(5), 3206–3216 (2014)
65. Landolt, E.: Extreme adaptation in the angiospermous hydrophytes. In: Anatomy of the Lemnaceae (Duckweeds). Borntraeger, Berlin (1998)
66. Sree, K.S., Appenroth, K.J., Institute, A.: Increase of starch accumulation in the duckweed Lemna minor under abiotic stress. Lbanian J. Agric. Sci. 11–14 (2014)

Chapter 2
Generic and Advanced Characterization Techniques

Doan Pham Minh, Philippe Accart, Céline Boachon, Rachel Calvet, Anthony Chesnaud, Sylvie Del Confetto, Jean-Louis Dirion, Jun Dong, Augustina Ephraim, Laurène Haurie, Nathalie Lyczko, Rajesh Munirathinam, Ange Nzihou, Séverine Patry, Christine Rolland, Lina María Romero Millán, Louise Roques, Abdoul Razac Sane, Rababe Sani, Elsa Weiss-Hortala and Claire E. White

Abstract Nowadays, the valorization of biomass, biowastes and by-products is among the key issue to be considered in the development of renewable energies from bioresources. Accurate analysis and characterization of these feedstocks is a crucial aspect in the understanding of their behaviour for further use. This chapter is focused on different characterization techniques which are commonly used up-to-date. They are classified in different categories: Sampling and storage; Proximate analysis; Ultimate analysis; Thermal analysis, Physical characterizations; Physico-chemical characterizations; Structural and textural characterizations; and Mechanical characterizations. For each of them, a general description of the technique is presented, followed by useful information on machines and experimental conditions such as sample preparation, sample pre-treatment, gas atmosphere, temperature program etc. Finally, examples, results treatment and exploitations will be provided to illustrate. This chapter provides an insight on generic and advanced characterization techniques for complex materials, such as biomass, biowastes and related bio-products, that will be again discussed along the handbook in the other chapters.

D. Pham Minh (✉) · P. Accart · C. Boachon · R. Calvet · S. Del Confetto · J.-L. Dirion · J. Dong · A. Ephraim · L. Haurie · N. Lyczko · R. Munirathinam · A. Nzihou · S. Patry · C. Rolland · L. M. Romero Millán · L. Roques · A. R. Sane · R. Sani · E. Weiss-Hortala
RAPSODEE CNRS, UMR-5302, Université de Toulouse, IMT Mines Albi, Campus Jarlard, 81013 Albi Cedex, 09, France
e-mail: doan.phamminh@mines-albi.fr

A. Chesnaud
MINES ParisTech, PSL Research University, CNRS, UMR 7633, MAT—Centre Des Matériaux, BP 87, 91003 Evry, France

C. E. White
Department of Civil and Environmental Engineering and the Andlinger Center for Energy and the Environment, Princeton University, Princeton, NJ 08544, USA

A. Nzihou (ed.), *Handbook on Characterization of Biomass, Biowaste and Related By-products*, https://doi.org/10.1007/978-3-030-35020-8_2

2.1 General Introduction

Up-to-date, world energy consumption continuously increases according to the increase of world population and societal development. Unfortunately, fossil fuels (coal, natural gas and oil) are still the main energy resources and represent around 85% of world energy consumption in 2016 [1]. To limit CO_2 emission, renewable energies with low environment impacts have to be developed in order to progressively replace fossil fuels. Different sources of renewable energies are available including mostly: hydraulic, solar, biomass, wind, marine and geothermal energies. These resources allow producing electricity and/or heat (hydraulic, solar, wind, marine and geothermal energies), and biofuels (hydrocarbons and hydrogen). For instance, biomass and biowastes are the most appropriate resources for the production of biofuels using biological and thermal processes. They are also widely valorized into useful chemicals materials in different fields such as construction materials, insulant materials for building sector etc. Research and development on the valorization of biomass and biowastes have been extremely active during the last decades.

Different techniques are available for the characterization of biomass and biowastes but there are increasing needs for new standards, protocols, methods and equipment to adapt to the complexity of biomass and biowastes nature. It is crucial to gather most of the techniques related to the characterization of biomass and biowastes in a unique handbook to help the community working with these resources to easily find a given appropriate characterization method and the most relevant way to exploit the results generated.

This chapter is devoted to the generic and advanced characterization techniques and is divided into following sections:

- Sampling and storage: This is very important for complex matrixes such as biomass and biowastes. Attention must be paid on this step in order to get meaningful results when working with these resources.
- Proximate analysis: This allows knowing the primary information about a biomass or biowaste: moisture content, volatile matter, ash content, and fixed carbon.
- Ultimate analysis: This gives the information about the elemental composition of sample.
- Thermal analysis: Thermal behavior of sample can be obtained by thermal analysis, which is useful for their valorization by thermal ways.
- Physical and physico-chemical characterizations: These allow the analysis of different parameters: particle size, porosity, specific surface area, surface functional groups etc.
- Structural and textural characterizations: These provide information about the crystalline structure and surface morphology.
- Mechanical characterizations: This section develops methods for measuring mechanical properties such as Young modulus, compressive and flexural resistance.

This chapter privileges the description of standards and methods available in the literature. But new protocols and methods answering current needs in research and development on biomass and biowaste are also discussed. For each characterization, a general description will be presented, followed by useful information on machines and experimental conditions such as sample preparation, sample pre-treatment, gas atmosphere, temperature program etc. Finally, examples will be illustrated with possible results treatment and exploitations.

Furthermore, Chap. 15 of the handbook is devoted to the in-depth characterization of solid residues resulted from biomass and biowastes processing. Figure 2.1 illustrates the relationship between these Chaps. 2 and 15.

2.2 Sampling and Storage

Biomass, biowastes and related by-products constitute large varieties of complex and heterogeneous media. The complexity can be understood by the fact that for a given variety of biomass, biowastes or related by-products, the composition could be very different. The previous chapter discussed and evidenced the abundance of different families of biomass. These materials exhibit also complex and heterogeneous structure. Thus, before any characterization, it is important to carefully ensure the steps of sampling and storage of samples for characterization, which might be meaningful for the system considered.

Because of the complexity of biomass, biowastes and related by-products varieties, significant number of standards and procedure for sampling, storage and preparation exit in the literature. Some of them are listed in Table 2.1 for various kinds of wastes. Before any sampling, storage and preparation of such a feedstock, it is recommended to refer to related standards in order to adapt to the most appropriate practices.

2.2.1 Sampling

In this part, only the standard EN 14899 entitled "Characterization of waste—Sampling of waste materials—Framework for the preparation and application of a Sampling Plan" is considered [2]. This standard describes the main steps for the preparation and application of a sampling procedure of a given waste. As stated in this standard, the sampling procedure describes the method of collection for laboratory samples necessary to satisfy test program objectives. The principles or fundamental rules described supply a framework for use in the following areas:

- preparation of standardized sampling procedure for use in normal or routine circumstances (i.e. preparation of derived standards covering well-defined sampling scenarios);

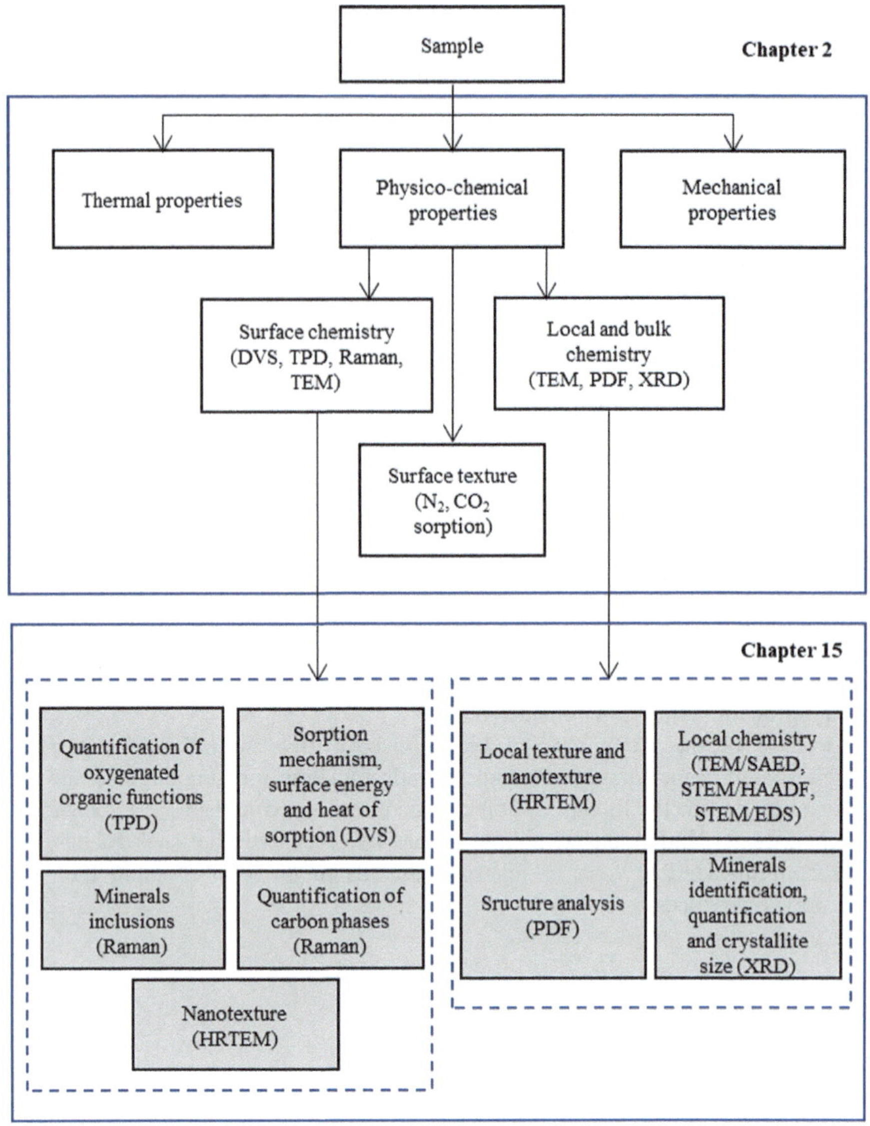

Fig. 2.1 Relationship between Chaps. 2 and 15

- incorporate specific sampling requirements in European and national legislation;
- design and prepare a "case by case" sampling procedure;

The standard EN 14899 includes two flow charts which summarize the essential procedure of the standard. The first flow chart defines 7 steps that make up the essential elements of the testing program, wherein is found the first step entitled: "Preparation of a sampling plan". This step is then detailed with another flow chart,

Table 2.1 Some standards and procedures related to sampling, storage and preparation of waste

Standard	Title	Ref.
EN 14899	Characterization of waste—sampling of waste materials—framework for the preparation and application of a sampling plan	[2]
FD CEN/TR 15310-1	Characterization of waste—sampling of waste materials—Part 1: Guidance on selection and application of criteria for sampling under various conditions	[3]
FD CEN/TR 15310-2	Characterization of waste—sampling of waste materials—Part 2: Guidance on sampling techniques	[4]
FD CEN/TR 15310-3	Characterization of waste—sampling of waste materials—Part 3: Guidance on procedures for sub-sampling in the field	[5]
FD CEN/TR 15310-4	Characterization of waste—sampling of waste materials—Part 4: Guidance on procedures for sample packaging, storage, preservation, transport and delivery	[6]
FD CEN/TR 15310-5	Characterization of waste—sampling of waste materials—Part 5: Guidance on the process of defining the sampling plan	[7]
ISO 13833	Stationary source emissions—determination of the ratio of biomass (biogenic) and fossil-derived carbon dioxide—radiocarbon sampling and determination	[8]
NF X 30-413	Déchets ménagers et assimilés—Constitution d'un échantillon de déchets ménagers et assimilés contenus dans une benne à ordures ménagères	[9]
NF X 30-437	Déchets ménagers et assimilés—Constitution et caractérisation, en entrée de centres de tri, d'un échantillon sur un lot de déchets ménagers et assimilés collectés sélectivement	[10]
EN 15442	Solid recovered fuels—methods for sampling	[11]
ASTM D7204-15	Standard practice for sampling waste streams on conveyors	[12]
EN 12255-8	Wastewater treatment plants—Part 8: Sludge treatment and storage	[13]
EN 15413	Solid recovered fuels—methods for the preparation of the test sample from the laboratory sample	[14]
ISO 6322-1	Storage of cereals and pulses—Part 1: General considerations for the keeping of cereals	[15]
ISO 20023	Solid biofuels—safety of solid biofuels pellets—safe handling and storage of wood pellets in residential and other small-scale applications	[16]
EN 15002	Characterization of waste—preparation of test portions from the laboratory sample	[17]
EN 15443	Solid recovered fuels—methods for the preparation of the laboratory sample	[18]
EN 16179	Sludge, treated biowaste and soil—guidance for sample pretreatment	[19]
ASTM D5231-92	Standard test method for determination of the composition of unprocessed municipal solid waste	[20]

which provides the key elements to produce a procedure [2]. Thus, before taking a sample, it is mandatory to collect all the information about the system considered. More details are available in the text of this standard [2].

2.2.2 Storage and Preparation

After mentioned sampling according to the procedure previously established, it is usually necessary to process other steps such as drying, particle size reduction, homogenization, sub-sampling and storage etc. for further use or characterization. Depending on the system considered, it is recommended to adapt to appropriate standard or procedure.

As an example, thereafter is shown the main steps from the standard EN 15413 "Solid recovered fuels—Methods for the preparation of the test sample from the laboratory sample" [18]. This standard concerns solid recovered fuels (SRF), which are also heterogeneous medium. It provides the sequence of operations intended to ensure the representativity of the test portions that have been taken according to the sampling procedure, prior to characterization of solid samples [18]. The main operations determined in this standard are as follows:

- Fraction separation: This is applicable for heterogeneous samples. The operation consists of the separation of different fractions into sub-samples which are directly weighed or indirectly measured. This allows a final weighed combination of different fractions' analysis results. This operation can be done manually or by sieving.
- Drying: This operation is necessary to remove water that can interfere with test portion preparation (e.g. during crushing or milling). There are different possibilities of drying: air drying at room temperature, oven drying at 40 °C or oven drying at 105 °C.
- Particle size reduction: This operation is needed in order to get the particle size required for the next step of characterization. It can be done by crushing/grinding, freeze crushing, milling, cutting and freeze cutting. Different cautions need to be considered because of the potential release of some elements such as mercury or contamination from equipment used.
- Homogenization: This is required before each operation that implies sub-sampling. This guarantees the homogeneity in properties and compositions for both the samples and sub-samples. This can be done manually or mechanically with adequate equipment.
- Sub-sampling: This operation is needed in order to divide a sample into different sub-samples for different characterizations. This operation can be done manually or mechanically.

In the case of SRF, after these operations, the storage can be done under the same conditions of its production for short term storage conditions before delivery to the lab, or at 4 °C for a long term storage condition before delivery to the lab.

2.2.3 *Example*

Below is described an example of sampling, pretreatment and conservation of a solid recovered fuel (SRF) from an industrial production line. This SRF production site transforms non-hazardous solid wastes principally containing woods, cardboards, papers, and plastics. Figure 2.2 shows raw solid wastes and the final SRF produced. The main steps of this SRF production are as follows:

- Delivery and reception of raw solid wastes.
- Mechanical sorting to separate different fractions for recycling purpose: metals, cardboards, wood, plastics.
- For the rejects from this mechanical sorting: coarse size reduction by grinding; sieving; then second sorting by density to recover again metals, cardboards, woods, plastics for recycling.
- For the rejects from this second sorting: size reduction to dozen mm by grinding to obtain the final SRF (Fig. 2.3).

2.3 Precision and Accuracy

Biomass, biowastes and related by-products are generally complex and heterogeneous mediums. To characterize such a medium, standardized sampling procedure must be respected as mentioned above to obtain representative samples. Then, the analysis of the samples must be performed with high precision and accuracy.

For a given measurement, it is recommended to repeat the measurement at least 3 times to evaluate the precision and the accuracy of the measurement. The precision indicates the level of variability of these repeated measurements. In other

Fig. 2.2 Non-hazardous solid waste before transformation (left) and SRF after sorting and grinding to chips below 28 mm (right)

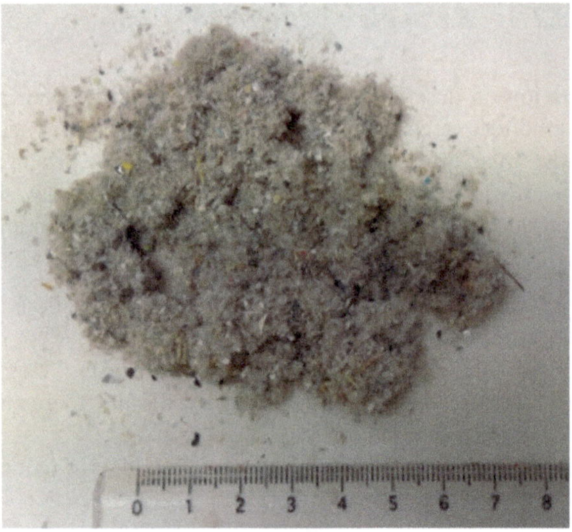

Fig. 2.3 Photo of the SRF after size reduction to particles smaller than 1 mm

words, the values obtained from measurements are close to each other under the same conditions. From scientific point of view, the precision of a result is expressed by its written format [21]. As example, the numbers 1.0 and 1.000 are mathematically equivalent. However, the precision is not the same. The number 1.0 is bounded by 0.95 and 1.05 while 1.000 is bounded by 0.9995 and 1.0005. Thus, writing 1.0000 means higher precision than writing 1.0.

However, a high precision is not enough to qualify a "good measurement". The precision reflects only the variability of measurements but does not show how the results of measurement are situated versus the goal. The accuracy describes the relationship between the mean value of several repeated measurements with the goal. High accuracy indicates that the mean value of the measurements is close to the goal [21].

It is recommended to reach both high precision and high accuracy regardless the type of measurement.

2.4 Proximate Analysis

2.4.1 Moisture Content Analysis

2.4.1.1 Introduction

Moisture content corresponds to the humidity (amount of water) in the sample. This parameter is expressed as the mass (or wt%) of water in the total mass of the sample (wet basis).

In a solid feedstock, the moisture content should first be quantified to have the mass distribution. In addition, the water content will help to define the suitable treatments that could be applied: grinding, dry or wet thermal treatment, biological treatment [22].

Moisture content could be determined using an oven method (standard ISO 18134) [23] and ASTM E790-15 [24] for solid biofuels, or ASTM D3173M-17a [25] for coke and coal or ASTM D4442-16 [26] for wood-based materials), a microwave oven method for particulate wood (ASTM E1358-97 [27]), using the distillation method [26] which is appropriate for chemically treated or impregnated material, or using an electric moisture meter. In this chapter, only oven drying method and chemical determination are described.

2.4.1.2 Standard Method: "Loss of Drying"

For a solid sample, the moisture content is basically evaluated using reference methods [23, 26] based on the mass loss during drying. This section is related to the "standard ISO 18134, Solid biofuels—Determination of moisture content—Oven dry method" [23]. In summary, the principle is to measure the mass variation due to the drying in an oven at 105 °C. This standard is divided into 3 parts and is related to solid biofuels (definitions in ISO 16559). The first part is related to "total moisture—reference method", the second one describes "total moisture—simplified method" while the third part addresses "moisture in general analysis sample".

Since the steps of the procedure are quite similar (Fig. 2.4), Table 2.2 presents the differences between the 3 standards and Table 2.3 describes the parameters for the calculations of the moisture content.

Part 1 is used to determine, with high precision, the total moisture content of solid biofuels by drying the sample in an oven. The main difference between

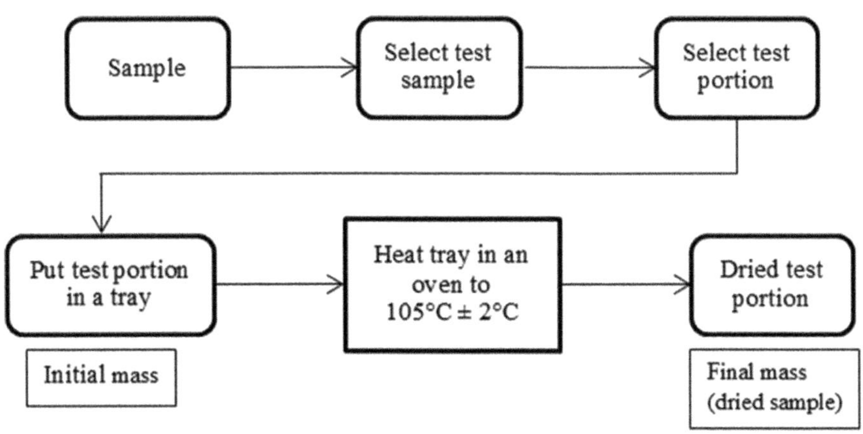

Fig. 2.4 Procedure used to determine moisture content

Table 2.2 Main steps of the procedure to determine moisture content using the 3 standards

Procedure	ISO 18134—part 1	ISO 18134—part 2	ISO 18134—part 3
Select a test sample[a] ISO 18135	Feedstock received in sealed air-tight container	Feedstock received in sealed air-tight container	Feedstock received in sealed air-tight container
Select a test portion[a] ISO 14780	Nominal top size < 31.5 mm	Nominal top size < 31.5 mm	Nominal top size < 1 mm
Mass of the test portion	Min 300 g ± 0.1 g[b]	Min 300 g ± 0.1 g[b]	Min 1 g ± 0.1 mg
Crucibles and trays	No lid Non-corrodible Heat resistant Clean surface to avoid adsorption/absorption	No lid Non-corrodible Heat resistant Clean surface to avoid adsorption/absorption	With lid Non-corrodible Heat resistant Clean surface to avoid adsorption/absorption
Layer of test portion (on crucibles and trays)	<1 g cm^{-2}	<1 g cm^{-2}	<0.2 g cm^{-2}
Temperature of the oven	105 ± 2 °C	105 ± 2 °C	105 ± 2 °C
Atmosphere	Air renewed 3–5 times per hour Flux adjusted to avoid particle entrainment	Air renewed 3–5 times per hour Flux adjusted to avoid particle entrainment	Air renewed 3–5 times per hour Flux adjusted to avoid particle entrainment Or nitrogen if material subject to oxidation (see ISO 11722) [28]
Drying time	Until constant mass (Δm < 0.2% of the initial mass within 1 h)—1 h at least and <24 h	Until constant mass (Δm < 0.2% of the initial mass within 1 h)—1 h at least and <24 h	Until constant mass (Δm < 1 mg within 1 h)—1–3 h[c]
Step after drying	Measure immediately the mass[d]	Measure immediately the mass[d]	Cool down to room temperature in a desiccator with desiccant[e]

[a]When sample or test sample is in a container where moisture might remain on the surface, shake the package before sampling test portion in order to allow the material to reabsorb the moisture

[b]For fine particulate solids, mass could be reduced to 200 ± 0.01 g or 100 ± 0.01 g

[c]Put uncovered tray filled with test portion and lid in the oven

[d]Remove the tray from the oven and measure the mass within 10–15 s to avoid moisture reabsorption

[e]Put the lid immediately on the tray once they have been removed from the oven. Solid samples are highly hygroscopic and the moisture sorption on this weak initial mass will strongly interfere with the moisture determination

Table 2.3 Parameters to calculate moisture content (wt% of moisture on wet basis) for the 3 standards

Procedure	ISO 18134 – part 1	ISO 18134 – part 2	ISO 18134 – part 3
Mass of empty container (m_1) at room temperature	Empty tray Mass ± 0.1 g	Empty tray Mass ± 0.1 g	Empty tray + lid Mass ± 0.1 mg
Mass of reference tray measured at room temperature (m_4)	Empty reference tray Mass ± 0.1 g	/	/
Mass of container + test portion before drying at room temperature (m_2)	Tray + test portion Mass ± 0.1 g	Tray + test portion Mass ± 0.1 g	Tray + lid + test portion Mass ± 0.1 mg
Mass of container after drying (m_3)	Tray + dried test portion, still hot[a] Mass ± 0.1 g	Tray + dried test portion, still hot[a] Mass ± 0.1 g	Tray + lid + dried test portion, room temperature Mass ± 0.1 mg
Mass of reference tray still hot (m_5)	Empty reference tray (1) Mass ± 0.1 g	/	/
Moisture content calculation (wt.% on wet basis)	$MC = \frac{(m_2 - m_3) - (m_4 - m_5)}{(m_2 - m_1)} \times 100$	$MC = \frac{(m_2 - m_3)}{(m_2 - m_1)} \times 100$	$MC = \frac{(m_2 - m_3)}{(m_2 - m_1)} \times 100$
Repeatability	Duplicate /	/	Duplicate Max 0.2 % absolute
Reproducibility	/	/	/
Reporting moisture content[b]	Mean value to the nearest 0.1%	Value to the nearest 0.1%	Mean value to the nearest 0.1%

a When measuring the mass of hot trays, use heat-insulating material on the balance pan to avoid direct contact.
b The result should be calculated to two decimal places and rounded to the nearest 0.1% for reporting

procedures from Part 1 and 2 is related to buoyancy effect. Indeed, the weight of the hot tray is lower than that of the cold one, and this mass difference would be weak compared to the test sample mass (300 g). Therefore, the procedure described in Part 2 would be suitable for routine control when high precision is not requested.

It should also be noticed that the size of the tray should adequately be selected to minimize buoyancy effect, meaning that the test portion layer should entirely cover the tray.

The third procedure (described in Part 3) is more suitable for experiments conducted at laboratory scale with small amounts of sample. Since the mass of the test portion requested to carry out Part 3 is low (1 g compared to 300 g in the 2 other standards) the effect of moisture reabsorption could not be neglected in the mass measurement. Therefore mass measurements in Part 3 should be done at room temperature (to avoid buoyancy effect) after a cooling step, and using a lid to prevent reabsorption of atmospheric moisture.

It could be noticed that dynamic equipment such as thermogravimetric analysis (TGA) can be used and refers to the standard ISO 18134–part 3, when the method is validated with reference test samples of biomass [23].

Note 1: volatiles

These standards have been developed for solid biofuels that contain small amount of volatiles. A special attention should be paid to biomass and bio-wastes that contain high amount of volatiles which are able to be volatilized at the drying temperature [29]. In that case, the water content will be overestimated.

Note 2: drying oven

To insure the drying of the test portion, the oven should not be overloaded, and the space between the trays should allow free flow of atmosphere and moisture.

Note 3: Scale

Scales (or balances) allow measurements at the nearest 0.2 g–0.01 mg. For the three standards described in the part, the precision of the scale requested is 0.1 g, 0.01 g or 0.1 mg. However a high precision scale presents a lower maximal weight. For the first 2 parts of this standard, the mass of the test portion is at least 300 g and the mass of the tray should be added. Analytical scale should meet the requirements, and thus selected according to its weight range and its precision.

To get accurate values from the scale, some precautions should be followed:

- Calibrate the scale when it is being installed for the first time, and then regularly or after any changes of locations.
- Center the bubble on the spirit level of the scale. The scale will have adjustable feet which you can turn individually to raise or lower a side. The device has to be even to be accurate.
- Close the door to avoid measurement perturbations due to airstream.
- Avoid pressing the surface of the table when recording the measure since vibrations will interfere with.
- Clean up the scale after use.

Note 4: repeatability and reproducibility

Since the heterogeneity of biomass and biowastes could be important, it is not possible to give a precision statement for these test methods. However moisture content determined with methods from Part 1 and 3 should be conducted in duplicate. It is also important to notice that both values obtained from Part 3 should not differ of more than 0.2% absolute to provide the mean value. In case the difference is higher, more test portions should be analyzed to meet the standard requirements.

Note 5: moisture content of pre-dried samples

In case the moisture content is too high for the cutting, the feedstock should be pre-dried. The total moisture content should be the sum of moisture lost during the pre-drying and the residual moisture.

Recommendation for test reports

The test report should include at least the following information:

- Identification of the laboratory performing the test and the date;
- Identification of the sample;
- Reference to the standard;
- Result of the test on wet basis;
- Any observation that might affect the result should be noted;
- Deviations from the established procedures should be described.

Alternative automatic method: Moisture analyzer

The moisture content determination could be also done quickly at lab scale using a moisture analyzer. This small equipment works according to the thermo-gravimetric principle, and consists of two components, a scale unit and a heating unit. The sample is loaded on the tray which is put on the scale unit. Once the analyzer is closed, a halogen radiator heats and dries the sample while the integrated scale continuously records the sample weight. When the sample no longer loses weight the instrument shuts off, and the moisture content is calculated based on the total mass loss. The sample absorbs the infrared radiation, and as a result heats up very quickly. Furthermore, the sample's temperature depends on its absorption characteristics and thus must not be the displayed temperature. This is different to the drying oven in which the sample is heated by convection and drying requires a long period of time. Halogen heating element consists of a glass pipe filled with halogen gas. As the mass of the halogen radiator is very low compared with that of a conventional infrared radiator, the maximum heating output can be quickly reached and outstanding controllability and even heat distribution is achieved. The two main parameters to be determined before starting a measurement are the drying temperature and the endpoint shutoff parameter (time or drying rate).

This kind of equipment is available in the market.

2.4.1.3 Chemical Determination

As explained in Note 1 from the standard method section, oven drying method might not be appropriate for sample containing numerous volatile compounds whose volatilization occurs at about 100 °C. Although the distillation method could be quite long and requires specific equipment to be accurate, there is a most adequate method to distinguish water from volatile compounds. At lab scale, one analytical solution arises: Karl Fischer titration with water extraction [30, 31].

Karl-Fischer titration principle:

Karl Fischer titration is a well-known technique used to measure water content, based on the back-titration of the acid formed during the Bunsen reaction (Eq. 2.1). However, it has been demonstrated that this reaction could be incomplete and the water content would be underestimated using pyridine as the base and methanol as the solvent.

$$SO_2 + I_2 + 2H_2O \rightarrow H_2SO_4 + 2HI \tag{2.1}$$

Recently, the reactive media has been improved since the solvent (methanol) participates in the reaction (increasing the sensitivity of the technique), pyridine (base) is not involved in the reaction and could be replaced by other bases, and the kinetics is improved in the pH range of 5.5–8. Finally, the reaction to be considered is the following:

$$ROH + SO_2 + I_2 + H_2O + 3R'N \rightarrow [R'NH]^+SO_4R^- + 2[R'NH]^+I^- \tag{2.2}$$

where ROH is the alcohol and R′N is the base. It could be noticed that imidazole is the most suitable base that offers the best pH range, and improves the titration. In addition methanol can easily be replaced by other alcohols: ethanol, methoxyethanol or diethylene glycol monoethylether, especially the latter improves the titer stability.

Karl-Fischer titration, end-point determination:

The water content is measured with an automatic Karl-Fischer titration. The system determines the iodine consumption until iodine excess detected with an electrode using bipotentiometric indication. A small, constant AC current—the polarization current "Ipol"—is applied to a double pin platinum electrode (sensor). As long as the iodine generated reacts with the water, there is no free iodine in the solution. A voltage of about 400–650 mV is necessary to maintain the polarization current between electrodes. At the end of water reaction, free iodine appears in the solution. In the presence of the electrode, iodine is electrochemically reduced following Eq. (2.3). The conductivity of the solution is thus greatly improved due to the presence of ions. The voltage between electrodes abruptly decreases to about 50–100 mV. This drop in voltage is used to indicate the end of the titration.

$$I_2 + 2e^- \rightarrow 2I^- \tag{2.3}$$

The Karl-Fischer titration is terminated when the potential drops below a specific predefined value. The end point value mainly depends on the polarization current (Ipol), but to a lesser extent the type of electrode (geometry, dimensions), the solvent and the stirring speed. As it will be explained in the next section, Karl-Fischer titration could be performed in two kinds of equipment, and the polarization current should be selected accordingly.

Karl-Fischer automatic equipment:

To provide accurate measurements, the Karl-Fischer reagent must be added precisely, accurately, and the titration cell must be as impervious as possible. The water adhering to the walls of the titration cell (inner water film) must be removed by swirling the titration vessel. The end point of titration is determined with the volumetric or the coulometric method (Table 2.4).

Note 1: generator electrode for coulometric method:

In coulometry, the iodine is electrochemically generated by anodic oxidation of iodine salts, following the half-electronic equation (Eq. 2.4). Basically, the coulometric cell is composed of both anode and cathode compartments.

$$2I^- \rightarrow I_2 + 2e^- \tag{2.4}$$

The amount of water titrated by coulometric technique is determined by the amount of charges (electrons) in the generator circuit. Basically Faraday's law (Eq. 2.5) expresses the relation between the number of charges in the circuit as function of the current intensity and duration time. At the same time, electrochemical reactions give the number of electrons exchanged (z) between the two-half electronic equations (see Fig. 2.5). Based on the Faraday constant, the mole number of electrons can be calculated (Eq. 2.6).

$$Q = It \tag{2.5}$$

$$Q = znF \tag{2.6}$$

where Q is in Coulombs (C), I in Amperes (A) and t in seconds (s), z the number of electrons, n in mol and F = 96485 C/mol.

In Karl-Fischer reaction two I^- ions are oxidized to one molecule of I_2, which then reacts with one molecule of water (Eq. 2.7). For the coulometric method, it is assumed that the reaction conversion is 100%.

$$n_{H_2O} = n_{I_2} = zn_{I^-} = \frac{Q}{F} \tag{2.7}$$

Table 2.4 Coulometric and volumetric Karl-Fischer titration (for water determination)

Method	Coulometric	Volumetric
Principle	I_2 from electrochemical generation	I_2 added by a burette
Range of water content	From 1 ppm to 5%	From 100 ppm to 100%
Sample amount	Low amount	Higher amount (to be adapted)
Sample type	Solid with oven Liquid Gas	Solid Liquid
Sample introduction and preparation	Direct introduction of solid sample (internal extraction) Direct introduction of liquid sample with syringe (external extraction of water from the solid) Direct introduction of gas from the oven (solid heated in the oven)	Direct introduction of solid sample (internal extraction) Direct introduction of liquid sample with syringe (external extraction of water from the solid)
Scheme	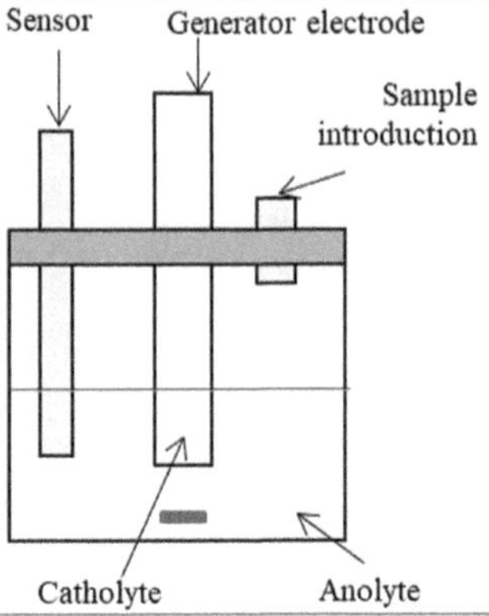	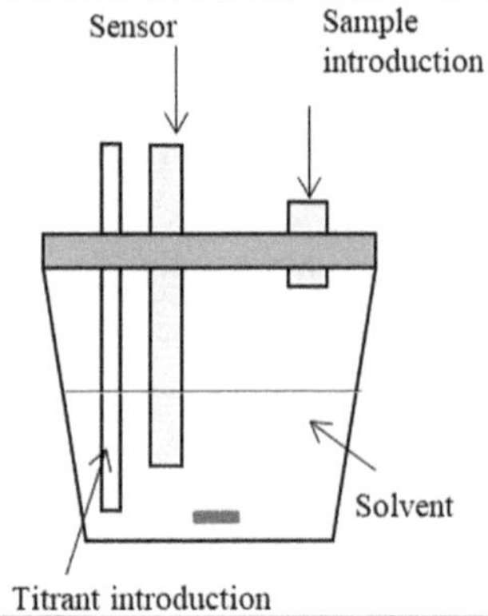
Anolyte/catholyte or titrant solution/ solvent	Anolyte: SO_2, imidazole and iodine salts (I^-) in methanol or ethanol (other solvents might be added such as chloroform, octanol, ethylene glycol …) Catholyte: the same as anolyte or specific reagent	One-component[a]: I_2, SO_2 and imidazole, in a suitable alcohol/methanol. *Titer less stable* Two-components[b]: I_2 in methanol/SO_2 and imidazole in methanol. *Titer stable.* Special reagent[c]: I_2, SO_2 and imidazole, in 2-methoxyethanol/2-chloroethanol and trichloromethane
Polarization current of the sensor	5 μA	24 μA
Calculation for end point titration	Charge requested for I_2 production (Faraday's law) $Q = It = znF$ Then $n_{H_2O} = \frac{Q}{zF}$	$[H_2O]_{sample} = \frac{1}{\alpha}\frac{[I_2]V_{eq}}{V_{sample}}$ Stoichiometry (α) Alcoholic solvents: $\alpha = 1$ Non-alcoholic solvents: $\alpha = 2$
Confidence	±0.5%	±0.5% (calibration requested)
Standard deviation	≈1%	≈1%

[a]The reagent can be stored for approximately two years

[b]A titration speed two or three times as high can be achieved with the two-component reagent. Both components are very stable in storage

[c]Aldehydes (R-CHO) and ketones (R-CO-R′) form acetals and ketals if titrated with standard methanol-containing reagents. As a result, additional water is produced and titrated at the same time, leading to higher water contents and a vanishing end point. Titration time is longer

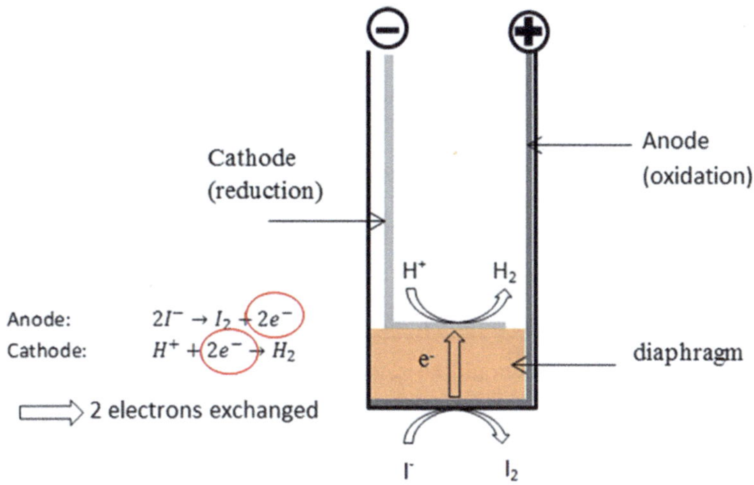

Fig. 2.5 Scheme of the generator electrode used in Karl-Fischer coulometric titration

Since current and time can be accurately measured, no standardization of the coulometric reagents is requested. Coulometry is therefore a reference method for the determination of water content. Iodine is generated by means of current pulses of 400, 300, 200 and 100 mA, for different pulse duration and frequency.

Originally, compartments were separated with a diaphragm (see Fig. 2.5) but some equipment are diaphragmless. The aim of the diaphragm is to avoid direct reduction of the I^- ions produced. Without diaphragm, the size of the cathode is smaller, and the electrode can be cleaned up easily and no residual contamination would appear (Table 2.5).

Note 2: stirring speed and dispersion of the volumetric Karl-Fischer titrant:

For the volumetric titration, iodine is gradually introduced by the burette. Thus a good mixing is essential for fast and accurate Karl-Fischer titrations. Mixing is influenced by the stirring speed, the position of the titrant addition and the shape of the vessel.

Experimentally, the optimum stirring speed is obtained when a distinct vortex is visible. A particular care is needed to avoid bubbles formation, since they have an impact on the bipotentiometric indication (electric conduction/resistance of the media). To improve the mixing of the titrant solution into the solvent, the location of the burette is a key issue. The titrant should be added where the turbulence is the greatest to ensure efficient mixing and far from the electrode to avoid direct measurement of iodine excess and let time to the reaction to be achieved. The conical shape of the beaker and the short stirring bar lead to turbulence at the point of titrant addition that effectively disperses the titrant. The turbulence is usually formed at the center of the beaker [30].

Table 2.5 Suitable uses of generator electrode with or without diaphragm (coulometric titration)

Generator electrode	With diaphragm	Without diaphragm
Water content determination for	Low water content Very accurate determination Nitro compounds Unsaturated hydrocarbons (easily reduced)	Hydrocarbons/chlorinated hydrocarbons Alcohols/phenols Ester/ethers Ketones/acetamides Petroleum oils and essence

Note 3: termination parameters for both methods:

Five termination parameters could be selected:

- Delay time: the titration is terminated when potential is below the endpoint value for 5–15 s. This is the most used method however the minimum rate must be adapted to the initial drift and to the titrant concentration.
- Absolute drift stop: the termination stop occurs when the drift became lower than the absolute drift value. Values of absolute drift should be optimized, and typical values are 3 and 10 μg/min for coulometric and volumetric methods respectively. The absolute drift should be higher than the initial drift (before titration) to reach the termination criterion.
- Relative drift stop: the titration is terminated as soon as the actual drift drops below the sum of the initial drift and the relative drift. Typical values are 2–5 μg/min and 5–15 μg/min for coulometric and volumetric methods respectively. This termination parameter is particularly interesting since it is independent on the initial drift, on the titrant concentration and on side reactions.
- Maximum titration time: titration stops after a certain elapsed time. This parameter is particularly suitable for titrations with the drying oven.
- Minimum titration time: titration could not stop before this elapsed time. This is particularly useful for samples releasing slowly water and it ensures a good repeatability.

Water determination in a solid using direct Karl-Fischer titration:

When the solid is soluble, water is easily released into the measurement cell, allowing a determination of the water content. This method is suitable for feedstocks such as cellulose, some kitchen wastes, sugar products … However, when the titration cell is opened to add the sample, about 50–100 μg water enter due to ambient humidity. Therefore only volumetric method should be used for direct titration of water in a solid sample. The sample should be quickly weighed (using back-weighing technique) and added to minimize air exposure.

The back-weighing technique is developed in 4 steps, to determine the exact mass added to the titration vessel:

- Tare the scale and weigh the boat (m_b);
- Weigh the sample (m_{app});
- Add the sample to the titration vessel;

- Weigh the "empty" boat (m_2, this mass might be higher than m_b);

The mass of the sample is then calculated with Eq. (2.8) below:

$$m_{sample} = m_{app} - m_2 \quad (2.8)$$

Water determination after extraction using Karl-Fischer titration:

As explained previously, direct determination of water in a solid requires the volumetric method. However for low concentrations requiring coulometric method, water should be extracted prior to the determination. It should be noted that only free water could be measured using Karl-Fischer titration. Three different extraction methods are described.

- Internal extraction:

Internal extraction is suitable for insoluble solids that release water quickly when crushed. The methodology is the following:

- Add the crushed samples into the titration vessel, using either methanol or methanol/ formamide mixture as the solvent;
- Use sufficient mixing time to extract water;
- For too long mixing time, use high-speed homogenizer.

- External extraction:

External extraction is suitable for insoluble solids that release water slowly when crushed as well as for samples with an extremely inhomogeneous water distribution. External extraction requires a solvent (containing low water content) such as methanol for insoluble organic solids, decanol/octanol for fatty and dairy products, formamide for natural substances, total water in sugar and starch, and chloroform for surface water in sugar.

The extraction should follow different steps:

- Determine water content of the solvent
- Crush the sample
- Mix the sample with the solvent
- Extract the water by shaking the mixture (manually, ultrasounds …)
- Take an aliquot and add to the Karl-Fischer vessel

The final moisture content includes the water content of the solvent that should be subtracted.

- Drying oven:

This method is suitable for solids and liquids, and is used when the water is released very slowly or when side reactions with titrant solution can occur. This system is composed of a Karl-Fischer titration cell and an oven. The oven is used to perform the drying step (volatilization of organic compounds and water), then the gas is

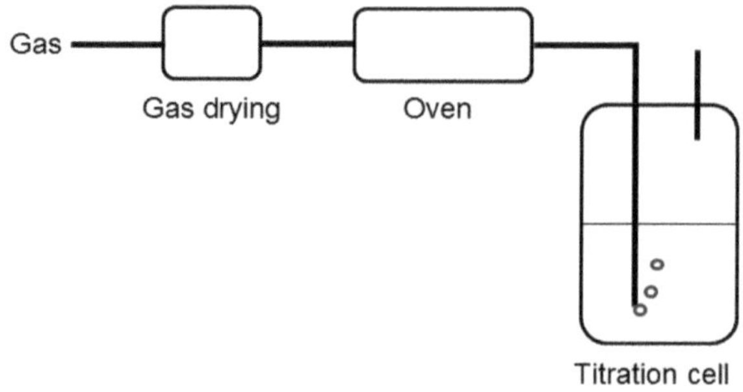

Fig. 2.6 Scheme of Karl-Fischer titration with oven

directly transferred to the Karl-Fischer vessel during a certain time (see Fig. 2.6). The inert gas should be as dry as possible to overcome bias in the water determination. The desiccants used are mainly molecular sieves and silica gel, and should be regenerated.

There are two procedures to use the drying oven for water titration:

- Continuous titration of water evolved: to perform this experiment, a short mixing time is required (20–60 s) as well as a maximum duration time. Both parameters are requested to avoid premature termination of the titration.
- Water is titrated after complete vaporization: water is vaporized and accumulated into the cell before starting the generator electrode. A termination parameter based on maximum time should be used.

As shown in Fig. 2.6, the steam bubbles into the vessel that causes a partial vaporization of the solvent or the anolyte. To minimize the anolyte vaporization, anolyte containing ethylene glycol is more suitable than that containing methanol, since the vaporization is less than 1 mL/h (at inert gas flow of 150–200 mL/min and 200 °C).

Karl-Fischer equipment (including oven) can be supplied by Mettler Toledo, Metrohm, Hach Lange or AnaLab.

Examples on the moisture content of different biomass and biowastes are presented in Sect. 2.4.5.

2.4.2 *Volatile Matter Analysis*

2.4.2.1 Introduction

Volatile matter content corresponds to the mass loss, less than that due to the moisture, when a solid is heated under standardized conditions mostly out of air contact (pyrolysis).

The volatile matter corresponds mostly to organic compounds that are released during the heating. This includes gases (H_2, CO, CO_2, CH_4) chemically bounded water, hydrocarbons, oxygenated organic compounds, tars, ammonia, sulfur organic compounds, some inorganic bounded to organic compounds … depending of the resource analyzed. This is different from volatile organic compounds (VOC) that are released at room temperature. The volatile matter determination is of great importance to define the quality of a biofuel, or the suitability of a thermal or biological treatment.

2.4.2.2 Measurement Method

The volatile matter content is determined using a partial pyrolysis method using restricted conditions in an oven or in a thermogravimetric analyzer. This is evaluated using reference methods: standard ISO 18123 [32] for solid biofuels, ASTM D1762-84 [33] for wood charcoal, or ASTM D3175-17 [34] for coal and coke, the principle is to measure the mass variation after ignition at a certain temperature for a short time. Each standard provides optimized parameters such as temperature (minimum 900 °C), and reaction time. The principle of the procedure is summarized in Fig. 2.7.

For the determination of the volatile matter, the key points are the oven, the thermocouples, the crucibles and the crucible stand. As defined previously, the determination requires pyrolysis conditions and should avoid air contact.

Recommendations for the oven:

To insure the pyrolysis of the test portion, the electrical oven should provide a uniform temperature (example: 900 ± 10 °C) and should have an appropriate heating capacity to match the desired temperature (example: 900 ± 10 °C) within 4 min when introducing the cold stand and crucibles. It is important for ovens to have flues so that the oven doors seal well. The flue should not be fitted with a butterfly valve to restrict air flow through the oven. The oven and the crucible stand (number of crucibles) should be selected accordingly. The temperature inside the oven is measured with a thermocouple. A sheathed thermocouple should permanently be installed inside the oven with its thermojunction as close as possible to the center of the heating chamber. A second thermocouple is used for calibration. The oven temperature readings should regularly be checked with the reference thermocouple. This reference thermocouple should be long enough to reach the zone of uniform temperature. An example of suitable oven used to determine volatile matter content can be found elsewhere [32].

Recommendations for the crucible:

The crucible should be cylindrical with a well-fitted lid. Both of them should be made of silica and the total mass should be between 10 and 14 g. For this target, the dimensions should be as follows: 21 mm of internal diameter and 38 mm of height. The fit of the lid on the crucible is critical, and the horizontal clearance does not exceed 0.5 mm. After selection, the crucible and the lid should be ground together to give smooth surfaces and a distinguish mark should be given. The perfect fit

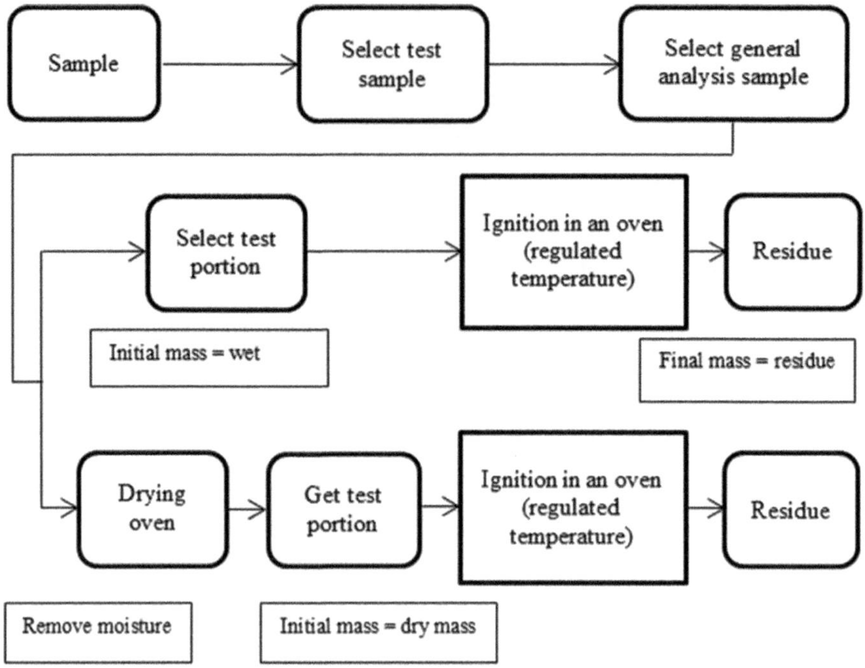

Fig. 2.7 Procedure used to determine volatile matter content

between crucible and lid provides a batch system where the solid is only in contact with air sealed, giving a partial pyrolysis. The fraction of volatilized matter is evacuated with the small overpressure induced. Details on the shape and dimension of a suitable silica crucible and lid used to determine volatile matter content can be found elsewhere [32].

Recommendations for the crucible stand:

The crucible stand is used to hold the crucible in the oven and should fit the requested heating time after introduction of cold crucible and stand (less than 4 min). The stand, a heat resistant steel wire, could welcome one or more crucibles and is designed accordingly. For example, it may consist of:

- For single determinations, a ring with ceramic discs 25 mm diameter and 2 mm thick, resting on the inner projection of its legs;
- For multiple determinations, a tray with ceramic plates 2 mm thick supporting the crucibles.

Details on the shape and dimension of a suitable crucible stand used to determine volatile matter content can be found elsewhere [32].

The steps from the "standard ISO 18123, Solid biofuels—Determination of the content of volatile matter" [32] is described in Table 2.6.

Table 2.6 Main steps of the procedure to determine the content of volatile matter [32]

Procedure	ISO 18123—wet sample	ISO 18123—dry sample
Select a test sample[a] ISO 18135	Feedstock received in sealed air-tight container	
Select a general analysis sample[a] ISO 14780	Nominal top size < 1 mm[b]	
Mass of the general analysis sample	1 g ± 0.1 g[c], record the mass at ± 0.1 mg	
Preparation of test portion	/	Drying oven method (ISO 18134—part 3)
Test portion	Analysis sample	Dry sample[d]
Test portion conditioning[e]	/	Dry test portion and crucible at 105 °C prior to weighing Cool the crucible in a desiccator with desiccant
Layer of test portion	Thick.[f] After filling crucibles, replace the lid and tap crucible to get an uniform surface	
Crucible and lid Crucible stand	Fused silica Heat-resistant steel wire	
Crucible and stand conditioning[g]	Heat the oven to 900 ± 10 °C Place the empty crucible(s) and lid(s) on the stand Place the stand in the oven Maintain at 900 ± 10 °C for 7 min ± 5 s Cool stand and crucibles to room temperature on a thermoresistant plate Store stand and crucibles in a desiccator with desiccant	
Oven procedure[h]	Heat the oven to 900 ± 10 °C Place the stand filled with test portion in hot oven (open and close the door as quick as possible) Maintain at 900 ± 10 °C for 7 min ± 5 s	
Step after drying[e]	Cool the crucible with ash on a heat resistant plate for 5–10 min Cool the crucible in a desiccator with desiccant to room temperature	

[a]When sample or test sample is in a package where moisture might remain on the surface, shake the package before sampling in order to allow the material to reabsorb the moisture
[b]Note that the nominal top size is the same as in standard ISO 18134–part 3 for the determination of moisture content. For some solids, the nominal top size should be reduced to lower values in order to improve the precision measurements
[c]The mass of the general analysis sample (and then test portion) should be sufficient for the measurement of the final mass with the same scale precision as for initial mass
[d]The dried sample should be kept dry using a closed container in a desiccator with desiccant
[e]Solid samples are highly hygroscopic and may reabsorb moisture. This procedure will provide high precision of the measurement of volatile matter content
[f]No value is precisely given, but the layer should be thick relatively to the crucible height
[g]The initial heating and cooling of crucible is done to minimize moisture absorption and thus to minimize experimental errors
[h]For multiple determinations, the stand should always be filled with the maximum number of crucibles, even empty

As seen in Table 2.6, the main difference between both procedures is the test portion used. Starting with a dry sample means that the moisture content as well as the volatile matter content have been determined on the same test portion. On the contrary, volatile matter and moisture contents are determined with different test portions (from the same test sample) using a wet sample. In case of high heterogeneity of the material, the successive determinations on the same test portion would be helpful for mass scale calculations. For the calculation of the volatile matter content (VM), percentage of volatile matter on dry basis, the following equations are applied:

$$VM(\%) = \frac{m_2 - m_3 - m_4}{m_2 - m_1 - m_4} \times 100 \tag{2.9}$$

$$VM(\%) = \left[\frac{100 \times (m_2 - m_3)}{m_2 - m_1} - MC\right] \times \frac{100}{100 - MC} \tag{2.10}$$

where: m_1 = mass of the empty crucible and lid (±0.1 mg) at room temperature (after conditioning step); m_2 = mass of the crucible and lid filled with test portion (±0.1 mg) at room temperature (either or not after test portion conditioning); m_3 = mass of the crucible and lid after heating (±0.1 mg) at room temperature; m_4 = mass of moisture determined with standard ISO 18134—part 3 (±0.1 mg); $m_4 = 0$ for dry sample; MC = moisture content (wt%) determined either on the current sample or on another test portion from the test sample.

The results should be calculated to 2 decimal digits and rounded to the nearest 0.1% for reporting.

The experiments should be done at least in duplicate for repeatability. The mean value of these duplicate determinations should not differ by more than 1% of the mean value. For reproducibility experiments done by two independent laboratories, the mean values from each laboratory should not differ of more than 3% of the mean value. It should be noticed that the reproducibility is also depending on the volume of air available in the crucible.

Recommendation for test reports:

The test report should include at least the following information:

- Identification of the laboratory performing the test and the date;
- Identification of the sample;
- Reference to the standard;
- Result of the test on dry basis;
- Any observation that might affect the result should be noticed;
- Deviations from the established procedures should be described.

Issues during the volatile matter content determination:

As expressed previously, different standards exist to determine the content of volatile matter. Some of them have previously been cited, showing that the procedure has been optimized to fit better with the nature of the solid analyzed. Here are explained some of the issues that could be encountered during the experiment in order to help readers to improve their own procedure.

- Particle size:

Regarding the nominal top size, the value should be less than 1 mm or lower depending on the type of solid biofuels as defined in the ISO 18122 standard. It could be noticed that the particle size is in the range 150-850 μm for the standard ASTM D1762-84, or lower than 250 μm for ASTM D3175-11.

- Volatilization conditions:

The standard test methods are sometimes adapted to fit the nature of the solid characterized, but the heating is always quite fast to high temperature: 900 °C [32] or 950 °C [33, 34], and the time should be short 7 or 10 min. Using TGA, basically N_2 is used and not batch air atmosphere, since the crucible is not covered. Some differences could appear with the different procedures as discussed by Singh et al. [35].

Notes for experimental procedure:

Note 1: desiccator and desiccant

Since the biomass, biowastes, and the residues as well are highly hygroscopic, the desiccant should frequently be controlled.

Note 2: scale

The precision of the scale requested for this test method is 0.1 mg. To get accurate values from the scale, some precautions should be carefully followed:

- Calibrate the scale when it is being installed for the first time, and then regularly or after any changes of locations.
- Center the bubble on the spirit level of the scale. The scale will have adjustable feet which you can turn individually to raise or lower a side. The device has to be even to be accurate.
- Close the door to avoid measurement perturbations due to airstream.
- Avoid pressing the surface of the table when recording the measure since vibrations will interfere with.
- Clean up the scale after use.

Examples on the volatile matter values of different biomass and biowastes are presented in Sect. 2.4.5.

2.4.3 Ash Content Analysis

2.4.3.1 Definition

Ash content corresponds to the total amount (mass or wt%) of minerals and metals in the sample (dry basis).

The ash content determines the weight of the non-combustible part of a solid that oxidizes during a combustion process but releases no energy. In a solid feedstock, the ash content is an important parameter that defines the quality of the biofuel. When biomass and biowaste are used for energy, the amount of ashes decreases the energetic value of the solid, and consequently has economic consequences. During thermoconversion processes, the ashes (fly or bottom ashes) needs to be removed and may contribute to slagging and corrosion in the plant.

2.4.3.2 Method

For a solid sample, the ash content is basically the solid residue remaining after combustion in an oven or in a thermogravimetric analyzer. This is evaluated using reference methods: standard ISO 18122:2015 [36] for solid biofuels, ASTM D3174-12 [37] for coal and coke from coal, ASTM D1762-84 [33] for wood charcoal, ASTM D1102-84 [38] for wood, ASTM E1755-01 for biomass [39], or ASTM E887-15 [40] for refused-derived fuel. The principle is to measure the mass variation after the combustion of organic compounds at a certain temperature. Each standard provides optimized parameters such as temperature (minimum 500 °C), heating rate, combustion time, or composition of the atmosphere. The principle of the procedure is summarized in Fig. 2.8.

The standard ISO 18122 [36] is described in Table 2.7.

As seen in Table 2.7, the main difference between these two procedures is the test portion used. Starting with a dry sample means that the moisture content as well as the ash content have been determined on the same test portion. On the contrary, ash and moisture contents are determined with different test portions (from the same test sample) using a wet sample. In case of high heterogeneity of the material, the successive determinations on the same test portion would be helpful for mass scale calculations. For the calculation of the ash content (A), percentage of ash on dry basis, the following equations are applied:

$$Ash(\%) = \frac{m_3 - m_1}{m_2 - m_1 - m_4} \times 100 \quad (2.11)$$

$$Ash(\%) = \frac{m_3 - m_1}{m_2 - m_1} \times 100 \times \frac{100}{100 - MC} \quad (2.12)$$

where: m_1 = mass of the empty crucible (±0.1 mg) at room temperature (after conditioning step); m_2 = mass of the crucible filled with test portion (±0.1 mg) at

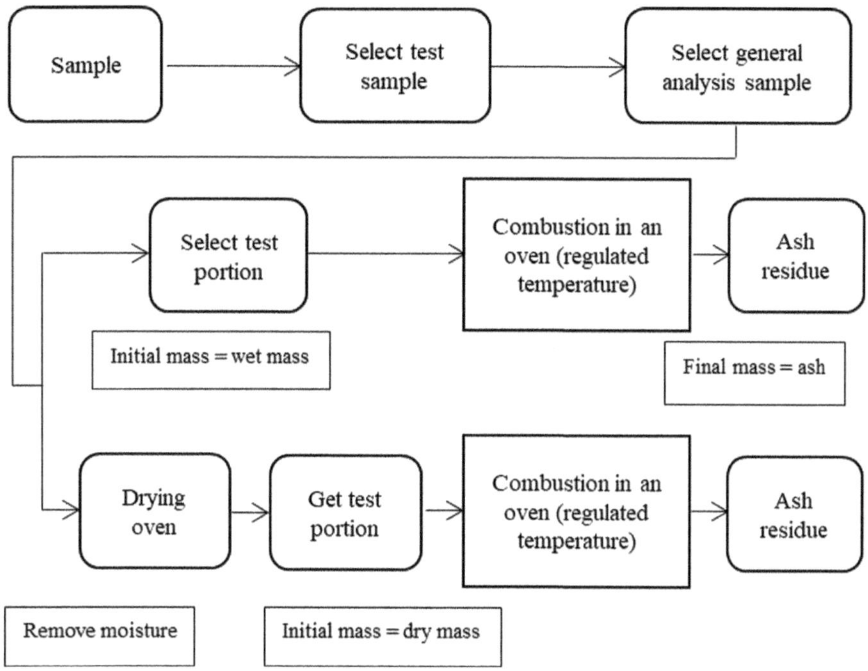

Fig. 2.8 Procedure used to determine ash content

room temperature (either or not after test portion conditioning); m_3 = mass of the crucible plus ashes (±0.1 mg) at room temperature; m_4 = mass of moisture determined with standard ISO 18134—part 3 (±0.1 mg); $m_4 = 0$ for dry sample; MC = moisture content (wt%) determined either on the current sample or on another test portion from the test sample.

The results should be calculated to 2 decimal digits and rounded to the nearest 0.1% for reporting.

The experiments should be done at least in duplicate for repeatability. When the calculated ash content (wt%) is smaller than 1%, the maximum acceptable difference of the repeatability and reproducibility must not be higher than 0.1 and 0.2% absolute, respectively. When the calculated ash content (wt%) is higher than 1%, the maximum acceptable difference of the repeatability and reproducibility must not be higher than 10 and 20% relative, respectively [36].

Note:

- Repeatability: to evaluate the repeatability of the procedure the same operator should prepare several test portion from the same test sample and perform the procedure for each test portion. Then, after calculating the mean value and the deviation, the latter should be compared to the value in the table (or in the standard).

Table 2.7 Main steps of the procedure to determine ash content

Procedure	ISO 18122 – wet sample	ISO 18122 – dry sample
Select a test sample[a] ISO 18135	Feedstock received in sealed air-tight container	
Select a general analysis sample[a] ISO 14780	Nominal top size < 1 mm[b]	
Mass of the general analysis sample	Min 1 g ± 0.1 mg[c]	
Preparation of test portion	/	Drying oven method (ISO 18134- part 3)
Test portion	General analysis sample	Dry general analysis sample[d]
Test portion conditioning[e]	/	Dry test portion and crucible at 105°C prior to weighing Cool the crucible in a desiccator with desiccant
Layer of test portion (on crucibles)	< 0.1 g cm^{-2} [f]	
Crucibles	Inert material (porcelain, silica or platinum)	
Crucible conditioning	Place the crucible in cold oven Heat to 550 ± 10°C Maintain oven at 550 ± 10°C for 60 min at least Cool the crucible on a heat resistant plate for 5-10 min Cool the crucible in a desiccator with desiccant to room temperature	
Oven programme for ash content determination	Place the crucible filled with test portion in cold oven Heat to 250°C, at heating rate of 4.5-7.5°C min^{-1} Maintain oven at 250°C for 60 min Heat to 550 ± 10°C, at heating rate of 10°C min^{-1} Maintain oven at 550 ± 10°C for 120 min at least (mass stabilization corresponds to Δm < 0.5 mg)	
Atmosphere	Air renewed 5 to 10 times per hour to insure sufficient oxygen content for combustion	
Step after drying[e]	Cool the crucible with ash on a heat resistant plate for 5-10 min Cool the crucible in a desiccator with desiccant to room temperature	

a When sample or test sample is in a package where moisture might remain on the surface, shake the package before sampling in order to allow the material to reabsorb the moisture.
b Note that the nominal top size is the same as in standard ISO 18134 – part 3 for the determination of moisture content. For some solids, the nominal top size should be reduced to lower values in order to improve the precision measurements.
c The mass of the general analysis sample (and then test portion) should be sufficient for the measurement of the ash mass (final mass) with the same scale precision as for initial mass.
d The dried sample should be kept dry using a closed container in a desiccator with desiccant.
e Solid samples are highly hygroscopic and may reabsorb moisture. This procedure will provide high precision of the ash content measurement.
f If the layer of test portion exceeds 0.1 g cm^{-2}, there is a risk of incomplete combustion and/or CO_2 absorption by calcium rich samples.

[a]When sample or test sample is in a package where moisture might remain on the surface, shake the package before sampling in order to allow the material to reabsorb the moisture

[b]Note that the nominal top size is the same as in standard ISO 18134 "part 3 for the determination of moisture content. For some solids, the nominal top size should be reduced to lower values in order to improve the precision measurements

[c]The mass of the general analysis sample (and then test portion) should be sufficient for the measurement of the ash mass (final mass) with the same scale precision as for initial mass

[d]The dried sample should be kept dry using a closed container in a desiccator with desiccant

[e]Solid samples are highly hygroscopic and may reabsorb moisture. This procedure will provide high precision of the ash content measurement

[f]If the layer of test portion exceeds 0.1 g cm^{-2}, there is a risk of incomplete combustion and/or CO_2 absorption by calcium rich samples

- Reproducibility: to evaluate the reproducibility of the procedure, at least 2 operators should prepare one test portion (from the same test sample) and perform the procedure themselves. Then, after calculating the mean value and the deviation, the latter should be compared to the value in the table (or in the standard).

Recommendation for test reports:

The test report should include at least the following information:

- Identification of the laboratory performing the test and the date.
- Identification of the sample.
- Reference to the standard.
- Result of the test on dry basis.
- Any observation that might affect the result should be noticed.
- Deviations from the established procedures should be described.

Issues During the Ash Content Determination

As expressed previously, different standards exist to determine ash content. Some of them have previously been cited, showing that the procedure has been optimized to fit better with the nature of the solid analyzed. Here are explained some of the issues that could be encountered during ashing in order to help readers to improve their own procedure.

- Particle size:

Regarding the nominal size, the value should be less than 1 mm or lower depending on the type of solid biofuels as defined in the ISO 18122 standard. It could be noticed that the particle size is in the range 150–850 μm for the standard ASTM D1762-84, or lower than 250 μm for ASTM D3174-12.

- Ashing temperature:

For biomass and biowastes, it is assumed that the remaining solid after oxidation at a temperature between 550 and 600 °C is the ash. The temperature programme has already been presented for the standard ISO 18122 [36], and the ashing temperature was set at 550 $\pm$ 10 °C. To insure a temperature over 550 °C, some other standards define the ashing temperature at 575 $\pm$ 25 °C for ASTM E1755-01 (developed for hard and soft wood as well as herbaceous materials …), and ASTM E830-87 (included in ASTM E887-15, developed for refuse-derived fuel), or between 580 and 600 °C for ASTM D1102-84 (developed for wood and wood products). For the test method ASTM D1762-84, which covers the determination of moisture, volatile matter, and ash in charcoal made from wood, ash content is determined at 750 °C (heating rate of 5 °C min^{-1}, for 6 h) since this is a residue from thermoconversion [41]. The standard test methods are sometimes adapted to fit the nature of the solid characterized, but the heating is always quite slow to avoid particle entrainment and ash volatilization. Indeed, in biomass and biowastes, a part of minerals and metals bounded to organic matter could be volatized during the decomposition of the organics. These volatile inorganics may rapidly recombine to be stabilized in the gaseous phase leading to a mass loss. At low heating rate, the recombination reactions are limited and the sequential volatilization or organic matter is favored [35].

As noticed in the ISO 18122 standard [36], the ash content of mineral biofuels is determined at 815 °C. A significant difference in ash content could be obtained, due to the decomposition of carbonates into CO_2, to the volatilization of inorganic compounds, or to an additional oxidation of ashes.

- Completion of ashing:

If there is any doubt on complete combustion, it is recommended in the ISO 18122 standard, to reload the crucible in the hot oven for additional 30 min periods until weigh stabilization. Another way would be to add some droplets of water or ammonium nitrate and to reload the crucible in the cold oven for additional 30 min periods until weigh stabilization. In ASTM E830-87 standard, it has been noted that particles of glass and sand tend to sinter at temperatures close to 675 °C. The ignition may be finished to constant weight at a temperature of 725 ± 25 °C.

Notes for Experimental Procedure:

Note 1: desiccator and desiccant

Since the biomass, biowastes, and the ash as well are highly hygroscopic, the desiccant should frequently be controlled.

Note 2: drying oven

To insure the combustion of the test portion, the oven should provide a uniform heat at the temperatures requested in the programme and should meet the heating rates defined in the standard.

Note 3: scale

The precision of the scale requested for this test method is 0.1 mg. To get accurate values from the scale, some precautions should be carefully followed:

- Calibrate the scale when it is being installed for the first time, and then regularly or after any changes of locations.
- Center the bubble on the spirit level of the scale. The scale will have adjustable feet which you can turn individually to raise or lower a side. The device has to be even to be accurate.
- Close the door to avoid measurement perturbations due to airstream.

- Avoid pressing the surface of the table when recording the measure since vibrations will interfere with.
- Clean up the scale after use.

Examples on the ash content of different biomass and biowastes are presented in Sect. 2.4.5.

2.4.4 Fixed Carbon Analysis

Fixed carbon is considered as the residue which is not volatile in a solid sample. For coal, coke, anthracene or carbonaceous structures with very high carbon content, fixed carbon is quasi exclusively composed of carbon. However dealing with biomass, residues, biochars or other heterogeneous materials, fixed carbon is taken as the non-volatile fraction. This is mainly composed of fused aromatic carbon structures. This parameter is expressed as the mass (or wt%) of fixed carbon in the total mass of the sample (dry basis).

The value of the fixed carbon content, or better the ratio of fixed carbon/volatile matter contents would help in selecting a treatment or an application. For example a solid with high fixed carbon content compared to the content of volatile matter will provide high char yield through thermochemical treatment, or might be refractory to biological treatment. In other words, a high content of fixed carbon is a synonym of carbon sequestration.

The proximate analysis of a solid sample could be expressed as the sum of moisture, ash, volatile matter and fixed carbon, as seen in Eq. (2.13). Basically standards are provided to measure moisture content, ash content and volatile matter content. The mass of fixed carbon is thus calculated by difference and the fixed carbon content is expressed as dry basis.

$$total\,mass = moisture + ash + volatile\,matter + fixed\ carbon \tag{2.13}$$

Examples on the fixed carbon value of different biomass and biowastes are presented in Sect. 2.4.5.

2.4.5 Examples of Proximate Analysis

The following examples have been taken from literature and values could be either obtained with ISO or ASTM standards. Table 2.8 presents proximate analysis of some woods while Table 2.9 shows the proximate analysis of some other resources (algae, grass, plants …). Moisture content is greatly related to the original ligno-cellulosic biomass or biowastes. Basically moisture content in dried wood used in laboratory experiments is less than 10 wt%. In all cases, the ash content depends on

Table 2.8 Ranges of proximate analysis for some woods

Feedstock	Moisture content (wt%)	Ash content (wt% db)	Volatile matter (wt% db)	Fixed carbon (wt% db)
Poplar [43]	8–59	0.4–4	72–86	12–28
Poplar [44]	≈54	1.05	83.99	14.96
Beech [44]	≈10	0.72	83.84	15.44
Spruce [44]	≈7	0.17	84.76	15.08
Pine [42]	3.87	0.60	80.00	19.40
Oak, red [42]	14.42	2.76	81.07	16.17
Willow [42]	43.50	1.59	83.40	15.01
Wood particle board [42]	8.10	0.60	77.70	21.70

Table 2.9 Ranges for proximate analysis of other biomass feedstocks

Feedstock	Moisture content (wt%)	Ash content (wt% db)	Volatile matter (wt% db)	Fixed carbon (wt% db)
Chlorella microalgae [42]	71.87	5.94	80.38	13.68
Peat [42]	29.40	6.30	68.10	25.60
Alfalfa stems [43]	9.29	5.27	78.92	15.81
Cotton stalk [43]	16.80	4.57	68.87	26.56
Miscanthus [43]	20–52	1–9	74–94	5–26
Switchgrass [43]	40–70	2–10	73–87	13–27
Almond shells [42]	7.85	3.04	76.10	20.86
Coconut shell [42]	5.60	0.50	76.10	23.40
Rice straw [43]	3–74	8–26	64–98	15–25
Rice husk [42]	11.20	20.60	54.40	25.00
Sugarcane bagasse [43]	10.39	2.44	85.61	11.95
Corn stover [43]	11–33	4–10	72–85	15–20
Wheat straw [43]	8–15	1–23	71–85	15–22
Wheat straw [44]	≈8	7.46	77.52	15.02

the origin and nature of lignocellulosic biomass. Virgin wood does not contain more than 5 wt% while rice straw is for example a feedstock with high ash content (around 20 wt% on dry basis). Lignocellulosic biomass has a high content of volatile matter, explaining the use of these feedstocks for bio or thermal conversion. Main values have been extracted from the Phyllis database [42] and the references to literature studies are provided for each resource on the dedicated page.

Proximate analysis is also provided for municipal solid waste (MSW) which is a mixture of food waste, paper, plastics, wood, textiles, glass and metals. The

Table 2.10 Ranges of proximate analysis for municipal solid wastes (MSW)

MSW origin	Moisture content (wt%)	Ash content (wt% db)	Volatile matter (wt% db)	Fixed carbon (wt% db)
Domestic MSW [45]	10–40	10–30	30–60	3–15
Huazhong [46]	8.80	5.93	82.28	11.79
Chongqing [47]	42.14	42.08	42.81	15.11
Hangzhou [48]	51.57	55.56	38.16	6.28
UK [49]	7.3	16.18	72.82	11.00

Table 2.11 Ranges of proximate analysis for biochars from lignocellulosic feedstock

Feedstock	Moisture content (wt%)	Ash content (wt% db)	Volatile matter (wt% db)	Fixed carbon (wt% db)
Wheat straw 550 °C [35]	3.59	20.19	16.41	63.40
Switchgrass 550 °C [35]	2.28	5.27	18.32	76.42
Rice husk 550 °C [35]	1.69	46.52	12.21	41.27
Miscanthus straw 550 °C [35]	1.85	11.97	16.96	71.08
Poplar 550 °C [50]	ND	7.25	20.37	72.38
Poplar 600 °C [50]	ND	7.17	17.81	75.01

proportion of each item greatly depends on the origin of the MSW (cities and Countries). Table 2.10 presents some values found in the literature.

When using the resources for thermal treatment, the volatile matter is consumed and the residual biochar contains a higher amount of fixed carbon and ashes, as seen in Table 2.11.

2.5 Ultimate Analysis

2.5.1 CHNSO Analysis

2.5.1.1 Introduction

Due to the diversity of biomass and biowaste, the accurate knowledge of the composition of the samples is of great importance to determine their possible valorization pathways. In particular, the quantification of the mass fractions of carbon, hydrogen, nitrogen and sulfur (CHNS analysis) is a very important step in the characterization of organic samples. Specific apparatus are available for these analyses.

2.5.1.2 Principle of the Measurement

CHNS analysis is usually performed via the complete combustion of the sample at high temperature ($\sim$1800 °C). The combustion process can be carried out under static or dynamic conditions (i.e. use of a fixed volume of oxygen or a constant flow of oxygen for a set time) according to the employed apparatus configuration. High purity oxygen (minimum 99.9995%) should be used in the combustion process to reduce the nitrogen and avoid measurement inaccuracies.

During the combustion process, the carbon present in the sample is converted into carbon dioxide, the hydrogen into water, and the nitrogen and sulphur to nitrogen and sulphur oxides. These combustion products are carried by a helium flow to a high purity cooper reactor to remove the remaining oxygen and to convert nitrogen oxides to nitrogen gas. Adsorbent traps are also used to remove additional combustion products and sulphur, if its determination is not intended [51]. Other inert gases as argon can be used as carrier gas instead of helium in some analyzers design.

Subsequently, the gas mixture containing CO_2, N_2, H_2O and SO_2 is analyzed to determine the concentration of each individual gas species. This step can be performed using two principal methods [52]:

- Gas species separation into a chromatographic column (GC) and subsequent quantification using a gas detector such as thermal conductivity detector (TCD).
- Individual gas species analysis with separate infra-red (IR detectors) and thermal conductivity cells.

Then, from the knowledge of the initial mass, a software provides the mass percentage of nitrogen, carbon, hydrogen and sulfur contained in the sample. The quantification of the elements requires a proper calibration of the analysis equipment using commercial standards.

2.5.1.3 Oxygen Quantification

CHNS analysis can also be completed by the determination of the oxygen content of the sample (CHNS/O analysis). In this case, the sample is pyrolyzed in a helium atmosphere at a temperature near 1000 °C. The pyrolysis gases containing oxygen are converted to carbon monoxide using a platinized carbon reagent, while water and other interference gases are removed using traps. Gas species concentration is then analyzed using one of the two presented methods. The oxygen concentration of the sample is determined form the CO concentration measured. As for the other elements, a software provides the mass percentage of oxygen from the knowledge of the initial mass of the sample.

A CHNS/O analyzer scheme and principle are presented in Fig. 2.9 for an apparatus using a chromatographic column for gas species separation and a TDC detector.

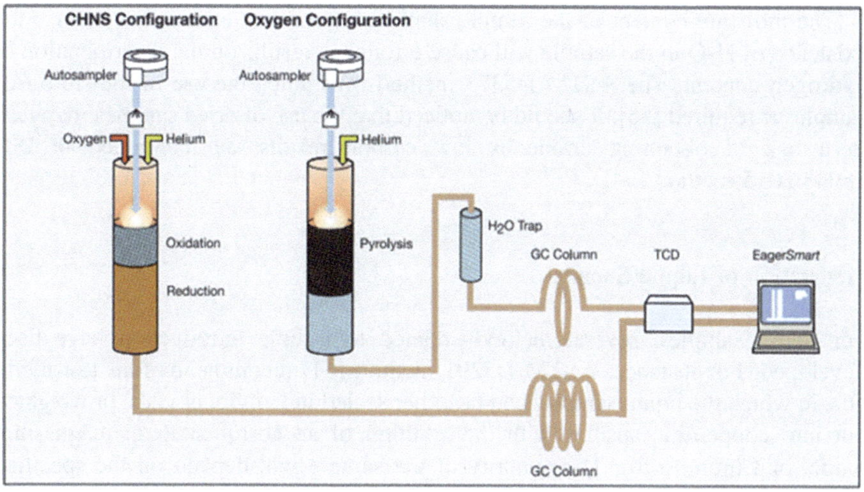

Fig. 2.9 CHNS/O analyzer schematic configuration. Reprinted by permission from Thermo Fisher Scientific. Copyright 2016 [53]

CHNS analysis can be performed on a wide variety of samples including solids, liquids and viscous samples. Accordingly, the sample preparation depends on the type of material being analyzed.

2.5.1.4 Sample Preparation

Most of the biomass and biowastes, such as wood, crops, seaweed, municipal solid waste, animal wastes, are found in solid state. However, there also exist many liquid-phase biomass and biowastes, for example bioethanol, biodiesel, tar, etc. The preparation of both solid and liquid samples for ultimate analysis will be introduced separately.

Preparation of Solid Samples

The sample generally needs to be crushed and homogenized before testing. The standard ISO 16948:2015 recommends a nominal top sample size of 1 mm or less [54]. A smaller sample size, e.g. 0.25 mm, may be required for some instruments to keep the desired precision. Or, an adequate sample size should be determined beforehand by validation experiments for unknown samples.

The quantity of the test sample may vary from 1–3 mg (microanalysis) to 100–500 mg (macroanalysis) [55], according to the appropriate type of instruments used.

The moisture content of the sample shall be determined concurrently, since the existence of H_2O in the sample will cause erroneous results on the determination of hydrogen content. The ASTM D5373 method mentioned the use of moisture-free samples if required [56]. It should be noticed that the use of dried samples may lead to a risk of obtaining erroneous low carbon results, as mentioned in ISO 16948:2015 method [54].

Preparation of Liquid Samples

For liquid samples, several methods related to sample introduction have been developed. For instance, ASTM D5291 method [51] recommends four test methods, in which the liquid samples can be either sealed in individual vials, or weighted out into copper/tin capsules with the addition of an absorbent (e.g. magnesium oxide, or Chromosorb). The quantity of the sample will depend on the specified instruments.

The viscous samples shall be warmed beforehand until being fluid, and shake for 5 s. A convenient, clean syringe or pipet shall be used to transfer the liquid samples into test capsules.

2.5.1.5 Sample Analysis

The prepared samples are analyzed using the specified instruments. Before the sample analysis, the apparatus shall be correctly calibrated. Standard reference materials and control samples are commonly used to check the performance of the instrument.

To start the analysis sequence, both ISO 16948:2015 [54] and ASTM D5291 [51] standards recommend running 2 blank samples to condition the instrument, 3–6 calibrated samples to check or carry out the calibration, and 1–3 standard samples to ensure the instrument performance of the actual type of samples, followed by the unknown samples.

The total carbon, hydrogen, nitrogen, and sulfur, can be obtained as a percentage by mass. The following Eqs. (2.14)–(2.17) allow calculating the results on dry basis:

For the carbon content:

$$\mathrm{C}_d = \mathrm{C}_{ad} \times \frac{100}{100 - M_{ad}} \tag{2.14}$$

For the nitrogen content:

$$N_d = N_{ad} \times \frac{100}{100 - M_{ad}} \tag{2.15}$$

For the sulfur content:

$$S_d = S_{ad} \times \frac{100}{100 - M_{ad}} \tag{2.16}$$

For the hydrogen content:

$$H_d = \left(H_{ad} - \frac{M_{ad}}{8.937}\right) \times \frac{100}{100 - M_{ad}} \tag{2.17}$$

where d represents the dry basis, ad is as determined, M_{ad} is the moisture content of the analyzed sample, and the constant 8.937 is the hydrogen concentration in the water that exists in the sample (the factor is calculated from the molar formula of H_2O taken into account the atomic weight of hydrogen at 1.008 and oxygen at 15.9994).

It should be highlighted that duplicated samples should be conducted simultaneously to ensure the robustness of the test results. The final report should be presented as the mean of duplicate determinations.

2.5.1.6 Examples

As an example, the CHONS characteristics of some typical biomass and biowastes are presented in Table 2.12.

Table 2.12 CHONS characteristics of several typical biomass and biowastes

Material	Composition in % mass (dry basis, ash free)				
	Carbon	Hydrogen	Nitrogen	Sulfur	Oxygen
Poplar wood	49.6	5.7	0.1	0.7	43.9
Peanut shell	48.2	5.9	0.8	0.0	45.1
Millet stem	42.9	5.7	0.8	0.0	50.6
Oil palm shells [57]	46.7	6.5	0.6	0	46.2
Bamboo guadua [57]	42.7	5.4	0.3	0	51.5
Municipal solid waste (EU average) [58]	59.6	9.5	2.0	0.3	28.6

2.5.2 *Elemental Analysis by Inductively Coupled Plasma Optical Emission Spectroscopy (ICP-OES)*

2.5.2.1 Measurement Principle

Taking into account the diversity of biomass and biowaste that can be valorized in energetic or non-energetic applications, the accurate elemental analysis of samples is an important step to determine possible treatments and valorization pathways.

Inductively Coupled Plasma Optical Emission Spectroscopy (ICP-OES) is an analytical technique for determining the inorganic elemental composition (e.g. alkaline and alkaline earth metals, transitions metals, rare earth metals) of a wide variety of samples. This technique uses an optical emission spectrometer to detect and quantify the radiation emitted by the sample atoms after being excited by an argon plasma at very high temperature (between 5000 K and 10,000 K) [59]. The elements present in the sample are identified according to the emitted rays wavelengths, which are characteristic of each particular atom. For its part, the measured emission intensities indicate the elements concentration, and are quantified using calibration standards. ICP-OES detection limit is in the ppb range for most elements, making this technique suitable for the determination of trace elements in samples.

ICP-OES has a wide analytical range and allows the identification of elements from lithium to uranium excepting halogens. Moreover, carbon, nitrogen, hydrogen and oxygen cannot be identified using this technique and are usually determined according to CHNS analysis presented in the previous section.

ICP-OES systems can be schematized as in Fig. 2.10. The nebulizer system atomizes the sample solution into the core of a high temperature plasma. Due to the very high temperature, atoms in the sample are then ionized and thermally excited, emitting a photon radiation. Photons are subsequently focused to an optical spectrometer, using a lens arrangement. Finally, the spectrometer separates the emitted

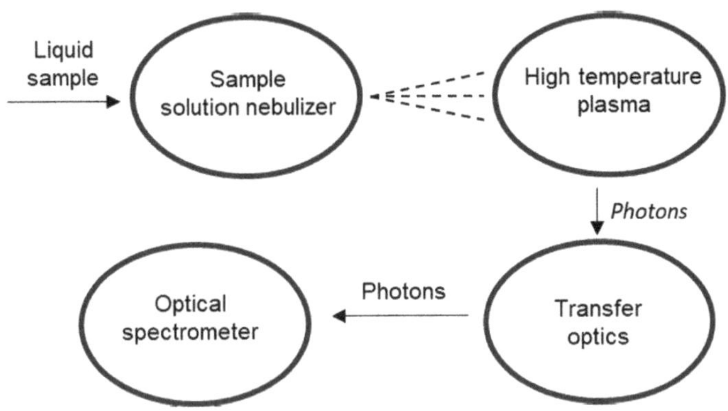

Fig. 2.10 Principle of the main steps of ICP-OES analysis

light according to its component wavelength. The elements present in the solution are then identified according to their characteristic wavelength emission. Further information about the theoretical aspects of this analytical technique can be found in dedicated ICP-OMS literature [59, 60].

2.5.2.2 Sample Preparation

As most of the biowastes and residues referred in this handbook are found in solid state, the sample preparation for ICP-OES analysis will be focused on solid samples. The correct preparation of samples will determine the quality of the obtained data.

Accordingly, it is important to say that one of the main challenges of this analytical technique is the complete digestion of solid samples prior to the analysis. Digestion is essential to ensure that the constitutive elements of the sample are correctly dissolved and can be measured. Biowastes are generally digested using a combination of mineral acids and other reagents, and high temperatures. The digestion conditions depend mainly on the type of material and the quantity to be analyzed. The most common acids and reagents used for solid samples digestion are presented in Table 2.13.

It is also important to say that solid samples should be ground before digestion in order to facilitate the dissolution process. It is recommended to use particles below 500 μm, as described by some sample preparation standards [62, 63] taking into account that the reduction in particle size increases the contact of the solid material with the digestion solution.

According to the sample and the elements to be analyzed, the sample digestion can be performed using open and closed systems.

Open Digestion

Open digestion allows the analysis of large sample quantities and use open vessels that are heated on a hot-plate. However, the digestion temperature is limited by the boiling point of the solution to avoid the loss of volatile elements present in the sample. Therefore, long digestion times and high acid consumption are required to obtain acceptable results. This technique is not recommended if the total

Table 2.13 Most common reagents used for solid samples digestion [61]

Compound	Formula	Boiling point (°C)
Hydrogen peroxide	H_2O_2	106
Nitric acid	HNO_3	122
Hydrochloric acid	HCl	110
Hydrofluoric acid	HF	112
Sulfuric acid	H_2SO_4	338

mineralization of the sample is required, as only a partial dissolution is often reached due to the temperature limitations of the procedure. Also, there are risks of sample contamination and volatile elements loss. Therefore, open digestion is generally used for partial digestion and leaching analysis of samples [61, 64].

Closed Digestion

The limitations of open digestion systems, in particular the low working temperature and the risks of volatilization of elements and sample contamination can be overcome using closed systems. This method allows higher temperatures for acid digestions, up to 300 °C, reducing considerably the acid consumption and the digestion time. Digestion process can be performed using heating blocks or microwave systems. Both of them allow a satisfactory digestion of most samples, suitable for further analysis using ICP-OES. Microwave heating systems allow extremely fast digestions compared to conventional heating blocks [65].

Digestion Vessels Materials

The employed containers for digestion solution preparation and storage must be naturally resistant to the used reagents. Glass, polyethylene (PE), polypropylene (PP) and polytetrafluoroethylene (PTFE) containers are generally used. However, it is important to mention that glass or quartz containers must be avoided when digestion process is performed using hydrofluoric acid (HF), as it reacts with siliceous compounds, contaminating the analysis solution.

For its part, the digestion vessels are usually made of TFM™-PTFE, as this material ensures temperature and chemical stability during the digestion process. TFM™-PTFE can be used at temperatures up to 250 °C, while PP and PE maximum working temperature is 130 and 60 °C respectively [66].

Sample Digestion for ICP-OES Analysis of Different Kinds of Samples

Several standards related with solid samples digestion have been developed, presenting general digestion conditions like acid mixtures, temperature and time. However, taking into account the great heterogeneity of biomass, solid biowastes and residues, standard methods could need to be adapted in some cases. Some existing standards for the digestion and ICP-OES analysis of different kind of materials are presented in Table 2.14.

Most organic samples can be completely dissolved using a mixture of H_2O_2 and HNO_3. Nitric acid is the most common digestion reagent used, as it does not generate spectral interferences and is compatible with ICP-OES instruments. However, inorganic content on biomass and biowastes could influence the optimal digestion conditions of samples. According to EN ISO 16967, the use of HF is

Table 2.14 Existing European standards for the digestion and ICP-OES analysis of various waste materials [62, 63, 67, 68]

Material	Standard
Municipal sludge	EN 16170—sludge, treated biowaste and soil—determination of elements using inductively coupled plasma optical emission spectrometry (ICP-OES)
Solid recovered fuels	EN 15410—methods for the determination of the content of major elements (Al, Ca, Fe, K, Mg, Na, P, Si, Ti)
Solid biofuels	EN 16967—biocombustibles solides—Détermination des éléments majeurs—Al, Ca, Fe, Mg, P, K, Si, Na et Ti
Waste (sewage sludge; incineration ash)	EN 13656—characterization of waste—microwave assisted digestion with hydrofluoric (HF), nitric (HNO_3) and hydrochloric (HCl) acid mixture for subsequent determination of elements

Table 2.15 Digestion methods for various biowastes using a close method and conventional heating

Sample	Weight	Reagent	Digestion system	T, °C	Time
Raw lignocellulosic biomass	250 mg	1.5 ml H_2O_2 (30%) 4 ml HNO_3 (65%) 0.5 ml HF (48%)	Closed (heating block)	220	1 h
Raw biomass (ash% >10)	150 mg	1.5 ml H_2O_2 (30%) 4 ml HNO_3 (65%) 0.5 ml HF (48%)	Closed (heating block)	220	4 h
Biochar	150 mg	1.5 ml H_2O_2 (30%) 4 ml HNO_3 (65%) 0.5 ml HF (48%)	Closed (heating block)	220	4 h
Solid recovered fuel	200 mg	3 ml H_2O_2 (30%) 3 ml HNO_3 (65%) 1 ml HF (48%)	Closed (heating block)	220	18 h
Biomass ash	50 mg	2 ml H_2O_2 (30%) 3 ml HNO_3 (65%) 2 ml HF (48%)	Closed (Tögl bomb)	220	4 h

recommended for the digestion of biofuels, as it allows the decomposition of silicates and aluminosilicates and the complete dissolution of associated alkaline and alkali earth metals [63, 69]. Biofuels with ash content above 10% will require higher quantities of hydrofluoric acid to ensure their total digestion and the use of H_3BO_3 for HF neutralization, in order to avoid damage to measurement systems and measurement interferences [68, 70]. HCl should be used when the analysis of precious metals is of interest [61].

The digestion procedure of biomass and biowaste depends on the nature of each sample. An overview of some experimental protocols for different kind of biowastes is presented in Table 2.15 using a closed digestion system and an electric heating block. In the case of microwave systems, digestion times could be significantly reduced. However, recommended sample weight and digestion reagents quantities remain the same.

The maximum sample quantity used in the digestion process depends mainly on the employed vessel capacity and should be carefully selected to avoid overpressure inside the digestion vessels. For raw biomass and biowaste, it is recommended to use between 150 and 500 mg. Considering that these materials are generally heterogeneous, higher amounts of sample could help to reduce the experimental variability. In this regard, it is also recommended to analyze each sample at least 3 times to verify the repeatability of results.

Digestion Method Development

Taking into account that biomass and biowastes are highly heterogeneous and come from different sources, it is often necessary to develop digestion methods for new materials. Depending on the sample nature, this process could take more or less time. Accordingly, the presented steps constitute a guideline for the digestion of new samples.

Preliminary information about digestion conditions of new samples can be extracted from relevant literature or standards related to similar materials. As already commented, the selection of the reagent mixture should be done according to the chemistry of the sample and the elements to be analyzed. Examples presented in Table 2.3 could be used as a starting point for the digestion program of solid biowastes.

It is recommended to begin digestion tests with lower sample quantities than reported in literature, in order to reduce overpressure risks, while maintaining the reagent quantities and temperature program. In the case of microwave heating, it is recommended to begin with low heating rates, usually between 10 and 20 °C/min, to avoid spontaneous reactions [71].

Acid mixtures and digestion temperatures and times should be optimized until the obtention of a clear solution. Solid particles in the solution indicate that the sample was not completely digested. In this case, the elemental composition analysis will not be accurate.

Finally, the sample amount can be gradually increased, in order to maximize the elements concentration in the digestion solution. However, it is important to take into account that too high sample amounts could cause overpressure in vessels.

2.5.2.3 Sample Analysis

When the total digestion of the sample is completed, the resulting solution is diluted volumetrically with demineralized water, according to the requirements of the analysis apparatus. Generally, the recommended final dilution volume is between 50 and 100 mL. The diluted solution can be then analyzed using the ICP-OES.

Table 2.16 ICP-OES analysis of a coconut shells and a solid recovered fuel

Coconut shells (CS)		CSR 28	
Major elements (mg/kg dry basis)		Major elements (% m dry basis)	
K	2807.8	Ca	3.35
P	396.7	Si	1.00
Ca	391.4	Al	0.99
Si	309.3	Fe	0.24
Al	261.6	Mg	0.22
Mg	170.4	K	0.14
Fe	159.8	Ti	0.14
Na	32.8	Minor elements (mg/kg dry basis)	
Ni	0.0	Sn	843
Zn	0.0	Na	394
Cu	0.0	P	200
		Zn	106
		Mn	78
		Cr	77
		Ni	46
		Cu	45
		Sb	45
		As	39
		V	28
		Pb	27
		Tl	20
		Cd	13
		Co	0
		Hg	<0.1

Before the sample analysis, the ICP-OES equipment should be correctly calibrated. To achieve this task, standard multielement solutions are commercially available at different concentrations and with different acid matrices. The expected element concentration in the analyzed samples should be in accordance with the calibration standard solution range and acid matrix, to achieve accurate results. It is recommended to have at least three calibration points at different concentrations for each element to be analyzed and to repeat the measurement twice or 3 times.

Also, for a reliable elemental analysis, it is recommended to perform an analytical background or blank test using only the reagent mixture used for the sample digestion, in order to identify possible contamination sources from reagents or sample preparation process.

The composition of the samples, in percentage can be calculated for each element according to Eq. (2.18).

$$m_{element} = \frac{C_{element} \cdot V_{solution}}{m_{sample}} \times 100 \quad (2.18)$$

where $m_{element}$ is the mass percentage composition of each element (%), $C_{element}$ is the element concentration in the analyzed solution (mg/L), $V_{solution}$ is the total solution volume (L) and m_{sample} is the mass of the sample used during the mineralization process (mg).

2.5.2.4 Examples

As an example, the inorganic composition of two different materials is presented in Table 2.16: Coconut Shells (CS), a tropical lignocellulosic residual biomass, and CSR 28, a solid recovered fuel.

It can be noted that ICP-OES is a straightforward technique for the inorganic elemental analysis of biowastes. The sample preparation is the most critical step for these materials, taking into account their great heterogeneity and variability. Accordingly, this section presented an insight to the general sample preparation methods and experimental protocols for ICP-OES elemental analysis of biomass and biowaste, and constitutes a guideline for users that start working with these kinds of materials.

2.5.3 *Elemental Analysis by X-ray Fluorescence (XRF)*

2.5.3.1 Introduction

X-ray fluorescence (XRF) spectrometry is a technique for elementary analysis which allows identifying and quantifying most of the chemical elements contained in a sample. The technique is fast, accurate and non-destructive and samples measured could be liquid, powder, compacted powder, and ceramic. The XRF spectrometer can be obtained from various suppliers, mainly Bruker [72], Thermo Fisher [73] or PANalytical [74].

Two type of XRF spectrometer exist which are energy dispersive systems (ED-XRF) and wavelength dispersive systems (WD-XRF). Elements that can be analysed are from sodium to uranium (Na to U) for ED-XRF and from beryllium to uranium (Be to U) for WD-XRF system. The main differences between the two XRF spectrometer principles are presented in the Table 2.17 [75].

X-rays are electromagnetic waves and their wavelength ranges from 0.001 to 10 nm, which corresponds to energy levels from 0.125 to 1250 keV. When X-rays interact with a sample, different phenomena occur (Fig. 2.11): a fraction of X-ray photons is transmitted and a fraction is absorbed; the fluorescence is the phenomenon in which the absorption of a radiation by a material produces the

Table 2.17 Differences between WDS-XRF and EDS-XRF

Operating characteristic	WDS (crystal dispersive)	EDS (silicon, energy diffraction)
Detectable elements	Variable, <30% Detects Z $\geq$ 4	~100% for 2–16 keV Detects Z $\geq$ 10 (Be Window) Detects Z $\geq$ 4 (windowless or thin window)
Resolution	Crystal-dependent (5 eV)	Energy-dependent (130 eV at 5.9 keV)
Maximum count rate (counts/s)	50,000 on an X-ray line	Resolution-dependent, <4000 over full spectrum for best resolution
Minimum useful probe size (nm)	~200	~5
Typical data—collection time	Tens of minutes	Minutes
Spectral artifacts	Rare	Major ones include escape peaks, pulse pileup, electron-beam scattering, peak overlap, and window absorption effects

Reproduced with permission from Springer Nature [75]

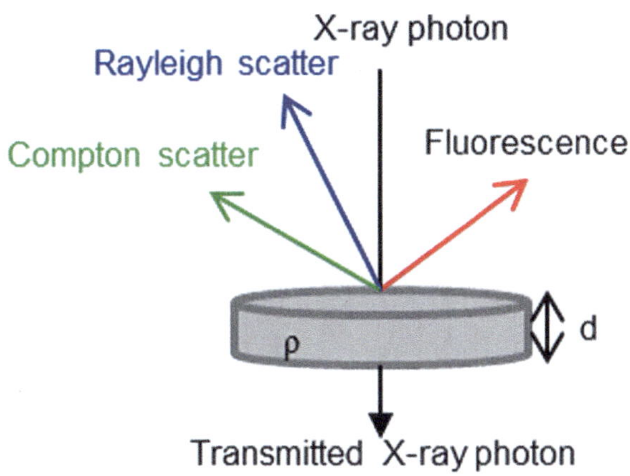

Fig. 2.11 X-ray interaction with matter

re-emission of a less energetic radiation characteristic of an electron transition from an excited state to a lower energy level. The two other phenomena corresponds to the Compton (with loss of energy) and Rayleigh (without loss of energy) scattering. The fluorescence and scattering depend on the thickness (d), the density (ρ), the chemical composition of the material and the energy of the incident X-rays.

2.5.3.2 Analysis Principle

The X-ray beam produced by an X-ray tube interacts with the matter; the radiation re-emitted by the sample is analysed. An atom consists of neutrons and protons around which the electrons gravitate. The electrons are on different levels of energy so called electron shells K, L, M and N as illustrated in Fig. 2.12. Each chemical element has its own energy levels that are listed in the periodic table of elements. For example, for lead (Pb) the energy levels (K_α, K_β, L_α, L_β etc.) for each shell can be extracted from the periodic table (Fig. 2.12).

Figure 2.13 represents the interaction between an incident X-ray beam and the electron shells of an atom divided among a number of energy levels. When an X-ray radiation hits an atom, an electron of an inner shell (E_o) is removed. The atom is then in an excited state. The de-excitation is accompanied by an electronic transition: an electron of higher layer (energy level E_1 or E_2) has to fill the void (left by the photoelectron, that is to say the electron ejected by photoelectric effect). This electronic transition releases the energy corresponding to the energy difference of the levels involved, either by the emission of an X-ray photon (X-ray fluorescence) of energy $(E_o - E_1)$ or $(E_1 - E_2)$, or by transmitting the energy to one of the atomic electrons (Auger emission), both processes being in competition. The energy of the emitted photon will be characteristic of the atom. Several lines can be emitted during these electron transitions (Fig. 2.14). For transitions from the L layer to the K layer (n = 2 to n = 1), we can observe the $K_{\alpha 1}$ (KLII) and $K_{\alpha 2}$ (KLIII) lines. For the transitions from the M layer to the K layer (n = 3 to n = 1), we can observe the $K_{\alpha 1}$ and $K_{\alpha 2}$ lines.

In Fig. 2.15 is presented an example of XRF spectrum. The emission peaks at 2.6, 3.7 and 4.5 keV are attributed to Cl, Ca and Ti sublayers, respectively. In the event the energy levels are very close, the $K_{\alpha 1}$ line cannot be distinguished from the

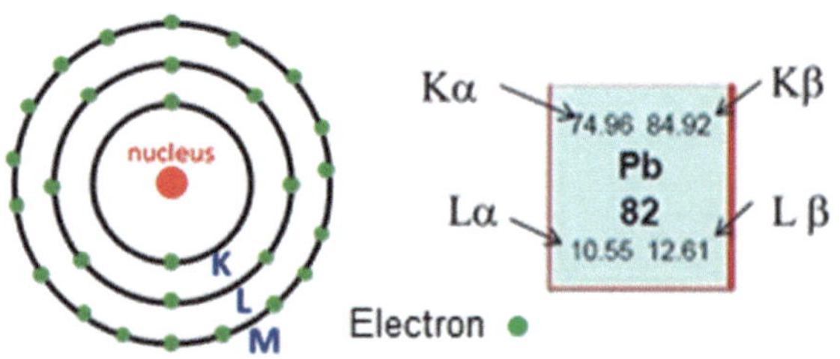

Fig. 2.12 Examples of energy levels

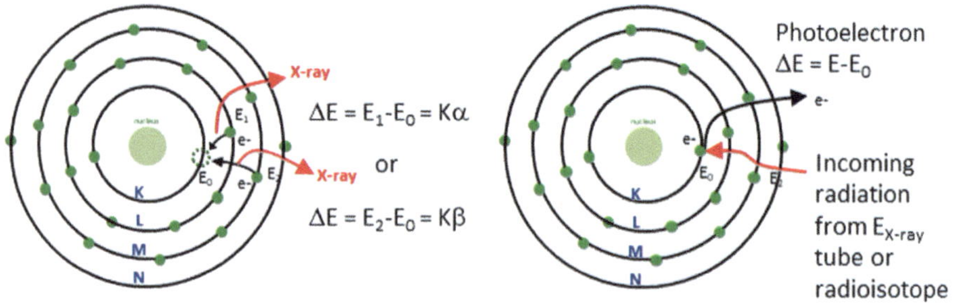

Fig. 2.13 Illustration of energy levels of electrons

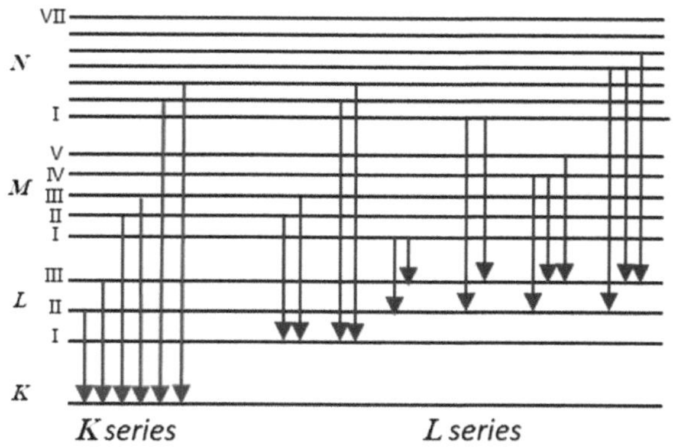

Fig. 2.14 Emission transitions for the K and L series

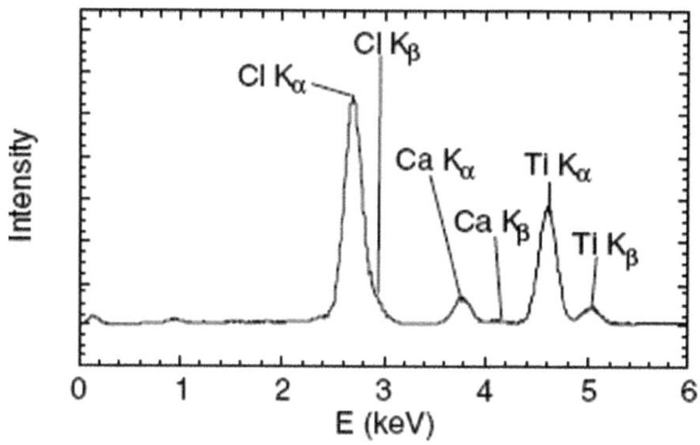

Fig. 2.15 Example of XRF spectrum. Reprinted with permission from Malvern Panalytical [74]

$K_{\alpha 2}$ and only K_{α} is measured. The X-ray fluorescence spectrum shows the intensity collected by the detector as a function of the different energies emitted by the sample. It consists of characteristic peaks, each corresponding to an electronic transition in an atom. The different elements are identified from the position of the peaks. For a line to appear on an X-ray fluorescence spectrum, the excitation radiation must be sufficiently energetic to remove an electron from the required level. For example, if the excitation radiation is an X-ray beam, in order for the K-type lines to be present on the spectrum, the energy of the incident X photons must be greater than that of its absorption threshold. As an example it is necessary that the energy is greater than 85 keV for lead.

X-ray fluorescence spectrometers use excitation X-rays. The beam is focalized on the sample (solid or liquid) which re-emits fluorescence X-ray radiations at an angle only limited by the absorption. In energy dispersive spectrometers (ED-XRF), the photons emitted by the sample are "sorted" into energy by a cooled semiconductor detector, then amplified and analyzed. In regards to X-ray photon analyses, it is important to take into account the "matrix effect". In X-ray fluorescence, the intensity of a line is not directly proportional to the concentration of the emitting element. The other atoms, constituting the sample, also modify the signal, which is called the matrix effect. The latter comprises two phenomena:

- X-ray absorption: the X-rays emitted by the tube are absorbed by the sample before reaching the target atom, this is the primary absorption. The fluorescent X-rays are then absorbed by the sample while emerging, it is the secondary absorption;
- Secondary fluorescence: the fluorescent X-rays emitted by an atom can in turn excite a neighboring atom. Therefore, the atoms receive X-rays from the tube and X-rays from other atoms. These phenomena that may occur are now taken into account in the calculations. Software models these two phenomena during calculations.

2.5.3.3 X-ray Fluorescence Intensity

For an isolated atom, the probability that an electron from an outer shell will take the vacant place while emitting an X-ray line is called the fluorescence yield (ω). This efficiency is close to unity for K-shell ionization and high Z-values. It is lower for L- and M-shell ionizations and decreases rapidly as Z decreases. The intensity of a line is therefore proportional to the yield ω which is the same for each series or sub-shells. It is therefore better to use the most intense lines for chemical analysis, i.e. the K_α lines for the K series and the L_α lines for the L series. The intensity measured for K lines are already higher than those of L lines. For example, in a given sample containing the same concentration of Fe (measured with the K lines) Pb (measured with the L lines), the intensity measured will be higher for Fe than for Pb.

As example, the model EPSILON 3 XL energy dispersive X-ray spectrometer (PANanalytical) has a rotating plate which can receive up to 10 samples. The optical path of this apparatus is represented in Fig. 2.16.

- X-ray tube (1): the anode is in Ag. Its power is 15 W for a voltage ranging from 4 to 50 kV and a current of 1 μA to 3 mA.

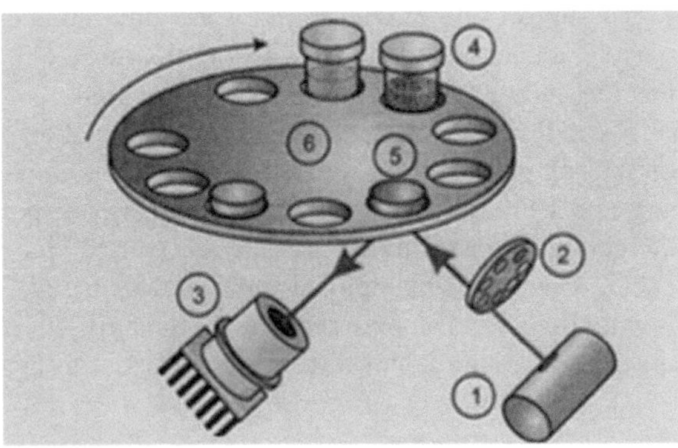

Fig. 2.16 Optical path of X-rays; 1: X-ray tube; 2: tube filters feeder (7); 3: detector; 4: samples; 5: measuring position; 6: autosampler. Reprinted with permission from Malvern Panalytical [74]

- Filters (2): there are 7: Cu-500, Tu, Ag-50, Al-200, Cu-300, Ag, white. As the detector analyses all energies at the same time, there is a risk of saturation. So, filters are used to attenuate the signal of elements to avoid saturation. The filters make it possible to cut the periodic table into 5 different analysis conditions, each corresponding to a series of elements.
- Detector (3): (SDD HR) High Resolution Silicon Drift Detector (SDD). It has a resolution ranging from 135 eV to 5.9 keV. It is controlled thermoelectrically by a Peltier system that stabilises a temperature below −17 °C.

The XRF-spectrometer is coupled with a specific software containing different programs as a function of a use as a qualitative or quantitative measurement. For quantitative measurements, standard samples are needed to determine a calibration curve.

2.5.3.4 Measurement Error

In X-ray fluorescence, the maximum count statistical error is the square root of the measured intensity. This is the total intensity that is to say the intensity measured in cps × time (s). To have a better precision it is necessary to increase the analysis time. If a light element or an element whose concentration is low is measured, increasing the measurement time makes it possible to reduce the statistical error of counting. For each type of element measured, the operating conditions as well as the gas atmosphere of the analysis will be different. These operating conditions used are presented in Table 2.18.

2.5.3.5 Sample Preparation

XRF allows the analysis of both liquid and solid samples. The sample preparation step plays a very important role.

For liquid samples, a cylinder made of inert ceramic materials is used (Fig. 2.17a). The dimension of this cylinder is of 32 mm in diameter and 26 mm in height in the case of a EPSILON 3 XL energy dispersive X-ray spectrometer

Table 2.18 Operating conditions measurements. Reproduced with permission from Malvern Panalytical [74]

Element conditions	kV	μA	Filter	Environment
<F–Si>	5	60	Non	He
<P–Cl>	10	30	Ti	He
<K–V>	12	25	Al-50	Air
<Cr–Co>	20	15	Al-200	Air
<Ni–Mo>	50	6	Ag	Air
<Rh–Sb>	50	6	Cu-500	Air
<default>	10	100	Non	Air

Fig. 2.17 Examples of sample preparation

(PANanalytical). A transparent film at the bottom of the cylinder makes possible the introduction of the sample. The type of the film used must be well-known. The characteristics of the commercially-available film are introduced in the software which are later taken into account for the calculation of analysis results. Mylar film is usually used with a thickness of 3.6 μm.

For solid samples, the measurement can be done with bulk powder, compact disc made by powder compression or fused bead made by powder melting technique. Bulk power measurement does not need any preparation. Powder is introduced into the sample chamber for the analysis. This technique could be used for the identification of elements present in a sample. It is not adapted for estimating the elemental concentration because of the inter-particular porosity of the sample. To reduce this porosity, powder can be compressed into compact disc (diameter of 32 mm and height of 1–3 mm (Fig. 2.17b). To prepare compact disc, a press is needed. In the example below, the press used is an Atlas Power Presses 15T which is a power assisted hydraulic press operating up to 15 tons/cm^2. The powder is introduced in the press chamber and a pressure varying between 5 and 15 tons is applied for at least 5 s. It is recommended to apply the same pressure and duration for a series of analysis. Disc prepared by this method has a plane surface with minimum of inter-particular porosity. Thus, it can be applied for semi-quantitative and quantitative analysis. Solid sample can also be prepared by melting technique, which allows completely eliminating the porosity of sample, as well as minimizing the matrix effect. Figure 2.17c illustrates an example of fused bead. For the preparation, the sample with a particle size smaller than 75 μm is mixed with a melting agent (fondant). The melting agent is usually lithium metaborate ($LiBO_2$) or lithium tetraborate ($Li_2B_4O_7$). The mixture is usually constituted of 10–20 wt% of sample. Then, the mixture is introduced into a platinum crucible and heated until 1100 °C in a muffle oven in order to melt the mixture. A platinum cover plate is also used to avoid volatilization of elements. During heating, the mixture is agitated by a movement of rotation or displacement. This melting mixture is used to prepared bead with a specific mold (3 mm of thickness and 32 mm of diameter).

For samples preparation, information like the matrix composition, the mass and the height of the sample have to be introduced in the software before the measurement. These data will be considered by the software. For semi-quantitative measurement, it is important to repeat at least threefold the measurement to ensure the reproducibility. For quantitative measurement, calibration curve using standards must be done prior to the measurement. The standards used must be prepared with the same "matrix" of the targeted sample to minimize the "matrix effect".

2.5.3.6 Semi-quantitative Measurement

This method allows calculating the elemental composition of a given sample using a specific software provided by the XRF-spectrometer constructor. Any calibration curve is needed. The accuracy of this method is generally of around $\pm 10\%$. For a liquid sample, the solvent composition has to be specified into the software. For a solid sample, the empirical chemical formula (i.e. CHNSO analysis for a woody biomass or its derivatives) of the matrix has to be determined using other methods (i.e. CHNSO, ICP-AES) which also must be specified into the software [76]. It has been previously demonstrated that using pure cellulose ($C_6H_{10}O_5$) as the default composition of the organic matrix gives comparable results to the use of an empirical formula for a broad range of biomass types [77]. The matrix composition is considered by the software for the calculation of the concentration of the element of interest from its XRF intensity.

Using the EPSILON 3 XL energy dispersive X-ray spectrometer (PANanalytical), the different steps for the semi-quantitative measurement are the following:

- In the software Toolbar go the "Measure" icon and select Omnian application;
- Specify the sample information:
  - Sample name;
  - Position in the apparatus (1 to 10);
  - Type: liquid, pellet or bead;
  - For liquid sample, it is necessary to input the diameter of the cup, the mass and height of the liquid and the type of film (Mylar, thickness of 3.6 μm) to the software;
  - For compact disc and fused bead sample, it is necessary to input the diameter, the mass and the height of the sample to the software.
- The analysis is carried out with the Omnian program which uses all operating conditions presented in Table 2.18.
- Close the cover plate of the apparatus and the analysis starts automatically;
- After the measurement, results of mass fraction obtained for elements or oxides are generated. At this step, the matrix composition has to be specified into the software which will be taken into account for the recalculation of the composition of the elements of interest.

Since the software considers only the global composition of the sample matrix, but does not go further to the effect of different components of the sample, the margin of errors from this method can be important (i.e. ±10%) [78, 79]. This problem can be fixed by periodically checking the instrument using well-know sample [78]. In addition, for biomass-based samples, the matrix is mainly composed of C, H and O, which are light elements and which mainly influence X-rays through energy absorption and scattering [77].

2.5.3.7 Quantitative Measurement

This method needs the determination of a calibration curve prior to the analysis. In all cases, standards for the preparation of calibration curve must have the same "matrix" than that of the sample to be analyzed.

For liquid sample, this criterion is easily satisfied because the standards can be quickly prepared using the same solvent (i.e. water, organic solvents) than the sample. Standards with various concentration of the element of interest are analyzed by a XRF spectrometer. Then, the calibration curve can be obtained by plotting the intensity as a function of the concentration. Linear calibration curve is generally preferred. In this case, at least 4 standards are needed to build a calibration curve.

For solid samples, compact disc or fused bead containing various concentration of the element of interest are prepared as described above. The preparation of fused beads standards having the same matrix is easy because the matrix is that of the melting agent. On the other hand, the preparation of compact disc standards having the same matrix is not easy, in particular for biomass and biowastes and their derivatives. In this case, a suitable solution is to grind properly the sample to be analyzed into fine powder (micrometric particles). Then, different well-defined amounts of the element of interest will be added to this fine powder in order to prepare different standards. These standards are measured and the calibration curve can be obtained by plotting the intensity as a function of the concentration. Figure 2.18 illustrates the principle of the quantitative analysis by this method. By this method, the intensity of the element of interest from the sample is always smaller than that of standards. Thus, its concentration is extrapolated from the calibration curve. The preparation of standards by this method practically eliminates the matrix effect.

The quantitative XRF analysis using an EPSILON 3 XL energy dispersive X-ray spectrometer (PANanalytical) includes the following steps:

- In the software Toolbar go to the "Application" icon and select "New Application";
- "New application" window appears where the name of the method can be added;
- In the new window, select the type of sample: liquid, pellet (compact disc) or bead. For fused bead, the following information must be specified: the formula of the melting agent, the initial weight of the melting agent used to make the bead, the sum of the weight of the melting agent plus that of the sample, and the value of L.O.I. (Loss On Ignition). For pellet sample, the weight has to be

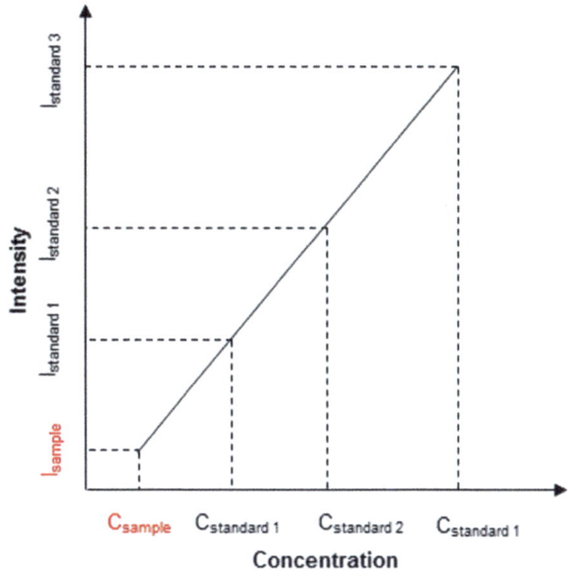

Fig. 2.18 Typical calibration curve obtained with standards prepared by adding well-defined amounts of the element of interest to the sample to be analyzed

specified. For liquid sample, the solvent formula, the initial volume of the sample and the final volume which corresponds to the total volume of the solvent and the sample must be specified;

- In the "Analysis conditions" window, select the element to be analyzed, the unit of the results (%wt. or ppm) and the operation conditions needed (see Table 2.18). For example, to analyze Ca and Cl, the operating conditions are <K-V> and <P-Cl>, respectively;
- Click on "Measurement optimization";
- Input the standards: For each standard prepared, specify the concentration of the element and its position in the apparatus. The measurement of these samples gives the intensity corresponding to each standard;
- Click on "Calibration right". A calibration curve is built showing the relationship between the concentration and the XRF intensity of the element of interest. This calibration curve is preferred to be linear, described by a mathematic equation of the first order;
- Sample analysis: From the XRF intensity of the element of interest in the sample to be analyzed, its concentration is calculated by the software from the calibration curve built above.

2.5.3.8 General Comments

For biomass, biowastes and their derivatives, different elemental analysis techniques are available:

- For light element, CHNSO analysis technique is adapted to determinate these major elements which constitute in general the matrix of sample.
- ICP-AES or ICP-OES allows analyzing most of the elements. But samples must be in liquid phase. A complete mineralization step must be performed for solid samples and this step is generally delicate for biomass, biowastes and their derivatives.
- XRF has advantage to be able to analyze both liquid and solid samples. However, the "matrix effect" must be carefully considered for semi-quantification or quantification analyses.
- For sulfur and halogens analysis, the combination of bomb calorimetry and ion chromatography is an alternative method [80–82].

2.5.3.9 Examples

Semi-quantitative Analysis

Results of XRF measurements obtained with a wood (poplar) and a refused-derived fuel (RDF), using the EPSILON 3 XL energy dispersive X-ray spectrometer from PANanalytical, are presented in Tables 2.19 and 2.20, as elemental or oxide concentration. The matrix of each biomass was determined by CHONS apparatus and then introduced in the XRF software. For the poplar, the matrix is composed by C, H, and O elements and represents 99% in weight of the sample. Concerning the refused-derived fuel, the matrix is composed of C, H, N O and S elements.

Another example concerns the analysis of chlorine (Cl) and bromine (Br) contents in rice husk (RH) using the EPSILON 3 XL energy dispersive X-ray spectrometer from PANanalytical. The sample was prepared with 1 g of RH, originally in powder form (250–500 μm particle size). The sample was pressed to form a pellet with a diameter of 32 mm and a thickness of 2 mm. To perform the test, it was necessary to input the matrix (empirical formula) of RH into the Epsilon 3 software. This matrix was determined to be $C_6H_{10}O_5$ based on the results of the

Table 2.19 Example of XRF analysis for a poplar wood

Element	%weight	Oxide	%weight
Mg	0.072	MgO	0.119
Si	0.111	SiO_2	0.237
P	0.047	P_2O_5	0.107
S	0.016	SO_3	0.039
Cl	0.01	K_2O	0.144
K	0.12	CaO	0.269
Ca	0.192	Fe_2O_3	0.001
Fe	0.001	ZnO	0.001
Zn	0.001	$C_4H_3O_6$	99.072
$C_4H_3O_6$	99.432	Cl	0.010

Table 2.20 Example of XRF analysis for a refused-derived fuel

Element	%weight	Oxyde	%weight
Na	0.076	Na_2O	0.105
Mg	0.126	MgO	0.214
Al	0.532	Al_2O_3	1.025
Si	1.311	SiO_2	2.841
P	0.032	P_2O_5	0.074
Cl	2.335	K_2O	0.391
K	0.342	CaO	16.152
Ca	12.182	TiO_2	0.924
Ti	0.585	Cr_2O_3	0.054
Cr	0.039	MnO	0.024
Mn	0.019	Fe_2O_3	0.769
Fe	0.568	NiO	0.009
Ni	0.008	CuO	0.124
Cu	0.105	ZnO	0.212
Zn	0.18	PbO	0.008
Pb	0.008	CH_2ONS	74.857
CH_2ONS	81.55	Cl	2.216

CHNSO analysis. The RH pellet was then placed in a holder in the XRF chamber and the concentrations of these halogens were calculated from the intensity of the emitted rays. Results obtained for the contents of Cl and Br in RH were of 0.046 and 0.001 wt%, respectively.

Analysis with a Calibrated Method

The aim of this test was to calibrate the XRF instrument in order to measure the Cl content of industrial wood waste. The samples used were pellets containing a mixture of 0, 5, 10, 100 wt% PVC in poplar wood (i.e. 0.01, 2.9, 5.7, 57 wt% Cl in pellet). These pellets of roughly 1 g and 2 mm thickness were prepared by first mixing PVC and poplar wood powders at the specified weight ratios, and then applying a pressure of 10 tons to the mixture using a laboratory press (Atlas Power Press 15 tons). Figure 2.19 displays the calibration curve obtained for Cl measurements in wood PVC-pellets. The linear regression performed on the data, yielded a coefficient of determination R2 close to 1.

To evaluate the calibration quality, the gradient of the calibration curve (i.e. 2412.9) was used to determine the chlorine content in a new set of samples containing 0, 1, 5 wt% PVC (i.e. 0.01, 0.57, 2.9 wt% Cl, calculated) in wood pellets. Figure 2.20 shows the curve obtained when the measured chlorine contents (using the calibration data) were compared to the known (calculated) quantities of chlorine in the pellets. The linear regression conducted on the data shows a high correlation between the measured and calculated chlorine contents.

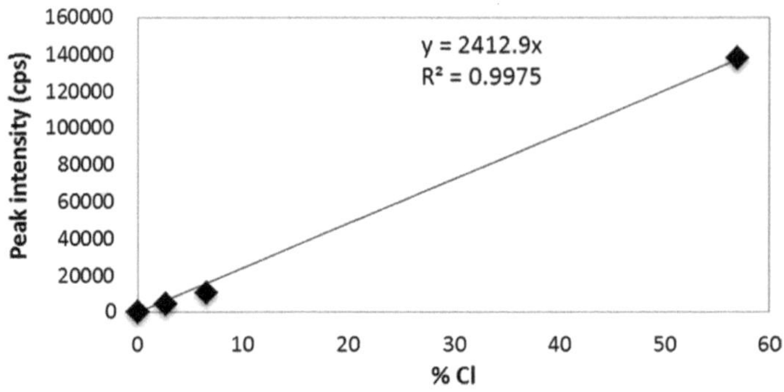

Fig. 2.19 Calibration curve of chlorine in PVC-wood pellets

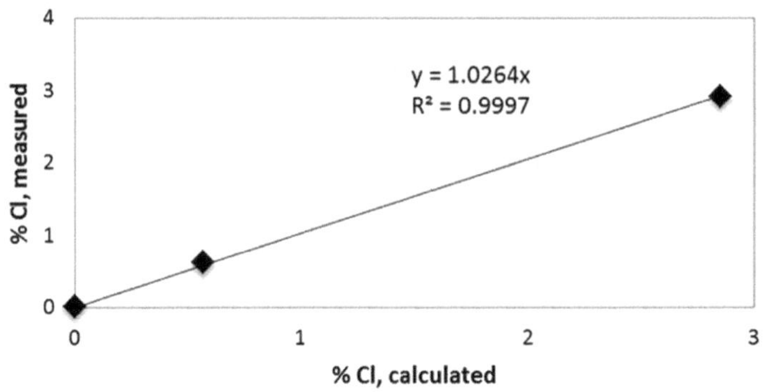

Fig. 2.20 Comparison of measured and theoretical chlorine in PVC-wood pellets

2.5.4 Total Organic Carbon Analysis

2.5.4.1 Introduction and Principle of Measurement

Total organic carbon (TOC) indicates the content of organic carbon in a liquid or solid sample. Carbon-containing compounds can be found in organic or inorganic form. Both of them can be determined by the method commonly named TOC analysis.

The principle of TOC analysis is similar for both liquid and solid samples. Firstly, total carbon (TC, organic carbon and inorganic carbon) is determined. Then, inorganic carbon (IC) is independently determined. Finally, total organic carbon (TOC) is obtained by the difference of TC and IC. Carbon-containing compounds are converted into gaseous CO_2, which can be detected and quantified by an IR detector.

2.5.4.2 Liquid Sample

For liquid sample, homogeneous solution is required. No particulate matter is admitted to do not clog the injection line.

Firstly, calibration curves are performed for both total carbon (TC) and inorganic carbon (IC). For TC calibration curve, potassium hydrogen phthalate (CAS 877-24-7, chemical formula $C_8H_5KO_4$) is usually used. For IC calibration curve, sodium hydrogen carbonate (CAS 144-55-8, chemical formula $NaHCO_3$) is usually used. Aqueous solutions containing different concentrations of potassium hydrogen phthalate or sodium hydrogen carbonate are prepared and used as standards. At least 4 standards are used to build a calibration curve. The concentration ranges of TC and IC must be adjusted in order to obtain linear calibration curves (usually <100 mg of C per liter).

For a given measurement, homogeneous sample is prepared for a required volume around 5 mL. This volume is enough to repeat three time the measurement of a sample. A fraction of sample is injected in a catalytic bed, which is composed of spherical Pt/Al_2O_3 particles and heated by an oven at 680 °C. Under a flux of synthetic air, all organic and inorganic carbons are converted into CO_2 gas. This CO_2 is detected by an IR detector. The concentration of total carbon (TC, usually in mg/L) can be calculated from the TC calibration curve. Then, another fraction of sample is used for IC analysis. This fraction is injected and mixed with 25 wt% H_3PO_4 acid solution, which converts all inorganic carbons (carbonates and hydrogen carbonates) into CO_2. This CO_2 is then detected by the IR detector. The value of IC concentration (usually in mg/L) can be obtained from the IC calibration curve. Both TC and IC determination steps are generally programmable and can be automatically executed by TOC machine. The difference of TC and IC gives TOC concentration (mg/L).

2.5.4.3 Solid Sample

Firstly, calibration curves are also performed for both TC and IC, using glucose and sodium hydrogen carbonate, respectively, for preparing standards. It is also important to work within the concentration ranges wherein TC and IC calibration curves are linear.

For solid sample, fine powder at the micrometric scale is in general required. A fraction of sample is placed in a ceramic nacelle, which is then pushed into an oven heated at 900 °C. It is worth noting that weighing precisely a sample is crucial for the calculation of the final result. At 900 °C and under synthetic air flux, all organic carbons are burned into CO_2. Also, all inorganic carbons are decomposed into CO_2. The value of TC (wt%) can be obtained from the TC calibration curve. Then, another fraction of sample is used for IC measurement. This fraction is wetted by a large amount of 25 wt% H_3PO_4 acid in a ceramic nacelle. Then, this nacelle is pushed into an oven heated at 200 °C. Inorganic carbons are decomposed into CO_2.

The value of IC (wt%) can be obtained from the IC calibration curve. The difference of TC and IC gives TOC concentration (wt%).

For the analysis of both solid and liquid samples, calibration curves can be validated for a long period (around 6 months for a research lab). However, it is recommended to periodically check calibration curves using standards.

2.5.4.4 Commercial Apparatus

Shimadzu is well known as manufacturer of TOC analyzer. Different series of machine exist such as TOC 5050A model, TOC V_{CSH} model, TOC-L model etc. [83]. Other manufacturers can be cited such as Mettler Toledo [84], General Electric Company [85] etc.

2.5.4.5 Examples

TOC is largely used for the determination of organic content in an aqueous solution, particularly in a given wastewater. For example, the value of TOC in olives mill wastewater can reach up to around 70 g L^{-1} [86].

2.5.5 Halogen Analysis by Calorimetric Bomb-Ionic Chromatography Coupling

2.5.5.1 Introduction and Principle of Measurement

Halogen contained in biomass and biowastes can produce corrosive and harmful components, such as halides or dioxins, when a thermochemical conversion is processed. The determination of chlorine, bromine and fluorine contents in feedstock is recommended prior to a thermal treatment.

Basically the determination of mineral composition could be performed with ICP-OES, however the technique demonstrates a low confidence regarding the quantification of the halogens. In fact, ICP-OES has low sensibility for halogen elements compared to other techniques. Therefore their detection and quantification is mostly performed with:

- Liquid ion chromatography [87], for the determination of chloride, bromide and fluoride;
- ICP-MS [88], for the determination of bromine;
- Silver nitrate titration [89], for the determination of chloride;
- Electrochemical probe [90], for the determination of fluoride.

Among them, the most preferred method is the quantification using liquid ion chromatography. For solid biomass or wastes analysis, a pre-treatment in order to transform halogen compounds present in the initial sample into a homogenous aqueous solution must be performed. Then, halogen anions will be analysed by ionic chromatography.

2.5.5.2 Ionic Chromatography

Liquid chromatography is an analytical technique for the separation of a mixture: components are eluted by a mobile phase (liquid) through a stationary phase (solid) and they are distributed according to their affinity with mobile and/or stationary phase and then detected and possibly quantified. Chromatographic systems are usually composed by at least: an injector for the sample introduction, a pump for the elution by the mobile phase, a chromatography column (stationary phase), a detector and an integrator for signal treatment (Fig. 2.21).

For ionic chromatography, the retention of ionic components in the sample is based on the ability to be exchanged with the counter-ion of the mobile phase. In most of the cases and for this analysis: the stationary phase is composed of an ion exchange resin, the liquid phase is an ionic aqueous eluent and the detection is performed with a conductivity detector. In addition an ion suppressor can be placed before conductivity cell to avoid the noise due to the conductance of eluent. The pump should be metal-free to avoid external contamination.

In other applications of ionic chromatography the liquid phase can be organic, and the detection can be operated by a UV-detector or an electrochemical detector.

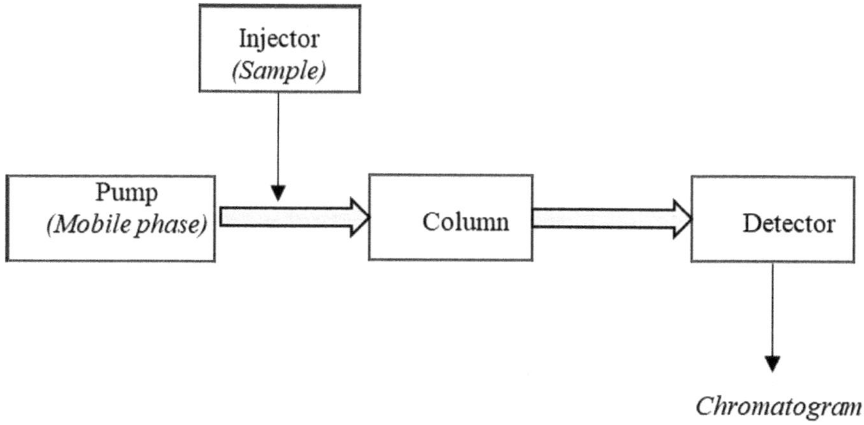

Fig. 2.21 Principle of a liquid chromatography system

2.5.5.3 Procedure of Calorimetric Bomb-Ionic Chromatography Coupling

Figure 2.22 shows the main procedure that could be followed to determine the halogen content in biomass and biowastes. The method consists in the combustion of the sample with pure oxygen in an adiabatic closed system (calorimetric bomb), which converts halogen elements of the sample into gaseous halogen compounds, the trapping of halides in an appropriate absorption solution, and finally the determination of halogens content by different methods. The standards EN 14582 [91], EN 15408 [92] and ISO 16994 [93] provide procedures for the extraction of halogen content in wastes, solid recovery fuels and solid biofuels respectively. As previously mentioned, the quantitative analysis could be performed with various methods. Oxygen combustion in a calorimetric bomb followed by a quantitative analysis with an ion chromatography is the preferred method that will be described.

Procedure for Combustion and Sample Preparation

This procedure is adapted from EN 15408 [92]. The calorimetric bomb system used to burn the biomass sample can be the same as the one used to measure the high heating value (HHV) of a fuel (Sect. 2.6.3). Hence, the procedure is similar and is summarized as follows:

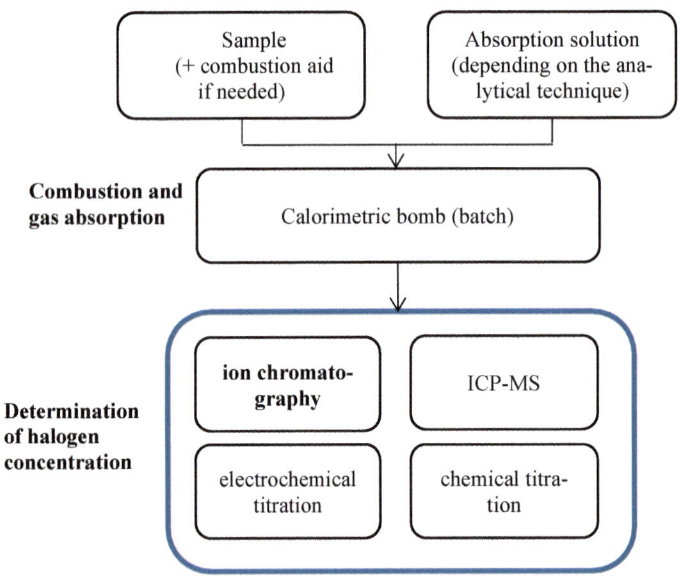

Fig. 2.22 Procedure for halogen quantification from biomass and biowastes

- Weigh about precisely 1 g of the sample (if a lower amount must be weighed, a particle size less than 1 mm is required in order to ensure the homogeneity of sample). The sample can be pressed to produce a pellet or placed and weighed directly in a specific capsule;
- Add a combustion aid to the sample if needed (e.g. cotton, benzoic acid, capsule or polyethylene bag);
- Add 10 mL of 0.2 mol/L KOH solution (the mobile phase for ion chromatography can also be used) to the decomposition vessel; fill the bomb with oxygen and set up the system according to the operator instructions. If the chlorine content is <1% or bromine is not to be determined, deionized water can be used instead of KOH solution;
- Start the combustion;
- After combustion, equilibrate at room temperature for at least 10 min. After opening the bomb, check the completeness of combustion, by observing the presence of solid residues. If unburned fractions are present, the experience must be redone. Quantitatively recover the absorbing solution in a 100 mL tarred bottle and fill up to the final volume with deionized water.

Note: It is recommended to process a blank control before each determination and between two determinations if the first one results in a high concentration of halogen elements.

Determination of Dissolved Anions Cl^-, Br^- and F^- by Liquid Ion Chromatography

Equipment

This is adapted from the standard ISO 10304-1 [87]. Samples, standard and calibration solutions must be stored in polyethylene vial before use. Moreover, they have to be filtered through a membrane filter with a 0.45 μm pore size before injection.

- Column: the stationary phase is an anion exchange resin. The working conditions and the nature of the column and mobile phase must be able to obtain a resolution (R) between two adjacent peaks higher than 1.3. Equation (2.19) shows the calculation of peak resolution R for two peaks namely 1 and 2, illustrated in Fig. 2.23; with t_{R1}: retention time (min) of the first peak 1; t_{R2}: retention time (min) of the second peak 2; $t_{R1} < t_{R2}$; w_1 and w_2: width (min) of the peak 1 and 2, respectively.

$$R_{2,1} = \frac{2(t_{R2} - t_{R1})}{w_2 + w_1} \tag{2.19}$$

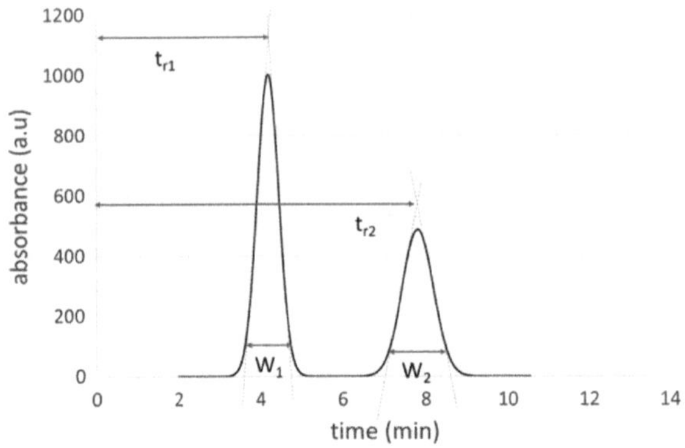

Fig. 2.23 Graphical representation of the parameters to calculate the peak resolution *R*

- A pre-column can be added to prevent the pollution of the column.
- Eluent: this is an aqueous solution of weak mono-and dibasic acids, the water for this preparation must be outgassed (to eliminate dissolved air), and the solution must be freshly prepared every three days.
- The choice of eluent depends on the column and the detector and it also contributes to obtain a good separation between two peaks on the chromatogram with the appropriate column.
- Detector: a conductivity cell with an ion suppressor is used. If the sensitivity must be increased, an UV detector can be used for the bromide determination.

Calibration

Stock (reference) solutions of chloride, bromide and fluoride, with a concentration of 1000 mg/L for each element are recommended. They can be commercially available or freshly prepared from salts dissolution (i.e. NaCl, NaBr and NaI). They are assumed to be stable for several months.

Standard solution: They are obtained by dilution of the stock solution to give intermediate concentration between stock and calibration solutions. They can be a mix of several stock solutions. They are assumed to be stable for one week (if stored in the dark under refrigerated conditions).

Calibration solutions: It is recommended to dilute standard solutions to obtain 5–10 calibration solutions to build a calibration curve. The concentration of calibration solutions must cover the expected work range. They must be freshly prepared the day of the analysis.

Analytical Series

- Blank control: A blank test containing the absorption solution or the mobile phase should be performed to evaluate the contribution of the entire process (combustion, absorption, injection and elution) to the halogen concentration. The composition of each halogen in this blank test is limited, depending on the concentrations measured in the sample. Indeed, the concentrations in blank tests should not exceed 10 or 30% when the measured concentration in the sample is above 0.5 mg/g or below 0.5 mg/g respectively.

In addition, a chromatography blank test can be prepared with deionised water to verify interferences due to eluent and chromatography system.

- Quality control: It is recommended to periodically control the apparatus, using an appropriate mixture which contains all the halogen elements with well-known contents.

The first kind of control must be processed at the same conditions as for the samples. It could help to calculate the trapping yield. This mixture could be a certified reference material (CRM) or a preparation from appropriate chemicals.

The second kind of control is necessary to check the validity of the calibration curve. Consequently, it must be processed under the same conditions as for the calibration solutions.

For both controls, it is recommended to prepare solutions having the concentration of the interesting element roughly in the middle of the calibration range. In the best case scenario, the control concentration of the interested element is close to that of sample.

- Calibration solutions: The solutions can be analysed after preparation as described above. Their injection allows identifying the peaks of interesting anions and building the calibration curve by plotting its intensity against its corresponding concentration. The analysis can be repeated to ensure its reproducibility.

If matrix interferences appear and trouble chromatogram, standard addition method can be applied, but it is not detailed in this document.

- Samples: It is recommended to at least repeat twice the analysis to obtain mean value of peak area. The identification is done by comparing retention time of peaks of the sample with those of standard solutions. The deviation of retention time must not exceed 10%.

Calculation

The results shall be calculated on dry basis, so the moisture content has to be determined (see Sect. 2.4.1).

The total chlorine, fluorine or bromine content expressed in dry weight is given by the following equation:

$$w = \frac{(C - C_0) \times V}{m_{dry}} \times 100 \tag{2.20}$$

where: C is the concentration of chloride, fluoride or bromide respectively in the solution (mg/l); C_o is the concentration of chloride, fluoride or bromide respectively in the blank solution (mg/L); V is the volume of the solution (L); m_{dry} is the dry mass of the test portion used (mg).

The concentration of the interesting element must be found within the calibration range. If it is not the case, the analysis must be redone with a new dilution of the sample.

2.5.5.4 Example

For this example, chlorine content in a solid recovered fuel (SRF) was determined using a C5000 Adiabatic Calorimeter (IKA®-Werke, Staufen, Germany) and Thermo Scientific™ Dionex™ ICS-3000 chromatography IC system, and following the procedure outlined in EN 15408 [92]. The SRF sample was initially ground to powder with a hammer mill and a 1.0 mm screen size. Next, the sample was dried according to EN 15414-3 [94] and then compressed to produce a tablet of roughly 300 mg.

Calorimetric bomb procedure:

The experimental procedure involved the following steps:

- The bomb calorimeter was calibrated using six benzoic acid tablets under 30 bar of oxygen at 24 °C;
- A 10 mL of 0.2 M KOH aqueous solution was added to the combustion bomb to capture any acid gases released during combustion;
- The SRF tablet of approximately 300 mg was weighed on a precision scale;
- The SRF tablet and the ignition cotton of known mass were placed in the crucible;
- The bomb, containing the crucible, was lowered in the calorimeter and the cover was then closed;
- The start button was pressed to begin the test;
- After combustion, the bomb was removed from the calorimeter and left to equilibrate for about 10 min;
- An additional step compared to the standard procedure has been performed to achieve the chlorine extraction. The bomb outlet was then connected, via a plastic tube and valve, to a 100 mL impinger bottle containing 50 mL of 0.2 M KOH. It took between 18 and 24 h to recover the solution in the bomb with minimal loss of undissolved chloride ions.

Ion chromatography analysis:

The analysis was performed with a Dionex C-3000 ion chromatography system. The main characteristic and the operational conditions for the analysis are as follows:

- Eluent: 15–30 mM KOH in deionized water;
- Eluent flowrate: 1.2 mL/min, isocratic flow;
- Guard column: IonPac A611-HC, 4 × 50 mm;
- Analytical column: IonPac AS-11-HC, 4 × 150 mm;
- Automated sampler: 25 μL sample injected in column using injection loop;
- Temperature of the column: 35 °C;
- Conductivity detector.

Calibration solutions were prepared using a commercial standard solution containing 1000 ppm of the anions, Cl^-, F^-, Br^-, SO_4^{2-}, PO_4^{3-} and NO^{3-}, in deionized water matrix. The chlorine content in the SRF sample was determined to be 0.36 wt% on dry basis.

2.6 Thermal Analysis

2.6.1 Thermogravimetric Analysis

2.6.1.1 Introduction

Thermal analysis includes different methods which have large applications in different fields. Various types of samples can be analysed: ceramics, composites, metals and alloys, mineral raw materials, polymers, resins, food products, pharmaceuticals and chemicals, nuclear materials, etc. In the field of energy, thermal analysis can be used for the characterization of petrochemical products, gas hydrate, coal, biomass and biowastes. This chapter will solely focus on biomass and biowastes.

When heated, a material can undergo different physico-chemical and mechanical transformations: decomposition, fusion, crystallisation, oxidation, etc. The main techniques of thermal analysis used are the following [95–97].

As shown in Table 2.21, TGA allows measuring the weight change of a material as a function of the temperature under controlled atmosphere. DTA and DSC indicate the endothermal or exothermal behaviour of a given physico-chemical transformation. DTA gives only temperature where a transformation happens compared to a reference, while DSC gives the precision on heat flow rate associated to the transformation. Finally, TMA measures dimension change under defined conditions of temperature, atmosphere and load applied on sample.

2.6.1.2 Choice of Thermal Analysis Technique

Table 2.22 gives a list of physico-chemical and mechanical transformations and the compatibility of thermal analysis techniques as a function of analysis target.

Table 2.21 List of the main techniques of thermal analysis

Analysis technic	Propriety measured	Abbreviation
Thermogravimetric analysis	Weight variation	TGA
Differential thermal analysis	Temperature difference	DTA
Differential scanning calorimetry	Heat flow rate difference	DSC
Thermomechanical analysis	Dimension change	TMA

Table 2.22 Adequacy of thermal analysis techniques versus physico-chemical and mechanical transformations; (●): the most appropriate technique; (▼): alternative solution; (–): not applicable

Analysis target	DSC	ATG	TMA
Specific heat	●	▼	–
Coefficient of expansion	–	–	●
Young's module	–	–	▼
Melting and crystallisation	●	–	●
Evaporation, sublimation	●	●	–
Glass transition	●	–	●
Solid-solid transition (polymorphism)	●	–	●
Purity analysis	●	▼	–
Decomposition, oxidation	▼	●	–
Kinetic	●	●	●
Enthalpy of reaction	●	●	▼
Crosslinking, vulcanization	●	–	▼

In this section, we discuss only TGA, TDA and DSC analyses. TMA technique will be discussed in another section.

2.6.1.3 General Description of TGA, TDA and DSC

A TGA device consists of a microbalance and a heating system (oven). The oven allows heating sample with a controlled temperature program, while the microbalance measures weight variation during the analysis. This technique allows obtaining the numeric values of temperature and weight variation as a function of time. In general, weight variation is calculated according to Eq. (2.21).

$$weight\ variation = \frac{m_o - m_t}{m_o} \times 100 \quad (2.21)$$

with m_0: initial weight; m_t: weight at a given time t.

The weight variation can be negative or positive. We can distinguish:

- Weight loss $(m_t < m_o)$: dehydration, desorption, decomposition (the breaking of chemical bonds), evaporation, reduction process etc.
- Weight gain $(m_t > m_o)$: adsorption, oxidation (interaction with an oxidizing atmosphere), carbonation etc.

For a measurement, two crucibles are used, one as reference and another containing sample. Different types of crucibles will be presented later.

A microbalance can have different configurations; the most commonly used are suspended, vertical, and horizontal. A given standard microbalance, as illustrated in Fig. 2.24, is composed of a beam whose center of gravity coincides with its axis of rotation articulated around a torsion band stretched between two springs. The

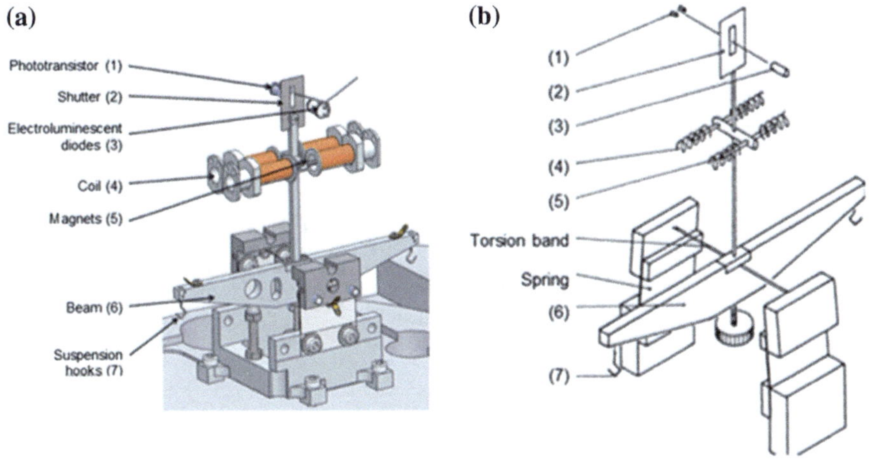

Fig. 2.24 Scheme of the electronic microbalance of B111 model from Setaram. Reproduced with permission from SETARAM

electromagnetic motor consists of a magnet and a coil. The detection of the position of the beam is carried out:

- either by a differential transformer (LVDT),
- or by an optical system based on a light emitting source (LED), detection cells (phototransistors) and a flap integrated with the beam.

When a TGA analysis is started, the microbalance is at the equilibrium position. When a weight variation occurs due to a transformation of sample, an electric current variation is applied to the coil in order to keep the microbalance at the equilibrium position. The intensity of the electric current variation is exploited to deduce corresponding weight variation. More information about TGA structure can be found at [98].

The DTA and DSC techniques also use two identical crucibles: the first contains sample to be studied and the second as a reference, containing (or not) an inert material such as alumina. They are subject to the same heating and/or cooling conditions, using in general the same oven.

The principle of a DTA is based on the recording of a difference of temperature ΔT between a sample and a reference for a given lap of time Δt (Fig. 2.25). To determinate this ΔT, two thermocouples are generally used for measuring the temperature of the two crucibles (reference and sample). The calorific data of the DTA only give qualitative information on the exothermicity or endothermicity of transformation. DTA is very often coupled with TGA as illustrated in Fig. 2.26. Thus, a microbalance is integrated inside the coupling which allows measuring weight variation (ATG) together with temperature difference (DTA).

A DSC allows measuring a difference of enthalpy ΔH between a sample and a reference within a lap of time Δt. By convention, the value of ΔH is negative for

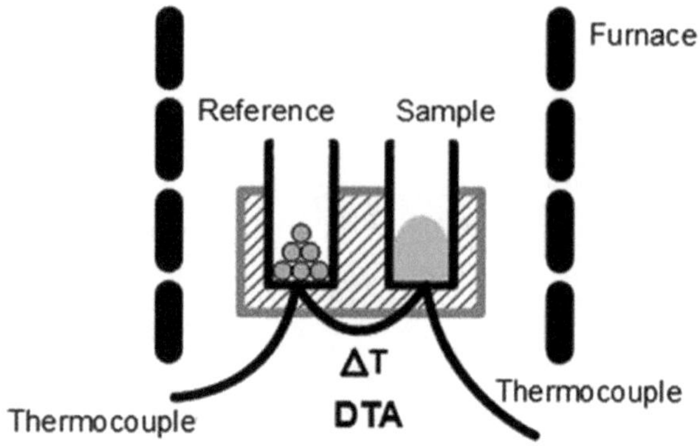

Fig. 2.25 Scheme of DTA system

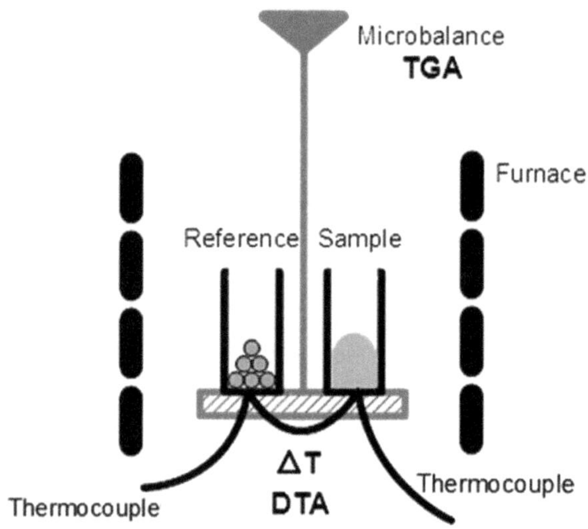

Fig. 2.26 Scheme of coupling TGA-DTA

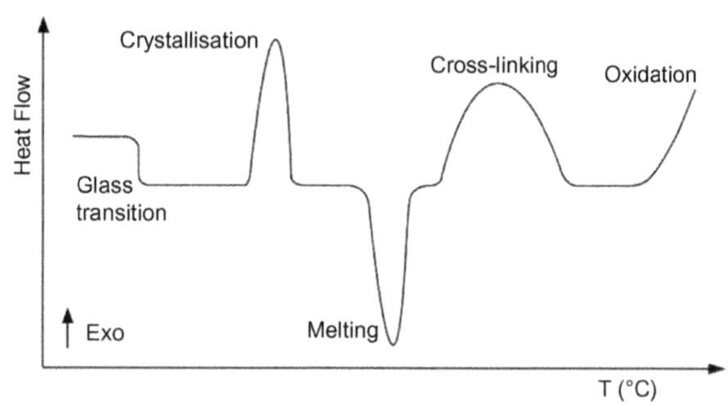

Fig. 2.27 Example of reactions recorded on the heat flow curves

exothermal phenomenon such as crystallization, condensation, oxidation, polymerization, etc.; and positive for endothermal phenomenon such as melting, dehydration, decomposition reactions, etc. Figure 2.27 illustrates some of the

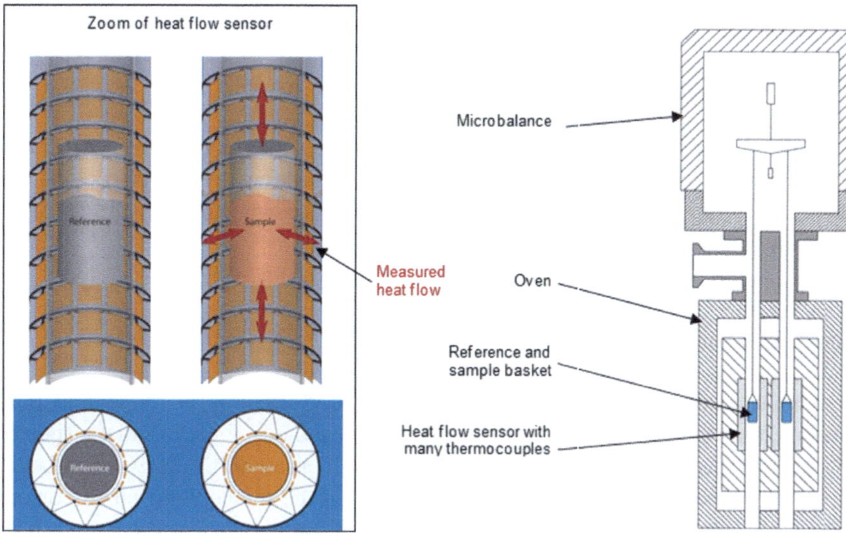

Fig. 2.28 Microbalance B111 coupling to a DSC 111 (SETARAM). Image on left-hand-side. Reproduced with permission from SETARAM. Image on right-hand-side: Reproduced with permission from Techiniques de l'Ingénieur [99]

transformations usually observed in thermal analysis. DSC apparatus measures the quantity of energy absorbed or released by a sample. We obtain heat flow (unit: mW or mJ/s) as a function of time or temperature. Heat flow is measured by a set of several thermocouples implemented around sample, as illustrated in Fig. 2.28. In this figure on the left-hand-side, the implementation of thermocouples of the B111 system (SETARAM) around two crucibles is schematized. On the right-hand-side, the scheme of the coupling TGA-DSC for the B111 system (SETARAM) is presented. The coupling TGA-DSC allows determining simultaneously weight variation as well as heat flow associated to the physico-chemical or mechanical transformation.

2.6.1.4 Commercial Apparatus

SETARAM is a reference constructor of thermal analysis apparatus such as for example SETSYS Evo or LABSYS Evo. There are many other manufacturers in the world like TA Instruments, NETZSCH or METLER TOLEDO.

2.6.1.5 Experimental Conditions

Device Installation

In order to avoid errors due to the vibrations of the surroundings of the device, it is very important that an ATG, ATD or DSC machine to be placed on an anti-vibration table. A thermal analysis device can give better results at night or on weekends (quiet periods) compared to working period.

Signals Correction

The temperature and heat flux signals must be corrected by using several pure metals such as indium, tin, zinc and aluminum as standards at the following speeds: 2, 5 and 10 °C/min. For these standards, a very high purity of 4N (99.99%) or better 5N (99.999%) is required. However, depending on the temperature range considered for the analysis, it is possible to use inorganic and organic products such as potassium nitrate, silver sulphate, quartz, for the correction step.

Alumina crucible is usually used for signals correction. However, it can also be used for samples analysis.

Thermal analysis apparatus is usually equipped with a software. This last allows calculating temperature correction coefficients and regression coefficients. After each correction, it is important to check that the new correction coefficients are good by performing a pure metal melting and calculating the uncertainty of this measurement (Eq. 2.22). The result must be less than or equal to the relative uncertainty given by the manufacturer (Table 2.23).

$$Uncertainty = \frac{Theoretical\ value - Measured\ value}{Theoretical\ value} \times 100 \tag{2.22}$$

The TG signal given by the microbalance must be very close to zero mg when two empty crucibles are installed. It must be stable over time at 30 °C. The TG signal should be checked regularly by doing a decomposition analysis of copper sulphate pentahydrate ($CuSO_4 \cdot 5H_2O$) for less than 600 °C or with calcium oxalate monohydrate ($CaC_2O_4 \cdot H_2O$) for more than 600 °C and then calculating the relative uncertainty of the observed mass losses. TG and the derivative DTG curves of the thermal decomposition of these two standards are given in Figs. 2.29 and 2.30.

Table 2.23 Characteristics of some metals as standards in thermal analysis [100, 101]

Pure metal	Indium	Tin	Lead	Zinc	Aluminium	Gold
Melting temperature (°C)	156.60	231.94	327.47	419.53	660.3	1064.18
Fusion enthalpy (J/g)	28.51 ± 0.19	60.55 ± 0.09	23.00 ± 0.06	108.62 ± 0.52	401.3 ± 1.9	64.5

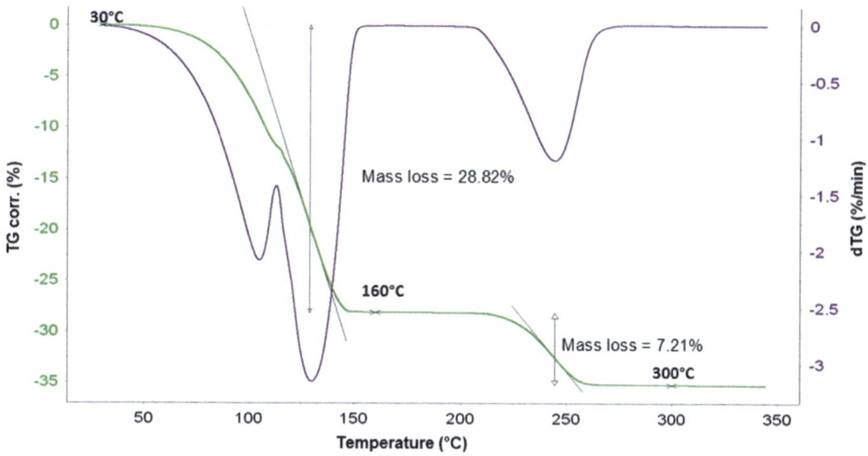

Fig. 2.29 Decomposition of copper sulfate pentahydrate

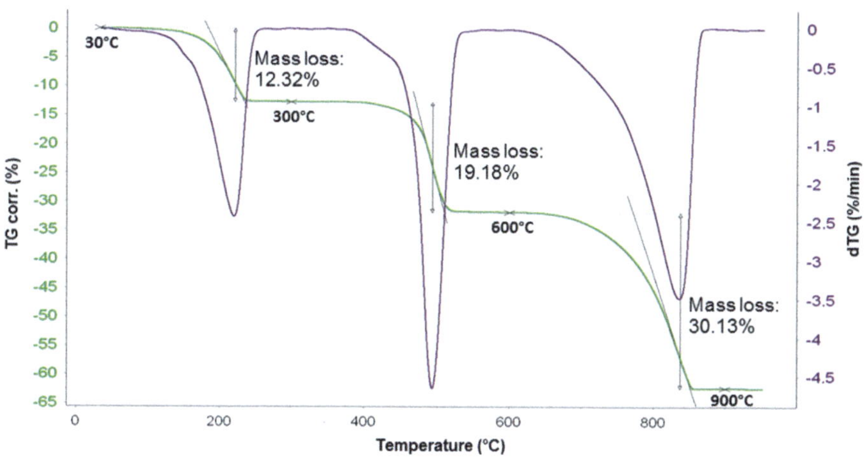

Fig. 2.30 Decarbonation of calcium oxalate monohydrate

Table 2.24 summarizes the main phenomena observed during the decomposition of copper sulphate pentahydrate for less than 600 °C or with calcium oxalate monohydrate.

Table 2.24 Main phenomena observed during the decomposition of copper sulphate pentahydrate for less than 600 °C or with calcium oxalate monohydrate

Working temperature	Standard	Temperature cycle	Theoretical results
Less than 600 °C	Copper sulphate pentahydrate $CuSO_4{\cdot}5H_2O$	Ramp from 30 °C to 600 °C at 10 °C/min under air atmosphere	The decomposition of copper sulphate pentahydrate is characterized by three weight losses: Around 70 °C: $CuSO_4{\cdot}5H_2O \rightarrow CuSO_4{\cdot}3H_2O + 2H_2O$ Around 120 °C: $CuSO_4{\cdot}3H_2O \rightarrow CuSO_4{\cdot}H_2O + 2H_2O$ (Weight loss between 30 and 160 °C: 28.82%) Around 220 °C: $CuSO_4{\cdot}H_2O \rightarrow CuSO_4 + H_2O$ (Weight loss: 7.21%)
More than 600 °C	Calcium oxalate monohydrate $CaC_2O_4{\cdot}H_2O$	Ramp from 30 to 900 °C at 10 °C/min under air atmosphere	The decarbonation of calcium oxalate monohydrate is characterized by three mass loss: Around 120 °C: $CaC_2O_4{\cdot}H_2O \rightarrow CaC_2O_4 + H_2O$ (Weight loss: 12.32%) Around 450 °C: $CaC_2O_4 \rightarrow CaCO_3 + CO$ (Weight loss: 19.18%) Around 700 °C: $CaCO_3 \rightarrow CaO + CO_2$ (Weight loss: 30.13%)

Atmosphere

Thermal analyses are carried out in different atmospheres (usually with a constant gas flow rate), or without gas flow. Table 2.25 shows the most used gases in thermal analysis and the associated objectives.

Crucible and Basket

Different crucibles and baskets are used depending on the applications and the temperature range. They can be made of aluminium, alumina, platinum, tungsten, silica and graphite (Fig. 2.31). Reference and sample crucibles must be identical. The crucible material must be inert in the temperature range studied. The melting point of the crucible must be higher than the maximum temperature of the cycle used. Figure 2.31 shows some examples of baskets and crucibles used in thermal analysis.

Aluminium crucibles: They are low cost and are designed for a simple use. The temperature is limited at 600 °C.

Table 2.25 Gases used in thermal analysis and associated objectives

Atmosphere	Gas	Examples of processes
Inert	N_2–Ar–He	Pyrolysis by studying the decomposition of the organic part of the sample, without oxidation from the atmosphere
Oxidant	Air–O_2	Combustion/oxidation processes
Reductive	H_2–CO–NH_3	Reduction reactions
Corrosive	SO_2–NH_3–NO_2	Chemical reactions
Wet gas	Vector gas + H_2O	Torrefaction, pyro-gasification, vapothermolysis, steam reforming
Under vacuum	none	Need of oven having a good tightness

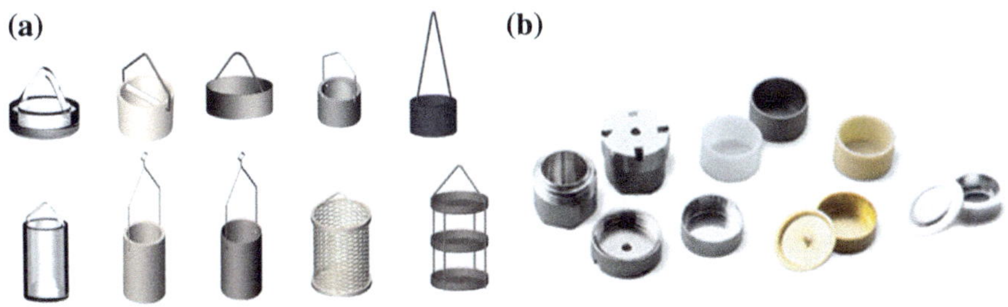

Fig. 2.31 Photo of different crucibles used in thermal analysis. Reproduced with permission from Setaram (a) and from Netzsch (b)

Alumina crucibles: They can be used for various types of samples: inorganic and organic products, metals, oxides etc. They are resistant to chemical attack such as acids, alkaline solutions, hydrogen and other reducing gases. They can work up to 1600 °C.

Platinum crucibles: They also can be used for various types of samples and can be used up to 1600 °C. They have also a very good resistance to the corrosion. They are non-porous and are very easy to be cleaned after each analysis. With high thermal conductivity, DSC results obtained with platinum crucibles are usually better compared to those obtained with alumina crucibles.

Caution: molten metals can form alloys with platinum crucibles, resulting in a damage of the crucible. Carbon black is also a "poison" for platinum crucibles under a non-oxidizing atmosphere. When the crucible holder is made of platinum, a platinum crucible can stick to this sample holder at around 1500 °C. The solution is to place a sapphire disk or alumina powder on the crucible holder surface which should avoid sticking problem.

After utilization, alumina or platinum crucibles are cleaned. In general, the cleaning of these crucibles includes: (1) immersion in a diluted acid solution (10 wt % hydrochloric or nitric acid) for at least 48 h, (2) sonification in an ultrasonic bath, (3) rinsing several time with pure water, (4) and heating to 1400 °C under air atmosphere. But the choice of these steps depends on samples analysed. Crucible containing traces of inorganic elements need acid washing while organic contaminants can be removed simply by air calcination at high temperature.

Temperature Program

For the stabilization of signals and a good purging of the oven, it is recommended to start a given analysis by an isotherm of 10 min under air or 20 min under other atmospheres at a controlled temperature. This last is usually equal or superior to 30 °C except when a cryogenic method is applied.

The heating rate is commonly between 2 and 20 °C/min. But there are commercial apparatus which can work with heating rate up to 100 °C/min. Table 2.26 shows the influence of the heating rate on the signals obtained in thermal analysis.

Importance of the Blank Curve

It is important to eliminate the interferences from instrument and the atmosphere of analysis (oven dilatation, effect of crucible, effect of gas flux …). Blank test is so needed. Blank test can be done by measuring signals without sample under the same experimental conditions. These signals are later subtracted from signals of the analysis with sample. The behaviour of sample can then be obtained properly. It is ideal to perform blank test before the analysis of each sample. But it is also possible to carry out blank test before each set of analysis for samples of similar nature [102].

2.6.1.6 Sample Preparation

Due to the low quantity of samples used for each TG-DSC or TG-DTA analysis (usually between 5 and 30 mg), the preparation and sampling are important steps

Table 2.26 Influence of the heating rate on the signals obtained

Parameter of analysis	High heating rate	Low heating rate
Resolution	Good	Excellence
Signal/noise report	Excellence	Good
Temperature gradient	High	Small
Sample size	Adapted for samples of small size	Adapted for samples of small to large size
Analysis time	More quickly	Longer
Weight loss and heat flux signals	Possibility of overlapping of peaks	Possibility to avoid overlapping of peaks

and need to be done with an adequate procedure which is adapted to sample. It is generally preferred to grind delicately the initial sample into fine powder to ensure that the analysis is representative.

It is necessary to tare a clean and empty crucible before each analysis by using an independent microbalance or directly with the microbalance of the apparatus.

TG-DSC or TG-DTA analyses are usually carried out with 5–30 mg of sample. This depends on the nature of sample, but also on each apparatus. For a set of samples, it is important to use similar weight for comparative study. This allows limiting artefacts due to size of samples.

For each analysis, it is necessary to distribute sample uniformly on the bottom of crucible, which must not be fulfilled. For viscous or liquid samples, it is recommended to use crucible equipped with a perforated cover.

Before analysis, it is necessary to control the position and the centering of crucible inside apparatus.

The treatment of data from TG-DSC or TG-DTA analysis is usually done with software provided with apparatus by the manufacturer. But it can be done with every spreadsheet such as Excel in Microsoftword. In general, derivative signals are more visible for observation, in particular for transformations of small amplitude.

2.6.1.7 Examples

Thereafter are presented TG-DSC analyses of two woods. The results are obtained after correction with blank test.

Figure 2.32 shows the results obtained from the TG-DSC analysis of a Scots pine [103]. The analysis was done with around 20 mg of a fine powder of Scots pin using a SETARAM LABSYS Evo apparatus. The following phenomena can be observed:

- Between 30 and 180 °C: The evaporation of water and possibly volatile maters is observed with a slight endothermic effect.
- Between 180 and 500 °C: A weight loss with an exothermic effect is observed. Three slopes of weight loss can be observed, which can be explained by the successive decomposition of hemicellulose, cellulose and lignin, which are the three main components of the wood.
- Finally, the weight and heat flow were stable above 500 °C. This corresponds to inorganic residues of this wood.

2.6.2 Kinetic of TG-DSC Analysis

The aim of this section is to present practical information for the adequate global kinetic analysis and modeling of biomasses and biowastes transformation processes from thermogravimetric data. Figure 2.33 shows the relationship between the

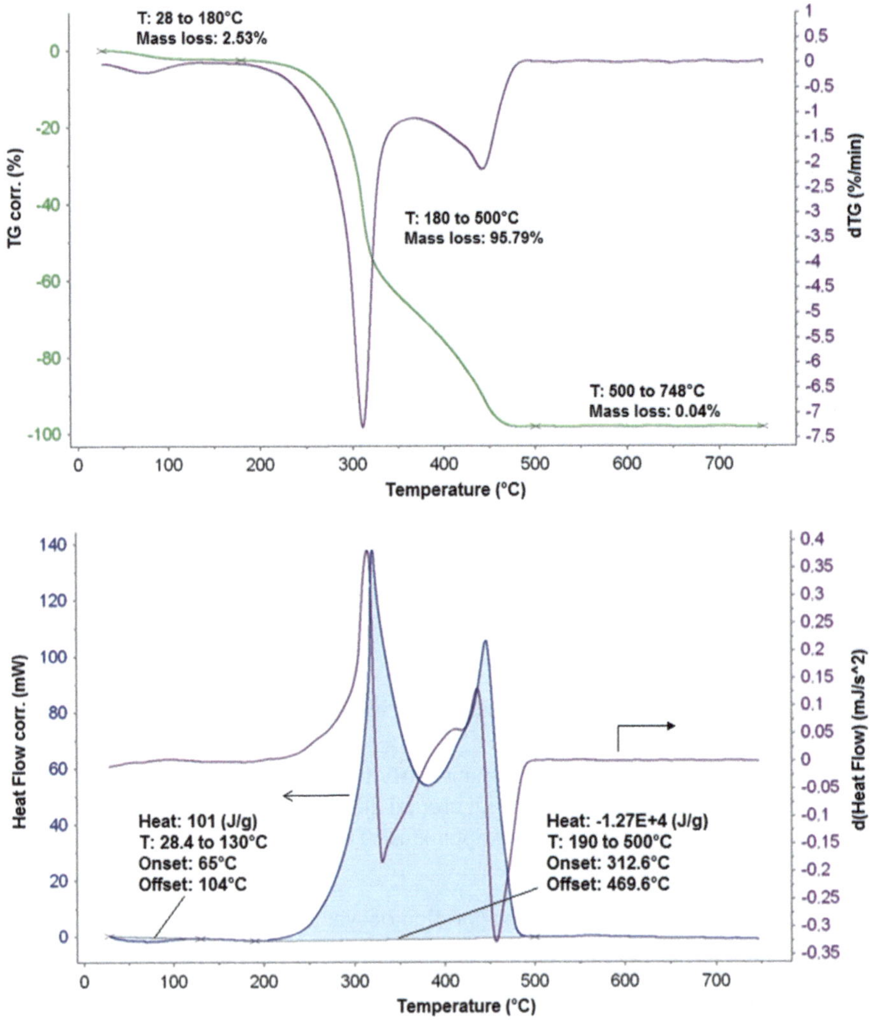

Fig. 2.32 TG-DSC analysis of a Scots pine under air atmosphere with the heating rate of 5 °C/min; Top: TG and derivative TG (or dTG) curves; Bottom: Heat flow and derivative heat flow curves

Sects. 2.6.1 and 2.6.2, and the two approaches of the kinetic modelling of thermogravimetric data.

It is important to note that global chemical kinetics is usually established for application purposes and describes the rate at which a process takes place. Three principal parameters need to be determined: the process activation energy E_a, the frequency factor A, and the reaction mechanism $f(\alpha)$. These parameters are also known as the kinetic triplet of a specific process, and are detailed in Sect. 2.6.2.1.

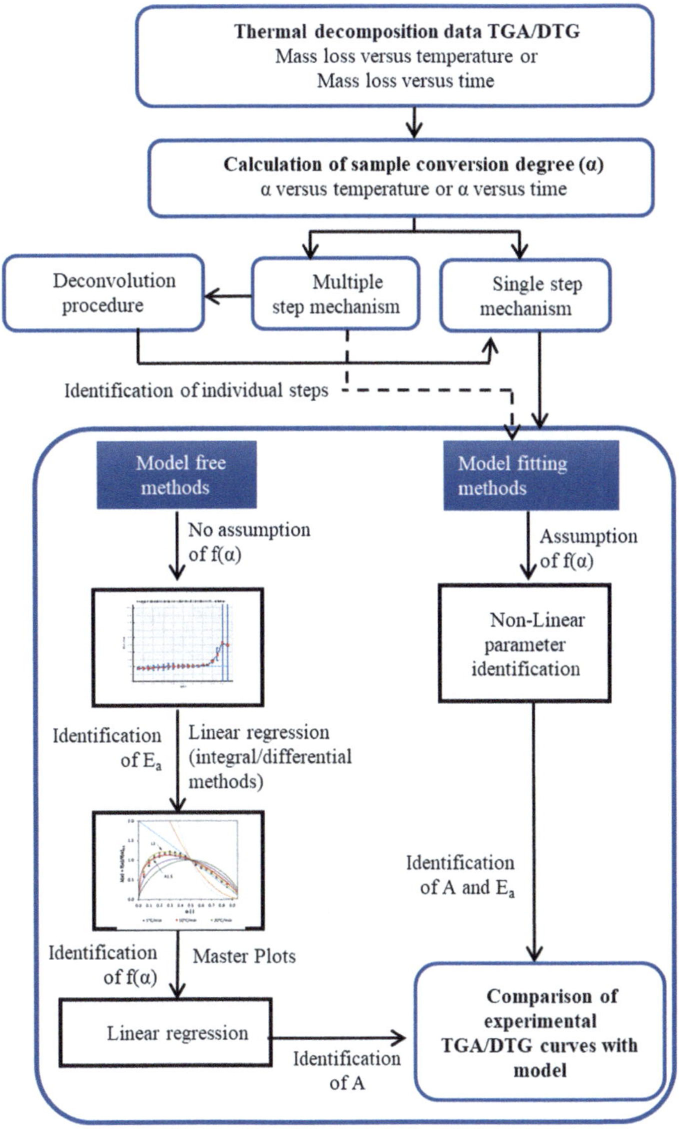

Fig. 2.33 Main steps of the kinetic modelling of thermogravimetric data

2.6.2.1 Chemical Kinetics Modeling

According to several authors, solid state kinetics can be described by the following Eq. (2.23) [104]:

$$\frac{d\alpha}{dt} = k(T)f(\alpha)h(P) \tag{2.23}$$

where α is the conversion degree of the sample during the analyzed process, $d\alpha/dt$ is the reaction rate, $k(T)$ is the rate constant representing the temperature dependence of the process, $h(P)$ the partial pressure dependence, and $f(\alpha)$ the function that describes the reaction mechanism. Generally, α can be calculated from TGA data according to Eq. (2.24).

$$\alpha = \frac{m_0 - m}{m_0 - m_f} \tag{2.24}$$

where m_0 is the initial mass of the dry sample, m_f the final mass, and m the current mass at a given temperature or time.

For its part, $k(T)$ can be described by the Arrhenius equation (Eq. 2.25):

$$k(T) = A\exp(-E_a/RT) \tag{2.25}$$

where A is the pre-exponential or frequency factor, E_a the apparent activation energy, T the absolute temperature, and R the universal gas constant. Equation (2.23) can be then rearranged as follows:

$$\frac{d\alpha}{dt} = A\exp(-E_a/RT)f(\alpha)h(P) \tag{2.26}$$

The $h(P)$ function is particularly important for gas-solid reactions, where the reactive gas atmosphere has an important impact on the process kinetics. This pressure dependence can be expressed as a power law or other different forms according to the process [105]. As most of the kinetic methods for thermal analysis consider that the reaction rate is only a function of T and α (i.e., processes under inert atmosphere like torrefaction, pyrolysis, etc.), the mathematical development of kinetic methods presented here is shown assuming no dependence on pressure. Nevertheless, the pressure dependence parameter can be included when necessary. Equation (2.26) can be then rewritten as:

$$\frac{d\alpha}{dt} = A\exp(-E_a/RT)f(\alpha) \tag{2.27}$$

As a reference, some of the most commonly used reaction models $f(\alpha)$ and their integral forms $g(\alpha)$ are presented in Table 2.27.

Different modeling approaches have been developed for the calculation of the global kinetic parameters of solid-state processes. They are generally classified in model-free and model fitting methods, according to the assumptions related with the reaction mechanism describing the process. Unlike model-free methods, the model fitting approach needs the previous assumption of a particular reaction model $f(\alpha)$

Table 2.27 Reaction mechanism used in solid-state kinetics [106]

Model		$f(\alpha)$	$g(\alpha)$
Order based	Or0—Zeroth order	1	α
	Or1—First order	$1-\alpha$	$-\ln(1-\alpha)$
	Or2—Second order	$(1-\alpha)^2$	$[1/(1-\alpha)]-1$
	Or3—Third order	$(1-\alpha)^3$	$(1/2)\left[\left[1/(1-\alpha)^2\right]-1\right]$
	Or n—nth order (n > 1)	$(1-\alpha)^n$	$[1/(n-1)]\left[(1-\alpha)^{(1-n)}-1\right]$
Diffusion	D1—One dimensional	$1/(2\alpha)$	α^2
	D2—Two dimensional	$[-\ln(1-\alpha)]^{-1}$	$\alpha+(1-\alpha)\ln(1-\alpha)$
	D3—Three dimensional	$(3/2)(1-\alpha)^{2/3}\left[1-(1-\alpha)^{1/3}\right]^{-1}$	$\left[1-(1-\alpha)^{1/3}\right]^2$
Power law	Pn—Power law	$n(\alpha)^{(n-1)/n}$	$\alpha^{1/n}$
Nucleation and growth	An—Avrami Erofeev	$n(1-\alpha)[-\ln(1-\alpha)]^{(n-1)/n}$	$[-ln(1-\alpha)]^{1/n}$
Geometrical contraction	R2—Contracting area	$2(1-\alpha)^{1/2}$	$1-(1-\alpha)^{1/2}$
	R3—Contracting volume	$3(1-\alpha)^{2/3}$	$1-(1-\alpha)^{1/3}$
Random scission	L2—Random scission L = 2	$2(\alpha^{1/2}-\alpha)$	–

for the estimation of the complete kinetic triplet of a process. Both approaches are described in Sects. 2.6.2.2 and 2.6.2.3.

2.6.2.2 Model-Free Methods—Isoconversional Methods

Isoconversional model-free methods are based on the hypothesis that the reaction rate at a constant conversion degree is only a function of temperature, and then, the apparent activation energy E_a can be calculated without the previous knowledge or assumption of the reaction mechanism $f(\alpha)$ [104, 107]. These methods can be applied to both isothermal and non-isothermal data.

As isoconversional methods are evaluated at fixed conversion degrees, the variation of E_a with α can give valuable information about the complexity of the process, in particular, about the existence of single-step or multiple-step kinetics. If the calculated E_a does not vary significantly with the conversion degree, it could be considered that the process can be described by single-step kinetics [106]. A maximum deviation of 20% of the mean activation energy value can be practically accepted.

From this evaluation, it is important to note that it is not possible to conclude that a process that can be represented by a single-step approach actually involves a single reaction. In general, chemical processes consist of several steps; however in some cases, one of them determines the overall kinetics and the process can be described as single-step one [104].

Isothermal Data

As already presented, isoconversional methods can be employed to both isothermal and non-isothermal data analysis. In the case of isothermal experiments, $d\alpha/dt$ expression can be rearranged as presented in Eq. (2.28), and the activation energy can be calculated from the slope of the plot $\ln(t)$ versus T^{-1}, for constant values of α.

$$\ln t = \frac{E_a}{RT} + \ln\left(\frac{g(\alpha)}{A}\right) \tag{2.28}$$

The value of $g(\alpha)/A$ can be then determined from the plot intercept. Therefore, the calculation of the pre-exponential factor depends on the previous knowledge or assumption of the reaction model $f(\alpha)$ and its integral form $g(\alpha)$. Some methods to correctly identify the reaction mechanism are presented in Section "Identification of the Reaction Mechanism f(α)".

Non-isothermal Data

For non-isothermal conditions, the temperature variation with time has to be taken into account. If the temperature change is linear with time, the reaction rate is expressed as in Eq. (2.29), where β is the heating rate.

$$\frac{d\alpha}{dT} = \frac{A}{\beta}\exp(-E_a/RT)f(\alpha) \tag{2.29}$$

Different mathematical approaches exist to solve this equation and calculate the kinetic parameters. They can be classified into integral and differential, according to the mathematical method employed for the evaluation of Eq. (2.29). Some of the most common ones used for solid-state reactions are summarized in Table 2.28.

Accordingly, from the presented methods, the first step to determine the kinetic triplet of a process is the calculation of the apparent activation energy, followed by the identification of the decomposition model $f(\alpha)$ and the estimation of the pre-exponential factor A. The E_a value at a constant conversion extent α can be determined from the slope of the isoconversional Arrhenius plots presented in Table 2.28 and the pre-exponential factor A from their intercept.

Table 2.28 Model-free isoconversional methods for non-isothermal kinetic analysis

Method	Expression	Isoconversional Arrhenius plot	Ref.
Differential methods			
Friedman	$\ln(d\alpha/dt) = \ln[Af(\alpha)] - E_a/RT$	$\ln(\beta d\alpha/dT)$ versus 1/T	[108]
Integral methods			
Kissinger-Akahira-Sunose (KAS)	$\ln(\beta/T^2) = \ln[(AR)/(E_a g(\alpha))] - E_a/RT$	$\ln(\beta/T^2)$ versus 1/T	[109]
Flynn-Wall-Ozawa (FWO)	$\log(\beta) = \log[(AE_a)/(Rg(\alpha))] - 2.315 - 0.4567E_a/RT$	$\log(\beta)$ versus 1/T	[110]
Starink	$\ln(\beta/T^{1.92}) = -1.0008E_a/RT + const$	$\ln(\beta/T^{1.92})$ versus 1/T	[111]

Identification of the Reaction Mechanism $f(\alpha)$

For both isothermal and non-isothermal data, the calculation of the pre-exponential factor needs the previous knowledge of the reaction model $f(\alpha)$. This can be a disadvantage of model-free methods, taking into account that many kinetic triplets can be calculated depending on the chosen $f(\alpha)$ function. In this regard, different methods have been developed for the accurate identification of the reaction model, valid for processes that can be represented by single step kinetics.

Generalized master plots approach

The use of master plots is a straightforward method to identify the kinetic model that best fit a certain process. Generalized master plots are reference theoretical curves dependent on the reaction model $f(\alpha)$ but independent on the E_a, A and the temperature program (isothermal or non-isothermal) [112]. Accordingly, the comparison of the experimental master plots to the theoretical ones allows the selection of the most appropriate kinetic model or model family.

Generalized master plots can be obtained from Eq. (2.27) by the introduction of the generalized time θ, representing the reaction time required to attain a certain α at infinite temperature [112, 113]:

$$\theta = \int_0^t \exp\left(\frac{-E_a}{RT}\right) dt \tag{2.30}$$

The differentiation of Eq. (2.30) and substitution in Eq. (2.27) leads to:

$$\frac{d\alpha}{d\theta} = Af(\alpha) = \frac{d\alpha}{dt} \exp\left(\frac{E_a}{RT}\right) \tag{2.31}$$

where $d\alpha/d\theta$ is the generalized reaction rate, and its integrated form is:

$$g(\alpha) = A\theta \tag{2.32}$$

The generalized master plots can be then obtained from the differential or the integral form of the generalized rate kinetic equation [114]. In the former case, taking 50% as the reference reaction extent, the data can be normalized as follows:

$$\frac{d\alpha/d\theta}{(d\alpha/d\theta)_{0.5}} = \frac{f(\alpha)}{f(\alpha)_{\alpha=0.5}} \tag{2.33}$$

Thus, the reduced-generalized reaction rate expression based on the differential form of the kinetic equation can be written as:

$$\lambda(\alpha) = \frac{f(\alpha)}{f(\alpha)_{\alpha=0.5}} = \frac{d\alpha/dt}{(d\alpha/dt)_{0.5}} \frac{\exp(E_a/RT)}{\exp(E_a/RT_{0.5})} \tag{2.34}$$

For isothermal programs, the experimental master plots can be obtained directly from an isothermal $d\alpha/dt$ curve.

For their part, the generalized master plots obtained from the integral form of the kinetic data can be calculated from Eq. (2.35) for isothermal data, and Eq. (2.36) for linear non-isothermal data.

Isothermal data:

$$\frac{g(\alpha)}{g(\alpha)_{\alpha=0.5}} = \frac{\theta}{\theta_{0.5}} = \frac{t}{t_{0.5}} \tag{2.35}$$

Linear non-isothermal data:

$$\frac{g(\alpha)}{g(\alpha)_{\alpha=0.5}} = \frac{\theta}{\theta_{0.5}} = \frac{p(x)}{p(x_{0.5})} \tag{2.36}$$

where

$$x = E_a/RT \tag{2.37}$$

$$p(x) = \frac{e^{-x}}{x}\left(\frac{x^3 + 18x^2 + 86x + 96}{x^4 + 20x^3 + 120x^2 + 240x + 120}\right) \tag{2.38}$$

Figure 2.34 show the generalized master plots in their differential and integral form for some of the most common reaction mechanisms used in solid state kinetics

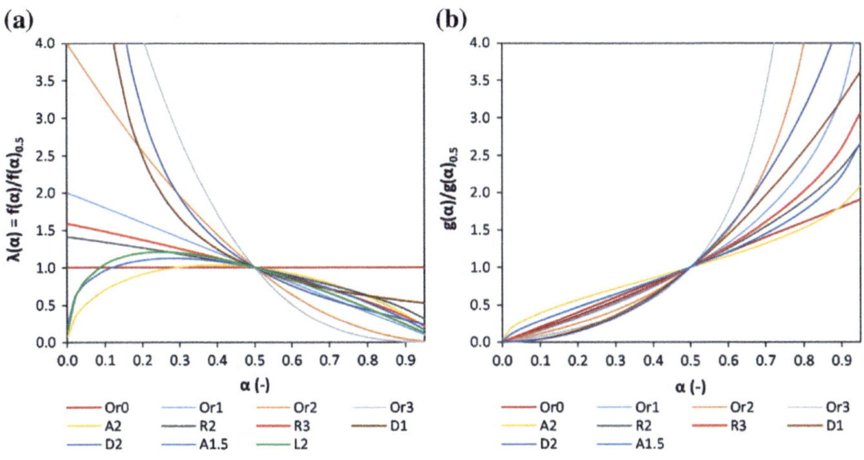

Fig. 2.34 Generalized master plots of the kinetic models calculated according to Eq. (2.34). a Differential form, b integral form

and presented in Table 2.27. As the referent reaction extent was 50%, all the functions coincide at $\alpha = 0.5$.

Compensation effect:

The reaction model E_a and the pre-exponential factor A can also be determined using the "compensation effect" observed when model-fitting methods are applied to a single experimental data set. If different reaction models $f(\alpha)_i$ are substituted into Eq. (2.27) or (2.29) and fitted to the experimental data, different pairs of Arrhenius parameters E_{ai} and $\ln E_{ai}$ are determined. Any inaccuracy in the selection of the reaction model is compensated by an inaccuracy in the process rate constant [115], and then, from a mathematical point of view, several Arrhenius pairs can describe the behavior of a specific process. These Arrhenius pairs have shown to be strongly correlated through the compensation effect as shown in Eq. (2.39), where a and b are the parameters of the compensation effect [115].

$$\ln A_i = aE_{a_i} + b \tag{2.39}$$

Then, the substitution of the E_a calculated using model-free methods into Eq. (2.39) allows the estimation of the pre-exponential factor A. For its part, the reaction model can be numerically reconstructed in its integral or differential form, according to Eqs. (2.27) and (2.29) for isothermal and non-isothermal data respectively [116]. The equation of the most suitable reaction mechanism of the analyzed process is then obtained from the comparison of the numerically calculated $g(\alpha)$ or $f(\alpha)$ with the theoretical mechanism equations in Table 2.27.

The main inconvenient of this method is the fact that model-fitting methods should be used to determine the compensation effect parameters, demanding more computations than the master plots approach. Nevertheless, both methods are suitable for the determination of the reaction model and pre-exponential factor of processes that can be represented by single step kinetics.

Isothermal or Non-Isothermal Experiments

The choice of the isothermal or non-isothermal approach depends on the type of process under study. For isothermal analysis, there is always a non-isothermal heat-up time that should be taken into account, since, during this period a partial decomposition of the sample can take place. Accordingly, α could not be zero at the beginning of the isothermal regime, and therefore this reaction extent should be taken into account in the kinetic calculations.

Isothermal approach may be used when the beginning of the reactions of interest can be controlled and started at a defined temperature, or when the decomposition during the heating-up stage is not of interest. Combustion or gasification experiments are examples of processes that are often analyzed using isothermal experiments. The beginning of the decomposition reactions can be controlled by switching the atmosphere from inert to reactive (e.g. CO_2, steam, air, etc.) after the isothermal regime is established.

In the case of thermal decomposition analysis of biowastes (e.g. torrefaction, pyrolysis), where the decomposition takes place over a wide range of temperature, the use of non-isothermal experiments is often a more straightforward way to avoid the problem of non-zero conversion at the beginning of the isothermal stage. Otherwise, the correct value of α at the beginning of the isothermal regime must be taken into account.

In both cases, it is recommended to perform model-free kinetic analysis from at least three different experimental conditions (e.g. temperature, heating rate, reactive gas partial pressure, etc.) in order to have adequate data to correctly determine the Arrhenius parameters from isoconversional plots.

An example of the use of model-free isoconversional methods for the determination of the kinetic parameters of two different thermochemical processes of biomasses or wastes is presented below for isothermal and non-isothermal data.

Non-isothermal data:

For non-isothermal analysis, the pyrolysis of commercial microcrystalline cellulose from MERK-Avicel® is presented here as example. The analysis of cellulose is of particular interest at it is one of the main components of lignocellulosic biowastes from agriculture and forestry. It is a long chain polymer representing the major structural component of cell walls in plants, and is usually denoted by the generic formula $(C_6H_{10}O_5)_n$ [117].

The experimental conditions used for the given example are presented in Table 2.29.

From the mass loss curve presented in Fig. 2.35a, it is possible to observe that cellulose decomposition under inert atmosphere takes place over a narrow temperature range, between 300 and 400 °C. The conversion degree α of the cellulose can be calculated according to Eq. (2.24). If the moisture release process is not of interest, it is recommended to correct the TGA data to work in a dry basis. Accordingly, the $d\alpha/dT$ curve is presented in Fig. 2.35b.

Any of the model-free methods presented in Table 2.28 can be used for the calculation of the kinetic triplet describing the microcrystalline cellulose pyrolysis.

Table 2.29 Experimental conditions for microcrystalline cellulose kinetic analysis

Sample	Microcrystalline cellulose MERK-Avicel®
Sample mass	~20 mg
Particle size	10–160 µm
Equipment	Seratam TG-DSC 111
Crucible	Alumina
Atmosphere	N_2—50 ml/min
Heating rate	2, 5, 10, 20 °C/min
Temperature range	30–700 °C

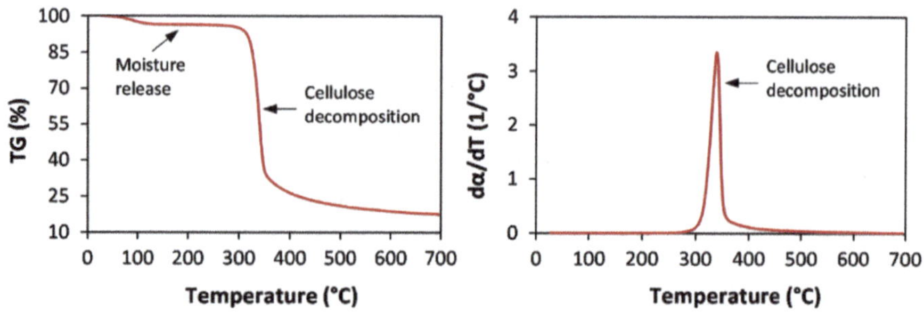

Fig. 2.35 TG and DTG curves of commercial microcrystalline cellulose from MERK-Avicel® pyrolysis at 10 °C/min

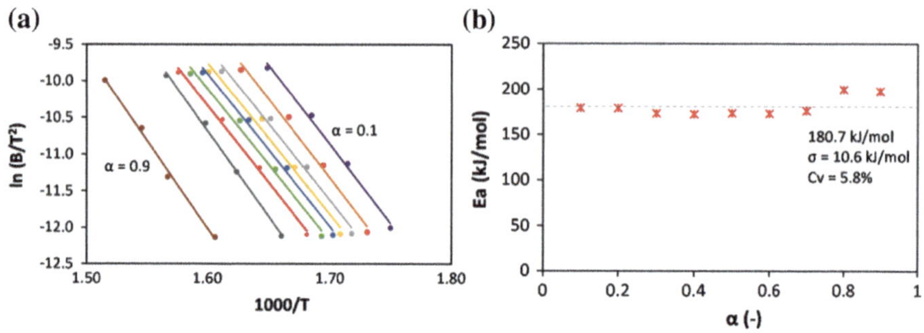

Fig. 2.36 a Kissinger-Akahira-Sunose isoconversional plots for microcrystalline cellulose pyrolysis. b Apparent activation energy values of microcrystalline cellulose pyrolysis as a function of α

Accordingly, the Arrhenius plots obtained using the Kissinger-Akahira-Sunose (KAS) isoconversional method are presented in Fig. 2.36a. From this method, the mean E_a value is 180.7 kJ/mol, in accordance with different reported kinetic studies for cellulose pyrolysis [118–121].

It is possible to notice from Fig. 2.36b that there is not a clear dependence between the apparent activation energy and the conversion degree from $\alpha = 0.1$ to $\alpha = 0.9$. Accordingly, it can be assumed that the thermal decomposition of the cellulose is described by single-step kinetics, and then, the most suitable reaction mechanism describing this process can be determined using the master plots approach.

The generalized reaction rate is calculated according to Eq. (2.34) over all the considered conversion range and compared to theoretical master plots. For cellulose, it can be observed from Fig. 2.37 that pyrolysis can be described by either an *A*1.5 or a *L*2 model. In this case, both models have very close master plots shape, and it is not easy to consistently distinguish between them. However, they have been largely reported in literature to correctly describe the thermal decomposition

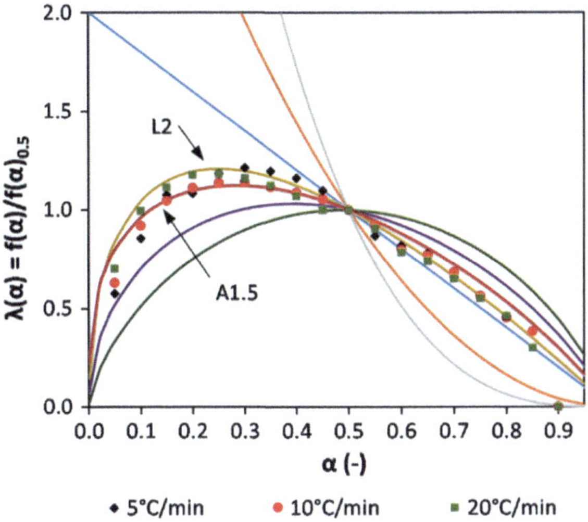

Fig. 2.37 Comparison between experimental and theoretical master plots (differential form) for microcrystalline cellulose pyrolysis

Table 2.30 Kinetic triplet calculated for the microcrystalline cellulose pyrolysis using the Kissinger-Akahira-Sunose isoconversional method

Apparent activation energy, E_a	Pre-exponential factor, A	Reaction model, $f(\alpha)$
180.7 kJ/mol	1.14E + 19 min^{-1}	A1.5, L2

behavior of different kind of polymers including cellulose [120]. Both models belong to the same model family (e.g. sigmoidal models) and are related to narrow reaction profiles as observed in Fig. 2.35. Therefore, from a mathematical point of view, they are suitable for the description of cellulose pyrolysis [57].

Finally, with the knowledge of the reaction model $f(\alpha)$, the pre-exponential factor A can be calculated from the intercept of each isoconversional Arrhenius plot (Fig. 2.36a). The kinetic triplet obtained for the example presented here using the KAS method is summarized in Table 2.30.

Isothermal data:

Coconut shells steam gasification is presented here as an example of isothermal kinetic analysis. The experimental conditions used for the given example are presented in Table 2.31.

From the experimental program presented in Fig. 2.38a, it can be observed that gasification reactions start after the isothermal regime is stablished, by switching the inert atmosphere to a reactive one (in this case N_2 + H_2O − 10 kPa H_2O partial pressure) [104].

For gas-solid reactions as gasification and combustion processes, the atmosphere partial pressure dependence can be generally described with a power law P^n [122–124]. The exponent n can be calculated from the slope of the plot $\ln(d\alpha/dt)$

Table 2.31 Experimental conditions for coconut shells steam gasification

Sample	Coconut shells
Sample mass	~20 mg
Particle size	100–200 µm
TG analyzer	Seratam TG-ATD 92
Steam generator	Wetsys humidity generator
Crucible	Alumina
Atmosphere	N_2, $N_2 + H_2O$
Steam partial pressure	3.7, 6.2 and 10 kPa
Total flow	67 ml/min
Heating rate	10 °C/min
Gasification temperature	750, 800 and 900 °C
Gasification time	5 h

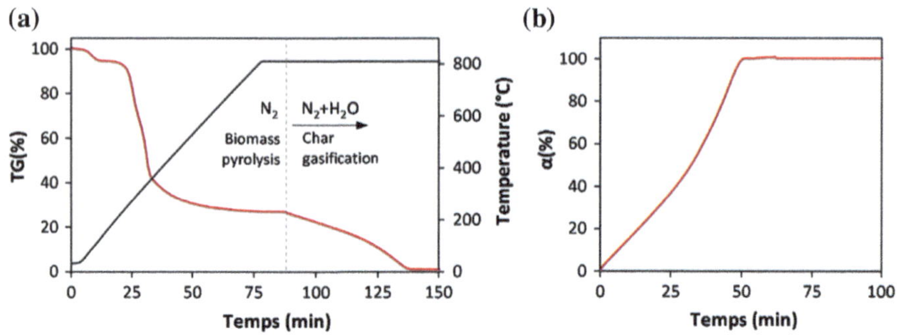

Fig. 2.38 a Experimental program for coconut shells steam gasification and mass loss curve. b Conversion degree during gasification stage

versus $\ln(P_{H_2O})$ for different α over all the conversion range, according to Eq. (2.26), and was found to be $n = 0.5 \pm 0.1$ in the case of the coconut shells steam gasification presented in this example.

As for non-isothermal experiments, the first step of the kinetic analysis is the calculation of α from the mass loss data, according to Eq. (2.24). In this case, m_o is the mass of the sample at the beginning of the isothermal stage, m_f is the final mass, and m is the current mass at a given time. Thus, the conversion profile of coconut shells steam gasification is presented in Fig. 2.38a.

The E_a values for constant α over all the conversion range are then calculated from the slope of the plot $\ln(t)$ versus T^{-1} as already described in Eq. (2.28). For this example, it is also possible to notice from Fig. 2.39a that the E_a is nearly constant over all the conversion range, and then, the process can be represented by a single kinetic triplet $E_a, f(\alpha)$ and A. The most suitable reaction model is determined using the master plots approach. It must be said that for isothermal analysis, the previous knowledge of the mean activation energy is not necessary for the calculation of $\lambda(\alpha)$ (see Eq. 2.34).

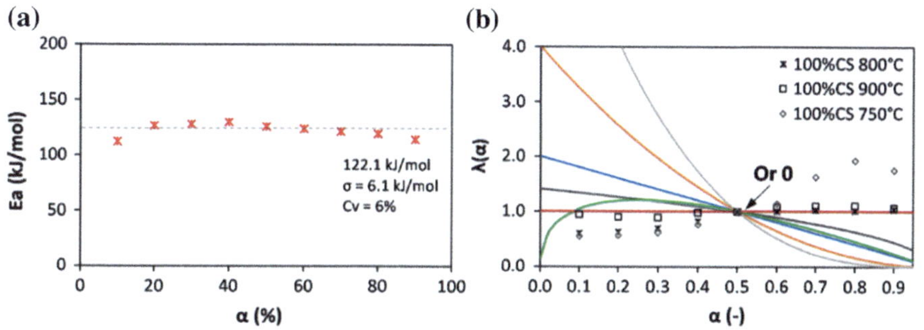

Fig. 2.39 a Apparent activation energy values of coconut shells steam gasification. b Comparison between experimental and theoretical master plots (differential form) coconut shells steam gasification

Table 2.32 Kinetic triplet calculated for coconut shells steam gasification and steam partial pressure dependence function

Apparent activation energy, E_a	Pre-exponential factor, A	Reaction model $f(\alpha)$	$h(P) = P^n$ n
122.1 kJ/mol	3.55E+3 min^{-1} $kPa^{-0.5}$	Or0	0.5

According to Fig. 2.39b, the most suitable reaction model describing the steam gasification of coconut shells is a zeroth-order model. Thus, with the knowledge of $f(\alpha)$, the pre-exponential factor A can be then determined from Eq. (2.28). The kinetic triplet for coconut shells gasification under the presented conditions is summarized in Table 2.32.

2.6.2.3 Model-Fitting Methods

A second class of methods, called *model-fitting methods*, consists in selecting a particular reaction model that is assumed to be able to well represent the evolution of the conversion degree with the temperature. The main interest with the model-fitting methods is that any models, even multi-step reaction models, can be chosen to describe simple or complex kinetics. However, like we will see below, some drawbacks can exist. The estimation of the kinetic parameters is based on the best fit of data. For that, a numerical criterion taking into account the discrepancy between the experimental data generated by the thermal analysis device and the values simulated with the model is defined. It is a least square criterion, i.e. computed from the sum of the squares of the residuals between the measured values and the calculated ones. In this paragraph, we limit our discussion to a TGA device measuring the variation of mass and also giving the derivative of the mass loss. For example, the two following least square criteria can be computed: the first one is based on the temperature derivative of the conversion degree and the second uses directly the conversion degree:

$$C_1 = \sum_i \left(\left(\frac{d\alpha}{dT}\right)^i_{model} - \left(\frac{d\alpha}{dT}\right)^i_{exp} \right)^2$$
$$C_2 = \sum_i \left((\alpha)^i_{model} - (\alpha)^i_{exp} \right)^2 \tag{2.40}$$

where i represents the index of data and associated computed values.

The numerical techniques to determine which values of the parameters minimize the chosen criterion are called least square regressions. Many numerical techniques exist. We only distinguish here the techniques used when the model is linear with respect to the parameters (*linear regression*) and the techniques when the model is non linear in respect to the parameters (*non linear regression*).

Selection of the Model

As previously mentioned, the first step is the selection of an appropriate model. This choice must be based on the a priori knowledge of the studied reactive system and the assumptions about the studied reaction mechanisms. The model can be single-step or multiple-step kinetics. However if a multiple-step kinetics model is selected, a strong attention must be paid to avoid a too much complex model. Indeed, because TGA device only allows measuring the global mass variation of a sample, it could be theoretically impossible to retrieve the "true" kinetic parameters of a model with too many constituents, reactions and parameters. The identifiability of a complex model is a property not easy to check in practice. In general with a non-identifiable model, many sets of kinetic parameters can be identified with the regression. Whatever the complexity of the selected model, the reliability in the identified parameters can be low if the used experimental data are only generated from a single experiment. This is the reason it is recommended to fit simultaneously to multiple data sets obtained under different temperature programs [125].

Linear Regression

A linear regression can be used if the model is linear with respect to the kinetic parameters. It's the case for a single-step reaction with a first order reaction model and with TGA data generated with non-isothermal conditions (constant heating rate β):

$$\frac{d\alpha}{dT} = \frac{A}{\beta}\exp(-E_a/RT)(1-\alpha) \tag{2.41}$$

A simple rearrangement of the equation allows writing:

$$\ln\left(\frac{d\alpha}{dT}/(1-\alpha)\right) = \ln\left(\frac{A}{\beta}\right) - E_a/RT \tag{2.42}$$

By linear regression, it is easy to compute the y-intercept and the slope of the linear relation and to deduce the values of A and E_a.

Non-linear Regression

A non-linear regression must be used if the model is non-linear with respect to the kinetic parameters. For example, this is the case for a nth order reaction model (where the numerical value of n has to be identified):

$$\frac{d\alpha}{dT} = \frac{A}{\beta}\exp(-E_a/RT)(1-\alpha)^n \tag{2.43}$$

The simultaneous identification of E_a, A and n implies to use a numerical method of optimization for minimizing a criterion, as those defined in Eq. (2.40).

The simultaneous identification of E_a, A and n implies to use a non-linear optimization method for minimizing a criterion, as those defined in Eq. (2.40). The non-linear optimization methods are iterative ones, i.e. the kinetic parameters values evolves step by step during iterative computations towards the best values of parameters.

One of the main consequences of using an iterative method is the necessity to initialize the value of kinetic parameters in order to start the computations. The choice of the initial values is an important step because they can influence the found solutions. Actually, the most used numerical methods are local methods—that means the algorithm will detect solutions as soon as some mathematical conditions will be checked on the derivatives of the criterion with respect to each kinetic parameter—and these detected solution can depend on the initialization of the kinetic parameters. A way for avoiding this drawback is to run several times the optimization process and compare the obtained solutions of the kinetic parameters and criterion before to select the definitive kinetic parameters. As the local methods of optimization are usually based on the derivatives of the criterion with respect to each kinetic parameter, a second consequence can appear when there is a large difference of magnitude in the computed derivatives; only the parameters associated to high derivatives will be well identified. For example the derivatives associated to the pre-exponential factors are generally smaller compared to those associated to the activation energies. A solution to this problem is to employ a reparametrization technique of the model before to run the optimization process in order to decrease the difference of magnitude between the derivatives of criterion with respect to each parameter. For example, the expression of the rate constant in Eq. (2.25) can be expressed under that formulation in order to reduce the magnitude between the derivative of criterion with respect to A and E_a:

$$k(T) = \exp(A') \exp\left(-E'_a/T\right) \tag{2.44}$$

with:

$$A = \log(A') \quad E'_a = E_a/T \tag{2.45}$$

Another advantage of the reparametrization is to reduce the high correlation existing between the estimates of the pre-exponential factor and the activation energy [126].

Application

The presented application concerns the thermal degradation of cardboard. Cardboard is a rather complex material, essentially composed of cellulose, hemicelluloses and lignin, and represents about 30% of municipal solid waste. The cardboard samples came from various forms of packaging. They were finely pulverized by cryogenic grinding. The sample mass was equal to 3.5 mg for all the experiments presented below. It is assumed that the thermal gradient inside the sample is small with such a mass.

Experiments were carried out with a SetaramTM 92-16.18 thermogravimetric analyzer. The thermogravimetric unit is composed of an electronic microbalance using a system of a balance beam and a furnace composed of a vertical tubular graphite tube. Cylindrical platinum crucibles were used. Experiments were conducted under a flow of an inert gas (nitrogen) at a constant flow-rate (33 ml min^{-1}) controlled with a mass flow controller. Because the buoyancy exerted on the sample decreases with increasing temperature, an apparent mass gain was observed (about 1 mg). So a preliminary "blank experiment", i.e. an experiment identical to the actual experiment except that an empty crucible is used, was carried out before each experiment. The evolution of measured mass of the blank experiment is subtracted from the recorded mass of the "actual experiment". Finally, in order to protect the balance from reaction with an incoming reactant gas stream, nitrogen was passed through the balance chamber for 4 h before each experiment. Three non-isothermal thermogravimetric experiments with the same range of temperature, between 450 and 850 K, and different heating rates (5, 10 and 15 K min^{-1}) were carried out for the determination of the kinetic parameters of the model given below. All the experimental conditions such as sample mass, atmosphere and gas flow rate were identical in all experiments.

The reaction scheme assumed here for the thermal degradation of cardboard was proposed by David [127]. The cardboard, noted C, is supposed to be degraded by two successive reactions. Gaseous products and an intermediary solid specie (I) are produced by the first reaction; a solid residue, called char (Ch), and other gases are the products of a second reaction from the intermediary solid specie. Mass stoichiometric coefficients a, a', b, b' were defined for representing the mass proportion between solid and gaseous products for each reaction. A first order Arrhenius kinetic model is used for the two reactions in the solid phase, then six parameters must be identified: the four Arrhenius parameters ($A1$, $E1$, $A2$, $E2$), a and b, the two mass stoichiometric coefficients. The global model for the solid species associated with the reaction scheme is given by the following system of algebra-differential equations:

$$\begin{aligned} \frac{dm_c}{dt} &= -k_1 m_c(t) = -A_1 \exp\left(\frac{-E_1}{RT}\right) m_c(t) \\ \frac{dm_I}{dt} &= a \cdot (k_1 m_c(t) - k_2 m_I(t)) \\ &= a \cdot \left(A_1 \exp\left(\frac{-E_1}{RT}\right) m_c(t) - A_2 \exp\left(\frac{-E_2}{RT}\right) m_I(t)\right) \\ \frac{dm_{Ch}}{dt} &= b \cdot k_2 m_I(t) = b \cdot A_2 \exp\left(\frac{-E_2}{RT}\right) m_I(t) \\ z &= m_c + m_I + m_{Ch} \end{aligned} \tag{2.46}$$

with the following initial conditions: $m_c(t = t_{init}) = 1, m_I(t = t_{init}) = 0$ *and* $m_{Ch}(t = t_{init}) = 0$.

m_c, m_I, m_{Ch} and z respectively represent the normalized mass of cardboard, solid intermediary, solid residue and the total normalized mass. t_{init} is the initial time corresponding to an initial temperature of 450 K. Because the temperature is controlled by a constant heating rate (β), the temperature evolution is given by a fourth differential equation with β respectively equal to 5, 10 or 15 K min^{-1}:

$$\frac{dT}{dt} = \beta$$

As previously explained, a reparametrization of the model has been realized (see Eqs. 2.44 and 2.45).

Because of the normalized mass are the variables used in this model, the following least square criterion has been defined:

$$C = \sum_{j=1}^{3} \sum_{i=1}^{2951} \left((z)^i_{j,model} - (z)^i_{j,exp}\right)^2 \tag{2.47}$$

where $(z)^i_{j,model}$ and $(z)^i_{j,exp}$ are respectively the ith computed and experimental total normalized mass in the jth experiment. A same number of experimental data, i.e. 2951, have been chosen for each experiment to give the same weight to each experiment in the regression.

The experimental derivatives of the mass (DTG) are not available in this example, so in addition to a numerical method for the non-linear least square regression, another numerical method for resolving the system of algebra-differential equations must be integrated. More precisely in each iteration, the differential equations (Eq. 2.46) are numerically resolved with the current kinetic parameters, next the criterion (Eq. 2.47) is computed and finally the parameters are modified according to the non-linear regression algorithm (here the Levenberg-Marquardt method). The operations are repeated until a solution minimizing the criterion defined in Eq. (2.51) is found. Many numerical libraries are available and can be found in Internet; here we used the software MATLAB and its optimization toolbox [128].

The values of the identified kinetic parameters are given in the Table 2.33. The experimental data for the experiment with a heating rate equal to 5 and 15 K min^{-1} (dashed lines) and the theoretical curves of the total normalized mass are respectively compared in Figs. 2.40 and 2.41. As a unique set of kinetic parameters has been identified in order to fit the three experiments with different heating rates, the experimental and simulated curves are not perfectly confounded. However, this set of kinetic parameters is able to well represent the thermal degradation of cardboard for many different experimental conditions. Thanks to the model, it is possible to compute and plot the evolutions of the individual normalized mass of each solid species (cardboard, intermediate solid and char). The simulated curves for a heating rate of 10 K min^{-1} are plotted in Fig. 2.42.

2.6.2.4 Deconvolution Procedures (Model-Free and Model Fitting Methods)

Until now, the methodology for the kinetic analysis of processes that can be suitable described by single step kinetics has been presented. However, this is not the case of most biomasses and biowastes transformation processes. If multiple steps take place during a transformation and can be identified from the understanding of the reaction mechanism, some mathematical approaches can be used to determine the associated kinetic parameters, assuming a parallel reaction model (PRM). Accordingly, the reaction rate can be written as presented in Eq. 2.48 [104]:

Table 2.33 Kinetic parameters for thermal degradation of cardboard based on three non-isothermal thermogravimetric experiments

Experimental heating rate (K min^{-1})	5/10/15
A1 (min^{-1})	7.20 × 1011
E1 (kJ mol^{-1})	142.01 × 103
A2 (min^{-1})	3.22 × 107
E2 (kJ mol^{-1})	103.33 × 103
a	0.37
b	0.31

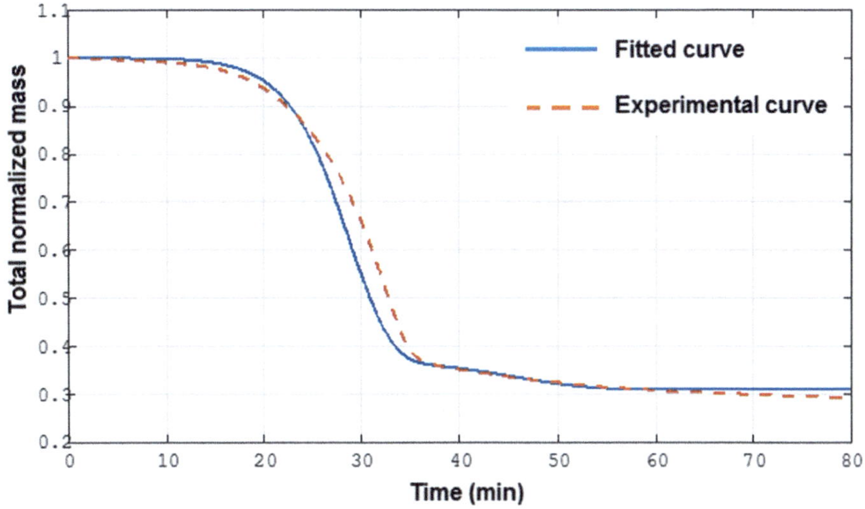

Fig. 2.40 Thermal decomposition of cardboard at 5 K min^{-1}. Comparison between the experimental curve and fitted curve with kinetic parameter from Table 2.33

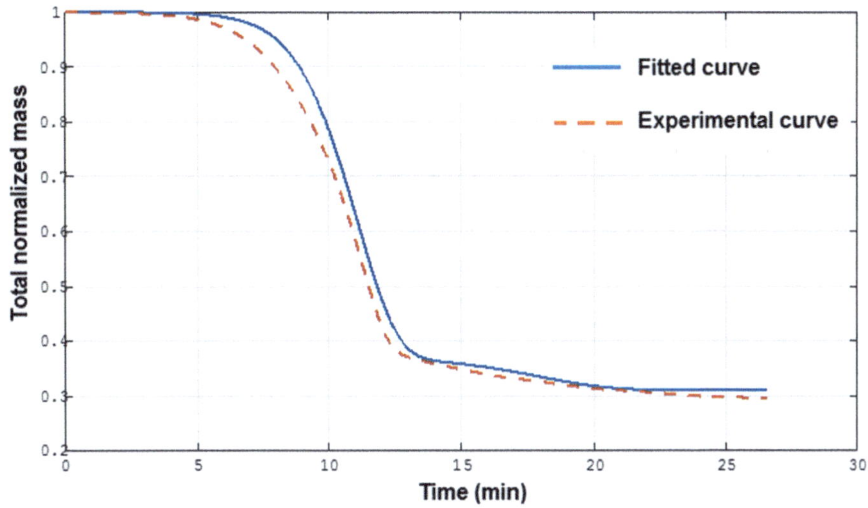

Fig. 2.41 Thermal decomposition of cardboard at 15 K min^{-1}. Comparison between the experimental curve and fitted curve with kinetic parameter from Table 2.33

$$\frac{d\alpha}{dt} = k_1(T)f_1(\alpha_1) + k_2(T)f_2(\alpha_2) + \cdots + k_n(T)f_n(\alpha_n) \tag{2.48}$$

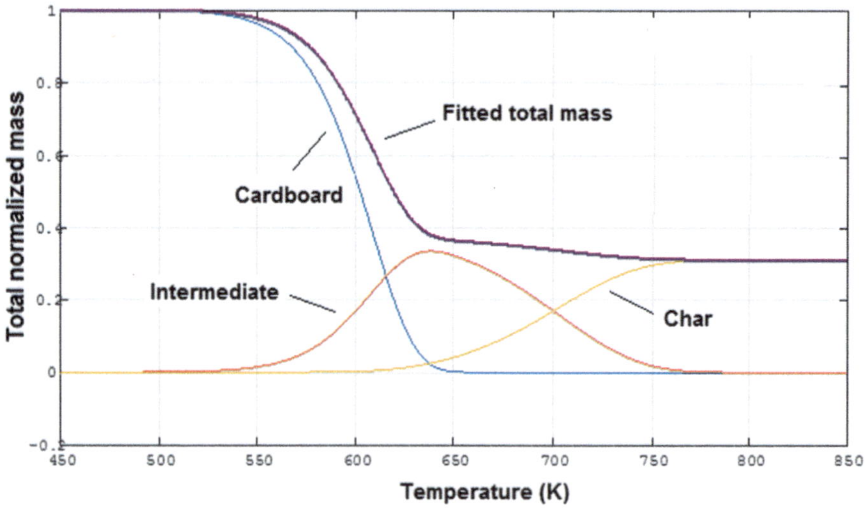

Fig. 2.42 Thermal decomposition of cardboard at 10 K min^{-1}. Time evolution of the fitted total normalized mass and simulated normalized mass of cardboard, Intermediate solid and char with kinetic parameters from Table 2.33

Thus, an alternative is the separation of each individual process by peak deconvolution of the experimentally obtained DTG data, using statistical functions. The global reaction rate is then expressed as the sum of the reaction rate of the individual parallel reactions i (from 1 to n). Once the individual processes have been identified, their kinetic analysis can be performed using one of the already presented approaches.

Several statistical functions can be used for peak deconvolution and are summarized in Table 2.34. The choice of the most suitable one depends mainly on the form of the experimental peak; in particular its width and asymmetry. Some functions do not allow the accommodation of asymmetrical peaks and cannot properly describe different processes.

Then, optimization methods are generally used to determine the function parameters that best fit a particular experimental curve. Usually, the fit error between the calculated and the experimental profile can be determined using Eq. (2.49).

$$Fit\ error(\%) = 100\left(\frac{\sqrt{\sum_{i=1}^{n}\left(\left(\frac{d\alpha}{dt}\right)_{i,exp} - \left(\frac{d\alpha}{dt}\right)_{i,calc}\right)^2}}{\left(\sqrt{N}\right)\left(\frac{d\alpha}{dt}\right)_{exp,max}}\right) \tag{2.49}$$

where $d\alpha/dt$ are the experimental and calculated values of the decomposition rate, and N is the total number of experimental points. The lowest is the fit error, the best is the agreement with the experimental data. It is must be noted that the

Table 2.34 Some statistical functions used for peak deconvolution in kinetic analysis

	Function	Parameters	Ref.
Gaussian	$\left(\frac{d\alpha}{dt}\right) = a\exp\left[-\frac{1}{2}\left(\frac{t-b}{c}\right)^2\right]$	*a*—amplitude *b*—peak temperature *c*—width	[129, 130]
Lorentzian (Cauchy)	$\left(\frac{d\alpha}{dt}\right) = \frac{a}{1+\left(\frac{t-b}{c}\right)^2}$	*a*—amplitude *b*—peak temperature *c*—width	[129, 130]
Weibull	$\left(\frac{d\alpha}{dt}\right) = \frac{a}{b}\left(\frac{t-c}{b}\right)^{a-1}\exp\left[-\left(\frac{t-c}{b}\right)^a\right]$	*a*—shape parameter > 0 *b*—amplitude > 0 *c*—peak temperature	[130–132]
Fraser-Suzuki	$\left(\frac{d\alpha}{dt}\right) = a\exp\left\{-\frac{\ln 2}{d^2}\ln\left[1+2d\left(\frac{t-b}{c}\right)\right]^2\right\}$	*a*—amplitude *b*—peak temperature *c*—half width *d*—asymmetry	[133]

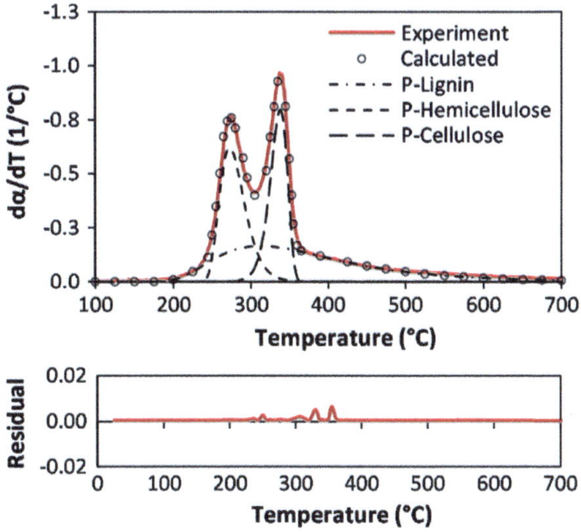

Fig. 2.43 Comparison between experimental data and deconvolution results for coconut shells pyrolysis at 10 °C/min. Adapted with permission from Springer Nature, copyright 2017 [57]

decomposition rate can also be expressed as a function of the temperature if required (e.g. non-isothermal data).

An example of the deconvolution of the pyrolysis DTG curve of coconut shells, a lignocellulosic agrowaste is presented in Fig. 2.43, using the Fraser-Suzuki function [105]. In this case, taking into account that lignocellulosic biomass is mainly composed by hemicellulose, cellulose and lignin, it is assumed that each component is decomposed independently, and then, 3 peaks are used for deconvolution (e.g. pseudo-hemicellulose, pseudo-cellulose, pseudo-lignin) [57, 105]. Fraser-Suzuki function was chosen for peak deconvolution as it allows the fitting of asymmetric curves and accurately describes the biomass decomposition behavior.

As observed in Fig. 2.43, the fit error between the calculated decomposition curve and the experimental data was 1%, showing a good description of the biomass decomposition behavior.

Though, it is important to say that the peak deconvolution procedure goes beyond the simple statistical fit and must be in accordance with the nature of the sample. For example, for some lignocellulosic residues a correct fit can be obtained with only two peaks, neglecting the fact that at least three components should be taken into account (hemicellulose, cellulose and lignin). Therefore, a fit with only two peaks does not represent the nature of the sample even if the fitting error is low.

From the deconvolution results, the analysis of individual peaks allows the determination of the kinetic parameters of each biomass pseudo-component, using any of the presented methods. Thus, the total reaction rate can be then described as in Eq. (2.48).

2.6.3 High and Low Heating Value Determination

2.6.3.1 Introduction

The heating value is the amount of energy released by the combustion of a unit quantity of a fuel. As a basis for determining the performance of an energy system, it is necessary to differentiate between higher heating value (HHV) or gross calorific value (GCV) and lower heating value (LHV) or net calorific value (NCV) [134].

Higher heating value (HHV) is defined as the amount of heat released by the unit mass or volume of fuel (initially at 25 °C) once it is completely combusted and the products have returned to a temperature of 25 °C. It includes the latent heat of vaporization of water. HHV is a useful indicator of biomass quality that depends on: (a) elemental composition, (b) moisture content, and (c) ash content [135]. The HHV can be experimentally determined in the laboratory with a bomb calorimeter. The HHV of different types of biomass are in the range of 14–23 MJ/kg (dry basis) [136–139]. The differences are mainly due to different carbon content (main energy source) and different ash content (non-combustible material).

The temperature of the exhaust flue gas of an industrial boiler is generally in the range of 120–180 °C. The combustion products are rarely cooled to the initial temperature of the fuel (≈25 °C). Hence, the water vapor in the flue gas does not condense; therefore, its latent heat of vaporization is not recovered. As a consequence, the effective heat available for use in the boiler is a lower amount [140]. We define a lower heating value (LHV) which is the amount of heat released by fully combusting a specified quantity of fuel less the heat of vaporization of water in the combustion product.

With regards to the LHV, existing methods cannot directly determine this energy: this value can be calculated from the HHV of the analyzed sample. The LHV requires information on the moisture and hydrogen content of the sample, which transform into water vapor during combustion of the sample. This will be further discussed in the following sections.

2.6.3.2 Bomb Calorimetry

Description

A bomb calorimeter system is based on measuring the heat released from the complete combustion of a fuel in an oxygen environment. A widely used bomb calorimeter is the IKA C5000. This system typically consists of a controller, measuring unit and cooling system [141].

Controller

The controller is the central control, operating, and display unit for measuring cell and cooling unit. It performs system calibration and stores experimental data and protocols. Furthermore the controller enables communication with peripheral devices such as a printer, analytical balance, sample rack, and an external PC.

Measuring Cell

This is where the combustion of fuel samples under precisely defined conditions takes place (Fig. 2.44). The bomb or decomposition vessel is immersed in a given amount of water, which is in turn placed in a thermally insulated jacket. A primary temperature transducer, placed inside the unit, records the change in the system temperature due to the combustion of the fuel in the bomb.

The measuring cell ensures that the following experimental conditions exist during determination of the heating values:

- The temperature of the fuel sample before combustion is 22 °C.
- An atmosphere comprising pure oxygen (99.5%) at a pressure of 30 bar exists during combustion.
- Adiabatic measuring method (DIN 51900-3 [142]): The temperature regulator adjusts the temperature of the water jacket to the temperature of the inner vessel. Hence, no heat exchange with the environment takes place during combustion.

To achieve these experimental conditions, the following modules are installed in the measuring cell:

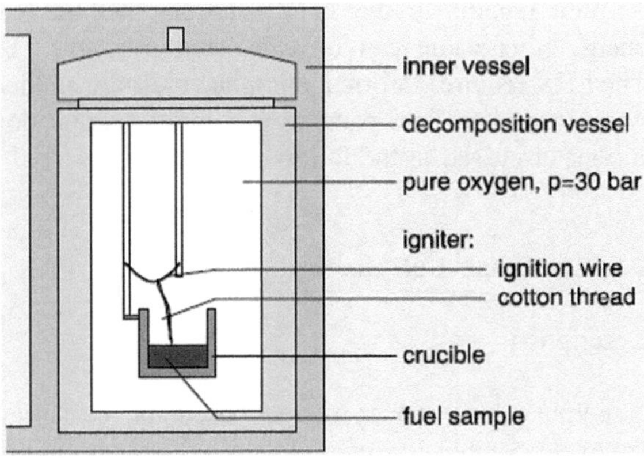

Fig. 2.44 Measuring cell of calorimeter system. Reproduced with permission from IKA [141]

- Water vessel with water jacket;
- Magnetic stirrer to ensure uniform heat distribution;
- Water circuit with pump, expansion container, and container for an external cooling unit;
- Heating system, temperature regulator;
- Oxygen filling station.

When the decomposition vessel is attached to the measuring cell, the experimental cycle can begin. When the experiment has been completed, the overpressure is released from the decomposition vessel, the inner vessel is drained, and the measuring cell cover is opened. The decomposition vessel can then be removed.

2.6.3.3 Cooling Unit

The system cools the water circuits of a measuring cell. It consists of the following subsystems:

- A heat exchanger, to provide the necessary cooling of the water circuit;
- A compressor with a condenser and evaporator to generate a sufficient cooling capacity for the measuring cell;
- A fan to carry away the heat generated by the measuring system using air.

2.6.3.4 Sample Preparation

Normally, solid biomass can be directly burnt in powder form in the crucible (Fig. 2.45). However, fast-burning of fine biomass particles may cause some particles to spatter into the solution in the decomposition vessel. In this case, it would be impossible to guarantee complete combustion. Furthermore, the inner wall of the

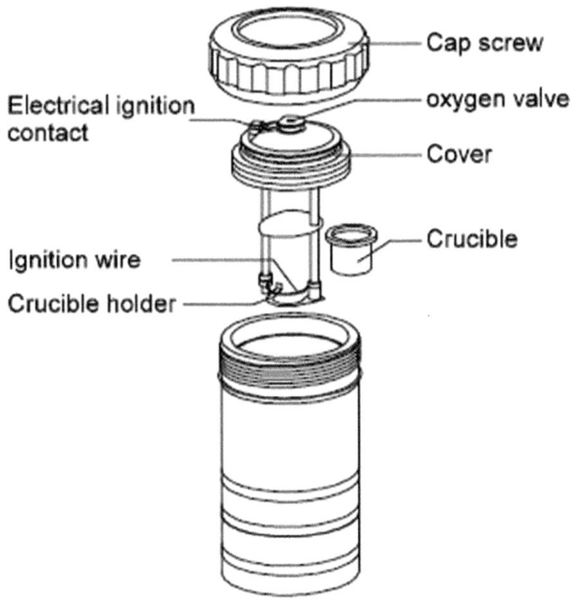

Fig. 2.45 Individual parts of the decomposition vessel; Reproduced with permission from IKA [141]

decomposition vessel could also be damaged. Hence, it is recommenced that, before combustion, the biomass particles be pelletized.

For liquid biomass such as bio-oil and black liquor, the samples can be poured directly into the crucible and weighed. If these liquids contain precipitating water, they must be dried or homogenized prior to weighing. The water content of these samples must be determined.

2.6.3.5 Bases of Expressing Biomass Heating Value

The heating value of biomass is often expressed on different bases or reference states depending on the situation. The following reference states of analysis are commonly used:

- As-received/supply state, Ar;
- Air dry state, ad;
- Dry and ash-free state, daf.

2.6.3.6 Correcting for Extraneous Energy

During the combustion experiment, heat is not only generated by the combustion of the biomass sample but also through extraneous energy from igniters, combustion aids and the formation of sulphuric and nitric acid (Fig. 2.46). This extraneous

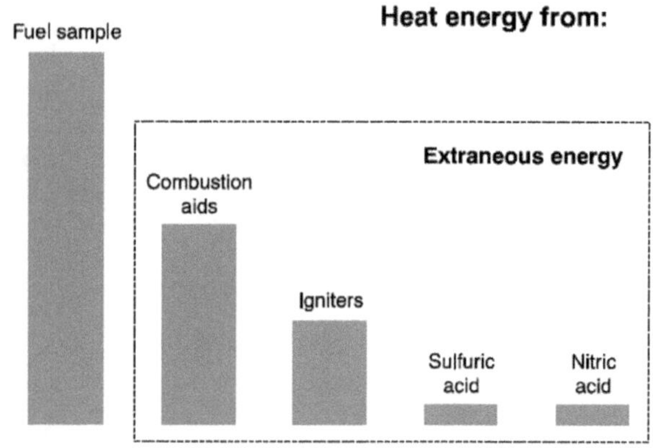

Fig. 2.46 Heat of combustion and extraneous energy. Reproduced with permission from IKA [141]

energy can vary significantly in relation to the heat of combustion of the biomass sample and must therefore be taken into account during calculations of the heating value. A schematic illustration of extraneous energy is available elsewhere [141].

2.6.3.7 Igniter and Combustion Aids

Cotton thread is often used to ignite the sample as most biomass are not easily inflammable and have relatively low calorific values. The heat of combustion of the cotton thread as well as the heat of electrical ignition would falsify the measurement. Hence, before cotton is used, it must be weighed since its weight (m_{ct}) together with its specific combustion heat (H_{ct}), is used to determine its HHV (Q_Z) as shown by Eq. (2.50).

$$Q_Z = m_{ct} \cdot H_{ct} \tag{2.50}$$

This HHV is included in the calculation as extraneous energy. The amount of igniter used should be kept as low as possible to minimize the error in measurements.

Gelatine or acetobutyrates capsules are normally used as combustion aids when the biomass sample in powder form cannot be molded into pellets, or is highly volatile. Similar to the case of using a cotton igniter, the higher heating value of the capsules must be known so that the combustion heat that they produce as extraneous energy can be taken into consideration.

2.6.3.8 Acid Correction

Biomass usually contains sulphur and nitrogen, which may form SO_2, SO_3 and NO_x when burnt in the decomposition vessel under the existing pressures and temperatures. When mixed with the water produced from combustion of the biomass sample, they form sulphuric and nitric acid and the heat of solution. According to DIN 51900 [142], this heat of solution must be taken into account when the gross calorific values are calculated.

Energy generated from formation of sulphuric acid:

$$Q_S = 94.62 \cdot m_S \cdot S_d \tag{2.51}$$

where S is the percentage of sulphur in the sample in the dry state (%).

Energy generated from formation of nitric acid:

$$Q_N = 43 \cdot m_N \cdot N_d \tag{2.52}$$

where N is the percentage of nitrogen in the sample in the dry state (%).

Thus, the sum of extraneous energies may be presented as:

$$\sum Q = Q_Z + Q_S + Q_N \tag{2.53}$$

2.6.3.9 Calibration of Bomb Calorimeter

General Description

The calorimeter system is calibrated at specific intervals for every operating condition used in order to ensure precise and reproducible measuring results. During calibration, the heat capacity of the calorimeter system is determined.

A specific amount of benzoic acid, a reference substance, is burnt in the decomposition vessel under experimental conditions (same oxygen pressure and amount of water in the decomposition vessel before combustion). The benzoic acid tablet is obtained from the National Bureau of Standards (NBS standard sample 39) with guaranteed HHV. Since the precise HHV of benzoic acid is already known, during combustion of this substance, it is also possible to calculate the heat capacity of the calorimeter system using the increase in temperature of the system.

Halogens

Corrosion by halogens released from biomass and waste on the surface of the decomposition vessel can be reduced if, prior to combustion, the vessel is filled with 5–10 mL of 0.5 M Na_2CO_3 depending on the halogen content in the biomass. It is therefore important to ensure that the system is calibrated beforehand with the same volume of 0.5 M Na_2CO_3 solution.

Calculations During Calibration

First law of thermodynamics applied to an oxygen bomb calorimeter (isochoric change with zero mechanical work) is expressed in the equation below:

$$\Delta U = q_{syst} \tag{2.54}$$

Given that the outer vessel is adiabatic, the combustion process will increase the temperature of the water surrounding the bomb by ΔT.

$$q_{syst} = \sum_{i=1}^{N} m_i \cdot H_i = C \cdot \Delta T \tag{2.55}$$

where C is the heat capacity of the calorimeter system, which can be determined by the calibration of the system by means of measuring the HHV of several benzoic acid tablets (C_6H_5COOH). The combustion reaction of the benzoic acid at 25 °C is defined as:

$$C_6H_5COOH(s) + 7.5O_2(g) \rightarrow 7CO_2(g) + 3H_2O(l) \tag{2.56}$$

The heat capacity of the calorimeter can therefore be calculated according to the following equation:

$$C = \frac{m_{ba} \cdot H_{ba} + Q_Z}{\Delta T_c} \tag{2.57}$$

where m_{ba} is the mass of the benzoic acid (g), H_{ba} is the calorific value of the benzoic acid tablet (J/g), Q_Z is the gross calorific value of the igniter and combustion aid (J], and ΔT_c is the observed change in temperature during the calibration experiment (°C).

A basic statistical analysis is performed on the results to determine the mean and standard deviation of the calorific values using the following equations:

$$\text{Average of } x = \bar{x} = \frac{1}{n} \sum_{J=1}^{n} x_j \tag{2.58}$$

$$\text{Standard deviation of the mean of } x = \sqrt{\frac{\sum_{j=1}^{n} n\left(x_j - \bar{x}\right)^2}{n(n-1)}} \tag{2.59}$$

where x_j are the calorific values and n is the number of measurements.

2.6.3.10 Determining Heating Value

Experimental Conditions

In a bomb calorimeter, combustion occurs under defined conditions. The sample is weighed and placed in a crucible, which in turn is enclosed in the decomposition vessel surrounded by water at 22 °C. Pure oxygen (99.95%) at 30 bar fills the decomposition vessel. The sample is then electrically ignited and the increase in temperature inside the calorimeter system is measured.

Calculations During Experiments

Having previously identified the HHV of the calorimeter during calibration *C*, the system automatically calculates the specific HHV of the fuel sample during analysis, which can be determined as follows:

$$HHV = \frac{C \cdot \Delta T_{fs} - Q_Z}{m_{fs}} \tag{2.60}$$

where ΔT_{fs} is the observed temperature increase of the water in the inner vessel of the measuring cell during combustion of the fuel sample (°C).

The lower heating value (LHV), also known as the net calorific value (NCV), is defined as the amount of heat released by fully combusting a specified quantity less the heat of vaporization of the water in the combustion product. The relationship between HHV and LHV is given by [140]:

$$LHV = HHV - h_g\left(\frac{9H}{100} - \frac{M}{100}\right) \tag{2.61}$$

where *LHV, HHV, H*, and *M* are lower heating value, higher heating value, hydrogen percentage, and moisture percentage, respectively, on an as-received basis. Here, h_g is the latent heat of steam in the same units as *HHV* (i.e., 970 BTU/lb, 2260 kJ/kg, or 540 kcal/kg).

2.6.3.11 Complete Combustion

Complete combustion of the sample is essential for the correct determination of the HHV and LHV. After each experiment, the crucible and all solid residues must be examined for signs of incomplete combustion. If slight soot deposits or unburned residue is found, the experiment must be repeated. If the biomass sample is in powder form, the sample could be molded into a pellet or inserted placed in a capsule before combustion, minimize the ejection of the sample particles from the crucible into the solution below, in the decomposition vessel.

2.6.3.12 Example 1: Agroforestry Species and Bio-based Industry Residues

Introduction

This example is based on the work of Gravalos et al. [143]. Samples of different agricultural residues and wastes (rice husks, apricot kernels, olive pits, sunflower husks, cotton stems, etc.), energy crops and wetland herbs (cardoon, switchgrass, common reed, narrow-leaf cattail), and forest residues (populus, fagus, pinus) were analyzed for their high heating values (HHV) or gross calorific values (GCV) based on CEN/TS 14918:2005 standard [144] and using an oxygen bomb calorimeter (Model C5000 Adiabatic Calorimeter, IKA®-Werke, Staufen, Germany).

Sample Preparation

The following procedure was used to prepare the samples of biomass residues:

- The biomass residues were first ground using a hammer mill and a 3.0 mm screen size;
- 0.5–1 g of each biomass powder was pressed to produce an unbreakable pellet (tablet) in order to limit the speed of combustion;
- For those biomass samples that were not easily pressed into pellets, capsules were used to hold and easily ignite the samples.

Analysis Procedure

The experimental procedure included the following steps:

- The instrument was calibrated and verified using a benzoic acid tablet;
- The biomass sample of 0.5–1.0 g was weighed on a precision balance;
- The powdered sample was inserted in the capsule and the capsule was pressed to compact the material;
- The capsule was carefully placed in the holder;
- The cotton thread was attached and the firing cotton was placed on top of the capsule;
- The bomb was lowered in the calorimeter and the cover was then closed;
- The start button was pressed to begin the test.

The C5000 Adiabatic Calorimeter automatically made the necessary calculations (Eqs. 2.50–2.60) to produce the gross calorific value (GCV) of the sample.

Results

Table 2.35 shows the minimum, maximum, range, mean and standard deviation of GCV of all the samples of agricultural residues and wastes. The results reveal that seeds and kernels have higher gross calorific values because of their higher unit mass and higher lipid content.

2.6.3.13 Example 2: HHV Analysis of Solid Recovered Fuel (SRF)

Introduction

In the analysis presented, the sample is a solid recovered fuel (SRF) composed of non-recyclable wood, plastics, paper and cardboard waste. The HHV and LHV of the sample was analyzed according to the standard EN 15400 [145] and using an oxygen bomb calorimeter (Model C5000 Adiabatic Calorimeter, IKA®-Werke, Staufen, Germany).

Sample Preparation

The SRF sample was initially ground to powder using with a hammer mill and a 1.0 mm screen size.

Next, the sample was dried according to EN 15414-3 [146] and then compressed to produce a tablet of roughly 300 mg.

Analysis Procedure

The experimental procedure involved the following steps:

- The bomb calorimeter was calibrated using six benzoic acid tablets under 30 bar of oxygen at 24 °C;
- A 10 mL solution of 0.5 M Na_2CO_3 (aq) was added to the combustion bomb to capture any acid gases released during combustion;
- The SRF tablet of approximately 300 mg was weighed on a precision balance;
- The tablet and the ignition cotton of known mass were placed in the crucible;
- The bomb, containing the crucible, was lowered in the calorimeter and the cover was then closed;
- The start button was pressed to begin the test.

Calculation of HHV

For this analysis, the sample was previously dried, hence only the hydrogen content of the sample was necessary to calculate the LHV. Information on the hydrogen

Table 2.35 Gross calorific value of agricultural residues and wastes

Agricultural residue and waste species	Minimum GCV (MJ kg^{-1})	Maximum GCV (MJ kg^{-1})	Range GCV (MJ kg^{-1})	Mean GCV (MJ kg^{-1})	Standard deviation
Rice husk with moisture content 8.30%	15.875	16.057	0.182	15.972	0.075
Rice husk without moisture (dried at 105 °C for 24 h)	16.546	16.748	0.202	16.643	0.078
Rice straw with moisture content 12.19%	15.012	15.130	0.118	15.092	0.049
Rice straw without moisture (dried at 105 °C for 24 h)	16.345	16.580	0.235	16.475	0.093
Pistachios shells	17.247	17.431	0.184	17.320	0.097
Leaves of pistachios trees	15.477	16.618	1.141	16.120*	0.585
Dark red cherry seeds	19.358	20.441	1.083	19.870*	0.544
Apricot kernels	18.508	18.642	0.134	18.562	0.071
Peach kernels	18.901	19.091	0.190	18.995	0.095
Watermelon fruit seeds	23.321	23.585	0.264	23.473	0.137
Grape seeds	20.038	20.839	0.801	20.388*	0.410
Olive pits	17.369	18.582	1.213	17.970*	0.537
Almont husks	18.007	18.318	0.311	18.176	0.112
Sunflower husks	18.444	18.831	0.387	18.674	0.173
Sunflower seeds cake with moisture content 12.72%	21.160	21.277	0.117	21.231	0.042
Rapeseed cake with moisture content 11.17%	21.442	21.844	0.402	21.569	0.158
Cotton (*Gossypium hirsutum* L.) plant root	17.671	17.768	0.097	17.707	0.036
Cotton plant main stem	17.645	17.799	0.154	17.733	0.054
Cotton plant terminal bud	16.105	16.643	0.538	16.399	0.160
Cotton plant vegetative branches	17.070	17.598	0.528	17.376	0.143
Cotton plant fruiting branches	17.065	17.505	0.440	17.368	0.143
Cotton plants leaves	15.955	16.126	0.171	16.059	0.048
Cotton plant (25 weeks/ harvest) bur	16.993	17.292	0.299	17.141	0.084
Cotton plant (25 weeks/ harvest) locks	16.543	16.870	0.327	16.679	0.099
Cotton plant seeds	22.750	23.078	0.328	22.933	0.111

Reproduced with permission from Scientific Research Publishing [143]
*Denotes significance ($p < 0.05$)

content in the sample is determined by the CHNS analyser. The LHV in MJ/kg is calculated using the following formula, according to standard EN 15400 [147]:

$$LHV = HHV - 0.206 \times \%H \tag{2.62}$$

where %H is the hydrogen content in the dry sample.

Results

The HHV of CSR measured by the bomb calorimeter was 25.8 MJ/kg. Given that the hydrogen content of CSR was 6.2 wt%, the LHV was calculated to be 24.5 MJ/kg (using Eq. 2.61).

2.6.4 *Heat Capacity Measurement*

2.6.4.1 Introduction

The isobaric specific heat capacity is defined as being the quantity of heat required to raise the temperature of 1 g of material by 1 K, a value expressed in J g^{-1} K^{-1}. At constant pressure, the isobaric specific heat capacity (C_p) is calculated using the following equation [148]:

$$C_p = \frac{1}{m}\frac{dH}{dT} \tag{2.63}$$

with C_p: isobaric specific heat capacity (J g^{-1} K^{-1}); m: mass (g); H: enthalpy (J); T: temperature (K).

The heat capacity of a sample (powder, ceramic, etc.) is a parameter allowing the calculation of the quantity of heat that it can absorb or accumulate (thermal insulation, energy-storage materials, etc.). This value enables the heat balance of materials to be established, which makes it possible to improve their cooling or insulating capacities.

When a material is heated, different physico-chemical phenomena can take place at a given temperature (e.g. quartz transition, glass transition, phase change, crystallization, fusion, or chemical reaction etc.). These transformations imply heat absorption or release. The measurement of C_p must exclude the energy due to these transformations.

It is also worth noting that C_p generally varies with the temperature. This variation depends on the nature of each material considered.

In this chapter, two methods of measurement are presented, which are often used for measuring C_p of a material.

2.6.4.2 Heat Capacity Measurement by a Calorimeter

Measurement Principle

Equation (2.63) can be also expressed as follows:

$$C_p = \frac{1}{m}\frac{dH}{dt}\frac{dt}{dT} \tag{2.64}$$

$$\frac{dt}{dT} = \frac{1}{V} \tag{2.65}$$

with V the linear heating rate during C_p measurement.

Thus, the following equation can be obtained:

$$\frac{dH}{dt} = m \cdot Cp \cdot V \tag{2.66}$$

with t: time(s); V: heating speed (K/s); $\frac{dH}{dt}$: measured signal that is the heat flow (mW).

This Eq. (2.66) is only applicable on a temperature range where the sample does not undergo any type of transformation, like crystallization, fusion, or chemical reaction, and which is outside the warm-up phase of the system (apparatus + crucible + sample).

Two measurements under the same experimental conditions are needed. The first one is carried out with the two empty measurement crucibles and the second one with the sample. The advantages of this method are that no standard reference is needed and it gives a C_p equation for a defined temperature range. The same heating rate must be used for both measurements, with and without the sample. Figure 2.47 illustrates the C_p measurement using a calorimeter.

Apparatus

An instrument that measures C_p is called a calorimeter. There are several suppliers, among whom is Setaram, with a well-known model, the "C80 calorimeter" [149]. This is a Calvet-type calorimeter, with an assembly of ring thermocouples. For the C80 model, there are a total of 9 rings surrounding the measuring crucible. Each ring has 38 thermocouples giving a total of 342 thermocouples, which ensure precise measurement "in 3D" of the heat flow absorbed or released by the sample during the test (Fig. 2.48).

Figure 2.49 shows a photo of the C_p measuring crucibles used for the C80 calorimeter, which are made of aluminium oxide. The specifications of the C80 calorimeter, as given by the supplier, are shown in Table 2.36.

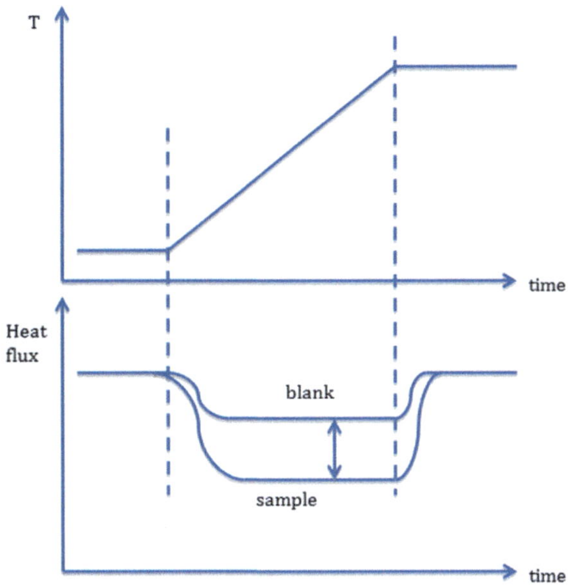

Fig. 2.47 Illustration of measurement of C_p with a calorimeter

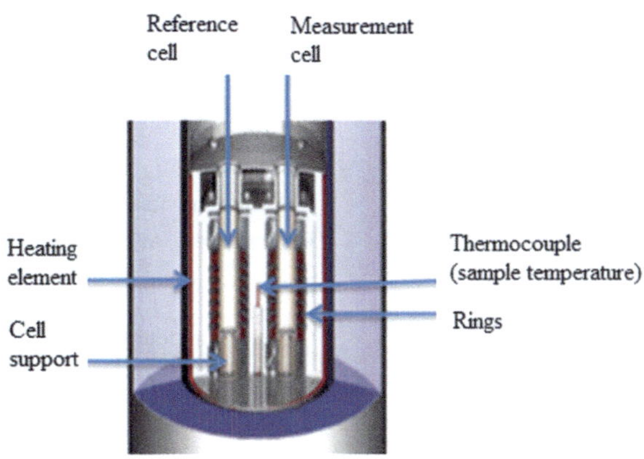

Fig. 2.48 Photo of the measurement crucibles of C80 apparatus. Reprinted with permission from Setaram Instrumentation

Indicative C_p values

Table 2.37 shows the C_p values of certain chemical products or materials.

Measurement and Precautions

Setting up the sample:

The following points need to be checked for any C_p measurement analysis:

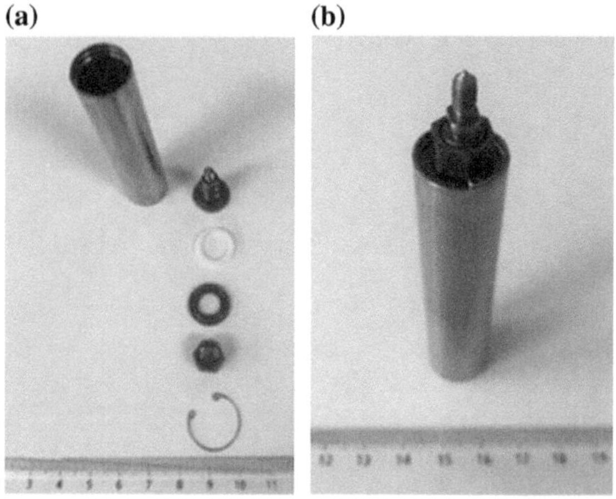

Fig. 2.49 Photo of the C_p measuring crucible used for the C80 calorimeter: (a) before and (b) after assembly

Table 2.36 Specifications of C80 calorimeter

Temperature range	T_{amb} to 300 °C
Accuracy of temperature	±0.1 °C
Precision of temperature	±0.05 °C
Programmable speed	±0.001 to 2 °C/min
Enthalpy accuracy	±1%
Calorimetric precision	±0.1%
RMS noise	1 μW
Resolution	0.10 μW
Sensitivity	35 μV/mW
Standard crucible volume	12.5 mL

Reproduced with permission from SETARAM Instrument

Table 2.37 C_p values of certain chemical products or materials

Nature of product	Indicative C_p value (J/g K)
Water	4.18
Organic compounds	1–2
Mineral compounds	0.7–1.2
Metals	0.2–0.6
Aluminium oxide	0.78

- Sample must not undergo any transformation in the temperature range under study;
- Sample must not lose weight;
- Sample should not interact with the crucible (table supplied by the manufacturer);
- Always use the same crucibles for measurements, empty-crucible blank and sample.

For solid samples:

- Sample must be completely dry so as not to cause a variation in the mass due to water (or solvent) loss during the analysis. Sample can be dried in an oven before analysis;
- Apply the same protocol to all the samples of the same series;
- The presence of humidity can cause variations in the C_p of sample;
- In the case of powder sample, it is requested to pack it down (or tap).

For liquid samples:

- A pressurized and sealed C_p crucible is recommended for liquid sample. To this end, Teflon seals can be used, with the crucible filled to 4/5 capacity, as shown in Fig. 2.50.

Programming of the Method

The main steps in measuring C_p are as follows:

- Plateau time of 120 min at initial temperature;
- Temperature ramp with a heating rate of ≤ 0.2 K/min;
- Plateau time of 120 min at the desired temperature;
- Before beginning the analysis, the system must be stabilized.

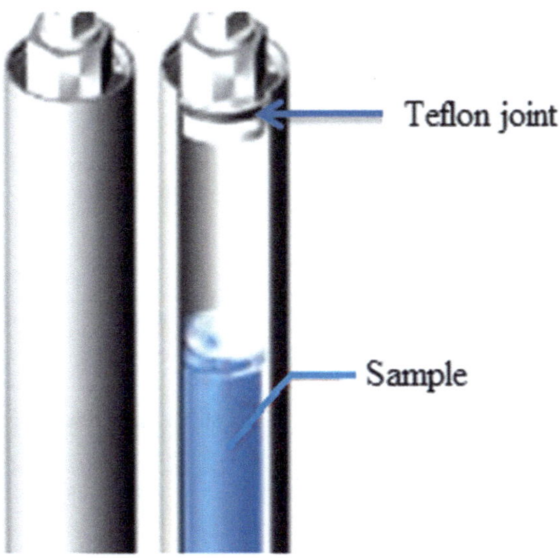

Fig. 2.50 Illustration of the use of a Teflon seal to ensure water-tightness of the measuring crucible for liquid samples. Reprinted with permission from Setaram Instrumentation

Example: Cp Measurement of a WWT (Waste Water Treatment) Dry Sludge

As an example, the C_p measurement of a WWT (waste water treatment) dry sludge is illustrated. The studied sludge is a non-homogeneous complex medium. Two measurements of C_p are presented for dry sludge in order to check the precision of the measurement. Another measurement was taken on the same dry sludge, but after grinding into fine particles.

The protocol of the measurement is outlined in Fig. 2.51. A first measurement was carried out with empty crucibles: the blank measurement Then, for each measurement, the sludge is placed in the crucible and the same protocol is applied. The sludge mass for each measurement is given in Table 2.38.

Figure 2.52 shows the results obtained for these three measurements. Firstly, we can see a very good repeatability of two measurement of dried sludge. Secondly, the C_p of ground sludge was lower than that of the initial dried sludge. Finally, in all cases, the C_p value slightly increased with the temperature within the temperature range studied.

In general, a mathematical equation expressing the evolution of C_p as a function of the temperature can be obtained with the software used for the apparatus.

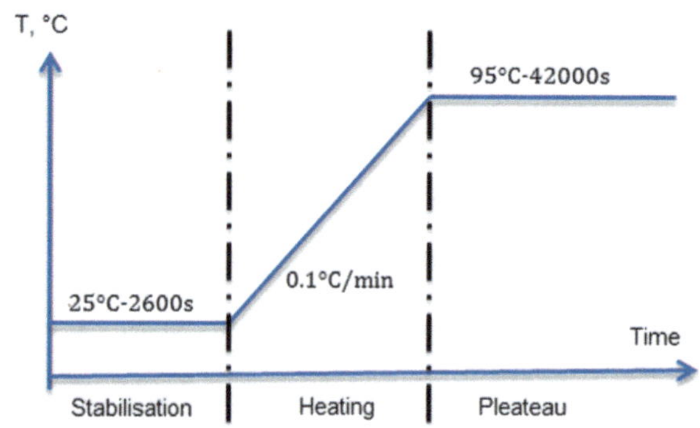

Fig. 2.51 Protocol of C_p measurement of a dry sludge

Table 2.38 Weight of sludge used for each measurement

Measurement	Weight (g)
C_p of dried sludge, 1st measurement	4.0725
C_p of dried sludge, 2nd measurement	3.8352
C_p of dried and ground sludge, 2nd measurement	7.7534

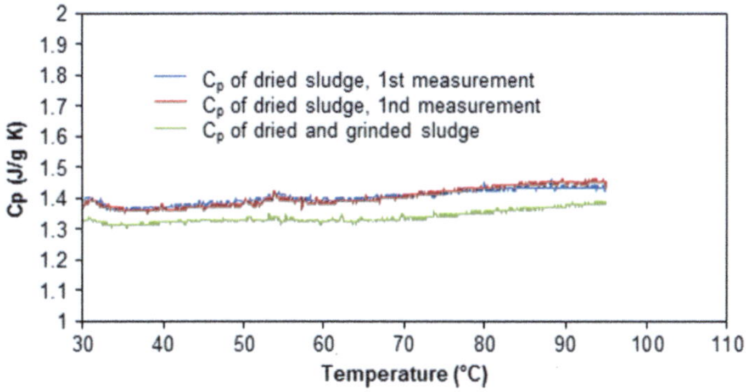

Fig. 2.52 Value of C_p of a sludge as a function of temperature from 35 to 95 °C

2.6.4.3 Heat Capacity Measurement with Modulated DSC

Introduction and Theory

As previously presented, differential scanning calorimetric analysis (DSC) measures heat flow absorbed or released by a sample, by its difference from the reference, as a function of a linear (or isothermal) temperature programming under a controlled atmosphere.

In modulated DCS (mDSC or TMDSC for Thermal Modulated Differential Scanning Calorimetry), heat flow absorbed or released by a sample (by its difference to the reference) is measured as a function of sinusoidal temperature programming around a linear (or isothermal) heat, for a given period and amplitude, under a controlled atmosphere and at a constant pressure. Figure 2.53 illustrates this sinusoidal temperature program.

It is worth noting that in DSC analysis the heating rate is linear (Eq. 2.67), while in mDSC, the temperature variation is sinusoidal (Eq. 2.67).

$$\text{DSC: } T = To + b \cdot t, \quad with \ \frac{dT}{dt} = b \tag{2.67}$$

$$\begin{aligned} \text{mDSC:} \quad & T = To + b \cdot t + A \cdot \sin(\omega \cdot t) \\ & \text{with } \omega = 2\pi f = 2\pi/P \\ & \text{and } \tfrac{dT}{dt} = b + A \cdot \omega \cdot \cos(\omega \cdot t) \\ & \text{and A: modulation amplitude; P: modulation period} \end{aligned} \tag{2.68}$$

Equation (2.69) comes into play:

$$\frac{dH}{dt} = C\frac{dT}{dt} + f(T,t) \tag{2.69}$$

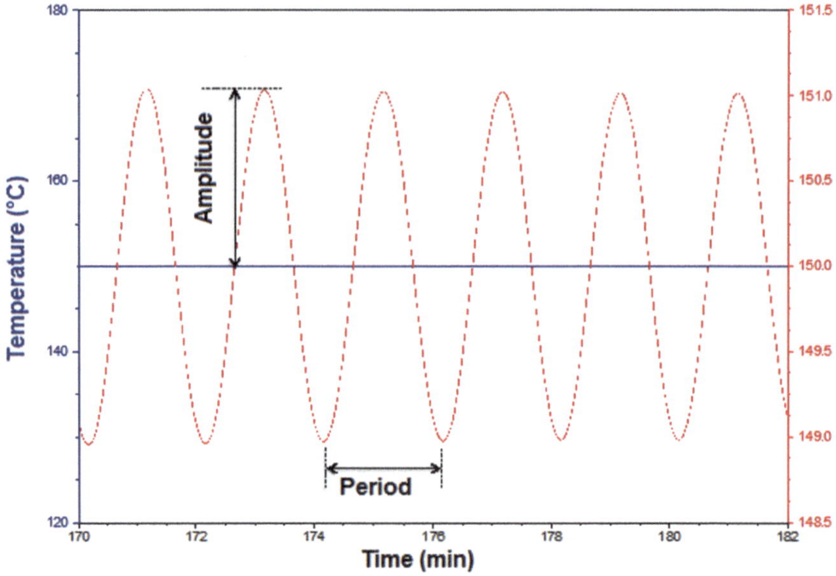

Fig. 2.53 Temperature modulation curve

with $\frac{dH}{dt}$ = Heat flow (W or J/s); C = calorific capacity (J/K); $\frac{dT}{dt}$ = heating rate (K/s); $f(T, t)$ = heat flow of the kinetic and enthalpy of the event at temperature T (W); and $C = C_p \times m$ (J/K).

Through the way it operates, modulated DSC allows the separation of the total heat flow $\frac{dH}{dt}$ from the heat flow caused by the heating rate $C\frac{dT}{dt}$, known as "reversing", and from the heat flow related to the absolute temperature of the event $f(T, t)$ known as "non-reversing".

For this reason, three signals are recorded during an analysis as illustrated in Fig. 2.54:

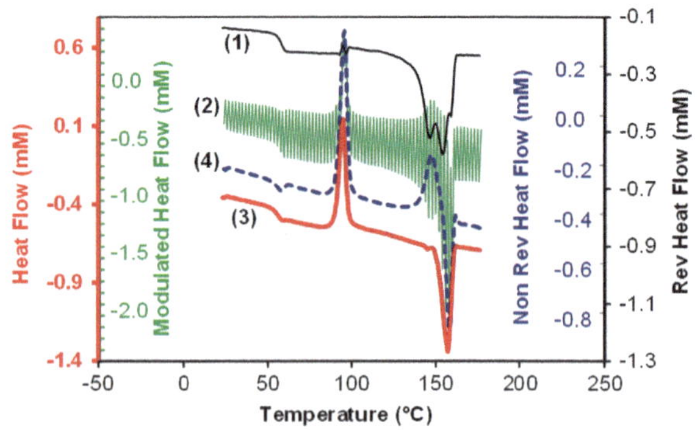

Fig. 2.54 Example of result that can be obtained by mDSC; (1) Rev Heat Flow (mM); (2) Modulated Heat Flow (mM); (3) Total Heat Flow (mM); (4) Non Rev Heat Flow (mM)

- The total heat flow signal, which corresponds to the heat flow obtained in conventional DSC;
- The reversing signal, linked to the C of the sample obtained by the modulation;
- The non-reversing signal, linked to a kinetic event, and which is obtained by the difference between the 2 preceding signals.

The advantage of mDSC compared to conventional DSC is that mDSC allows separating events linked to a kinetic component from events related to a C component (Table 2.39). Thus, it is possible to obtain the calorific capacity of a sample directly (Fig. 2.54).

Apparatus

The apparatus used for mDSC is the same as for conventional DSC with an option included to manage modulation. Not all suppliers of differential scanning calorimeters offer this option.

The apparatus consists of an oven in which an empty crucible, acting as a reference, is placed next to a crucible containing the sample. The oven is scanned by an inert dry gas, usually nitrogen of 99.99% purity.

In the DSC Q200 model, made by TA Instruments, there is a T_{zero} temperature sensor between the two platforms of the reference and sample crucibles (Fig. 2.55).

Table 2.39 Separation of thermal events in mDSC as a function of the calorific or kinetic capacity component

C component	Kinetic component
Reversing signal	Non-reversing signal
Glass transition	Evaporation
Most melting	Crystallization Thermoset cure Protein denaturation Starch gelatinization Decomposition Some melting

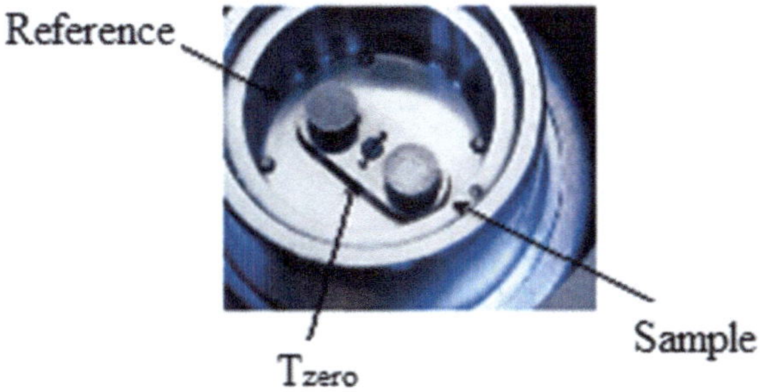

Fig. 2.55 Illustration of a DSC Q200 oven

This sensor measures the temperature difference between T_{zero} and the sample temperature, and between T_{zero} and the reference temperature. This allows measurement of a heat flow caused by the sample and the system (T_{zero}) as well as a heat flow caused by the reference and the system (T_{zero}). From this difference between the two heat flows, the heat flow of the sample can be deduced. The oven is cooled continuously by a cooling system, allowing several consecutive measurement cycles to be carried out.

Analysis Conditions

In order to carry out a heat capacity measurement, the sample analysed must not show any thermal event or mass change over the temperature range used. To verify this, it is preferable to carry out a thermogravimetric analysis beforehand on the same temperature range.

Preparation of the Sample

The sample is prepared in a crucible suited to the apparatus being used, and fitted with a lid. The type of crucible used will be chosen for its chemical compatibility with the sample and the analysis temperature. For temperatures below 400 °C, aluminum crucibles are the most widely used. Whatever type is used, the reference and sample crucibles must be identical in type, size and mass.

For powder sample, it is recommended to pack it into the bottom of the crucible, to improve the heat transfer.

If the sample is monolithic (a block), it must be cut into appropriate dimensions to fit into the crucible with the lid in place.

The DSC crucible being generally of low volume (<100 μL), it is preferable to fill it to its maximum capacity so that the quantity of sample analysed is representative of the amount collected and the heat flow obtained is large enough to be measured. The recommended mass of sample should be about 15–20 mg.

A pre-treatment of sample can be carried out in an oven at 100 °C in order to eliminate any adsorbed humidity. This humidity can give rise to a loss of mass during the measurement, which would be a source of error in determining heat capacity.

The quantity of sample must be weighed on a precision balance after the pre-treatment.

Analysis

Before proceeding to the determination of the C_p of sample, the apparatus must be calibrated for heat capacity by means of a sapphire disc. This calibration is in addition to the heat-flow calibration carried out in traditional DSC, which is usually done with Indium.

mDSC calibration:

Calibration for mDSC serves the purpose of determining a heat capacity constant, *K*, and is carried out using a sapphire disc. *K* is a multiplication factor allowing measurement of the heat capacity by mDSC. Before calibration, *K* is equal to 1.

$$C = \frac{Variation\ of\ modulated\ heat\ flow}{Variation\ of\ modulated\ heating\ rate} \cdot K \tag{2.70}$$

The sapphire disc is placed in a crucible of the same type as that used for the analysis of sample, with the lid on. In the oven, an empty crucible with its lid on is placed on the reference side, while a crucible containing the sapphire, again with its lid on, is positioned on the sample side. Calibration must be carried out by applying the same analysis conditions as for the sample. The thermogram recorded allows the C signal to be plotted as a function of temperature. The heat capacity constant corresponds to the mean value of the ratio of theoretical C to measured C at each temperature.

$$K = \frac{Theoretical\ C_p}{Measured\ C_p} \tag{2.71}$$

with: Theoretical C_p = theoretical heat capacity of sapphire; Measured C_p = measured heat capacity of sapphire.

In Table 2.40, an example of calibration with the sapphire is given.

Table 2.41 lists the theoretical heat capacity values of the sapphire for a given temperature. The value of *K* obtained must be entered in the software, which will apply this multiplication factor to the results, according to Eq. (2.70).

Method of Analysis

The analysis is carried out under a flow of inert gas, usually 99.99% pure nitrogen, with a flow rate of 50 mL/min. If pre-treatment in an oven has not been done, it is recommended to engage a heat cycle before measuring the specific heat. The sample mass used for measuring the C_p will then correspond to the mass after the heat cycle.

Temperature

The temperature at which the C_p must be measured depends on the study and the envisaged application, as well as on the type of sample. The sample must be thermally stable and not show any thermal event at the chosen temperatures.

Table 2.40 Example of C_p calibration

Temperature, °C	Rev measured C_p, J/(g °C)	Rev theoretical C_p, J/(g °C)	*K*
−83	0.4364	0.4659	1.0676
−73	0.4673	0.5014	1.0730
−63	0.4995	0.5356	1.0723
−53	0.5295	0.5684	1.0735
−43	0.5588	0.5996	1.0730
−33	0.586	0.6294	1.0741
−23	0.6169	0.6579	1.0665
−13	0.6449	0.6848	1.0619
−3	0.6726	0.7103	1.0561
0	0.6806	0.7180	1.0550
6	0.697	0.7343	1.0535
16	0.7248	0.7572	1.0447
26	0.7515	0.7788	1.0363
36	0.7756	0.7994	1.0307
46	0.7989	0.8188	1.0249
56	0.8219	0.8373	1.0187
66	0.8446	0.8548	1.0121
76	0.8657	0.8713	1.0065
		Mean *K*	1.0500

Table 2.41 Sapphire theoretical heat capacity

Temperature (°C)	Theoretical C_p (J/g/°C)	Temperature (°C)	Theoretical C_p (J/g/°C)	Temperature (°C)	Theoretical C_p (J/g/°C)
−93.15	0.4291	66.85	0.8548	236.85	1.0484
−83.15	0.4659	76.85	0.8713	246.85	1.0557
−73.15	0.5014	86.85	0.8871	256.85	1.0627
−63.15	0.5356	96.85	0.902	266.85	1.0692
−53.15	0.5684	106.85	0.9161	276.85	1.0756
−43.15	0.5996	116.85	0.9296	286.85	1.0817
−33.15	0.6294	126.85	0.9423	296.85	1.0876
−23.15	0.6579	136.85	0.9545	306.85	1.0932
−13.15	0.6848	146.85	0.966	316.85	1.0987
−3.15	0.7103	156.85	0.977	326.85	1.1038
0.00	0.718	166.85	0.9875	336.85	1.1089
6.85	0.7343	176.85	0.9975	346.85	1.1137
16.85	0.7572	186.85	1.007	356.85	1.1183
26.85	0.7788	196.85	1.0161	366.85	1.1228
36.85	0.7994	206.85	1.0247	376.85	1.1271
46.85	0.8188	216.85	1.033	386.85	1.1313
56.85	0.8373	226.85	1.0409	396.85	1.1353
				406.85	1.1393

Reproduced with permission from TA Instruments

Measurement is made in isotherms, but several successive levels of different temperatures can be achieved.

Modulation

The modulation conditions applied are:

- Amplitude of + or −1 °C;
- Period length of 100 or 120 s.

Duration of the Isotherm

The duration of the isotherm on which the modulation will take place must be a minimum of 30 min. This isotherm can be repeated several times. 8–10 min should be allowed for the modulation to stabilize on the isotherm. The temperature points during this stabilization phase must not be taken into account in the calculation of the C_p.

Repeatability

It is advisable to repeat the measurements. Repeat measurements on the same crucible can be used to evaluate measurement error.

Repeatability on the sample is evaluated by taking at least three different extractions from the same sample.

Recommendation

It is advisable to verify the heat capacity constant K obtained though the calibration with the sapphire disc, by carrying out an analysis in the same analysis conditions on the same sapphire disc. The C_p results obtained should be very close to the theoretical results in Table 2.41.

Results Treatment

The measurement results are presented in the form of a graph plotting the C_p and the modulated flow as a function of time (Fig. 2.56).

In order to conform to the requirements of the study, in the example shown in Fig. 2.56, C_p measurements have been done at 3 temperatures, 30, 50 and 180 °C. At the beginning of each isotherm, the modulated heat flow is unstable, so that zone is not used for the treatment of the results.

The heat capacity result at a given temperature corresponds to the mean value of the C_p obtained on the isotherm in the stable zone of the modulated heat flow.

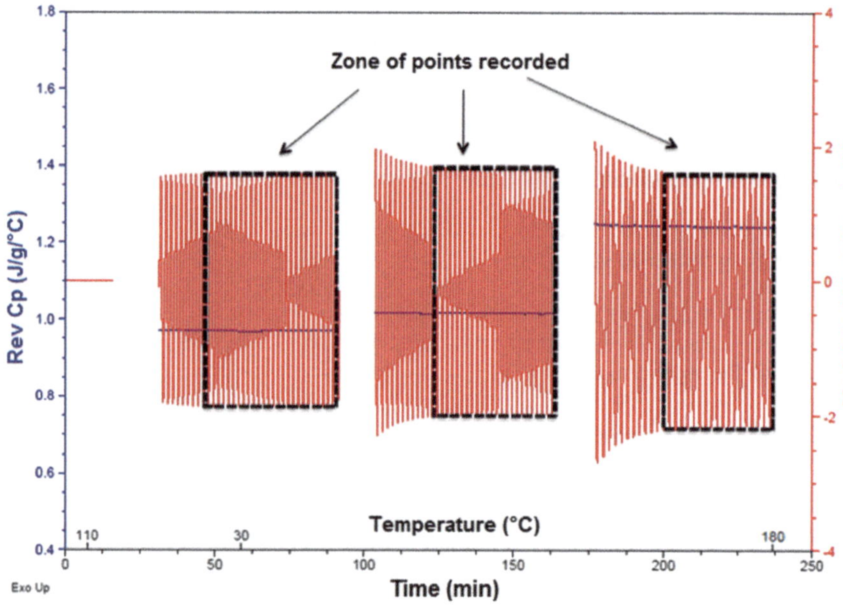

Fig. 2.56 Example of measurement analysis of C_p obtained for a mixture of clay and olive pit flour, after heat treatment

Calculation of the C_p is done by means of Excel-type calculation software after exporting the points.

Examples of Results

The heat capacity values obtained for a mixture of clay + 4 wt% olive pit flour are shown in Table 2.42. The sample was analysed with and without oven pre-treatment at 80 °C.

- Without pre-treatment: measurement repeatability was carried out;
- With oven pre-treatment at 80 °C: measurement repeatability was carried out on the sample with 3 extractions.

The heat capacity values obtained with oven pre-treatment at 80 °C are between 26 and 21% higher than those obtained without pre-treatment. This gap tends to narrow when the measurement temperature increases. This can be explained by the fact that oven pre-treatment eliminates all traces of residues and/or gases such as air or carbon dioxide, and any humidity that may influence the measurement. Thus it is clearly necessary to carry out pre-treatment or perform a heat cycle before taking measurements.

Table 2.42 C_p results obtained from mixture of clay with 4 wt% of olive pit flour

OPF 4%				
Modulation condition				
Amplitude		Period	Duration of isotherm	
1 °C		100 s	60 min	
Results				
Test	Pre-treatment	Cp @ 30 °C, J/(g °C)	Cp @ 50 °C, J/(g °C)	Cp @ 180 °C, J/(g °C)
Test 1	/	0.6677	0.703	0.9099
Repeatability Test 1	/	0.6619	0.6976	0.9151
Test 2	Oven-baked at 80 °C	0.8877	0.9276	1.1367
Test 3	Oven-baked at 80 °C	0.8555	0.8926	1.111
Test 4	Oven-baked at 80 °C	0.9376	0.9804	1.2017
Average	*Baked*	*0.8936*	*0.9335*	*1.1498*
Standard deviation	*Baked*	*0.0414*	*0.0442*	*0.0467*

Repeatability analysis carried out on the same crucible (test 1 and repeat test 1) gives a difference between measurements inferior to 1%. This confirms the thermal stability of the sample (at the measured temperatures) and good measurement accuracy.

Repeatability analysis on the sample (tests 2, 3 and 4) gives a measurement difference of 7–9%. This means that the sample has a good homogeneity.

For information, Table 2.43 shows the values of some different samples which are usually found in materials field.

Table 2.43 Example of C_p values for different sample types

Samples	C_p at 25 °C (J/g °C)
Glass	0.75
Brick	0.84
Gypsum	1.09
Cellulose and cotton	1.3–1.5
Charcoal	0.84
Granite	0.79
Sand, quartz	0.83
Peat	1.9
Wood	1.3–2.4

2.7 Physical Characterizations

2.7.1 *Particle Size Distribution Determination Using Laser Scattering*

2.7.1.1 Introduction

Laser granulometry is a technique used to measure the particle size distribution (PSD) of various pulverulent materials. The optical signals resulting from the interaction between an incident laser beam and particles (grains, aggregates, agglomerates, globules etc.) provide a granulometric distribution (particle size distribution) ranging from few nanometers to few millimeters.

2.7.1.2 Principle of the Technique

This technique is based on the diffraction and the diffusion of a laser beam. A micron-focused laser beam passes through particles dispersed in the liquid or gas phase, and creates an extinction and a backscattering characteristics of their size and number. In fact, a laser beam crosses through a continuous liquid or gas phase until it meets an element with a different refraction and/or absorption index. This change causes the beam to deviate around the particle (Fig. 2.57) and generates a diffraction pattern specific to the size of the observed object, the peak intensity of which will strongly depend on their number.

A detector, consisting of multiple photodiodes, records angle and intensity of emitted radiation to build a diffraction pattern characteristic of the scattering medium. The biggest particles (or elements) diffract the beam at smallest angles,

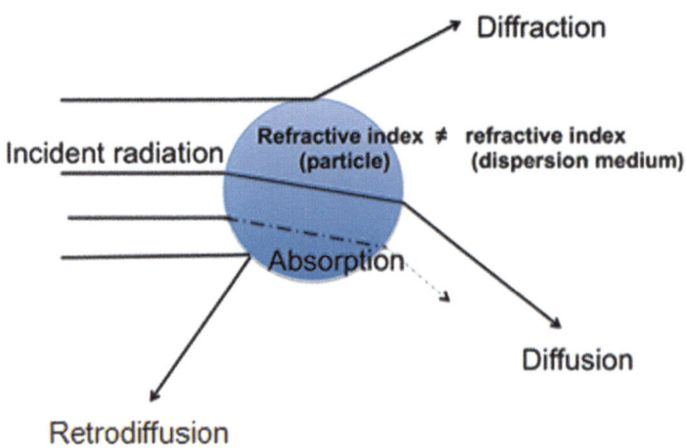

Fig. 2.57 Incident laser beam deviation from a particle

while the smallest diffuse it at largest angles, but with a lower light intensity. A population of given particle in size will give rise to a particular diffraction pattern that is correlated to the measurement method used.

2.7.1.3 Theory

The Lorenz-Mie theory [150], or Mie solution is a solution of the Maxwell equations that describe the coherent diffusion of light by a spherical particle [151, 152] characterised by its equivalent diameter and refractive index. However, this theory is applicable under the condition that the equivalent diameter of particles belongs to a class of sizes varying between 0.1 and 10 times the wavelength of the incident beam. This limitation explains why numerous granulometer manufacturers propose apparatus including several lasers with different operating frequencies which also enable to measure the size of very small particles of the order of a few tens of nanometers. To do that, the optical properties (refraction and absorption indexes) of particles and of the medium within which they are dispersed must be indicated into the software. Based on these optical data, the analysis software can solve the complex equations, from the signal measured, i.e. the diffusion angles and intensities, in order to determine the size of different suspended particles. The sizes calculated are displayed using a histogram by number frequency or by volume; it is also possible to obtain, by calculation, results by number or expressed in other ways. The results by volume can be correlated with a histogram in mass which may be obtained by sieving [153].

The Fraunhofer's theory is an approximation of the Mie's theory. It only considers surface optical phenomena [154, 155]. The Fraunhofer's theory can be applied only to particles larger than 30 μm. Below this size, the scattered intensity is two low to be detected as illustrated in Fig. 2.58 (left-hand-side). Under the same conditions, Mie's theory allows obtained higher scattered intensity and so can be used for this class of particles (Fig. 2.58 on right-hand-side).

2.7.1.4 Sphere of Equivalent Volume

A laser granulometer measures the diameter of a sphere having the same volume as a given particle. It is possible to calculate a granulometric distribution by number in order to obtain other information. While a distribution by volume highlights the biggest particles, a distribution by number shows the smallest elements according to the following principle (Fig. 2.59):

The granulometric analysis in Fig. 2.60, carried out on big particles onto which finer ones were coated, demonstrates the interest of displaying the particle size distribution by number (right-hand-side) instead of by volume (left-hand-side). Indeed, the representation of the particle size distribution by number (right-hand-side) clearly reveals the population of smallest particles.

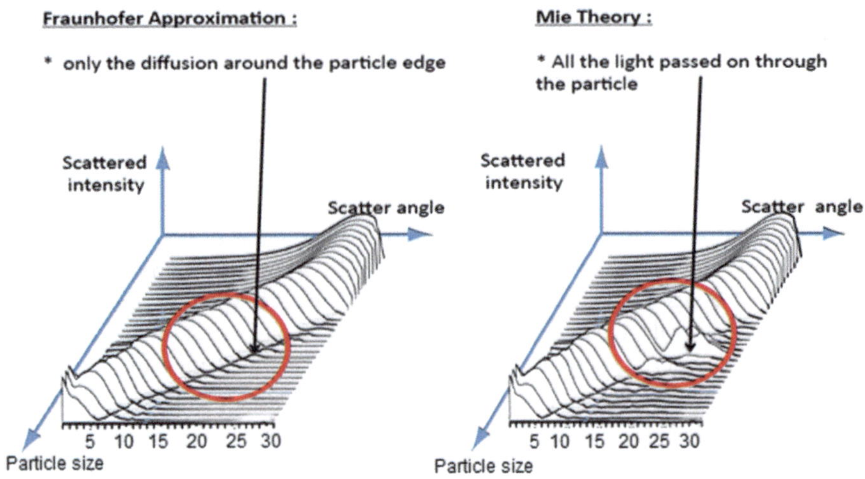

Fig. 2.58 Optical phenomena occuring when particles are smaller than 30 μm—Malvern Instruments—practical guide to laser granulometry. Reproduced with permission from Malvern Panalytical SAS

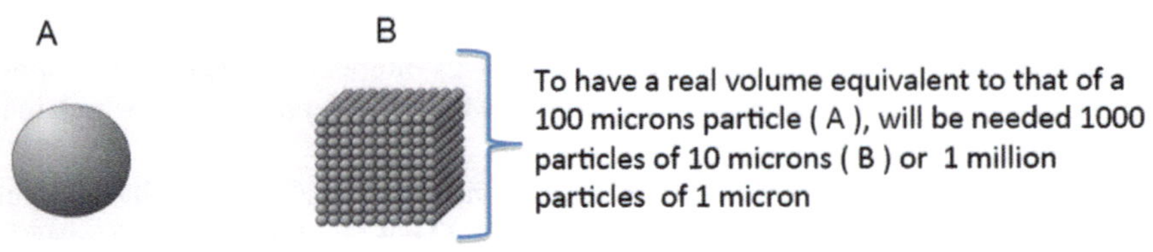

Fig. 2.59 Volume and number, two different pieces of information

The two results of this same analysis show the percentage of particles expressed by volume or number depending on their size in microns. The dispersion has been carried out in air at different pressures. The representation of the particle size distribution by volume (Fig. 2.60, left-hand-side) indicates the presence of a population of particles having an equivalent diameter of about 100 μm, i.e. the "biggest" particles in the sample, however it hides the existence of the smaller ones (size below 10 μm). In other words, the representation by volume does not allow revealing the smallest particles in the dispersed media. From the representation in number (Fig. 2.60, right-hand-side), it is clear that the greatest numbers of particles are the smallest, i.e. which have the size inferior or equal to 10 μm.

Results show that this coating of fine particles of less than 10 μm on particles of 100 μm resists to a dispersion submitted to a pressure up to 0.5 bar: no fine particles are detected and both representations in volume or number are identical. Above 1.5 bar, the finest particles are liberated from the biggest ones and, since the latter only represent a tiny fraction compared to the numerous fine particles. So this

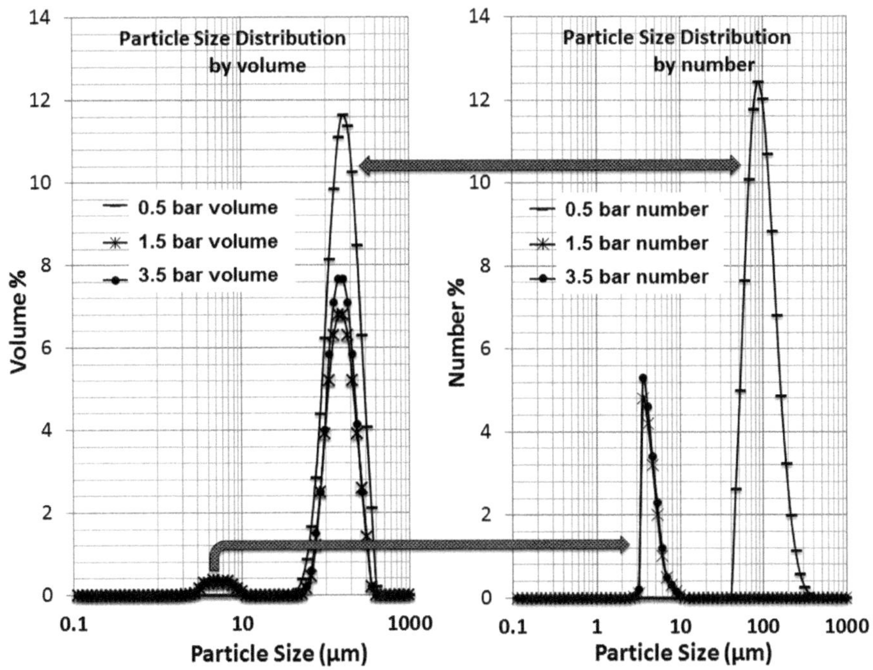

Fig. 2.60 Display of an analysis by number or volume

example shows the interest of using both representations that are complementary dependant on the information expected.

2.7.1.5 Laser Granulometer

Several companies develop and supply laser granulometers. They all have specific technical features that enable to solve a wide variety of analytical requirements. The main differences between granulometers are: the number and the position of detectors, the number of laser sources proposed to reach better resolution (2 or 3 wavelengths), the size of optical bench (long or short), the regulation of resonance frequency of ultrasonic bath in which sample is dispersed, the dispersion modules usable with different Venturi chambers for dry and/or fragile samples or, on the contrary a gravity dispersion module without air supply to measure particle size distribution of agglomerates. For dispersions in a liquid, measurement cells of different volumes can be used, with an ultrasound probe for optimal dispersion of the pulverulent material. All laser granulometers work on the same basic principle as illustrated in Fig. 2.61.

A laser source (A) emits a laser beam which impacts a measurement cell (C). The presence of elements whose the size should be measured will deviate the

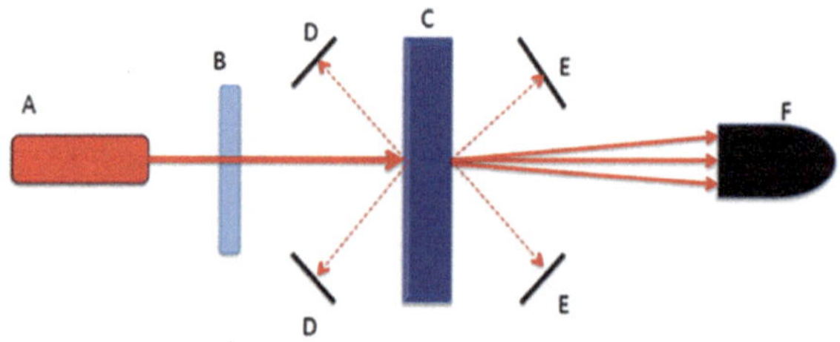

Fig. 2.61 Basic principle of laser granulometer; A: red laser; B: Fourier lens; C: measurement cell; D: back scatter detectors; E: side scatter detectors; F: focal plane detector

incident beam onto detectors, localized in (E) and (D), for the smallest particles, and onto a focal plane detector (F) for the largest particles. The difference in laser intensity with and without particles enables the obscuration to be measured, which is directly linked to the quantity of elements encountered by the laser. In order to improve the resolution of the granulometer for measuring smaller particle sizes, many suppliers add one or more light sources and dedicated detectors as illustrated in Fig. 2.62.

Various options are available for a laser granulometer. It can be coupled with other technique like particle's shape measurement (by CILAS). Some models are equipped with a system to optimize the dispersion of the elements to be measured (i.e. apparatus of SYMPATEC GmbH).

A number of variants are available for detection, dispersion (dry or wet) or for combining with complementary techniques. This great diversity enables each user to find adequate apparatus. Malvern Instruments MS 3000 laser granulometer is well adapted for wet and dry dispersion. Other manufacturers include

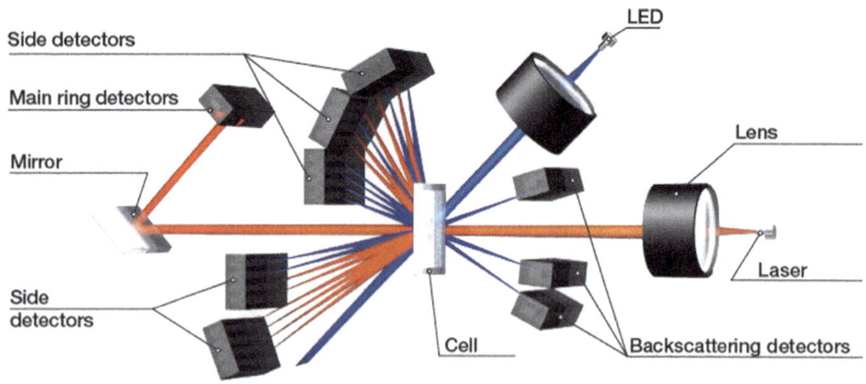

Fig. 2.62 Laser granulometer with a red laser and a blue LED—Horiba—LA960. Reproduction with permission from HORIBA

(non-exhaustive list) Shimadzu, Fritsch, Sympatec, Microtrac, Horiba, Malvern Instruments, Cilas, Beckman Coulter, Angstrom Advanced Inc, Bettersize.

2.7.1.6 Sample Preparation and Analysis

Two methods are generally applied:

- Dry method: dispersion of pulverulent sample in a flow of dry air at controlled pressure.
- Wet method: dispersion of sample in a non-dissolving dispersing liquid.

The sample to be measured could be a powder, a liquid or a paste. It is possible to measure particle size distribution of liquid samples, paste or dry powder by simply dispersing them in a liquid in the case of wet measurements or directly by dry analysis if they are dry powder.

The most classical and most frequent representation of the particle size distributions is the logarithm diagram for which the size is plotted on the x-axis in log base scale (usually in µm) and the y-axis is displayed according to an arithmetic scale from 0 to 100%. The y-axis can represent the proportion by volume or by frequency of particles that have a given diameter or belong to a size class. Some log-normal representations are also used; the x-axis is in log and the y-axis in normal Gaussian distribution.

Type of Dispersion

It is preferable to analyze liquid or paste samples by a wet method and dry samples using a dry method. This allows remaining close to the initial state in which the sample is received. Drying can, for example, create agglomerates that are difficult to break up in an air flow and putting a dry sample into a liquid may create binding forces between the sample and the dispersant, which will have an impact on the size measured.

Particular cases: for safety reasons, it is imperative that a dry explosive and/or nanometric powder is measured using the wet method. A suspension with an unidentified dispersant liquid must be dried before the measurement.

Sample Preparation Depending on Type of Dispersion

Dry method

The sample must be perfectly dry. In this measurement configuration, all the particles of sample pass only once through the measurement cell before being removed by the vacuum system. The whole volume of the sample is analysed by programming a measurement time greater than the time needed for the sample to pass through the cell. An initial sample is tested in order to adjust the apparatus parameters. This adjustment should be carried out while ensuring a level of laser obscuration perfectly identical so that the dispersion energy per unit of volume dispersed is comparable. Note that the obscuration reflects the equivalent optical concentration of dispersion. Medium without sample addition corresponds to an obscuration of zero. An obscuration of 1–5% is a good reference figure for the dry method, however a minimum volume of 20 mL of powder to be analysed is recommended. Several samples of identical volume are thus required from the same batch for the repeatability of the measurement. Each sample must be representative of the whole and must have been taken into account with good sampling practice. Several analyses are carried out in order to obtain repeatable measurements. According to the ISO 13320-1 [156] standard, when using a laser granulometer, five analyses are generally required, carried out under identical operating conditions and in which the characteristic sizes do not vary by more than:

- 6% for $d_{50} < 10$ μm and 10% for d_{10} and d_{90};
- 3% for $d_{50} > 10$ μm and 5% for d_{10} and d_{90}.

The dispersion pressure is the most important factor (Fig. 2.63): if it is too low, it will not disperse aggregate particles while high dispersion pressure could also generate attrition and modified the particle size distribution.

In order to define optimum dispersion pressure, it is necessary to measure several representative batches while gradually increasing the pressure, for example in

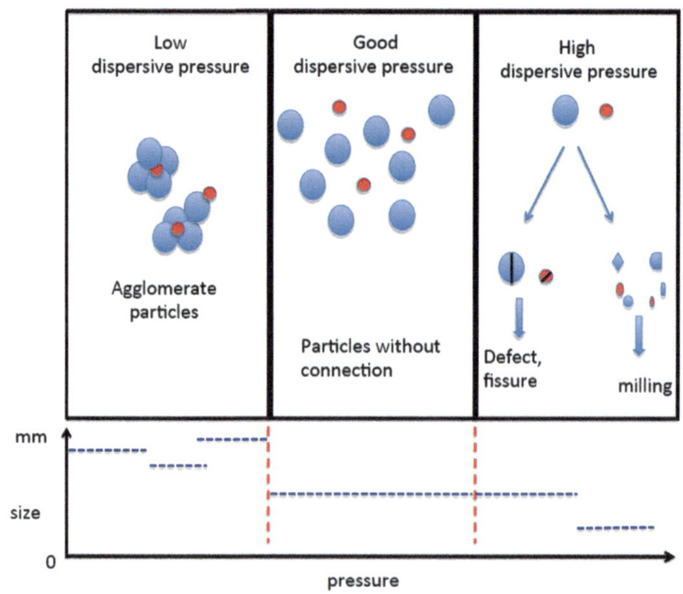

Fig. 2.63 Effect of increasing dispersion pressure on the size measured

increments of 0.5 bar. A correct dispersion pressure is characterized by stable size measurements. Ideally, the dispersion pressure that is considered as correct should be validated by another analytical technique, such as microscopy.

Liquid dispersion

In the case of a liquid dispersion, the batch obtained after having dispersed the sample is either a paste or a dispersion. A sample of the batch is placed in the measurement cell. Thus, it is important to have mixed the whole batch beforehand; otherwise it could have undergone segregation, decantation or phase separation. Unlike dry dispersion, measurements using the wet method take place in a "closed" circuit. Each particle will pass through the measurement cell several times. So, several factors will have a direct impact on the quality of the measurement:

- the liquid in the cell must be a good dispersant;
- the stirring speed and the measurement time must be well-adapted to the sample;
- the obscuration.

Observations:

The liquid in the cell must be a good dispersant:

The liquid in the measurement cell is used either to disperse or to transport the sample within the apparatus. A test should be carried out before by introducing the sample into the measurement cell:

Sample ⇨ beaker with the liquid supposed to be a good dispersant ⇨ manual mixing.

- If the solid remains at the surface, it is then advised to add a surfactant that should likely modify the surface tension for better wettability;
- If the solid remains agglomerated at the bottom of the beaker, the liquid should then be changed for a better dispersion;
- If the solid shows clearly distinct particles or a cloudy suspension without big lumps (flocs, agregates, agglomerates etc.) at the bottom of the beaker, the liquid is suitable for the dispersion. The analysis using the laser granulometer could then be carried out. If the dissolution of fine particles takes place, the liquid must be changed to avoid this dissolution effect.

The stirring speed and measurement time must be adapted to the sample:

The stirring speed regulates the flow of liquid circulating through the measurement cell. When the particles are very dense, the stirring speed should be gradually increased until the particles are suspended. As an example, Fig. 2.64 shows the relationship of the size (top) and obscuration (bottom) as a function of the stiring speed. Above 3000 rpm, the size becomes stable. So, the analysis can be performed with a stirring speed above 3000 rpm.

Depending on the objective of the measurement, the duration allocated to each measurement could be adapted. Short measurement time generally allows observing agglomerates. On the other hand, long measurement time allows observing primary

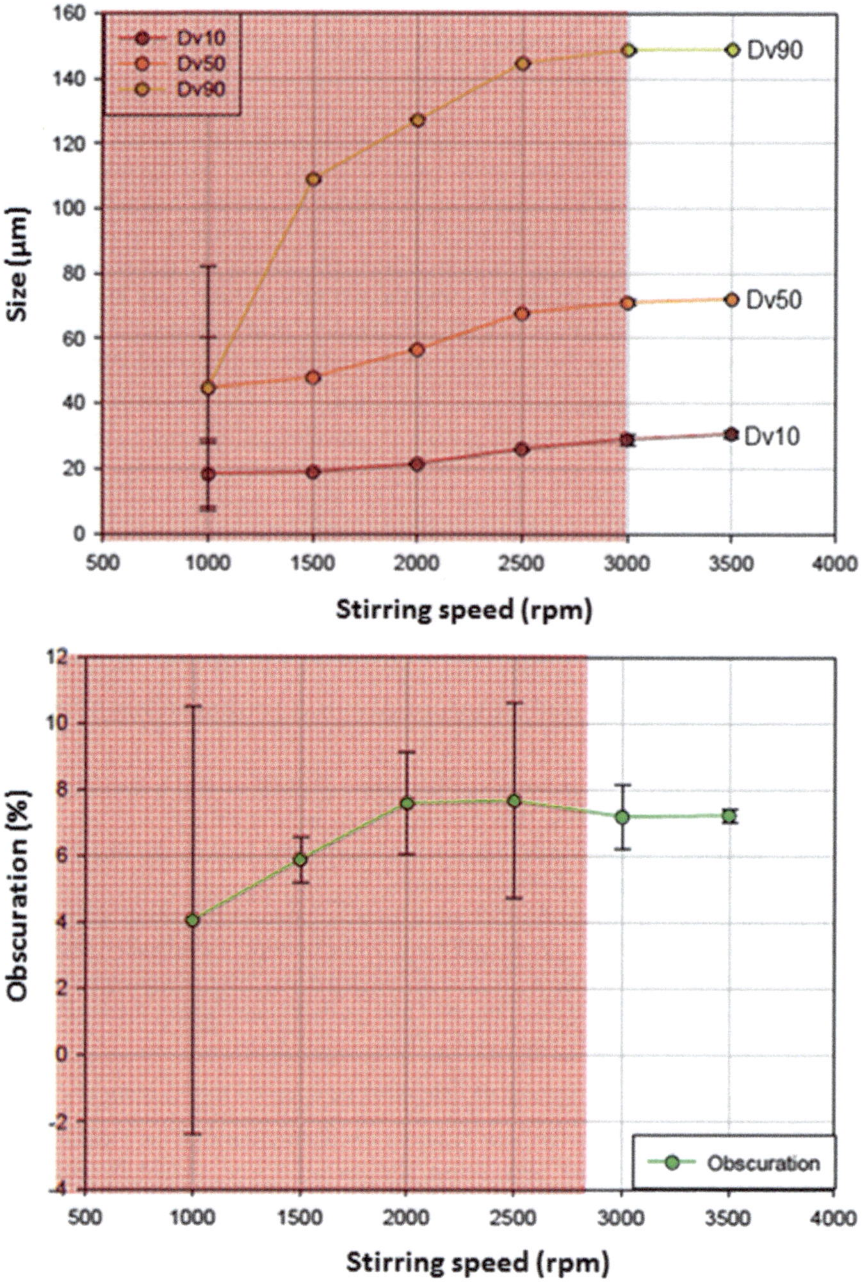

Fig. 2.64 Relationship between size (top) and obscuration (bottom) to the stirring speed of a copper powder of density 8.92 g/cm^3 [74]. Reproduced with permission from Malvern Panalytical

particles which constitute sample. Sample should be measured as soon as it has been introduced into the cell. Then, further measurements should be carried out at regular intervals. This allows observing the initial state of the sample as well as the evolution of the sample up to its stabilisation. When sample is stabilized, primary particles (or agglomerates which could not be further dispersed) are observed. The total time could range from few seconds to several minutes. The analysis time could be defined if a microscopic observation has been performed prior to laser granulometric analysis.

It is possible to pre-disperse sample in a beaker by stirring with the addition of a surfactant or a dispersant or even by using ultrasound. The sample will then be placed in the measurement cell of the granulometer once this pre-dispersion has been carried out. This method can be applied when only the size of primary particles is targeted, because this method leads to a loss of information about the initial state of agglomeration.

Obscuration:

As previously indicated, obscuration is directly linked to the number of elements encountered by the laser. However, obscuration values must be respected. In liquid dispersion, for a homogeneous sample, an obscuration of 1–10% enables high-quality measurements to be carried out. For a heterogeneous sample, such as a sludge, it is recommended to increase the obscuration in order to analyze the greatest possible number of elements. An obscuration of 15–25% should be targeted. However, a particular attention must be taken to not exceed the limit of 30% obscuration so as to avoid multi-diffraction phenomena as shown in Fig. 2.65. When there are too many particles (i.e. obscuration is higher than 25%), the laser beam is deviated several times. This phenomenon is observed by the activation of the wide-angle detectors. The laser granulometer interprets it as resulting from the presence of fine particles, which is obviously not the case.

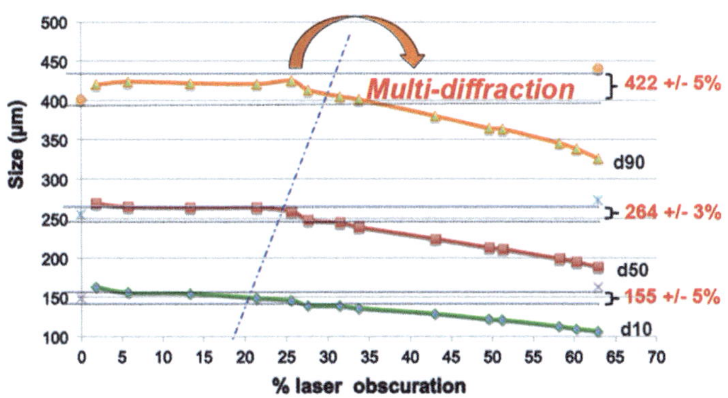

Fig. 2.65 Example of the multi-diffraction: size change as a function of obscuration. Measurement of glass beads

Particular case with the wet method: ultrasound

The cells used for the wet method are often equipped with an ultrasound source. It is thus possible for a sample to be dispersed by the action of ultrasound. It should be noted that:

- Ultrasound may have no effect on the dispersion of the sample;
- Ultrasound may disperse then re-agglomerate the separated particles if too much energy is applied. Cycles of dispersion/re-agglomeration are sometimes observed;
- The sample may be heated up, with the possible sample degradation;
- Some liquids (e.g. ethanol) cannot be used with ultrasound because these liquids modify laser obscuration under ultrasound effect; although ultrasound can help with dispersion, it should be used with precaution. The suspension becomes cloudy which the laser granulometer could account for particles. Preliminary test will be necessary to define the power and emission time.

2.7.1.7 Making Accurate Measurements with a Laser Granulometer

In the previous section, various operating conditions were defined:

For dry dispersion:

Several batches should be correctly sampled, measurement time defined in relation to the volume, obscuration adjusted from 1 to 5%, dispersion pressure fixed.

For liquid dispersion:

Mixing of the batch for homogenization, then taking a sample from the batch and introducing it into the measurement cell (directly into the cell or dispersed outside the cell); the liquid in the cell is a good dispersant; the stirring speed is sufficient to entrain all the particles; the duration and frequency of measurement is defined; the use of ultrasound is possible but not mandatory. These operating conditions are necessary for accurate analysis.

Whether dry or liquid dispersion is used, all analyses should follow the same protocol:

- Data to be supplied for the analysis: Measurement time, frequency of measurement, stirring speed, dispersion pressure.
- Data to be defined for size calculation: Refraction and absorption indices for the particles to be analysed, refraction index of the dispersion medium, density for the calculation of the specific surface area if necessary.
- Bringing the cell up to working pressure for dry dispersion or introduction of the dispersant liquid into the cell for liquid dispersion: The introduction of liquid into the circuit should be firstly performed without circulation then with several variations of the flow rate of the pump. This operation enables any bubbles to be

removed from the measurement cell as illustrated in Fig. 2.66. If the bubbles are not entrained, it will be necessary to clean the optics of the measurement cell (emptying, cleaning and refilling of the cell). The use of a surfactant can improve the dispersion of the sample. However, foam can be formed. In this case, the surfactant concentration as well as the stirring speed must be adjusted to eliminate foam. If this is not possible, another dispersion method should be adapted.

- Alignment of the laser:

The laser beam should be focused on the central detector (detector 0)—as shown in Fig. 2.67 using an automatic alignment. This detector receives the undeviated laser ray if there are no pollutants or other elements in the measurement cell which could impede the incident beam pathway. The laser power is thus maximal on detectors (0) and (1).

Fig. 2.66 Presence of bubbles in the measurement cell in liquid dispersion

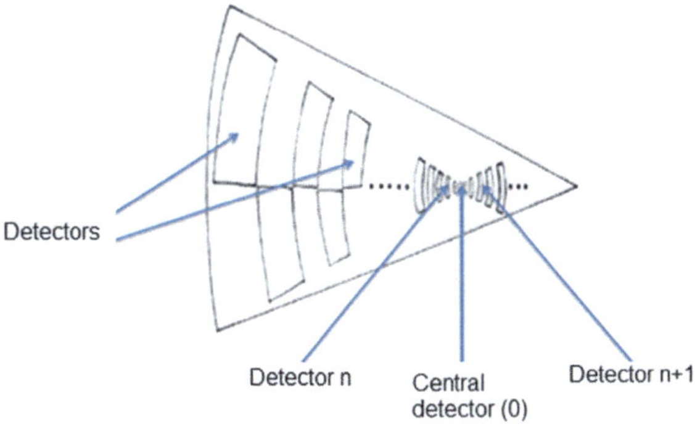

Fig. 2.67 Plan of a logarithmic detector—every detector is numbered. Reproduced with permission from Malvern Panalytical

Measurement of background noise:

The signals that impact each detector are recorded. They will be subtracted from the signals recorded during analyses. The aspect of the background noise is important: it should exhibit an exponential declining shape as displayed in Fig. 2.68. The presence of pollution (impurities, bubbles, scratched optics etc.) will produce an abnormal spectrum then requiring a cleaning of the optical surfaces as shown in Fig. 2.69. The signal that impacts the central detector must have a certain level of intensity for the analysis to be validated. This level depends on the laser used.

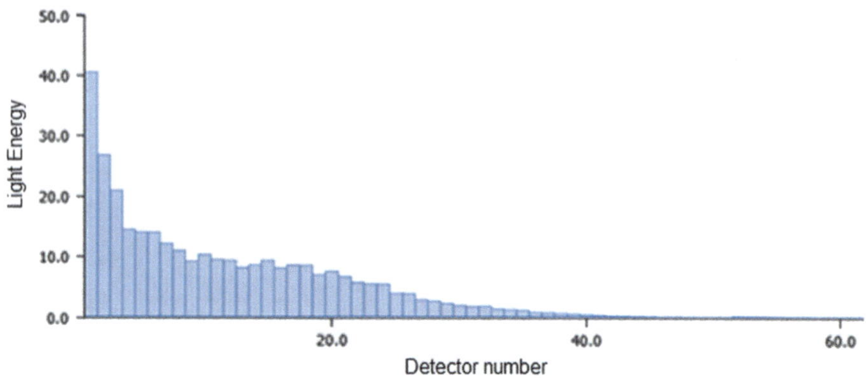

Fig. 2.68 Exponential decay: the granulometer is ready for an analysis. Reproduced with permission from Malvern Panalytical

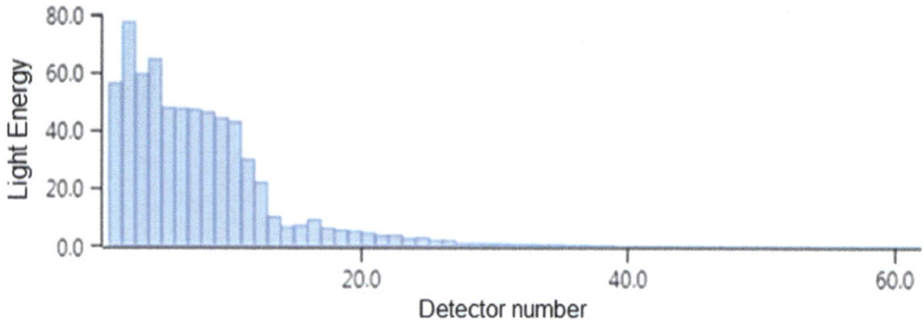

Fig. 2.69 Abnormal spectrum and the energy recorded at detector (0) is not the highest; the granulometer must be cleaned. Reproduced with permission from Malvern Panalytical

Verification before introduction of the sample:

When the previous steps have been validated, the laser power level recommended by the manufacturer must be checked. This level must not be lower than that of this reference value. The laser obscuration (=light diffracted) must have the lowest value possible, equal to or less than 0.1%. The laser granulometer is ready to perform a sample measurement.

Check the quality of the analysis that has been carried out by observing:

- Data and background noise: in the example presented in Fig. 2.70, the background noise is represented by the decreasing exponential shape of the line and the signal at detector 0 on the left has the maximum energy. The signal displayed in the form of "bars" corresponds to the measurement with the sample minus the initial background noise without particles. There are no negative data points, the corrected signal is significantly higher than the background noise in the main detection zone (detector 40). So the data acquisition can be validated.
- The residual and the fit: To treat the raw data obtained above, mathematical treatment is necessary. It is crucial to select the suitable optical indexes which are characteristic for sample and medium studied. The objective is to minimize the difference or residual between the data corrected by the optical matrix and the theoretical image (the fit) of the distribution. As examples, Figs. 2.71 and 2.72 show respectively an appropriate data treatment and a wrong one.

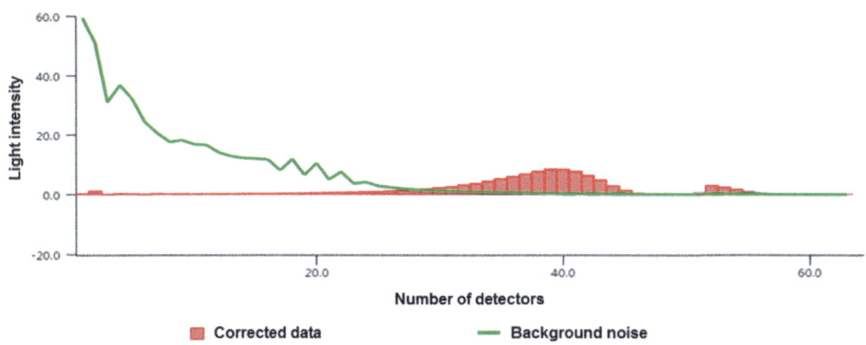

Non	Bruit de fond	Signal	Données	Non	Bruit de fond	Signal	Données	Non	Bruit de fond	Signal	Données	Non	Bruit de fond	Signal	Données	Non	Bruit de fond	Signal	Données
1	59.35	56.72	0.11	14	12.25	11.87	0.19	27	1.86	3.03	1.26	40	0.23	8.46	8.24	53	0.14	2.43	2.29
2	50.90	49.63	1.09	15	12.04	11.66	0.18	28	1.55	2.98	1.49	41	0.18	7.68	7.51	54	0.09	1.60	1.51
3	31.01	29.49	-0.08	16	11.74	11.36	0.17	29	1.32	3.02	1.76	42	0.13	6.37	6.25	55	0.06	0.82	0.77
4	36.72	35.22	0.20	17	8.07	7.93	0.23	30	1.12	3.17	2.10	43	0.10	4.84	4.74	56	0.03	0.34	0.31
5	31.98	30.59	0.09	18	11.89	11.63	0.28	31	1.01	3.47	2.51	44	0.07	2.85	2.78	57	0.02	0.19	0.17
6	24.40	23.22	-0.05	19	6.46	6.48	0.31	32	0.86	3.83	3.01	45	0.04	1.27	1.23	58	0.01	0.17	0.16
7	20.56	19.78	0.17	20	10.55	10.44	0.39	33	0.75	4.33	3.61	46	0.02	0.37	0.35	59	0.007	0.06	0.06
8	17.66	16.89	0.05	21	4.89	5.12	0.45	34	0.71	5.04	4.36	47	0.01	0.18	0.17	60	0.005	0.009	0.004
9	18.27	17.53	0.11	22	7.45	7.65	0.54	35	0.59	5.70	5.14	48	0.01	0.01	0.004	61	0.006	0.009	0.003
10	16.93	16.18	0.04	23	3.57	4.04	0.63	36	0.54	6.42	5.91	49	0.01	0.02	0.004	62	0.007	0.01	0.006
11	16.71	16.07	0.13	24	4.11	4.67	0.75	37	0.45	7.14	6.71	50	0.01	0.02	0.01	63	0.00	0.00	0.15
12	14.14	13.62	0.14	25	2.74	3.50	0.89	38	0.39	7.91	7.54	51	0.00	0.00	0.37				
13	12.86	12.42	0.15	26	2.32	3.27	1.05	39	0.29	8.55	8.27	52	0.15	2.95	2.81				

Fig. 2.70 Examples of optical signals with and without the sample

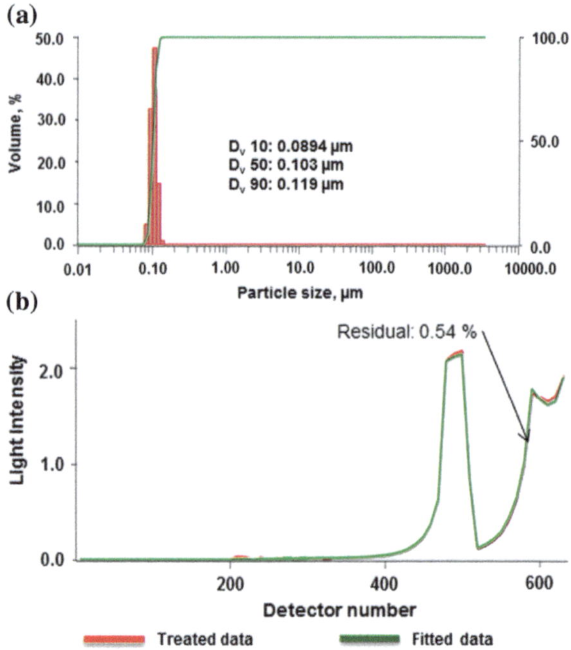

Fig. 2.71 Illustration of the importance of optical data: good optical data (1.59/0/1.33) leading to low residual value

2.7.1.8 Results Presentation

The results obtained from a laser granulometer analysis provide at the very least a percentage by volume as a function of the particle size in microns. Analyses are generally presented in the form of a results sheet (Fig. 2.73) showing the information necessary for the analysis, the characteristic values obtained, a graphical representation and a table showing the percentage of particles in each size-class.

The results may be presented graphically in different ways (Fig. 2.74). The first representation is a histogram displaying the percentage by volume in each class of size. These classes are displayed according to a linear scale. The second representation is identical to the first but with the addition of a green line representing the cumulative percentage of passes. This line gives the values for the different percentages (10, 50, 90%) having a defined size. The third representation shows the frequency (red line) and cumulative refusals (green line) on a x-axis logarithmic scale. The choice between expressing the results in refusals or in passes must correspond to the value of the percentages displayed (e.g. 10% refusal = 90% passes). The logarithmic representation is recommended because it gives more information about the finest particles. Table 2.44 indicates some characteristic values. Other results may be presented depending on the specifications required by each user. It is, moreover, common to export only the values that enable the granulometric curves to be retraced for easier treatment of data and interpretation of results.

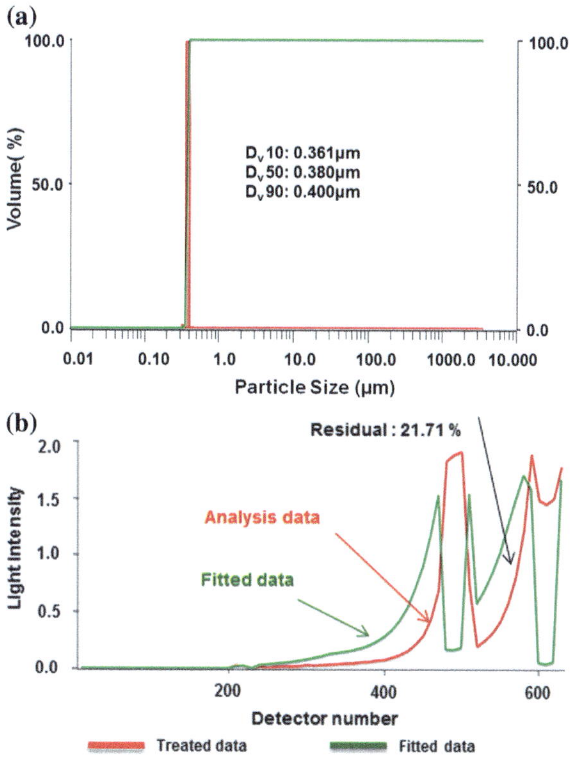

Fig. 2.72 Illustration of importance of optical data: bad optical data (1.603/0.1/1.33) leading to high residual value

2.7.1.9 Examples of Results

Liquid Dispersion: Characterisation of Sewage Sludge Before and After Digestion

An original procedure for sewage sludge characterisation was developed in order to specifically analyse the particle size distribution [157].

<u>*Operating protocol*</u>:

The operating protocol was established by taking into account the variety of elements containing sewage sludge. In fact, there are neither standards nor standardized operating protocols enabling precise characterisation of the granulometry of such a sample. Water is used as the dispersant: its refraction index is 1.33. The sludge is so heterogeneous that it is impossible to define a refraction or absorption index for it. A mean value of refraction index of 1.52 and an absorption index fixed at 0.1 were used. The residuals obtained during the analyses were correct (1%) thus validating the choice of these coefficients. For the measurement, a volume of 100 mL is placed in the dispersion cell. The alignment and the optical background noise are recorded. The sludge was first mixed manually when possible in order to

Measurement Details

Sample Name	exemple	**Measurement Date Time**	12/04/2017 13:16:58
Operator Name	Administrateur	**Analysis Date Time**	19/12/2017 11:50:20
SOP File Name	PVAROS3K02.msop	**Result Source**	Edited

Analysis

Particle Name	Glass Beads (typical)	**Particle Refractive Index**	1.520
Dispersant Name	Dry dispersion	**Dispersant Refractive Index**	1.000
Particle Absorption Index	0.000	**Laser Obscuration**	4.16 %
Weighted Residual	0.36 %	**Scattering Model**	Mie
Analysis Model	General Purpose	**Analysis Sensitivity**	Enhanced

Result

Concentration	0.0077 %	**Span**	0.928
Uniformity	0.285	**Result Units**	Volume
Specific Surface Area	42.92 m²/kg	**Dv 10**	38.3 μm
D [3,2]	57.1 μm	**Dv 50**	60.8 μm
D [4,3]	64.1 μm	**Dv 90**	94.8 μm

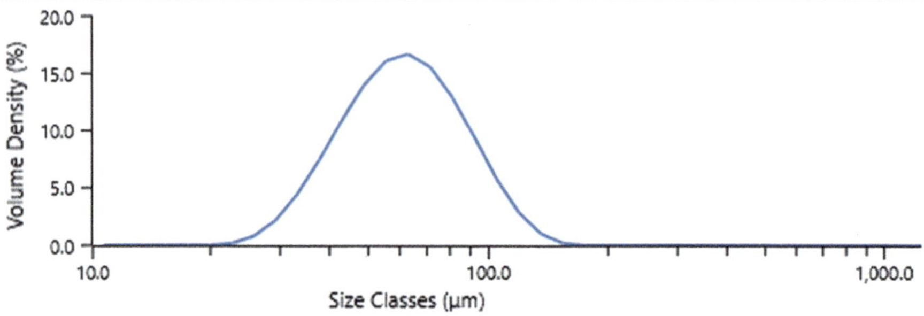

Size (μm)	% Volume In	Size (μm)	% Volume In	Size (μm)	% Volume In
7.64	0.00	40.1	9.03	211	0.00
8.68	0.00	45.6	11.67	240	0.00
9.86	0.00	51.8	13.49	272	0.00
11.2	0.00	58.9	14.02	310	0.00
12.7	0.00	66.9	13.05	352	0.00
14.5	0.00	76.0	10.80	400	0.00
16.4	0.00	86.4	7.79	454	0.00
18.7	0.00	98.1	4.74	516	0.00
21.2	0.11	111	2.28	586	0.00
24.1	0.61	127	0.75	666	0.00
27.4	1.74	144	0.11	756	0.00
31.1	3.63	163	0.00	859	0.00
35.3	6.18	186	0.00	976	0.00

Fig. 2.73 Example of results sheet

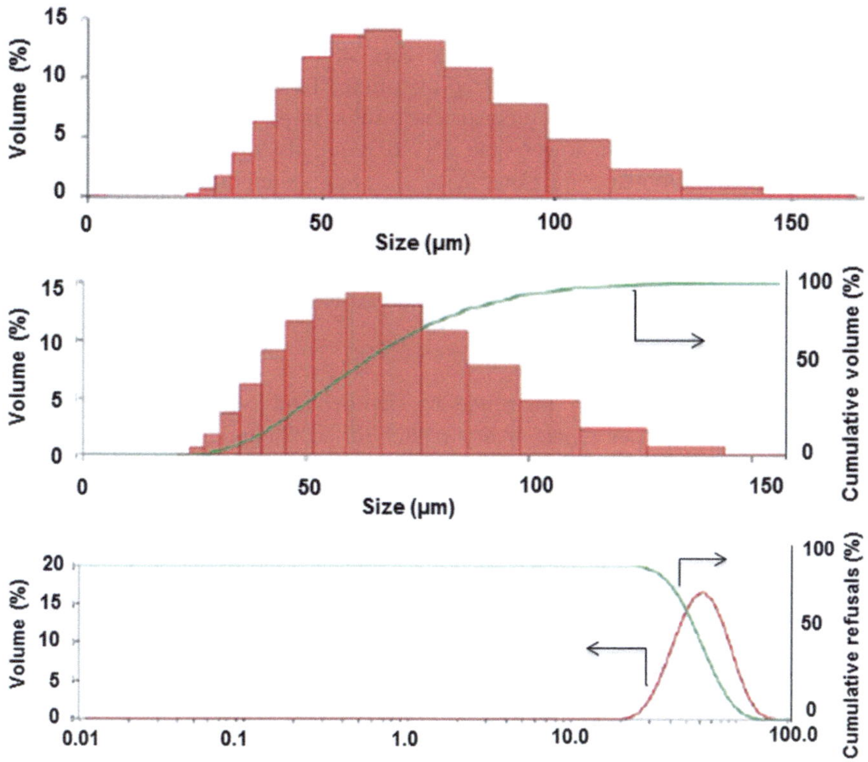

Fig. 2.74 Different graphical representations of an analysis

Table 2.44 The main expression of diameters in particle size distribution analysis with a laser granulometer

Parameter	Expression	Note
Percentiles D_{vx} (x% of the population resides below this size)	D_{v10}, D_{v50}, D_{v90}	$SPAN = \frac{D_{v90}-D_{v10}}{D_{v50}}$
Volume moment mean (De Brouckere mean diameter)	$D[4,3] = \frac{\sum D_i^4 n_i}{\sum D_i^3 n_i}$	Sensitive to the presence of large particles
Surface area moment mean (Sauter mean diameter)	$D[3,2] = \frac{\sum D_i^3 n_i}{\sum D_i^2 n_i}$	Sensitive to the presence of fine particles

ensure a good homogenization of the sample and to be sure that the sample taken for analysis is representative of the whole. This sample is taken in order to yield an obscuration ranging between 15 and 25%. These high levels of obscuration enable a large quantity of particles to be analyzed without multi-diffraction phenomena occurring. The circulation of the liquid in the measurement cell is fixed at 75% of the maximum pump speed, i.e. 2600 rpm; the heaviest particles are thus entrained during the stirring. The analysis begins when the obscuration stabilizes; in the study presented, optimal dispersion was obtained instantaneously. Five measurements were then made on the same sample for very short periods. For some sludges, dispersion/agglomeration were observed over time. To validate the measurements, at least two samples must lead to consistent results. In particular, the difference on the D [3, 4] must remain lower than 5%. If required, further analyses must be carried out. Granulometric measurements of sewage sludges often require many samples because of the heterogeneity of the initial medium and the difficulty in obtaining a sample that is representative of the whole batch. Analyses of sludges before entering the digester (undigested sludge) and digested one at different concentrations of flocculant show changes in the granulometric distribution.

Results:

The results in Fig. 2.75 are presented by number and by volume for different masses of flocculant per tonne of dry matter (dried sludge). For the undigested sludge (Fig. 2.75, top), the number distribution indicates that the concentration of flocculant below 24 kg/tDM did not allow flocculation of finest particles.

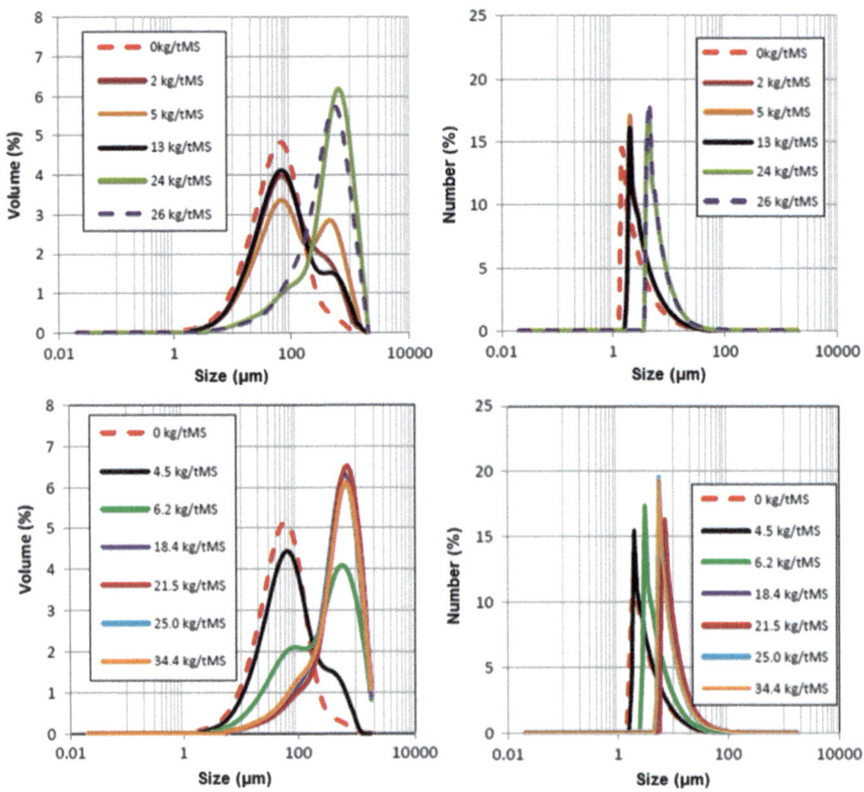

Fig. 2.75 Influence of the concentration of flocculant on the particle size distribution by volume and by number of the two sludges. Reproduction with permission from Charlou [157]

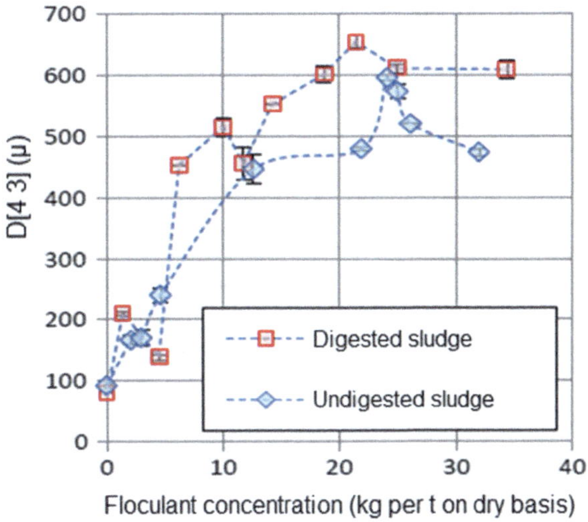

Fig. 2.76 Influence of the concentration of flocculant on the D [3, 4] of sludge before and after digestion Reproduction with permission from Charlou [157]

The volume distribution shows, moreover, that without flocculant (0 kg/tDM), the distribution is monomodal with a peak near the finest particles. The gradual addition of flocculant moves the granulometric distribution towards the bigger sizes. Digested sludge (Fig. 2.75, bottom) shows the same trend but for a concentration of 21.5 kg/tDM. Figure 2.76 shows the values of D [3, 4] of these granulometric analyses in relation to the concentration of flocculant. When the two concentrations defined above are reached, the addition of further flocculant has no effect on the size because the De Brouckere diameter (Equations in Table 2.44) remains stable.

Solid Dispersion: Characterisation of Agricultural Residues in the Manufacture of Terracotta Products

Laser granulometry has been used to characterise different granulometric fractions of a wheat straw, wheat flour and olive kernel which were previously ground to fine powders [158]. These dry materials were used in manufacturing industrial processes.

Operating protocol

The different materials to be characterised were fluids, without problems of sticking. The first step consists in the determination of the working pressure which must be enough for the dispersion of particles. Different pressures were tested: 0.5; 1.5; 2.5; and 3.5 bar. Air was used as the dispersant: its refraction index is 1. The working pressure chosen was 3.5 bar. At this pressure, the sizes observed were the smallest, corresponding to microscopic observation. For these materials, it is impossible to define a refraction or absorption index due to the lack of data. A mean value of refraction index of 1.52 and an absorption index fixed at 0.1 were applied.

The residuals obtained during the analyses were correct (1%) which validated the choice of these coefficients. For each pressure, a volume of 20–50 mL was placed in the dispersion cell. The alignment was done and the optical background noise was recorded at the tested pressure. When the sequence of measurements started, ten analyses were programmed on the same batch. The hopper was manually adjusted to give an obscuration of 1–5%. These low obscuration values enable the same level of energy to be applied to the same quantity of material being analysed in a repeatable manner. In order to validate the measurements, obscuration values below 1% and above 5% were eliminated. The results obtained at 0.5 bar were not repeatable, likely because of the difference in energy yield applied. The other pressures tested were validated because they led to at least 5 results which were close to each other. It should be noted that the analyses carried out on the batches with the largest particles (millimetric) required several samples to be taken in order to validate the results (limit of resolution, elements to big, poor matter transportation etc.).

Results:

The results in Fig. 2.77 are presented by number and by volume for each of the powders. The analysis by number enables the estimation of the population presenting the highest number of particles, the latter having an influence on the porosity of the finished product.

These different fractions correspond to powders obtained by sieving. In order to validate the influence of the size factor independently of the nature of the particles, two other powders (olive kernels and wheat flour) were characterised (Fig. 2.78).

Each of these powders was added at the level of 4% to the initial mixture in the manufacturing process. Different tests were then able to be carried out on the finished product.

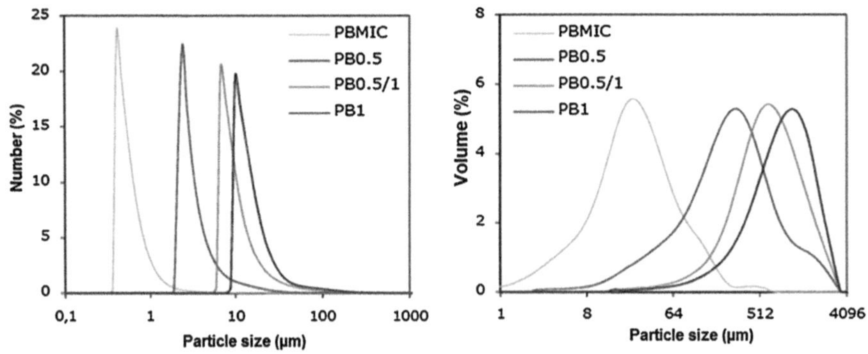

Fig. 2.77 Particle size distribution by number and volume of the different granulometric classes of wheat straw (PB). Reproduced with permission from Nigay [158]

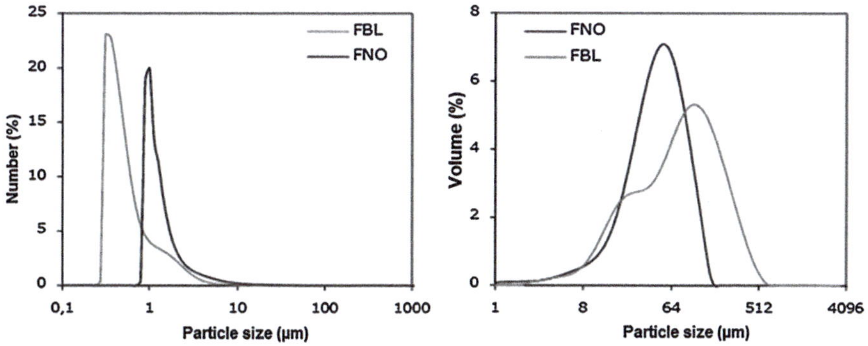

Fig. 2.78 Particle size distributions by number and volume of wheat flour (FBL) and olive stones (FNO). Reproduced with permission from Nigay [158]

2.7.2 *Density Analysis*

2.7.2.1 Introduction

A solid can be characterised by its density (ρ) which is the ratio of its mass to its volume (Eq. 2.72). In the international system (SI), density is expressed in "kg m^{-3}" but the unit "g cm^{-3}" is more often used.

$$\rho = \frac{m}{V} \tag{2.72}$$

where: m = mass of the solid in kg, V = volume occupied by the solid in m^3.

The true density of a solid is obtained by measuring its true volume, i.e. the volume occupied by the solid minus the volume of open pores (Fig. 2.79b). Closed pores are not taken into account by the measurement because they are inaccessible.

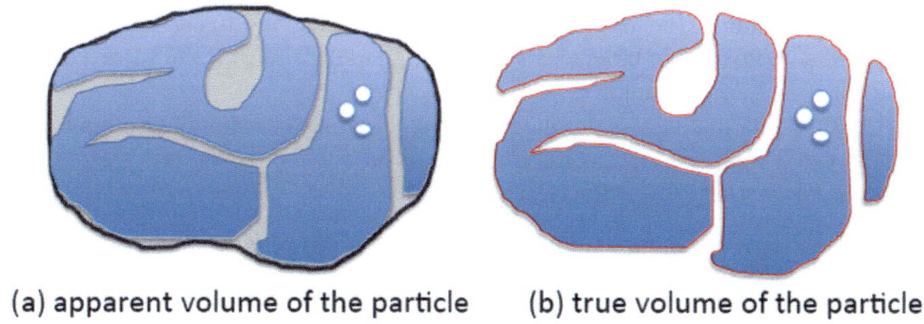

Fig. 2.79 Diagram of a porous particle a with the apparent volume of the particle, b with the true volume of the particle

The term apparent density is obtained with the apparent volume of the solid, i.e. the volume occupied by the solid plus the volume of its pores (Fig. 2.79a). The next sections describe the two methods for the determination of the apparent density.

2.7.2.2 Tap-Density

Definition

Tap-density of a powder is the ratio of the mass of the powder to the volume occupied by the powder after it has been mechanically tapped for a defined period of time. Tapping step allows minimizing the extraparticulate porosity. Tap-density represents the random dense packing of a powder sample. It can be calculated using Eq. (2.73).

$$\rho_{tap} = \frac{m}{V_t} \tag{2.73}$$

where: m: mass of sample (g); V_t: tap volume of sample (cm^3); ρ_{tap}: tap-density (g/cm^3).

Tap-density values strongly depend on the shape of the powders particles. The values are higher for more regularly shaped particles (i.e. sphere) compared to irregularly shaped particles. It is a useful parameter for several engineering applications (i.e. agrofood, pharmaceutical etc.). The packing properties of a powder can affect operations of solid dosage manufacturing, including bulk storage, powder flow and compaction.

Apparatus and Principle of Measurement

As example, the model SVM 22 from ERWEKA is described thereafter. The apparatus consists of the following parts:

- A graduated glass cylinder (100, 250 or 500 mL);
- Two test stations (for this model) where the cylinders are fixed. The apparatus described here can simultaneously measure 2 tap-densities for the same samples or for two different samples;
- A settling unit that allows producing 250 $\pm$ 15 strokes per minute from a height of 3 $\pm$ 0.2 mm;
- A digital interface for setting test runs time.

The procedure used to measure the tap-density of powders, granules and similar products includes the following steps [159]:

- Choose an appropriate volume of glass cylinder to adapt to the quantity of sample;
- Introduce sample in the glass cylinder up to a given volume. Note the volume (V_0) and the mass used (m_d);
- Secure the cylinder on the test station;
- Carry out 10, 500 and 1250 taps on the sample and read the corresponding volumes. It is recommended to name them V_{10}, V_{500} and V_{1250}.
- If the difference between V_{500} and V_{1250} is less than or equal to 2 mL, V_{1250} can be considered as the tapped volume (V_t). If the difference between V_{500} and V_{1250} is higher than 2 mL, repeat in increments such as 1250 taps, until the difference between succeeding measurements is less than or equal to 2 mL.
- Finally note V_t and calculate the tapped density by using Eq. (2.73). Do not forget to note the final number (N) of taps used. It can be a reference for others tests or for reproducibility tests.
- Repeat this procedure at least 3 times for the same sample to check the repeatability.

2.7.2.3 Triple-Weighing

Triple-weighing allows also measuring bulk density of a sample. This non-destructive technique consists in measuring three times the weight of a sample under different conditions as follows:

- In atmospheric air when the sample is dry (m_s);
- When the sample is immersed in water (m_h);
- When the sample is impregnated (m_w).

This technique is generally used in ceramics manufacturing field (i.e. fired-clay products, monolithic ceramics for energy storage etc.).

An apparatus for triple-weighing measurement includes the following parts:

- Cutting and polishing equipments for samples preparation in order to get specific dimensions;
- Device for the samples impregnation and saturation. Figure 2.80 shows a photo of an experimental set-up for impregnation and saturation steps. It consists of a vacuum pump, a desiccator with adapted dimension and a pipeline allowing the impregnation of sample with water. The vacuum pump allows reducing the pressure to at least 2 kPa; the later can be read by a pressure gauge.
- Measurement equipment including a sliding caliper (1/50 mm) or other standardized device for dimension measurement and a precision balance adapted to the mass of sample to be measured. This balance allows evaluating the hydrostatic weight as illustrated in Fig. 2.81 [160].

The procedure used to measure the apparent density by triple-weighing technique includes the following steps [161]:

Fig. 2.80 Example of an experimental set-up for the impregnation of samples: a vacuum pump, b desiccator, c pipeline allowing the impregnation of sample with water

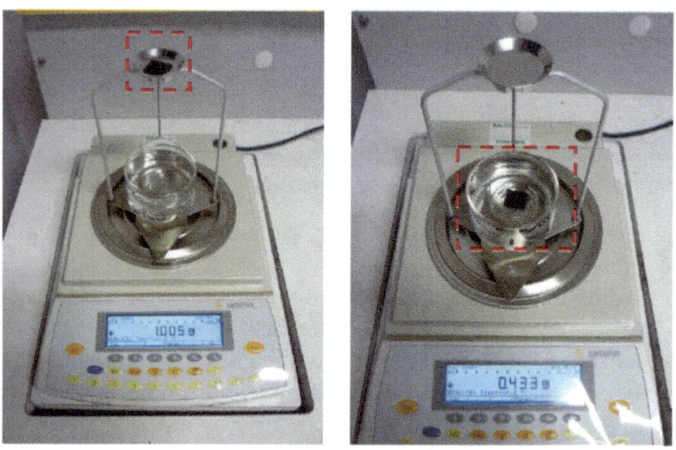

(a) Weighing in the air of a sample (b) Weighing in the water of a sample

Fig. 2.81 Example of balance and equipment to weight the sample. Reproduced with permission from Farrugia [160]

- Prepare sample (i.e. cubic shape) using the process mentioned in the previous section. In all cases, the sample must be carefully polished in order to have regular surfaces;
- Measure the dimensions of sample by using a sliding caliper. It is recommended to do the measurement at least 3 times for each sample;
- Measure the dried sample weight (m_s);
- Put the sample in the desiccator;
- Reduce the pressure inside the desiccator to below 2 kPa for at least 4 h;
- Introduce water for a partial immersion of the sample and keep the desiccator below 2 kPa for at least 2 h;
- Add water to completely immerse the sample and keep the desiccator below 2 kPa for at least 24 h;
- Reset the atmospheric pressure inside the desiccator and keep the sample immersed in water for at least 24 h;
- Weigh the sample under water (m_h);
- Take the sample out of the water and quickly wipe and weigh it (m_w).

The calculation of the apparent density (ρ_d) by triple-weighing can be carried out by using Eq. (2.74):

$$\rho_d = \frac{m_s}{V_t} \tag{2.74}$$

where m_s is the dry weight and V_t the apparent volume of the sample. V_t could be calculated from the dimension of the sample determined before but it is recommended to determine V_t by Eq. (2.75) where q_w is the water density (0.998 g/cm^3 at 20 °C) [161].

$$V_t = \frac{m_w - m_h}{q_w} \tag{2.75}$$

From the data obtained with the triple-weighing method, it is also possible to determine three other physical characteristics of a monolith as follows:

- The porosity (ε) by using Eq. (2.76), where ρ_s is the true density of the sample obtained by another technique (helium pycnometer for example):

$$\varepsilon = 1 - \frac{\rho_d}{\rho_s} \tag{2.76}$$

- The connected porosity (n_c) by using Eq. (2.77):

$$n_c = \frac{m_w - m_s}{m_w - m_h} \tag{2.77}$$

- And the closed porosity (n_o) by using Eq. (2.78):

$$n_o = \varepsilon - n_c \tag{2.78}$$

2.7.2.4 True Density

Introduction

As previously presented, the true density of a solid is obtained by measuring its mass and true volume, i.e. the volume occupied by the solid minus the volume of open pores (Fig. 2.79b). Closed pores are not taken into account by the measurement because they are inaccessible. Helium pycnometry allows determining the true volume of a known mass of a solid. Thus, the true density can be determined. Helium is used thanks to its ability to penetrate pores. This technique is non-destructive.

Principle of Measurement

The method involves measurement of pressure. Helium is introduced into the measurement cell containing the sample at pressure P_1 at ambient temperature (Fig. 2.82). The gas occupies the volume of the measurement cell minus the volume occupied by the solid. The gas is then released into an expansion chamber of volume V_{exp}. Thus, the pressure falls from P_1 to P_2. By application of Mariotte's law, the true volume of the sample is calculated by Eq. (2.79). These steps form a measurement cycle which should be repeated several times for an analysis.

$$V_{sample} = V_{cel} - \frac{V_{exp}}{\frac{P_1 - P_{atm}}{P_2 - P_{atm}} - 1} \tag{2.79}$$

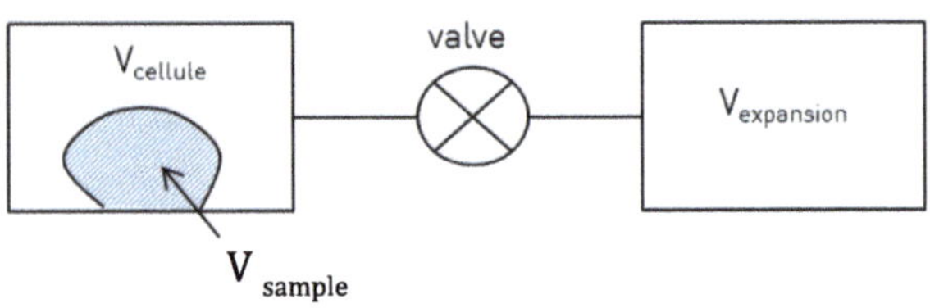

Fig. 2.82 Simplified diagram of a helium pycnometer (Accupyc model). Reproduced with permission from MICROMERITICS

where: V_{sample}: true volume of the sample; V_{cel}: volume of the measurement cell; V_{exp}: volume of the expansion chamber; P_1: pressure of gas in the measurement cell; P_2: pressure of gas after release into expansion chamber; P_{atm}: atmospheric pressure.

Equipment

There are several equipment manufacturers on the market, i.e. Micromeritics (model ACCUPYC II 1340) and Quantachrom (models ULTRAFOAM 1200 and PENTAFOAM 5200). The next part of this section is adapted to the model ACCUPYC II 1340 from Micromeritics.

To make measurements under optimum conditions, the pycnometer must be coupled with:

- a thermostatic bath in order to maintain the temperature of the cell at 20 °C;
- a vacuum pump in order to be able to "degas" the sample and carry out purges;
- a bottle of helium that is at least 99.995% pure.
- There should be a balance with a precision of 0.1 mg near the instrument.

Conditions for Analysis

The analysis is carried out in two stages:

- Preparation of the sample, including weighing and "degassing".
- Purging and analysis of the sample.

Sample Preparation

Samples such as clay, sand, soil or wood must be dried at 105 °C for 24 h in an oven. The quantity of sample must be adapted to the volume of measurement cell. The cell should be filled to at least two thirds of its volume. It is recommended to use measurement cell having high volume (i.e. typically 1–10 cm^3) to obtain representative result. Both monolithic or powder samples can be analysed.

Prior to the measurement, it is also preferable to eliminate any molecules such as carbon dioxide, water, nitrogen etc. that are absorbed on the surface of sample. Thus, a primary vacuum (at least 10^{-2} mbar) is applied for around 5 min but this duration depends on the sample and the quantity used. In order to avoid eventual pollution of sample and analyzer, the instrument should not be opened between the degassing and the analysis steps.

Sample must be weighed before and after degassing step since the degassing can lead to a mass loss. In general, the mass after degassing step is taken.

Purge and Analysis of the Sample

It is necessary to regularly calibrate measurement cells with standards (metal spheres) provided by manufacturer. The volume of standards is precisely known.

The density of a sample depends on the temperature and pressure of measurement. Generally, the density is determined at 20 °C (regulation with a thermostatic bath) and atmospheric pressure.

Moisture content of certain samples (i.d. woods, biowastes etc.) can modify density values. In this case, it should be necessary to increase the equilibrium level of the instrument from 0.005 psi/min (34.5 Pa/min) to a value above 0.5 psi/min (3447 Pa/min).

In addition to the "degassing" performed under primary vacuum, the instrument is programmed to carry out purge cycles of the sample with an inert gas flow to completely clean the sample and expansion chambers before starting the analysis. In most of cases, 10 purges at 19.5 psi (134 447 Pa) are enough.

Several cycles of purges and measurements are carried out consecutively. At least 10 purges at 19.5 psig are needed for a measurement. For fine particles at micrometric scale, the number of purges should be at least 20 in order to obtain a stable bed of particles. After purging step, the analysis takes place. The result given by the instrument corresponds to the mean of the densities obtained for the number of cycles carried out.

The granulometry, the porosity and the homogeneity of the sample are important characteristics for a good level of repeatability. At least 10 cycles of measurement should be performed for one sample. The standard deviation must be below 10^{-3}. If it is not the case, the measurement must be restarted (without opening the measurement cell). Figure 2.83 shows the main steps to perform a measurement.

Examples

Example 1: Determination of the True Density of a Rice Sample

Equipment used: Micromeritics ACCUPYC 1330 Helium Pycnometer.
Parameters: Temperature: 20.4 °C; volume of cell: 11.3756 cm^3; duration of primary degassing: 4 min; number of purges: 10; number of measurements: 10; equilibrium level: 0.005 psig/min.
Steps of measurement: The calibration of the instrument was checked. The apparent volume of sample was measured using a graduated test tube. Its mass was determined.
Results: The results are presented in the tables below (Tables 2.45, 2.46 and 2.47). The mean density of the rice being studied is 1.46 g/cm^3.

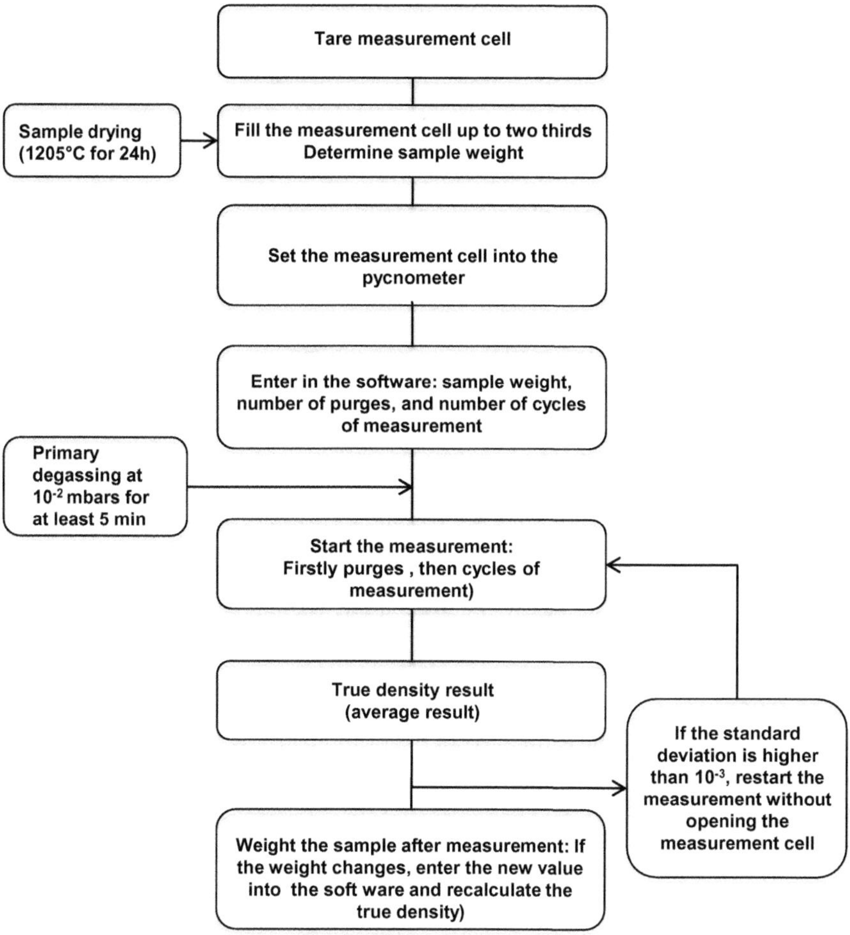

Fig. 2.83 Main steps of true density measurement

Table 2.45 True density analysis of a sample of a rice

Reference	Mass (g)	Apparent volume (cm^3)	True volume (cm^3)	Density (g/cm^3)	Standard deviation (g/cm^3)
Rice n°1	7.9723	8.9	5.4507	1.4626	0.0001
Rice n°2	8.1433	9.0	5.5662	1.4630	0.0001
Rice n°3	8.4719	9.3	5.7878	1.4637	0.0001

Table 2.46 True density analysis of a coffee powder

Reference	Mass (in g)	True volume (in cm^3)	Density (in g/cm^3)	Standard deviation (in g/cm^3)
Coffee	1.2322	1.2334	0.9990	0.0032

Table 2.47 True density analysis of a coffee powder

Reference	Mass, g	True volume, cm^3	Density, g/cm^3	Standard deviation
Raw sample n°1	2.0193	1.4040	1.4382	0.0021
Raw sample n°2	2.0457	1.4254	1.4352	0.0018
Crushed sample n°1	2.3857	1.6331	1.4608	0.0006
Crushed sample n°2	2.3574	1.6134	1.4611	0.0005
Crushed sample n°3	2.3893	1.6342	1.4620	0.0004

Example 2: Determination of the True Density of a Coffee Powder

Equipment used: Micromeritics ACCUPYC 1330 Helium Pycnometer.
Parameters: temperature: 20.1 °C; volume of cell: 6.0911 cm^3; duration of primary degassing: 4 min; number of purges: 10; number of measurements: 10; equilibrium level: 0.005 psig/min.
Steps of measurement: The calibration of the instrument was checked. The apparent volume of sample was measured using a graduated test tube. Its mass was determined.
Results: The density of the coffee is 0.99 g/cm^3.

Example 3: Determination of the Density of Maritime Pine Wood Chips

Sample: Two types of wood were analyzed: (1) Raw maritime pine wood chips (without grinding) with irregular and elongated shape (length of around 5 mm). (2) Grinded wood which are more homogenous compared to raw sample, and has particle size around 1 mm.

Equipment used: Micromeritics ACCUPYC 1330 Helium Pycnometer.

Parameters: temperature: 19.9 °C; volume of cell: 11.5998 cm^3; duration of manual degassing: 4 min; number of purges: 10; number of measurements: 10; equilibrium level: 0.005 psig/min.

Results: The densities of raw sample and crushed sample were respectively 1.44 and 1.46 g/cm^3. These results show that the granulometry and irregularity of samples has an impact on the true density.

2.8 Physico-chemical Characterizations

The surface reactivity of materials is related to their physico-chemical properties. On the one hand, the texture and the porosity are important to develop an interface between the solid and the second "reactive" phase that could be liquid or gaseous. On the other hand, the specific functions belonging to the surface are of great importance for chemical reactions. Especially, carbon materials can be composed of organic functions, including the carbon matrix, and minerals that could be tuned for specific applications [162]. Reaching the surface, and more specifically the interface, all crystalline materials will miss a stacking layer providing a certain surface energy. This energy is directly related to the electronic destabilization of the structure. For amorphous materials, the functions arising at the surface are not stabilized with the same strong or weak bonds as in the bulk. Therefore heterogeneous interactions appear accounting for a surface reactivity. From biomass and biowaste and their related products, many applications could be selected such as soil amendment [163], sorbent [164], catalysts [162, 165], catalyst support [166] or electrode/supercapacitors [167]. All of these applications require information on surface properties and surface groups. These can be obtained by different techniques, i.e. Boehm titration for acidic or basic properties; infra-red spectroscopy, ultraviolet—visible spectroscopy, and Raman spectroscopy for chemical organic functions, Temperature Programme Desorption for oxygenated organic functions, Dynamic Vapor Sorption for surface energy etc.

2.8.1 Infrared and Ultraviolet—Visible Spectroscopies

The main objective of this section is focused on the application of infrared (IR) and ultraviolet—visible (UV-Vis) spectroscopies for the analysis of biomass, bio-wastes and related products. A selection of some of the techniques will be developed in this section. For more details, several books are available in the literature [168–175].

2.8.1.1 Introduction to the IR and UV-Vis Spectroscopic Methods

Spectroscopic methods of analysis deal with interactions between electromagnetic radiations and matter which can be solutions or pure molecules, solids, liquids or gases depending on the techniques.

Electromagnetic Radiation

All lights (including IR and UV-Vis) can be considered as an electromagnetic radiation, or the combination of an electric and a magnetic field; both of them oscillate perpendicularly from each to other and perpendicularly to the propagation direction of the light as a sinusoidal law.

An electromagnetic radiation can be characterized by:

- its frequency (ν), with Hertz (Hz) as unit.
- its wavelength (λ), with length units (nm, μm etc.).
- its wavenumber ($\bar{\nu}$), which is the inverse of the wavelength.
- The relationship between these different terms is as follow:

$$\nu = \frac{c}{\lambda} = \bar{\nu}c \tag{2.80}$$

with c is the light speed = 2.99712×10^8 m/s in the air, λ: the wavelength and $\bar{\nu}$ the wavenumber of the radiation.

The wavenumber $\bar{\nu}$ is usually used in IR spectroscopy, expressed in cm^{-1} whereas in UV-Vis, the electromagnetic radiation is characterized by the wavelength λ expressed in nm.

An electromagnetic radiation can be considered as an energy carrier and relates to a beam of photons. The energy of a photon is proportional to the radiation frequency:

$$E = h\nu = h\frac{c}{\lambda} = hc\bar{\nu} \tag{2.81}$$

with h the Planck constant = 6.626×10^{-34} J s.

Figure 2.84 shows the electromagnetic spectrum which extends from radio waves to gamma rays [168]. The IR radiations are between micro-waves and UV-Vis, and UV-Vis between IR and X-rays.

IR and UV-Vis areas are sub-divided according to their closeness to the visible area. IR spectroscopy is sub-divided into near IR ($\bar{\nu}$: 14,000–4000 cm^{-1}); mid IR ($\bar{\nu}$: 4000–400 cm^{-1}) and far IR ($\bar{\nu}$: 400–200 cm^{-1}). Generally, IR spectroscopy is referred to mid IR. UV-Vis spectroscopy is sub-divided into near ultraviolet region ($\bar{\nu}$: 200–400 nm), far ultraviolet region ($\bar{\nu}$: 10–200 nm) and visible region ($\bar{\nu}$: 400–800 nm). In UV-Vis spectroscopy, the common devices operate between 190 and 800 nm.

Absorption of Radiation

To understand the interaction of an electromagnetic radiation with matter, it is necessary to firstly understand the quantum mechanics of atoms and molecules [168, 169], but it is not the aim of this section. To simplify, it is known that each elementary particle (atom, ion, molecule …) possesses several energetic states. The lowest energy state is called the fundamental state. A particle can have a transition from the fundamental state to a higher energy state, called excited state.

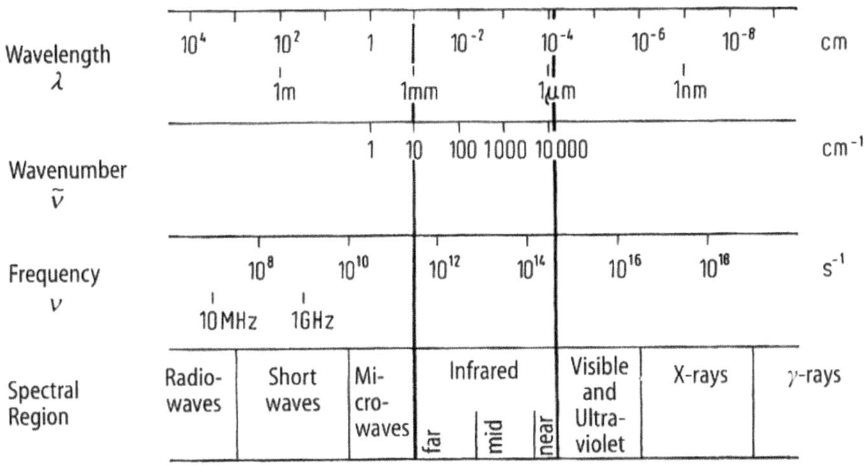

Fig. 2.84 The electromagnetic spectrum [168]. Copyright Wiley-VCH Verlag GmbH & Co. KGaA. Reproduced with permission [168]

This transition can be generated by an electromagnetic radiation, when electromagnetic radiation energy is absorbed by the particle. However, this absorption only happens when the energy of the corresponding photon is exactly equal to the energy gap ΔE between two states involved in the transition (Eq. 2.82), the absorption is quantized (Fig. 2.85):

$$\nu = \frac{\Delta E}{h} = \frac{E_2 - E_1}{h} \tag{2.82}$$

where ν is the frequency radiation, ΔE the energy gap between the two levels involved in the transition and h the Planck constant.

The lifetime of excited state is very short (around 10^{-9}–10^{-6} s). Excited particle returns quickly to fundamental state by liberating the energy previously absorbed under heat or radiation forms. The absorption of electromagnetic radiation induces various transitions which depend on the radiation frequency. Actually the total

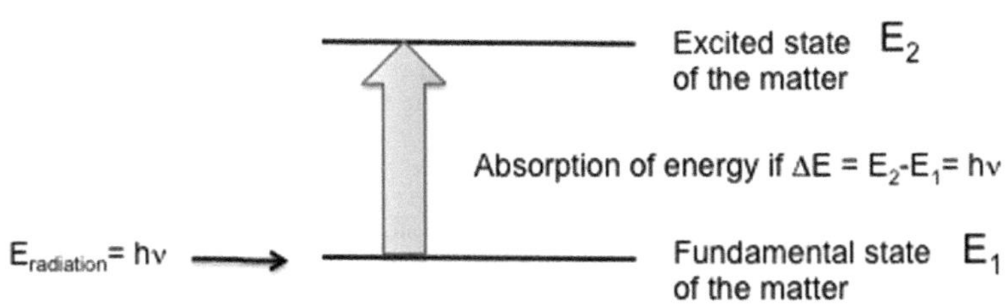

Fig. 2.85 Transition from a fundamental to an excited state as the result of an electromagnetic radiation absorption

mechanical energy of a molecule is the sum of three terms, the electronic (E_{el}), the rotational (E_{rot}) and the vibrational (E_{vib}) ones (Eq. 2.83):

$$E_r = E_{el} + E_{vib} + E_{rot} \tag{2.83}$$

where E_{el} is the energy of valence electrons which form bonds on molecules; E_{vib} is the energy associated to vibration of bonds in a molecule; E_{rot} is the energy of rotation of the molecule as a whole about an axis.

Each energetic electronic, vibrational or rotational term, is associated to different energetic levels, a fundamental and several excited levels (Fig. 2.86).

If a molecule absorbs a radiation in the far IR or in the microwave region, only the molecule rotation is affected and only the rotational term in Eq. (2.83) is modified. Transitions occur between a fundamental rotational level to an excited rotational level.

The energy of a mid IR radiation corresponds to the energy difference between two vibrational levels in a molecule. As the vibrational energy levels are quantified, the IR spectrum is expected to exhibit different discrete lines. Each line corresponds to a vibrational transition between basic and excited vibrational levels of the transition. However, in IR field, each vibrational level is sub-divided into several rotational levels, each one being quantized. So a vibrational transition is the result of numerous rotational transitions leading to a band. So, an IR spectrum is then

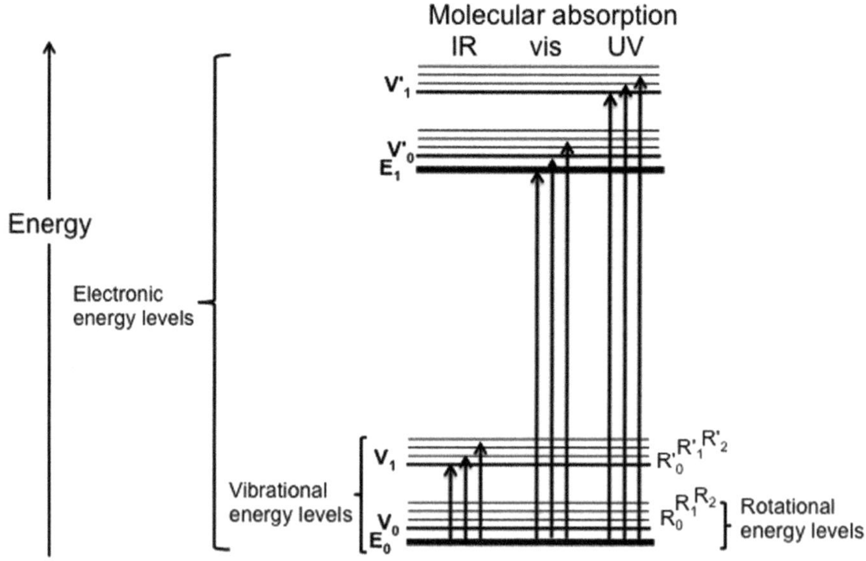

Fig. 2.86 Energy levels of a molecule (E_0, E_1, are respectively a fundamental and an excited electronic levels, V_0 a fundamental vibrational level, V_1, V'_0, V'_1 excited vibrational levels, R_0 a fundamental rotational level, R_1, R_2, R'_0, R'_1, R'_2 excited rotational levels)

composed with bands, but not discrete lines. An IR spectrum is also called a vibrational-rotational spectrum.

If the radiation belongs to the UV-Vis region, the transitions occur between electronic levels. Transitions of valence electrons from a bonding or non-bonding orbital to an anti-bonding orbital occur, called electronic transitions. Each electronic level, basic and excited states, is sub-divided in several vibrational levels, each one subdivided in several rotational levels. Then, an electronic transition is the result of a whole of transitions, from the fundamental electronic level to one of the numerous rotational/vibrational levels. This is associated to an excited electronic level. The energetic difference associated to these different transitions is so weak that, instead to appear as a discrete line, a UV-Vis transition appears as a broad absorption band.

In summary, IR radiations lead to modifications of vibrational and rotational motions of bonds in molecules and UV-Vis radiations induce electronic transitions of valence electrons involved in bonds. The energy level implied in UV-Vis radiations is higher than that implied in IR radiations. Actually, the gap between two vibrational or rotational states is lower than the gap between two electronic states. In other words, IR radiation is not enough energetic to induce electronic transitions as UV-Vis radiations. In practice, UV-Vis radiations can lead to chemical reactions with bonds breaking which is not the case with IR radiations.

Absorption Spectrum

The IR or UV-Vis spectrum of a sample is a plot of absorbed or transmitted energy by the sample, expressed by the absorbance or the transmittance, against the frequency or the wavelength of the absorbed radiation. The absorbance A is defined as follow:

$$A = \log \frac{I_o}{I} \tag{2.84}$$

with I_o: intensity of incident radiation; and I: intensity of transmitted radiation.

The absorption depends on the substance concentration by the Beer-Lambert law (Eq. 2.85). This allows performing quantitative study in IR and UV-Vis spectroscopy:

$$A = \log \frac{I_o}{I} = \varepsilon l c \tag{2.85}$$

with ε: the molar absorptivity or molar extinction coefficient, which is characteristic for a substance and for a wavelength; l: the pathlength of the beam of light through the material sample; c: the concentration of substance studied. The concentration c is usually expressed in mol L^{-1}; l in cm and ε in mol^{-1} L cm^{-1}.

The absorbance is commonly used in UV-Vis spectroscopy as the percent transmittance, %T, (Eq. 2.86) is preferred in IR spectroscopy, leading to bands upward pointing in UV-Vis spectroscopy and downward pointing in IR spectroscopy.

$$\%T = \frac{I}{I_o} \times 100 \tag{2.86}$$

2.8.1.2 IR Spectroscopy

Interest of IR Spectroscopy

The IR spectroscopy is commonly used to provide information about the presence or absence of hydrocarbon bonds (C–H, C=C, C≡C) and specific functional groups (O–H, N–H, C=O etc.). In the 1500–400 cm^{-1} range, there is a complex area with many bands, more or less strong, frequently overlapping each other. It is called the fingerprint region. It is generally difficult to identify specific functional groups in this area, but it can be used to compare samples. If two samples display the same IR spectrum, it is a proof that it is the same molecule. IR spectroscopy is not the suitable technique to identify the chemical structure of a molecule. It provides information on molecular fragments or specific functional groups. It must be completed by other techniques such as NMR or mass spectroscopy, chromatography.

Different Vibrational Modes

In IR spectroscopy, a molecule can be considered as an aggregate of spheres, representing the atoms linked to each other by bonds. This can be compared to a harmonic oscillator. The IR radiation absorption will lead to vibrations modifying the distance between atoms and bond angles (Fig. 2.87). Covalent bonds can vibrate in two modes: stretching and bending.

The stretching mode:

A stretching vibration corresponds to a change in the bond length. Two kinds of stretching vibrations are distinguished, the symmetrical stretching and the asymmetrical stretching.

The bending mode:

A bending vibration corresponds to a change in the angle between two bonds. Four bending vibration exist, scissoring, rocking, wagging and twisting.

Each vibration mode absorbs specific frequencies. The bands corresponding to stretching vibrations appear at higher frequencies than bands corresponding to bending vibrations.

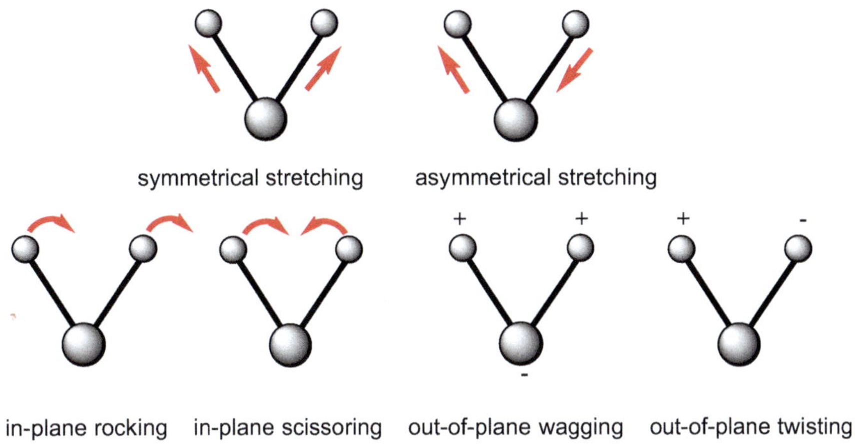

Fig. 2.87 Illustration of vibrational modes

Infrared Active Bonds

Only polar bonds absorb in IR spectroscopy such as C–H, O–H, C=O, etc. The absorption is the result of the interaction between the electrical component of the electromagnetic radiation and the dipole moment of the bond. The intensity of the bands depends on the magnitude of the dipole moment associated with the bond, the greater the polarity of the bond, the stronger its absorption band. Some molecules are IR inactive. For example the O_2, N_2 gaz molecules of air don't absorb in IR.

Factors Affecting the Vibrational Frequencies in IR

If a bond between two atoms is compared to the model of two balls related by a spring (a harmonic oscillator), the system can be described by Hooke's law (Eq. 2.87) that relates frequency of vibration, mass of atoms and force constant:

$$\bar{v} = \frac{1}{2\pi c}\sqrt{\frac{k}{\mu}} \tag{2.87}$$

where

μ: the reduced mass (the product of the two masses of atoms divided by their sum).
k: the force constant of the bond (depending on the bond strength, the stronger the bond, the larger the k).

According to Hooke's law, the frequency of vibration depends on:

- The strength of the bond: stretching vibration of a triple bond appears at higher frequency than that of a double bond, the later appears at higher frequency than that of a single bond:

Bond	C≡C	C=C	C–C
Frequency of vibration	2150 cm^{-1}	1650 cm^{-1}	1200 cm^{-1}

- The atoms electronegativity: the higher the electronegativity difference is, the higher the frequency vibration is:

Bond	C=O	C=N	C=C
Frequency of vibration	1720 cm^{-1}	1650 cm^{-1}	1620 cm^{-1}

- The masses of the atoms linked by the bond: the lighter the atoms linked by the bond are, the faster the bond vibrates:

Bond	C–H	C–C	C–O	C–Cl
Frequency of vibration	3000 cm^{-1}	1200 cm^{-1}	1100 cm^{-1}	750 cm^{-1}

- The frequency vibration also depends on the environment of the bond:

Bond	$C{=}O_{carboxylic\ acid}$	$C{=}O_{ester}$	$C{=}O_{aldehyde}$	$C{=}O_{ketone}$
Frequency of vibration	1760 cm^{-1}	1750 cm^{-1}	1730 cm^{-1}	1720 cm^{-1}

The IR Absorption Range

The IR spectrum can be divided in three main areas:

- The area 4000–1500 cm^{-1} corresponds to the stretching vibrations of the main bonds: N–H, O–H, C–H, C≡C, C≡N, C=C, C=O, C=N, etc.
- The area 1500–1000 cm^{-1} corresponds to the bending vibrations, also called the fingerprint of the compound, it is a complex area showing many bands but characteristic of a given molecule. In this region, also appears the stretching band of the C–O bond of an alcohol or an ester.
- The area 1000–400 cm^{-1} is useful to identify ethylenic or aromatic structures.
- To help the identification of functional groups in a spectrum, IR correlation charts are available in literature, giving absorption frequencies, typically reported in wavenumber, for common molecular bonds and functional groups (Table 2.48) [176].

Figure 2.88 shows an example of FTIR spectra obtained with a Perkin-Elmer RXI spectrometer for hydroxyapatite (HA), hydroxyapatite-modified lignocellulose (LH),

Table 2.48 An example of IR correlation chart

Functional group	Bond	Wavenumber (cm^{-1})	Band description
Alcohols, phenols	O–H, "free"	3640–3610	Strong, sharp
Alcohols, phenols	O–H, hydrogen-bonded	3550–3200	Strong, broad
Amides, amines[a]	N–H	3500–3200	Medium
Alkynes	C–H	3330–3270	Strong, narrow
Carboxylic acids	O–H	3300–2500	Medium, broad
Arenes	C–H	3100–3000	Strong
Alkenes	C–H	3100–3000	Medium
Alkanes	C–H	2950–2850	Medium
Aldehydes	C–H	2830–2695	Medium
Nitriles	C≡N	2260–2210	Medium
Alkynes	C≡C	2260–2100	Weak
Carbonyls, carboxylic acids, esters	C=O	1780–1680	Strong
Amides	C=O	1690–1630	Strong
Alkenes	C=C	1680–1610	Medium
Alcohols, carboxylic acids, esters, ethers	C–O	1320–1000	Strong
Chloroalkanes	C–Cl	800–700	Medium
Bromoalkanes	C–Br	690–515	Medium

Understanding advanced organic and analytical chemistry: the learner's approach, Chan and Tan [176]; Copyright@2011

[a]For both primary and secondary amides or amines

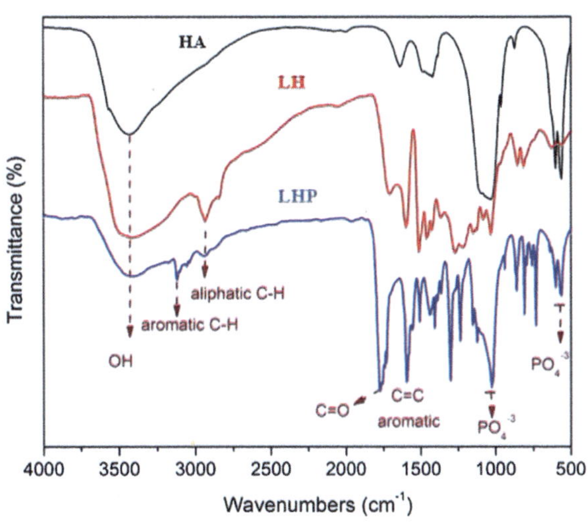

Fig. 2.88 FTIR spectra of hydroxyapatite (HA), hydroxyapatite-modified lignocellulose (LH), and LH grafted Perylene-3,4,9,10 tetracarboxylic dianhydride (LHP). Reproduced with permission from Elsevier [177]

and LH grafted Perylene-3,4,9,10 tetracarboxylic dianhydride (LHP) [177]. Vibration peaks are plotted as a function of the wavenumbers. Other examples of FTIR application will be presented and commented in the next sections.

Apparatus

IR spectrometers can be classified into two categories: dispersive Infrared spectrometer and Fourier Transform Infrared spectrometer (FTIR). They are different from their optical systems as shown in Fig. 2.89.

In the case of two-beam dispersive IR spectrometer, IR radiation is divided into two similar beams, one passes through reference cell and another one passes through sample cell. The monochromator is composed of a set of slots and a light dispersive system, which allow sending alternatively the light transmitted from reference and sample cells to the detector at each wavenumber. This concept has several drawbacks, mostly the mechanical slippage of optical system, slow scan speed, mediocre resolution, accuracy and sensitivity, and narrow frequency of analysis.

FTIR spectrometer is equipped with an interferometer which allows a wavenumber-dependent radiation modulation. In general, a Michelson interferometer, containing a fixed mirror, a movable mirror and a beam splitter, is used in which the two-beam interferences are converted by mathematical Fourier Transform into classical IR spectrum. Compared to dispersive technique, FTIR is faster, more sensitive, with simple mechanical design and high resolution.

For each concept, different types can be found for IR radiation source, IR detector, cell (for gas, liquid, or solid sample and for study under a specific condition of temperature or pressure), and holder for pellets and film etc. [168].

Sample Techniques

IR Spectroscopy Measurement by Transmission Mode

For the measurement of gas sample, gas cell is used for sampling. The optical pathlength can vary from some cm up to around 40 m depending on the type and

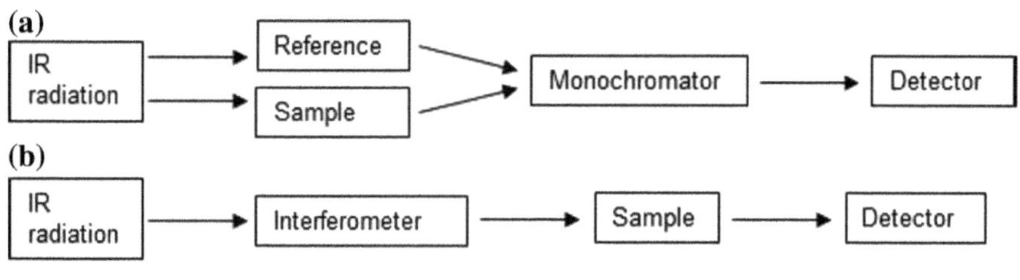

Fig. 2.89 Scheme of a two-beam dispersive IR spectrometer; and b Fourier transform IR spectrometer

the concentration of gas analyzed. Cell needs to be purged several times to avoid contamination.

Liquid sample can be analyzed under pure compound form or homogenous solution using a solvent. The most used solvent is CCl_4 (with the limit that it's suspected of causing cancer), which has only an absorbance band at 750–800 cm^{-1}. Liquid sample can be held by two discs of different materials or by a cell having a defined pathlength (25–1000 μm) [171] (Table 2.49). The most common materials used are KBr and KCl because they are cheap, or ZnSe and CaF_2 for aqueous or wet samples. The wavenumber range depends on the material of cell. Quantitative analysis is only possible when using cell having a defined pathlength for measurement. Figure 2.90 shows an example of demountable and sealed cells for analysis of liquids.

Solid sample can be analyzed by different ways as listed below:

- Dissolution in a solvent.
- Dispersion in organic oil called Nujol.
- Production of thin film.
- Dispersion of solid powder in KBr: A hydraulic press at around 5–8 t/cm^2 is used to prepare thin disc from the mixture of KBr and sample powders. This disc is then ready for analysis.

Table 2.49 Selected infrared transmitting material

Material	Wavenumber (cm^{-1})	Comments
NaCl	40,000–625	Common, low cost, hygroscopic
KBr	40,000–385	Common, low cost, very hygroscopic
BaF_2	50,000–870	Water insoluble, easily cracked
CaF_2	50,000–1100	Water insoluble, good at high pressures
ZnS	10,000–715	Water insoluble
ZnSe	10,000–515	Water insoluble

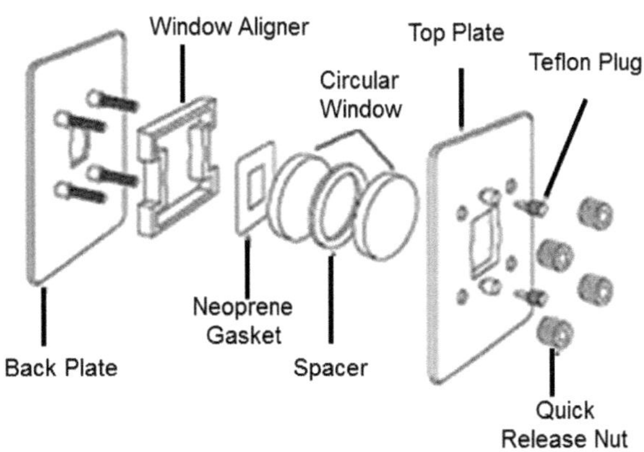

Fig. 2.90 Illustration of demountable cell diagram. Reproduced with permission from Perkin-Elmer [178]

IR Spectroscopy Measurement by Reflection

IR spectroscopy measurement by reflection has advantages to do not need any sample preparation, but it is only for solid or liquid samples because reflection takes place at surface.

When an IR radiation strikes a surface, it can be reflected, transmitted or absorbed. It depends on the refractive indices of the two media (which usually corresponds to air and solid sample). The most common techniques of measurement by reflection are the followings:

- Specular reflection: it is used for samples which reflect less (varnish, painting, polymer film …). The accessory allows studying the reflectance or the reflected light which depends on the refraction indices of each compound at a given wavenumber [168].
- Attenuated total reflectance (ATR): This is also known as internal reflection spectroscopy (IRS) or multiple internal reflectance (MIR). ATR allows obtaining infrared spectrum of the surface of a liquid or solid sample which either is too thick or has too strongly absorption to be analyzed by standard transmission spectroscopy. For this technique, sample is placed on the surface of a transparent crystal called internal reflection element (IRE) which is usually made from a material with a high refractive index (Ge with n = 4, ZnSe with n = 2.4, KRS-5 with n = 2.4, diamond with n = 2.4). This IRE is radiated with an incidence angle which must exceed the critical angle (θ_c). This last one depends on the nature of IRE used and sample. The incident light is totally reflected, generally several times, and the sample interacts with the evanescent wave. Absorption spectrum can be conveniently obtained by measuring this interaction [168].
- Diffuse reflectance: This technique is usually used for solids powder having a rough surface. When incident light is sent to sample surface, specular and diffuse reflectance can happen together. The apparatus for diffuse reflectance measurement allows collecting diffusive light from sample. For substance dissolved in a solvent, KBr is impregnated with this solution then the solvent is evaporated to obtain a solid sample. By comparing the diffuse reflectance of the mixture sample/KBr with that of pure KBr powder, one can obtain a spectrum which is comparable to that from transmission mode. For this technique, important parameters are: particle size, packing density and the homogeneity of the mixture [168].

Example of Infrared Spectroscopy Application

Lehto et al. [172] recently studied the effect of different pretreatments of industrial silver/white birch (*Betula pendula/B. pubescens*) and Scots pine (*Pinus sylvestris*) chips by hot water and dilute sulfuric acid having pH of 3 using attenuated total reflectance (ATR) infrared spectroscopy. Wood chips with the maximum thickness of 7 mm, maximum width of 13 mm, and minimum width 7 mm were used. The pretreatment was performed with a 1.25 L rotating stainless steel autoclaves heated in an oil bath, at 130 and 150 °C for 30–120 min. The liquid-to-wood ratio was 5 L/kg. Pretreated wood chips were separated from the hydrolysates by using filtration bags. Chips were then washed with tap water. For ATR infrared spectroscopy analysis, wood chips were grounded and sieved into particles smaller than 0.5 mm, using a Fritsch Pulverisette analytical grinder (Fritsch GmbH, Idar-Oberstein, Germany). These fine particles were used to make wood pellets with a Bruker Alpha FTIR spectroscopic device (Bruker Optics, Billerica, MA, USA) which were fitted with a Platinum ATR single reflection diamond ATR module. The ATR measurement was carried out in the 4000–400 cm^{-1} wavenumber range. Final spectra were averaged from 64 scans at a spectral resolution of 2 cm^{-1}. Figure 2.91 shows an example of ATR infrared spectra obtained for the untreated woods and woods after pretreatment under different conditions of solvent, temperature and duration [172].

In this work, ATR infrared spectra allowed determining the characteristic bands of various functional groups such as OH stretching in hydroxyl groups; CH

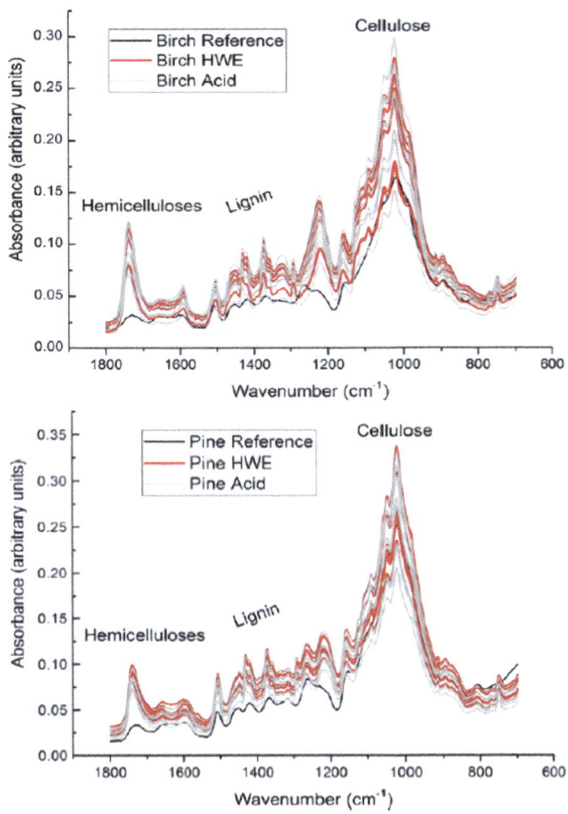

Fig. 2.91 ATR infrared spectra of the fingerprint regions for hot-water extracted (red) and acid extracted (grey) birch (left) and pine (right) wood samples treated at varying pretreatment temperatures and times. Reproduced with permission from Elsevier [172]

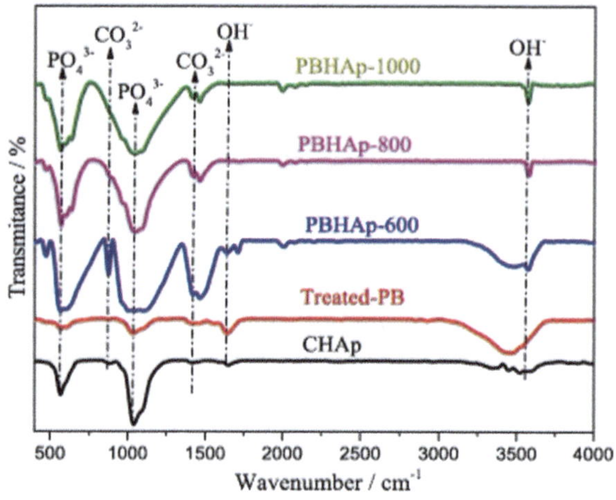

Fig. 2.92 FTIR results of synthesized HAp from pig bone at different calcination temperatures. For comparison, treated pig bone and commercial HAp spectra is also [179]. Reproduced with permission from Elsevier

stretching in methyl and methylene groups; C=O stretching in unconjugated ketone, carbonyl, and aliphatic groups (xylan); C=O vibration of esters (in lignin), ketones, and aldehydes in hemicelluloses; C=C aromatic cycle; and many others. By affecting the functional groups to the main component of woody materials, the author could evidence the increase in hemicellulose content and the decrease in the lignin content by pretreatment processes, as shown in Fig. 2.91.

Ofudje et al. [179] presented a new synthesis of hydroxyapatite powder (HAp) from pig bone waste via heat treatment in a muffle oven in air between 600 and 1000 °C, the interest was to reduce environmental pollution and adding values to waste as HAp is a material of great interest in biomedical applications. The FTIR spectroscopy was carried out to identify the functional groups contained in the HAp powder (Fig. 2.92). The FTIR device was a tensor 27 (Bruker, Germany). The analyses were made by transmission on pellets. A mixture made of 1:99 of HAp powder and KBr was mixed in a mortar, pestle and then compressed at about a load of 5 tons to form a pellet of 2 mm diameter. All spectra were recorded in the range of 4000–400 cm^{-1} with resolution of 4 cm^{-1} and 64 scans. The analyses showed the presence of phosphate, carbonate and hydroxyl functional groups in the pig bone derived HAp at different temperatures, testifying to the synthesis of pure HAp in conformity with standard HAp powder from Sigma Aldrich.

2.8.1.3 UV-Vis Spectroscopy

Introduction to UV-Vis Spectroscopy

If IR spectroscopy provides information on molecular structure, UV-Vis spectroscopy is commonly used to quantitative analysis applying the Beer-Lambert law. The UV-Vis spectrum is recorded in the region extended from 190 to 900 nm.

The absorption of a radiation on the UV-Vis region induces transition of a valence electron from a fundamental to an excited energetic level. These transitions are called electronic transitions. A UV-Vis spectrum is the plot of the intensity of absorption in terms of absorbance (without unit) against wavelengths of absorbed radiations expressed in nm (Fig. 2.93). Each transition in the UV-Vis region leads to a broad band in the spectrum. The wavelength of maximum absorbance is a characteristic value, designated as λ_{max} and associated to a ε_{max}.

Electronic Transitions

According to molecular orbital theory, the absorption of a UV-Vis radiation involves transition of a valence electron from a bonding or non-bonding orbital to an anti-bonding orbital. The bonding orbitals are called σ or π, the non-bonding n and anti-bonding σ^* and π^*. Figure 2.94 presents the relative energies implemented by the four electronic transitions.

- $\sigma \rightarrow \sigma^*$ transition

The transition of an electron from a bonding σ orbital to the associated anti-bonding orbital σ^* is designated $\sigma \rightarrow \sigma^*$ transition and requires so high energy that this transition is not observed in UV-Vis region commonly used in spectroscopy (190–900 nm). This is related to the fact that σ bonds are very strong. To observe this transition, it requires an evacuated spectrophotometer, since oxygen in the air absorbs strongly under 190 nm and specific cells, quartz cells can be used only up to 190 nm. With a common spectrophotometer, saturated hydrocarbons as hexane or cyclohexane with only single bonds C–C or C–H are transparent and can be used as solvents.

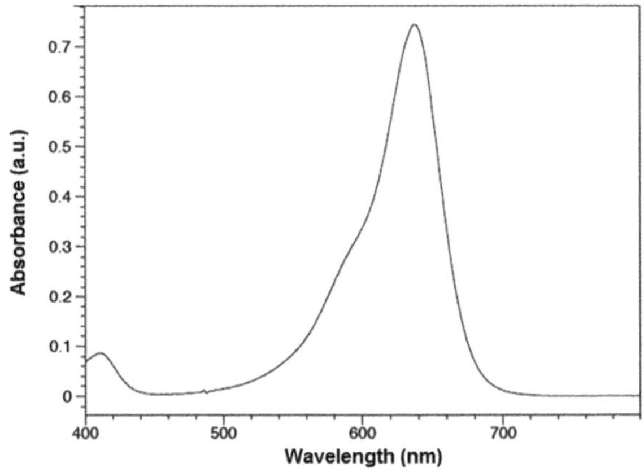

Fig. 2.93 VIS spectrum of patent blue V sodium salt

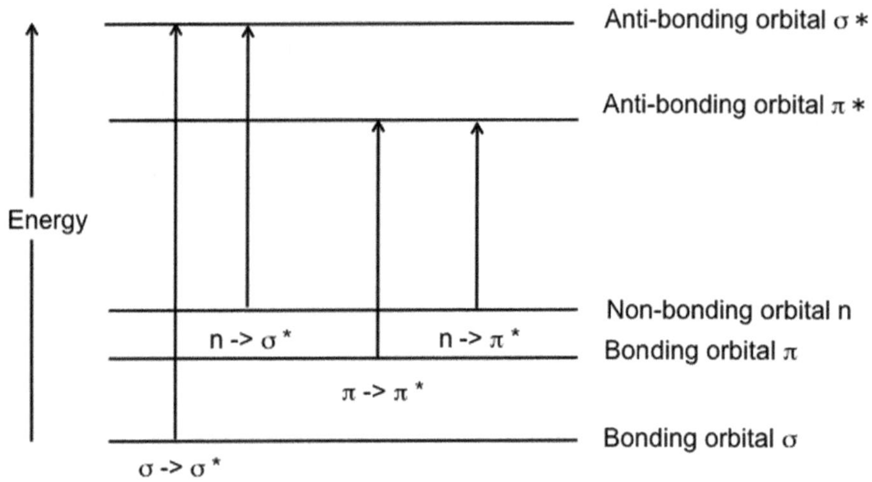

Fig. 2.94 Electronic transitions between different molecular orbitals

- $\pi \rightarrow \pi^*$ transition

The transition of an electron from a bonding π orbital to the associated anti-bonding π^* orbital is designated $\pi \rightarrow \pi^*$ transition. These transitions appear in compounds containing one or more unsaturated groups such as C=C, C=O, C=N, NO_2, S=O, etc.

The molar absorptivities ε associated to $\pi \rightarrow \pi^*$ transition are higher than 10,000 L mol^{-1} cm^{-1}. The unsaturated functional groups which absorb in UV-Vis region are called chromophore. The increase of conjugated double bonds shifts the absorption maximum to higher wavelengths. An extensively conjugated π-electrons system leads to colored compounds.

- $n \rightarrow \sigma^*$ or $n \rightarrow \pi^*$ transitions

The transitions of an electron from a non-bonding n orbital to an anti-bonding σ^* orbital or π^* orbital are designated $n \rightarrow \sigma^*$ or $n \rightarrow \pi^*$ transitions respectively. The molar absorptivity ε associated to $n \rightarrow \sigma^*$ transition is lower than 100 L mol^{-1} cm^{-1}, leading to bands with low intensity.

The $n \rightarrow \pi^*$ transition is observed in molecules containing one or more unsaturated groups involving an heteroatom like C=O, N=O, NO_2, C=S. Such groups show two bands, one corresponding to the $\pi \rightarrow \pi^*$ transition and a second one at a lower energy and a higher wavelength corresponding to the $n \rightarrow \pi^*$ transition.

The $n \rightarrow \pi^*$ transition is observed in molecules containing non-bonding electrons on a heteroatom involved in a saturated system like alcohols, amines, alkyl halides or ether oxides. This transition requires a lower energy than the transition or $\sigma \rightarrow \sigma^*$ but a higher than the $n \rightarrow \pi^*$ one. In many cases (alcohols, alkyl chlorides), the λ_{max} of the band is lower than 190 nm so it cannot be observed with a regular spectrophotometer. In a C–Y bond, the energy required for $n \rightarrow \sigma^*$

Table 2.50 Characteristics of some common unconjugated chromophore

Chromophore	Example	λ_{max} (nm)	ε_{max}	Transition	Solvent
>C=C<	Ethylene	171	15,530	$\pi \rightarrow \pi^*$	Vapor
—C≡C—	Acetylene	150	~10,000	$\pi \rightarrow \pi^*$	Hexane
		173	6000	$\pi \rightarrow \pi^*$	Vapor
>C=O	Acetaldehyde	160	20,000	$n \rightarrow \sigma^*$	Vapor
		180	10,000	$\pi \rightarrow \pi^*$	Vapor
		290	17	$n \rightarrow \sigma^*$	Hexane
	Acetone	166	16,000	$n \rightarrow \sigma^*$	Vapor
		188	900	$\pi \rightarrow \pi^*$	Hexane
		279	15	$n \rightarrow \sigma^*$	Hexane
–COOH	Acetic acid	204	60	$n \rightarrow \sigma^*$	Water
$-CONH_2$	Acetamide	178	9500	$\pi \rightarrow \pi^*$	Hexane
		220	63	$n \rightarrow \sigma^*$	Water
–COOR	Ethyl acetate	211	57	$n \rightarrow \sigma^*$	Ethanol
$-NO_2$	Nitromethane	201	5000	$\pi \rightarrow \pi^*$	Methanol
		274	17	$n \rightarrow \sigma^*$	Methanol
>C=N—	Acetoxime	190	5000	$n \rightarrow \sigma^*$	Water
–C≡N	Acetonitrile	167	Week	$\pi \rightarrow \pi^*$	Vapor
–N=N–	Azomethane	338	4	$n \rightarrow \sigma^*$	Ethanol

Reproduced with permission from Springer Nature [180]

transition increases with the electronegativity of the atom Y. When the electronegativity increases, the excitation of non-bonding n electrons becomes more difficult and the probability of transition decreases. The molar absorptivity ε increases with the probability of the transition. Thus, amines show both higher λ_{max} and ε_{max} than alcohols, according the lower electronegativity of nitrogen than oxygen, e.g. trimethylamine λ_{max} 227 nm, ε_{max} 900 and methyl alcohol λ_{max} 183 nm, ε_{max} 150 [179]. Thus alkanes, methanol and ethanol are good solvents for recording a UV-Visible spectrum because they do not absorb under 200 nm.

Table 2.50 shows few examples of λ_{max} and ε_{max} for some electronic transitions.

Effect of Solvents on a UV-Vis Spectrum

The UV-Vis spectrum of a given compound is measured in very dilute solutions. The wavelength of maximum absorbance must be lower than two absorbance units. Above this value, the Beer-Lambert law cannot be applied because the evolution of the absorbance is not further linear with the concentration. The most important criterion in the choice of solvent is that it must be transparent within the wavelength

Table 2.51 Maximum of lambda (λ_{max}) of different solvents

Solvent	λ_{max} (nm)
Acetonitrile	190
Cyclohexane	212
Diethyl ether	205
Dimethylformamide	271
Dimethyl sulfoxide	270
Dioxan	216
Ethanol	207
Hexane	199
Heptane	200
Methanol	210
Propan-2-ol	209
Propanone (Acetone)	331
Tetrahydrofuran	238
Toluene	335
Trichloromethane (Chloroform)	247

range being examined. The most common solvents used are water, ethanol and hexane. Table 2.51 gives examples of solvents used in UV-Vis spectroscopy with their lower wavelength cut off limits.

Polar solvents can affect the position (λ_{max}), the intensity and the form of absorption bands.

- A shift of an absorption maximum towards longer wavelength or lower energy is called a bathochromic shift.
- A shift of an absorption maximum towards shorter wavelength or higher energy is called a hypsochromic shift.

So it is important to give the nature of the solvent when expressing the λ_{max} and ε_{max} of a chromophore.

The substituents that themselves do not absorb UV-Vis radiations but induce shifts on the absorption maximum of transitions bands are called auxochromes. The substituents like hydroxyl, alkoxy, halogen, amino groups are some examples of auxochromes.

Conjugated Systems

The *Woodward-Fieser* rules are empirical rules to predict the λ_{max} of conjugated dienes and polyenes. The method of calculation is based on the λ_{max} of a parent chromophore: an acyclic, homoannular or heteroannular diene, then modified by the presence of different groups (alkyls, alkoxy, etc.) or double bonds (Table 2.52).

Table 2.52 Woodward Fieser rules (values from [180])

Base values for acyclic or heteroannular diene	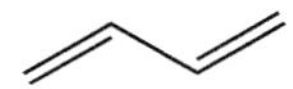 s-trans 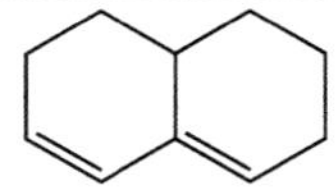 heteroannular diene (transoid) 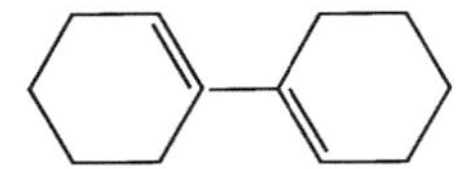	214 nm
Base values for homonuclear diene	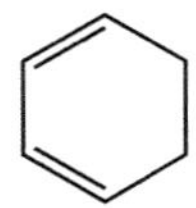 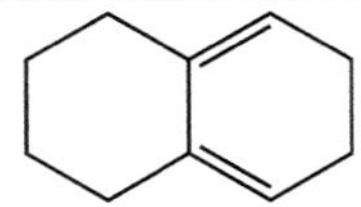 homoannular diene (cisoid)	253 nm
Increments:		
An additional conjugated double bond		+30 nm
An exocyclic double bond		+5 nm
An alkyl group or ring residue		+5 nm
An halogen: Cl,Br		+5 nm
–OR		+6 nm
–SR		+30 nm
–NR2		+60 nm
–Ph		+60 nm

In a homoannular diene, the same ring contains conjugated double bonds and cisoid means s-cis configuration (s for a single bond between the two double bonds).

In a heteroannular diene, the double bonds are not contained in the same ring and they have a transoid configuration.

Some examples of experimental λ_{max} compared to λ_{max} calculated with Woodward-Fieser rules can be found in [180].

Measurement

To measure the absorption of a solution, a cell is carried out. Different optical lengths (0.001–10 cm) are commercially available (Fig. 2.95). The most common cell has an optical length of 1 cm for routine measurements. Two sides crossed by the electromagnetic radiation should not be touched by hands to not interfere with the sample absorption. The cells are made of different materials. The most common are plastic, optical glass and quartz.

The transparency domain for the glass cell is roughly the visible region [174] and it is almost the same transparency domain for plastic cells. But glass and plastic cannot be used at wavelengths under 280 and 320 nm respectively, because their absorbance becomes too high. The great interest of plastic cells is that they are very cheap, but the optical quality is lower than for optical glass. The quartz cell presents wavelength range larger than plastic and glass ones: 190–2500 nm, but they are very expensive. Below 190 nm, quartz cell absorbs the light and must be replaced by synthetic silica, sapphire or magnesium fluoride cells which allow expanding the range to 180, 170 and 120 nm respectively [174]. For accurate measurements, a pair of cells can be used with the same material and the same optical length, they are called matching pair. The pair of matching cells can be used for a dual beam spectrophotometer [174].

In practice, if the sample absorbs in the UV range, the spectrum must be recorded with a quartz cell. The glass and plastic cells can only be used with a colored sample, on a wavelength range corresponding to the visible region.

Manufacturers can provide accessories for thermostat and stir the cells.

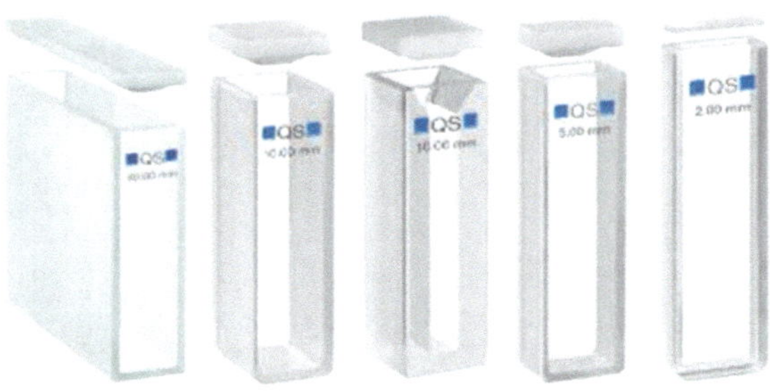

Fig. 2.95 UV-Vis cells of different optical lengths. Reproduced with permission from Hellma, VWR Company

Quantitative Analysis

UV-Vis spectroscopy is usually carried out to make quantitative analysis. The first step consists in recording the spectrum of the sample of interest in solution, the variation of the absorbance against the wavelength at a fixed concentration of the sample. The interest is to determine the λ_{max} of the molecule depending on the solvent and in some cases on the pH. Measuring the absorbance at λ_{max} leads to a better accuracy of the measurement. The second step consists in the preparation of a set of standards containing a known amount of the sample and to determinate the spectra. The variation of the absorbance at λ_{max} is plotted against the concentration of the sample for the different standards. According to the Beer-Lambert law, a linear relationship between the absorbance and the concentration is obtained. This step is called calibration and is essential to make quantitative analysis. The Beer-Lambert straight line or calibration straight line allows determining the concentration of a sample from its absorbance at λ_{max}.

The Beer-Lambert law is non-linear at higher concentration (Fig. 2.96). The most important reason is electrostatic interactions between molecules in close proximity.

For standards and the sample, the measured absorbance should not exceed 2 absorbance units (AU). So the solution has to be enough diluted. Typical concentrations of solutions can vary between 10^{-2} and 10^{-5} mol L^{-1} for ε varying from 100 to 100,000 L mol^{-1} cm^{-1} for an absorbance around 1 AU according to the Beer-Lambert law.

UV-Vis analysis is also useful to follow kinetic of a reaction by measuring the variation of absorbance over time at a fixed wavelength due to the linear relationship between absorbance and concentration.

Apparatus

Typically, a common spectrophotometer is composed of four main parts, a radiation source generating a light, a monochromator yielding the desired range of

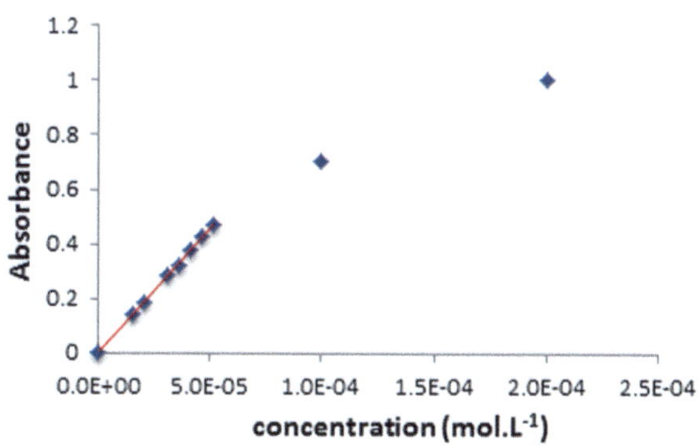

Fig. 2.96 Non linearity of Beer-Lambert law at higher concentrations

wavelengths, a sample container, and a detector [175]. It is possible to have two or even three monochromators to improve resolution and accuracy of the analysis. The sample compartment can be placed in front of or behind the monochromator (Fig. 2.97).

In the diode array based instrument with a dispersion of wavelengths post-sample, the full intensity of the light is sent on the sample. To prevent the photo-degradation of sample, the irradiation should be rapid (approximately few seconds). If the monochromator is in front of the sample, the sample is not subject to photodecomposition, the light going through the sample is weaker after dispersion. The device can be a single or a dual beam. In a single beam device, it is necessary to acquire a background with the cell containing the solvent before acquiring the sample spectrum. In a dual beam configuration, the light is split into two beams, one passing through the sample and the other through the reference, the cell containing only the solvent (Fig. 2.97c). The absorption of the cell and the solvent is subtracted to the one of the sample.

Light Sources

In the UV range between 180 and 350 nm, a deuterium lamp is frequently employed. It can also be used in far UV range (<180 nm), but another problems arise like the absorption of oxygen below 190 nm and the absorption of common quartz cells. The deuterium lamp is associated with a tungsten filament lamp for the visible and near infrared range between 330 and 900 nm. The red limit depends on the material of the photo-cathode, 840 nm for the most common one and 1000 nm for few other ones. High intensity xenon arcs allow covering a large wavelength region from 175 up to 1000 nm [175].

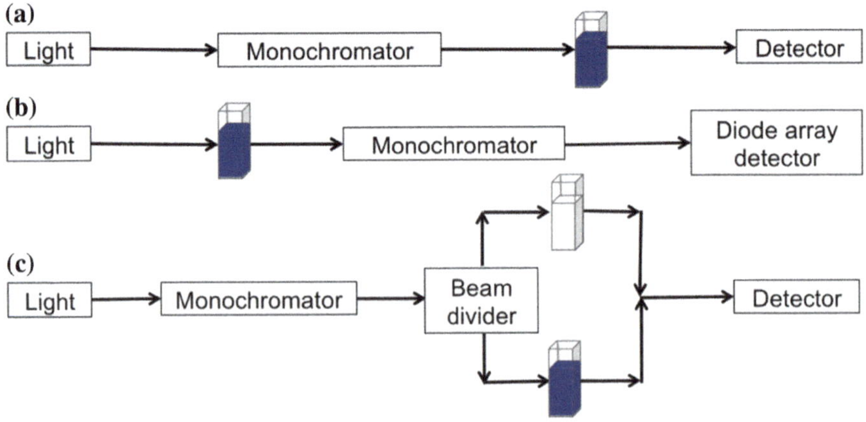

Fig. 2.97 Three spectrophotometer configurations. a Single-beam, b single-beam reverse optics and c dual beam

Monochromator

Diffracting gratings or prisms are employed as monochromators. Their rotation makes varying the wavelength of light coming out from exit slit. The size of the slits is an important variable, and with the dispersion of the monochromator's optics, they determine the spectral band-width (SBW). This last one determines the resolution and the measured molar absorptivity ε of the band. Smaller the SBW is, higher the resolution is.

Sample Container

A single beam device has only one sample container. In this case, a cell containing the solvent (reference) is firstly measured, followed by the measurement of the cell containing the sample. On the other hand, a dual beam has two sample containers. In this case, both the reference and sample can simultaneously be measured.

Detectors

The detector converts a light intensity to an electrical signal, energy of incident photons into photoelectrons in order to generate a spectrum. Photomultiplier tubes (PMT) and silicon photodiodes are common detectors used in a UV-Visible spectrophotometer. The photomultiplier tube is composed of a photocathode and a series of dynodes in an evacuated glass enclosure. Its principe is based on photoelectrons discharged from a photoelectric surface when it's struck by photons of radiation. The photodiode array is a multichannel photon detector which collects the entire spectrum simultaneously thanks to many hundreds of diodes.

In a conventional scanning UV-Vis spectrophotometer with a PMT, a single wavelength is recorded at any given point of time whereas the diode array detector can measure the absorption for all dispersed radiations simultaneously. With the diode array detector, the spectrum can be recorded in around one second while scanning UV-VIS instrument takes between 20 and 60 s to perform the same scan.

Example of UV-Vis Spectroscopy Application

UV-Vis spectroscopy is a convenient method to quantify a sample or to follow a reaction because it is a quick, easy and cheap technique.

Lee et al. [181] applied UV-Vis spectroscopy to quantify solubilized lignin generated from steam treatment of different biomass: willow, softwood, aspen, hemp, and canary grass. The spectra, between 200 and 800 nm, were acquired on a Cary 100 bio UV-Vis spectrometer, in quartz cells. All standards were dissolved in 0.035 mol L^{-1} NaOH. Lignin is a photosensitive compound so vial containing the sample was coated with aluminium until the analysis. The UV-Vis measurements on standards and samples were realized at 212, 225, 237, 27, 280 and 287 nm. Reference solution was consistent with NaOH solutions used for the sample. Lignin presents two characteristic bands at 200–230 and 260–280 nm. But acidic degradation products from the carbohydrate fraction of biomass with absorbance maximums at 205 and 280 nm, overlap with lignin absorbances. A limitation of UV-Vis spectroscopy is the study of mixtures when the spectra of different products overlap. Lee et al. showed that with a concentration of hemicellulose below 75%, acidic degradation products do not affect significantly the UV-Vis response of the lignin, the lowest standard deviation in the results with increasing hemicellulose concentration until 75% was obtained at 270 nm (Fig. 2.98). This technique was applied by authors to quantify lignin in broths produced from one-step steam process and correlate successfully with gravimetric method.

UV-Vis spectroscopy was also carried out by Giannakopoulos et al. to study oxidation of municipal wastewater (WW) by complexation with natural polyphenols like gallic acid in alkaline pH, under ambient oxygen and temperature [182]. Gallic acid (GA) is the most efficient antioxidant in plants and possesses a high radical scavenging activity. The measurements, between 200 and 1000 nm were made with a Hitachi U-190 single beam UV/Vis instrument with 800 nm/min wavelength scan speed. Before UV-Vis analysis, the samples were filtered using prewashed 0.45 μm fiber filter. Because of the pH influence on the studied organic acids, the measurements were realized at the same pH. Figure 2.99 shows the UV-Vis spectra of GA before and after oxidation by atmospheric oxygen. Three bands at 220, 280 and 332 nm were observed in GA and attributed to the π system of the benzene ring. After oxidation, these absorptions were decreased, shifted and broadened to horizontal plateaux among to 260–290 nm and 330–430 nm associated by authors with the n-π^* transition due to the C–C and C–O–C coupling,

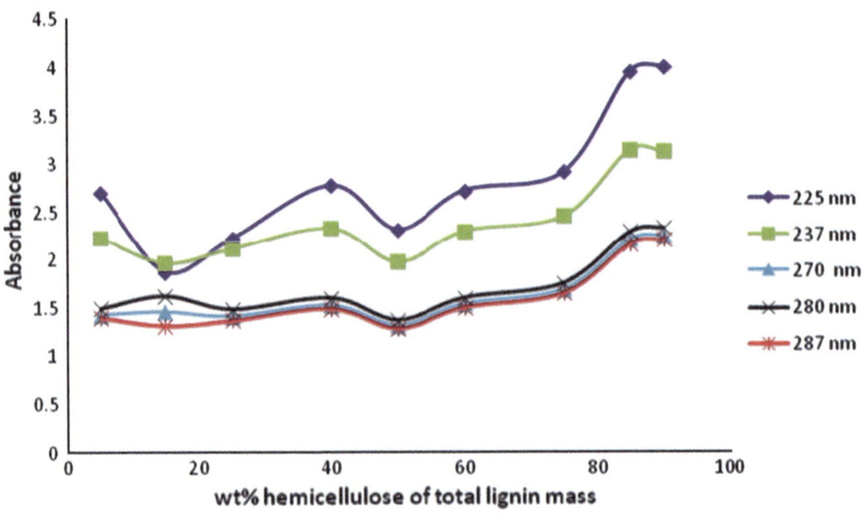

Fig. 2.98 Various wavelengths and the effect of hemicellulose concentration on the absorbance of a 0.05 g/L reed canary grass lignin (pure). Reproduced with permission from Elsevier [181]

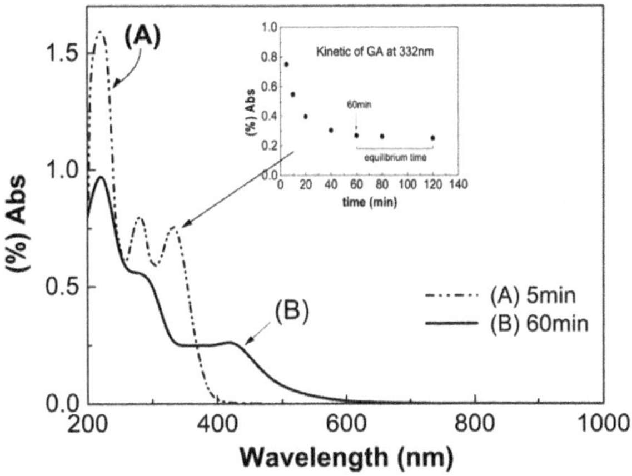

Fig. 2.99 UV/Vis spectra of GA at pH > 7 in the presence of O_2 at (A) 5 min for short incubation times (initial time) and (B) 60 min for long incubation times (equilibrium time). Reproduced with permission from Elsevier [182]

reflecting a polymerization reaction with increased conjugation length and a reduction in energy required for the electron transitions.

In order to observe the oxidation reaction of GA in presence of WW, an amount of diluted GA was added to WW solution. Authors also studied the UV-Vis spectra of WW compounds, GA-GA polymer and GA-WW complex at pH > 7 (Fig. 2.100). The spectra of GA-GA and GA-WW complexes were similar, suggesting the presence of similar chromophores and/or similar structural analogies. The authors suggested that at pH > 7, free phenolic radicals of GA interact with amino groups of proteins to form polymer products (GA-WW).

UV-Vis spectrophotometry was also used by Sirajuddin et al. [183] to determine ultra-trace of hydroquinone (HQ) in waste photographic solutions using $KMnO_4$ as oxidizing agent as well as signal enhancer. The analysis was realized with a Lambda 2 UV-Vis spectrometer of Perkin-Elmer company and thermostated cells.

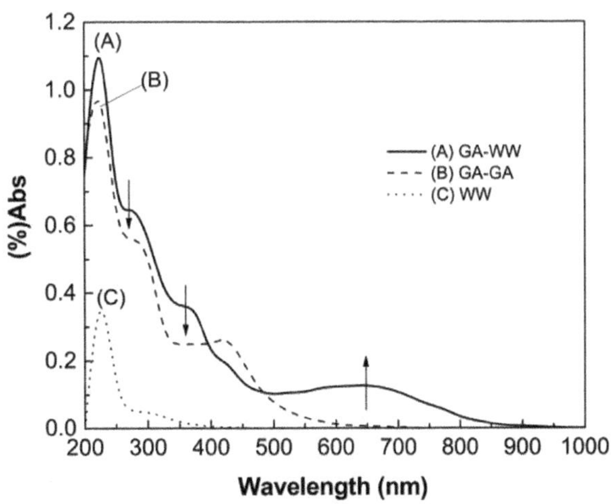

Fig. 2.100 UV/Vis spectra of (A) GA polymerization in the presence of WW, (B) GA polymerization in the absence of WW and (C) WW, at pH > 7. The arrows show the evolution of the bands. Reproduced with permission from Elsevier [182]

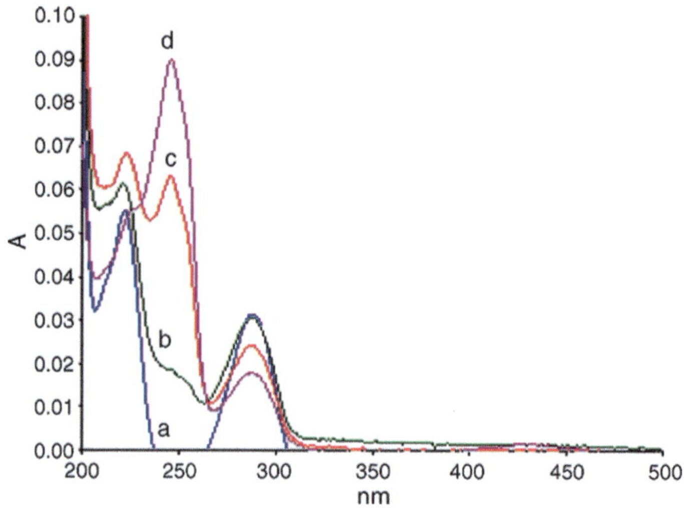

Fig. 2.101 UV spectra showing formation of: (a) HQ (288 nm) and SQ (222 nm) in water without $KMnO_4$ and (b), (c) and (d) BQ (245.5 nm) in presence of 0.1, 0.5 and 1.0 μM $KMnO_4$, respectively. Reproduced with permission from Elsevier [183]

Figure 2.101 describes the conversion of 1 mg ml^{-1} aqueous HQ (λ_{max}, 288 nm) into semiquinone (SQ) (λ_{max}, 222 nm) and benzoquinone (BQ) (λ_{max}, 245.5 nm) in the absence and presence of $KMnO_4$ solution. After dissolution of HQ in water, the SQ appears due to photo-oxidation of HQ. With 1 mM $KMnO_4$ solution, it leads to catalytic oxidation of HQ, and HQ and SQ are nearly completely transformed into BQ.

2.8.1.4 Raman Spectroscopy

Introduction

Raman spectroscopy is a non-destructive technique for observation and characterization of the molecular composition and external structure of a material. It allows to study vibration modes of chemical bonds of molecules in gaseous, liquid (pure or in solvent) or solid (crystalline or amorphous) phases. The sample is subjected to an incident beam (UV-visible light) and, after interacting with it, the excited molecular bonds emit a photon with a specific energy and frequency dependent on their nature. A Raman spectrum is then a graph where the intensity of the vibration mode is plotted as function of its frequency. This resulting graph is a reconstructed chemical image of the sample [184]. The Raman spectrum is acquired using a confocal microscope coupled with the Raman spectroscopy to record a chemical composition image of the sample with a high spatial resolution. This technique provides topological or map analyses with an excellent quality while Infra-Red

(IR), which is a complementary analysis tool, addresses chemical information from the bulk [185]. Raman imaging has gained momentum in the analysis of solid residues. Raman imaging enabled the assessment of carbon molecule, the study of the distribution of different components in solid residues and their quantitative distribution analysis at the vicinity of a surface. The interest for this technique can be explained by the simultaneous access to a visual representation of the sample chemistry, using spatial information, and a quantification using spectral information [186]. This technique has also the advantage of being a non-invasive and solvent free technique. Nevertheless, Raman imaging presents some limitations, for example the fluorescence which can be a limit for some applications [184–186]. With this technique, only a light sample preparation is required for solid residues (please see more details in Chap. 15).

Theory

Prior to explain the confocal Raman spectroscopy theory, a short introduction about the confocal Microscopy is presented [186]. Raman spectroscopy experienced a renewed interest when the optical microscope was introduced since the use of the latter allows focusing the energy of the laser on a small surface of the sample. Thus, micro-Raman equipment improves the analysis resolution and the dissemination of the technique. Basically, an optical microscope collects the light emitted by all illuminated points and the observed image contains high definition points (from the focus plane) and blurred regions (out of the focus). This technique has a low resolution and does not give 3D information. On the contrary, the confocal microscope provides images with a short depth of field, and only the emitted rays from the focus plane are collected. A lens or an objective is used to focus a point-like light source onto a sample. The focused spot has a spatial extension (radius of the spot), which is determined by the wavelength. The emitted rays are focused through one or two lens(es) (two lenses in Fig. 2.102) which are intended for defocusing and refocusing. Finally, only the fraction of the emitted rays from the focus plane will succeed to go through the pinhole in front of a detector. The size of the pinhole is very important since the detector should only collect the central part of the focus that has passed through it. The detectors used are photo-multiplicators tubes (PMT) or charge-coupled device (CCD) camera with a high sensitivity. Depending on the beam position on the sample, the light intensity is measured and digitized to obtained numerical images.

As seen in Fig. 2.102, only the rays coming from the focus plane (and thus from the sample) will succeed to pass through the pinhole and will be recorded. This characteristic is of great importance since the reconstructed image is obtained with a good contrast. In addition, when moving the focal plane in the depth of the sample, 3D images can be obtained. When the incident beam reaches the sample, different phenomena appear including elastic and inelastic scattering (see Fig. 2.103). Raman spectroscopy is based on inelastic scattering. Each material has a specific

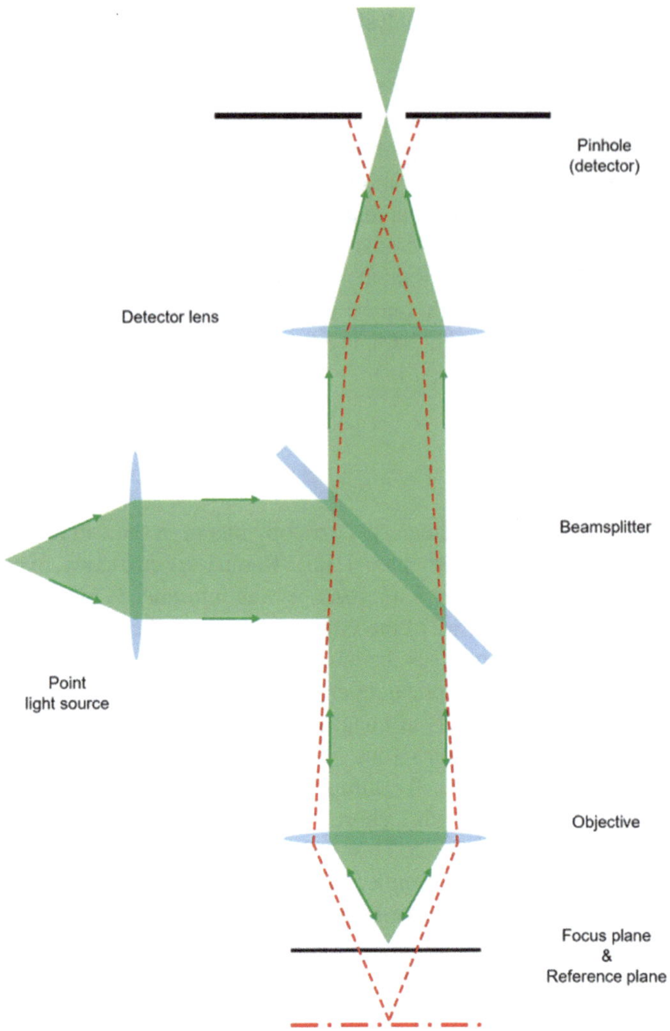

Fig. 2.102 Scheme of a confocal microscope. Reproduced with permission from Witec [187]

and unique Raman spectral footprint, which is characterized by the position and intensity of Raman bands. The spectrum includes data related to bond vibration and thus it is characteristic of the chemical composition and structure of the material analyzed. The "Raman effect" is the phenomenon that consists in having a frequency-shifted spectrum in the light scattering when the sample is submitted to a monochromatic light. When a material is illuminated with a monochromatic light (with an electromagnetic excitatory radiation at a certain frequency), the following phenomena happen:

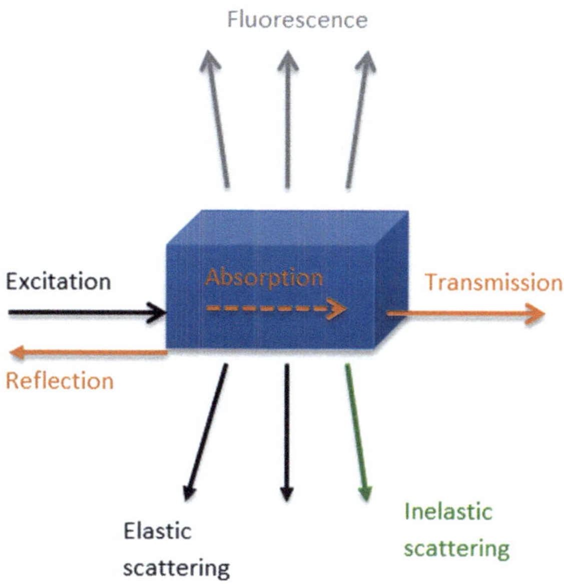

Fig. 2.103 Interaction of visible light with matter

- a part of the photons from this radiation is reflected or absorbed;
- a small part is scattered in all directions. Elastic and inelastic scatterings appear. For less than 1/1000 of these photons, a frequency shift is observed which corresponds to the Raman effect (inelastic scattering).

Both elastic and inelastic scattering phenomena are presented in an energy level diagram (Fig. 2.104). Once the incident beam interacts with the matter, the electrons absorb the energy to move towards an excited state (virtual state). When the electrons go back to the fundamental state, the energy of the emitted photon is similar to the energy absorbed. This transition is thus elastic and so-called the Rayleigh scattering. When the energy of the emitted photon is different from the energy absorbed, the phenomenon is inelastic and is called the "Raman effect". When the energy of the photon is lower than the incident beam, there is a Stokes Raman effect. On the contrary, a higher energy corresponds to the anti-Stokes Raman effect. It should be noted that fluorescence (Fig. 2.103) may hide the Raman effect. Indeed, fluorescence corresponds to the photon absorption and its spontaneous emission with a fluorescent light. When the fluorescence and Raman effect occur both at the same electronic states, the fluorescence will hide the Raman band. On the contrary, when the fluorescence occurs from excited states of lower energy, both signals are easily separated.

The anti-Stokes Raman peaks have negative wavenumber while Stokes Raman peaks have positive one (Fig. 2.105). Generally, the Stokes peaks are only used. As explained previously, the Raman spectrum is plotted as a function of the Raman shift which corresponds to the shift of frequency from the Rayleigh scattering.

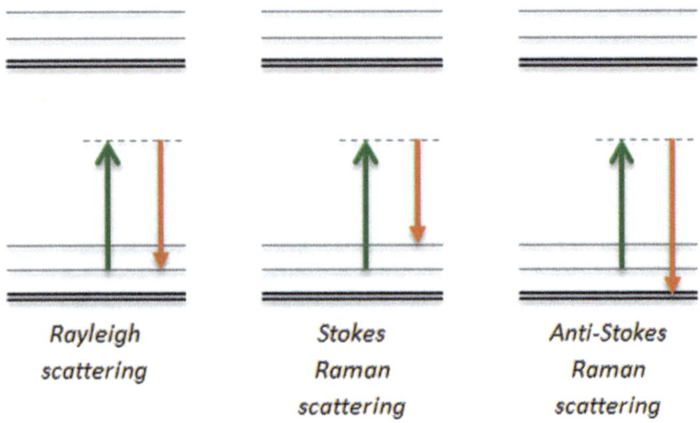

Fig. 2.104 Energy level diagram

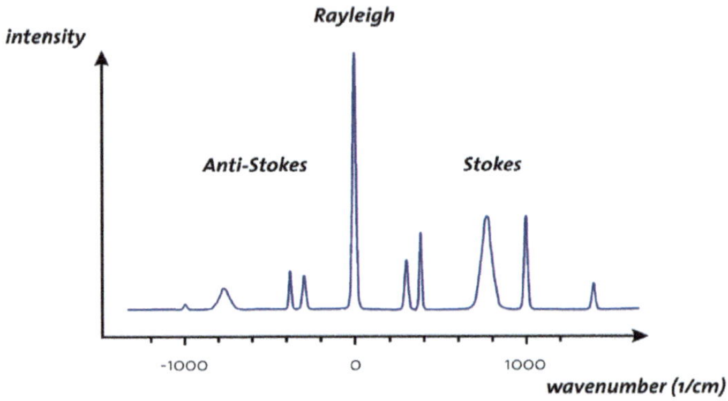

Fig. 2.105 Representation of Rayleigh and Raman peaks. Reproduction with permission from Witec [187]

From the Raman spectra, lots of information about the chemistry of the sample can be obtained. Each peak is characterized by:

- Its spectral position (which can be related to the frequency of the vibrational mode);
- Its intensity (which is related to the number of diffusing molecules and the vibrational mode);
- Its polarization state which give information about the symmetrical mode;
- Its profile (study of the movements or interactions in condensed phase, or in gas phase depending on the temperature considered).

The Raman spectrum is characteristic of the sample as the wavelength offset depends on the bonds in the molecule and is unique (strongly related to the crystal or molecule vibration). So, the offset analysis allows characterizing chemical composition and crystalline structure of the sample.

Raman Analysis Method

To analyze the sample with Raman Confocal Microscopy, the method consists in focusing, through a lens, the monochromatic light beam on the surface of the sample and then to analyze the scattered light. This light goes through another lens and is collected in a monochromator and its intensity measured with a detector. To have a proper focus on the sample and to be able to collect qualitative images, an appropriate objective and pinhole size should be used. The choice will depend on the sample and the kind of observation expected. The choice of the laser beam is also very important. Depending on the sample and the laser beam wavelength, the fluorescence emission can overlap and hide the Raman spectrum. Table 2.53 gives useful information in order to calculate the resolution of the microscope (the smallest distance it can separate, in lateral and in axial position) as a function of the wavelength and the lens aperture (based on Eqs. 2.88 and 2.89) [186]. Finally, Table 2.53 shows that the best resolution (lowest distances) is obtained with the lowest wavelength and the highest diameter of the aperture of the objective.

$$\Delta X_{lateral} \approx 0.61 \frac{\lambda}{NA} \tag{2.88}$$

$$\Delta X_{axial} \approx 1.4 \cdot n^2 \frac{\lambda}{NA^2} \tag{2.89}$$

with λ the wavelength of the light beam (in nm), NA the numerical aperture of the objective (dimensionless) and n the refractive index of the sample.

After choosing the appropriate laser beam and lens, and prior to start the analyses, the light and the laser beam characterization should have been achieved. In order to have a good laser beam alignment, a reference sample should be used (silicon sample for example) and the light should be in the center of the area as

Table 2.53 Information on resolution calculation

Wavelength (nm)	NA	$\Delta X_{lateral}$ (nm)	ΔX_{axial} (μm)
532	0.90	360	0.92
532	0.75	433	1.3
532	0.40	811	4.7
633	0.90	429	1.1
633	0.75	515	1.6
633	0.40	965	5.5
785	0.90	532	1.4
785	0.75	639	1.9
785	0.40	1197	6.9

illustrated in Fig. 2.106. The procedure for alignment may change depending on the Raman confocal-microscope used. The main idea is to check the intensity of the first vibrational peak of the reference sample (silicon) using the oscilloscope. For this purpose, the highest intensity of the peak should be reached by adjusting first the Z lens position and then the X and Y lens positions.

To carry out an analysis, the optical focus, which will involve the laser focus, should be performed. The analysis should be started with a "large" lens (X10 for example) to focus on the sample, or on the area expected using upright (Z) and horizontal movements (X and Y). To have a better focus, the lens should be changed for an accurate one (50X or 100X) and focus should be refined (Fig. 2.107).

Once the focus is properly made, the type of analysis should be selected: single point analysis, line scan, surface area scan, depth scan or stack mode. For all these methods, different parameters should be selected: the point, line or area of analysis, the integration time (recording time on each point or line to collect data) and the number of accumulations on each point. For line, area or depth scan, the number of analysis points per line and the number of lines in the area should be selected. This will have an impact on the area covered and the quality of the image. At the end, all these collected spectra will be reduced to a unique spectrum. Table 2.54 summarizes the different kinds of available analyses.

From the raw spectra, basic and advanced post-treatments are available. With both area modes (surface and depth) scan, the chemical distribution on horizontal or vertical surface is obtained, resulting in a 2D chemical distribution after advanced post-treatment. With a 3D software, a 3D image of the sample can be reconstructed from stack mode data (do refer to the two examples displayed in Fig. 2.108). All these advanced post-analyses will be detailed in Chap. 15. In the current chapter, only "basic" post-analyses will be described.

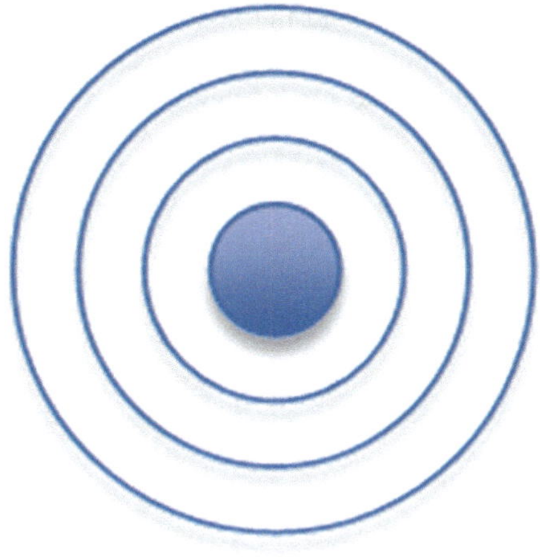

Fig. 2.106 Good laser beam alignment

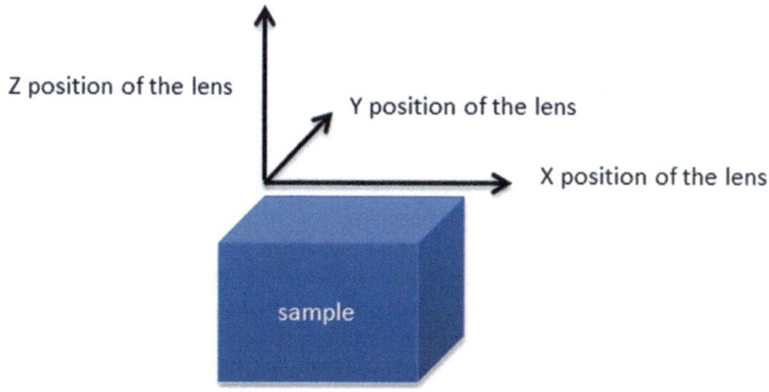

Fig. 2.107 Positions of the lens

Raw Raman spectra should be post-analyzed to obtain accurate and useful information. The first step is the cosmic ray removal to suppress the noise. This is often an available option in the software of the Raman equipment. Cosmic rays are particle from outer space which can interact with atoms or molecules from Earth. When these particles hit a CCD detector, a false signal is generated in the spectrum (very sharp and long peak) that is not a Raman peak. To remove the spectral cosmic rays, the spectrum in the spectral domain must be filtered. During the filtration, each pixel is compared to its neighboring pixels and if the difference exceeds a limit, then it is considered as a cosmic ray and the corresponding peak is subtracted. Then, the background noise is removed prior to apply a baseline correction (level to zero). This is also often an available option in the Raman software. Each spectrum in the spectral images is corrected from the baseline using an order polynomial fit (order chosen accordingly) in order to remove the background. For example, Fig. 2.109 presents a raw spectrum recorded on an oxygenated pyrolysis biochar obtained from a biomass mixture (wood waste, flocculation sludge and food waste, namely BPUBFDA) and its corresponding post-analyzed spectrum. For this analysis, Raman spectra of the model substance were collected using a Witec Alpha300 micro-spectroscope (Witec GmbH, Ulm, Germany) equipped with a 532 nm green laser, and Witec Project® software for post-analysis. The studied spectral range was 150–3800 cm^{-1}. In Chap. 15, the method implemented to assign peaks to a specific chemical structure will be detailed. Raman spectral images were collected through a 50X air objective with a 0.1 s spectral collection time for each spectrum and 167 nm steps between each spectral collection point (10 accumulations/point). The spectral image was 8 × 7 μm in size.

The principle of the procedure, explained above, is summarized in Fig. 2.110.

The Raman analysis and "basic" post-analysis can usually be carried out with a standard Raman microscope. Some options can be added to make easier the advanced post-analysis (like 3D imaging and volume visualizations, cluster analysis and graph demixer). These options will be explained in more details in Chap. 15 but a brief explanation is given in this chapter. With cluster analysis option, similar

Table 2.54 Summary of the analyses available for the Confocal-Raman equipment

Type of analysis	Single point	Line scan	Surface area scan	Depth area scan	Stack mode
Scheme	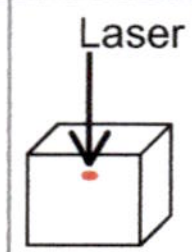	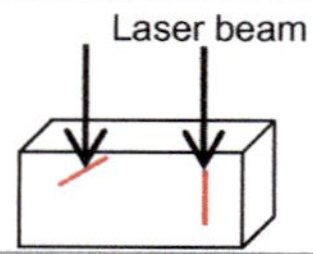		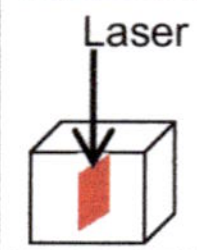	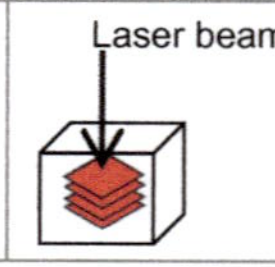
Parameters					
Position (X, Y, Z)	1 position	Initial and final	Initial and final	Initial and final	Initial and final
Accumulation number	X	X	X	X	X
Integration time	X	X	X	X	X
Number of analyses per line		X	X	X	X
Number of lines per image			X	X	X
Spectra	1 spectrum	1 spectrum per analysis point 1 average spectrum	1 spectrum per analysis point 1 average spectrum	1 spectrum per analysis point 1 average spectrum	1 spectrum per analysis point 1 average spectrum
Comments	Suitable for homogeneous samples	Investigate chemical homogeneity/heterogeneity	Investigate chemical homogeneity/heterogeneity	Investigate chemical homogeneity/heterogeneity	Investigate chemical homogeneity/heterogeneity

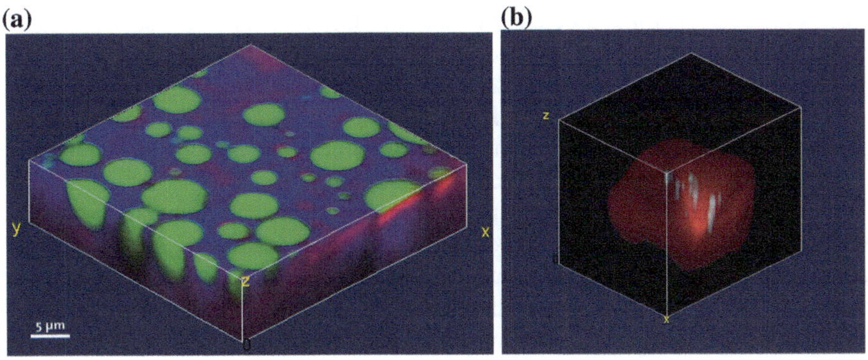

Fig. 2.108 Example of 3D images recorded with the Raman stack mode. a 3D image of an oil-alkane-water emulsion (green: oil, red: alkane, blue: water) and b 3D image of diamond (red) included in a quartz matrix. Reproduced with permission from Witec [188]

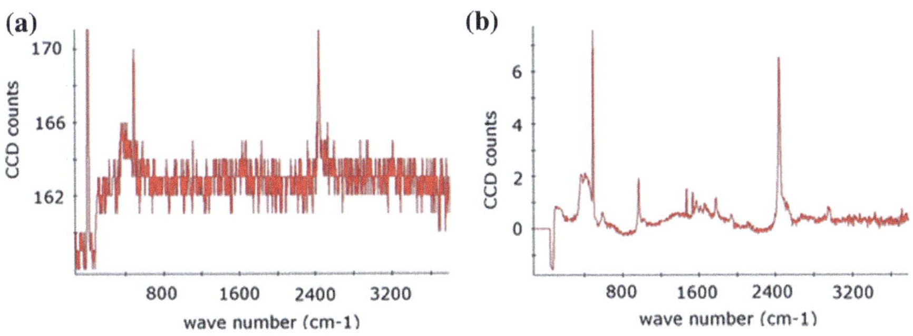

Fig. 2.109 Raw Raman spectra (a) and post-analyzed spectra (b) of a biochar sample (BPUBFDA)

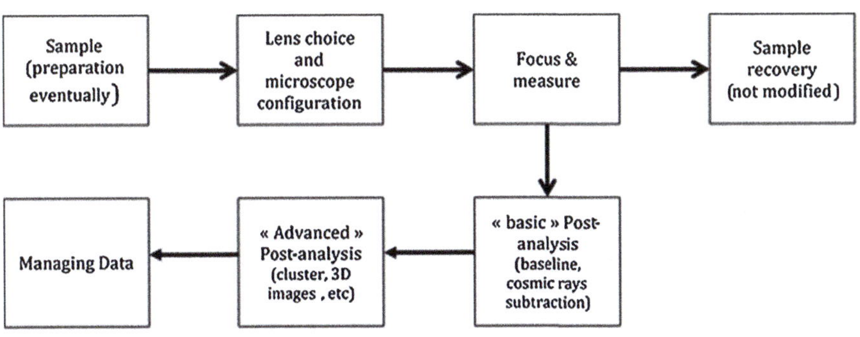

Fig. 2.110 Procedure to perform a Raman spectrum

spectra are classified into a defined number of clusters and a color-coded image is obtained on the basis of the selected clusters. With graph demixer option, spectral information can be subtracted or added from/to other spectra. Some supplementary units can also be added to the standard system like real surface. This is a patented option for Witec equipment which enables confocal Raman imaging guided by surface topography. In confocal microscopy, owing to the presence of the pinhole, the light from outside the focal plane is strongly reduced. With this option, whatever the sample's topography, its surface is kept in focus during the entire measurement procedure. Other useful options can be added to the standard system, like multi-laser coupler (when several lasers can be used to adapt the wavelength on sensitive sample) or stages for heating and cooling (better temperature control of the sample). Just for information, in this chapter, a Raman spectroscopy/confocal microscope from Witec GmbH has been mentioned but several suppliers also exist: Thermofisher Scientific, Horiba scientific, Renishaw and Bruker.

2.8.2 Zeta Potential Determination

2.8.2.1 Introduction

A dispersed system can be a colloidal suspension which corresponds to suspensions of small particles ranging between few nanometers and few micrometers as shown in Fig. 2.111 and can be classified in different ranges [189]. Three main categories of colloidal suspensions are defined as follows [190]:

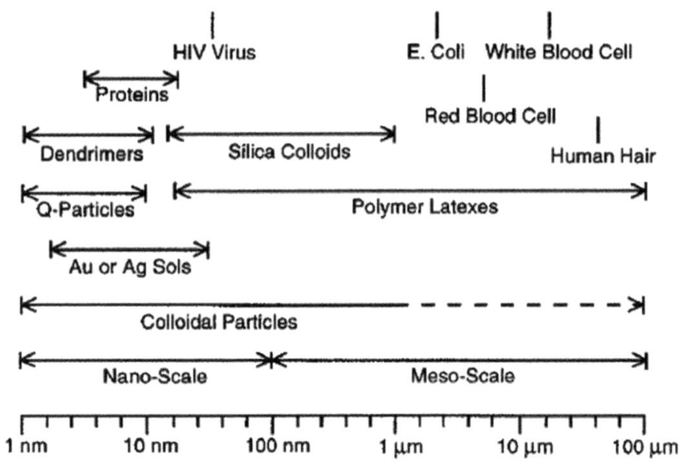

Fig. 2.111 Size ranging of colloidal suspensions. Reproduced with permission from John Wiley and Sons [189]

- Solid particles in liquid (suspension);
- Small liquid drop in another liquid (emulsion);
- Gas in a liquid (foam).

This section will be focused on the suspension of solid particles dispersed in a liquid. The stability of the suspension depends on the attractive and repulsive forces between particles. For example, attractive forces can generate flocculation which is a reversible phenomenon or coagulation which is an irreversible phenomenon. The basic theory used to explain the stability of a suspension is the DLVO (Derjaguin, Verwey, Landau and Overbeek) stability [191, 192]. DLVO theory considers that the total potential, V_T (2.90) between two particles is the sum of attractive potential (Van Der Waals), V_A (Eq. 2.91) and repulsive of the double layer V_R (Eq. 2.92):

$$V_T = V_A + V_R \tag{2.90}$$

where V_A corresponds to the Van Der Waals potential and V_R to repulsive potential

$$V_A = \frac{-A}{12 \times \pi \times D^2} \tag{2.91}$$

where A is the Hamaker constant which depends on dielectrical properties of particles in aqueous solution [193, 194], and D is the particles distance.

$$V_R = 2 \times \pi \times \varepsilon \times a \times \zeta^2 \times exp(-K \times D) \tag{2.92}$$

where: a is the particle radius, ε is the permittivity medium, K is a function of the ionic composition and ζ is the zeta potential.

When two particles with charged surface approach and their electrical double layers begin to overlap, a repulsive force is induced. DLVO theory suggests that electrical double layer (Fig. 2.112) repulsion will stabilize suspension when solid particle is in aqueous phase.

2.8.2.2 Principle of Measurement of the Zeta Potential

The zeta potential represents the charge that a particle acquires through the ions that surround it when in solution. So, the measurement of zeta potential allows explaining the dispersion, flocculation or aggregation phenomenon which can occur in suspension. The knowledge of zeta potential can be used to improve stability and the storage time of a formulation, to control flocculation and decantation process or to control feedstock quality. The zeta potential is related to the surface charge of solid particles which is a property that all solid materials possess when suspended in a liquid phase. This surface charge presented in the Fig. 2.112 is called the electronic double layer.

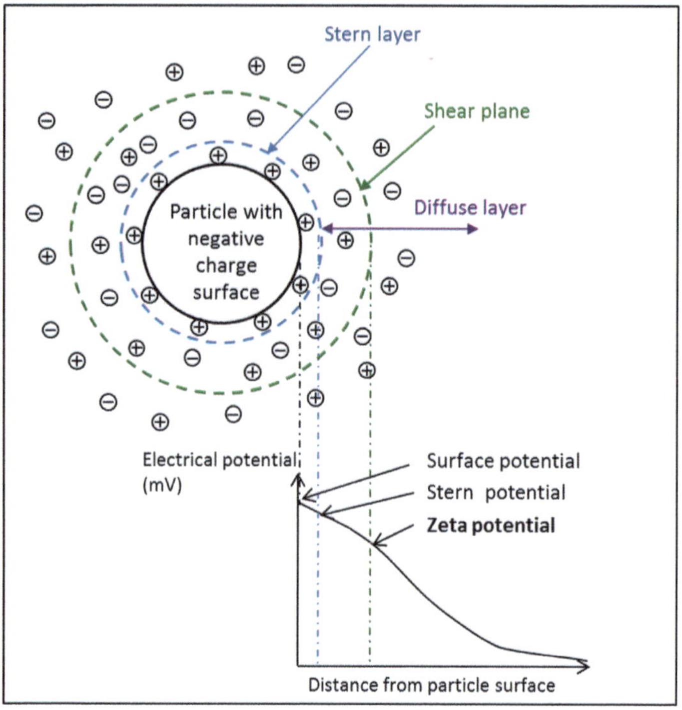

Fig. 2.112 Electronic double layer. Reproduced with permission from Malvern Panalytical

Electronic Double Layer

When a solid particle is in contact with a liquid, the solid particle acquires spontaneously an electrical surface charge called electric double layer as presented in the Fig. 2.112 [195]. The charging mechanisms can be assigned to electron affinity differences of two phases, ionization of surface groups, and differential ion adsorption from electrolyte solution and differential ion dissolution from solid particle.

The first model to describe the electronic double layer was developed by Helmholtz in 1879 which consider ions layer adsorbed on the surface [196]. This model was completed then by Perrin [197]. After Gouy and Chapman considered the double layer is not compact as in the theory of Helmholtz and introduce the diffuse layer theory [198, 199]. In 1924, Stern developed a model based in the two previous models in which he takes into account the specific size of adsorbed ions [200]. This model considers the Helmoltz's layer (compact layer) and the Gouy-Chapam's layer (diffuse layer) called electronic double layer of Gouy-Chapman-Stern. Other model appears later [201] but the Stern's model is the most used. Approximations considered in the Gouy-Chapman-Stern's model are: ions are considered as point charges, only Coulomb interactions are significant, the electrical permittivity is constant inside the double layer and the solvent is uniform at the atomic scale.

As shown above the theory of electric double layer is complex and was discussed by Parson [202]. The description of the electric double layer is presented in Fig. 2.112. The two layers considered are a layer closed to the surface charge (either positive or negative) comprises ions adsorbed directly onto the particle (Stern layer) that is considered motionless. The second layer is a diffuse layer where occurs diffusion of ions which are distributed according to the influence of electrical forces composed of free ions attracted to the surface charge via Coulomb force. These two layers are separated by the shear plane.

Zeta Potential Measurement

When a charged particle moves in a liquid under extern electric fields it is called electrophoresis and the Stern layer is moving in a part of the diffuse layer. Fluid between particle surface and the shear layer have a rate equal to zero. The plane out has a rate, so the speed of the particle is related to the electric potential of the shear plane and corresponds to the zeta potential as shown in Fig. 2.113. The zeta potential (ζ) corresponds to the measure of the electrical charge of particles suspended in a liquid and the measured is in mV.

Factors influencing zeta potential are the pH, the conductivity (salt concentration) and an additive (e.g. a polymer). The measurement of the zeta potential as a function of pH gives information on the stability of the suspension as shown in Fig. 2.113. If acid is added, positives charges increase and the suspension is stable. If the pH increases, the charges will be neutralized and we reach the isoelectrical point. It is the unstable zone of the suspension. The addition of basis allows increasing negative charges. The isoelectrical point is also called the zeta potential zero and corresponds to the pH at which the surface average charge of the solid

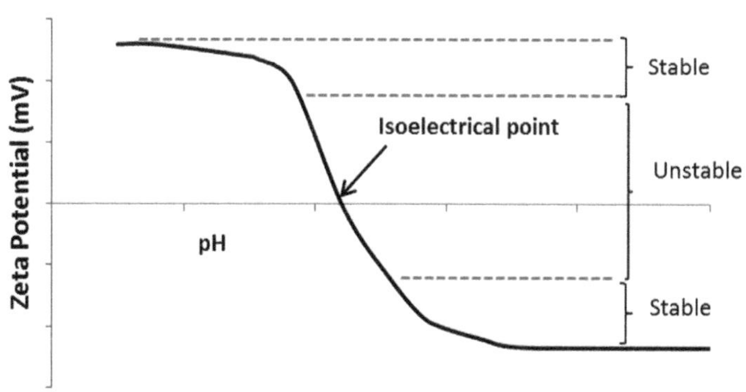

Fig. 2.113 Zeta potential as a function of pH

particle is neutral [203, 204]. At the isolelectrical point, the balance of sorbed and desorbed protons H^+ is zero.

Electrophoresis

Zeta potential allows acquiring information about this electrical charge by electrophoresis. Under the effect of an electric field between two electrodes (+) and (−), the particles will migrate in one direction or the other as a function of their electrical surface charge. The zeta potential value gives information about repulsion forces between particles and on the stability.

Zeta potential is measured by electrophoresis analysis and calculated from theoretical models based on the electrophoresis mobility. The measure is based on the rate at which a charged particle moves under an electric field. The particle migration rate is proportional to the zeta potential. The speed is measured with an anemometer Doppler laser and calculation is based on the Marian Smoluchowski theory [205]. Several books are available in which a description of the electrophoresis and theories used are well described [206, 207].

When an electric field is applied to a suspension, charged particles are attracted to the opposite charging electrode. The moving speed of particles is expressed by the following equation [208]:

$$\mu = V/E \tag{2.93}$$

where μ is the electrophoretic mobility (in: (μm • cm)/(V • s)), V is the speed (in: μm/s) and E is the electric field applied (in: V/cm).

As shown in Fig. 2.114, under the electrical filed, different forces exist and are expressed by the following equations:

- Electric force (solvent drive by counter ions):

$$\overrightarrow{F_E} = qE \tag{2.94}$$

where q is the charge of the particle.

- Hydrodynamic friction force (opposite direction of the particle movement) gives by the Stokes's law:

$$\overrightarrow{F_H} = 6\pi\eta a_0 U \tag{2.95}$$

where a_o is the radius of the particle, η is the fluid viscosity and U the particle speed.

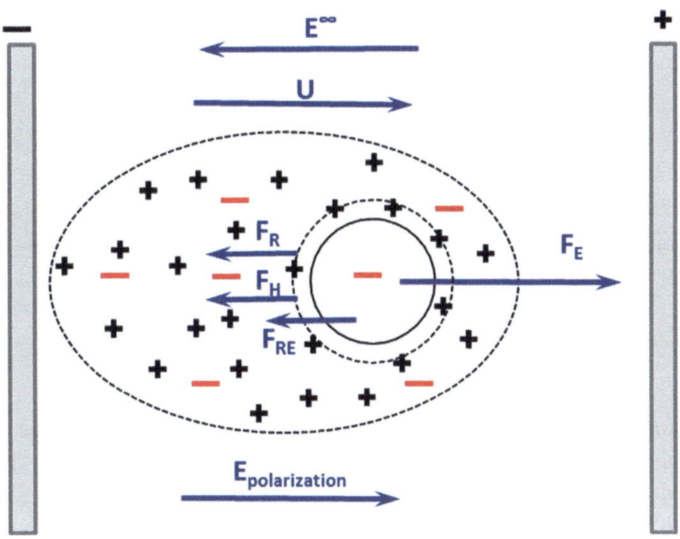

Fig. 2.114 Electric double layer deformation under electric field

From Eqs. (2.94) and (2.95), the relation (2.93) can be rewritten as following:

$$\mu = q/6\pi\eta a_0 \tag{2.96}$$

Under an electrical field, the counter-ions located in the diffuse layer which are subjected to electrostatic attractive force tend to move in the opposite direction to this one. The result is a distortion of the double layer as is presented in Fig. 2.114 where two forces appear, an electrostatic force $(\overrightarrow{F_{RE}})$ and a relaxation force due to the breaking of charges around the moving particle $(\overrightarrow{F_R})$ [209]. The shear plane position depends on the physico chemical properties of the medium [210, 211].

Different authors have developed models to connect the electrophoretic mobility of a particle and its surface potential [205, 212].

First, the Henry's model describes the electrophoresis of a spherical particle by the following equation [212]:

$$\mu = \frac{2\varepsilon\zeta f(\kappa a)}{3\eta} \tag{2.97}$$

where μ is the electrophoretic mobility, ε is the dielectric constant, ζ is the zeta potential, η is the viscosity, a the particle radius and $f(\kappa a)$ the Henry's function.

The Henry's function is a monotonic function of the product (κa) ranging between 1 and 1.5 for a variation of (κa) from 0 to infinity [210]. In literature tables are available and give values of $f(\kappa a)$ as a function of (κa) [209, 210, 212–214].

Some approximations was performed by Von Smoluchowski and Huckel as a function of the values of $f(\kappa a)$.

If $f(\kappa a)$ is less than 1, the double layer is large and the Henry's equation tends to 1. So, Eq. (2.98) is obtained as follows:

$$\mu = \frac{2\varepsilon\zeta}{3\eta} \tag{2.98}$$

If $f(\kappa a)$ is higher than 1, the double layer is small, the Von Smoluchowski's approximation is used and the Henry's equation tends to 1.5. So, the Von Smoluchowski's equation is as follows:

$$\mu = \frac{\varepsilon\zeta}{\eta} \tag{2.99}$$

If the solid particle is dispersed in water at 25 °C, Eq. (2.99) can be simplified with ε equal to 79 and η equal to 0.8872. So, the zeta potential can be expressed by the following equation:

$$\zeta = 11.23\mu \tag{2.100}$$

2.8.2.3 Application

Apparatus and Sample Preparation

Different apparatus are commercialized by Malvern Panalytical [215], Horiba [216] or Occhio Instruments [217] among other. As example, the model Zetasizer Nano ZS of Malvern Panalytical allows measuring the particle size distribution in the range of 0.3–10 μm, the density and the zeta potential. The principle is based on the light diffusion in an ultra-diluted or concentrated medium. The measurement method is the capillarity electrophoresis which consists in applying an electric field on the particle surface before measuring the speed of particles.

The zeta potential is measured by Doppler laser effect, diffused light which corresponds to the shifting of the frequency of a wave between the measurements on transmission and on reception, when the distance between the transmitter and the receiver varies over time.

A beam passes through the scattered particles (scattering beam) and another (reference beam) passes through the sample or bypass the cell. The measurement is obtained by comparing the difference in frequency between the incident beam (reference beam) and the scattered beam. The mobility of the particles under the influence of an electric field can then be determined. In this apparatus, only the laser beam passes through the capillary and the reference beam is diverted outside the cell and has the same frequency and phase as the laser. The corresponding optical configuration is presented in Fig. 2.115. When the particles are dispersed in the cell they scatter the light as they pass the laser beam. This scattered wave has a shift in frequency (and phase) which is a function of the speed of the particles. When a

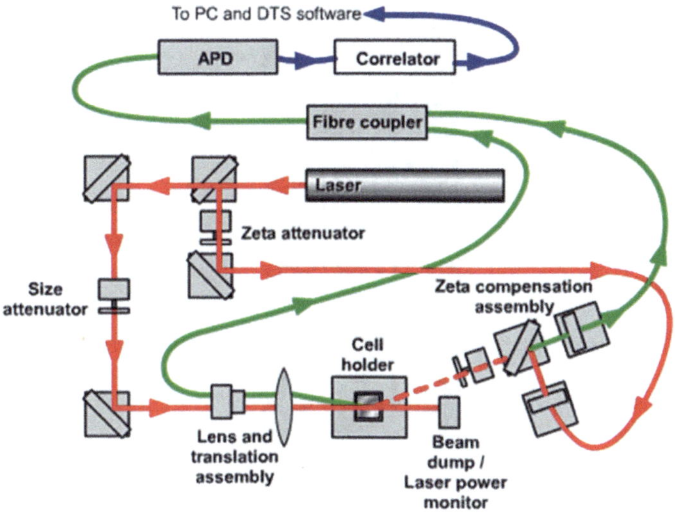

Fig. 2.115 Optical configuration of the apparatus Zetasizer Nano ZS from Malvern Panalytical. Reproduced with permission from Malvern Panalytical [215]

particle moves, the diffused light with a Doppler effect corresponds to the frequency at the front higher than the frequency at the back.

Measurement Cell

Two types of cells (capillary and dip) are available as illustrated in Fig. 2.116. The capillary cell (left-hand-side) can be used under the following conditions: for dielectric constant >10, aqueous medium, low sample quantity. It is usually disposable and is used with the titrator. The dip cell (right-hand-side) can be used with aqueous or non-aqueous medium for larger range of zeta potential compared to capillary cell. It can be disposable.

Experimental Protocol for Zeta Potential Measurement by the Titration Method

The suspension prepared is introduced in the tube as illustrated in Fig. 2.117a. Around 10–20 mg of solid particles with a mean diameter smaller than 10 μm are mixed with 10–15 mL of distilled water. It is important to have a suspension and avoid any sedimentation that would distort the final result. The mass of solid particles can be adjusted as a function of potential measurements. The acid and basic solutions are added automatically by the software as a function of the value of the zeta potential.

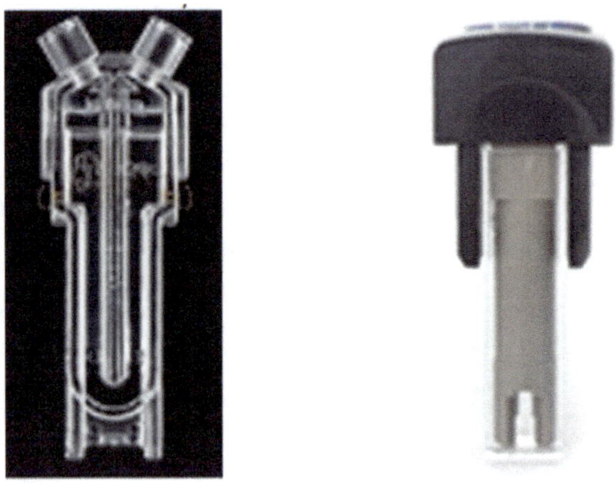

Fig. 2.116 Measurement cells for Zetasizer Nano ZS apparatus. Reproduced with permssion from Malvern Panalytical

If the zeta potential is measured as a function of the concentration of an additive, the volume of the initial solution should be of about 1.5 mL.

For measurement, the cell is firstly introduced in the apparatus as illustrated in Fig. 2.117b. The volume minimum for measurement is 2.5 mL. Usually, the volume of the solvent used is in the range of 10–15 mL. The following steps are suggested for zeta potential measurement by the titration method:

1. Introduce the cell used (capillary or drip)
2. Place three tubes of solutions: HCl = 0.25 M; HCl = 0.01 M and NaCl = 0.1 M

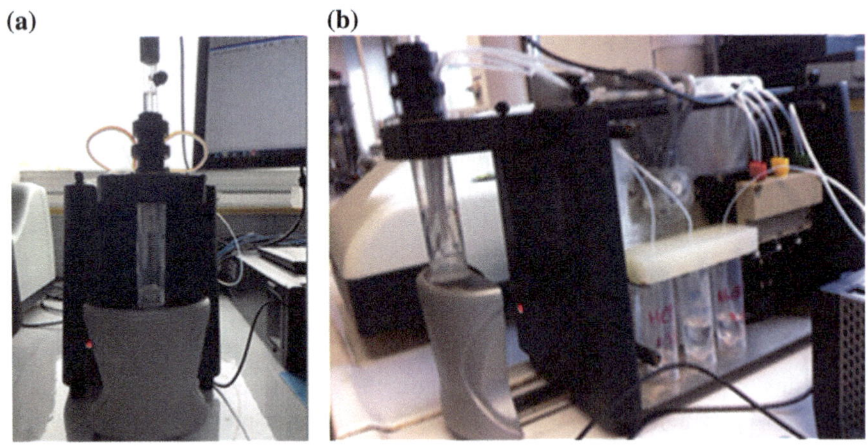

Fig. 2.117 Components of Zetasizer Nano ZS apparatus. Reproduced with permssion from Malvern Panalytical

3. To prepare the suspension: introduce 10 mL of water and 20 mg of solid particles. Particles have to remain in suspension in the solution under magnetic stirring. The maximum particle size accepted for analysis is of 10 μm.
4. In the software:

   - To create the file: File → Open → Measurement
   - To open an analysis method: File → Open → SOP → Titrator
   - Indicative informations:

     Sample: sample name
     Material: choice of particle type RI and absorbance
     Dispersant: choice of liquid used
     Model: Smoluchowski for aqueous medium
     T °C: temperature of the measurement in °C
     Choice of the cell: Cell DTS1070 (for example)
     Zeta potentiel: automatic
     Titration: To choose the acid and basis concentrations used
     Sequence: to choose "start a current pH"
     Indicate the solution volume introduced

5. Tools → Instrument → MPT-2 Manual Control
   Then, wait for the stability of the suspension pH.
6. Start *SOP*

The measurement of zeta potential as a function of pH will stop automatically.

It is possible to verify the repeatability of the measurement. For that, the number of counts has to be in the range of 150,000–300,000. As example, Fig. 2.118 shows a good analysis, wherein the phase as a function of time is stable and presents only one peak well defined. The total counts in this case reached more than 200,000 counts. On the other hand, Fig. 2.119 shows an incorrect analysis. Phase plot was not stable with several peaks. This type of results indicates that the concentration is not within the measurable range.

Example of Measurement

Figure 2.120 shows the results obtained for the zeta potential measurement of two alum sludges obtained from two drinking water treatment plants. Samples preparation was made with 10 mL of distilled water and 15 mg of sludge with a mean diameter of 14 μm and the measurement was made at 22 °C. These sludges are different from their mineral composition. Consequently, the zeta potential (or point of zero charge—PZC) is of 3.29 and 4.33 for sludge 1 and sludge 2, respectively. This isoelectric point corresponds to the pH where the average surface charge of solid particle is neutral. For pH higher than the PZC, the surface charge is negative. For pH smaller than the PZC, the surface charge is positive.

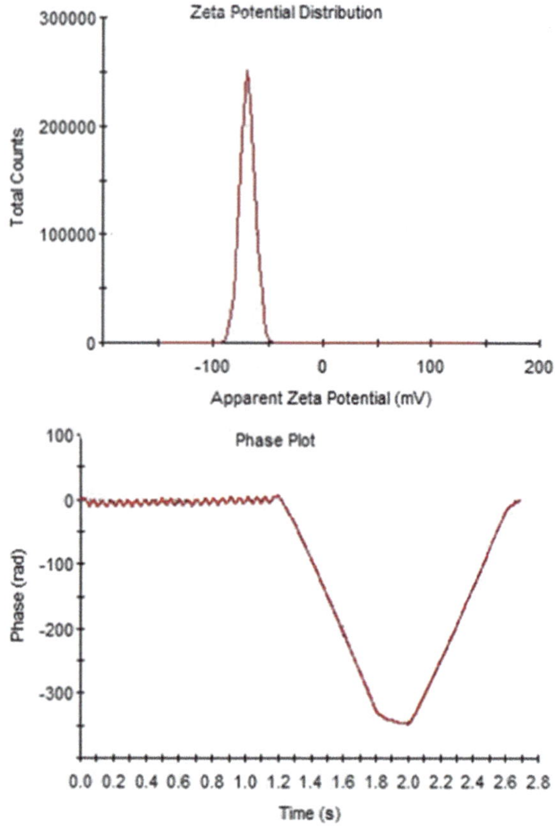

Fig. 2.118 Example of a good zeta potential measurement. Reproduced with permssion from Malvern Panalytical

2.8.3 *Cation Exchange Capacity Determination*

2.8.3.1 Introduction

Cation Exchange Capacity (CEC) measures the ability of a solid sample to fix cations from the liquid phase wherein the solid is dispersed. The value is expressed as centimole (or millimole) of positive charges per kilogram of solid ($cmol^+ kg^{-1}$ or $mmol^+ kg^{-1}$) at a specified pH.

In the field of biomass and biowaste, CEC is mainly determined for by-products used in soil application, and the CEC measures the ability of clay and organic matter of the soil to fix cations, which are nutrients, to avoid their leachability. This could be an indication of the soil fertility [163, 218]. Solids with high CEC values are convenient to be used in soil applications. The cations considered in the CEC are Ca^{2+}, Na^+, K^+, NH_4^+, Al^{3+} and Mg^{2+}, and their adsorption on the negative charges (permanent and pH-dependent) of the solid are of great importance to maintain an overall electrical neutrality. It should be noted that some negative

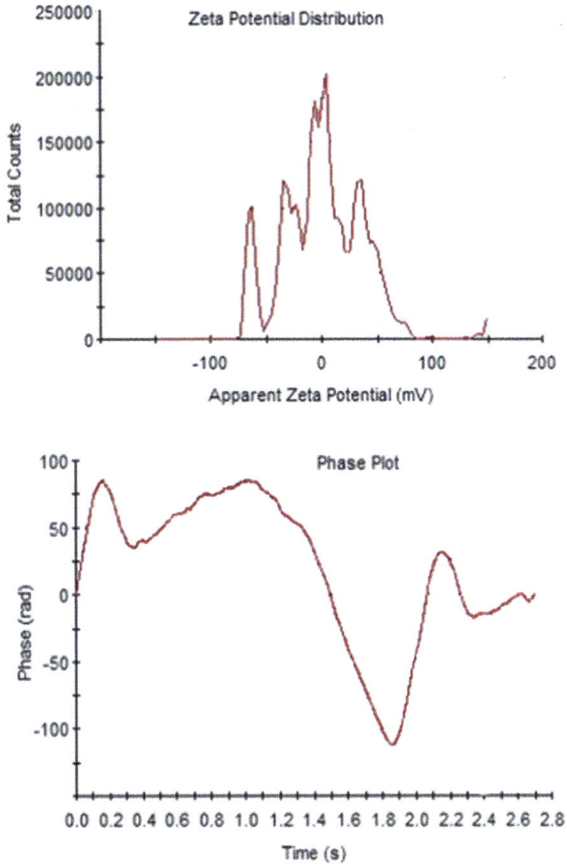

Fig. 2.119 Example of an incorrect zeta potential measurement. Reproduced with permssion from Malvern Panalytical

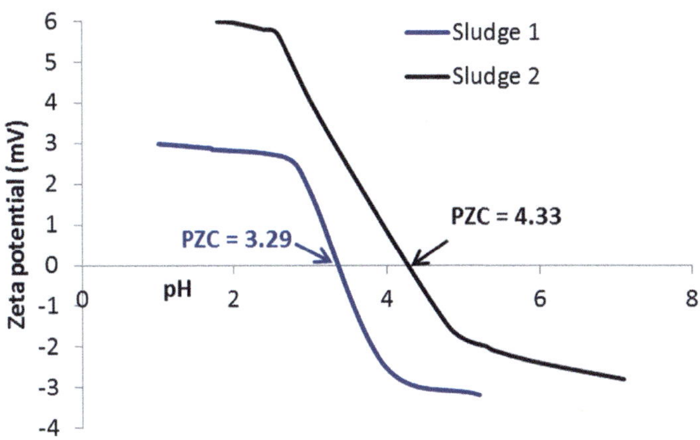

Fig. 2.120 Zeta potential measurement as a function of pH

charges are pH-dependent, thus the CEC value will depend on the amount of this kind of negative charges. Therefore the value of CEC determined for a solid should refer to the pH used for the determination. Permanent negative charges are basically due to the isomorphic substitution within the lattice of phyllosilicates. Variable negative charges are related to hydroxyl groups in carboxylic, lactonic and phenolic groups. In case the amount of variable negative charges is high in a solid, the CEC values will increase with the pH [218]. It is important to keep in mind that the CEC value is not intrinsic property of the solid and is greatly related to the method used [218].

2.8.3.2 Methods

Different methods exist to measure the CEC of a solid sample. The principle is based on the affinity of different cations to cover/saturate the active negative sites at the surface. To determine experimentally the CEC, different principles could be applied:

- Summation: CEC is calculated by summing the quantity of each cation released after leaching with an index cation (will not be developed).
- Direct displacement: the solid is saturated with index cation and the latter is then displaced with another cation. In this method a small fraction of liquid from index cation solution is retained by the solid. The CEC is determined from the released index cation.
- Displacing following washing: the basic principle is the same as direct displacement method; however the displacing is done after washing the solid to remove retained solution.
- Radioactive tracer: the exchangeable cations are eluted with Cs 137 tracer (will not be developed).

It is assumed that the cation exchange selectivity depends on the valence and the effective radius (size of the crystalline and hydrated ions), and is classified as follows [218]:

$$Al^{3+} > Ca^{2+} > Mg^{2+} > K^{+} \approx NH_4{}^{+} > Na^{+}$$

Therefore, the different techniques are based on the use of a specific cation to saturate the surface (exchange step), followed by either a direct or indirect quantification. In this document, four standard methods will be developed:

- Method using ammonium acetate (from NF X31-130 [219]);
- Method using sodium acetate (from Laird and Fleming [220]);
- Method using hexamminecobalt trichloride (from NF X31-130 [219] and ISO 23470 [221, 222]);
- Method using barium chloride (from ISO 11260 [223, 224]).

All the standards developed along this section are based on the CEC determination of soil and related materials. As explained previously, the CEC is not an intrinsic parameter of the solid sample and its determination may suffer from different sources of errors. For example, the following situations may result in underestimation of CEC value [218]:

- Index cation insufficiently strong;
- Minerals dissolved in the saturation solution;
- Index cation removed by proton during washing steps;
- Fine particles lost during centrifugation or filtration;
- Heterogeneity of the solid sample;
- Longer equilibration time.

The following situations may result in overestimation of CEC value [218]:

- Incomplete washing;
- Leaching of non-exchangeable cations;
- Presence of highly soluble minerals;
- Heterogeneity of the solid sample.

Method Using Ammonium Acetate

The test sample should be air-dried (40 °C) and pretreated according to the standard ISO 11464 [225]. The principle of the procedure is summarized in Fig. 2.121.

The procedure of the standard NF X31-130 [219] is described in Table 2.55.

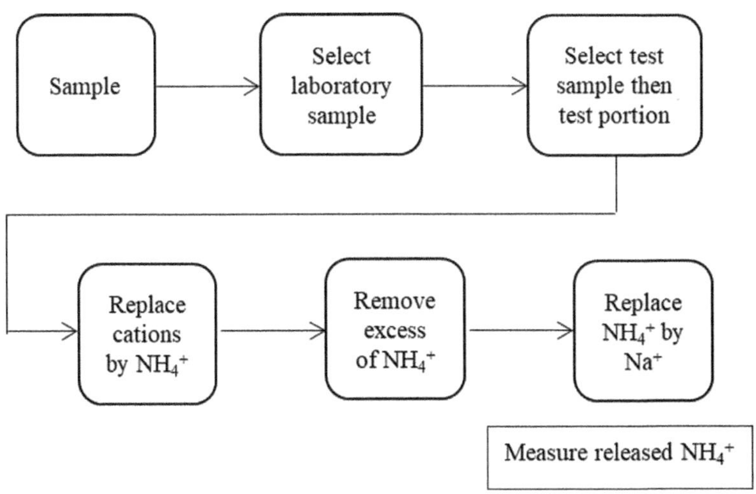

Fig. 2.121 Scheme of the procedure used to determine CEC using ammonium acetate

Table 2.55 Procedure to determine CEC using ammonium acetate procedure

Procedure	Ammonium acetate procedure
Select a test sample ISO 11464 [225]	Air dried feedstock Particle size <2 mm
Test portion	2.50 g at 0.1% of test sample
Materials	Percolation column + percolation media[a] Ammonium acetate 1 mol L^{-1}, pH = 7[b]
Exchange by percolation (negative sites saturated with NH_4^+)	Put 2.5 g of solid in the percolation column Do percolation: fill the column (4 times) with ammonium acetate 1 M (total volume is about 75 mL) Rinse with about 75 mL of ethanol (95°) to remove excess reagent[c] Dry sample + percolation media 24 h at room temperature
Extraction[d] (NH_4^+ released and replaced by Na^+)	Put sample + percolation media in a stirred container Add 50 mL of sodium chloride 1 mol L^{-1} Shake during 1 h Filter the solution[e] Store filtered solution in a sealed container no more than 24 h
Analytical measurements[f]	Analysis of ammonium in the filtered solution: Distillation and titration (similar to Kjeldahl) [219] Absorptiometry with continuous flow analyzer [219] Absorptiometry with flow injection analyzer [218]
Data collection	For spectroscopy methods, a calibration curve should be done Measurement of the NH_4^+ concentration (cations released) is determined with the calibration curve
CEC calculation (concentration of positive charges)	NH_4^+ = 1 positive charge per molecule $CEC(\text{cmol}^+\text{kg}^{-1}) = \frac{[NH_4^+] \times V_{NaCl}}{m_{drysample}} \times 100$ With: $[NH_4^+]$, concentration of the filtrate (mol L^{-1}) V_{NaCl}, volume of the filtrate (mL) $m_{dry\ sample}$ in g

[a]The percolation equipment should be a packed-bed column filled with cellulose for about 15 mm. The quality of the cellulose should be selected to avoid mineral leaching during percolation

[b]The ammonium acetate solution should be prepared by dissolving ammonium acetate (CH_3COONH_4) into distilled water (conductivity <0.5 μS/cm) and the pH adjusted to 7.00 ± 0.05 using solutions of ammonium hydroxide (NH_4OH) and acetic acid (CH_3COOH). In the standard method, the mass of solid is 77 g and water volume is adjusted to 1 L

[c]In the standard, the completion of the rinsing step is checked with Nessler reagent. However it is assumed that a total ethanol volume of 75 mL is sufficient [226]

[d]The extraction step might be done with a mechanical extractor [218]

[e]The filter should be ashless and should not contain leachable sodium

[f]Methods using flow analysis are automating wet chemical procedures and are therefore particularly suitable for the processing of large sample series at a high analysis frequency (up to 100 samples per hour). Both methods (FIA and CFA) consist of the automatic dosage of the sample introduced into a flow system (manifold) in which the sample analytes react with the reagent solutions on their way through the manifold. The sample preparation may be integrated into the manifold. The reaction product is measured in a flow detector

Ammonium analysis:

Distillation and titration:

This technique is not developed in the present section but the basic principles include: (i) the extract solution is distillated with soda addition; (ii) and the distillate is recovered in a boric solution and then ammonium is titrated with a sulfuric acid solution.

Absorptiometry with injection flow analyzer:

The sample containing ammonium is injected into a continuous carrier stream and mixed with an alkaline solution. The ammonia formed is separated from the solution in a diffusion cell over a hydrophobic semipermeable membrane and taken up by a streaming recipient flow containing a pH indicator. The color change of the indicator is continuously measured with the flow photometer [227].

This kind of equipment could be supplied by Hach (QuickChem FIA+), OI Analytical (OI Analytical Flow Solution) or Global FIA (Flo Pro Chemstack), for example.

Absorptiometry with continuous flow analyzer:

The titration of ammonium is based on the Berthelot reaction. In a continuous stream, ammonium in the sample reacts in alkaline solution with hypochlorite (ClO^-). Then the chloramine reacts with salicylate at a temperature of 37–50 °C, using nitroprusside catalyst, to form a blue-green indophenol. The intensity of the color is quantitatively measured in a flow photometer at 640–660 nm [227].

Skalar (San++) or Systea (FlowSys III Continuous Flow Analyser) for instance could supply the equipment needed.

Calibration solutions for absorptiometry with continuous flow analyzer [219]:

The standard solution should correspond to 0.1 $mol^+ L^{-1}$. For this purpose, dissolve 6.607 g of ammonium sulfate ($(NH_4)_2SO_4$) in 1 L volumetric flask and make the volume with a sodium chloride solution (1 mol L^{-1}). The 5 calibration solutions, having [NH_4^+] equal to 0.000, 0.005, 0.010, 0.015 and 0.020 $mol^+ L^{-1}$, should be prepared from the standard solution and the volume in the volumetric flask should be adjusted with the sodium chloride solution (1 mol L^{-1}). The analyzer should be cleaned after each analyses.

Method Using Sodium Acetate

The sodium acetate method has been proposed by Laird and Fleming [220] for the characterization of soil samples. However, it has been also used by some authors for the analysis of biochar samples from different kind of feedstocks [228]. The sodium acetate method involves the saturation of the samples with Na^+ ions that replace the exchangeable cations in the sample surface, followed by the subsequent displacement of Na^+ ions by a replacement solution. This method is recommended for samples that could contain carbonates, and in consequence, the washing solution pH is adjusted to 8.2 to minimize the carbonate dissolution during the Na saturating procedure (Fig. 2.122).

The procedure of the sodium acetate method proposed by Laird and Fleming is described in Table 2.56.

Method Using Hexammine Cobalt Trichloride

The test sample should be air-dried (40 °C) and pretreated according to the standard ISO 11464 [225]. The principle of the procedure is summarized in Fig. 2.123.

As the pH of a soil suspension in the hexamminecobalt trichloride solution is close to the pH of the suspension in water, this method is considered to give the effective CEC, i.e. the CEC at the soil pH.

The size of hexamminecobalt ions, used for exchange reaction, is larger than that of ammonium ions. Therefore hexamminecobalt ions may not enter into micropores, and the CEC may be smaller.

The procedure of the standard ISO 23470:2011 [221] is described in Table 2.57. It should be noticed that this present section develops different standard methods used for CEC determination. However each standard contains also procedure for the determination of some exchangeable cations. In this context, a calcite saturated hexamminecobalt trichloride solution is specified particularly for determination of exchangeable Ca in solids containing calcium carbonate. This method was introduced to avoid erroneous (inflated) Ca values.

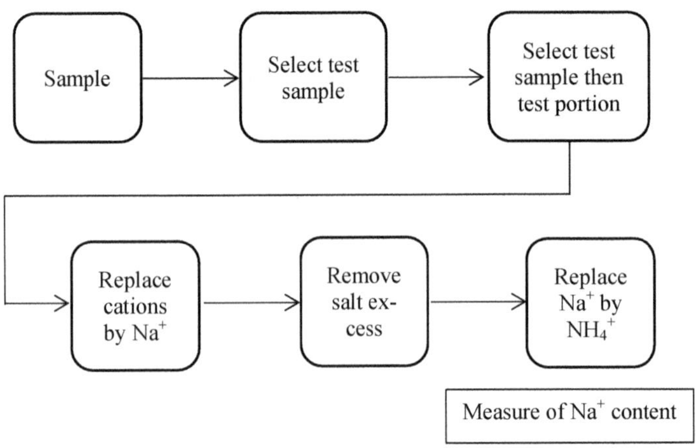

Fig. 2.122 Procedure used to determine CEC using sodium acetate

Table 2.56 Procedure to determine CEC using the sodium acetate method

Procedure	Sodium acetate procedure
Select a test sample ISO 11464 [225]	Air dried feedstock
Test portion	0.2 g of sample 0.5 g to determine the water content of the air dried sample (Heat in an oven at 200 °C for 2 h)
Materials	Saturating solution: 1 M NaOAc pH = 8.2. Adjust the pH to 8.2 by adding acetic acid (CH_3COOH) or sodium hydroxide (NaOH)[a] First rinsing solution: 80% isopropyl alcohol (IPA) Second rinsing solution: 100% isopropyl alcohol Replacing solution: 0.1 M NH_4Cl[b,c] Centrifuge and centrifuge tubes Conductivity meter
Saturating step	Put the test portion in a centrifuge tube Add 15 ml of saturating solution to the tube. Carefully vortex the sample and then centrifuge[d] Decant and discard the supernatant Repeat 2 more times
First rinsing step	Add 15 mL of the first rinsing solution[e] Carefully vortex the sample and then centrifuge Measure the conductivity of the supernatant and discard it If the conductivity of the supernatant is above 6 µS cm^{-1}, repeat the rinsing steps with the first rinsing solution If the conductivity of the supernatant is below 6 µS cm^{-1}, switch to the second rinsing solution
Second rinsing step	Add 15 mL of the second rinsing solution Carefully vortex the sample and then centrifuge Measure the conductivity of the supernatant and discard it If the conductivity of the supernatant is above 1 µS cm^{-1}, repeat the rinsing steps Stop rinsing when the measured supernatant conductivity is below 1 µS cm^{-1} (slightly above the conductivity of 100% IPA rinsing solution)
Drying step	After the final rinse step, allow the sample to air-dry by opening the sample container
Replacing step	Add 15 mL of the replacing solution[c] Carefully vortex the sample and then centrifuge Decant and save the supernatant in a 100 ml container Repeat this step 7 more times, adding the supernatant to the same 100 ml container Bring the volumetric flask to 100 mL with deionized water
Analytical measurements	Analyze the sodium content by inductively coupled plasma-atomic emission spectrometry (ICP-OES). Before analysis, standard solutions should be used in order to determine a calibration curve

(continued)

Table 2.56 (continued)

Procedure	Sodium acetate procedure
CEC calculation	CEC of the sample is calculated according to the following equation: $CEC(\text{cmol kg}^{-1}) = \frac{(0.435 \cdot C)}{(W \cdot F)}$ where C—Na concentration in the 100 ml volumetric flask (mg L^{-1}) W—Weight of air-dry sample added to centrifuge tube (g) F—Water correction factor (–) $F = \frac{weight\ of\ oven\ dried\ sample(g)}{weight\ of\ air\ dried\ sample(g)}$

[a]The sodium acetate solution should be prepared by dissolving 136.08 g of NaOAc $3H_2O$ in 750 mL of distilled, deionized water. The solution pH is adjusted to 8.2 by adding acetic acid (CH_3COOH) or sodium hydroxide (NaOH). Then, the water volume is adjusted to 1 L

[b]The ammonium chloride solution should be prepared by dissolving NH_4Cl in distilled, deionized water. In the procedure, the mass of solid is 5.35 g of NH_4Cl and water volume is adjusted to 1 L

[c]Some authors use NH_4OAc instead of NH_4Cl as replacing solution in the sodium acetate method [229]. In the procedure described here, NH_4Cl is recommended because the Na contamination of this reagent is very small compared to other salts

[d]According to sodium acetate method described by Laird and Fleming [220], it is recommended to centrifuge at 3000 rpm for 5 min. However, if the sample is not sufficiently consolidated after centrifugation, a higher speed and time can be used

[e]As the sodium acetate is not soluble in isopropyl alcohol, a mixture of water-isopropyl alcohol should be used for the first rinses

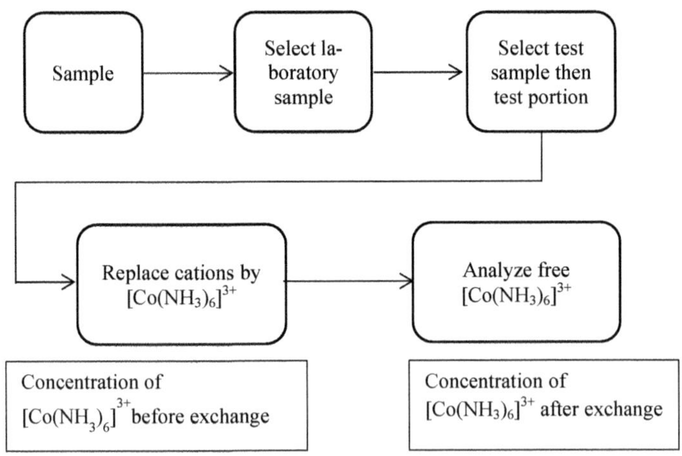

Fig. 2.123 Procedure used to determine CEC using hexammine cobalt trichloride

Mass of the test portion:

In Table 2.57, the mass of the test portion is not given. Indeed, the centimoles of positive charges carried by the test portion should lie between Q/10 and Q/3 inclusive, giving that Q is the number of centimoles of positive charges provided by a given volume of hexamminecobalt solution. Table 2.58 gives suitable mass of test

Table 2.57 Procedure to determine CEC using hexammine cobalt trichloride

Procedure	Hexammine cobalt trichloride procedure
Select a test sample ISO 11464 [225]	Air dried feedstock Particle size < 2 mm
Test portion	1.25–10 g at 0.1% of test sample
Materials	Container for shaking[a] Exchange solution (C_o, initial concentration): [Co $(NH_3)_6Cl_3$] = 0.0166 mol L^{-1}[b] or calcite saturated $Co(NH_3)_6Cl_3$[c] End-over-end shaker[d] Centrifuge in relevant centrifuge vials[e]
Exchange reaction (negative sites saturated with $[Co(NH_3)_6]^{3+}$)	Put test portion in a container for shaking Add 50 mL of $Co(NH_3)_6Cl_3$ solution (either or not calcite saturated) Shake for 60 ± 5 min Centrifuge the mixture and collect the clear supernatant[f] Store clear supernatant for no more than 24 h
Blank test[g]	*Do a blank test with the same procedure as for exchange reaction, but without test portion*
Analytical measurements	Analysis of ammonium concentration (C_f, final concentration): Distillation and titration (similar to Kjeldahl) [219, 221] Spectrophotometric determination, λ = 475 nm [219, 221] Analysis of total cobalt concentration: Spectrometric determination (ICP-AES or FAAS) [221]
Data collection	For spectrophotometric method, a calibration curve should be done Measurement of the ions concentration is determined with the calibration curve
CEC calculation (concentration of positive charges)	$[Co(NH_3)_6]^{3+}$ = 3 positive charges per molecule $CEC(\text{cmol}^+\,\text{kg}^{-1}) = \frac{3\times\left[(Co(NH_3)_6)^{3+}\right]_{ex}\times V_{ex}}{m_{drysample}} \times 100$ With: $\left[(Co(NH_3)_6)^{3+}\right]_{ex}$, exchanged concentration (C_o-C_f), in mol L^{-1} V_{ex}, volume of $Co(NH_3)_6Cl_3$ solution, in mL $m_{dry\ sample}$ in g

[a]Containers for shaking should be tightly stoppered, with rigid or non-rigid wall, and should have a volume between 75 and 100 mL

[b]The hexamminecobalt trichloride solution should be prepared by dissolving hexammine cobalt trichloride (Co $(NH_3)_6Cl_3$) into distillated water (conductivity <0.5 µS/cm). In the standard method, the mass of solid is 4.458 g ± 1 mg and water volume is adjusted to 1 L

[c]Put 1 L of the hexamminecobalt trichloride solution previously prepared in a 2 L beaker and add 2 g of fine-grained calcite. Put the mixture in ultrasonic bath for 30 min, and then shake the solution with a magnetic stirrer for 30 min. Protect the beaker by a plastic foil on top, and let undissolved calcite settle overnight. The solution may be decanted or filtered, however addition of undissolved calcite for exchange reaction should be strictly avoided

[d]End-over-end shaker should allow the permanent suspension of the solid/extraction solution mixture, in ambient air at 20 ± 2 °C

[e]Centrifugation time and speed depend on the type of centrifuge and are selected with a view of securing a clear supernatant. 4000 rpm for 20 min are generally sufficient but special care should be taken due to the high content of fixed hexamminecobalt ions on colloidal particles

[f]It is also possible to do filtration instead of centrifugation. In this case, the blank test will be used to correct the quantity of cations exchanged since the filter may retain a small fraction of the hexamminecobalt ions. For details regarding the calculation, please see [221]

[g]The blank test is used for the ammonia nitrogen determination by distillation

Table 2.58 Suitable mass of test portion (50 mL solution) for a range of expected CEC values, from [219, 221, 222]

Expected CEC ($cmol^{+}$ kg^{-1})	<5	5–10	10–32	32–64
Mass of test portion (g)	10	5	2.5	1.25

portion for a range of expected CEC values. Evaluation of plausible CEC value may be performed using the diagram type called "Carbonate and Sulphate Field Model (CSF model)" [222, 230].

Exchange solution:

Hexamminecobalt trichloride is recommended as an exchange solution for non-calcareous soils with a pH value <6.5. This method is considered to give the effective CEC, since pH of the soil in the solution is quite similar to the pH of soil suspended in water.

A calcite saturated hexamminecobalt trichloride solution is recommended for calcareous soils with a pH value >6.5. The pH value of the calcite saturated hexamminecobalt trichloride solution being approximately 9, the results are comparable to the determination of potential CEC.

Spectrophotometric determination of ammonium:

Issues with soluble organic matter in spectrophotometric determination:

The CEC calculation is based on the measurement of hexamminecobalt ions in the solution before and after exchange reaction. The difference in concentration will give the quantity of ions exchanged at the surface. However, the presence of soluble organic matter might interfere with the spectrophotometric determination. The soluble organic matter is coloring the solution and absorbs light at 475 nm. In case of high solubilized quantities of organic matter, a second determination of absorbance should be done at a second wavelength (λ = 380 nm). Considering R_1 and R_2, the ratios between the optical absorption of the pure hexamminecobalt solutions at both wavelengths (R_1), and the solutions with dissolved organic matter at both wavelengths (R_2), the optical absorption of the hexamminecobalt solution could be recalculated using Eq. (2.101) [221]. The new value of the absorbance is then used to calculate the concentration with the calibration curve.

$$DO = \frac{(X_1 - R_2 X_2) R_1}{R_1 - R_2} \tag{2.101}$$

where:

- DO is the optical absorption of the hexamminecobalt solution;
- $R_1 = \frac{DO(475\,\text{nm})}{DO(380\,\text{nm})}$ is the ratio between the optical absorption of the pure hexamminecobalt solutions at both wavelengths, the value of 8.13 is given in the standard method;

- $R_2 = \frac{DO(475\text{ nm})}{DO(380\text{ nm})}$ is the ratio between the optical absorption of the hexamminecobalt solutions containing soluble organic matter at both wavelengths, the value of 0.2 is given in the standard method;
- X_1 and X_2 are the optical absorption of the test solution at 475 and 380 nm respectively.

Calibration solutions for spectrophotometric determination:

The standard solution should be prepared by dissolving 4.458 g of hexamminecobalt trichloride ($Co(NH_3)_6Cl_3$) in 1 L volumetric flask and make the volume with distilled water. The 4 calibration solutions should be prepared from the standard solution diluted with water at ratios 1/5, 2/5, 3/5 and 4/5. The calibration curve should be established with 5 points.

Method Using Barium Chloride Solution

The test sample should be air-dried (40 °C) and pretreated according to the standard ISO 11464:2006 [225]. The principle of the procedure is summarized in Fig. 2.124.

Using this procedure, the CEC is determined at the pH of the soil and at low total ionic strength (0.01 mol L^{-1}). The procedure of the ISO 11260:2011 [223] is described in Table 2.59.

Determination of magnesium concentration and CEC:

Issues with soluble organic matter:

Organic matter contributes to the CEC. In case organic matter is dissolved in barium chloride solution, the number of negative sites will decrease and then the

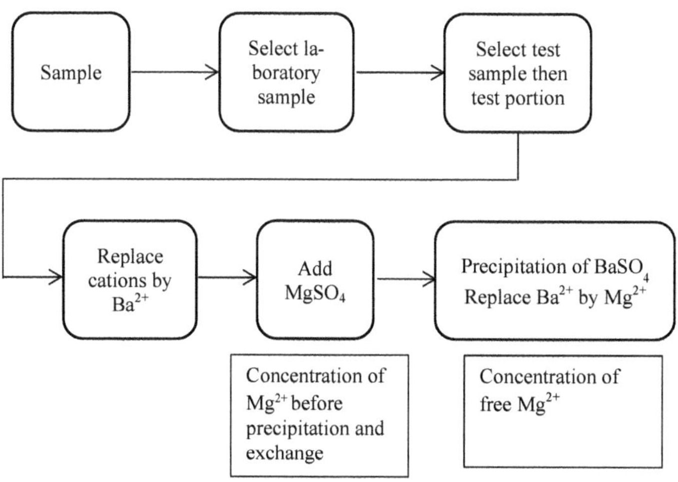

Fig. 2.124 Scheme of the procedure used to determine CEC using barium chloride

Table 2.59 Procedure to determine CEC using barium chloride

Procedure	Barium chloride procedure
Select a test sample ISO 11464 [225]	Air dried feedstock Particle size <2 mm
Test portion	2.50 g at 0.1% of test sample
Materials	Stoppered polyethylene centrifuge tube 50 mL Exchange solution: $[BaCl_2]$ = 0.1 mol L^{-1}[a] Stabilization solution: $[BaCl_2]$ = 0.0025 mol L^{-1}[b] Precipitation solution: $[MgSO_4]$ = 0.020 mol L^{-1}[c]
Exchange reaction (negative sites saturated with Ba^{2+})	Put test portion in a centrifuge tube (m_1) Add 30 mL of $BaCl_2$ solution (0.1 mol L^{-1}) Shake for 1 h Centrifuge the mixture and remove the supernatant[d] Repeat addition of $BaCl_2$, shaking and centrifugation twice more for the same test portion
Stabilization	Add 30 mL of $BaCl_2$ solution (0.0025 mol L^{-1}) to the cake Shake overnight Centrifuge the mixture and remove the supernatant[d] Record the mass of the tube (m_2)
Precipitation and exchange (negative sites saturated with Mg^{2+})	Add 30 mL of $MgSO_4$ solution (0.02 mol L^{-1}) Shake overnight Centrifuge the mixture[d] Decant the supernatant through a coarse filter paper Store filtered solution
Blank test[e]	*Do a blank test with the same procedure (each step), but without test portion*
Analytical measurements	Analysis of magnesium concentration: Spectrometric determination (FAAS [223, 224] or ICP-AES [224])
Data collection	A calibration curve should be done Measurement of the ions concentration is determined with the calibration curve
CEC calculation (concentration of positive charges)	Mg^{2+} = 2 positive charges per molecule $CEC\left(\mathrm{cmol}^+\,\mathrm{kg}^{-1}\right) = \frac{2\times\left[\mathrm{Mg}^{2+}\right]_{ex,corr}\times V_{\mathrm{MgSO_4}}}{m_{drysample}}\times 100$ With $\left[\mathrm{Mg}^{2+}\right]_{ex,corr}$, exchanged concentration corrected, in mol L^{-1}[e] V_{MgSO4}, volume of $MgSO_4$ solution, in mL $m_{dry\ sample}$ in g

[a]The barium chloride solution should be prepared by dissolving barium chloride dihydrate ($BaCl_2$, $2H_2O$) into distilled water (conductivity <0.5 μS/cm). In the standard method, the mass of solid is 24.43 g ± 1 mg and water volume is adjusted to 1 L

[b]The barium chloride solution should be prepared by diluting 25 mL of $BaCl_2$ solution at 0.1 mol L^{-1} in water (volumetric flask of 1 L)

[c]The magnesium sulfate solution should be prepared by dissolving magnesium sulfate heptahydrate ($MgSO_4$, $7H_2O$) into distilled water (conductivity <0.5 μS/cm). In the standard method, the mass of solid is 4.93 ± 0.01 g and water volume is adjusted to 1 L. Please note that magnesium sulfate heptahydrate may lose water of crystallization, and the reagent should be standardized by titration with EDTA at pH 10 using Eriochrome Black T indicator

[d]Centrifugation is performed at 3000 g for 10 min. Note that the relative centrifugal force (RCF) is given in the standard, but microcentrifuges have only settings for speed (revolutions per minute, RPM). A relation between both parameter exists: RCF = $(1.118 \times 10^{-5}) \times R \times speed^2$, where R is the radius of the rotor in cm

[e]The blank test is used to correct the quantity of cations exchanged since the filter may retain a small fraction of the magnesium ions. See next section for calculation details

CEC value will be underestimated. A brown-yellow color of the supernatant collected after barium chloride leaching indicates the presence of soluble organic matter.

Calibration solutions for spectrometric determination:

To prevent the formation of refractory compounds of magnesium during the spectrometry determination, an acidified lanthanum solution is added to sample before analysis. The diluted solutions are prepared as follows:

- Put 0.2 mL of sample in a 100 mL volumetric flask;
- Add 10 mL of lanthanum acidified solution (put 15.6 mg of lanthanum nitrate hexahydrate + 42 mL of HCl + water to make the volume to 500 mL);
- Make the volume with water to 100 mL, then the dilution factor is equal to 500;
- The value of CEC should not exceed 40 $cmol^+ kg^{-1}$.

CEC calculation from concentrations of diluted solutions:

The calculation of CEC presented in the standard is reported in Eq. (2.102) [223]. This equation contains the concentrations of dilute solutions, as well as the dilution factor of 500 (see Eq. (2.102)).

$$CEC\left(cmol^{+} kg^{-1}\right) = \frac{2 \times \Delta\left[Mg^{2+}\right]_{dil,corr} \times V_{MgSO_4}}{m_{drysample}} \times \frac{500}{10} \tag{2.102}$$

where the concentration is in mmol L^{-1} and the mass in g. Be careful that the concentration units are different between this equation and the one in Table 2.59.

Corrected magnesium concentrations for CEC calculation:

As expressed in Eq. (2.102), the CEC is given by the quantity of magnesium ions exchanged. This value is actually the difference in magnesium concentration between the blank (quite similar to the initial concentration) and the final filtered solutions (concentration measured with spectrometry) (see Eq. 2.103). However, the solid sample is subjected to many leaching solutions. When centrifugation is performed, the solid sample is not dried and contains a residual volume of liquid ($BaCl_2$). Therefore, the concentrations should be expressed based on the same volumes to be able to calculate the difference. The actual concentration of magnesium in single solutions is then the concentration of magnesium measured multiplied by the total volume (volume added + remaining liquid) and divided by the volume of $MgSO_4$ (same volume as in the blank).

$$\left[Mg^{2+}\right]_{ex} = \left[Mg^{2+}\right]_{blank} - \left[Mg^{2+}\right]_{sample} \tag{2.103}$$

$$\left[Mg^{2+}\right]_{sample,corr} = \frac{\left[Mg^{2+}\right]_{sample} \times V_{total}}{V_{MgSO_4}} \tag{2.104}$$

In the standard, it is advised to calculate the total volume as the sum of wet mass of the solid plus the volume of $MgSO_4$ solution. The wet mass is measured by the difference in mass between m_2 (mass of wet sample + tube, see Table 2.59) and m_1 (mass of air dried sample + tube, see Table 2.59).

2.8.3.3 Typical Values of CEC

The CEC determination is of great importance for characterizing the soils. From literature, it could be seen that the values of CEC for various soils varied depending on the composition, location and depth. These values are influenced by minerals and organic matter. Clay particles have generally the highest cation exchange capacity followed by silt and sand (Table 2.60). It should be noted that the clay materials have structural and textural properties (tetrahedral, octahedral sheets and the distance between the sheets) that improves the CEC of the solid (see more details in dedicated literature). In addition, organic matter content of soils has also an important influence on the CEC since it varies between 100 and 300 $cmol^+ kg^{-1}$ [231]. This is one of the reasons why the biochars (residues from thermoconversion) have been investigated to improve soil fertility. A large number of samples have been studied and the CEC values of biochars are varying between 4 and 15 $cmol^+ kg^{-1}$ [218], the values are also increased to 20–45 $cmol^+ kg^{-1}$ with the presence of minerals as for straw and rice straw derived biochars [232]. The CEC values from sodium acetate method measured for steam gasification chars from different lignocellulosic agrowastes are in accordance with literature reported data. In particular, using the sodium acetate method, coconut shells and oil palm shells chars showed a CEC of 16 and 11.5 $cmol^+ kg^{-1}$ respectively, while bamboo guadua chars, with a high mineral content (>30% ash content), showed a CEC value of 44 $cmol^+ kg^{-1}$. Gasification chars were produced at 850 °C with a 30% steam/70% N_2 reacting atmosphere.

Table 2.60 Typical values of CEC

Type of material	CEC range ($cmol^+ kg^{-1}$)
Sand	1–5
Fine sandy loam	5–10
Loams and silt loam	5–15
Clay loam	15–30
Clay: Allophane Montmorillonite Chlorite, illite Kaolinite	>30 (100–150) (60–100) (20–40) (2–16)

2.8.3.4 Comparison of the Standard Methods for CEC Determination and Examples

An interesting work has already been published by Cieselski et al. [226] on the comparison of the three methods described in standards for 48 different soils. The authors measured the CEC of each soil using the three techniques and draw 2D graphs to compare two methods (see Fig. 2.125). From their work, it could be seen that:

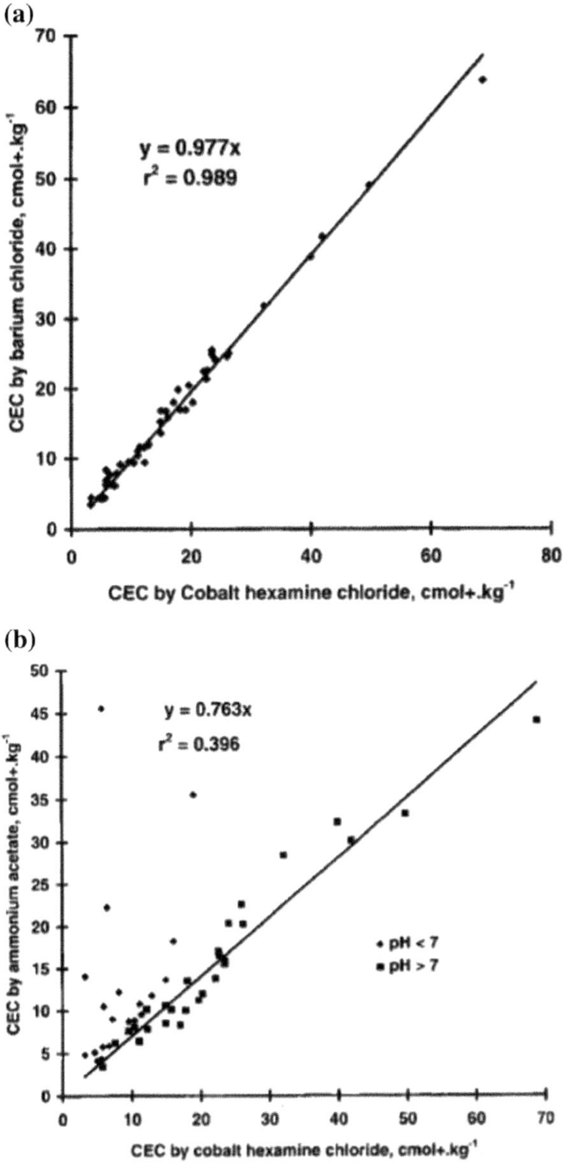

Fig. 2.125 Comparison of CEC values obtained with a barium chloride and hexamminecobalt trichloride methods and b ammonium acetate and hexamminecobalt trichloride methods. Adapted with the permission from Agronomy for Sustainable Development [226]

Barium chloride and hexamminecobalt trichloride methods are giving very similar values. Both methods are carried out in unbuffered solutions.

A great variation exists between ammonium acetate and hexamminecobalt trichloride methods, especially at acidic pH. Ammonium acetate method is carried out in buffered solution. A significant difference has been also observed with both techniques applied to volcano soils in Japan [221].

Barium chloride and hexamminecobalt trichloride methods are considered as equivalent. Regarding the ammonium acetate method, it has been shown that this method is mainly overestimating the values obtained by other methods (Fig. 2.125b). The buffering capacity of the exchange solution in case of ammonium acetate will influence the pH-dependent charges of the soil. The authors observed that the pH of the solution was quite similar when suspending soil in water or soil in hexamminecobalt [226]. On the contrary they showed that the pH of the mixture soil/ammonium acetate solution was quite stable (buffered solution at 7) and then was far from the resulting values in unbuffered solution. Based on these results, the authors decided to recalculate the CEC values from ammonium acetate method, taking into account organic carbon content and water pH (see the equations in [226]). The new CEC values from ammonium acetate methods are compared again with those obtained with hexamminecobalt trichloride method. Figure 2.126 shows a good agreement between the corrected values and the values obtained with the two other methods. The authors finally concluded that the 3 methods were convenient, but the values from ammonium acetate method should be corrected. As seen in the previous sections, the three methods are quite time consuming and need successive steps. In a practical point of view the authors suggest the use of the hexamminecobalt trichloride method since there is only one exchange step [226].

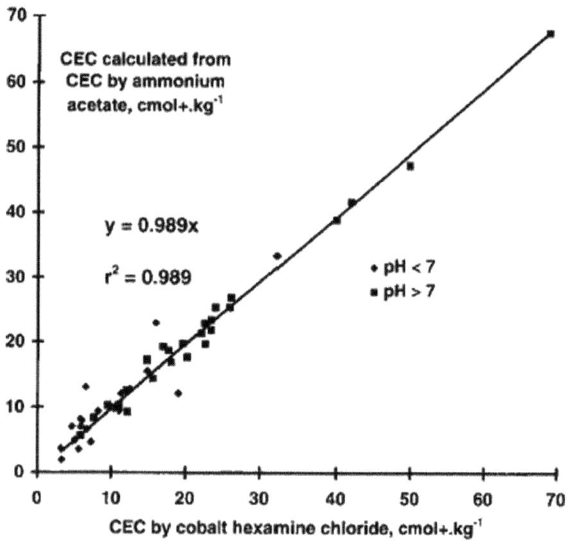

Fig. 2.126 Comparison of corrected CEC values obtained with ammonium acetate and hexamminecobalt trichloride methods. Adapted with the permission from Agronomy for Sustainable Development [226]

Based on the discussion above, it appears that the organic matter could influence the determination of the CEC from ammonium acetate method. It has been also discussed that soluble salts may interfere with the CEC determination using this method [218]. Indeed they compared the CEC values of same resources (biochars) provided with the same method (ammonium acetate) by two different laboratories. They observed some differences they attributed to:

- The experimental procedure, a laboratory washed the biochar before experimental determination, resulting in release of soluble salts;
- The biochar mass reference in the calculation, since one laboratory expressed the CEC as mass of dry biochar or as mass of dry washed biochar;
- The number of percolation steps (see Table 2.55);
- The technique used for ammonium analysis;
- The retention of index cation in the filter.

The experimental procedure of the standard methods is quite time consuming due to the number of steps requested. It has also been suggested to evaluate the CEC value as the contribution of clay and organic matter [231]:

$$CEC_{estimated} = w_{clay} CEC_{clay} + w_{OM} CEC_{OM} \tag{2.105}$$

where:

w is the percentage in mass of clay or organic matter in the solid;
CEC is an average value of clay or organic matter (it is suggested $\approx$ 80 $cmol^{+}\ kg^{-1}$ and 200 $cmol^{+}\ kg^{-1}$ respectively for a soil from Alberta Peace Region) [231].

2.8.4 *Surface Acido-basic Composition Determination Using Boehm Titration*

2.8.4.1 Introduction

Acidic functions at the surface of materials are related to oxygenated functional groups such as carboxylic, lactonic and phenolic groups [163, 233, 234]. The method developed for this quantification, namely Boehm titration [235], determines the concentration of each oxygenated functional groups families in mmol g^{-1}. Voll and Boehm also studied the quantification of basic surface functions derived from γ-pyrone structures [236]. However acidic rather than basic functions are mainly characterized since their amounts are significantly higher (5–25 times more) [233, 237].

Carbon materials could be potential candidates for applications such as soil amendment [163], sorbent [164], catalysts [165, 238–242] or support for catalyst [166], where heterogeneous interactions between the solid and a gaseous or liquid phase are involved. Acidic or basic functions are developing the acidic/basic

Table 2.61 Acidic organic functions determined with Boehm titration

Acidic functions	Formula	pK_a range	Base used
Carboxylic	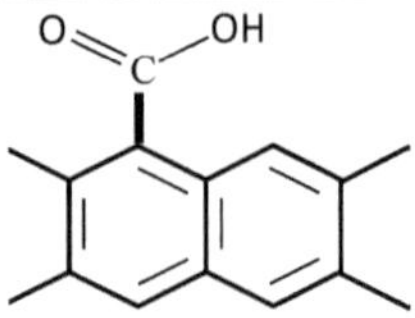	5.0–6.4	$NaHCO_3$, Na_2CO_3 and NaOH
Lactonic	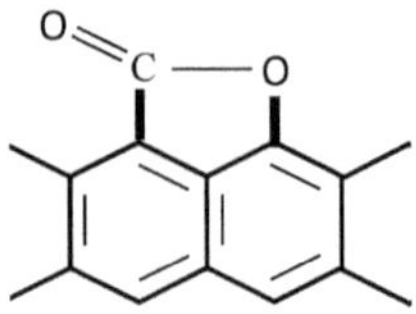	6.4–10.3	Na_2CO_3 and NaOH
Phenolic	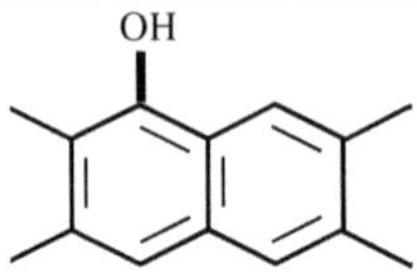	10.3–13	Only NaOH

character of the surface materials, and a deeper knowledge will help for a better understanding of reactivity mechanisms: for example in selective sorption or acid-basic catalysis [243–245].

Boehm titration was originally developed for carbon black to evaluate their quality [237]. The principle is based on the selective neutralization of acids functions with different base solutions. Basically carboxylic acids are considered as weak acids having pK_a lower than 6.4. A second family corresponds to lactonic groups in the range of pK_a 6.4–10.3. Finally the phenolic groups have a pK_a value in the range of 10.3–13. The functions and pK_a are summarized in Table 2.61. To fit with these pK_a ranges, the strength of the base should be adapted. Basically three base solutions are considered: $NaHCO_3$ (for carboxylic group neutralization), Na_2CO_3 (for carboxylic and lactonic groups neutralization) and the strongest base NaOH (for the three organic functions).

This method has been used for other carbon materials such as biochars. Although well established, Boehm titration is not a standard. This section will somehow address some researches using this method for biochar surface functions analysis [235, 246–249].

2.8.4.2 Boehm Titration Method and Principle

The different steps of the Boehm titration are shown in Fig. 2.127, and the reactions involved are also summarized in Table 2.62. In summary, the acidic functions at the surface are neutralized with the base solutions (3 different experiments). After

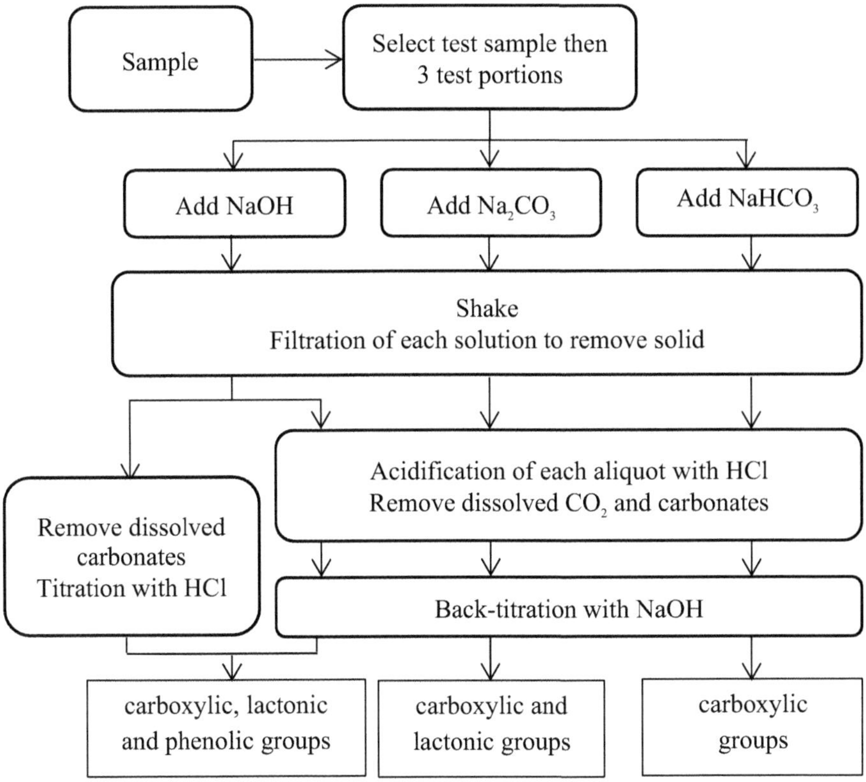

Fig. 2.127 Scheme of the Boehm titration procedure

reaction between the base solution and the acidic surface functions, the solid is separated from the liquid phase. The addition of a strong acid (HCl) to this aliquot of basic solution consumes the remaining base, and then the excess of HCl is titrated by NaOH. As seen in Table 2.62, the number of acidic sites at the solid surface is equal to the quantity of Na^+ consumed, and a stoichiometric factor should be taken into account with Na_2CO_3. The experimental procedure is described in Table 2.63, based on the values from Goertzen et al. [246]. Each procedure should be triplicated (3 subsamples x 3 base solutions). Using adequate calculations (see Sect. 2.8.4.3) the amount of each acidic function is then calculated. It should be noted that using the NaOH strong base as the neutralizing agent, a direct titration with HCl is also suggested.

2.8.4.3 Boehm Titration: Calculation of Acidic Sites

As the acidic functions are neutralized by base solution, the amount of acidic site is expressed as follows:

Table 2.62 Reactions between acidic functions and the base solutions

	Reactions with the base
$NaHCO_3$	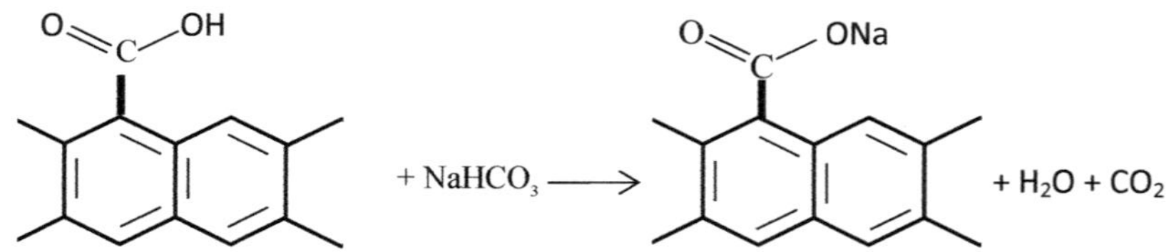
Na_2CO_3	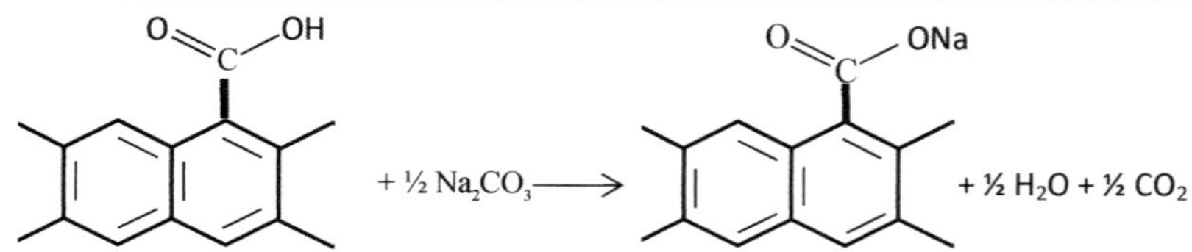
	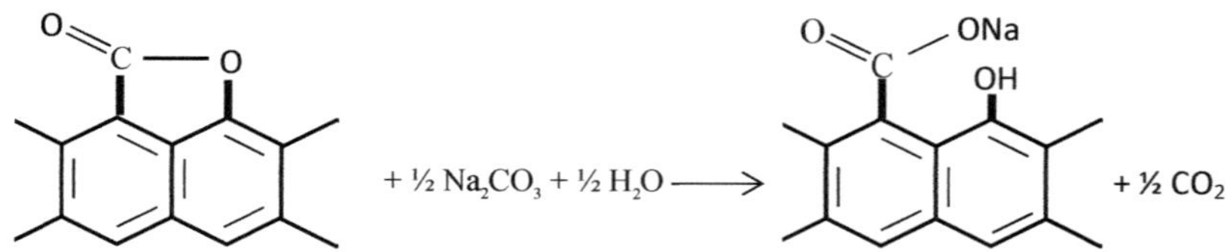

(continued)

Table 2.62 (continued)

	Reactions with the base
NaOH	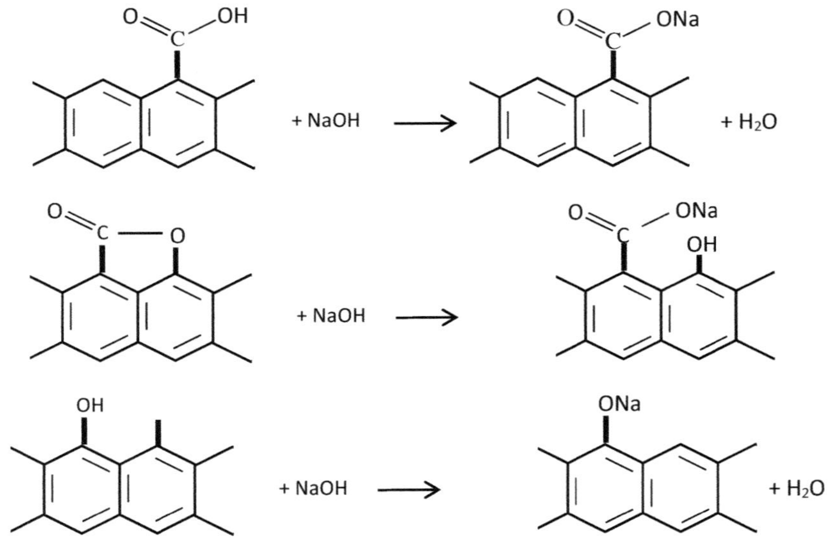

$$n_{acid} = n_{\mathrm{NaOH}} = n_{\mathrm{NaHCO_3}} = 2n_{\mathrm{Na_2CO_3}} \tag{2.106}$$

The stoichiometric coefficient differs when Na_2CO_3 is used:

$$\propto = \frac{n_{acid}}{n_{\mathrm{Na_2CO_3}}} = 2.$$

The quantity of acidic sites could be also written as the difference between initial and final amounts of base:

$$n_{acid} = \propto \left(n^0_{base} - n^1_{base}\right) \tag{2.107}$$

An aliquot of solution 1 is further used for either direct or back-titration. Then:

$$n^1_{base} = n^{aliquot}_{base} \times \frac{V_{base}}{V_{aliquot}} \tag{2.108}$$

The remaining base in the aliquot reacts with HCl. For this reaction, the stoichiometric factor using Na_2CO_3 is the same as the α defined previously. A certain amount of HCl has not been consumed:

$$n^{remaining}_{\mathrm{HCl}} = n^0_{\mathrm{HCl}} - \propto n^{aliquot}_{base} \tag{2.109}$$

And finally the remaining HCl is titrated with NaOH:

$$n^{remaining}_{\mathrm{HCl}} = n^{tit}_{\mathrm{NaOH}} = [\mathrm{NaOH}]^{tit} V_{\mathrm{NaOH},equi} \tag{2.110}$$

Combining Eqs. (2.106)–(2.110), the quantity of acidic functions is calculated as follows:

$$n_{acid} = \propto [base]^0 V_{base} - \frac{V_{base}}{V_{aliquot}} \times \left([\mathrm{HCl}]^0 V_{\mathrm{HCl}} - [\mathrm{NaOH}]^{tit} V_{\mathrm{NaOH},equi}\right) \tag{2.111}$$

For aliquots from NaOH base solution, a direct titration of the remaining quantity of NaOH could also be done:

$$n^{aliquot}_{NaOH} = n^{tit}_{HCl} = [HCl]^{tit} V_{HCl,equi} \tag{2.112}$$

The quantity of NaOH in the aliquot is expressed as in Eq. (2.108). Combining both equations, the quantity of acidic functions is given in Eq. (2.113):

$$n_{acid} = [\text{NaOH}]^0 V_{\text{NaOH}} - \frac{V_{\text{NaOH}}}{V_{aliquot}} \times [\text{HCl}]^{tit} V_{\text{HCl},equi} \quad (2.113)$$

The quantity of acidic functions should be calculated for dry solid (typically as mmol g^{-1}), and the value obtained in Eq. (2.111) or Eq. (2.113) should be divided by the mass of dry solid. The average values of acidic sites should be calculated from triplicate test portions.

The use of the 3 base solutions results in 3 values of acidic sites: $n_{acid}(\text{NaOH})$, $n_{acid}(\text{NaHCO}_3)$, and $n_{acid}(\text{Na}_2\text{CO}_3)$. Based on the ability of each base to react with carboxylic, lactonic and phenolic functions, the final quantities are expressed as follows:

$$n_{carboxylic} = n_{acid}(\text{NaHCO}_3) \quad (2.114)$$

$$n_{lactonic} = n_{acid}(\text{Na}_2\text{CO}_3) - n_{acid}(\text{NaHCO}_3) \quad (2.115)$$

$$n_{phenolic} = n_{acid}(\text{NaOH}) - n_{acid}(\text{Na}_2\text{CO}_3) - n_{acid}(\text{NaHCO}_3) \quad (2.116)$$

2.8.4.4 Issues During Boehm Titration: CO_2 Dissolving and Release

Basically, the carbonate system in water shows the following equilibrium equations:

$$\text{CO}_{2(g)} + \text{H}_2\text{O} \rightarrow \text{H}_2\text{CO}_3 \quad \text{Henry constant} \quad (2.117)$$

$$\text{H}_2\text{CO}_3 + \text{H}_2\text{O} \rightarrow \text{HCO}_3^- + \text{H}^+ \quad \text{pK}_a = 6.4 \quad (2.118)$$

$$\text{HCO}_3^- + \text{H}_2\text{O} \rightarrow \text{CO}_3^{2-} + \text{H}^+ \quad \text{pK}_a = 10.3 \quad (2.119)$$

In an open system, the Henry constant determines the concentration of H_2CO_3 in equilibrium with the gaseous phase. Once the pH of the solution increases, the dissociation of H_2CO_3 to its conjugated bases increases as well, and this shift is compensated with an increase of dissolving CO_2. Therefore, the Boehm base solutions, as well as the NaOH titration solution are clearly subjected to dissolve CO_2 from the atmosphere. To overcome the bias induced by the atmospheric CO_2, the solutions should be freshly prepared prior to be used, and stored in sealed containers as well. In addition, when acidification step occurs, bases are neutralized and CO_2 is expected to be released at this low pH. Then the objective of the back-titration is to provide a direct quantification of protons (H^+) from HCl using NaOH. Since HCl and NaOH are strong acid and base compounds, the equilibrium will be reached at pH = 7. As demonstrated by Goertzen et al. [246], the presence of dissolved CO_2 in the solution provides a second inflexion point and shifts the equilibrium pH to higher values. The authors studied in detail the solution degassing and the impact on the resulting calculations. Particularly they noticed that a variation of 0.5 mL of titration solution will increase the quantity of acidic sites of about 90 μmol. They finally recommend to degass the acidified solutions for 2 h with N_2 or Ar prior to perform the titration, and during the titration as well.

Regarding the endpoint determination, indicator mixture was originally used: methyl red for direct titration (from basic to acidic solution), phenolphthalein or methyl blue for indirect titration (from acidic to basic solution). Potentiometric determination is also suggested to follow the pH of the solution along the titration and to determine the endpoint from the first derivative curve. Potentiometric method is more time consuming than a single point determination with indicator, and gives higher precision. However both methods give close results, this could be explained by experimentalist and burette errors.

2.8.4.5 Adaptations of Boehm Titration Procedure for Biochars and Related Pyrocarbons

Boehm titration was designed for carbon black characterization [233, 236], which are highly carbonaceous materials. The method assumed that only organic insoluble acidic sites react with the base solution. However, biochars and related pyrocarbons are not carbon black materials, since the carbonaceous structures and the ash contents are different [243, 250]. Certain acidic species such as carboxylic acids or phenols which do not compose the aromatic carbon network may be dissolved in the base solution. In addition, inorganic acidic species (such as silica, alumina) and inorganic basic species (carbonates, oxides or hydroxides) may also be dissolved using the base solution [234, 249]. Especially the acidic species are introducing bias in the Boehm titration since they will be protonated and precipitated during the acidification step. To overcome an inaccurate quantification of acidic sites, adaptations to the original Boehm methods have been suggested [234, 246–249].

Solid Sample Pretreatment: Removal of Base-Soluble Acidic Compounds and Ashes

This pretreatment aimed at removing base-soluble acidic compounds, and thus considering that they are not part of the solid matrix [234, 249]. In addition inorganic species are also removed to overcome bias in titration of organic functions. For this method (described in Fig. 2.128), the modifications are applied to the test sample and the resulting pre-treated test portions are further used in the conventional procedure described in Table 2.63. It should be noted that the acidic organic sites of interest for Boehm titration are firstly deprotonated with NaOH addition then protonated again after HCl addition.

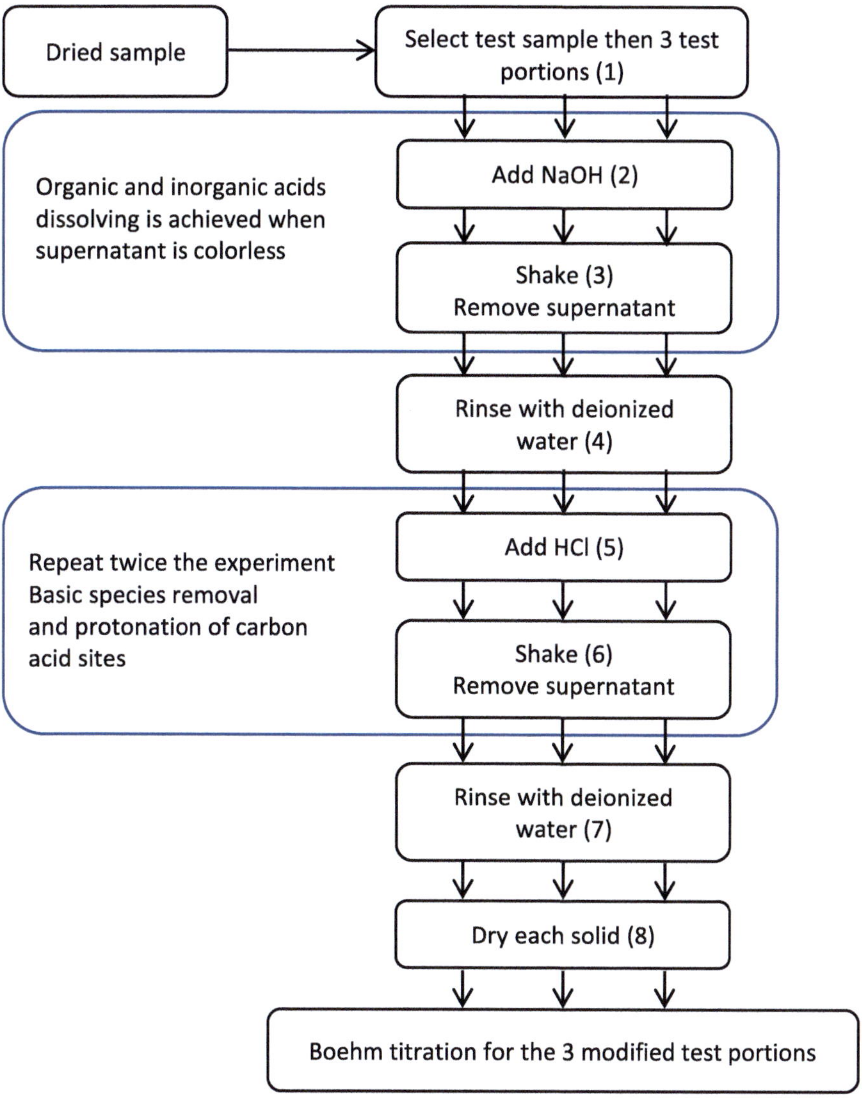

(1) For this procedure the initial mass is 1.5 g (each subsample of test portion) and then, after removal of soluble organic and inorganic species, the mass of each test portion should be recorded.

(2) About 40 mL of NaOH solution (0.05 mol L^{-1}) is added to the sample.

(3) The mixture is shaken for 24 h and then centrifuged and filtered to separate the solid sample. The procedure is achieved once the supernatant is colorless or light-yellow colored. Repeat (2) and (3) until the supernatant is colorless.

(4) Rinsing with deionized water is achieved once the solution reaches a conductivity of 30 $\mu S\ cm^{-1}$.

(5) About 40 mL of HCl solution (0.05 mol L^{-1}) is added to the sample.

(6) The mixture is shaken for 24 h and then filtered to separate the solid sample. Repeat (5) and (6) twice.

(7) Rinsing with deionized water is achieved once the solution reaches a conductivity < 10 $\mu S\ cm^{-1}$.

(8) Drying is performed at 50°C for 48 h.

Fig. 2.128 Pre-treatment procedure for solid samples

Table 2.63 Procedure to determine acidic functions using Boehm titration

Procedure	$pK_a < 6.4$	$pK_a < 10.3$	$pK_a < 13.0$
Select a test sample	Dried solid for calculations on dry mass basis		
Mass of test portion subsamples	1.50 g	1.50 g	1.50 g
Base solution	$[NaHCO_3]° = 0.05$ mol L^{-1}	$[Na_2CO_3]° = 0.05$ mol L^{-1}	$[NaOH]° = 0.05$ mol L^{-1}
Reaction with base solution	Put 1.5 g of solid in a container for shaking Add 50.00 mL of $NaHCO_3$ solution Shake for 24 h[a] in sealed tubes	Put 1.5 g of solid in a container for shaking Add 50.00 mL of Na_2CO_3 solution Shake for 24 h[a] in sealed tubes	Put 1.5 g of solid in a container for shaking Add 50.00 mL of NaOH solution Shake for 24 h[a] in sealed tubes
Separation[b]	Filter the solution Immediate analysis	Filter the solution Immediate analysis	Filter the solution Immediate analysis
Data collection	Back-titration: Take 10.00 mL of filtered solution Add 20.00 mL of HCl at 0.05 mol L^{-1} Remove CO_2 [d] Titration with NaOH 0.05 mol L^{-1} [e]	Back-titration: Take 10.00 mL of filtered solution Add 30.00 mL[c] of HCl at 0.05 mol L^{-1} Remove CO_2 [d] Titration with NaOH 0.05 mol L^{-1} [e]	Back-titration: Take 10.00 mL of filtered solution Add 20.00 mL of HCl at 0.05 mol L^{-1} Remove CO_2 [d] Titration with NaOH 0.05 mol L^{-1} [e] Direct titration: Take 10.00 mL of filtered solution Remove CO_2 [d] Titration with HCl at 0.05 mol L^{-1}

[a]As described by Oickle et al. [247], the best agitation method is shaking since it reduces the damages on the solid particles compared to sonication or stirring. The duration of the agitation should be sufficient for base diffusion into the porous structure. A time of 24 h is suggested in this procedure, but 72 h could be possible. A larger duration might be requested for very high porous materials. Although CO_2 is produced during the reaction with Boehm bases (Table 2.62), the pH of the solution, as well as the amount of CO_2 molecules produced should not result in CO_2 overpressure into the tube

[b]Filtration was the original method selected for the separation. However a discussion aroused since the cellulose/paper filters may contain surface groups that will interfere with the solution. As shown by Oickle et al. [247], a common filter paper will not interfere with the titration. A cellulose filter (0.45 μm) is often selected in the literature [234, 248]. If the immediate analysis after filtration is not possible, do not store in refrigerated atmosphere for more than 24 h

[c]HCl is used to acidify the filtered solutions, meaning that the proton (H^+) should react with the remaining base. Using Na_2CO_3, 2 protons are requested to neutralize the base, and then the acidification is performed with 30 mL of HCl solution

[d]This point will be discussed in detail in the next section (*Issues during Boehm titration: CO_2 dissolving and release*)

[e]NaOH solutions should be standardized prior to perform the titration, since carbonates might be dissolved and NaOH pellets are highly hygroscopic as well

Solid Sample Pretreatment: Barium Procedure

The present adapted method considers base-soluble acidic compounds as part of the solid matrix. The aim of this method is to remove ashes and to limit the interferences induced by base-soluble compounds during acidification prior to back-titration [248]. First carbonates, acetate salts and other inorganic species are dissolved using an acidic solution. Second $CaCl_2$ is used for precipitation of soluble organic compounds (Fig. 2.129).

2.8.4.6 Experimental Recommendations

Standardization of NaOH Solution

The NaOH solution should be standardized to minimize errors in calculating the quantity of acidic sites. NaOH is highly hygroscopic and can dissolve atmospheric CO_2. To ensure the accurate concentration of the solution made with NaOH pellets, the solution is titrated with a potassium hydrogen phthalate solution (about 0.2 g ± 0.01 mg of dried solid introduced in 25 mL of deionized water). The titration is conducted with phenolphthalein indicator [246]. This NaOH solution is then used to standardize the HCl solution.

Centrifugation:

The relative centrifugal force (RCF = 9600 g) is given in the procedure, but microcentrifuges have only settings for speed (revolutions per minute, RPM). A relation between both parameters exists:

$RCF = (1.118 \times 10^{-5}) \cdot R \cdot speed^2$, where R is the radius of the rotor in cm.

Samples Containing High Amounts of Ashes

For samples containing high amount of ashes, both methods may fail in giving the "right" values since the Si, Al and Fe oxides will not be dissolved during pretreatments. Their presence may act as acidic functional groups resulting in an overestimation of acidic groups [234, 248].

2.8.4.7 Examples

Some examples are provided with biochars. Figure 2.130 shows values obtained by Fidel et al. [248] using the barium chloride technique with corn stover pyrolysis biochars. The values of each acidic family are in the range of 0.15–1.24 mmol g^{-1}.

Figure 2.131 shows values obtained by Suliman et al. [251] with hybrid poplar wood biochars prepared at different temperatures. They followed the standard

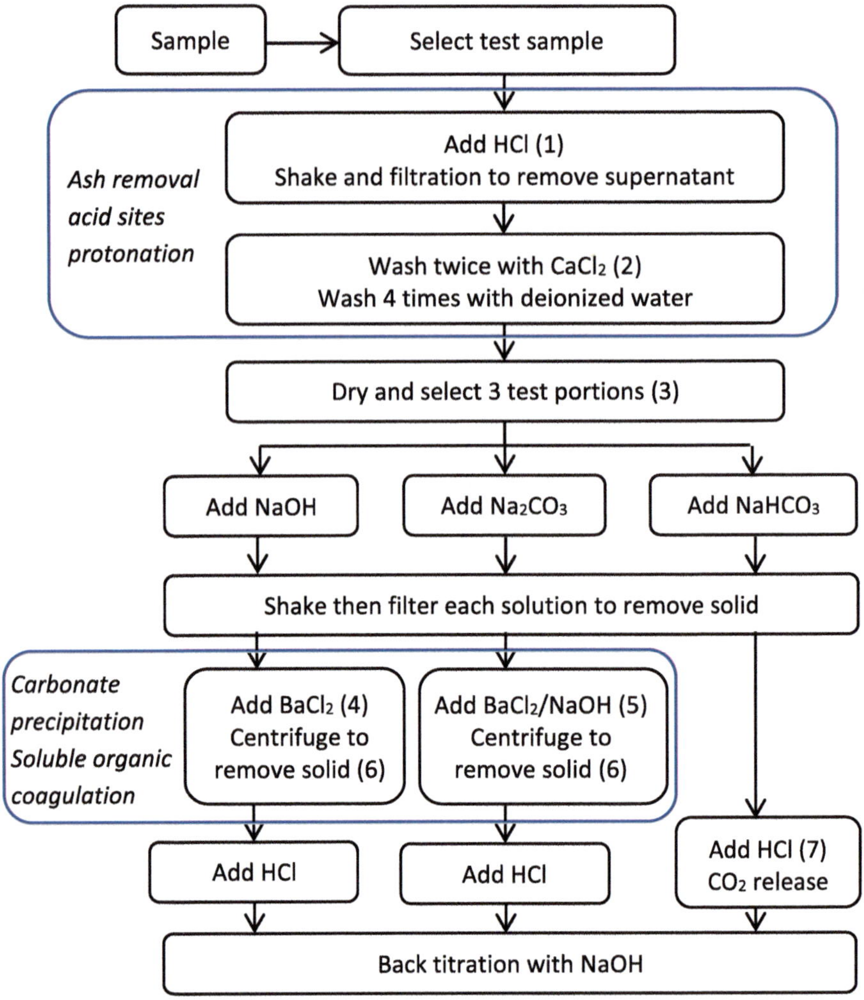

(1) 500 mL of HCl 0.05 M are added to about 10 g of solid. The mixture is shaken for 24 h, the slurry pH should be between 1 and 2.
(2) Acidification dissolves ashes but modify the solid pH before Boehm titration. $CaCl_2$ solution (500 mL, 1 M) is shaken with the solid for 15 min and filtered to ensure neutralization and heavy metal leaching. Solid on the filter is then washed with deionized water (500 mL) to remove remaining chlorine. Washing is achieved when pH of the water outflow is close to 5 (measure the pH) and chlorine is removed (measure the conductivity or check AgCl precipitation by adding drops of $AgNO_3$).
(3) Drying is performed at 50°C for 48 h (or more than 60 h). Record the ash-free solid mass. Then the Boehm procedure is started.
(4) 10 mL of NaOH extract is mixed with 10 mL of $BaCl_2$ solution (1 M). $BaCl_2$ is used to precipitate carbonates.
(5) 10 mL of Na_2CO_3 extract is mixed with 10 mL of $BaCl_2$ solution (1 M) and 10 mL of NaOH (0.05 M). $BaCl_2$ is used to precipitate carbonates, the NaOH solution will insure a sufficient pH to turn all carbonates to CO_3^{2-}.
(4) and (5) The mixture is centrifuged at 9600 g for 10 min. A clear supernatant is required for further steps (acidification and back-titration).
(6) In case the supernatant is not clear enough, a filtration could be done (0.02 μm) to remove fine colloidal.
(7) Carbonate interferences in this extract are simply avoided using CO_2 degassing with N_2 prior (2 h) and during titration.

Fig. 2.129 Scheme of the barium procedure

Boehm titration without any pretreatment. The values of each acidic family are in the range of 0–0.06 mmol g^{-1}. These values are lower than that from Fig. 2.130, meaning that the resource and the process have some influence.

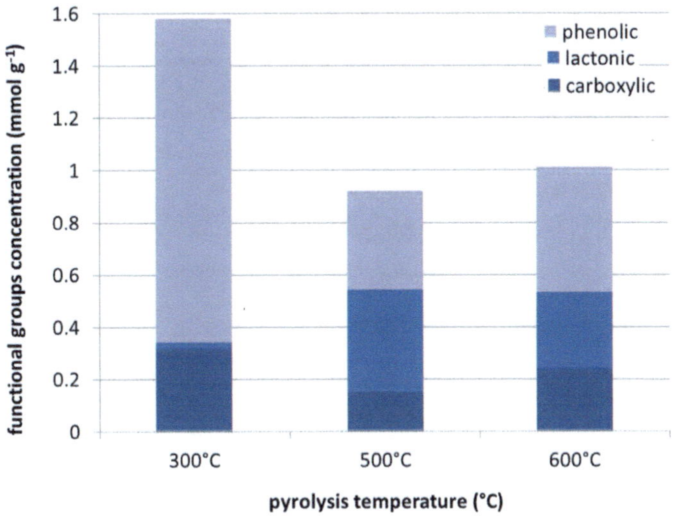

Fig. 2.130 Functional group concentration of corn stover biochars

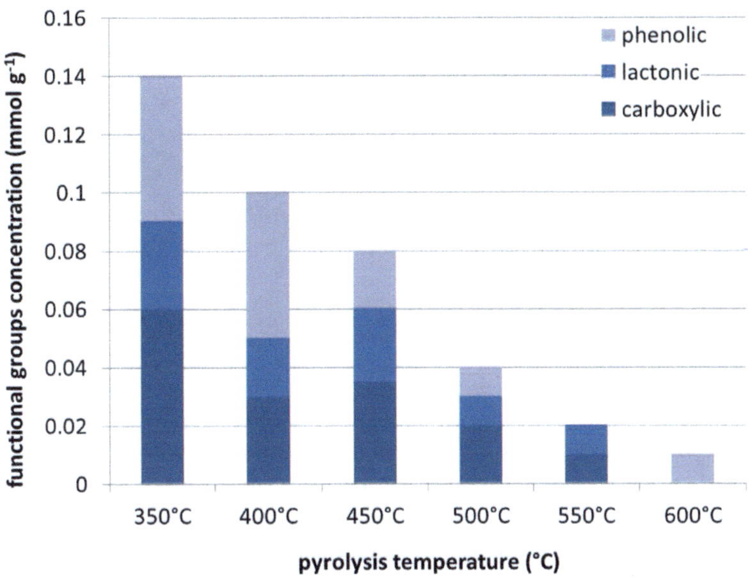

Fig. 2.131 Functional group concentration of poplar wood biochars. Reproduced with permission from Elsevier [251]

Although Boehm titration (related to organic functions) and Cation Exchange Capacity (exchangeable cations) are quantifying different kinds of positive charges, some authors have compared the values, as presented in Fig. 2.132 [252].

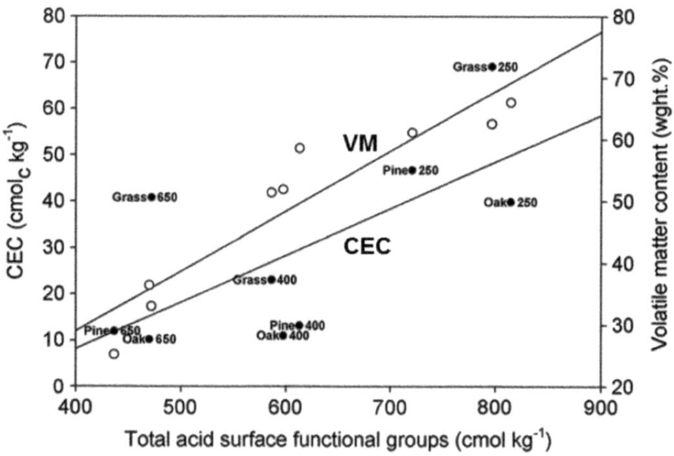

Fig. 2.132 Functional group concentration and CEC of various pyrolysis chars. Reproduced with permission from Elsevier [252]

However a special care should be taken for the methods used (here the CEC value has been measured with KCl solution at pH = 7, which is not a standardized method, and the acidic groups were determined with the original Boehm titration without any pretreatment), and the ash content of the solid considered (for example biochars from grass have more than 10 wt% of ashes).

2.8.5 Determination of Polycyclic Aromatic Hydrocarbons (PAH) Content

2.8.5.1 Introduction

Polycyclic Aromatic Hydrocarbons (PAHs) are molecules containing carbon and hydrogen atoms, organized in fused aromatic rings where some non-aromatic rings may also be included. PAHs basic structures are naphthalene (2 fused aromatic rings) and anthracene/phenanthrene (3 fused aromatic rings) as presented in Fig. 2.133. The term "PAHs" refers to a large number of chemicals that could be harmful, and are likely found as complex mixtures [253, 254]. PAHs and their derivatives are generated by incomplete combustion of organic material, such as biomass, biowastes, coal, fossil fuels. Therefore PAHs are one of the most widespread organic pollutants since several polycyclic aromatic hydrocarbons (PAHs) are known to be mutagenic and/or carcinogenic [253]. PAHs are individually quantified and the total PAH content is calculated in mg kg^{-1} on dry basis.

PAHs are non-polar and lipophilic molecules. Due to their chemical structure, their volatility and solubility in water decrease with the number of fused-rings.

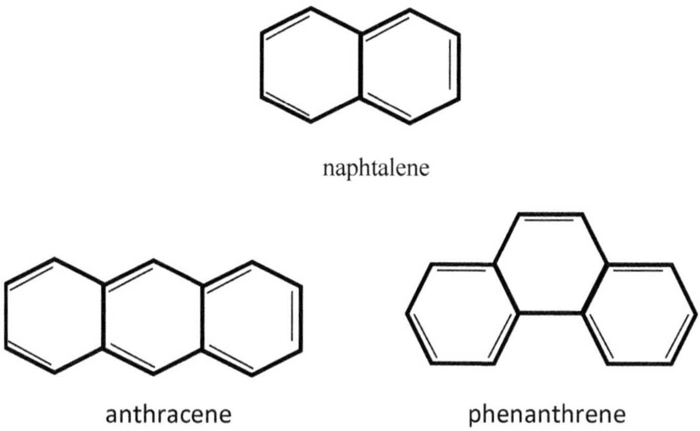

Fig. 2.133 Basic structures of PAHs

On the contrary, their solubility in oily substances, such as in human body, are facilitated with the increasing number of fused-rings.

Light PAH, such as naphthalene, is volatile and could be dissolved in water. PAHs containing 3 and 4 fused-rings, such as anthracene, phenanthrene and pyrene, are semi-volatile, and could be dissolved to a lesser extent in water. Heavier PAH exist in solid phases, which might be the residues from thermochemical processes, or soils and sediments as well [254]. Further description of PAHs contaminants: distribution, occurrence, concentrations human health effects etc. can be found elsewhere [254].

Many PAHs are attracting attention from European Commission (EU), US Environmental Protection Agency (EPA) and EU Scientific Committee for Food (SCF) points of view to be monitored in the environment (air, water, soils, sediments and wastes) and from industries [255–260]. The most common PAHs found in the incomplete combustion of biomass and biowastes, in addition to those presented in Fig. 2.133, are depicted in Fig. 2.134. Among them, benz[a]anthracene is one of the utmost PAH monitored due to its toxicity.

2.8.5.2 Measurement Methodology

PAHs are individually quantified by standard methods and the total PAHs content is also calculated in mg of PAH kg^{-1}. The bioavailability of PAH is also of great interest and a procedure will be described.

The standard method EN 16181 [261] is described in this section. This standard is affordable for sludge, biowaste and soil. This method is focused on the quantification of the 16 PAHs which are recommended to be monitored by EU, EPA and SCF (see Table 2.64).

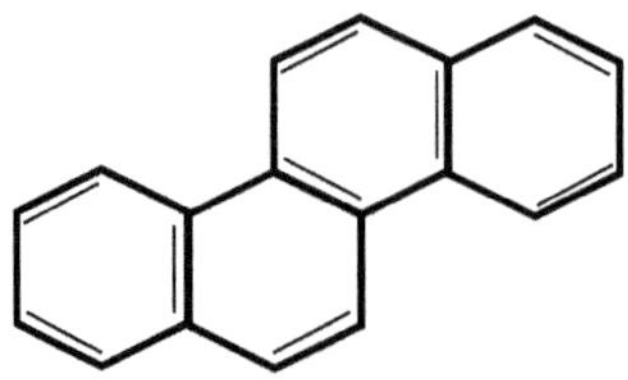

chrysene

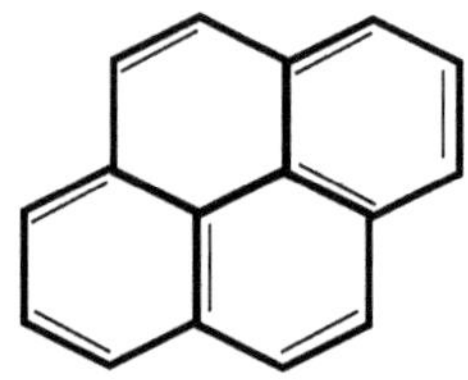

pyrene

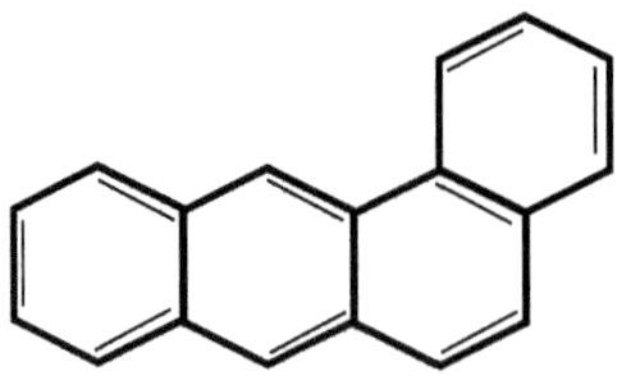

benzo[a]anthracene

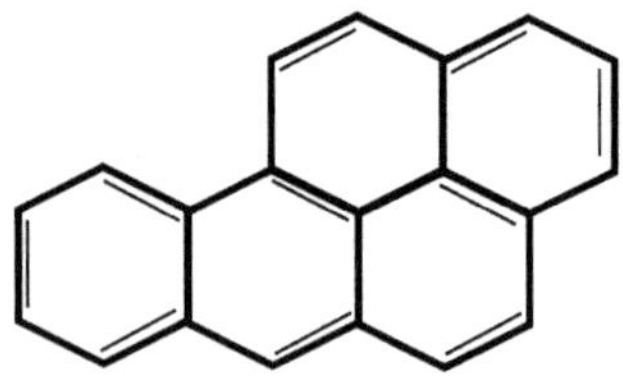

benzo[a]pyrene

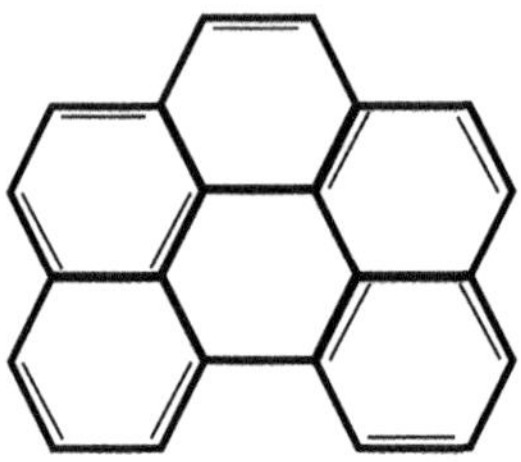

benzo[ghi]perylene

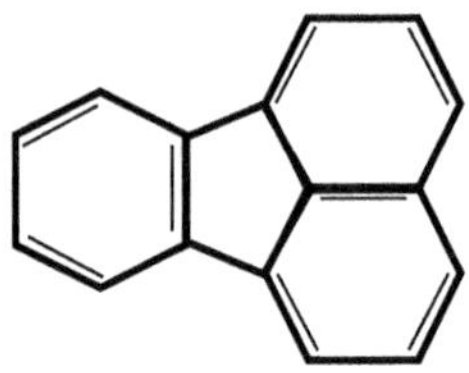

fluoranthene

Fig. 2.134 Main PAHs structures

In summary, PAHs are extracted with solvents from the pretreated solid residue and the resulting solution should be diluted or concentrated to improve the PAH detection in chromatographic techniques (see Fig. 2.135). This dilution/concentration step also serves for solvent adjustment. In some cases an additional cleaning is required prior to chromatographic analysis, in order to limit the number of peaks related to chemicals out of concern. The methods for detection and quantification of PAHs are performed with Gas Chromatography-Mass Spectroscopy (GC-MS) or High-Performance Liquid Chromatography with Ultra Violet-Diode Array Detector and/or Fluorescence Detector (HPLC-UV-DAD/FLD).

A blank test following extraction, dilution or concentration, cleaning if necessary and analysis, should be done with the same amount of reagents in order to evaluate external contamination.

Test portion selection should be carried out using ISO 18135 [262] and ISO 14780 [263] (standard methods applied to solid biofuels), or according to EN 16179 [19] (standard method applied to sludge, treated biowaste and soil). Obviously, all

Table 2.64 List of the 16 PAHs measured in the standard method

PAH name	CAS number
Naphthalene (see Fig. 2.133)	91-20-3
Acenaphthene	83-32-9
Acenaphthylene	208-96-8
Fluorene	86-73-7
Anthracene (see Fig. 2.133)	120-12-7
Phenanthrene (see Fig. 2.133)	85-01-8
Fluoranthene (see Fig. 2.134)	206-44-0
Pyrene (see Fig. 2.134)	129-00-0
Benzo[a]anthracene (see Fig. 2.134)	56-55-3
Chrysene (see Fig. 2.134)	218-01-9
Benzo[b]fluoranthene	205-99-2
Benzo[k]fluoranthene	207-08-9
Benzo[a]pyrene (see Fig. 2.134)	50-32-8
Indeno[1,2,3-cd]pyrene	193-39-5
Dibenz[a,h]anthracene	53-70-3
Benzo[ghi]perylene (see Fig. 2.134)	191-24-2

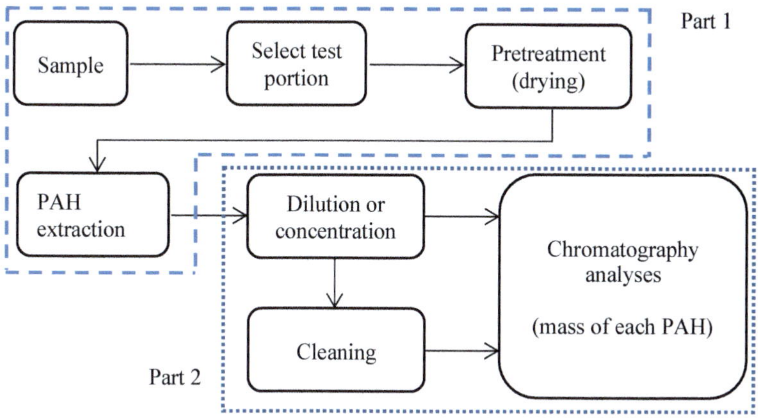

Fig. 2.135 Procedure used to determine PAHs content

samples (laboratory samples, test samples and test portions) should be stored in steel, aluminum or glass containers to avoid PAH contamination from plastic containers. The test portion is then dried to improve the homogeneity and the extraction efficiency (EN 16179 [19]).

Depending on the feedstock, 2 drying techniques are available: freeze drying, or Na_2SO_4 drying for sludge-like samples. Some samples with very low moisture content could be used without drying if freshly produced, otherwise after storage, drying should be performed. It should be noted that some of the PAHs are volatile and an oven drying at 105 °C should be avoided.

Extraction procedure should be selected among the 3 available techniques depending on the moisture and the matrix. This procedure will not use the same kind of solvents and will influence the further steps of concentration, dilution, and cleaning depending on the compatibility with GC or HPLC eluent. Therefore the following sections detail separately the first part (including extraction) and the second one (after extraction).

Part 1: Extraction of PAHs

The first steps of the standard EN 16181 [261] are described in Table 2.65.

Extraction procedures are able to extract up to 20 g of dry sample. If the sample has a low density or if the sample is homogeneous, fewer amounts can be used. Basically, for soil: 10–20 g; for sludge: 2–10 g; for compost: 5–20 g; for biowaste: 2–20 g. The amount of sample shall be weighed with an accuracy of at least 1%.

For sake of clarity, the names of the extraction methods (1, 2 and 3) (Table 2.65) are reported as named in the standard. However, methods 1 and 3 being quite similar, they are presented first.

Extraction Method 1 (Acetone and Agitation) for Dry Samples

This extraction procedure is strongly recommended if aggregates should be broken in the sample to reach the PAHs. This is especially important with soil samples containing clay particles. The principle of the method is to dissolve the PAHs into a hexane-like solvent. The experimental procedure is described in Table 2.66.

Table 2.65 Part 1 of the procedure to determine PAH content

Steps	PAH content procedure (EN 16181:2018)		
Type of sample	Dry	Dry, highly contaminated	Wet
Select a test portion EN 16179 [19]	Mass > 15 g, no particle size restriction 2 < mass < 15 g, particle size ≤ 1 mm 0.2 < mass < 2 g, particle size ≤ 500 µm		
Drying EN 16179 [19]	Freeze drying (EN 16720) Anhydrous Na_2SO_4		No drying for high moisture content
Extraction	Method 1 (acetone/ agitation)	Method 2 (toluene/ Soxhlet)	Method 3 (acetone/ agitation)

Table 2.66 Procedure 1 to extract PAHs from dry solids

Steps	Procedure for extraction method 1
Break aggregates	Put a defined mass of dried test portion in a flask *Add internal standard solution for GC or surrogate for HPLC*[a] Add 50 mL of acetone Shake 30 min in shaking device[b]
PAH extraction	Add 50 mL of hexane-like solvent[c] Shake 12 h in shaking device[b] Decant the mixture: store the supernatant (extract 1) and put the solid in a new flask
Wash the solid	Add 50 mL of hexane-like solvent[c] Shake in shaking device[b] Decant the mixture: store the supernatant (extract 2) and store the solid
Remove acetone	Put extract 1 and extract 2 in a separating funnel Add 400 mL of water[d] Shake and remove aqueous phase Proceed to water rinsing twice Collect the extract
Drying	Dry the extract with anhydrous Na_2SO_4 Collect Na_2SO_4 and rinse it with hexane-like solvent[c] Collect the rinsing phase and add it to the extract
Final extract	*Final extract of about 100 mL does contain PAHs into hexane-like solvent and potentially some impurities*

[a]For GC analysis, a defined volume of internal standard solution (see Sect. "Analysis with GC-MS") should be added before acetone addition for internal calibration. However, pay attention that the volume (or mass) added will insure a sufficient concentration of these standards in the final extract (about 0.1 μg/mL, or adapted to the measurement range). For HPLC analysis, a surrogate such as 6-methyl chrysene should be added before extraction for performance evaluation (recovery)

[b]Shaking device should have horizontal movement (200 strokes to 300 strokes per min)

[c]Hexane-like solvent has a boiling range comprised between 40 and 60 °C

[d]In the standard procedure, tap water is allowed for this step

Extraction Method 3 (Acetone and Agitation) for Wet Samples

This method has been designed for the PAH extraction of wet samples. The main differences with method 1 are: (i) the use of NaCl as promoter of liquid-liquid extraction, and (ii) the total water volume should be 50 mL. If the mass of sample and its moisture content does not match with 50 mL of water in the flask, a defined volume of water should be added. The experimental procedure is described in Table 2.67.

Extraction Method 2 (Toluene and Soxhlet) for Dry Samples

This extraction procedure is strongly recommended for highly contaminated dry solids and sludges. The principle of the method is to dissolve the PAHs into toluene using the Soxhlet apparatus. This apparatus and the experimental procedure are

Table 2.67 Procedure 3 to extract PAHs from wet solids

Steps	Procedure for extraction method 3
PAH extraction	• Put a defined mass of wet test portion in an Erlenmeyer flask (1 L) • *Add internal standard solution for GC or surrogate for HPLC*[a] • Add 40 g of NaCl[b] • Add 100 mL of acetone • Add 50 mL of hexane-like solvent[c] • Shake at least 12 h in shaking device[d] • Decant the mixture and collect the supernatant[e]
Remove acetone	• Put extract in a separating funnel • Add 150 mL of water[f] • Shake and remove aqueous phase • Proceed to water rinsing twice • Collect the extract
Final extract	*Final extract of about 50 mL does contain PAHs into hexane-like solvent and potentially some impurities*

[a]For GC analysis, a defined volume of internal standard solution (see Sect. "Analysis with GC-MS") should be added before acetone addition for internal calibration. However, pay attention that the volume (or mass) added will insure a sufficient concentration of these standards in the final extract (about 0.1 μg/mL or adapted to the measurement range). For HPLC analysis, a surrogate such as 6-methyl chrysene should be added before extraction for performance evaluation (recovery)
[b]NaCl is added in order to improve the phase separation, since it reduces the emulsification character of the water/organics mixtures
[c]Hexane-like solvent has a boiling range comprised between 40 and 60 °C
[d]Shaking device should have horizontal movement (200 strokes to 300 strokes per min)
[e]Decantation could be performed with either separating funnel or centrifugation
[f]In the standard procedure, tap water is allowed for this step

depicted in Fig. 2.136. For this extraction method, Pressurized Liquid Extraction (PLE) can also be used.

The Soxhlet extraction apparatus consists of a round bottom flask, e.g. 100 mL, Soxhlet extractor and Soxhlet thimble, e.g. 27 mm × 100 mm, vertical condensers, e.g. 300 mm, and heating device. The solvent is feed into the flask while the solid is put into the thimble. For this standard method, it is recommended to use 70 mL of toluene. By heating the solvent, the vapors are produced and then condensed at the top of the apparatus. The condensed solvent comes then across the solid sample, such as a cross-current solid/liquid extraction. The duration of the extraction should be estimated to fit with a minimum of 100 extraction cycles.

Part 2: Identification and Quantification of PAHs

After the extraction procedures, the extract should properly be prepared for injection in the chromatographic columns. On the one hand, the extracts could be prepared for injection without cleaning procedure (see Fig. 2.135), using concentration or dilution to fit with the lower and upper detection limits. During this step, the solvent should also meet the requirements to perform the HPLC analyses

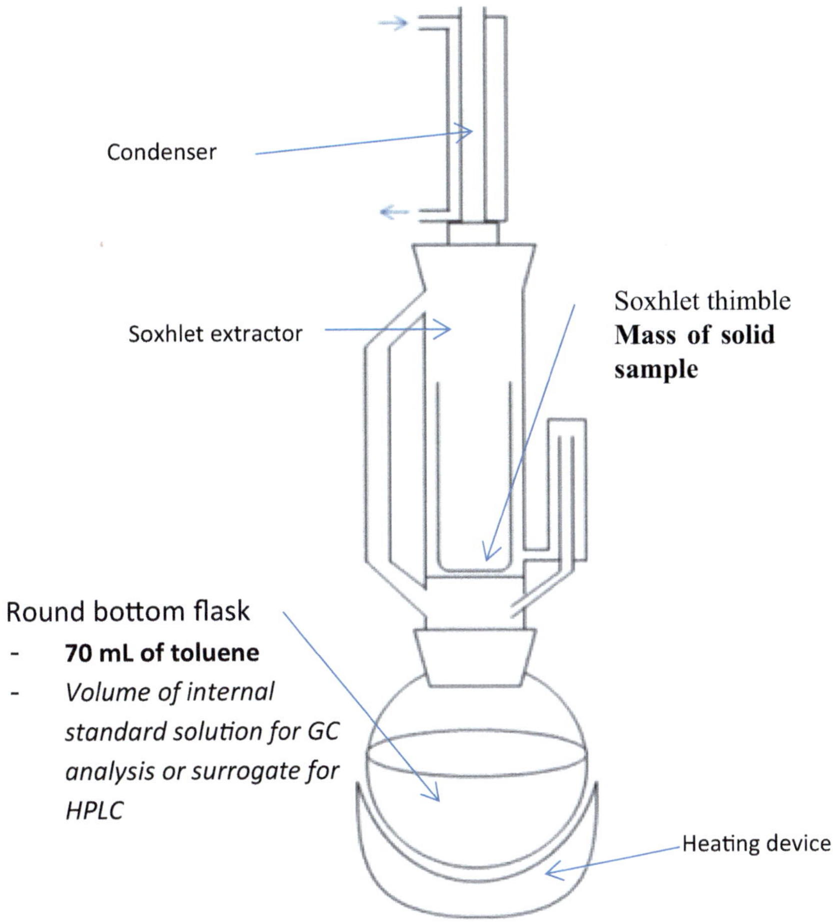

Fig. 2.136 Soxhlet apparatus for PAH extraction (method 2)

(eluent) and GC-MS analysis (boiling point) as well. On the other hand, the extracts could be cleaned-up after concentration and before injection (see Fig. 2.135) to remove impurities and spike peaks. As mentioned previously, solvent adjustments are also performed after the cleaning. This section presents also the requirements for gas and liquid chromatography for identification and quantification of PAHs.

Direct Preparation of the Extracts

For GC-MS analysis both solvents are suitable for the injection in the apparatus. However extracts should not be evaporated to dryness because the loss of the 2- or 3-ring compounds may occur. Therefore a parent solvent such as cyclohexane (boiling point at 80.75 °C) or toluene (boiling point at 110.60 °C) should be added as keeper in the concentrated extract [261, 264]. A keeper is not requested for extracts from method 2 since toluene has been used as the solvent. An injection standard is added to the final extract. This is used to check the recovery of the internal standard solution. Table 2.68 shows the steps for concentration/keeper addition (and optional dilution) of the extracts for further direct GC-MS analysis.

HPLC analysis will be further performed with acetonitrile eluent. Hexane-like solvent (extraction methods 1 and 3) should be replaced by acetonitrile. Toluene used as extraction solvent in method 2 can interfere with the detection of naphthalene and 3-ring PAHs. Therefore, extracts could be used as such when these substances are not present, or if a dilution is necessary. On the contrary, for volatile and semi-volatile PAHs quantification, and after a concentration, solvent replacement is requested. Table 2.69 shows the steps for concentration, solvent replacement (and optional dilution) of the extracts for further direct HPLC analysis.

Preparation of the Extracts with Cleaning Procedures

The purpose of the cleaning is to remove compounds that can interfere with the PAHs of interest in the GC-MS or HPLC chromatograms or if they influence the measurement procedure (i.e. contamination of injector system, column and

Table 2.68 Direct preparation of extracts for GC-MS analysis of PAH

Main steps	Detailed procedure		
Type of extract	Extraction 1 ≈100 mL extract (in hexane-like solvent)	Extraction 3 ≈50 mL extract (in hexane-like solvent)	Extraction 2 ≈70 mL extract (in toluene)
Concentration	Concentrate the extract to 10 mL using concentrator or rotary evaporator[a] Put the extract in calibrated test tube and carefully concentrate to 1 mL under slow nitrogen stream at room temperature[b]		*Optional concentration to a certain volume using concentrator or rotary evaporator*[a]
Add keeper substance	Add 100 μL of cyclohexane or toluene		
Optional: dilution	*Add defined volume of cyclohexane*		
Final extract	Select defined volume of the extract Add defined volume of injection standard[c]		

[a]Pay attention to the temperature to avoid the loss of volatile PAH
[b]Pay attention to the gas flow rate to avoid the loss of volatile PAH
[c]Injection standard solution is used to check the recovery of internal standard solution (see Section "Analysis with GC-MS"). The volume should be equivalent to a mass of 1 μg and the concentrations of PAHs in the injection standard solution should be in line with the concentrations in the internal standard solution to calculate the performance

Table 2.69 Direct preparation of extracts for HPLC analysis of PAH

Main steps	Detailed procedure		
Type of extract	Extraction 1 ≈100 mL extract (in hexane-like solvent)	Extraction 3 ≈50 mL extract (in hexane-like solvent)	Extraction 2 ≈70 mL extract (in toluene)[a]
Concentration	Concentrate the extract to 10 mL using concentrator or rotary evaporator[b] Put the extract in calibrated test tube and concentrate to 1 mL (or more) under slow nitrogen stream at room temperature[c]		
Replace solvent and optional dilution[d]	Add 0.8 mL of acetonitrile in the calibrated test tube Concentrate to 0.8 mL using an inert gas at room temperature Add acetonitrile to 1 mL		

[a]The final extract of toluene can be diluted with acetonitrile in accordance with the expected contamination level prior to HPLC analysis
[b]Pay attention to the temperature to avoid the loss of volatile PAH
[c]Pay attention to the gas flow rate to avoid the loss of volatile PAH. Do not evaporate the extracts to dryness, as losses of the 2- or 3-ring compounds may occur
[d]After concentration, the volume of extract in hexane-like solvent is targeted to 1 mL. However, for highly concentrated solutions, the concentration could be stopped for a higher volume. Then, only an aliquot of 1 mL will be further used for solvent replacing. The final volume of 0.8 mL refers to the added volume of acetonitrile, since the evaporation of hexane-like solvent should be achieved

detection). If no or negligible interfering substances are present, cleaning is not necessary. However, before using a cleaning to real samples, attention should be paid to ensure that recoveries after the cleaning are at least 80% for all relevant PAHs using a spiking standard. According to [261], different cleaning methods may be used including: (i) "alumina" method for acidic polar compounds; (ii) "silica" method for basic polar compounds; (iii) "gel permeation" method for high molecular compounds, lipids and oils; and (iv) "DMF/cyclohexane" method for aliphatic hydrocarbons, lipids and oils.

The cleaning should be done after the concentration step (see Fig. 2.135), meaning that the concentrated extracts of about 1 mL refer to the solutions obtained after concentration in Tables 2.68 and 2.69.

Alumina and silica methods:

Both methods are based on adsorption chromatography in a simple column to remove polar compounds. The different steps of the procedure are detailed in Table 2.70. Alumina used for this procedure should be basic or neutral, with a specific surface area of 200 $m^2\ g^{-1}$. Silica used should be silica gel 60, with a particle size of 63–200 μm. For both methods eluent is hexane-like solvent, even for extract from method 2 where solvent is toluene. For the extract from method 2, the final concentration will remove hexane-like solvent in priority compared to toluene which will be recovered in the final extract.

Table 2.70 Procedures for alumina and silica cleaning methods

Steps	Alumina method	Silica method
Column preparation	Take a glass chromatography tube Put small amount of quartz wool Put 2.0 ± 0.1 g of deactivated Al_2O_3 [a]	Take a glass chromatography tube Pack glass wool and deactivated SiO_2 (10 g) Put it into the column Add 1 cm layer of anhydrous Na_2SO_4 Add 20 ml of hexane-like solvent
Extraction	Transfer extract with a pipette Rinse test tube twice with 1 mL of hexane-like solvent and transfer rinsing solution to the column Elute with about 10 mL of hexane-like solvent	Put extract quantitatively into the column when the level of solvent is drained to approx. 0.5 cm above packing Elute with a total volume of 100 mL of hexane-like solvent
Final extract	Take the eluate *Add keeper substance*[c] Concentrate at T ≤ 40 °C to 0.5 mL[d]	Take the eluate *Add keeper substance*[c] Concentrate to desired volume[d]

[a]Preparation of deactivated aluminum oxide: add 10 g of water to 90 g of aluminum oxide (0 wt% of moisture content). Shake until all lumps have disappeared. Allow the aluminum oxide to equilibrate before use for some 16 h, sealed from air. Use it for a maximum of 2 weeks

[b]Preparation of silica gel 60 with 10 wt% of water content: heat silica gel 60 for at least 3 h at 450 °C, cool down and store in a desiccator containing magnesium perchlorate or another suitable drying agent. Before use, heat it once again at least for 5 h at 130 °C in a drying oven, and cool in a desiccator to ambient temperature. Transfer the silica gel to a flask and add 10% water (w/w). Shake for 5 min intensively by hand until all lumps have disappeared and then for 2 h in a shaking machine. Store the deactivated silica gel sealed from air and use it for maximum of one week

[c]100 μL of the keeper substance (cyclohexane or toluene) is only added to solutions from extraction methods 1 and 3, since toluene from method 2 will be the main solvent of the final concentrated extract

[d]The eluate is concentrated to desired volume following the procedure detailed in Tables 2.68 and 2.69

Gel permeation chromatography (GPC):

This cleaning procedure is designed to remove high molecular compounds, lipids and oils from the extract. The concentrated extract should be reduced to a solid residue under nitrogen flow, and immediately dissolved in 15 mL of elution mixture (cyclohexane and ethyl acetate 1:1 in volume). 5 mL of this dissolved residue is put into the GPC column (porous styrene divinylbenzene resin). A typical GPC system setting is depicted in Table 2.71, giving recommended values for first fraction, PAHs elution and last fraction [261].

Table 2.71 Typical GPC system settings

Parameter	Values
Flow rate	5 mL min^{-1}
Volume of the sample loop	5 mL
First fraction	120 mL, 24 min
PAHs elution	155 mL, 31 min
Last fraction	20 mL, 4 min

The eluate is collected in a glass vessel, a keeper substance is added (if requested as explained previously), and the eluate is concentrated to desired volume (following the procedure detailed in Tables 2.68 and 2.69).

DMF/cyclohexane partitioning:

This cleaning procedure to remove aliphatic hydrocarbons, lipids and oils is especially recommended for GC-MS measurements. The detailed procedure is described in Table 2.72. Dimethylformamide (DMF)/water mixture is first selected to dissolve PAHs and aliphatic hydrocarbons as well. When cyclohexane is added, aliphatic hydrocarbons are preferentially partitioned to cyclohexane phase. The PAH enriched phase of DMF/water could also contain oils and lipids. The second extraction serves the back-extraction of PAHs to cyclohexane due to high water content in the mixture.

General Comments About the Analytical Techniques

Chromatography technique selection:

It is worth noticing that GC-MS used to be the technique for PAH identification and quantification. HPLC techniques have recently been developed and are suitable for PAHs analysis. However the analytic technique should be selected according to the purpose of the study:

- The separation of dibenzo[a,h]anthracene and indeno[1,2,3-cd]pyrene by GC-MS is critical since they have very similar retention time. However, mass for diagnostic ions could overcome this difficulty.

Table 2.72 DMF/cyclohexane partitioning cleaning method

Steps	Detailed procedure
Extraction of aliphatic hydrocarbons to cyclohexane	Take a separatory funnel of 100 mL Dissolve extract (1 mL) in 10 mL of DMF/water 9:1 Put mixture in separatory funnel Extract with 10 mL of cyclohexane, repeat twice Store cyclohexane phase
Extraction of PAH from DMF	Put DMF/water phase in a separatory funnel of 500 mL Add 100 mL of pure water Extract with 10 mL of cyclohexane, repeat once
Final extract	Take final cyclohexane extract Add keeper substance[a] Concentrate to desired volume[b]

[a]100 μL of the keeper substance (toluene) is added before concentration
[b]The solution is concentrated to desired volume following the procedure detailed in Tables 2.68 and 2.69

- GC-MS separation of benzo[b]fluoranthene and benzo[k]fluoranthene is also critical since they have similar retention time and diagnostic ions as well.
- For HPLC analysis, a special care should be taken on the extraction solvent elimination since toluene has a broad peak that will interfere with naphthalene and 3-ring PAHs.
- HPLC with fluorescence detector could not identify and quantify acenaphthylene.
- Fluorescence detector is also strongly influenced by oxygen dissolved in the extract and the mobile phase.
- Ultraviolet detectors are providing bigger linear ranges compared to fluorescence detectors, meaning that UV is more suitable for a large variability of concentrations in the extracts.
- Fluorescence detectors are more selective and more sensitive, and thus required for trace elements quantification.

Standard solutions:

Due to the close responses of the PAH with both chromatographic techniques, the use of commercially available certified solutions are strongly recommended for the preparation of standard solutions.

Standard solutions for PAH identifications should be prepared with the suitable solvent for subsequent analysis: acetonitrile for HPLC and cyclohexane or toluene for GC. Each single substance stock solution should be prepared to achieve a mass concentration of 10 μg/mL. These solutions will serve to identify each of PAHs from the extract.

Multiple substances stock solution is also prepared in suitable solvent at a mass concentration of 10 μg/mL to serve as the preparation of calibration solutions.

The stability of all these solutions should regularly be checked.

Notices:

- Pay attention to equilibrate the chromatography system before measuring samples or calibration solution and check regularly the baseline noise.
- Reproducible retention times are usually achieved after two or three injections of a reference or standard solution.
- If the response of a PAH in the analysis extract is out of the working calibration range, dilute the sample and inject again.

Analysis with GC-MS

In gas chromatography, internal calibration method is used and different solutions should be prepared to identify and quantify the PAHs in the extract:

- Internal standard solution: solution containing at least 5 selected deuterated PAHs (see Table 2.73) in suitable solvent, and used for quantification of PAHs in the extracts. This solution should be added to the sample before extraction. The aim of this solution is to overcome the injection volume errors and matrix

Table 2.73 List of the deuterated PAHs

PAH name	CAS number
Naphthalene-d8	1146-65-2
Acenaphtene-d10	15067-26-2
Acenaphthylene-d8	93951-97-4
Fluorene-d10	81103-79-9
Anthracene-d10	1719-06-8
Phenanthrene-d10	1517-22-2
Fluoranthene-d10	93951-69-0
Pyrene-d10	1718-52-1
Benzo[a]anthracene-d12	1718-53-2
Chrysene-d12	1719-03-5
Benzo[b]fluoranthene-d12	93951-98-5
Benzo[k]fluoranthene-d12	93952-01-3
Benzo[a]pyrene-d12	63466-71-7
Indeno[1,2,3-cd]pyrene-d12	203578-33-0
Dibenz[a,h]anthracene-d14	13250-98-1
Benzo[ghi]perylene-d12	93951-66-7

effect by using PAH which are not present in the feedstock. It is recommended to select a sufficient number of deuterated PAHs in order to have their retention times close to those of the native PAHs.

- Calibration standard solutions: solutions of cyclohexane containing the 16 PAHs to be quantified as well as the deuterated PAHs selected for the internal standard solution. This solution will give response coefficients of each PAH to deuterated PAHs (internal calibration).
- Injection standard solution: solution containing deuterated benzo[e]pyrene or $^{13}C_{12}$-labelled PAH in suitable solvent to be added to the final extract before injection in the GC. The aim of this solution is to monitor the variability of the instrument response. It should be noticed that the injection standard solution comes with an extraction standard solution in another standard method [264].
- Performance standard solution: solution containing an equal volume of an internal standard solution and the injection standard solution (it may be one of the calibration standard).

As mentioned in the definitions of the different solutions, deuterated PAHs are used for the internal calibration since they are not endogenous PAHs in the feedstock. Table 2.73 presents the list of the deuterated PAH that can be used for these solutions.

GC-MS settings:

The GC should be equipped with a non-discriminating injection system, a capillary column and a mass spectrometric detector. The carrier gas is helium or hydrogen. Table 2.74 presents an example of typical conditions that should be optimized by

Table 2.74 Typical GC-MS settings from

Module	Settings
Column	• Capillary column • Stationary phase: 5% of phenyl-methyl silicone coated onto fused silica • Chromatographic resolution >0.8 for benzofluoranthenes • Length: minimum of 30 m • Internal diameter: 0.25 mm • Film thickness: 0.2 μm
Oven temperature program	60 °C, 2 min 60–120 °C at 30 °C min^{-1} 120–300 °C at 5 °C min^{-1} 300 °C, 15 min
Injector temperature	260 °C
Splitless injection	1 μL, keep the split 1.8 min closed
Carrier gas	Helium 0.8 to 1 mL min^{-1}
MS	Full scan or ion monitoring/recording mode (SIM/SIR) Use mass numbers in ISO 22892 [265] for quantitative analysis in selected ion monitoring mode (SIM)

each laboratory [261]. It should be noticed that some differences in the settings could be seen between the two standards [261, 264].

Calibration using the internal standard method:

The first step of this calibration method is to determine the response coefficients of each PAH relatively to the internal standard. A known volume of the internal standard solution (deuterated PAH) is added to the multiple substances stock solution (16 PAHs). This stock solution is then diluted to prepare at least 5 calibration solutions. The concentrations should be equally distributed between the solvent and the highest concentration. The calibration solutions are injected and the area of each peak is recorded. For linear calibration method, the relative response coefficient of each PAH is calculated as follows:

$$\frac{A^{ith\ solution}_{jth\ PAH}}{A^{ith\ solution}_{kth\ deuterated\ PAH}} = s_{jth\ PAH}\frac{C^{ith\ solution}_{jth\ PAH}}{C^{ith\ solution}_{kth\ deuterated\ PAH}} + b_{jth\ PAH} \tag{2.120}$$

where A corresponds to the peak area of each PAH or deuterated PAH in each calibration solution; C is the corresponding mass concentration in the calibration solution (in μg L^{-1} or μg mL^{-1}); s is the slope of the linear calibration function and b the intercept with the ordinate.

A non-linear calibration is also possible, but a daily calibration is requested.

The procedure described above corresponds to an initial calibration that should be done for the first time or after maintenance of the GC-MS. A daily check of the calibration should also be done. For a linear calibration, only 2 of the above 5 calibration solutions should be injected. These 2 solutions are selected to be at about 20 ± 10% and 80 ± 10% of the established linear range. A maximal deviation of 10% is accepted for the resulting straight line compared to that of the initial calibration. Otherwise, the initial calibration described above should be performed again. For a non-linear calibration all the 5 calibration solutions should daily be injected.

Identification of the PAHs:

The identification of the PAHs in GC-MS should follow the standard ISO 22892 [265]. Three steps are described:

- Step 1: relative or absolute retention time of peaks should be the same as those obtained from the calibration solution. Allowed deviations in relative retention time (t_r') of the sample compared to values from the calibration solution are:
  - For $t_r' < 500$ s: maximum deviation of 1 s;
  - For $500 < t_r' < 5000$ s: deviation of 0.2%;
  - For $t_r' > 5000$ s: maximum deviation of 6 s.
- Step 2: each PAH gives 3 diagnostic ions (see Table 2.75) [261, 265] with a certain relative intensity that should be calculated in the standard solution (Irstd) and in the sample solution (Irs). The relative intensities calculated for each diagnostic ion in the sample should not deviate by more than ±(0.1 × Irstd + 10)%. Every diagnostic ion gives 1 identification point. At least 3 points must be obtained in order to achieve identification.
- Step 3: if only 2 diagnostic ions fit the requirements above, additional experimental information are requested to achieve a third identification point.

Quantification of the PAHs:

For GC-MS analyses, the final extract contains the PAH extracted from the feedstock, the deuterated PAHs from the internal standard solution (this solution has been added before extraction procedure) and the deuterated benzo[e]pyrene or $^{13}C_{12}$-labelled PAH from the injection solution. The sample is used right after adding appropriate volume of the injection standard solution as mentioned in Table 2.68. The volume should be equivalent to a mass of 1 μg [264]. The total volume of the sample (V_s, in mL) is then measured for further calculations. For the extract, Eq. (2.120) could be rewritten as follows (including volume concentrations):

$$\frac{A_{jth\ PAH}^{extract}}{A_{kth\ deuterated\ PAH}^{extract}} = s_{jth\ PAH} \frac{C_{jth\ PAH}^{extract}}{C_{kth\ deuterated\ PAH}^{extract}} + b_{jth\ PAH} \qquad (2.121)$$

Finally, the mass concentration of each PAH (w, in mg kg^{-1}) in dry basis is calculated using Eq. (2.122) and should be rounded to two significant digits:

Table 2.75 Diagnostic ions for GC-MS analysis: m/z values and relative intensities in bracket

PAH name	Diagnostic ion 1 (m/z)	Diagnostic ion 2 (m/z)	Diagnostic ion 3 (m/z)
Naphthalene	128 (100)	102 (11)	–
Acenaphtene	154 (70)	153 (100)	76 (10)
Acenaphthylene	152 (100)	150 (3)	*76 (0)*
Fluorene	166 (81)	165 (100)	*139 (4)*
Anthracene	178 (100)	152 (12)	*76 (0)*
Phenanthrene	178 (100)	152 (9)	*76 (3)*
Fluoranthene	202 (100)	200 (31)	*100 (3)*
Pyrene	202 (100)	200 (2)	*101 (4)*
Benzo[a]anthracene	228 (100)	226 (3)	*114 (2)*
Chrysene	228 (100)	226 (6)	*113 (4)*
Benzo[b]fluoranthene	252 (100)	250 (22)	126 (5)
Benzo[k]fluoranthene	252 (100)	250 (22)	126 (5)
Benzo[a]pyrene	252 (100)	250 (18)	*113 (11)*
Indeno[1,2,3-cd] pyrene	276 (100)	138 (12)	*274 (4)*
Dibenz[a,h] anthracene	278 (100)	139 (9)	*276 (5)*
Benzo[ghi]perylene	276 (100)	138 (12)	*274 (4)*

$$w_{jth\ PAH} = \frac{\left(\frac{A_{jth\ PAH}^{extract}}{A_{kth\ deuterated\ PAH}^{extract}}\right) - b_{jth\ PAH}}{s_{jth\ PAH} \times m \times d_s} \times C_{kth\ deuterated\ PAH}^{extract} \times V_s \tag{2.122}$$

where A is the peak area of the measured PAHs and the corresponding deuterated PAH, b and s are related to the calibration function (see Eq. 2.120), V_s is in mL, C is the concentration of the selected internal standard in µg/mL, m is the mass of the feedstock used for the extraction (g) and d_s its dry matter fraction.

Performance of the method:

The performance of the method is evaluated using PAHs from the internal standard solution (deuterated PAH) and those from the injection standard solution (deuterated benzo[e]pyrene or $^{13}C_{12}$-labelled PAH). The recovery rate of each PAH (U, %) is calculated using the performance standard solution and following Eq. (2.123):

$$U_{jth\ PAH}(\%) = \frac{A_{jth\ PAH\ internal}^{extract}}{A_{PAH\ injection}^{extract}} \times \frac{A_{PAH\ injection}^{performance}}{A_{jth\ PAH\ internal}^{performance}} \times 100 \tag{2.123}$$

The recovery for each PAH should be at least 60%. The average ratio of all PAH should be at least 75% of the ratio in the standard.

Analysis with HPLC

Using HPLC, external calibration method is used and the following solutions should be prepared:

Calibration standard solutions: solutions of acetonitrile containing the 16 PAHs to be quantified;

Reference solution: this solution serves the identification of retention times;

Surrogate: this specific substance (6-methylchrysene for example) is added to the solution before extraction to check for the recovery of this substance throughout the whole procedure.

HPLC settings:

The HPLC should be equipped with a degassing assembly, analytical pumps suitable for binary gradient elution, column thermostat maintaining a constant temperature with a precision of ±0.5 °C, a reverse phase column, and a multi-wavelength ultraviolet detector as well as a fluorescence detector capable of programming at least six pairs of wavelengths, including damping/amplification, preferably equipped with monochromator(s). Table 2.76 presents some typical conditions detailed in the standard method [261]. However recent developments have been made for the PAH quantification and quicker methods are now available (see application notes of various manufacturers).

Calibration using the external standard method:

The multiple substances stock solution (16 PAHs, each at 10 μg/mL) is diluted to prepare at least 5 calibration solutions in acetonitrile. For external calibration, the linear regression should be done for each PAH by plotting concentration in the calibration solution (μg/mL) versus peak area.

Table 2.76 Typical HPLC-UV-DAD or HPLC-UV-DAD/FLD settings

Module	Settings
Column	Reverse phase column Stationary phase: spherical silica gel 5 μm, C18 with specific bonding Chromatographic resolution >0.8
Mobile phase	Acetonitrile (AN) Water (W)
Gradient	0–5 min: 50% AN, 50% W 5–25 min: 50–90% AN, 50–10% W 25–45 min: 90% AN, 10% W
Temperature	25 °C
Injection volume	5 μL, with loop
UV-DAD	Diode array detector over UV domain
FLD	Select at least 6 couples of excitation/emission wavelengths (see standard method for optimal couples [261]). Only acenaphthylene could not be detected

A reference solution should be prepared from the stock solution by diluting 50 μL in acetonitrile (total volume of 5 mL). The concentration of each PAH in this reference solution is 100 pg/μL. This solution serves as a reference chromatogram before initial calibration to determine retention times.

The procedure described above corresponds to an initial calibration that should be done for the first time or after maintenance. A daily check of the calibration should also be done. Every ten samples, a standard solution with concentration in the working range should be injected. The concentration of each PAH should not deviate by more than 10%. Otherwise, the initial calibration described above should be performed again.

Identification of the PAHs:

The identification of the PAHs with HPLC method is achieved when:

- The retention time of the substance is equal to the retention time in the reference solution (tolerance ±1%, maximum 10 s), using the same analytical conditions;
- And the presence of PAHs is confirmed at a second wavelength or using both detectors.

Quantification of the PAHs:

The final extract contains the PAH extracted from the feedstock and the surrogate. From the calibration regressions, the mass concentration of each PAH (w, in mg kg^{-1}) in dry basis is calculated using Eq. (2.124) and should be rounded to two significant digits:

$$w_{jth\ PAH} = \frac{A_{jth\ PAH}^{extract} \times a_{jth\ PAH}}{m \times d_s} \times V_s \tag{2.124}$$

where A is the peak area of the measured PAHs, a is the slope of the linear regression, V_s is the volume of the extract in mL, C is the concentration of the selected internal standard in μg mL^{-1}, m is the mass of the feedstock used for the extraction (g) and d_s its dry matter fraction.

Performance of the method:

The performance of the method is evaluated using recovery of surrogate by comparing its experimental concentration to its theoretical concentration. This recovery should be between 70 and 110%.

2.8.5.3 Example of Single and Total PAH Contents

From the detailed procedure, the mass concentration of each PAH is obtained on dry basis (mg kg^{-1}). The sum of all the single PAHs gives the total PAH content of the feedstock. An example of single PAH and total content is presented in Table 2.77.

Table 2.77 Example of single PAH content and total PAH content for a soil and a soil containing 1.16 wt% of biochar using GC-MS analysis

PAHs	Soil (mg kg^{-1})	Soil + char (mg kg^{-1})
Naphthalene	0.0098 ± 0.0002	0.0263 ± 0.0046
Acenaphtene	nd	nd
Acenaphthylene	nd	nd
Fluorene	0.0023 ± 0.0008	0.0033 ± 0.0006
Anthracene	0.0003 ± 0.0002	0.0014 ± 0.0014
Phenanthrene	0.0118 ± 0.0036	0.0212 ± 0.0063
Fluoranthene	0.0035 ± 0.0010	0.0075 ± 0.0030
Pyrene	0.0031 ± 0.0007	0.0069 ± 0.0020
Benzo[a]anthracene	0.0039 ± 0.0007	0.0057 ± 0.0009
Chrysene	0.0007 ± 0.0003	0.0014 ± 0.0010
Benzo[b]fluoranthene	0.0067 ± 0.0014	0.0091 ± 0.0029
Benzo[k]fluoranthene	0.0005 ± 0.0001	0.0014 ± 0.0003
Benzo[a]pyrene	0.0001 ± 0.0002	0.0019 ± 0.0009
Indeno[1,2,3-cd]pyrene	0.0023 ± 0.0008	0.0040 ± 0.0022
Dibenz[a,h]anthracene	0.0009 ± 0.0002	0.0014 ± 0.0004
Benzo[ghi]perylene	0.0046 ± 0.0011	0.0070 ± 0.0013
Total	0.506 ± 0.017	0.0986 ± 0.0190

Reproduced with permission from Elsevier [266]

2.8.5.4 PAH Bioavailability

Understanding the PAH bioavailability is of great importance when speaking about the use of biochar as soil amendment. As explained previously, PAHs are partitioned between hydrophilic and hydrophobic media. If PAHs are dissolved in water, they are bioavailable for living substances and will increase the toxicity. Therefore the bioavailability is measured by using the partitioning of PAHs between a passive sampler and the aqueous media. The bioavailable PAHs content is the amount of dissolved PAH in the aqueous media $\left(w_{jth\ PAH}^{water}\right)$. The procedure is explained in Fig. 2.137 [267]. PAHs contained in the biochar are expected to be partially released to the sampler or to the aqueous phase. By shaking during a sufficient time, equilibrium is reached between both media. After this step, PAHs adsorbed by the sampler are extracted to be quantified. Recommended passive sampler plastics are polyoxymethylene (POM), silicon and low-density polyethylene. These passive samplers are calibrated, meaning that the partitioning coefficients of each PAH for the sampler/water media is known. By measuring the PAH concentration in the sampler, the PAH concentration in water is deduced (Eq. 2.125).

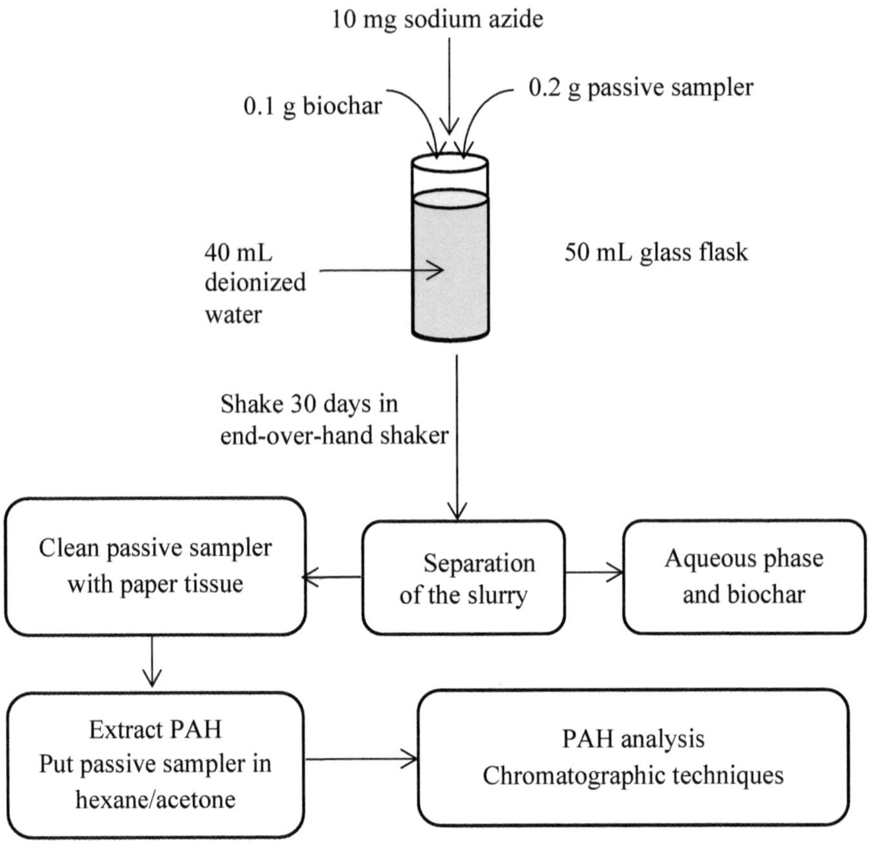

Fig. 2.137 Procedure used to determine PAHs bioavailability

$$w_{jth\ PAH}^{water} = \frac{w_{jth\ PAH}^{sampler}}{K_{jth\ PAH}^{sampler}} \tag{2.125}$$

where w is the mass concentration (mg kg^{-1}) of each PAH either in water or in sampler, and K is the partitioning coefficient (Table 2.78).

2.8.6 Surface Area Determination Using Vapor Sorption

2.8.6.1 Introduction

Surface characterization of solid materials is of great importance to investigate and to understand their behavior and reactivity. For example, the properties of materials

Table 2.78 Examples of total PAH contents and bioavailable PAH percentages for biochar

Biochar	PAHs content ($mg\ kg^{-1}$)	Bioavailable PAH (%)
Wheat straw 550 °C	0.057	0.6
Wheat straw 700 °C	0.006	6.9
Switchgrass 400 °C	0.042	0.9
Switchgrass 550 °C	0.044	0.9
Miscanthus straw 550 °C	0.321	0.1
Miscanthus straw 700 °C	0.054	0.7
Mixed softwood 700 °C	0.010	3.8
Rice straw 450 °C	0.207	0.1

Reproduced from Singh et al. [267] with permission from CSIRO Publishing

under humid conditions have to be well known (water uptake and release) for their application and storage issues. Carbon material could be used as sorbent [164], catalysts [162] or support for catalyst [166], where heterogeneous reactions between the solid and a gaseous or liquid phase could occur. The physico-chemical properties of the surface involved are the textural characteristics (porosity and specific surface area) [243, 268], the mineral content [165, 250], the oxygenated organic groups [239, 240, 242] and the structure of the carbon matrix [165, 238]. Dynamic Vapor Sorption is useful to investigate the active surface area rather than the total specific area.

The concept of active surface area (ASA) was initially introduced by Laine et al. [269] on graphitized carbon black and was applied to the study of carbon reactivity [270]. This ASA technique is based on the determination of the amount of oxygen chemisorbed on the active sites by temperature programmed desorption (TPD). However, the active surface will essentially represent polar sites while graphitic network (basal planes) corresponds to non-polar groups. For example, polar active surface area (ASA) at various burn-off of Graphon samples represented only 0.2 to 2.1% of the total surface area (TSA = S_p) [269]. Carbon materials from biomass and related products do not contain only graphitic network, organics are also identified in the structure and at the surface [164, 250] resulting in a modification of the number and polarity of active sites [271].

Dynamic vapor sorption (DVS) is largely used to describe surface properties of powder in the field of pharmacy [272], food processing [273] or drying [274]. DVS is a gravimetric method where the sample (a solid, a powder, a fiber, or a film) is exposed to known conditions of temperature and relative humidity. Basically DVS is performed using water vapor, since it is strongly adsorbed by solid materials as a condensable vapor. This means that water vapor will effectively interact with any solid materials whether they are organic, inorganic or biological.

Compared to gas sorption (Ar or N_2), the sample is not pretreated (no degassing) and the adsorption of the vapor is performed at ambient temperature or above, typically from 10 to 85 °C. In addition, this dynamic technique is based on the gravimetric measurement of water or solvent adsorbed until equilibrium state that results in an adsorption isotherm at temperature range mentioned above.

The Dynamic Vapor Sorption equipment is designed to accurately measure a sample's change in mass as it sorbs precisely controlled concentrations of water or organic vapors in dry air/N_2 carrier gas. The DVS gravimetric method uses a Cahn microbalance for the accurate determination of mass variations. The mass of vapor absorbed on the solid is measured as function of the relative humidity (i.e. vapor pressure) and is compared to a reference sample. The relative humidity is controlled by bubbling dry nitrogen through the water or the solvent. The mass of the sample is also continually measured that provide dynamic information of the adsorption phenomenon.

2.8.6.2 Principles of DVS Measurement

The DVS equipment described was provided by Surface Measurement Systems (DVS2, SMS). Figure 2.138 presents the scheme of this DVS instrument. The system is made of a chamber working at controlled temperature of 20–45 °C. The sample is placed in a pan and a similar pan is hanged on the other side of the Cahn microbalance (see Fig. 2.138). This DVS equipment contains two separate chambers where the pans are placed. The incubator is feed with the mixture of dry nitrogen and solvent (water or organic solvent) saturated nitrogen, where the solvent saturated nitrogen mixture is gradually introduced to a certain relative humidity. Targeted values of relative humidity are achieved using precise mass flow controllers to provide a mixture of dry and saturated vapors with specific ratio. The temperature and humidity control are provided by sensors located below the sample and the reference pan. The main advantage of this equipment is the stability of the

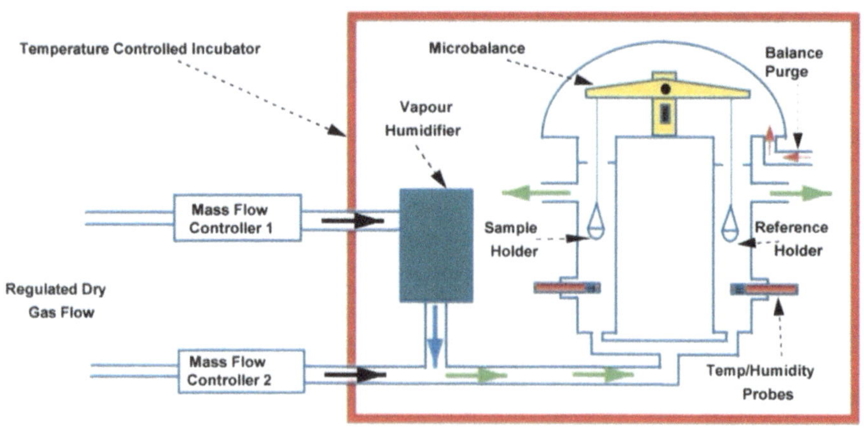

Fig. 2.138 Scheme of the DVS2 instrument SMS®. Reproduced with permission from SMS-Surface Measurement System [275]

microbalance. Indeed, dry nitrogen (manually controlled) is introduced in the microbalance chamber to avoid humidity accumulation that would degrade the quality of the microbalance system and thus its baseline.

It should be noted that using organic solvents in DVS2 requires the removal of video camera and temperature/humidity sensors. Therefore the partial pressure of the solvent vapor in the mixture is only determined by the mass flows.

Figure 2.139 presents a scheme of the DVS Advantage equipment from SMS. This system is also composed of a microbalance where the sample and the reference have been hanged. An Infra-Red or Raman spectrometer could optionally be added to follow the chemical modifications over the experiment. The main difference with the previous equipment is found in the determination and the control of the relative humidity. For this equipment, the partial pressure of water and vapors is controlled with a real-time sensor (optical detection). This system is thus able to control the partial pressure of organic vapor solvents in the incubator. As said previously, DVS equipment offers moderate working temperatures. An additional preheater could be added in the sample chamber to warm up to 150 °C or 300 °C.

DVS instruments have been designed to work continuously for laboratory experiments, and the shut off is not necessary after each analysis.

Figure 2.140 summarizes the procedure used to perform an experiment. The key issues are the parameters used in defining the conditions for acquisition (method). Basically the sample is placed in a pan and brought to desired temperature in a dry atmosphere and is kept to reach the equilibrium. The stabilized mass corresponds to the dry mass of the solid. Increments of relative humidity (RH) were managed by a computer program by steps or using a ramp. The equilibrium criteria were defined as the duration of the step (minimum and maximum values) and the adsorption kinetics (mass variation over time, and duration time of the plateau). For the latter,

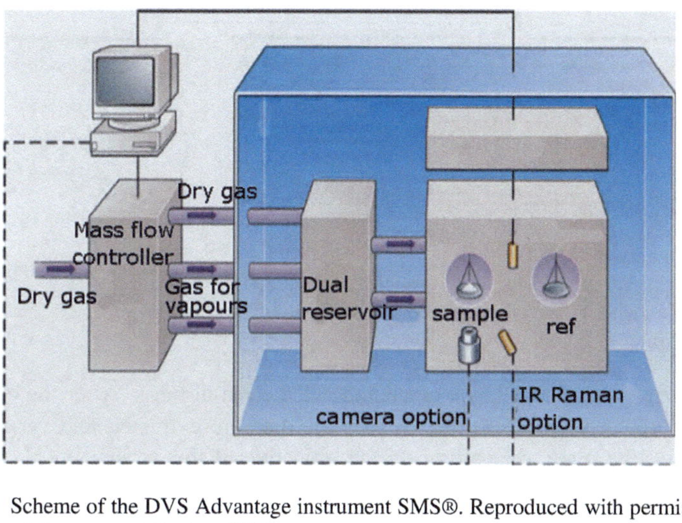

Fig. 2.139 Scheme of the DVS Advantage instrument SMS®. Reproduced with permission from SMS-Surface Measurement System [276]

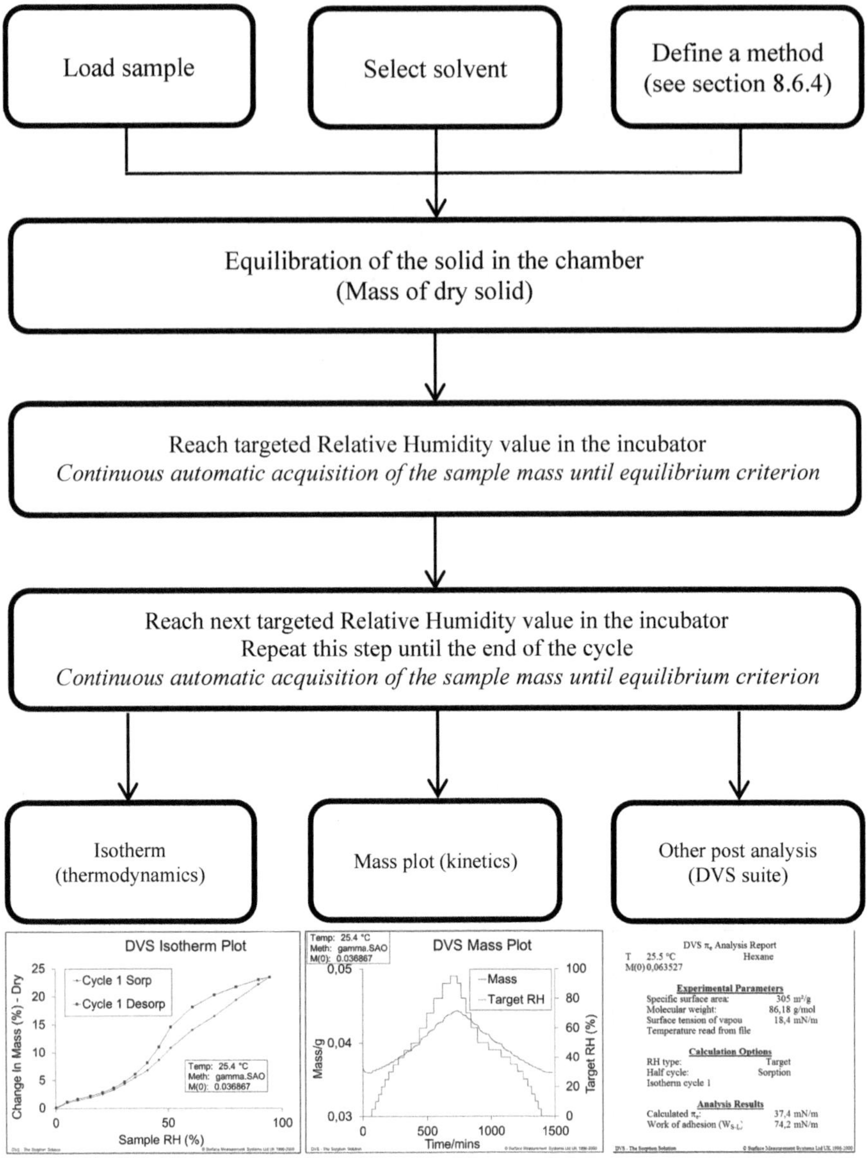

Fig. 2.140 Scheme of the procedure used to perform a DVS experiment

the parameter is expressed as a percentage of the initial mass. After the experiment (one or more sorption/desorption cycles) the data are extracted and various calculations could be made depending on the objective of the experiment.

2.8.6.3 Experimental Recommendations to Perform an Experiment

Figures 2.138, 2.139 and 2.140 show some basics about the DVS technique. The aim of this part is to give practical information to start an experiment, based on DVS2 (SMS) and DVS Advantage (SMS) instruments. The elements of the DVS instrument are summarized in Table 2.79.

Prior to the measurement, the calibration of the microbalance and humidity generators should be performed. In case of unexpected shutdown, the calibrations should be checked. Methods for verification or calibration of humidity generators described in ASTM E 2551 [277] could be used. For the microbalance calibration, specific calibration weight samples should be used, and the calibration should be checked at each working temperature. A summary of the procedure is provided below.

Prior to start an experiment, the following items should be turned on:

- Cahn microbalance;
- DVS control unit;
- Computer;
- Oven (incubator);
- Carrier gas supply (dry nitrogen at constant pressure of 1–2 bars);
- DVS software.

After an experiment, this is not necessary to shut-off the instrument. When a shut-off is requested, the following items should be turned off:

- DVS software;
- Carrier gas;
- Computer;
- DVS control unit;
- Cahn microbalance (note that the microbalance should be recalibrated after shut-off);
- Oven (incubator).

Sample Pan

Quartz pans are generally suitable for standard materials characterization. They should be used as a pair (one for the reference position and one for the sample position) as their matched weights will minimize buoyancy drift. Please note that a quartz pan is fragile and a particular care should be taken during its manipulation.

Stainless steel pans and stainless steel meshes are also available. For certain instruments, the microbalance might be adapted to measure up to 5 g. For this specific purpose glass pans are provided. Figure 2.141 presents different size and types of pans.

For accurate and reliable mass measurements in DVS equipment, sample pan should be as clean as possible, and the static from the sample should be removed.

Table 2.79 Elements of the Dynamic Vapor Sorption instrument

Elements	Type and range
Pan material	Stainless steel Quartz Glass
Mass control (microbalance) and precision	Mass of 1 mg–1 g ± 1 μg (DVS Advantage) Mass of 1–10 g ± 5 μg (DVS2)
Amplitude of the mass change (including the sample and the pan masses)	150 mg for 1 mg–1 g 1 g for 1–10 g
Mass flow controller	Mass Flo® controller MKS N_2 requirement: 0–500 sccm[a] Total mass flow (water)
Total mass flow with water Total mass flow with organic solvent	350–500 sccm[a] 150–300 sccm[a]
Vapor	Water, relative humidity (RH) of 0–96% ± 0.5% (DVS2) Water, RH of 0–98% ± 0.5% (DVS Advantage) Organic solvents, P/P_o of 0–90% ± 0.7% (DVS Advantage)
Control of humidity (DVS2) Control of partial pressure (DVS Advantage)	Rotronic sensor Unique-in-line transducer technology
Pressure	1 atm
Temperature control	10–45 °C ± 0.1 °C Thermocouples Pt-100 type K
Temperature of pre-heater (DVS Advantage)	Up to 150 °C or 300 °C

[a]sccm = standard cubic centimeter per minute

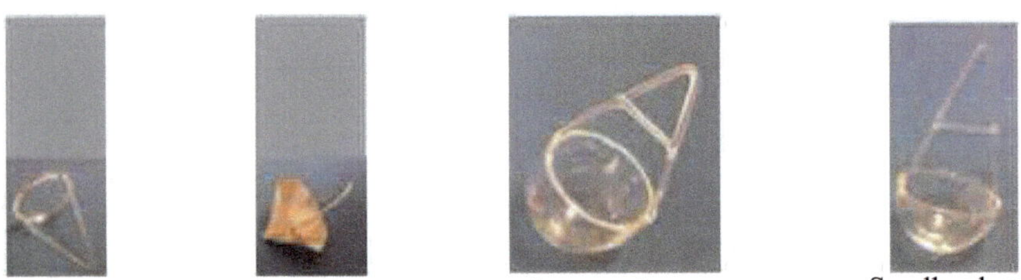

Small tray Perforated cone Large volume Small volume

Fig. 2.141 Examples of sample pans

For the cleaning of the pans, the basic procedure is the following:

- Clean with pure or deionized water;
- Rinse with a clean pure solvent such as ethanol (ethyl alcohol) or isopropyl alcohol (IPA);
- Dry in air (naturally).

If residual contamination remains after cleaning procedure, acceptable alternatives are to use a different solvent or a mild short-term ultrasonic cleaning. Please note that ultrasonic cleaning may occasionally result in pan breakage.

Please note that the following cleaning strategies are not accurate: flame (induces micro-cracks), detergents (not entirely removed they will induce a baseline shift) or the use of tissues to wrap them (this induces a static build-up).

The cleaning procedure would affect the baseline. In order to achieve a satisfactory baseline, the two important parameters below must be checked:

- The pan is dry, and therefore the baseline under these conditions (absence of water vapor) is stable. Basically after spending 5–10 min in 100% dry nitrogen the pan drying should be achieved and should provide a stable baseline;
- The sample pans have no static effect. To remove static effect, the pan should be maintained for 5 min in the manifold at a target moisture (P/P_o = 95%).

Microbalance

The calibration of the microbalance should be made regularly and at least after changing the temperature in the operating conditions, a stop, or when the DVS equipment is moved. The calibration is performed by starting the "calibration" procedure from the software with an empty microbalance, and then a certified calibration weight of 100 mg should be used (stabilization time is of about 5 min). The accuracy of the change in mass is limited to 150 mg (from 0 to 150 mg absolute values) to maintain a precision of ±5 μg of the measurements. In other words, the total mass should be measured in the range of 0–150 mg along the experiment. Therefore the mass sample should be chosen accordingly.

For higher sample mass (up to 1 g, upper limit of the instrument described), the use of counterweights is requested. The idea is to maintain the range 0 (tare) to 150 mg as targeted region for measuring the mass during the experiment. For this purpose selected calibration weight (m_{CW}) should be hanged to the microbalance (sample and reference chambers) and the balance should be tared to 0. Simultaneously, the value of the counterweight (m_{CW}) should be added to the control software for displaying the correct value of the mass.

Vapor Solvent

Water and organic solvent could be used in the instrument. The exit vapor gas stream should be vented to avoid emissions of toxic solvents in the room. For DVS SMS instrument the recommended organic solvents are summarized in Table 2.80. Only solvents of high purity should be used and HPLC-grade is recommended. It should be noted that the suitable organic solvents should be in vapor state at the experimental conditions, meaning that they should be selected according to their saturation vapor tension. Solvents with high flammability should not be used in the instrument.

Table 2.80 Suitable organic solvents for DVS instrument

Family of compound	Solvents
Alkane	Heptane, octane, nonane, decane, undecane, dodecane, cyclohexane
Alcohols	Methanol, ethanol, 1-butanol, isopropanol, 1-propanol
Aromatic	Toluene
Ester	Ethyl acetate, butyl acetate
Ketone	Acetone, 2-butanone
Other	1,4-dioxane, dichloro-methane, acetonitrile

Partial Pressure Measurement for DVS Advantage

The vapor pressure of organic and inorganic solvents is measured using a real time, in-line optical detector. This optical sensor determines the temperature at which a liquid droplet film forms on a special optical mirror. This dew point temperature is then converted into the solvent partial pressure using the sample temperature and standard vapor pressure tables for the solvent. The optical sensor is mounted at the rear of the DVS stand and in-line with the vapor flow stream. The sensitivity of the detector is factory-set for the most volatile vapors to be used. The DVS control software incorporates a number of special control features to maintain the cleanliness and performance of the optical mirror.

In order to maintain the accuracy of the Optical Sensor, it is recommended to perform a Dynamic Contamination Correction cycle (DCC) prior to running a partial pressure method. Since the measurements of partial pressure are based on the condensation over the mirror, its surface should be perfectly clean. Indeed any contaminants on the mirror may modify the temperature of film formation, and thus the value of the partial pressure.

2.8.6.4 Defining a Method to Run an Experiment

Once the instrument is stable (all the items turned on as explained in the previous section), the run could start. First, the mass of the sample should be carefully selected to fit with the upper limit (150 mg), to limit the experimental total duration time and to adjust the equilibrium criteria when expressed as dm/dt. If the change in mass is expected to be less than 0.5%, a sample mass of 50–100 mg should be chosen to get a more acceptable signal-to-noise ratio. On the contrary, for greater changes in mass, a range of 10–20 mg is advised to fit the mass range and to get suitable total experimental time as well. For very hydrophilic materials, a mass up to 10 mg will be preferred to limit the total experimental time. Then the equilibrium criteria should be defined in the software: duration time of each stage (for example 30 min for sorption, or 3000 min for drying) and the dm/dt parameters. This is important to properly define the equilibrium criterion for dm/dt method. Default settings considers the last 5 min of the plateau for dm/dt calculation as a percentage of mass variation over the time. For DVS2 instrument, this value could be changed (for example 10 min) and a maximum and minimum time of the plateau should be added, basically values used are 20 and 360 min respectively. Suitable values of dm/dt would be of 0.002% min^{-1} or 0.0005% min^{-1} for sample mass of 10–20 mg or 100 mg, respectively. The value of the equilibrium criterion should be lowered for dense materials (pellets) where the solvent diffusion will be lowered by the compact structure. Another way to define the dm/dt value is to consider a maximal mass variation equal to the precision of the balance over the duration time of the plateau, following Eq. (2.126).

$$\frac{dm}{dt} = \frac{precision\ of\ the\ balance / initial\ mass\ of\ the\ sample}{duration\ time\ of\ the\ plateau} \times 100 \tag{2.126}$$

where "*precision of the balance*" and "*initial mass of the sample*" are expressed in mg, and "*duration time of the plateau*" in minute.

Once the mass is chosen, the pan should be filled with the powder or the pellet. The powder should be spread as oddly as possible to improve the contact surface with the vapor. Once the pan is filled, it is hanged on the microbalance. A certain stabilization time is requested to ensure a total drying of the sample in N_2 atmosphere, and then the initial mass of the sample would be determined (m_0). For some samples, the drying might be done at higher temperature than that used for the experiment, when pre-heater is available. In case of drying experiment, the initial mass m_0 should be recorded once the pan has been hanged and stabilized (no oscillation) on the microbalance. The conditions that should be defined prior to perform an experiment are summarized in Table 2.81 (for DVS2 instrument).

In terms of the software, there is different software depending on the equipment (DVS2 and DVS Advantage). For DVS Advantage instrument, a method defines a type of analysis while a sequence is composed of successive methods that will automatically be run. Among the methods, two kinds of procedures could be selected: preheat or partial pressure method. For preheat method, the input data required are start and stop temperatures as well as the duration time. The user could define variable staged preheat methods to form a final preheat sequence. For a multiple stage sequence, stages may be isothermal (similar start and stop temperature up to 300 °C) or as a ramp.

2.8.6.5 Example of an Experiment

This section provides the first raw data extracted from an experiment: the mass plot over time and the isotherm plot. This experiment has been performed on a biochar from treated wood and the conditions of the sequence are summarized in Table 2.82.

As seen in Fig. 2.142, the drying method let the sample be carefully dried before starting the partial pressure method. The blue curve shows the targeted partial pressure (or relative humidity for water) as defined in the different half-cycles in

Table 2.81 Parameters to be defined to plot an isotherm and some recommendations for values (DVS2)

Parameters of the method	Range of values and comments
Equilibrium criterion of each stage	Time of the stage (for example 30 min) dm/dt value in % min^{-1} and the corresponding time of the plateau (for example 0.002 or 0.0005% min^{-1} over 10 min, depending on the sample mass, but a maximum of 9.99% min^{-1} is allowed), add the minimum and maximum values of the plateau if required
Start and stop relative humidity (%)	RH values in % (at least 5% difference since the accuracy is of ± 0.5% with water)
Temperature (°C)	T = 25 °C in general (values in the range 10–45 °C)
Isotherm edition	Temperature (see above) Equilibrium criterion (see above) Cycle determination: half, full or multiple Half cycle determination: start and stop RH (of the cycle, example 0–50%) + step size (RH increment, example 5%) Full cycle determination: same as half cycle (example increase of RH) + the symmetric opposite variation (decrease of RH with the same step value) Multiple cycles: reproduce the full cycle a certain number of times
Total gas flow (sccm)	In general 350 sccm with water or 200 sccm with organic solvents
Sampling rate (time)	time step for each data acquisition (in general 20 s)
Save data rate (time)	time step for each saving of the file (in general 1 min)
Select the solvent	Water or other organic solvent included in the software (see list in Table 2.80)

Table 2.82 Sequence used for the experiment example

Method	Values
Stabilization (drying)	Mass of sample in the stainless steel pan: about 37 mg Time = 30 min T = 25 °C State the initial dry mass m_0
Isotherm method	T = 25 °C Solvent: water Half-cycle 1: P/P° = 0–50% every 5% Half-cycle 2: P/P° = 50–90% every 10% Half-cycle 3: P/P° = 95% Half-cycle 4: P/P° = 90–50% every 10% Half-cycle 5: P/P° = 50–0% every 5% Equilibrium criterion = dm/dt < 0.002% min^{-1} for 10 min Total gas flow = 350 sccm Sampling rate = 20 s

Table 2.82. The red curve corresponds to the mass of the sample pan over time. The profile of the water uptake is basically a rapid increase followed by a slower increase of the mass, as seen in Fig. 2.143. With biomass resources and especially

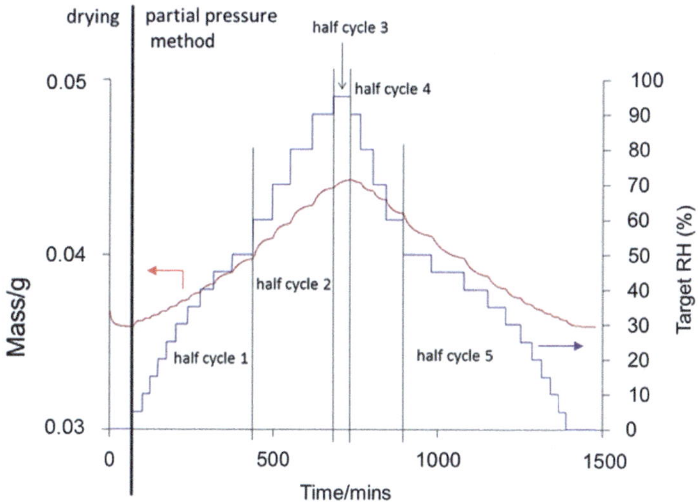

Fig. 2.142 Example of mass plot using DVS2 (SMS) for a biochar at 25 °C (water)

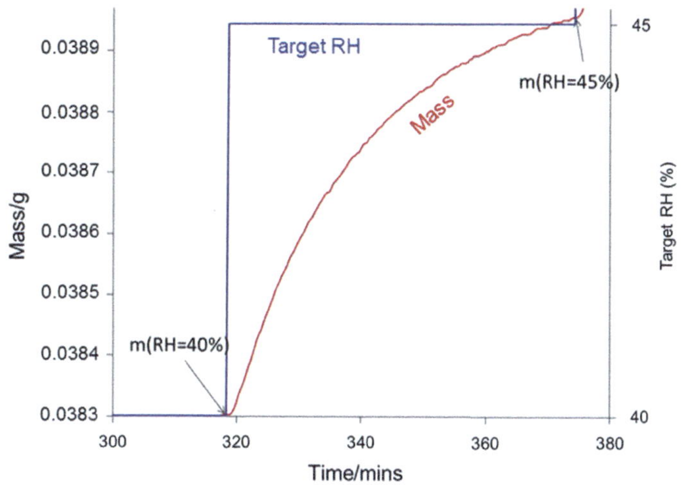

Fig. 2.143 Enlargement of a region of the mass plot using DVS2 (SMS) for a biochar at 25 °C (water)

biochars, this profile is the most commonly found in the literature for water uptake [278, 279].

The values of mass at the end of each partial pressure step (for example 40 and 45% in Fig. 2.143) correspond to the equilibrium ones, requested for the isotherm plot. The isotherm corresponds to the mass of water adsorbed (or the change in mass as in Fig. 2.144) when the sample is in equilibrium with the atmosphere at a

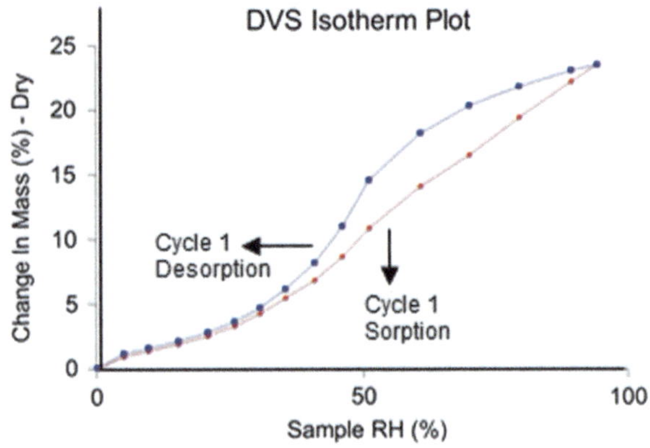

Fig. 2.144 Example of isotherm plot using DVS2 (SMS) for a biochar at 25 °C (water)

specific partial pressure. Figure 2.144 shows the sorption (half-cycles 1 and 2) and the desorption (half-cycles 4 and 5) isotherms, plotted thanks to all the equilibrium mass values extracted from Fig. 2.142. Although these isotherms are used for water adsorption, the shape of the isotherm should also be referred to the IUPAC classification [280] for interpretation. The shape of the hysteresis is also giving lots of information related to the sorption mechanism or the textural properties. From the mass plot and the isotherm, different post-analyses might be done depending on the research objectives such as active surface area (like BET equation, see gas adsorption for more detail), kinetics of the sorption (this specific feature will be explained in Chap. 15), type of pores (based on the shape of the desorption isotherm, when ONLY physisorption occurs)... The next section will explain the possibilities offered by DVS technique to characterize the solids. For example, a BET specific surface area (value of surface area from the BET equation) is calculated from the post-analysis of the isotherm. A value of about 100 $m^2\ g^{-1}$ has been calculated in the region of 5–21% of partial pressure. Using N_2 adsorption, a BET specific surface area of about 600 $m^2\ g^{-1}$ has been obtained. Finally, this experiment shows that only 1/6 of the total available surface is active for water sorption at room temperature for the biochar.

2.8.6.6 Characterization Options

Table 2.83 gives some investigations that are possible with DVS equipment. For more details related to practical performing the experiment, application notes from SMS Company would help the reader. The specific properties such as sorption kinetics and surface energy will be detailed in Chap. 15.

Table 2.83 Physico-chemical characterizations using DVS

Property of the sample	DVS method principle	Comments
Adsorption mechanism [281, 282]	Isotherm plot (sorption and desorption) using water or organic solvent	Chemi versus Physisorption using hysteresis and/or multicycle
Adsorption mechanism [283]	Isotherm plot (sorption and desorption) using water or organic solvent at different temperatures	Chemi versus Physisorption using heat of sorption
Adsorption mechanism (smooth surface or porous surface) [284]	Isotherm plot (sorption and desorption) using water or organic solvent	Chemi versus Physisorption Monolayer, multilayer or cluster mechanism
Texture of the surface [281]	Isotherm plot (sorption and desorption) using water or organic solvent	Use the hysteresis to define non-porous surfaces, presence of mesopores, micropores
Active surface area [285]	Isotherm plot (sorption and desorption) using water or organic solvent	Post-analysis with isotherm models
Sorption kinetics	Mass plot using water or organic solvent (small size steps)	Post-analysis with kinetics models
Storage behavior (moisture uptake) [286]	Mass plot using water	Post-analysis with kinetics models
Drying properties	Mass plot (desorption)	Post-analysis with kinetics models
Diffusion coefficient (bulk absorption) [287]	Mass plot using water or solvent: large size step to fill the pores	Post-analysis (Fick law)
Heat of sorption [283]	Isotherm plot (sorption and desorption) using water or organic solvent at different temperatures	Post-analysis (Clausius-Clapeyron equation)
Surface energy [288]	Isotherm plot (sorption) using polar and non-polar solvent	Post-analysis (Young equation)
Chemistry (amorphous content with calibration curve) [289]	Isotherm plot (sorption) using water or organic solvent of compound with various amorphous content	Post-analysis (calibration curve)
Chemistry (solvate, hydrates) [281, 290]	Isotherm plot (sorption and desorption) using water or organic solvent	Analysis of the hysteresis Post-analysis in case of calibration curve

2.8.6.7 Other DVS Instruments

Different equipments exist depending on the operating conditions and the type of post-analyses. Table 2.84 proposes a summary of instruments that might currently be found with some specific features.

Table 2.84 Dynamic vapor sorption instruments

DVS instrument	Specific features
SMS instruments: Common characteristics for most of the instruments Partial pressure range 0–98% Temperature range 5–85 °C Ultra-precision microbalance (±0.1 μg) Preheat up to 300 °C 1 sample Options: video, Raman and IR spectroscopy	
SMS DVS Adventure	Achieve the true 0% of partial pressure, only water
SMS DVS Resolution	Achieve the true 0% of partial pressure, water and organic solvents Ultrasonic sensor for accurate partial pressure measurement Option for 5 simultaneous samples
SMS DVS Endeavour	5 simultaneous samples, only water Ultrasonic sensor for accurate partial pressure measurement
SMS DVS Vacuum	Temperature range 20–70 °C Pressure range from 0.01 to 1000 Torr or 0.005–10 Torr Gases, organic solvents, water
SMS DVS Intrinsic	Temperature range 20–40 °C, compact system (no options) Only water
TA instruments Q5000 SA	Quite the same common characteristics as for SMS DVS instruments Automatic sampling for 10 sample pans Only water
TA instruments VTI-SA+	Quite the same common characteristics as for SMS DVS instruments Water and solvents Option: camera, Raman
Hidden Isochema XEMIS	High pressure: 100 mbar to 200 bar Low and high temperature: 196–500 °C Gases (organic, mineral)
Quantachrome Aquadyne	Only water Temperature range: 10–85 °C Partial pressure range 2–98%
Quantachrome Vistar	Water and organic solvents Temperature range: −20–100 °C Up to 4 simultaneous samples
Mettler Toledo SPS products	Only water Temperature range: 5–60 °C Partial pressure range 0–98% 11 or 23 simultaneous samples Options: Camera, Raman, NIR

2.8.7 *Specific Surface Area and Pore Size Determination by Gas Adsorption*

2.8.7.1 General Introduction to Gas Adsorption

Adsorption is a spontaneous phenomenon where atoms in a fluid, which can be a gas or a liquid, adhere and cover the surface of a solid. Depending on the level of energy involved between the adsorbed atoms, called the adsorbate, and the solid, called the adsorbent, two types of adsorption are encountered.

Chemisorption: when the energy involved is high, above 100 kJ mol^{-1}. This level of energy is explained by the creation of a covalent bond between the adsorbate and the adsorbent. The properties of the solid are therefore modified and the phenomenon is irreversible.

Physisorption: the level of energy between the adsorbate and the adsorbent is low (20–40 kJ mol^{-1}). In this case, the phenomenon is reversible, as the adsorbate could desorb under specific conditions.

It is important to do not confuse adsorption, a phenomenon happening at the surface of a solid, with absorption, where the molecules penetrate into the volume of solid (or liquid).

The gas adsorption, in particular the physisorption is largely used for the investigation of textural properties of divided solids. Whatever the field of application, i.e. the catalysis, the gas separation, or the gas purification etc., the investigation of surface properties is inevitable.

The quantity of gas retained at the surface of a solid will depend on the extent of this solid surface, gas pressure and temperature. The characteristics of the solid such as the specific surface area, the pore size distribution and pore volume could be determined when a gas such as nitrogen or carbon dioxide is used. Consequently, it is important to precise the experimental conditions used, including the nature of the gas, for each analysis result.

Extensive literature is available on the adsorption theory and techniques [291–300].

Definition of Specific Surface Area

The specific surface area is defined as the total surface area of a solid divided by its mass and expressed in m^2/g. It is calculated using Eq. (2.127).

$$a = \left(\frac{n_m^a}{m^s}\right) N_A \sigma_m \tag{2.127}$$

where n_m^a is the quantity of adsorbate (mol) to form a monomolecular layer on the surface of the solid; m^s is the mass of the solid (g); N_A is the Avogadro number

(6.023 E[23] per mol); and σ_m is the surface of one molecule of adsorbate (σ_m is equal to $0.162 \cdot 10^{-18}$ m^2 for N_2 at 77 K).

Divided solids, also called pulverulent materials or just simply 'powders', can be in various shapes: sphere, needle, plate, and so on. They can be variable in diameter, from few nanometers to few microns. These powders can be in the form of agglomerates, have surface irregularities (smooth or rough) or porosity (open or closed). These parameters (size, shape and porosity) influence measurement.

Definition of Porosity

A pore is a cavity that is deeper than it is wide. There are two types of porosities: open porosity with an open access at the surface of the solid and closed porosity which by definition has no access from the surface of material. Open porosity and intergranular of both aggregates and particles can be measured.

The porosity yield is determined as follows (Eq. 2.128).

$$\epsilon = \frac{V_{pt}}{V_{pt} + V_s} \tag{2.128}$$

where V_{pt} is the total porous volume and V_s is the solid volume.

The International Union of Pure and Applied Chemistry (IUPAC) classifies pores into 3 categories according to size:

- macropores: pore diameter greater than 50 nm;
- mesopores: pore diameter between 2 and 50 nm;
- micropores: pore diameter smaller than 2 nm.

As shown in Fig. 2.145, pores in different forms can be encountered. Experimental measurements can be carried out for the determination of open porosity as well as the total volume of the closed and open porosity. The difference between these two measurements gives the volume of closed porosity. In general, the closed porosity of a material is very low. The porosity of a given material can be studied by scanning electron microscopy (SEM), transmission electron microscopy (TEM), gas adsorption, and mercury porosimetry. However, it is difficult to get a

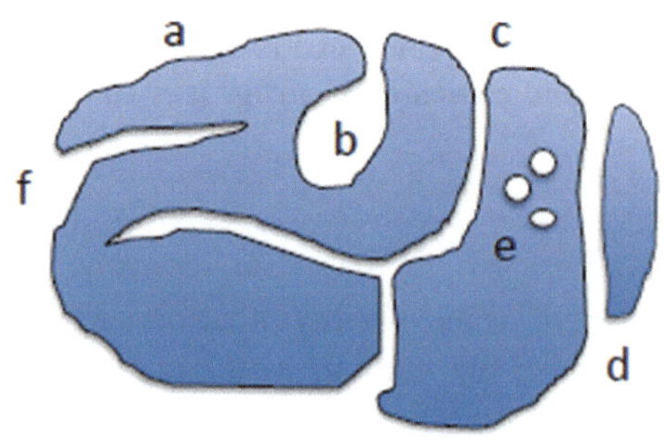

Fig. 2.145 Illustration of different pore forms (a) surface roughness; (b) bottle-shaped pores (pore entry narrower than pore bottom); (c) interconnected pores; (d) cylindrical pores; (e) closed pores; (f) finger pores

statistical result with SEM and TEM. Gas adsorption is appropriate for micropores and mesopores analysis while mercury porosimetry is used for macropores analysis (see Sect. 2.8.8).

Adsorption Isotherm

An adsorption isotherm corresponds to a set of equilibrium states of the adsorbate at the surface of the solid and its gas phase contained in the atmosphere at a given temperature.

These equilibrium states are represented graphically by tracing the quantity of gas adsorbed per gram of adsorbent as a function of the relative equilibrium pressure, i.e. relative to the equilibrium pressure of the adsorbed gas on the saturation vapour pressure of this gas. This relative equilibrium pressure is written as P/P_o.

Generally, the isotherm is characteristic of an adsorbate-adsorbent pair.

The notion of equilibrium here is linked to a state of thermodynamic equilibrium, and the adsorption isotherm is the result of experimental measurement. The isotherm therefore depends on experimental conditions (times, pressure and temperature) as well as on the capacities of the apparatus used.

Figure 2.146 shows a representation of the adsorption process of a gas on the surface of a solid as a function of the change in relative pressure.

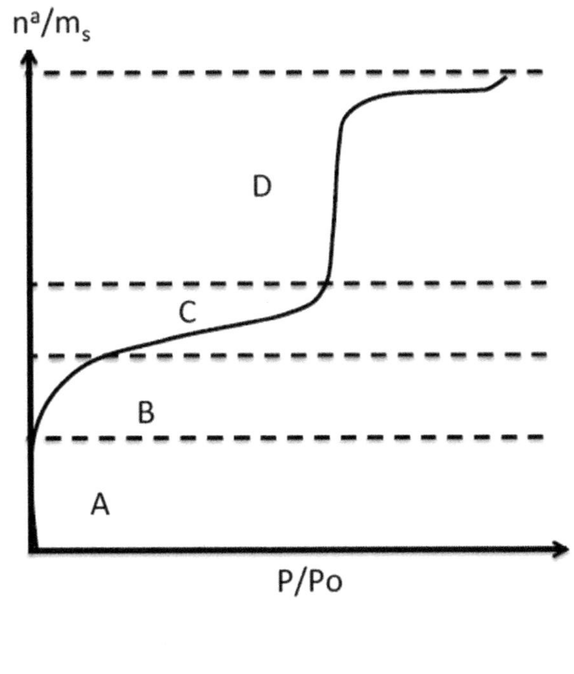

Fig. 2.146 Example of an adsorption isotherm with different stages of adsorption

It is observed, at low relative pressures corresponding to domain A, that the sites with the highest interaction energy are filled. These sites correspond to a surface in the narrowest pores, named micropores.

Then, the sites with lower energy will be filled (domain B). This corresponds to the domain where there is formation of an adsorbed gas monolayer on the surface of the adsorbent.

In domain C, the surface of the adsorbent is much filled with the gas resulting to a multilayer structure.

Adsorption ends close to the adsorbate saturation pressure. This corresponds to domain D which implies the largest pores and a capillary condensation phenomenon at the surface of the solid.

The appearance of the isotherm is characteristic of adsorbate-adsorbent interactions that reveal a surface structure. As an aid to interpretation, IUPAC determined a classification for the most commonly isotherms. This classification, established in 1985, had previously been described by Brunauer et al. [301]. It has recently been revised by IUPAC to include new isotherm types obtained from new materials as shown in Fig. 2.147.

This classification describes 6 types of isotherm.

- Type I isotherms are obtained for microporous materials, such as active charcoal or zeolites. High adsorption at low relative pressures can be observed, representing filling of micropores followed by a plateau that reflects saturation of pore. There are two forms of Type I isotherm. Type I(a) is obtained for very narrow micropores (width below 1 nm) referred as ultramicropores. Type I(b) is obtained with materials having pore size distributions over a broader range including wider micropores and possibly narrow mesopores (width greater than or equal to 2.5 nm) called supermicropores;
- Type II isotherms are obtained for nonporous or macroporous materials on the surface of which the adsorbed gas layer becomes thicker as relative pressure increases. This is called multilayer adsorption. Point B is identified as being the end of the monolayer;
- Type III isotherms differ from Type II isotherms at low relative pressures. It is obtained for materials with no interaction with the adsorbed molecule;
- Type IV isotherms are characteristic of materials that exhibit mesoporosity. This porosity is confirmed by the presence of a saturation plateau at the highest relative pressures. In some cases, this saturation plateau can be reduced to an inflection point. The phenomenon of mesoporosity capillary condensation leads to formation of hysteresis that may or may not overlap the adsorption;
- Type V isotherms are a compilation of Type III and Type IV isotherms. This type of isotherm is rarely encountered;
- Type VI isotherms are observed for solids with energetically homogenous surfaces on which adsorption layers form one after another.

When the adsorption and desorption isotherm is studied, a hysteresis loop caused by the capillary condensation phenomenon is observed. The hysteresis form is

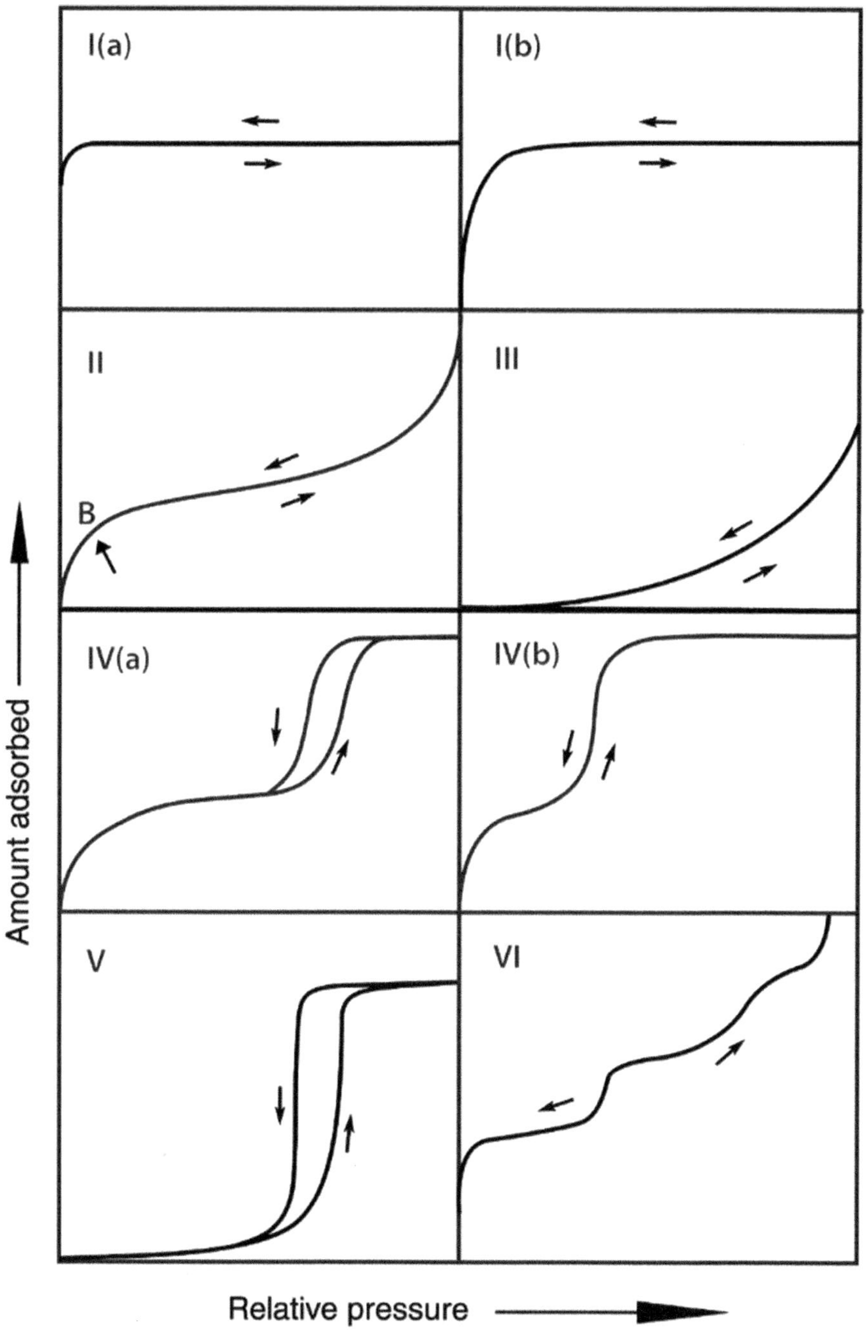

Fig. 2.147 IUPAC classification of adsorption isotherms. Reproduced with permission from De Gruyter [295]

directly linked to pore form. IUPAC has classified this hysteresis into 6 classes, as shown in Fig. 2.148.

- Hysteresis H1 is obtained for type IV isotherms (characteristic of mesoporous samples) with a saturation plateau at high relative pressures. This loop presents parallel adsorption and desorption branches that are almost vertical. This appearance is characteristic of an adsorbent with mesopore that are narrow and probably cylindrical, with equal pore entry and exit diameters;
- Hysteresis H2 is obtained, as for H1, for type IV isotherms. The desorption is delayed compared to adsorption. This type of hysteresis is characteristic of an adsorbent with complex mesoporosity, with pores that are interconnected or "*ink-bottle-shaped*" (where pore entry diameter is different from pore bottom diameter);
- Hysteresis H3 is generally observed for adsorbents forming aggregates or with a leaf-shaped structure. This appearance is characteristic of an unstable structure created by capillary condensation that has the effect of expanding the structure, in the manner that can be compared to a dry sponge and the same sponge that expands after being wetted. This type of hysteresis is also observed if the pore size distribution is very wide (mesoporosity, even a degree of macroporosity). Pores are not completely saturated by the condensed gas: there is no saturation plateau;

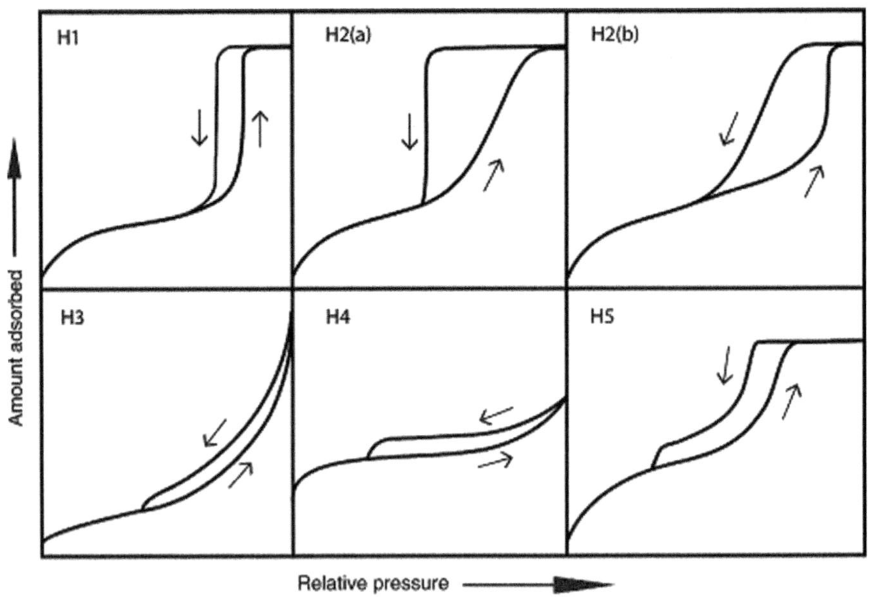

Fig. 2.148 IUPAC classification of hysteresis loops. Reproduced with permission from De Gruyter [295]

- Hysteresis H4 is similar to H3. The only difference is at low relative pressures where the filling of micropores occurs;
- Hysteresis H5 is encountered for certain porous structures that may present cavitation.

More details on adsorption isotherms and hysteresis classification can be found at [295]. As the IUPAC classification was established using model materials, it is commonplace when making analyses to encounter so-called "*composite adsorption isotherms*" that may embody several types of isotherm in the classification.

Applied Models

It is from these adsorption-desorption isotherms obtained that mathematical models will be applied to assign quantitative values to specific surface area, porous volume and pore size distribution.

The difficulty lies in understanding these more or less empirical models, recognizing their respective hypotheses and evaluating whether they can be applied to the isotherm obtained. In this section, the theories utilised and the hypotheses underpinning each model used will not be detailed as this is not the primary aim of this handbook. They can be found elsewhere [291–295, 302].

The management softwares provided by the majority of apparatus manufacturers provide a list of models, such as BET (Brunauer, Emmett and Teller), BJH (Barrett, Joyner and Halenda), HK (Horvath and Kawazoe), which are the best-known and most commonly used models.

Less frequently used models such as Dubinin, Freundlich, and DFT (Density Functional Theory) etc., are also available but they are not discussed in this document.

Table 2.85 lists the models most commonly encountered. It helps to identify the suitable model to be used to obtain information from a given isotherm type.

For example, for a type I isotherm, characteristic of a microporous solid (pores below 2 nm), the BET or Langmuir equations (in certain cases) can be applied to obtain specific surface area. The HK (Horvath and Kawazoe) or *t*-plot models can be applied for the determination of microporous size and volume, but the BJH model cannot be applied for this isotherm, as this is specific to mesopores.

Isotherms are traditionally measured using nitrogen at 77 K (temperature of liquid nitrogen). For more specific measurements, other gases can be adsorbed such as argon, krypton or carbon dioxide at the temperature of liquid nitrogen (77 K) or liquid argon (87 K). In the case of carbon dioxide, the isotherm is created at 273 K or 298 K.

Table 2.85 Models applied to determine specific surface area, pore volume and size distribution as a function of the isotherm type

Model		BET	Langmuir	t-plot	BJH	HK
		Brunauer Emmett Teller	Langmuir	t-plot (Harkins and Jura)	Barrett, Joyner Halenda	Horvath and Kawazoe
Mathematical equation		$\frac{n^a}{n^a_m} = \frac{C_x}{(1-x)(1-x+C_x)}$	$\frac{P/P_o}{n_a} = \frac{1}{n^a_m b} + \frac{P/P_o}{n^a_m}$	$t = \left(\frac{0.1399}{0.034-\log(P/P_o)}\right)^{0.5}$	$\delta V_{p,n} = \left(\frac{\bar{r}_{p,n}}{\bar{r}_{p,n}-t_n}\right)^2 \left[\delta V^l_n - \delta t_n \sum_{i=1}^{n-1} \delta a_{p,i} + \delta t_n t_n \sum_{i=1}^{n-1} \frac{\delta a_{p,i}}{\bar{r}_{p,i}}\right]$	$\Phi(z)_{pore} = k\left[\left(\frac{\sigma}{z}\right)^{10} - \left(\frac{\sigma}{z}\right)^4 + \left(\frac{\sigma}{W-z}\right)^{10} - \left(\frac{\sigma}{W-z}\right)^4\right]$
Isotherm type	Information	Specific surface area	Specific surface area	Microporous volume	Mesoporous volume and size distribution	Microporous volume and size distribution
I(a)	Micropore d < 1 nm	• *	• *	•	X	•
I(b)	Micropore + narrow mesopores d < 2.5 nm	• *	• *	•	X	•
II	Nonporous or macropores	•	X	•	X	X
III	Nonporous	•	X	X	X	X
IV(a)	Mesopores >4 nm Hysteresis	•	X	•	•	•
IV(b)	Mesopores <4 nm	•	X	•	•	•
V	Weak interactions adsorbent-adsorbate	•	X	X	•	X
VI	Filling layer by layer	•	X	•	•	•

• Model can be applied to this type of isotherm
X Model not suited to this type of isotherm
* Can be applied under certain conditions

2.8.7.2 Apparatus

Models and Suppliers

The best-knowns apparatus on the market are from QUANTACHROM, BELL JAPAN and MICROMERITICS. The principle of these equipments is based on the manometric measurement (Fig. 2.149). The quantity of gas adsorbed is determined by measuring gas pressure at a known temperature.

Other apparatus function according to the gravimetric measurement principle. These apparatus will not be discussed in this chapter.

The apparatus used for the measurements provided is a 3Flex from Micromeritics.

Measurement Principle

The principle of a manometric measurement apparatus is to measure gas pressure in a known volume. Figure 2.149 shows a diagram of the apparatus in simplified form. It comprises four parts connected together via a chamber calibrated by the apparatus manufacturer, known as the manifold. Into the manifold, which is maintained at a constant known temperature, the adsorbate is introduced at pressure P_1. Knowing the volume, the quantity of gas contained in the manifold is deduced.

The gas is then introduced into the measuring cell containing the sample held in a Dewar flask of liquid nitrogen at 77 K. The gas adsorbed occupies the volume of the manifold plus the volume of the measuring cell containing the sample.

Once pressure is stable, equilibrium is considered to be attained and pressure P_2 (lower than P_1) is recorded. The quantity of non adsorbed gas is calculated based on

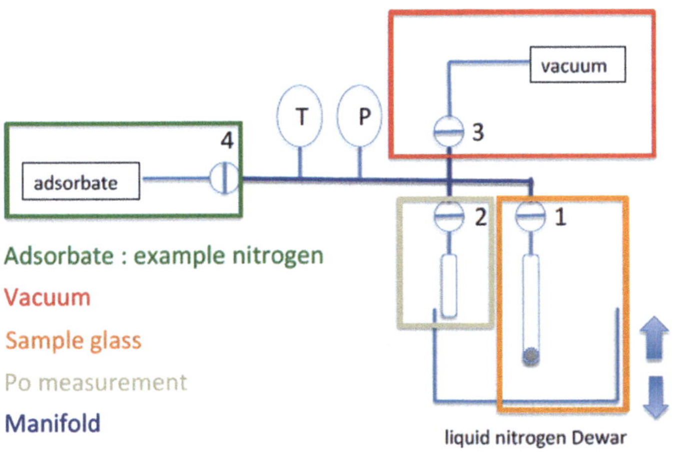

Fig. 2.149 Simplified diagram of manometric adsorption measuring apparatus

pressure P_2. This quantity of gas is subtracted from the quantity of gas initially introduced into the manifold. The quantity of gas adsorbed, n_a, is then deduced. The first plot on the adsorption isotherm has now been obtained. This operation is then repeated until a pressure is obtained that is equal to the saturation pressure of the gas adsorbed at liquid nitrogen temperature, i.e. until attaining a relative pressure of $P/P_o = 1$. The adsorption isotherm is built step-by-step using gas increments.

The quantity of gas adsorbed, n_a, can be determined after each increment providing that the vacuum volume in the measuring cell containing the sample, called the dead volume, is measured beforehand.

The concept of measuring the dead volume will be covered later on, in the section on analysis conditions.

Desorption is the opposite phenomenon to adsorption. It is obtained by measuring the quantity of gas desorbed whilst relative pressure falls. In no case, can the desorption branch pass under the adsorption branch. This would indicate that the quantity desorbed was greater than the quantity adsorbed.

Desorption provides a better understanding of the adsorption: the determination of the shape of pores for example can be obtained.

2.8.7.3 Preparation of the Sample and the Measuring Cell

Sample preparation is a very important stage, and yet it is one of the frequently overlooked. Preparation impacts dead volume measurement results and the quantity of gas adsorbed, which determines the quality of the adsorption isotherm. This step is carried out in 2 stages:

- preparation of the measuring cell: includes taking and weighing the sample;
- degassing the sample: depending on temperature conditions and duration of degassing.

In order to be analysed, the sample must be introduced into a measuring cell which is specific to the apparatus used. The same measuring cell is used for the degassing and analysis phases. The main difference between measuring cells provided by the various manufacturers is their shape and size. They are made of glass and have a sealed stopper to prevent atmospheric contamination of the sample after degassing.

The sample quantity placed in the cell will depend, among other factors, on the specific surface area of the sample and its porosity as well as its shape (monolith or powder) and particles size distribution.

Equation (2.129) enables a calculation of the geometric surface developed by nonporous spherical particles. The surface is inversely proportional to the particles size distribution of the sample. Using this equation, the surface developed by the sample can be evaluated in order to deduce the quantity of the sample to be taken. Even if the sample is porous and absolutely not spherical, this equation still provides an initial assessment. It is advised to make another analysis after adjusting the sample mass.

$$a = \frac{6}{\rho \cdot d} \quad (2.129)$$

where, a is geometric surface in $m^2\ g^{-1}$; ρ, solid density in $g\ m^{-3}$ and d, particle diameter in m.

For accuracy reasons of measurement, it is recommended to have enough materials in the measuring cell, which corresponds to a total surface of at least 10 m^2. In order to follow this recommendation, quantities of samples to be taken have been calculated in Table 2.86 for three examples of solids with specific surface areas of 1–600 $m^2\ g^{-1}$. For solid A, with a very low specific surface area, a sample mass of 1000 mg in the measuring cell is needed. If the sample is a powder, there will be a very thick layer of it in the measuring cell which may present gas diffusion problems during degassing and analysis, which may well be a source of error in later results. Conversely, solid C has a very large specific surface area of around 600 m^2/g, and so will require a very small quantity of the sample in the measuring cell, i.e. 16 mg. In this case, this raises the question of the representivity of the sample relative to the original sample. A greater quantity of the sample will be used to avoid this problem, and a surface larger than 10 m^2 will therefore be measured in the measuring cell.

Because the volume of sample used for adsorption measurement is generally small (<1 cm^3), the sampling step to get a representative sample is of crucial importance. This is all the most important if the sample is a divided solid that is heterogeneous in terms of particle size and shape (needle, sphere, agglomerate, etc.). If it is too difficult to take a sample, granulometric fractioning may be used with the aid of a sieve.

When introducing a solid into the measuring cell, the operator must be careful to use a measuring cell that is clean and in good condition with no cracks. The joints on the stopper used to seal the measuring cell must be in good condition. The sample should be introduced into the cell without contaminating the inner surface of the cell with fine particles that may impact the quality of the analysis and be a source of error when measuring. This phenomenon is a frequent occurrence where the sample is an electrostatic powder. It can be avoided or limited by using a static eliminator device, such as the AD-1683 from Delta-Labo [303]. It is sufficient to position the device in front of the sample for a few minutes prior to using it in order to eliminate electrostatic charges and make the solid easier to handle.

This problem is often encountered with processed wood chars and fine powders.

Table 2.86 Comparison of sample masses to place in the measuring cell as a function of the specific surface area developed by the samples

Sample	S (m^2/g)	Sample weight (mg)	Comment
A	1 (Kr)	1000	Gas diffusion risk
B	50 (N_2)	200	Good
C	600 (N_2)	16	Sampling and representivity are problematic

Weighing accurately the sample is also of crucial importance for the sample preparation step. The results obtained will be compared to the sample weight, therefore, weighing is an important stage and it must be carefully carried out. Every measuring cell is weighed three times: empty; with the sample before degassing; and, finally, with the sample after degassing. It is recommended to weigh measuring cells using suitable calibrated balances.

It is the sample mass after degassing that is used for the calculation. Knowing the sample mass before degassing makes possible the calculation of the mass loss by degassing, which provides additional information about the sample.

2.8.7.4 Sample Degassing

Degassing consists in freeing the sample surface of gas species. It allows eliminating physisorbed molecules at the surface of the solid, such as atmospheric humidity, carbon dioxide or other residual post-process molecules in order to free access to the pores and liberate the surface. Adsorption cannot occur before this stage. Degassing must take place according to a properly defined procedure established by the operator and suited to the sample type so that the results obtained are reproducible and comparable.

The apparatus manufacturers offer degassing in a gas stream (Bell Japan) or at temperature and in a vacuum (Micromeritics and Quantachrom). This stage can be carried out directly within the analysis machine or using a separate unit that operates independently. In the examples given below, degassing was performed using a separate unit, a Smart VacPrep from Micromeritics.

Three degassing parameters are to be considered: temperature, duration and vacuum attained.

Degassing Temperature

In most cases, degassing occurs in a vacuum and the sample is heated to accelerate the desorption of molecules. The chosen temperature for heating the sample must not modify its surface and structure.

It is therefore important to properly understand the sample before subjecting it to heat treatment, by using other analysis techniques such as thermogravimetric analysis, for example. This analysis makes possible track changes in sample mass as a function of temperature in a controlled atmosphere. The mass loss observed indicates the release of adsorbed molecules or a chemical transformation in the sample caused by the temperature. It is therefore important to know the temperatures at which these phenomena occurred, as well as what lies at the origin of mass loss, to avoid deterioration of the sample in the degassing system.

Organic compounds will, in general, be treated at lower temperatures than mineral compounds.

In cases where the sample can be heated: the recommendation is to heat in 2 stages. For the first stage, the sample is heated around 90 °C for 1 h until a vacuum value of around 10 mmHg is reached. A second stage at the desired degassing temperature is then carried out. The temperature must be increased, not more than 5 °C per minute, to avoid modifying the porous sample structure through gas release.

In cases where the sample cannot be heated, i.e. above 90 °C: heat the sample at isotherm temperature. Degassing duration is longer.

If case of doubt, it is advised to avoid heating at too high temperature and to use lengthy degassing times.

Degassing Duration

Degassing duration depends on the porous structure of the materials to be analysed. The more complex and narrow the sample's porous network, the longer the degassing time. The same applies to a sample that is sensitive to temperature.

For materials such as alumina and silica, the recommended degassing times according to ISO 18757 standard [304] are in the order of 1–3 h.

For materials sourced from biomass, degassing times are longer, in the order of 10–24 h. In cases where the sample cannot be heated, this time can comfortably exceed 24 h.

It is essential to carry out several tests. Optimal degassing times correspond to the isotherm for which the maximum quantity is adsorbed.

Required Vacuum

The level of vacuum to attain is a function of the analysis to be carried out. For analysis of samples presenting microporosity, the vacuum to attain is of the order to 10^{-4} mmHg. But for a non-microporous sample, vacuum in the order of 10^{-3} mmHg is sufficient. The system used for degassing the sample must be fitted with a pump enabling this level of vacuum to be attained. The speed of vacuum release is set if the degassing control software permits. This speed must be slow, in the order of 5–10 mmHg/min to avoid altering the structure of the sample. The degassing stage is completed once the vacuum requested is attained and the determined degassing duration is reached. After the degassing stage, the sample must be reweighed. This mass of the sample is the mass to be used in the analysis. As the specific surface area of a solid is not an intrinsic value to the material but a value that relates to degassing and analysis conditions, it is advised to indicate the degassing conditions used along with the results obtained. It is further advised to establish an identical degassing protocol for all samples so that the results can be properly compared.

Analysis Conditions

Before the analysis, it is necessary to take the following precautions:

- check the system tightness by the verification of the stability of the pressure;
- use dry adsorbate and helium (to measure dead volume) with a purity of at least 99.99%;
- measure saturation pressure (P_o) of the adsorbate at the temperature of liquid nitrogen at regular intervals throughout the analysis process;
- maintain the sample at the temperature of liquid nitrogen throughout the analysis process by using an insulating jacket around the measuring cell or any other equivalent system provided by the manufacturer;
- optimise adsorbed gas increments on the acquisition isotherm. Because the apparatus works by taking manometric measurements per gas increment, all errors are cumulative. They repeat, as many times as there are points on an adsorption isotherm. This means that particular care should be paid to the point acquisition zones. These depend on the type of isotherm.

Other factors, such as equilibrium times corresponding to waiting time between pressure stabilisation and acquisition point recording, measuring dead volume or the range of relative pressures used, all vary as a function of the type of sample.

Two types of sample are identified for bio-sourced materials characterization by gas adsorption:

- microporous samples, with porosity below 2 nm.
- non-microporous samples, with porosity above 2 nm or non-porous samples. Remember that macropores with pore diameter >50 nm cannot be characterised by gas adsorption. Mercury porosity measurement can be used in this case which is separately presented later in this chapter (Sect. 2.8.8).

Microporous Sample

It has been previously shown how micropores fill at very low relative pressure ($P/P_0 < 0.01$). In order to reach the smallest pores, it is necessary to attain a pressure of up to 10^{-7} mmHg. Obtaining these pressures requires an apparatus with appropriately sensitive pressure sensors as well as the verification of the stability of the pressure.

Gas Increments

Filling of micropores demands extremely small quantities of gas. In order to dose the gas as accurately as possible, it is preferable to work by incremental volumes (dose in cm^3/g of sample at standard conditions for temperature and pressure). These gas volume increments are generally in the order of 3–5 cm^3/g depending on the porosity of the sample. Starting from P/P_o close to 0.01, adsorbate increments continue by pressure increment until $P/P_o = 1$.

Number of Acquisition Points

In the case of a type I isotherm, it is recommended to have enough experimental points covering from the beginning to the end of the formation of monolayer. For the rest of the isotherms, a handful of additional points are plotted in order to construct the isotherm's plateau.

For the desorption, the points at the beginning (close to $P/P_o = 1$) and at the end of the desorption process must be collected to build the whole isotherm. Desorption curve can overlap adsorption curve, but can never be below the adsorption curve.

Equilibrium Time

The more the pores are narrow and/or the pore network is complex, the longer the time necessary to reach the adsorption equilibrium is needed. This is due to the diffusion phenomena of adsorbate molecules inside the porosity. Thus, the equilibrium time between each point must be enough long to be stabilized. But it is important to also take into account the limitations of the apparatus used to maintain the analysis as a function of the capacity of the Dewar flask containing liquid nitrogen. A minimum equilibrium time of 30 s should be done. For a microporous sample, an analysis time between 24 and 48 h is usually needed.

Dead Volume

Measurement of the dead volume consists of measuring the volume in the measuring cell which is not occupied by the sample. This corresponds to the volume of vacuum in the measuring cell. This therefore depends on the volume of the measuring cell, the quantity of sample introduced into it and the sample's porosity.

The quantity of gas introduced into the measuring cell depends on the temperature of the cell. The cell is at the temperature of liquid nitrogen (at the bottom of the cell) and at room temperature at the top. Because of its vertical position in the apparatus, it is only partially immersed in the liquid nitrogen. Dead volume is therefore measured at the temperature of liquid nitrogen and at room temperature. The software takes into account the change in the temperature gradient during the analysis to determine the quantity of gas adsorbed.

Dead volume is measured by gas expansion, usually using helium. Helium is considered not to adsorb on the sample surface. However, this is not the case with microporous solids. There are two possibilities for avoiding the problem of helium adsorption with a microporous solid:

- either, the apparatus offers the possibility of determining dead volume after analysis. In which case, helium adsorption will not be a source of error when measuring the quantity of the chosen adsorbate.
- or, the dead volume is calculated based on the measurement of the volume of the empty measuring cell and the volume occupied by the sample calculated using its true density [305]. In this case, the calculations must be made at room temperature (Eq. 2.130) and at the temperature of liquid nitrogen (Eq. 2.131). The values will then be entered in the sample's analysis file by the operator.

$$V_{ws} = V_{wm} - \left(\frac{m_s}{\rho_s} \times \frac{T_{std}}{T_{amb}}\right) \tag{2.130}$$

where V_{ws} is the free space calculated with sample in cm^3 at room temperature; V_{wm}, free space measured for the empty tube in cm^3 at room temperature; m_s, mass of sample to be analysed in g; ρ_s, sample true density measured with helium pycnometer in g/cm^3; T_{std}, standard temperature is 273.15 K; T_{amb}, ambient temperature in K.

$$V_{cs} = V_{cm} - \left(\frac{m_s}{\rho_s} \times \frac{T_{std}}{T_{bath}}\right) \tag{2.131}$$

where V_{cs} is the free space calculated with sample in cm^3 at liqid nitrogen temperature; V_{cm}, free space measured for the empty tube cm^3 at liqid nitrogen temperature; m_s, mass of sample to be analysed in g; ρ_s, sample true density measured with helium pycnometer in g/cm^3; T_{std}, standard temperature is 273.15 K; T_{bath}, bath temperature in K.

The quantity of gas adsorbed on the sample surface is calculated using this dead volume.

Non-microporous Sample

Non-microporous sample includes non-porous, mesoporous and macroporous materials. The gas adsorption is suitable for the analysis of non-porous and mesoporous samples. On the other hand, it is not adequate to macroporous materials because their pore diameters are too large to be identified by this technique.

Analysis of a non-microporous sample (i.e. starting feedstock) is simpler and quicker than for microporous samples (i.e. biochars from pyro-gasification of biomass). In reality, non-microporous samples are less frequently encountered.

In most cases, nitrogen at the temperature of liquid nitrogen is used to build adsorption-desorption isotherms. If the specific surface area is smaller or equal to

1 m^2/g, the use of krypton as adsorbate is recommended. In this case, the full isotherm is not completed and the measurement with the value of P/P_o up to 0.3 is enough for the determination of the specific surface area.

Gas Increments

Gas increments to build the adsorption isotherm are determined using pressure measurements in accordance with a calibration table of gas doses established by the operator. The table of pressures where the acquisition points are to be recorded is constructed so as to prioritise certain regions of the isotherm such as:

- the zone where a monomolecular layer is formed;
- the region close to $P/P_o = 1$ for the adsorption and desorption branches, if a full isotherm is plotted in order to confirm, or otherwise, the presence of the characteristic saturation plateau of mesopores;
- the region where the hysteresis loop closes: close to $P/P_o = 0.42$ for nitrogen adsorption at 77 K, and $P/P_o = 0.28$ argon adsorption at 77 K;

For these non-microporous materials, the hysteresis loop is important and its measurement must be carefully done. From the data of hysteresis loop measurement, the porous volume and the pores size distribution can be determined.

For materials having the specific surface area smaller than 5 m^2/g, the measurement up to P/P_o of 0.3 is enough in order to determinate the specific surface area. For these samples, an adsorption with krypton at 77 K gives better results than nitrogen adsorption.

Equilibrium Time

An equilibrium time of 30 s is recommended.

Dead Volume

Helium expansion is used at the beginning of the analysis to measure dead volume.

In cases where the surface developed by the sample is very small, a glass insert may be placed in the measuring cell with the sample. This will reduce the dead volume and limit measurement errors that may result.

Recommendations

The information below may help to successfully plot the isotherms:

- Check that the apparatus used must be able to provide the desired measurements;
- isotherm measurements must be repeated several times in order to optimise analysis conditions and assess sample heterogeneity.

Table 2.87 Compilation of common errors, their causes and possible solutions

Errors encountered	Causes	Possible solutions
Negative isotherm	Poor estimation of dead volume Adsorption of helium Insufficient degassing	Measure external dead volume Measure dead volume after analysis Modify degassing conditions
	Very little surface	Unsuitable adsorbed gas Unsuitable pressure range
S-shaped isotherm	Lack of equilibrium Helium adsorption	Increase equilibrium time Measure external dead volume Measure dead volume after analysis
Desorption passes under adsorption	Cumulative error Sample heating Leakage problems	Do not put too many points Liquid nitrogen level, analysis too long Check stopper seals and electro valves
The loop hysteresis is closed below the limit value	Lack of equilibrium Sample heating Cumulative error Leakage problems	Increase equilibrium time Liquid nitrogen level, analysis too long Do not put too many points Check stopper seals and electro valves
Specific surface area too small $<1\ m^2/g$		Increase the quantity of sample in the cell Check experimental error by conducting dummy test with an empty cell: possible adsorption on measuring cell walls

Table 2.87 provides a non-exhaustive list of commonly encountered errors, their causes and possible solutions.

2.8.7.5 Exploitation of the Results

In general, gas adsorption/desorption is not applicable for raw biomass, biowastes and co-products because they are non porous materials and are in most case not stable under the analysis conditions. On the other hand, when these feedstocks are treated by a thermal process, i.e. pyrolysis, gasification, the resulting solid residues, which are usually carbonaceous materials, can be characterized by this technique.

This paragraph looks at the models most commonly used to process isotherms. The treatment of raw data depends on the isotherm obtained and the expected information. The treatment of gas adsorption/desorption data consists in the application of models built for different kinds of materials. In this chapter, only the most commonly models used will be presented. It is also recommended to read the references available in the literature for each theory and model that will not further developed in this book.

For non-porous and macroporous materials, specific surface area is the principal information obtained by gas adsorption/desorption. For microporous (porosity smaller than 2 nm) and mesoporous materials (porosity in the range of 2–50 nm), pores size distribution can be obtained, in addition to the specific surface area. For microporous materials, the measurement should be carefully performed, because of the delicate penetration of adsorbate molecules inside micropores.

The next sections provide a "guide" for: (i) the graphical construction of adsorption/desorption isotherm; (ii) the method for obtaining the specific surface area; (iii) and the method for the determination of the porous volume and pores size distribution. Most of the examples shown in these sections have been obtained from nitrogen adsorption-desorption. Another section is dedicated to carbon dioxide adsorption-desorption for microporosity analysis (Sect. 2.8.7.12).

2.8.7.6 Graphical Construction of Adsorption/Desorption Isotherm

As previously presented in Figs. 2.147 and 2.148, the construction of adsorption/desorption isotherm consists in plotting the adsorbed quantity of adsorbate on the surface of the solid as a function of the relative pressure. The determination of the adsorbed quantity of adsorbate must be done with both adsorption and desorption steps to build full isotherm.

The isotherm obtained for a microporous sample is of type I according to the IUPAC classification (Fig. 2.147). As example, Fig. 2.150 shows the adsorption isotherm obtained for a solid microporous sample of char derived from natural biomass (coconut shell) processed using steam at 750 °C. The adsorption branch rises steeply at low relative pressures ($P/P_o < 0.1$), corresponding to filling the micropores, then there is a saturation plateau around 190 cm^3/g STP (Standard Temperature and Pressure). Desorption superimposes almost perfectly onto adsorption.

The log scale isotherm (Fig. 2.151) provides greater detail about micropore filling. In this example, micropore filling occurs in a single stage. The isotherm corresponds to one experimental measurement of the sample. Different models can be applied to determinate the values of different parameters. These values are obtained under well-defined operational conditions. So, it is vital to work in strictly identical conditions in terms of preparing and analysing samples for each series of samples.

For macroporous or nonporous samples, it is not necessary to build a complete isotherm with adsorption and desorption. This will not deliver any additional information. Analysis can be halted at $P/P_o = 0.3$, where the BET law applies for the determination of the specific surface area (shown later). The specific surface area obtained is generally small for these samples.

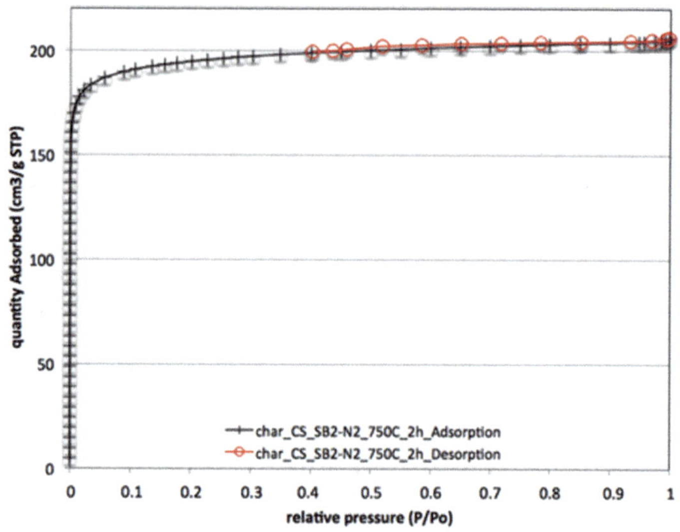

Fig. 2.150 Nitrogen adsorption and desorption isotherm at 77 K of a solid microporous sample (coconut shell char obtained by gasification at 750 °C for 2 h under N_2)

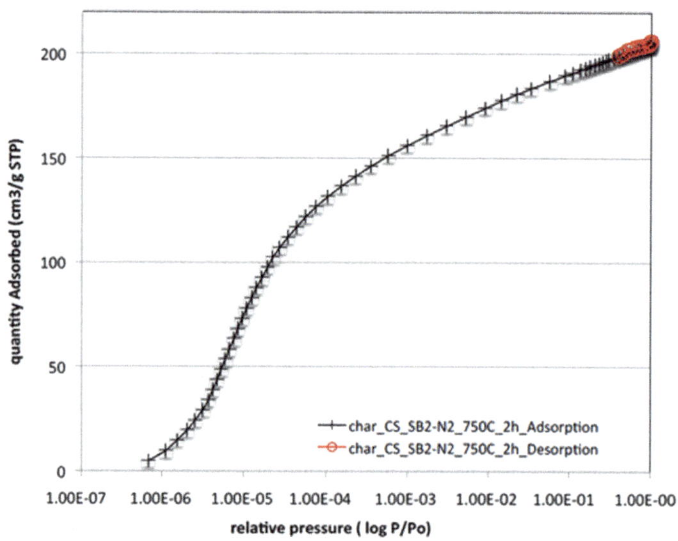

Fig. 2.151 Log scale representation of nitrogen adsorption isotherm at 77 K of a solid microporous sample (coconut shell char obtained by gasification at 750 °C for 2 h under N_2)

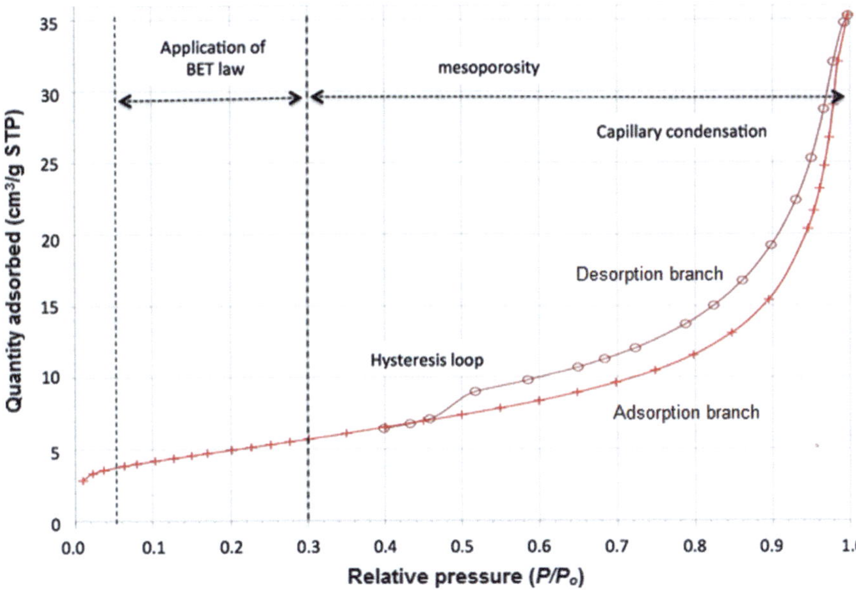

Fig. 2.152 Example of the adsorption isotherm of a mesoporous sample (solid recovered fuel)

For mesoporous samples, the complete isotherm (adsorption-desorption) is useful in determining pore size distribution as well as defining pore shape. This is because the hysteresis obtained is used to deduce pore shape (Fig. 2.148).

As example, Fig. 2.152 shows the adsorption isotherm for a solid recovered fuel—SRF. The isotherm obtained is of type II or IV because when P/P_o is close to 1, the curve tends toward a probable plateau. The presence of this probable saturation plateau implies that:

- either, the sample has significant mesoporosity (check by applying the BET method);
- or, the sample is macroporous.

2.8.7.7 Determination of the Specific Surface Area

The determination of the specific surface area (m^2/g) is among the most important applications of gas adsorption/desorption. Equation (2.127) allows calculating the specific surface area of a solid. This last depends on the molecular crowding of the adsorbate (σ_m, or the surface of one molecule of adsorbate) as well as on the quantity of the adsorbate necessary to form a monolayer on the surface of the solid (n_m^a). So, according to Eq. (2.127), n_m^a should be obtained in order to determinate the specific surface area.

Langmuir Model

In the literature, the chemisorption of a gas on the surface of a solid (generally at higher temperature than that of physisorption) leads to the formation of a monolayer of adsorbate. In this case, Langmuir model can be applied to calculate n_m^a. Equations (2.132) and (2.133) mathematically describe this model.

$$\theta = \frac{b\frac{P}{P_o}}{1 + b\frac{P}{P_o}} \tag{2.132}$$

$$\theta = \frac{n_a}{n_m^a} \tag{2.133}$$

with θ the fractional occupancy of adsorption sites (Eq. 2.133) at the working temperature; n_a the quantity of adsorbate which is adsorbed on the surface of the solid at the relative pressure P/P_o; b the constant of the Langmuir model, which is related to the activation energy of the considered adsoption.

When the physisorption takes place with the formation of isotherm of the type I, Langmuir model can be applied to determinate n_m^a. The combination of Eqs. (2.132) and (2.133) leads to the following linear relationship between $\frac{1}{n_a}$ and $\frac{1}{P}$ (Eq. 2.134).

$$\frac{1}{n_a} = \frac{1}{n_m^a} + \frac{1}{bn_m^a}\frac{P_o}{P} \tag{2.134}$$

Thus, a linear cover should be obtained by plotting $\frac{1}{n_a}$ versus $\frac{1}{P}$ and the value of $\frac{1}{n_m^a}$ is equal to the ordinate at the origine.

BET Model

BET model (Brunauer, Emmett and Teller model) is the most used for the determination of the value of n_m^a, and so the specific surface area of a solid material. This model has been built on the hypothesis of the multilayer adsorption, as illustrated in Fig. 2.153. This signifies that the adsorbate can form multimolecular layers on the

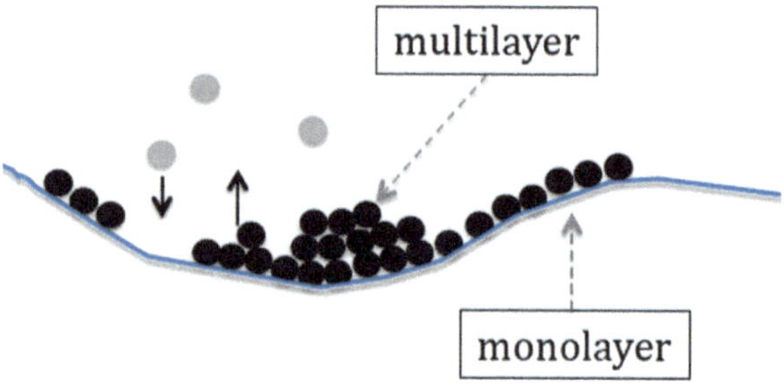

Fig. 2.153 Illustration of multilayer physisorption of a gas on the surface of a solid

surface of the solid by increasing the relative pressure. Equation (2.135) mathematically describes this model:

$$\frac{n^a}{n^a_m} = \left(\frac{Cx}{1-x}\right)\left(\frac{1-(N+1)x^N + Nx^{N+1}}{1+(C+1)x - Cx^{N+1}}\right) \tag{2.135}$$

$$C = \exp\left(\frac{E_1 - E_l}{RT}\right) \tag{2.136}$$

where n^a: molar quantity of adsorbate; n^a_m: molar quantity of adsorbate necessary to cover the surface of the solid by a monoalayer; C: BET constant; E_1: the adsorption energy of the 1st monolayer; E_l: the liquefaction energy of the adsorbate for other monolayer; x: relative pressure P/P_o; N: number of layers adsorbed; R: ideal gas constant.

The BET equation (Eq. 2.135) can be transformed as follows:

$$\frac{1}{n^a(P_o/P-1)} = \frac{1}{n^a_m C} + \left(\frac{C-1}{n^a_m C}\right)P/P_o \tag{2.137}$$

Thus, Eq. (2.137) is an equation of the first order by plotting $\frac{1}{n^a(P_o/P-1)}$ as a function of P/P_o. We note that n_a and P are experimentally measured which allow to calculate the values of $\frac{1}{n^a(P_o/P-1)}$ and P/P_o. In general, Eq. (2.137) is generally linear in the relative pressure range (P/P_o) of 0.05–0.3. An example of this linear dependency is shown in Fig. 2.154.

Plotting $\frac{1}{n^a(P_o/P-1)}$ as a function of P/P_o can conduct to two following possibilities:

- Case 1: The experimental points are linear in the relative pressure P/P_o used (0.05–0.3). In this case, the correlation coefficient R^2 must be close to 1. The

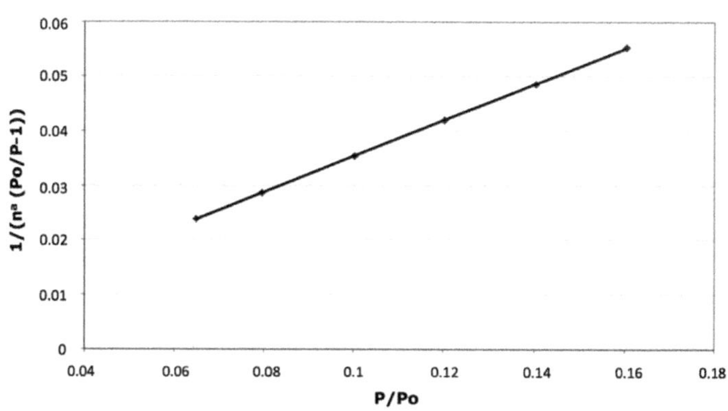

Fig. 2.154 Example of the linear transformation of BET model

specific surface area can be determined. In this case, the term $\left(\frac{C-1}{n_m^a C}\right)$ is the slope of the linear curve, and $\frac{1}{n_m^a C}$ is the ordinate value at the origine (Eq. 2.137). Thus, two equations containing two variables are obtained:

$$\left(\frac{C-1}{n_m^a C}\right) = slope \tag{2.138}$$

$$\frac{1}{n_m^a C} = ordinate\,at\,the\,origine \tag{2.139}$$

From these Eqs. (2.138) and (2.139), C and n_m^a can be determinated. The BET content C must not be negative. The specific surface area can be obtained by applying the value of n_m^a to Eq. (2.127). This specific surface area is usually noted as "S_{BET}".

- Case 2: The experimental points are not linearly plotted. So the specific surface area cannot be calculated. This can be explained either by experimental error or the relative pressure range selected for applying BET model is not good. To this last, a specific verification procedure (Rouquerol transformation) must be done as shown in the next section.

Rouquerol transformation

Rouquerol transformation [306] is a procedure to determinate the relative pressure range which is validated for applying BET model. The following steps need to be done:

- From the experimental values of n_a and P, calculate the value of the term $n^a\left(1-\frac{P}{Po}\right)$;
- Plot $n^a\left(1-\frac{P}{Po}\right)$ as a function of $\left(\frac{P}{Po}\right)$;
- Only the points in the zone wherein $n^a\left(1-\frac{P}{Po}\right)$ increases with $\left(\frac{P}{Po}\right)$ could be used for BET model application.

Figure 2.155 shows an example of the Rouquerol transformation of a char obtained by gasifying a coconut shell at 750 °C for 2 h under N_2. The term $n^a\left(1-\frac{P}{Po}\right)$ increases with $\left(\frac{P}{Po}\right)$ and reaches a maxium. Then it decreases. In this case, only the points below the maximum of the term $n^a\left(1-\frac{P}{Po}\right)$ can be used for the application of the BET model.

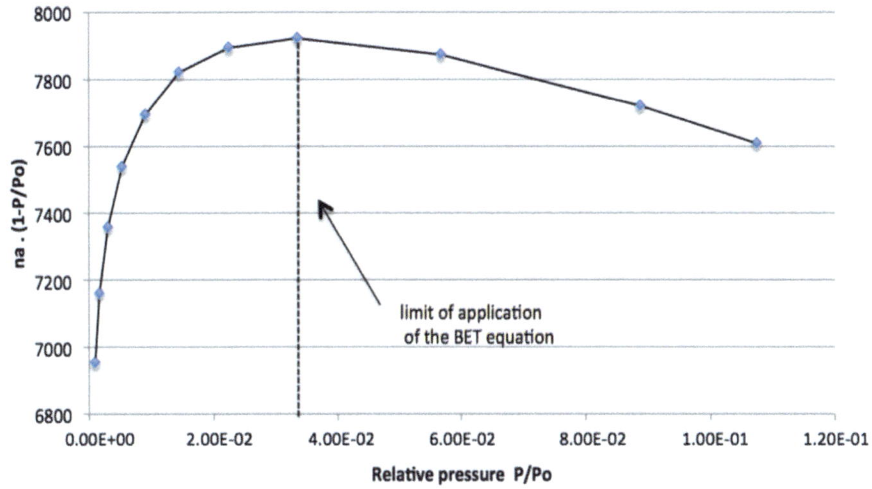

Fig. 2.155 Example of Rouquerol transformation curve of a char obtained from the gasification of a coconut shell at 750 °C for 2 h under N_2

Application of the BET Model to Microporous Materials

As mentioned above, the BET equation is in general linear within the relative pressure range of 0.05–0.3. In the case of microporous solids with a type I isotherm and a pronounced saturation plateau, multi-molecular adsorption has not occurred and applying the BET model in the relative pressure range of 0.05–0.3 makes no sense. This is reflected in a negative value of C constant in Eq. (2.137). For this type of isotherms, Rouquerol transformation must be performed to determinate the relative pressure range applicable for the BET model.

Figure 2.156 shows the linear curve of the BET model obtained for a microporous sample in the relative pressure range of up to 0.12. After applying the Rouquerol transformation as mentioned above, three zones can be separated in Fig. 2.156. Zone A of the graph corresponds to the points which are validated by the Rouquerol transformation for the application of the BET model. Thus, S_{BET} is calculated using only these points. In this zone A, the relative pressure values are outside the domain of application generally used (P/P_o between 0.05 and 0.3). It is worth noting that, to calculate the specific surface area, 5–8 points are needed.

Section B of the graph corresponds to the relative pressures for which the points superimpose onto the BET line but their inclusion in the calculation of specific surface area produces a negative C constant. They cannot therefore be used.

Section C of the graph corresponds to pressures at which the points begin to diverge from the BET line. The BET equation is no longer applicable.

In the case of microporous materials, the BET model is applied for the first points of the gas adsorption. Within this zone (Zone A in Fig. 2.156), the surface of the solid has not been covered by a monolayer of adsorbate yet. Thus, the BET surface obtained does not correspond to a true value for specific surface area, which

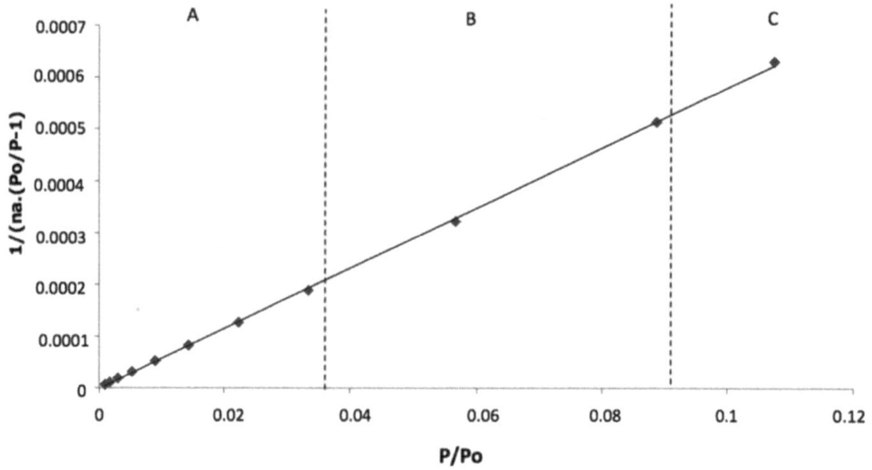

Fig. 2.156 Linear BET transformation from a microporous solid (char obtained from the gasification of a coconut shell at 750 °C for 2 h under N_2)

must be calculated from the quantity of adsorbate to form a monolayer on the micropore surface (n^a_m). We refer to the "equivalent S_{BET}".

For microporous materials, it is also noted that:

- The C constant is very high for microporous solids, greater than 1000;
- The adsorbate quantity to form the monolayer (under STP conditions) calculated using the BET equation is close to that obtained at the isotherm plateau.

Application of the BET Model to Non-microporous Materials

Non-microporous materials include non-porous, mesoporous and macroporous materials. Their specific surface area is determined by application of the BET law in the suitable relative pressure range, generally from 0.05 to 0.3. It is nonetheless advisable to check the validity of BET law:

- the C constant is positive.
- the relative pressure range selected is validated by using Rouquerol transformation.
- the correlation coefficient of the linear BET equation is close to 1.

2.8.7.8 Determination of Porous Volume and Pore Size Distribution of Microporous Materials

Adsorption isotherm of microporous materials is of the type I. If this isotherm has a saturation plateau clearly identified as illustrated in Fig. 2.147, t-Plot and Horvath-Kawazoe models can be used to get microporous volume and pore size distribution. Generally, they can be applied until the relative pressure where the saturation plateau take place.

t-Plot and Horvath-Kawazoe are the models most used to determine microporous and micropore size distribution.

t-Plot Method

Basis of t-Plot method

This method, developed by De Boer et al. [307], evaluates the thickness of a multi-molecular layer formed on the "external surface". Remider that in general, the term external surface is the surface of non-porous material. Here, this term "external surface" is extended to non-porous, macroporous and mesoporous materials.

For non-porous, macroporous and mesoporous materials, the "external surface" includes the surfaces of pores, on condition that their size allows the multilayer adsorption. In these cases, the external surface (a_t) is equal to the BET surface.

For materials containing micropores together with other larger pores, the external surface (a_t) is different to the *equivalent BET surface* (equivalent S_{BET}). This difference corresponds to the surface developed by the micropores.

De Boer et al. [302, 307] proposed that the thickness t of the multilayer adsorption is equal to the number of monomolecular single layers N multiplied by the thickness of a monomolecular layer e.

$$t = e \times N \tag{2.140}$$

To develop the method, other hypotheses have been considered as follows:

The adsorbed layer per gramme of solid is considered as liquid which occupies a volume:

$$V^l = (n^a/m^s)(M/\rho^l) \tag{2.141}$$

with V^l: liquid volume of adsorbate; m^s: solid mass; M: molecular mass of adsorbate; ρ^l: density of the liquid adsorbate.

The external surface, a_t of one gramme of solid, covered by the adsorbed layer of the thickness t is calculated as follows:

$$a_t = V^l / t \tag{2.142}$$

Each monomolecular layer has the same thickness, *e*. For liquid nitrogen at 77 K, *e* is equal to 0.354 nm.

The number of monomolecular layers of adsorbate which corresponds to the thickness *t* can be calculated as follows:

$$N = n^a / n^a_m \tag{2.143}$$

So, for liquid nitrogen at 77 K, the thickness *t* is as follows:

$$t(\text{nm}) = 0.354 n^a / n^a_m \tag{2.144}$$

By ploting the thickness, *t*, as a function of the relative pressure, P/P_o, it has been observed that the isotherms obtained with various non-porous materials was almost the same. This curve $t = f(P/P_o)$ was called "universal curve" (Fig. 2.157 [302]).

By plotting $\boldsymbol{n^a/m^s}$ as a function of the thickness, ***t***, the "transformed curve ***t***" was obtained as illustrated in Fig. 2.158. Thus, three cases can happen:

- For non-porous materials (isotherm of the type II), the "transformed curve *t*" is linear in the full relative pressure range. The slope, $s(t)$, of this linear curve is linked with the external surface a_t, by Eq. (2.145). So, this equation allows determining the external surface a_t, when knowing the value of the slope and the

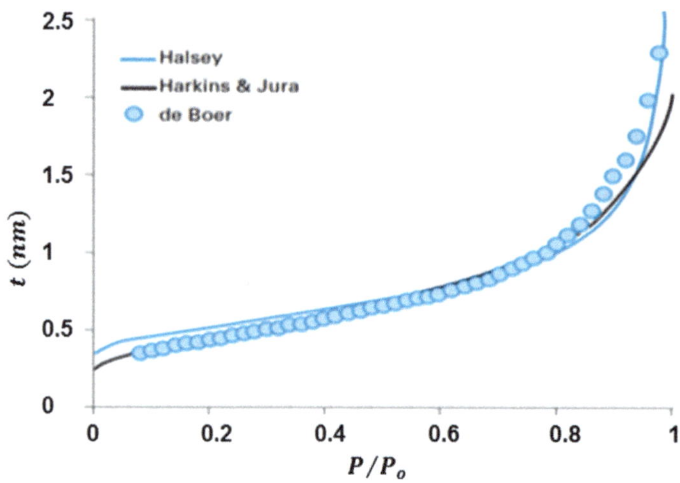

Fig. 2.157 «De Boer's universal curve» expressing the dependency of the thickness ***t*** as a function of the relative pressure $\boldsymbol{P/P_o}$ for non-porous materials. Reproduced with permission from Techniques de l'Ingénieur [302]

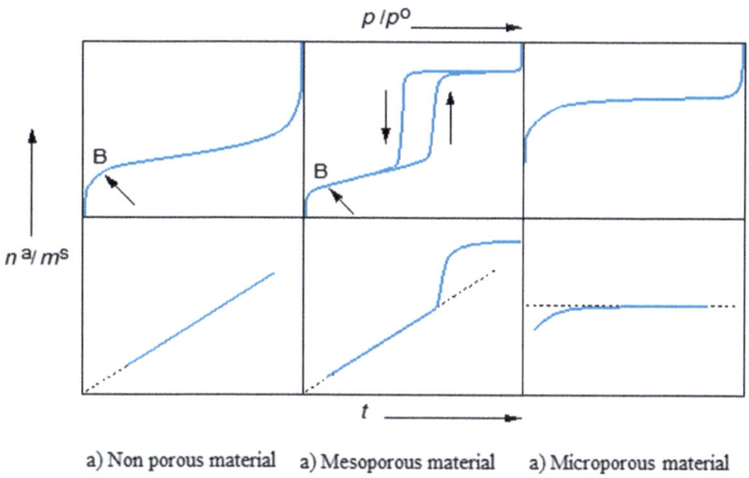

Fig. 2.158 De Boer's transformed curves *t*. Reproduced with permission from Techniques de l'Ingénieur [294]

nature of the adsorbate (for liquid nitrogen at 77 K as adsorbate, ρ^l is equal to 0.809 g/cm^3 and M is equal to 28 g/mol).

$$s(t) = (n^a/m^s)/t = a_t(M/\rho^l) \tag{2.145}$$

- For mesoporous materials (isotherm of the type IV), the "transformed curve *t*" gives a linear part that passes through the origin and a curved part corresponding to capillary condensation.
- For microporous materials (isotherm of the type I), the "transformed curve *t*" is not a line while micropores are filling. The curve becomes horizontal once the micropores are saturated.

Calculation of the Thickness t

Different empirical formula which express the relationship between the thickness *t*, and the relative pressure have been developed.

Harkins and Jura have proposed Eq. (2.146), which directly links the value of *t* with the relative pressure P/P_o up to its limited value of 0.8:

$$t(\text{nm}) = \left(\frac{0.1399}{0.034 - \log(P/P_o)}\right)^{0.5} \tag{2.146}$$

Frenkel, Halsey and Hill have also developed another equation linking t with the relative pressure P/P_o, called equation FHH or equation Halsey:

$$t(\text{nm}) = 0.354\left(\frac{-5}{\ln(P/P_o)}\right)^{0.333} \quad (2.147)$$

The validity of these two equations has been confirmed as shown in Fig. 2.157 [302]. The experimental curve of De Boer was well represented by Eqs. (2.146) and (2.147) up to the limited value of the relative pressure of 0.8.

Harkins and Jura's equation is usually used to calculate the external surface, a_t, and the microporous volume of a solid. For that, the following steps are applied:

- Perform the adsorption/desorption experiment to collect the experimental data (n_a and P/P_o);
- Select the points of the adsorption isotherm in order to determinate the microporous surface. These points must correspond to the relative pressure below 0.8;
- Use the Harkins and Jura's equation (Eq. 2.146) to calculate the thickness t;
- Plot n_a/m_s versus t to build the "transformed curve t";
- From this "transformed curve t", select the points which are linear which must have the value of t superior to 0.354 nm if liquid nitrogen is used (this is the thickness of a liquid nitrogen monolayer). An example of these selected points is shown in Fig. 2.159. Build this linear curve. The slope of this linear curve is used to calculate the external surface, a_t.
- The microporous surface is then calculated by the difference between the specific surface area (using BET model) and the external surface, $\boldsymbol{a_t}$.
- From the "transformed curve $\boldsymbol{t}$", the value of the ordinate at the origine gives the microporous volume of the studied sample.

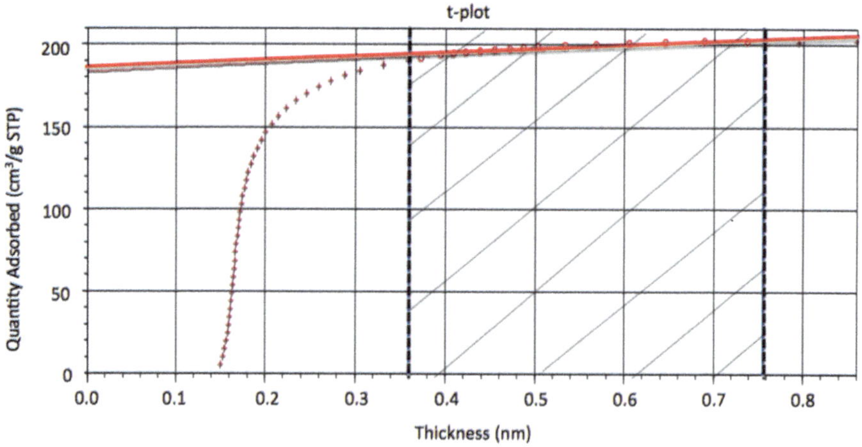

Fig. 2.159 Transformed curve $\boldsymbol{t}$; the points selected to build the linear curve was within the thickness range of 0.36 to 0.76 nm; the external surface $\boldsymbol{a_t}$ can be calculated from the slope of this linear curve

Examples of Application of the t-Plot Method

The following examples illustrate the calculation of the external surface, a_t, of a microporous and a mesoporous materials.

Figure 2.150 shows the nitrogen adsorption isotherm at 77 K obtained with a carbonaceous microporous material. This isotherm can be considered as a "composite" of the type I and type II, which indicates the presence of both micropores and mesopores. The equivalent specific surface area obtained for this material by the BET model was of 777 m^2/g. The "transformed curve *t*" is presented in Fig. 2.159. This curve did not pass through the origine which is characteristic for the presence of micropore. The linear part of this curve could be built with the value of thickness t above 0.6. The slope of this linear part allowed calculating the external surface of 42 m^2/g. So the specific surface area due to micropores was

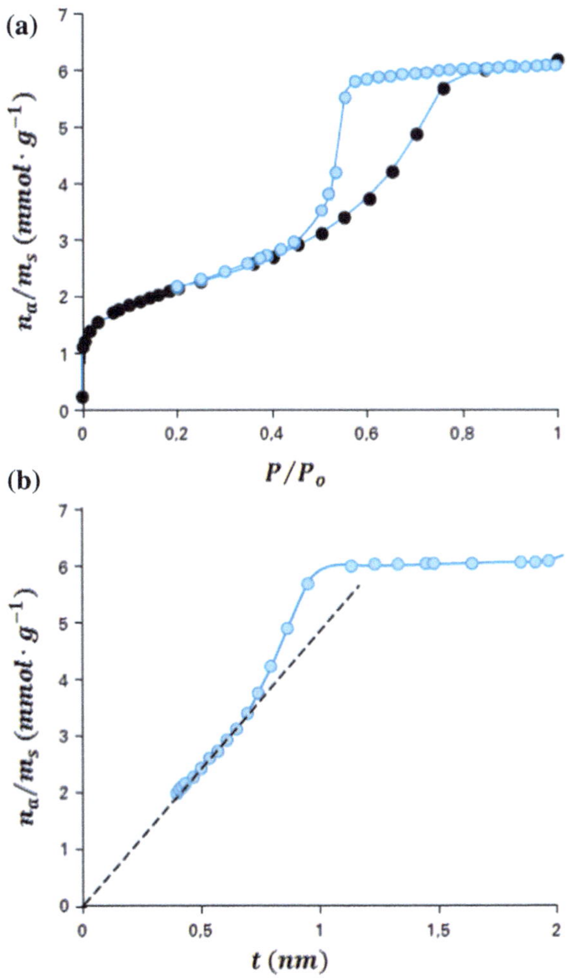

Fig. 2.160 Example of nitrogen adsorption isotherm on a mesoporous glass Vycor®: a adsorption/desorption isotherm; b transformed curve *t*. Reproduced with permission from Techniques de l'Ingéieur [302]

calculated from the difference between the specific surface area and the external surface (735 m^2/g). So, the specific surface area of this material was mainly caused by micropores.

Figure 2.160a [302] shows the nitrogen adsorption isotherm at 77 K obtained with a mesoporous glass (Vycor®), which is of the type IV. Using the BET model, the specific surface area of this material was calculated and reached 148.9 m^2/g. The "transformed curve *t*" (Fig. 2.160b) shows a linear part which passed through the origine as expected for a mesoporous material. From the slope of this linear part, the external surface, a_t, could be calculated which reached 148.9 m^2/g, a value close to the one obtained with the BET model. So, this material does not contain micropores because the specific surface area is equal to the specific surface caused by mesopores.

Figure 2.161a [302] shows the nitrogen adsorption isotherm at 77 K obtained with a non-porous black carbon (*Vulcan*), which is of the type II. The "transformed curve *t*" (Fig. 2.161b) shows a linear part up to the thickness of 1.2 nm. Above this thickness where the relative pressure is superior to 0.82, the intergranular condensation took place which explaining the exponential increase of the "transformed curve *t*" in Fig. 2.161b. For this sample, the external surface and the BET surface was found to be similar to each other.

Limit of the t-Plot Method

The universal curve established by De Boer is validated for the relative pressure above 0.15. Below this value, this curve changes from each to other material. In fact, below the relative pressure of 0.15, the adsorption implies specific interactions of each couple of adsorbent/adsorbate. These interactions cannot be expressed by a universal curve (or a general equation). Thus, the "transformed curve *t*" sometimes does not pass through the origine (i.e. for microporous materials), while it is expected.

Note: *From the literature, other methods exist, i.e. method Sing* [302], *method Dubinin-Radushkevich-Stoeckli* [308], *for studying microporous materials. They are not discussed in this chapter.*

Horvath and Kawazoe Model

Basis of Horvath and Kawazoe Model

The Horvath and Kawazoe (HK) [309] method provides an estimate of microporous volume and pore size distribution for pores smaller than 2 nm.

It is based on the hypothesis that the relative pressure necessary to fill a pore of a give pore size and form directly depends on the energy of adsorbate/adsorbent interaction, as illustrated by the equation below.

$$RT\ln(P/P_o)_{pore} = f\left(\Phi_{(r)pore}\right) \tag{2.148}$$

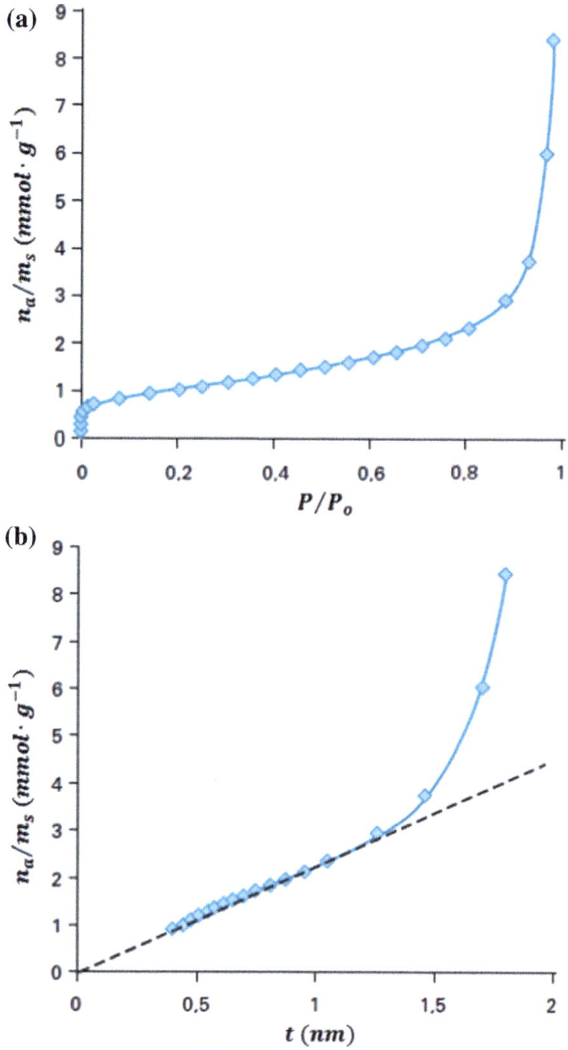

Fig. 2.161 Example of nitrogen adsorption isotherm on a non-porous black carbon (*Vulcan*): a adsorption/desorption isotherm; b transformed curve ***t***. Reproduced with permission from Techniques de l'Ingéieur [302]

where $\Phi_{(r)pore}$ is the Lennard-Jones interaction potential of a gas molecule at the distance r from the surface of the adsorbent, which is characteristic for the couple adsorbate/adsorbent used.

As example, when nitrogen is adsorbed on the surface of carbonaceous molecular sieve containing cylindrical microporous of the diameter W (nm), the Eq. (2.148) has been proposed as follows [302]:

$$\ln(P/P_o) = \frac{61.23}{(W-0.64)}\left[\frac{1.895\times10^{-3}}{(W-0.32)^3} - \frac{2.709\times10^{-7}}{(W-0.32)^9} - 0.05014\right] \quad (2.149)$$

To develop this model, the following assumptions have been considered:

- At a given relative pressure, the micropores are either fully filled or empty;
- The adsorbate reacts as an ideal gas;
- The contribution of entropy in the free adsorption energy is negliged before that of the internal energy.

The HK model was initially established for carbon-based materials with slit-shaped micropores. Then, it was extended by Saito-Foley and Cheng-Yang to take into account samples with cylindrical or spherical pores [310, 311]. Two other hypotheses have been considered:

- The pores are cylindrical with infinite length;
- The internal pores wall is constituted of an single atom layer which ensures the continuous interaction potential.

For more information on the basis and the development of this model, the readers can refer to the articles available in the literature [309–311].

Procedure for the Application of HK Model

As a rule, organic samples will be processed using the HK model [309] and inorganic samples using the Saito-Foley model [310, 311].

In order to be able to apply these models, it is vital to have plotted an isotherm at low relative pressures ($P/P_o < 10^{-4}$) and to have a certain knowledges on the nature of the sample (i.e. graphite carbon, aluminophosphate, aluminosilicate, zeolite etc.) and eventually the form of pores (i.e. cylindrical, spherical, split). These informations are considered by the software during the data treatment.

Applying this method can lead to a peak close to 0.4 nm of the pore size when using nitrogen as adsorbate. This is generally an artefact, created by the model and the analysis conditions.

The following steps are generally needed for the determination of the micropores size distribution by the HK model:

- Obtaining the isotherm at low relative pressure ($P/P_o < 10^{-4}$);
- On this isotherm, selecting the points up to the saturation plateau within the microporous range;
- Selecting the geometry of pores (split, cylinder, sphere) of the sample on the software;
- Selecting the nature of the adsorbate and that of the adsorbent. The software automatically gives the physical properties of this couple adsorbate/adsorbent;
- The software calculates the pores volume of a given width for each relative pressure. Micropores size distribution can be obtained by plotting the pores volume as a function of the pores width;
- The total volume of micropores and the mean diameter of pores can also be calculated for the relative pressure limit selected from the isotherm.

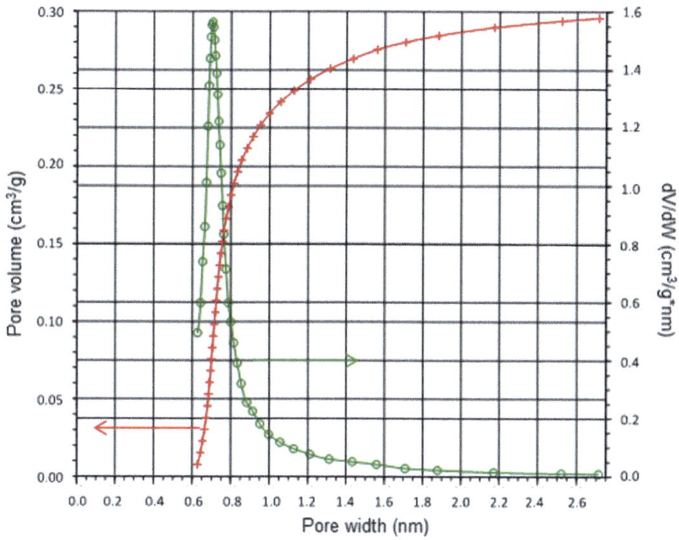

Fig. 2.162 Pore size distribution with the HK model obtained with a char from woody biomass gasification under N_2 at 750 °C: slit-shaped pores

As example, Fig. 2.162 shows the micropores size distribution of a char obtained by the gasification of a woody biomass under N_2 at 750 °C. Other examples of results are presented in Table 2.88.

2.8.7.9 Determination of Porous Volume and Pore Size Distribution of Non-microporous Materials

Introduction

Non-microporous materials include non-porous, mesoporous and macroporous materials. Non-porous materials have any porosity to be determined. For macroporous materials, gas adsorption/desorption is not appropriate for the determination of the porous volume and pores size distribution and mercrury porosimetry is used to study these materials. For mesoporous materials, the Barret, Joyner and Halenda (BJH) model has been developed for the investigation of their properties [312–316].

This model is based on the following hypotheses:

- The Kelvin equation can be applied for the full mesoporous range (Eq. 2.150). This equation allows determining the pressure limit at which the capillary condensation happens.

Table 2.88 Examples of results obtained from gas adsorption analysis for some organic and inorganic samples; SRF: solid refused fuel

Origin			Organic			Mineral
Samples			coconut_N_2_750	coconut_H_2O_750	syngas_activated_char H_2S	Mixt clay + SRF_600 °C
Weight (g)			0.6488	0.8705	0.5515	0.5569
Degasification						
Preparation		Heat (°C)	150	150	30	110
		Rate of heat, °C/min	5	5	–	5
		Vacuum, mmHg	1.00E−03	1.00E−03	1.00E−02	1.00E−02
		Time (min)	600	600	1000	600
Analysis conditions						
Analysis		Adsorptive	N_2	N_2	N_2	N_2
		Gas dose, cm^3/g STP	5	5	–	–
		Equilibrium time (s)	30	30	30	30
		Free space A = after B = before	A	A	A	B
		Isotherm	I	I	II	IV
Results						
Specific surface area	BET	S m^2/g	776.98	640.73	40.96	17.67
		C	4148.43	6820	129.28	
		<P/Po<	9.0E−4 < P/Po < 3.3E −2	3.0E−4 < P/Po < 3; 4E −2	0.06 < P/Po < 0.22	0.037 < P/Po < 0.20
	Langmuir	S m^2/g	NA	NA	NA	NA
		<P/P_o<	NA	NA	NA	NA

(continued)

Table 2.88 (continued)

Origin			Organic			Mineral
Microporosity	t-plot	V, cm^3/g	0.2829	0.224	6.60E-04	–
		S_{microp}, m^2/g	735.11	588.52	1.93	–
		S_{ext}, m^2/g	41.86	52.2	39.02	–
		$<P/P_o<$	$0.10 < P/P_o < 0.59$	$0.56 < P/P_o < 0.39$	$0.1 < P/P_o < 0.49$	–
	Harvath and Kawazoe	Pore geometry	Slip	Slip	–	–
		Adsorbent	Carbon-graphite	Carbon-graphite	–	–
		V, cm^3/g	0.2938	0.2461	–	–
		$d_{p,}$ Å	7.5	7.6	–	–
		$<P/P_o<$	$6.74E\text{-}6 < P/P_o < 0.10$	$4.1E\text{-}7 < P/P_o < 0.15$	–	–
Mesoporosity	BJH_{desorp}	V, cm^3/g	–	–	–	0.0448
		S, m^2/g	–	–	–	8.82
		$d_{por,}$ nm	–	–	–	10.17
		$<P/P_o<$	–	–	–	$0.68 < P/P_o < 0.99$

$$\ln \frac{P}{P_o} = \frac{-2\gamma V_l}{r_K RT} \cos\theta \tag{2.150}$$

where γ and V_l: the surface tension, the molar adsorbate volume in liquid form at adsorption temperature (8.85 mN m^{-1} and 34.6 cm^3 mol^{-1} for liquid nitrogen, respectively); r_K: the Kelvin radius; R: ideal gas constant; T: temperature and θ: liquid to surface contact angle which is considered as an ideal wetting, thus $\cos\theta = 1$;

- The capillary condensation takes place: the adsorbed gas is condensed inside the pores by increasing the relative pressure;
- There are only mesopores which are of well-defined form with rigid structure, and their filling does not depend on their localisation inside the porous network;
- The multimolecular adsorption in the pores happens likely on a plane surface.

In most of cases, the BJH model is applied for the desorption branch which represents well the equilibrium between the adsorbate in the gas phase and under the condensed state inside the capillary.

Principle

The principle of the BJH method has well been mentioned elsewhere [322]. Briefly, when the relative pressure is decreased from $(P/P_o)_i$ to $(P/P_o)_j$, a desorption takes place and its volume, ΔV_{ij}, can be measured. It is assumed that this volume corresponds to two phenomena:

- emptying pores of its capillary condensate; these pores having their critical radius related to $(P/P_o)_j$ according to the Kelvin law;
- reducing in thickness of the physically adsorbed layer of adsorbate in the pores previously emptied of its condensate, Δt.

This is schematized in the equation below:

$$\Delta V_{desorbed} = V_{capillary\ condensate}^{pores\ of\ size\ r_p} + \Delta t \cdot A_{internal\ pores\ of\ size\ r_{p+1} > r_p} \tag{2.151}$$

For each relative pressure, the pore radius r_p can be calculated by the Kelvin low and the thickness of the adsorbed layer in the emptied pores having the radius $r_{p+1} > r_p$ can be calculated by Harkins and Jura law. Thus, it is possible to calculate the the mesoporous volume and surface of each category of the corresponding mesopores.

The desorption points on the isotherm used for application of the BJH law are taken at relative pressures between 0.42 and 1 (or close to 1) for nitrogen desorption at 77 K.

More information about the development of different steps of the BJH model is available elsewhere [302, 312–314].

Procedure for the Analysis of a Mesoporous Materials Using the BJH Model

This section presents the main steps for the analysis of a mesoporous sample using a 3flex apparatus (Micromeritics).

- Perform the adsorption/desorption to obtain the isotherm;
- From this isotherm, select the points for BJH model application; these points are generally selected from the desorption branch;
- Calculate the thickness *t*, preferently by Harkins and Jura model;
- Apply the BJH model within a defined range of mesopores (generally from 1.8 to 150 nm) with the software. A plot of porous volume versus pore size can be obtained;
- The results obtained include the cumulative specific surface area due to mesopores, the total mesoporous volume, the mean diameter of pores, and the pore size distribution.

Figure 2.163 shows the pores size distribution of a sample composed of natural clays and fine powders of a solid recovered fuel (SRF) after calcination at 600 °C. A widespread distribution with macroporosity (pore diameters above 50 nm) is observed.

The application of the BJH law can result in formation of a peak around 2 nm, as shown in Fig. 2.164. This peak is an artefact that does not reflect any real porosity. The form of hysteresis causes this peak. In this case, it is preferable to use the adsorption branch.

The BET surface obtained corresponds to the total surface (micropores + mesopores) developed by the sample. Under no circumstances can the BJH surface (the sum of the surfaces of all the categories of pores determined by the BJH model) be larger than the BET surface. If this is the case, it will be necessary to check the isotherm obtained and to confirm the existence of mesoporosity by the presence of a plateau or pseudo-plateau where P/P_o is close to 1.

If the surface calculated by BJH is smaller than the BET surface, this is probably indicative of the existence of microporosity. The isotherm obtained is a type I + IV composite type.

The application of the BJH law poses a problem in cases where the pore structure is not rigid. The structure expands under the effect of capillary condensation. This is reflected in hysteresis of the H3 type (see Fig. 2.148). The difficulty presented by this type of samples encountered lies in confirming the presence or absence of mesoporosity. One solution is to use other characterization techniques such as scanning electron microscopy, thermoporometry, etc. Changing the adsorbate or swapping nitrogen for argon is not always enough to resolve this problem.

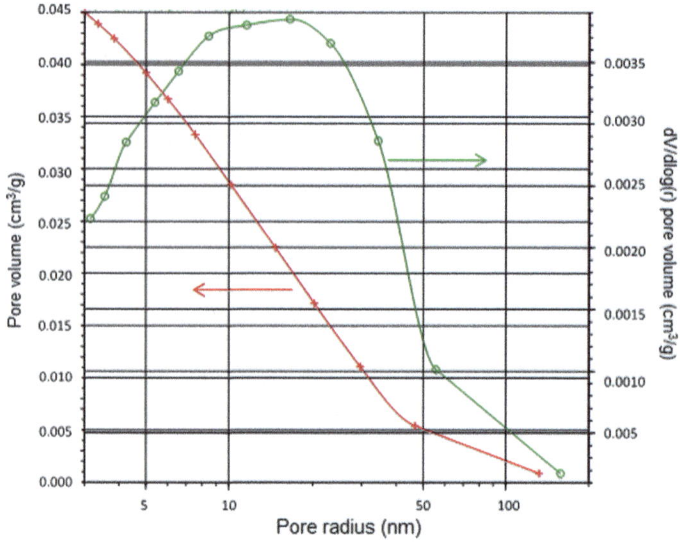

Fig. 2.163 Cumulative (left-hand-side axis) and differential pore size distribution (right-hand-side axis) by application of the BJH law to a mixture of natural clay and solid recovered fuel processed at 600 °C

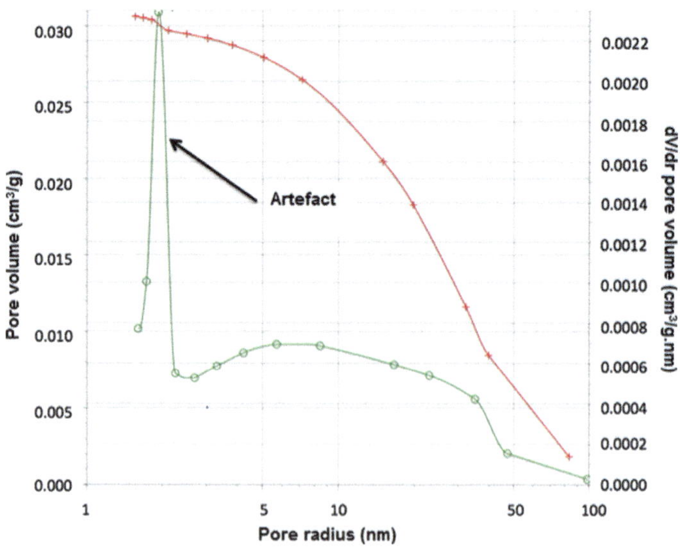

Fig. 2.164 Illustration of artefact when application of the BJH law for the pore size distribution of a char sample from wood

2.8.7.10 Summary on Gas Adsorption Analysis

To conclude the section on the exploitation of the results, and to assist operators to identify the various stages to carry out before obtaining a result, Fig. 2.165 schematically summarises the different steps of an adsorption analysis. This diagram lists the key stages and the order in which they must be performed. It is intended as a guide for operators. Operators must refer to the conditions for preparation, analysis, processing and for applying the different models that are covered in this article.

2.8.7.11 Examples

Table 2.88 shows the results obtained from some biomass-based materials. The samples were analysed and processed using the protocols described in the sections above. Nitrogen has been used to establish adsorption-desorption isotherms.

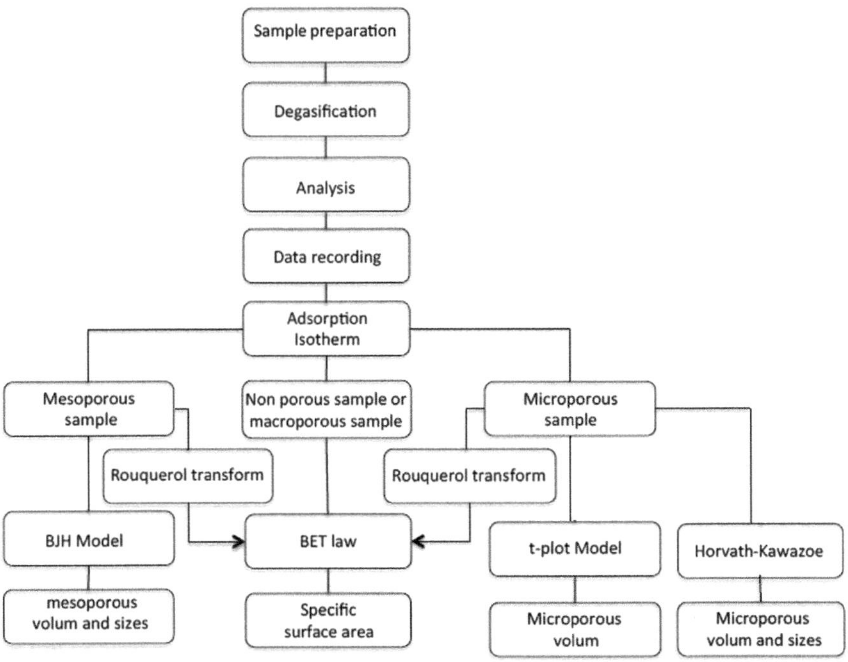

Fig. 2.165 Diagram of the analysis and processing protocol

The examples concern samples of mineral and organic origins:

- Char from the thermo-conversion of coconut under a stream of nitrogen or water vapor at 750 °C (coconut_N_2_750 or coconut_H_2O_750);
- Char from the thermo-conversion of a mixture of coagulation-flocculation sludge from wastewater treatmet plant and food waste;
- Ceramic obtained by the firing of a mixture of clay with solid recovered fuel (SRF) at 600 °C.

2.8.7.12 Microporosity Determination Using Carbon Dioxide Adsorption

Abbreviations:

Langmuir theory	Monolayer model of adsorption
BET	Brunauer-Emmett-Teller theory, multilayer model for physisorption;
HK	Horwath and Kawazoe theory;
GCMC	Grand Canonical Monte Carlo;
DFT, NLDFT	Density Functional Theory, Non-local Density Functional Theory;
$\Omega[\rho(r)]$	Grand potential of the adsorbate molecule;
$F[\rho(r)]$	Functional representing Helmholtz free energy, this is the sum of the ideal term, excess hard sphere repulsion term, and the attractive term (for 2 molecules at position r and r') calculated in a mean-field fashion;
$\rho(r)$	Number density profile in the pore, meaning the number of adsorbate molecules in a pore having a radius r;
$V_{ext}(r)$	External potential;
μ	Chemical potential of the adsorbate;
w, w_{cc}	Pore width, distance between carbons centers of the pore walls;
$N_{NLDFT}(p/p_0, w)$	Amount adsorbed per unit pore area for a given partial pressure (P/P°) and pore width (w);
$\rho(z, p/p_0, w)$	Number density distribution of the adsorbate;
$Q(T, V, \mu)$:	Partition function for N-particles system;
$N_{\mu VT}(s^N)$:	Number of molecules adsorbed;
$U(s^N)$:	Potential energy function;
s^N:	Reduced vector of the coordinates of all molecules;
β:	Reciprocal temperature ($1/k_BT$);
Λ:	Thermal Broglie wavelength;

$N_{exp}(p/p_0)$: Experimental amount adsorbed for a given partial pressure (P/P°);
$V_{diff}(w)$: Differential pore volume distribution;
$V_{cum}(w)$: Cumulative pore volume distribution;

Although the literature extensively uses the gas adsorption at very low temperature (−186 or −196 °C) using Ar or N_2 inert gases, the use of CO_2 at ambient temperature (mainly between 0 and 25 °C) is gaining attention to characterize microporosity.

BET theory uses the same approach as Langmuir theory but for a series of discrete layers. This involves vertical interaction between molecules, but lateral interaction is ignored. BET theory assumes that the surface is energetically homogeneous; i.e., all adsorption sites on the bare surface have the same adsorption energy (internal and external surface). However this theory would not be valid for certain types of materials; and CO_2 adsorption is additionally used. Basically CO_2 adsorption allows the characterization of pore size below 0.4–0.7 nm [317–319]. Although the molecular size of N_2 molecules is lower than that of CO_2 molecules, N_2 molecules are kinetically limited to access these very small pores (low temperature). In addition, nitrogen is not an optimal adsorbate for micropore analysis, mainly because of its quadrupole moment, which strongly affects the micropore filling pressure [320]. On the other hand, CO_2 adsorption is not a reliable technique for pore size analysis of solids with polar surfaces, because of the specific interactions caused by the large quadrupole moment [320].

The CO_2 adsorption isotherms, and the subsequent treatments, are obtained following the procedures described in Fig. 2.166.

As detailed in previous sections on gas adsorption, the outgassing procedure should carefully be carried out at temperature that does not affect the properties of the material. However the outgassing temperature should accurately be selected to

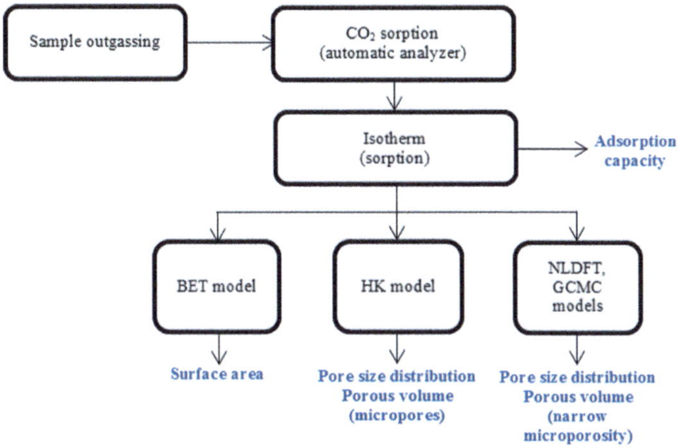

Fig. 2.166 Determination of micropores using CO_2 gas adsorption

remove bounded water, and to let the gas access the narrow micropores. A recent research proposes a mathematical correlation highlighting the relationship between BET values of specific surface area and the outgassing temperature, and the authors recommended outgassing at temperature as close as possible to 105 °C [318].

Using CO_2 adsorption, only low relative pressures could be reached (for example P/P° = 0.018 or 0.033 at T = 25 and 0 °C), and the adsorption isotherm could not fully be acquired as with N_2. Therefore the isotherm only refers to the microporosity and does not contain mesoporosity information. For type I isotherm adsorption branch, BET equations may be applied for specific surface area calculation (Eq. 2.137). Using N_2 or Ar adsorption, Dubinin-based equations, HK and t-plot models are applied to the micropores determination (pore size and volume) at a partial pressure P/P° < 0.5 (see Sects. 2.8.7.8 and 2.8.7.9). However, these models suffer from limitations to accurately determine the micropore size distribution and microporous volume [320–322]. Therefore, CO_2 adsorption could only be modeled using HK equation, which is basically a semi-empirical relation between the partial pressure and the gas-solid interaction. In the narrow micropores, this relation is quite limited since interactions between walls of the pore are not taken into account. Therefore, improvements in isotherm modeling (especially at very low P/P°) aroused with the development of molecular modeling such as Grand Canonical Monte Carlo simulations (GCMC) and Non-local Density Functional Theory (NLDFT) calculations to better understand the interactions between adsorbed species and porous carbon materials.

For NLDFT and GCMC methods, the basic principles are the calculation of local isotherms for a range of different pore size. This is basically included in the software for specified probe molecule, solid material and analysis temperature. From the experimental isotherm, the effective pore-size distribution is then calculated as a histogram of the molecular density versus the pore diameter.

The following sections present briefly the principles of NLDFT and GCMC for local isotherms calculation (in the pore) and the pore size distribution.

Isotherm Calculation from Density Functional Theory (DFT)

Density Functional Theory has been developed to better describe, from a theoretical point of view, the fluid behavior in non-homogeneous systems. Particular interest was focused on phase changes in fluids confined by the presence of walls, capillaries, and slits. In other words, this theory assumes that the fluid has a specific electronic density that will change due to interactions with the pore wall, other fluid molecules in the gas, and other molecules adsorbed at the opposite wall in the very narrow pores. Considering one fluid component (gas) in an external potential V_{ext} (perturbation due to the surface), the grand potential Ω of the adsorbate molecule is a functional of the number density profile in the pore $(\rho(r))$ at a given temperature and volume [320–325]. $\rho(r)$ could be expressed as the number of adsorbate molecules in a pore having a radius r.

$$\Omega[\rho(r)] = F[\rho(r)] + \int \rho(r)(V_{ext}(r) - \mu)dr \tag{2.152}$$

In highly simplified approach, Eq. (2.152) contains 2 terms. The first one is the functional representing Helmholtz free energy. The last term is a functional due to the external potential, and in other words, corresponds to the interaction between the adsorbate and the surface. This is an integral over *dr*, an elemental volume around the position selected (*r* is the position vector). The term related to the Helmholtz free energy is expressed as a sum of the ideal term, excess hard sphere repulsion term, and the attractive term (for 2 molecules at position r and r') calculated in a mean-field fashion. Please refer to literature for detailed equations.

Regarding the excess hard sphere repulsion term, different approaches would be used [322, 324], but the most popular is the smooth density approximation (SDA) of Tarazona et al. [323] (see literature for more details on the equations [321–325]). Indeed this term could be solved either or not locally. LDFT and NLDFT have been applied by Tarazona et al. on adsorption isotherms, and they concluded that the NLDFT is better representing the interactions at the surface since the model better fit the experimental isotherms [323]. It is important to note that this term is crucial for microporosity since it includes carbon–fluid interaction potential inside a slit-shaped or cylindrical micropore based on the overlapping potentials of the opposing walls [322, 323].

The fluid-fluid interactions involved in the attractive term are modeled based on the Lennard-Jones potential and the Weeks-Chandler-Andersen perturbation scheme, see literature for more details [321]. It means that the kernel model selected in software includes proper interaction parameters. For CO_2 gas adsorption, the intramolecular energy potential is not negligible and the model is based on 3 Lennard-Jones sites with charges centered on each site [321, 322].

Finally, the set of isotherms is determined. First the equilibrium density profiles $(\rho(z))$ is calculated for each value of the chemical potential (μ) by minimization of the grand potential (Ω, Eq. 2.152). Then the amount adsorbed per unit pore area (N_{NLDFT}) for a given partial pressure (P/P°) and pore width (w) is calculated (Eq. 2.153), from Avogadro number (N_A) and by integration of the number density distribution of the adsorbate, $\rho(z, p/p_0, w)$, over z. More precisely the number density distribution depends on z, the position into the pore, partial pressure, and w the diameter of the pore. The pore diameter is illustrated in Fig. 2.167.

$$N_{NLDFT}(p/p_0, w) = \frac{1}{2N_A} \int_0^{wcc} \rho(z, p/p_0, w)dz \tag{2.153}$$

Local Isotherms Calculation with Grand Canonical Monte-Carlo Model (GCMC)

The calculations of adsorption in pores could also be performed in the Grand Canonical Monte-Carlo (GCMC) ensemble, an open system, where the influential

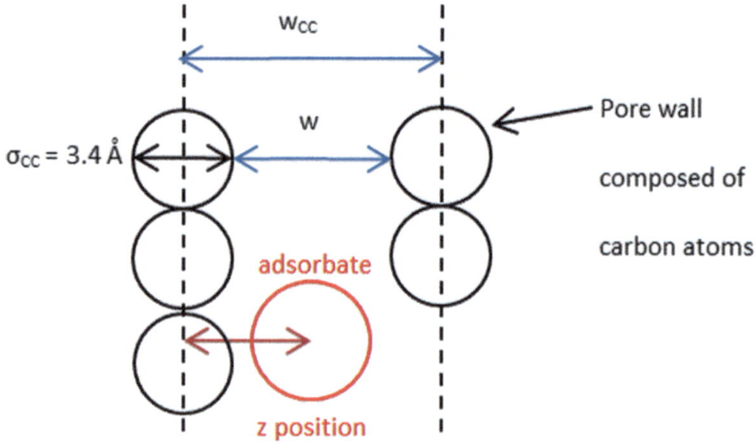

Fig. 2.167 Representation of pore width and the distance between carbon-carbon centers

parameters are temperature, pore volume and chemical potential. This leads to fluctuations in both the energy and number of particles, which provides a way to determine the density. Finally the simulation gives the number of molecules at a given chemical potential. For easier comparison with experimental isotherms, the chemical potential should be expressed as function of the partial pressure [322]. In the Grand Canonical Ensemble, the partition function Q for N-particles system is expressed as in Eq. (2.154) (see literature for more detail [326]).

$$Q(T, V, \mu) \equiv \sum_{N=0}^{\infty} \frac{e^{\beta N \mu} V^N}{\Lambda^{3N} N!} \int \mathrm{e}^{-\beta U(s^N)} ds^N \tag{2.154}$$

where $U(s^N)$ is the potential energy function, s^N is the reduced vector of the coordinates of all molecules, $\boldsymbol{\beta}$ is the reciprocal temperature ($1/k_BT$), Λ is the thermal Broglie wavelength.

The starting point of Monte-Carlo simulation is the use of Metropolis method [326], where the sampling scheme computes thermal averages of functions that depend explicitly on the coordinates of the molecules in the system. Thus, this technique could not be used to determine the integral in Eq. (2.154). However, the Metropolis method can be used to measure the difference in free energy between two possible states of a system. Therefore the principle of GCMC simulation is to calculate the difference between the potential energy of the system at a certain state and after a perturbation (displacement, removal or insertion of molecules) and to evaluate the probability density of this perturbation. It should be noted that this technique is limited when the molecule is large and complex, especially for insertion perturbation. Then this new state should be accepted or rejected, and the simulation continues with the potential energy. The corresponding probability density is expressed using Eq. (2.155). For each set of perturbations at a selected relative pressure the potential energy of the system is calculated for N molecules. Since the GCMC calculations contain the position vector of all molecules, the number of molecules adsorbed could be calculated and the isotherm could be drawn.

$$N_{\mu VT}\left(s^N\right) \propto \frac{e^{\beta N\mu} V^N}{\Lambda^{3N} N!} e^{-\beta U(s^N)} \tag{2.155}$$

Pore size Distribution

From the NLDFT or GCMC calculations, a set of isotherms for pores of different width is available and an effective pore-size distribution would be calculated. It should be noticed that carbon solids do not exhibit a perfect pore shape as in the models (slit-shaped or cylindrical) and the graphene defects (layers and edges) are not included [322]. The pore area distribution (S_{diff}) could be calculated using Eq. (2.156) from the experimental values (N_{exp}) and the calculated density distribution function $\rho(z, P/P_0, w)$, included in N_{NLDFT} or N_{GCMC}. This follows a common procedure of inverting the integral adsorption equation by the Tikhonov regularization method (see [325] for more detail).

$$N_{exp}(p/p_0) = \int_0^\infty N_{NLDFT}(p/p_0, w) S_{diff}(w) dw \tag{2.156}$$

The distribution is then presented as specific cumulative (V_{cum}) and differential (V_{diff}) pore volume distributions.

$$V_{diff}(w) = \frac{w}{2} S_{diff}(w) \tag{2.157}$$

$$V_{cum}(w) = \int_0^w V_{diff}(w) dw \tag{2.158}$$

Examples and Experimental Recommendations

Based on the previous explanations, the selection of the gas probe should be linked to the objective of the study. To illustrate the use of CO_2 and the NLDFT model for the determination of pore size distribution, wood biochar is taken as an example. Data have been acquired using gas adsorption on Micromeritics 3Flex analyzer, using the 3Flex V3.01 software. Different models are available in the software and CO_2/Carbon model has been used for this example. This model has been developed for carbon materials based on graphitic structures. Therefore the model is not convenient for solid having larger amorphous phase content. To determine the pore size distribution, after selecting a model from NLDFT suite, the regularization value should manually be adjusted to better fit the isotherm.

Fig. 2.168 Adsorption isotherms on a spent wood biochar using CO_2 at 0 °C and CO_2 at 25 °C (biochar 1, see description in the main text)

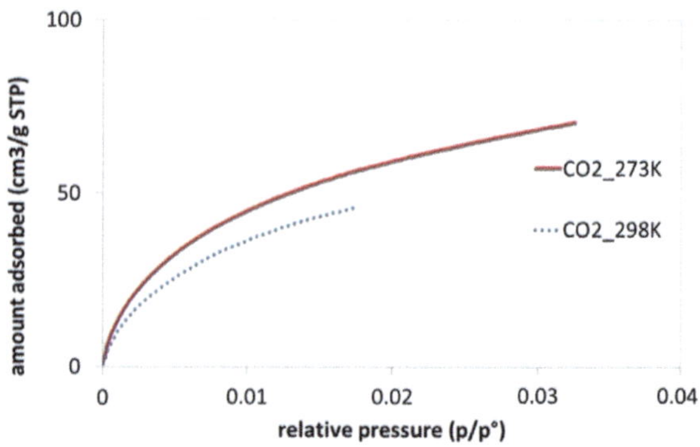

Table 2.89 Specific surface area and microporous volume, values determined with slit-shaped pores, for 2 wood biochars

Parameter	Probe	Amount adsorbed	Specific surface area		HK model	NLDFT model
Sample	Gas/K	q_{max}, cm^3/g (STP)	S_{BET}, m^2/g	$S_{Langmuir}$, m^2/g	V_p, cm^3/g (d_p in nm)	V_p, cm^3/g (d_p in nm)
Biochar 1	CO_2_273	70.4	391	414	0.148 ($d_p < 0.9$)	0.048 ($d_p \leq 0.45$)
Biochar 2	CO_2_273	70.2	353	387	0.147 ($d_p < 0.9$)	0.047 ($d_p \leq 0.45$)

As explained previously, the experimental CO_2 adsorption isotherm is acquired in a shorter range of partial pressure. Figure 2.168 shows the isotherms for CO_2 adsorption at 0 and 25 °C on a spent wood biochar. This biochar has been obtained by the pyrolysis of a wood at 700 °C during 30 min. Then, the biochar has been activated at 850 °C under steam for 80 min. Finally, the activated biochar has been used for H_2S removal tests at room temperature until saturation. The adsorption tests have been performed on this spent biochar. This biochar is named "Biochar 1".

Table 2.89 presents results acquired on two biochar samples. The first one is the biochar 1 described above. The second one is a Biochar 2 which has been produced from steam gasification of bamboo guadua at 750 °C during 1 h.

From the experimental isotherms, and using HK and NLDFT models, the pore size distribution has been determined with a slit-shaped pore. For NLDFT, the model used in the software corresponded to NLDFT for CO_2/Carbon. Figure 2.169 shows the pore size distributions of Biochar 1 using both models. For a same biochar, Table 2.89 shows that the micropore volume was calculated for pore size lower than 0.45 nm using NLDFT model, while HK model does the calculation up to 0.9 nm. Figure 2.170 also shows a good agreement between NLDFT and GCMC models for the cumulative pore volume using gas adsorption on carbon fibers.

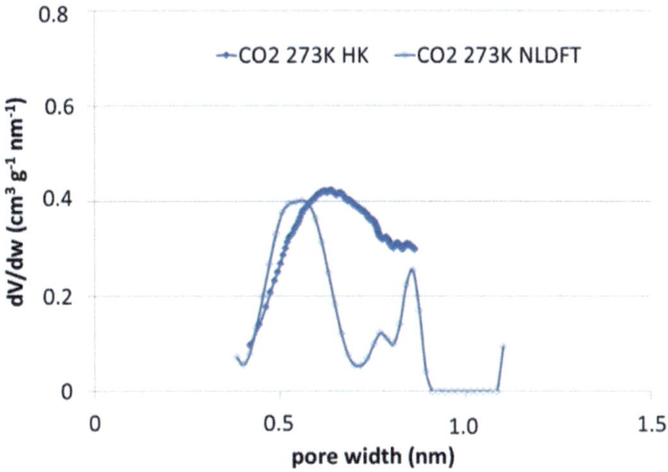

Fig. 2.169 Pore size distributions for CO_2 adsorption using HK and NLDFT models, results from a wood biochar (namely Biochar 1)

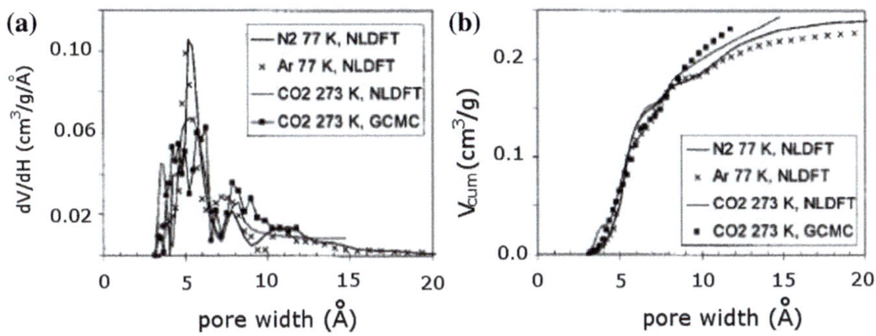

Fig. 2.170 Comparison of NLDFT and GCMC models for N_2 and CO_2 adsorption on carbon fibers, from [321]. Reprinted (adapted) with permission from American Chemical Society [321]

2.8.8 Determination of Macropores by Mercury Porosimetry

2.8.8.1 Principle of the Measurement

Mercury intrusion porosimetry (MIP) is a powerful technique utilized for the evaluation of porosity, pore size distribution, and pore volume (among others) to characterize a wide variety of solid and powder materials.

The technique consists in forcing mercury to penetrate into the pores by applying pressure (Fig. 2.171). The instrument has two pressure outlets: a low-pressure outlet that works in the range of 0.003–0.25 MPa and a high-pressure outlet working in the range of 0.025 MPa–200 or 400 MPa depending on the model.

A known quantity of the sample is placed in a measurement cell called a penetrometer (Fig. 2.172). The volume of the empty penetrometer is first determined by a calibration. The quantity of mercury in the penetrometer during the calibration or during the analysis is determined by a measurement of capacitance carried out at the closure system of the penetrometer.

The analysis starts with the low-pressure outlet where a vacuum is created in the penetrometer in order to degas the sample, thus liberating access to the pore space. Mercury is then introduced into the penetrometer via the capillary tube and will then fill the space left by the sample. This is the only time during the analysis when mercury is added to the measurement cell. The increase in pressure moves the mercury from the tube towards the intergranular space and/or the biggest pores.

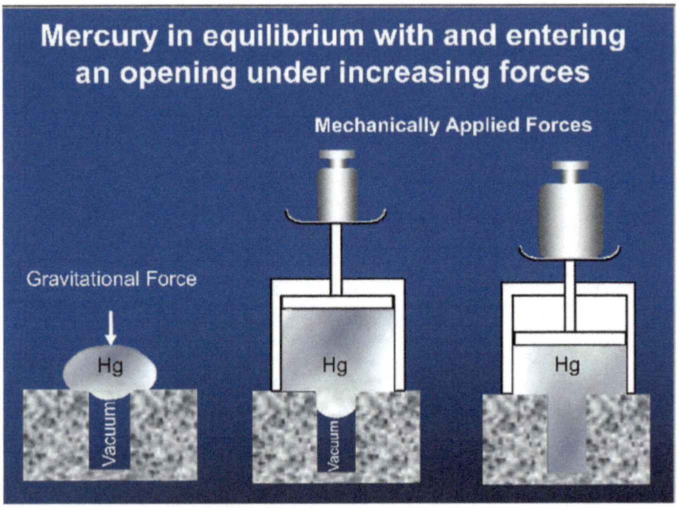

Fig. 2.171 Illustration of application of external pressure to the sample for introducing mercury into the pores [327]. Reproduced with permission from MICROMERITICS

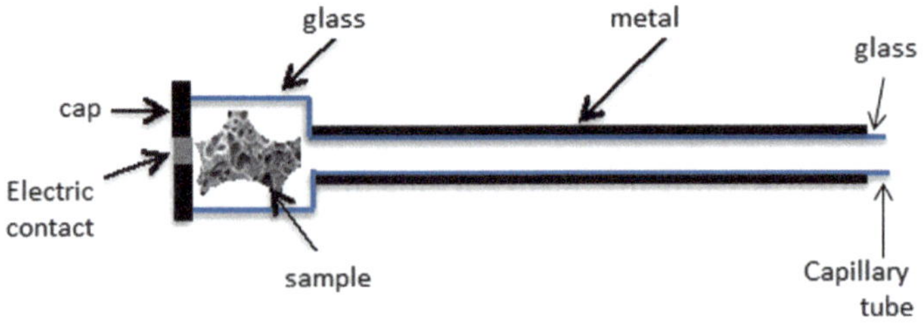

Fig. 2.172 Measurement cell (or penetrometer)

The analysis continues using the high-pressure outlet, after weighing the penetrometer containing the sample with the mercury. This stage is useful for calculating (using the software that runs the instrument) the volume occupied by the sample (volume of the solid plus the volume of the pores). Since the mass of the introduced sample is known, the apparent density of the sample can be deduced.

On the high-pressure outlet, the applied pressure forces the mercury to penetrate the pore space. The pressure applied is inversely proportional to the pore diameter (Eq. 2.159). The quantity of mercury having gradually penetrated the pore space is measured by capacitance. Between each pressure increment, a set time is left in order to allow the mercury to penetrate the pore space. The volume of mercury introduced corresponds to the volume of the pores. The hypothesis of cylindrical pores enables the distribution of pore volume and the distribution of pore size to be obtained.

The Washburn's equation (Eq. 2.159) describes the mechanism of intrusion by connecting a pressure applied to a pore radius:

$$r_p = \frac{-2\gamma_{LV}\cos\theta}{P} \tag{2.159}$$

where: r_p is the pore radius (μm), γ_{LV} is the surface tension of mercury (liquid-vapor) in N/cm, θ is the contact angle (°) et P is the pressure applied (MPa).

The Washburn's equation is based on the following assumptions:

- γ_{LV} and θ are constant regardless of the temperature and shape of the pore;
- Presence of only open pores having cylindrical shapes and being non-interconnected;
- No presence of closed pores;
- The sample is incompressible and non-expandable under the effect of the pressure.

As previously explained, the quantity of mercury is quantified via calibration curves at each pressure. From Eq. (2.159), pore radius can be calculated for each pressure. Thus, the final results can be expressed either by the cumulative intrusion or percentage of pore as a function of pore radius.

However, the following assumptions must be taken into account when performing the analysis and interpretation of the results:

- γ_{LV} varies little with temperature and strongly depends on the purity of the mercury used (the average value is above 480–485 $\times$ 10^{-5} N/cm in the case of the use of tridistilled mercury);
- θ varies depending on the nature of the solid and the shape of the pores (the average value is 130° and 140°);
- The pressure applied to the sample is above 0.03 and 400 MPa (maxi) could compress the sample or destroy the porous network.

2.8.8.2 Penetrometer

The penetrometer corresponds to the measurement cell used on the mercury porosimeter.

It consists of (Fig. 2.172):

a glass "head" where the sample is placed
a glass stem with a capillary tube, covered in platinum.
a cover fitted with an attachment where the capacitance measurement is carried out.

The seal between the "head" and the cover is made using a specific kind of grease.

Each penetrometer is sold with a certificate indicating the volume of the capillary tube and its dielectric constant (given in μL/pF).

The penetrometer containing mercury acts like a condenser. The glass of the penetrometer acts as an insulator while the mercury and the external metallic coating act as a conductor. For this reason, it is essential that the penetrometer used for the analysis should be clean and that the metallic coating of the capillary tube should be in good condition in order to ensure correct measurement of the capacitance.

There is a whole range of penetrometers which vary in terms of their volume and/or their shape. It is essential to adapt the penetrometer to the sample.

Choice of Penetrometer

Two shapes of penetrometer are available (Fig. 2.173), one specific to the analysis of powder where the mercury is injected above the bed of powder by the cone and the other more specific to the analysis of a "solid", monolithic sample.

For a given shape of penetrometer, the length and the diameter of the head can vary, thus giving a wide range of penetrometer volumes that can accept samples of various sizes.

The choice of penetrometer volume is also made depending on the porosity of the sample. An initial analysis is often carried out in order to determine whether the penetrometer is suitable before carrying out a second analysis whose results will actually be used.

To help choosing the penetrometer volume and in order to achieve the most accurate measurements, the manufacturer recommends using between 20 and 80% of the volume of the capillary tube.

If the volume of mercury in the chosen capillary tube is below 20%, it is preferable either to increase the quantity of sample or to use a penetrometer with a smaller volume.

If the volume of mercury in the chosen capillary tube is above 80% it is preferable either to reduce the quantity of sample analysed or to use a penetrometer with a greater volume.

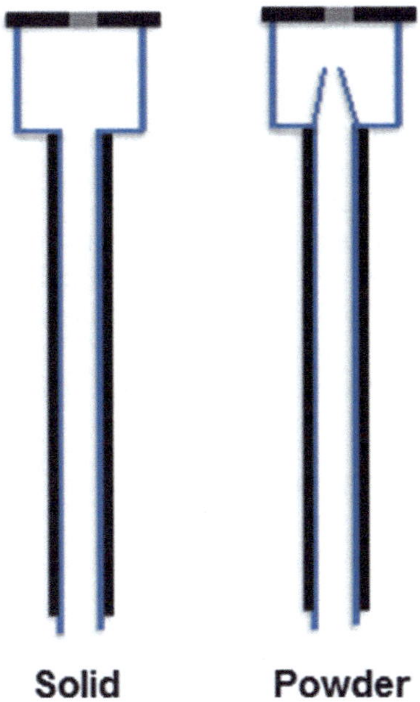

Fig. 2.173 Available penetrometer shapes

Calibration of Penetrometer

Before carrying out the analysis, it is essential to calibrate the penetrometer. This serves to determine its volume. This value is used by the software to determine the volume occupied by the sample in order to calculate its apparent density and the porosity.

This measurement is made only at the low-pressure outlet where the mercury is introduced into the penetrometer. Since the latter has been weighed before and after being used with the equipment, the mass of mercury introduced may be calculated.

Since the density of mercury, 13.5335 g/mL, is known, the volume introduced, which corresponds to the volume of the empty penetrometer, may be calculated. The value thus obtained must be entered into the penetrometer characteristics used by the equipment software.

Mercury

Mercury is used because of its incompressibility and the fact that it is a non-wetting liquid. It is nevertheless a toxic heavy metal which implies conditions of manipulation and treatment that are expensive and difficult to manage. For these reasons, many laboratories have stopped using this technique which, however, for the moment remains practically irreplaceable.

Before any manipulation it is recommended that the operator should refer to the mercury safety datasheet, in order to identify the personal and collective protective equipments necessary to the use of this substance. The mercury used for the analyses must be at least of 99.9% of purity.

After the analysis, the penetrometer will contain the sample, polluted by the mercury contained in the pore network, as well as liquid mercury. It is recommended when cleaning the penetrometer to separate the polluted solid sample from the liquid mercury.

The polluted solid sample should be stored prior to the treatment by a specialist waste treatment company whereas the liquid mercury should be recovered so that it can be distilled by a specialised company for further reuse.

To clean the penetrometer after an analysis, a vacuum cleaning system may be used, consisting of a pump connected to a container and a pipe.

2.8.8.3 Analysis

Sample Preparation

If the sample to be analysed contains residual water or solvent it must be dried before analysis, since the presence of such substances can influence the measurement. Furthermore, the sample must be metal-free, to avoid the formation of an amalgam with the mercury.

The protocol for preparing the sample consists essentially in several weighing steps:

- weigh the empty penetrometer (with the specific grease that makes the seal between the penetrometer and the cover)
- weigh the quantity of sample introduced into the penetrometer: this quantity should be adjusted for a second analysis so that the volume of mercury in the capillary tube is between 20 and 80% of the volume of the penetrometer.
- finally, weigh the penetrometer + the sample + the mercury before using the high-pressure outlet. This weighing determines the quantity of mercury introduced into the penetrometer.

The precision of the weighing is a non-negligible parameter in terms of the quality of the results, on the one hand to avoid an accumulation of errors during repeated weighing of the penetrometer and on the other hand because the volume of mercury introduced will expressed in relation to the mass of the sample.

The quantity of sample used should be representative of the whole batch.

Pressure Increments

During the analysis, the sample undergoes significant variations in pressure. It may be compressed and/or densified. This densification results from crushing of the pore network if it is not rigid enough to resist the applied pressures. In order to limit such phenomena, it is recommended that the pressure increments be configured so that the mercury penetrates the pores slowly. In concrete terms, this means using small increments and long set times (minimum 30 s) to leave the mercury time to penetrate.

The measuring instruments offer three possible modes for incrementally increasing the pressure:

- Pressure scanning: this consists of scanning the whole range of pressures by introducing the mercury continuously without set times. This mode is highly destructive but gives an idea of how the sample will behave in relation to pressure.
- Fixed pressure increments: the pressure increments and the resting time between each increment are fixed. This mode suits samples with a wide range of pore sizes. However, several measurements are required in order to obtain the correct set time and pressure increment.
- Increments in volume: these are made at fixed pressure. This mode enables the operator to go more slowly in very porous zones and thus to do less damage to the sample.

Figure 2.174 shows the influence of this mode of intrusion on the volume of mercury introduced. Line 1 corresponds to pressure scanning whereas line 5 results from the volume increment mercury intrusion mode. For the volume of mercury introduced, line 1 shows that the mercury has reached smaller pore radii than those under the conditions shown by line 5, which confirms crushing and/or modification of the pore structure.

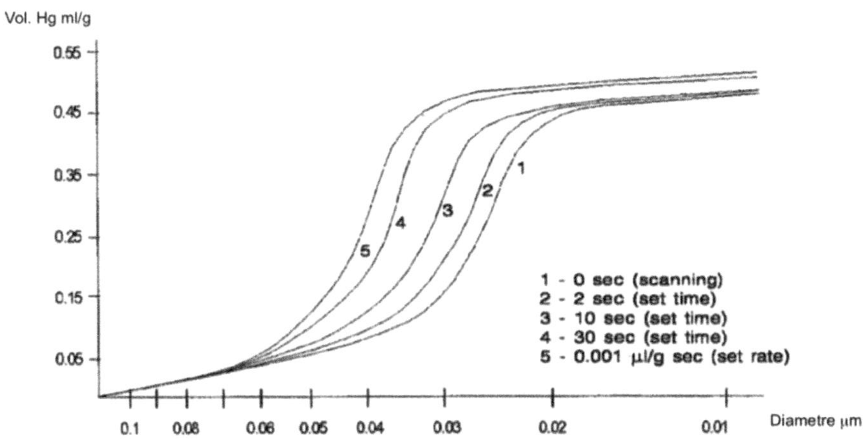

Fig. 2.174 Influence of the intrusion mode on an analysis of an alumina [328]. Reproduced with permission from MICROMERITICS

Program

The results are displayed graphically on a programme (Fig. 2.175). This is obtained by plotting the volume of mercury introduced against the pressure applied (or the pore diameter). The intrusion of the mercury (by increasing the pressure) and its extrusion (by a reduction in the pressure during which the mercury leaves the pore network) can then be observed.

The resulting hysteresis loop is never closed, because mercury remains trapped in the network.

Several intrusion-extrusion cycles may be carried out if the pore network is rigid enough to withstand the pressure constraints.

Extrusion of mercury should never occur during intrusion. If this happens it means either that the set time is too short or that there is a leak of mercury from the penetrometer.

Horizontal extrusion indicates irreversible modification of the pore structure.

A hump that forms at the start of an extrusion is due to a heating of the system called the "thermometer effect". Correction with an empty measurement cell is then necessary.

A perfectly linear intrusion of mercury can indicate compression of the sample.

Figure 2.175 represents two intrusion cycles of mercury and one extrusion cycle.

- The first mercury intrusion cycle consists of two sharp upward curves;
- The first curve starts at low pressure: the amount of mercury introduced indicates a rearrangement of the particles in the penetrometer followed by intrusion into the intergranular pore space;

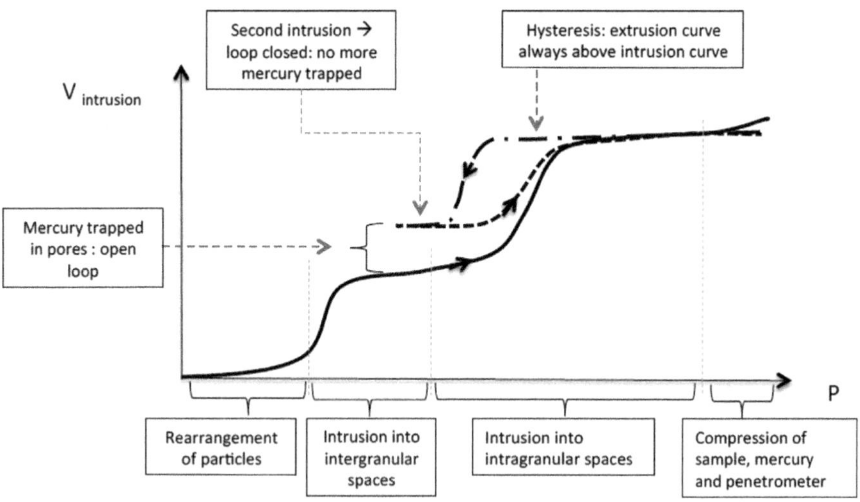

Fig. 2.175 Programme obtained after two mercury intrusion cycles

- The second curve is obtained with the high-pressure outlet and indicates an intrusion of mercury into the pore network.

The pressure applied during this first cycle very often modifies the pore structure. The mercury extrusion cycle is always above the intrusion cycle on the graph. The shape of the hysteresis helps with the interpretation of the shape of the pores. The hysteresis loop remains open because mercury remains trapped in the pore network.

The second intrusion cycle, which has only one sharp upward curve, enables the real volume of mercury introduced to be determined and the hysteresis loop to be closed.

An increase in the volume of mercury introduced during the analysis does not necessarily indicate an intrusion of mercury into the pores. It can also result from crushing of the pore network or from densification of the sample. Visual inspection of the sample is often necessary after the analysis to help with the interpretation.

2.8.8.4 Example

Macropore distribution of a ceramic material made from clay, sand and a refused derivative fuel (RDF, 8 wt%), called thereafter "*ceramic 8%.wt RDF*". The method of sample preparation has been previously described in the document. Refused derivative fuels (RDF) from municipal solid wastes contain mainly organic matters (cardboard, plastics, papers …) with some inorganic fractions. They can be valorized in different processes such as porous agents for producing fired clay brick with low thermal conductivity (high thermally-insulant character). This ceramic was prepared by adding RDF under fin powder form to a fabrication mixture of clay, sand and water. After shaping and drying, the monoliths were fired under controlled conditions of temperature program which allowed decomposing organic compounds initially present in RDF.

Figure 2.176 shows the pore size distribution of the ceramic fired at 800 °C. The results obtained are presented either by the cumulative intrusion or the percentage of pores as a function of the pore diameter. Pores smaller 1 μm represented around 65% (zone A) while pores between 1 and 10 μm were of around 30% and pores larger than 10 μm around 5%.

2.8.9 Temperature Programmed Techniques

2.8.9.1 Introduction

Temperature programmed (TP) techniques are generally used to study or monitor a chemical reaction of a probe molecule (for e.g. H_2, CO_2, H_2S, NH_3, etc.) with a substrate by linearly increasing the reaction temperature with time [329]. Reactions

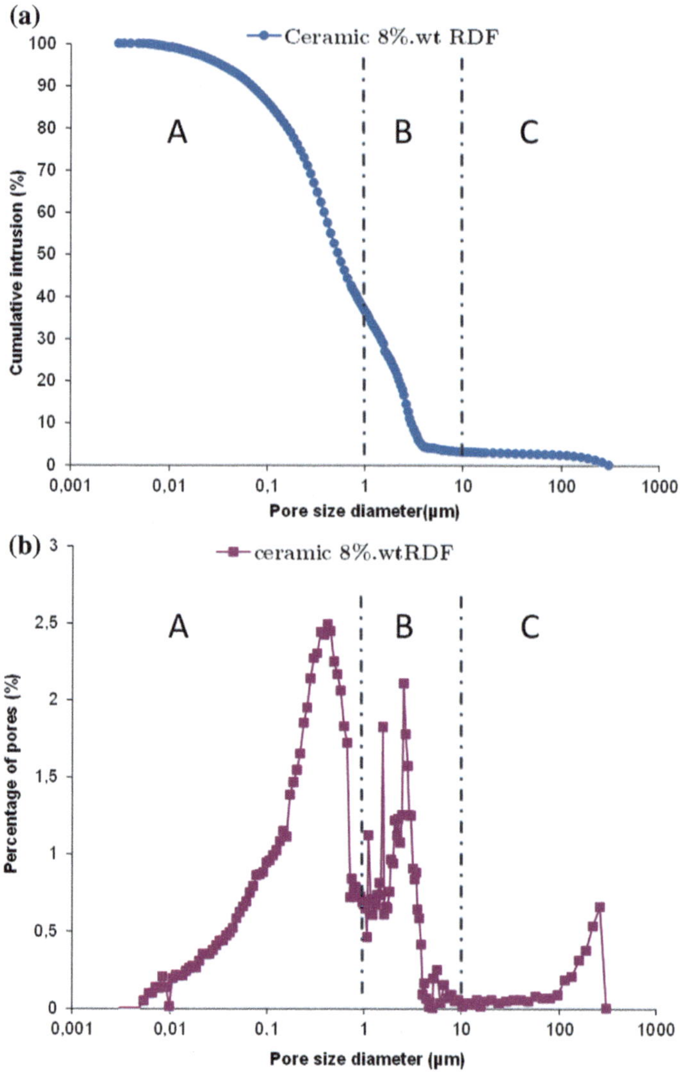

Fig. 2.176 Pore size distribution of a RDF: a cumulative intrusions as a function of pore diameter size; b pore percentage as a function of pore size

such as reduction with H_2, oxidation with O_2, and sulfidation with H_2S happening on the sample under study can be investigated using analytical techniques that are named as temperature programmed reduction (TPR), temperature programmed oxidation (TPO), and temperature programmed sulfidation (TPS), respectively. When the surface process investigated on the sample is desorption of a probe molecule, then the technique is termed as temperature programmed desorption (TPD) wherein the solid is previously saturated with probe molecule. If the probe molecule adsorbs irreversibly on the sample via surface reaction, then is called a temperature programmed reaction spectroscopy (TPRS), which can yield information related to reaction kinetic and insights into surface reaction mechanisms. Generally, in all these methods, temperature (T) is varied as a linear function of time (t), i.e. the heating rate ($\beta = dT/dt$) is a constant value. In general, TP techniques are widely applied in the field of heterogeneous catalysis for investigating catalysts supported on porous materials or well-defined crystalline solids. More detailed information about these techniques can be found in literature [329] and references therein. In brief, TPR and TPO methods help to determine the reducibility of the species present on the surface of a solid and to examine the extent to which the solid can be oxidized (which was previously reduced), respectively. From TPR and TPO results, adequate temperature of reduction and oxidation of a given material can be precisely determined. TPD analysis is very useful in determining the acido-basic nature; number, type, and strength of active sites available on the surface of a solid; and also obtain kinetic and thermodynamic parameters of the desorption process or decomposition reactions by measuring the rate of desorption of adsorbed molecules as a function of temperature [330].

In a typical TPD experiment [330, 331], a small amount of sample (100-200 mg), which is loaded in a reactor (usually U-shaped quartz reactor), can be heated linearly to a given temperature as a function of time in a oven maintained at atmospheric pressure. In order to remove any physisorbed molecules (for e.g. water from moisture), the sample is first heated up to a set pretreatment temperature for a defined period of time under inert gas (He) and then the temperature is brought back to room temperature (can be higher or sub-ambient temperature, if the adsorption of probe molecule is temperature dependent). Next, the probe molecule (NH_3, CO_2, etc.) diluted in inert gas flows over the sample for a predetermined time under isothermal conditions to fully saturate the surface area of the sample. The probe molecules such as NH_3, pyridine, CO_2 and SO_2 are used to titrate acidic or basic surface sites. Later, the excess and physisorbed probe gas molecules are removed under inert gas, following this step, the sample is heated linearly with time to a predetermined temperature. During the increase in temperature of the sample, the chemisorbed probe molecules start to desorb and the detector present in the downstream measures the change in the inert gas flow based on which the amount of desorbed molecules are quantified. Usually TPD instrument is equipped with thermal conductivity detector (TCD), and it does not distinguish different types of molecules. Therefore, when there is more than one molecule in the desorbed gas stream, coupling mass spectrometer to TPD instrument allows to quantitatively measure the desorbed gas composition.

In a typical measurement of TPR or TPO, a small amount of sample (100–200 mg) is also used. The sample is placed in a reactor which is usually of U-shaped quartz reactor, heated by an electrical oven. The sample can be pretreated, depending on the objective of the analysis. For example, in the elaboration of solid catalyst or adsorbent, a pretreatment by air calcination can be applied before the reduction step. In this case, TPR experiment can be performed after air calcination step.

TPR experiment consists of the heating of the sample under a flux of reducer gas, which is usually hydrogen diluted in an inert gas such as He. At the outlet of the reactor, a detector is used to analyze the consumption of hydrogen during the experiment. Different type of detector can be used such as TCD, online mass spectroscopy, or $\mu - GC$.

TPO experiment is similar to TPR experiment, but the probe gas used is an oxidant gas (such as oxygen diluted in inert gas) instead of a reducing gas. A detector is also used to analyze the consumption of oxidant gas during sample heating.

2.8.9.2 Instruments

The instrument schematic for TP techniques from Micromeritics, AutoChem II 2920 [332] is depicted in Fig. 2.177. The sample tube (sample holder/reactor) is charged with the substrate and could be heated in an oven in a controllable manner to the set temperature. A thermal conductivity detector is used to detect the probe molecule before and during the reaction. A cold trap is very handy to trap condensable molecules produced during the reaction (for e.g. water formed during the reduction of metal oxides under H_2). This AutoChem II 2920 model allows performing different TP techniques: TPR, TPO, TPD, and even chemisorption and BET measurement [332]. Other commercial equipment, which is equivalent to AutoChem II 2920, is ChemBET Pulsar model from Quantachrome Instruments [333].

Commercial equipment such as AutoChem II 2920 and ChemBET Pulsar need high initial investment costs. For the measurement of TPR, TPO and TPD, hand-made equipment can be obtained by associating the following principal devices:

- A tubular electrical oven;
- A U-shaped or tubular quartz reactor;
- A TCD detector (or any online gas analyzer);
- Gas bottles with associated gas pressure regulation systems.

Figure 2.178a shows the scheme of this hand-made equipment, which is still used in laboratory. Figure 2.178b shows an example of TPR results obtained with different TiO_2 powder impregnated with ruthenium precursors using such hand-made TP equipment [334]. The different peaks observed in Fig. 2.178b correspond to the reduction of ruthenium precursor (Ru^{3+}) into metallic ruthenium (Ru°).

The two main steps involved in TPD analysis are, (i) flow of targeted probe molecule onto the sample during which the adsorption of probe molecules takes place and (ii) heating of the sample in a controlled fashion allows the desorption of probe species into the gas phase that are detected and quantified using one of the detectors such as thermal conductivity detector, flame ionization detector, or mass

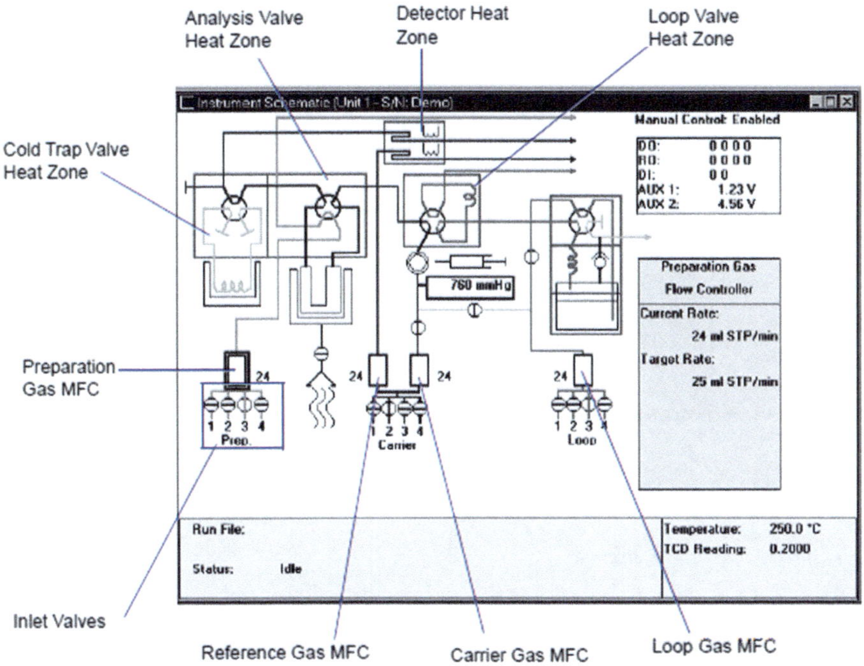

Fig. 2.177 Instrument schematic of AutoChem II 2920. Taken from the Micromeritics operator's manual for AutoChem II 2920. Reproduced with permission from Micromeritics [332]

spectrometer. Typical TPD profile is depicted in Fig. 2.179. The area under the TPD profile is proportional to the amount of probe molecules chemisorbed on the sample. The position of the peak maximum (T_{max}) provides information related to the activation energy of desorption. Since the adsorption process is spontaneous process, the activation energy for desorption represents the heat (enthalpy) of adsorption. In general, the higher value of T_{max} suggests that desorption is more difficult, which is an indication of strong interaction between the probe species and the active site on the sample's surface. When there is more than one binding state for the probe molecule on the sample's surface, there will be multiple peaks in the TPD profile, consequently, different activation energies for desorption. Simultaneous desorption of probe molecules from different adsorption sites on the sample's surface leads to overlapping peaks in the TPD spectrum (Fig. 2.179). For more detailed information to extract reaction mechanisms, kinetic date, measuring surface areas of adsorption from the TPD spectra can be found in the literature [330, 331]. In order to get a reliable and quantitative information from TPD analysis, the mass transfer properties such as inter-particle and intra-particle diffusion effects must be properly taken into consideration [335].

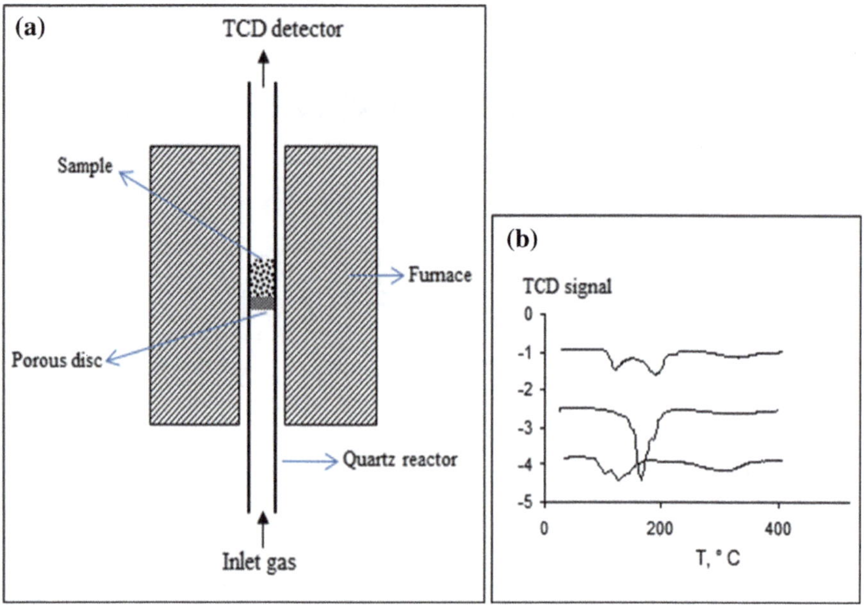

Fig. 2.178 Scheme of an equipment made with an electrical oven: a tubular quartz reactor and a TCD detector; b example of TCD signals of TPR experiments for different TiO_2 powder impregnated with ruthenium precursor. Adapted with permission from Pham Minh [334]

Fig. 2.179 Typical TPD profile of NH_3 obtained on zeolite (H-Mordenite) [336]. Reprinted with permission from American Chemical Society

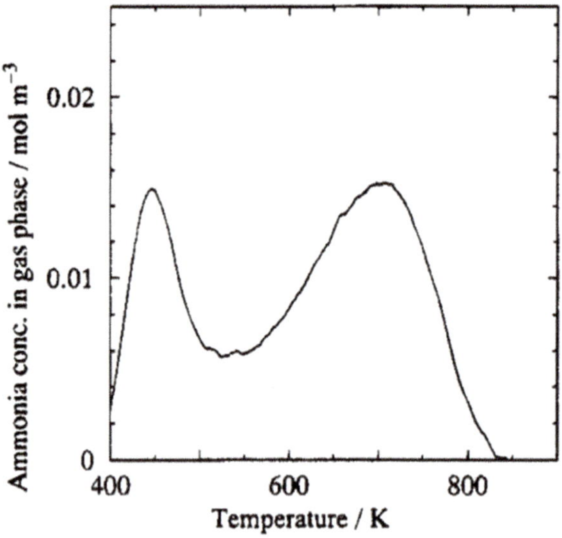

2.9 Structural and Textural Characterization

2.9.1 X-Rays Diffraction (XRD)

X-ray powder diffraction (XRPD) is a non-destructive technique widely used to characterize polycrystalline powders. The best known use of XRPD is the determination of the phase composition of a mixture including (poly)crystalline and amorphous phase identification and quantification. For any given crystalline phase, XRPD may also be used to determine the crystallite size, under the condition that the sample is crystallized, as well as the crystal structure and the microstrain.

2.9.1.1 Crystallography

Before going into the details of X-ray diffraction, it seems necessary to remind some fundamentals of crystallography. A solid can be classified as crystalline, polymer or amorphous.

A solid is called crystalline when atoms or molecules are in an ordered arrangement in a repeating three dimensional system so called the crystal lattice. For amorphous solids, there is no ordered structure at short distance.

Crystalline materials are characterized by the long-range ordered periodic arrangements of atoms. The basic repeating unit that defines the crystal system is named the unit cell; it is defined by dimensional (a, b and c) and angular (α, β and γ) lattice parameters, which describes its geometry and size (Fig. 2.180a). From symmetry elements, crystalline classes are defined and they are 32 in total (Fig. 2.181). The 14 Bravais systems along with elements of symmetry give rise to 230 space groups. All description and definition of the different crystal systems and space groups are described in International Tables For Crystallography [337].

From these different lattices, set of points, so called nodes, are defined and localized at the intersection of unit cells. The nodes are identified in the system by the extremity of a, b and c vectors as illustrated in the Fig. 2.180b. Any plane

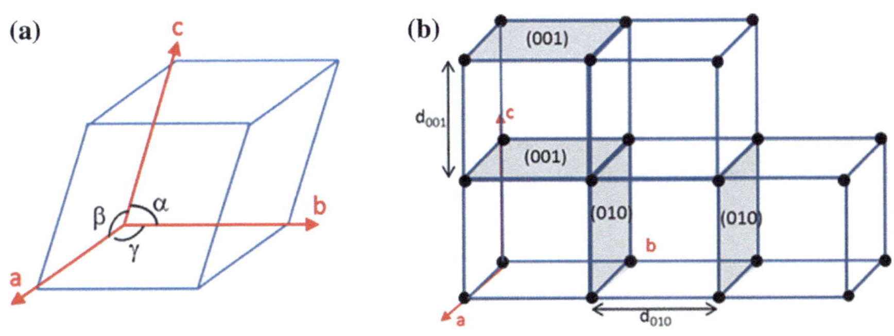

Fig. 2.180 Illustration of crystallographic parameters

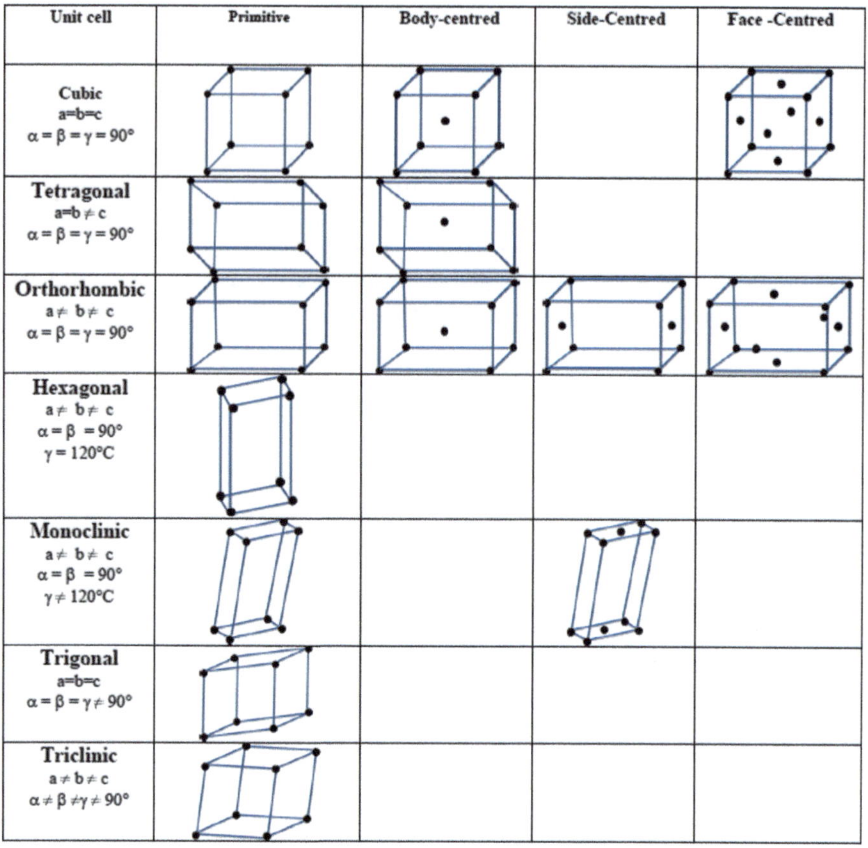

Fig. 2.181 Illustration of Bravais lattices

passing through 3 non-colinear points is a reticular plane defined by the indices of Miller (*h*, *k*, and *l*) which are always whole numbers. It contains an infinity of nodes that form a 2D network and is part of a set of parallel, equidistant planes that pass through all nodes of the network, no reticular plan being empty. The distance between inter-reticular plans is noted d_{hkl}. The Miller indices (*hkl*) define the intersection of the plane of atoms with the axes of the elementary unit cell. The corresponding plane intersects the axis in a/h, b/k and c/l. Examples of reticular planes (001) and (010) are presented in the Fig. 2.180b. Owing to the periodicity of the lattice, there is an infinity of identical plans.

2.9.1.2 Powder XRD Principle

XRD Production

X-rays were discovered in 1896 by Wilhelm Röngten and as a reward he received the Nobel prize of Physics. The wave nature of X-rays was established in 1913 by first experiments of Max Von Laue.

X-rays are electromagnetic radiations of same nature of light, however with a shorter wavelength (Fig. 2.182). The energy is quantified in the form of photons, each of them carrying an energy expressed by the following equation:

$$E = \frac{hc}{\lambda} = h\nu \tag{2.160}$$

where "h" is the Planck constant ($6.626 \cdot 10^{-34}$ J s) and "c" the speed of light ($3 \cdot 10^{8}$ m s^{-1}). The wavelength of X-rays is between 0.1 and 100 Å which corresponds to an energy between 0.1 and 100 keV.

The X-rays are produced within a X-ray tube, as presented in the Fig. 2.183 when electrons emitted by a heated cathode and accelerated by an electrical field in a vacuum chamber, impact a metallic target as referred to an anode and called anti-cathode for history reason. When a current passes through the tungsten filament (cathode), electrons are thermally emitted and they are accelerated towards the anode by the high potential difference between the anode and the cathode. While electrons strike the anode, X-rays are continuously produced and they come out of the tube through thin beryllium windows.

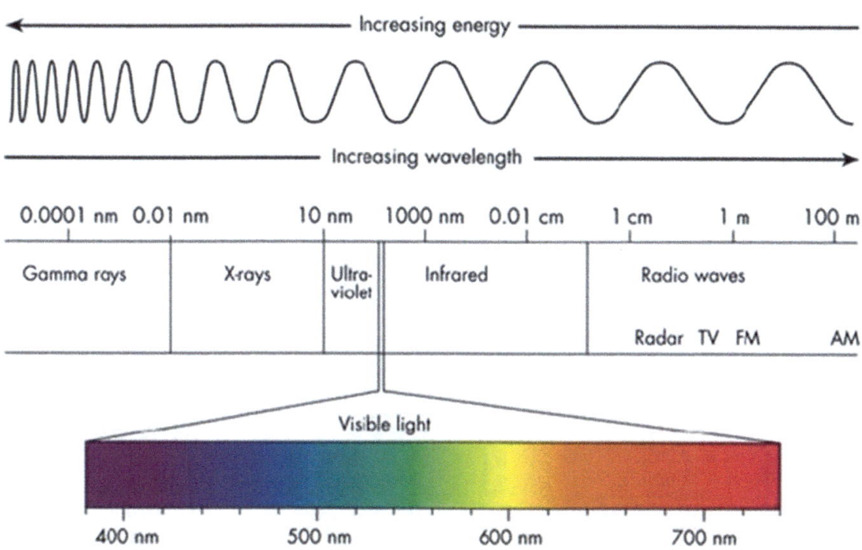

Fig. 2.182 Electromagnetic radiations (Philip Ronan, Gringer)

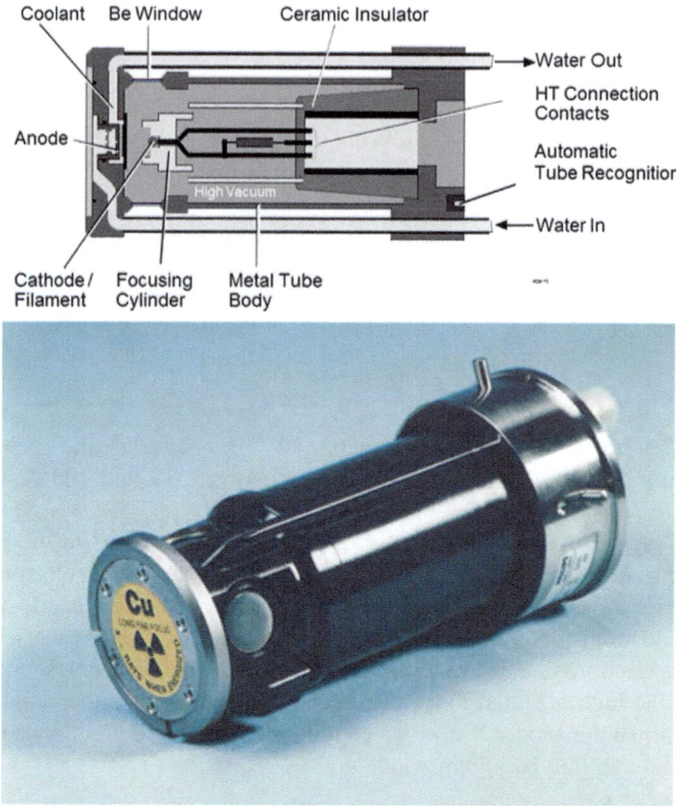

Fig. 2.183 Example of X-ray tube using Cu as an anti-cathode. Schematic of its longitudinal cross-section (on the top) and photograph (on the bottom) of the tube. Reprinted with permission from Malvern Panalytical [74]

The wavelength of the characteristic radiation from an X-ray tube depends on the nature of the anode (anti-cathode) material used. Table 2.90 presents the available anode materials and their characteristic wavelengths. The most used X-ray tube is the Copper (Cu) anode.

Table 2.90 X-rays tube characteristics

Anode Material	Wavelength, $K_{\alpha 1}$ (Å)	Wavelength, $K_{\alpha 2}$ (Å)	Wavelength, K_{α} (Å)
Mo	0.70930	0.71359	0.71073
Cu	1.54056	1.54439	1.54184
Co	1.78897	1.79285	1.79026
Fe	1.93604	1.93998	1.93735
Cr	2.28970	2.29361	2.29100

Diffraction

When a crystal is irradiated by an X-ray beam with a wavelength λ, a X-ray diffraction is observed (Fig. 2.184) at specific angles 2θ linked to the inter-reticular distances d_{hkl} (crystal lattice constant) according to the Bragg's law (Eq. 2.161):

$$2d_{hkl}\sin(\theta) = n\lambda \tag{2.161}$$

where n (a whole number) corresponds to the order of reflation; λ (in Å) is the wavelength of the X-ray radiation used (Cu, Co, Mn, Cr, …), d_{hkl} (in Å) is the inter reticular distance of the (hkl) diffracting plane and θ is the half angle formed between the incident beam and the diffracted rays measured by a goniometer.

Instrumentation

The XRPD technique consists in recording the intensity of the diffracted X-rays after interaction with the matter as a function of the Bragg angle 2θ, the so-called diffractogram or X-ray diffraction pattern. Different diffractometers, coupled with specific softwares to collect, extract, compil and analyze experimental datas, are marketed by PANalytical X'Pert Pro Multipurpose Powder Diffractometer [74], Rigaku SmartLab Multipurpose Diffractometer [338] and Bruker [72].

Experimental setups implement different goniometer configuration as Bragg-Brentano, Debye-Scherrer or Guinier which are well described by Pecharsky and Zavalij [339].

A powder X-ray diffractometer is composed of a X-ray tube (source), a goniometer on which is placed the sample, some slits, a detector. Powder diffractometers typically use the Bragg-Brentano geometry (Fig. 2.185). The incident angle corresponds to the angle formed between the X-ray and the focal plan of the goniometer. The diffracted angle is the angle formed between the diffracted beam and the detector. To that extent, two configurations are possible, $\theta - \theta$ and $\theta - 2\theta$. In the first case, the sample is fixed and the X-ray tube and detector rotate at

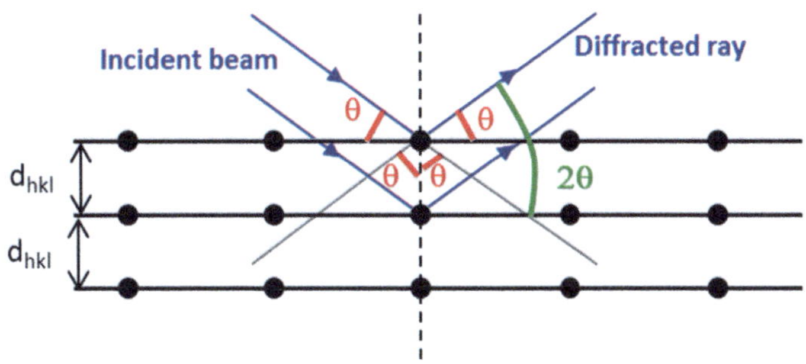

Fig. 2.184 Demonstration of the Bragg's law

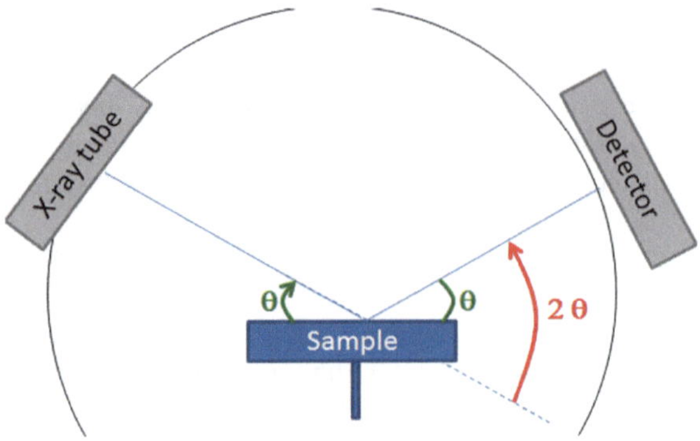

Fig. 2.185 Schematic of the Bragg-Brentano geometry

constant speed ($\theta°/\text{min}$). The sample can rotate around itself to improve statistical measurements with a Spinner. In the second case, the X-ray tube is fixed. The sample rotates with a speed of $\theta°/\text{min}$ and the detector moves twice as fast as this first, then with a speed of $2\theta°/\text{min}$.

The X-ray tube used can integrate different metal anode listed in the Table 2.90. A crystal monochromator or a filter can be placed upstream or downstream from the sample in order to obtain a monochromatic X-ray radiation. When the X-ray incident beam interacts with the inter reticular planes of the crystalline material, scattering occurs and diffracted rays appear, under the conditions that the interference between them is constructive; these diffracted rays are characteristics of the crystal structure of compounds contained in the sample. The detector records the intensity of diffracted rays at an angle of 2θ. Different detectors exist as scintillation counters, solid state detector, gas proportional or charged coupled device (CCD). Description of XRD principle and instrumentation including details about the path traveled by the X-ray beam in the different optic parts of the device are deeply explained and described in specialized books [339–342].

2.9.1.3 Sample Preparation

The sample preparation is the most important step for X-ray diffraction to guarantee reliable analyses. During the preparation, the sample can be contaminated by the presence of dust or other materials remaining from a previous experiment. In some specific experimental conditions, the sample can be analyzed up to a depth of 1 or 2 mm, but in most cases the penetration depth of X-ray in the matter can't exceed one hundred microns. If composition gradients in depth exist in the sample, the measurement will not be representative of the latter. An example of a kit of sample preparation provided by PANalytical is shown in the Fig. 2.186.

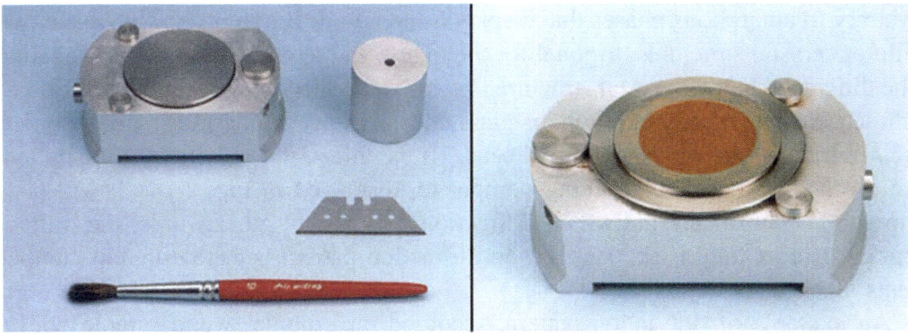

Fig. 2.186 Kit of sample preparation. Reprinted with permission from Malvern Panalytical [74]

2.9.1.4 XRD Powder Diffractograms

The intensity of diffracted X-rays is measured as a function of the Bragg angle (2θ). All crystalline phases have a distinct diffractogram; it is the "identity card" of a compound.

The position and intensity of the diffraction peaks in an X-ray diffraction pattern are determined by the crystal structure (atom types and positions) of a material while their width and shape are mostly a function of instrument and microstructural parameters. Therefore, the diffractogram is composed of diffraction peaks the 2θ positions of which is related to inter-reticular distances according to the Bragg's law, so to the lattice parameters of the compound. The intensity of the diffraction peaks are in particular related to the nature and position of atoms which diffract. So the intensity of the *k-th* diffraction line is expressed by the following equation:

$$l_k = SM_kL_pF_k^2 \tag{2.162}$$

where:

S is the scale factor, it is a numerical constant;
M_k is the multiplicity of the *k-th* diffraction line and corresponds to the number of diffraction lines being at the same inter-reticular distance and having the same structure factor;
L_p: Lorentz Polarisation factor is the product of 3 factors, the Lorentz factor, the polarisation factor and the geometric factor. These factors can be calculate but it necessary to know the instrumental XRD and its instrumental resolution;
F_k: Structure factor of the *k-th* diffraction line.

These different factors are very well described in the book of Waseda et al. [343].

In Fig. 2.187a are presented the diffractograms recorded on amorphous and cristallized SiO_2, using copper anticathode. Crystalline materials like quartz and cristobalite produce X-ray diffraction patterns with well-resolved diffraction peaks contrary to amorphous phases that display a spread-out Bragg peak. They have two different crystal structures, trigonal for the quartz and tetragonal for the cristobalite. The difference in their crystal structure is clearly reflected by the relative position of diffraction peaks. As the amorphous glass does not have a long-range atomic order, their diffraction patterns show broad peaks. If the three SiO_2 polymorphs are mixed, the diffractogram of the resulting composite, displayed in Fig. 2.187 b will be a superimposition of the individual diffraction patterns. In other words, the diffractogram of a mixture is the sum of the diffraction patterns of the different components taken separately.

An example of diffraction pattern, recorded on a poplar wood sample with a Phillips Panalytical X'pert Pro MPD diffractometer, is presented in Fig. 2.188. The X-ray radiation source used is the copper (Cu) whose the average K_α wavelength is

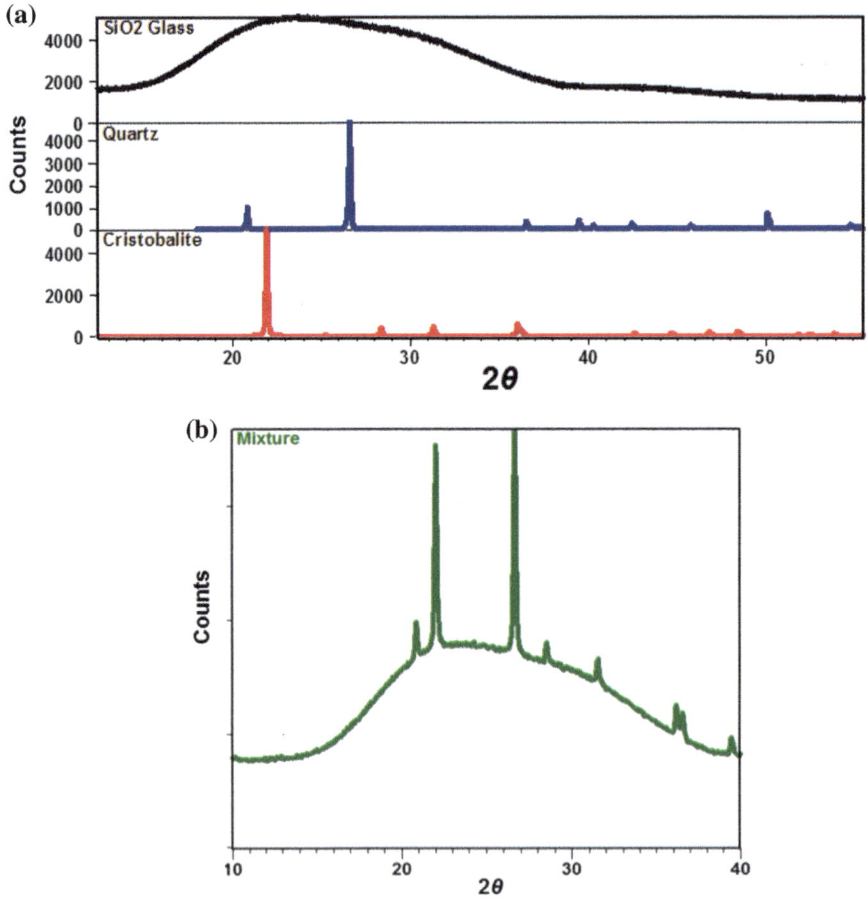

Fig. 2.187 a Diffractograms of three individual SiO_2 polymorphs; b diffractogram of a mixture of three SiO_2 polymorphs [344]

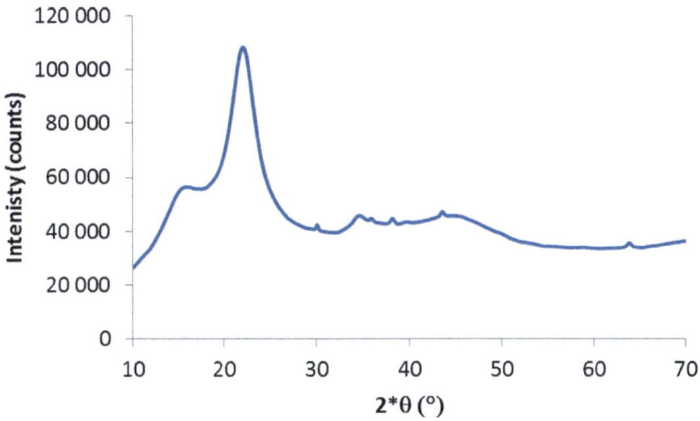

Fig. 2.188 Diffractogram of a raw poplar wood

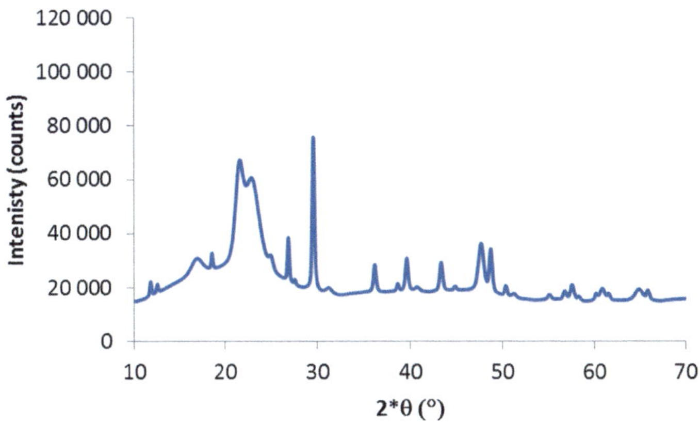

Fig. 2.189 Diffractogram of a refused-derived fuel

1.543 Å. The $\theta - \theta$ configuration in the Bragg-Brentano geometry is used. A nickel filter is used (up stream or down stream) to remove the Cu K_β radiation. The apparatus operates at a voltage of 45 kV and an intensity of 40 mA. The diffractogram was recorded in a 2θ range from 10 to 75° with a scan speed of 0.027°/s. The phase identification was carried out using the JCPDS and COD databases. The diffractogram displays a broad diffraction peak between 20 and 30° which can be assigned to an amorphous carbon ((002) peak) phase. Small diffraction peaks are also observable and corresponds to some crystalline phases. Another example of a X-ray powder diffraction pattern recorded on a refused derived fuel is presented in Fig. 2.189. This sample consists in a mixture of crystalline phases ($Mg_3Si_2O_5(OH)_4$, $CaCO_3$, $Al(OH)_3$, …) which may explain the presence of a large number of diffraction peaks. The sample also contains an amorphous phase which can be detected by the presence of a peak spread between 20 and 25°. The method used for the identification and quantification of phases as well as the determination of crystalline size is further described in Chap. 15.

2.9.2 *Pair Distribution Function (PDF) Analysis*

2.9.2.1 Introduction

As mentioned in the previous section, X-ray (and neutron) powder diffraction (XRPD) is a well-established scattering technique for identifying and analyzing the composition of materials [345]. Data are collected in reciprocal space in the form of a diffraction pattern, as previously outlined in Section "Instrumentation". Sharp Bragg peaks in a diffraction pattern arise due to the presence of one or more micron-sized crystalline phases, while the presence of a nanocrystalline phase manifests itself in the data via broadening of these Bragg peaks [346]. For an amorphous phase, diffuse peaks that span several degrees in 2θ are visible in the diffraction pattern instead of well-resolved Bragg reflections. Typically, these diffuse peaks are not recognized during data analysis processes using classical software, and are systematically incorporated into the "background" [347, 348]. Nevertheless, there may be important information about the atomic structure of disordered crystalline, glassy and amorphous phases (e.g., atom-atom distances, coordination numbers, extent of ordering at the nanoscale) hidden within the diffuse scattering [349], which could be extracted via a careful data collection and analysis.

2.9.2.2 Basics of an Elastic Scattering Experiment

A typical elastic scattering experiment involves measuring the constructive and destructive interference arising from the scattered waves from a sample. These waves (typically X-rays or neutrons) are scattered from the atoms in the sample, where an incoming collimated parallel beam of radiation (i.e., a planewave) is spherically scattered by an individual atom, and the resultant scattering from an array of atoms leads to constructive and destructive interference of the waves being measured at the detector location. The relationship between the atomic structure of a crystalline sample and the measured intensity at the detector can be explained using Bragg's Law (Eq. (2.163). The geometry of scattering experiment is given in Fig. 2.190 [345].

$$n\lambda = 2d \sin\theta \tag{2.163}$$

In Eq. (2.163), λ is the wavelength (in Å) of the incoming X-ray radiation, d is the distance (in Å) between planes of atoms in the crystal, n is the order of reflection, and θ is the angle formed between the direction of the incident X-ray beam and that of the radiation diffracted by the crystal.

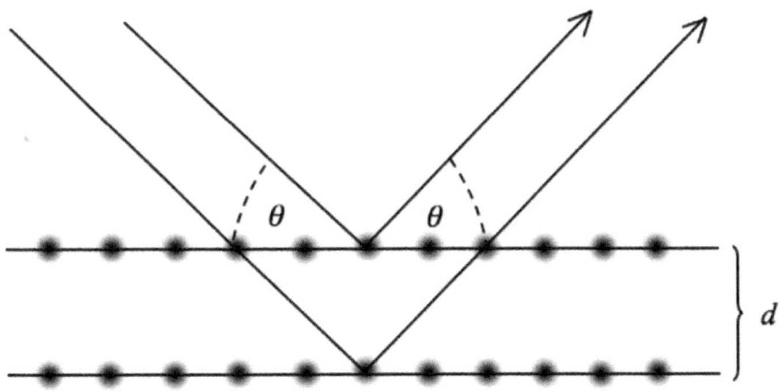

Fig. 2.190 Schematic representation of the Bragg's law. The incident parallel X-ray radiation diffracted by the crystallographic planes (horizontal lines) at an angle θ

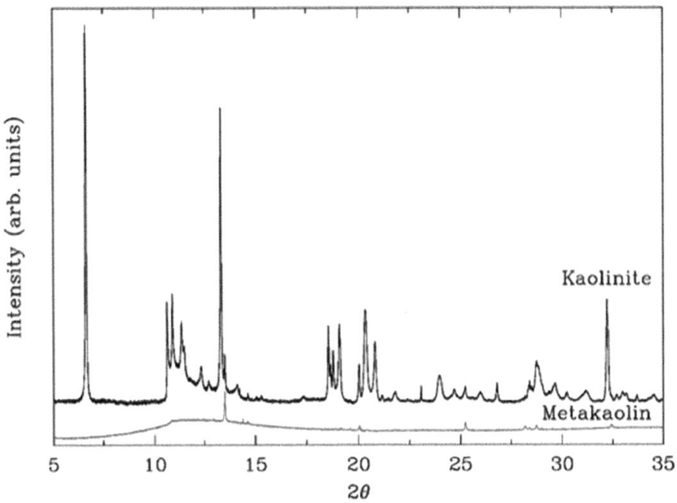

Fig. 2.191 High-resolution X-ray diffraction pattern of high purity kaolinite (crystalline phase) and metakaolin (amorphous phase) obtained on the Powder Diffraction beamline at the Australian Synchrotron using a wavelength of 0.82631 Å. The only Bragg reflections visible for metakaolin are attributed to the anatase/rutile impurities. The amorphous nature of metakaolin is visible via the broad diffuse peak located at ~10–17° in 2θ

Plotting the constructive/destructive interference (intensity) as a function of the Bragg angle 2θ for a powder crystalline material results in a diffraction pattern where the sharp intense Bragg peaks provide information about the crystal structure (s) present in the sample. For an amorphous material, such Bragg peaks will not be present, but constructive/destructive interference still occurs owing to scattering from the atoms in the sample, leading to the existence of broad diffuse peaks in the diffraction pattern. As an example, Fig. 2.191 displays diffraction patterns for crystalline (black curve) and amorphous (light grey curve) samples.

Note: the sharp-rise-slow-fall (that is a technical term used for stacking faults in clay minerals and the accompanying behavior in the diffraction pattern) observed for the kaolinite (crystalline phase) in Fig. 2.191 *between 11 and 13° in 2θ is due to the stacking faults inherent to this clay mineral.*

In real space, the atomic structure of a material can be represented by an atomic distribution function, which is closely related to its autocorrelation function (ACF). As an example, Fig. 2.192 displays the pair distribution function (PDF) for graphite and the corresponding atom-atom distances in the graphene layer. In a PDF, the intensities of the atom-atom distances are weighted according to the scattering strength of the respective atoms, which is not the case for the ACF. The PDF $G(r)$ of an atomic structure is defined in Eq. (2.164) and gives the probability of finding two atoms i and j separated by a distance r_{ij} in real space, weighted by the scattering lengths b_i and b_j (described below) and averaged over all pairs of atoms in the sample (b is the compositional average of the scattering length for the

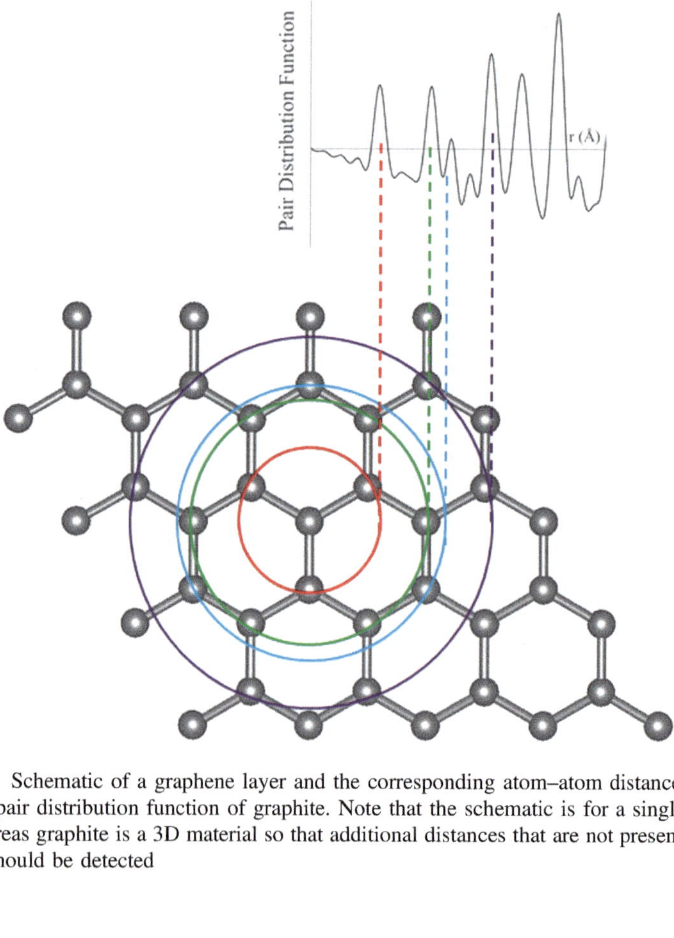

Fig. 2.192 Schematic of a graphene layer and the corresponding atom–atom distances found in the X-ray pair distribution function of graphite. Note that the schematic is for a single graphene layer, whereas graphite is a 3D material so that additional distances that are not present in the 2D structure should be detected

sample) [350]. The atomic number density of the sample, ρ_o, is also included in the PDF.

$$G(r) = \frac{1}{r}\sum_{i,j}\frac{b_i b_j}{b^2}\delta(r - r_{ij}) - 4\pi r\rho_o \tag{2.164}$$

The relationship between the PDF and the radial distribution function (RDF, denoted by $g(r)$) for an atomic structure is given in Eq. (2.165) for a monatomic sample (sample consisting of one element type).

$$g(r) = \frac{G(r)}{4\pi r\rho_o} + 1 \tag{2.165}$$

Finally, it is important to note that although Fig. 2.192 depicts a central atom from which the atom-atom distances are calculated, the PDF (and RDF/ACF) is a histogram of all atom-atom distances in the material.

Since X-rays are scattered from the electron cloud in an atom, the scattering strength (sometimes referred to as length) of an element for X-ray radiation is proportional to the atomic number (i.e., the number of electrons in an atom). On the other hand, neutrons are scattered from the nucleus of an atom, and the neutron scattering strengths of elements are tabulated and can be extracted from the literature (known as the coherent neutron scattering lengths that have non-monotonic behavior with atomic number). Neutrons are advantageous to use for materials containing both light and heavy elements since in such materials X-rays tend to be insensitive to the atom-atom correlations involving the light elements. Furthermore, for materials containing elements that are neighbors in the periodic table, and therefore very similar in electron density, neutrons may provide enhanced contrast between these elements compared to X-rays. However, care must be taken to limit the amount of incoherent neutron scattering from a sample when performing neutron PDF analysis, since there is no structural information associated with this form of scattering. The incoherent (and coherent) scattering lengths of all elements are found in tabulated data in the literature, e.g. [351].

2.9.2.3 Data Collection

Although the diffuse (and Bragg) scattering for an amorphous/disordered material can be analyzed in reciprocal space, it is more straightforward to analyze the data in real space, which is obtained via a sine Fourier transform [350] of the normalized scattering data. The process used to obtain the final dataset in real space is provided in Fig. 2.193, and is described in detail below. For a given scattering experiment, the scattered intensity $I(Q)$ is measured as a function of momentum transfer Q, where Q is related to the scattering angle θ as shown in Eq. (2.166). λ is the

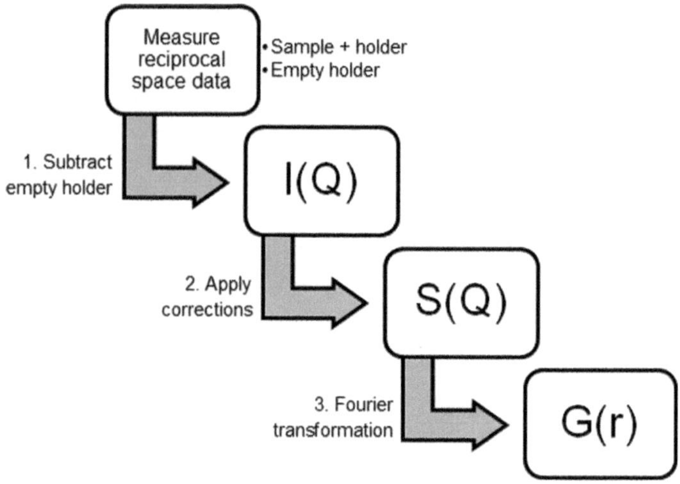

Fig. 2.193 Scheme of the PDF data collection and subsequent data processing for X-ray scattering measurements. Note that once data of the reciprocal space are measured, all steps (1–3) are routinely carried out by a single software program, such as PDFgetX3

wavelength of the incident radiation source (typically X-rays, but can use neutrons and even electrons can also be used).

$$Q\left(\text{Å}^{-1}\right) = \frac{4\pi \sin\theta}{\lambda} \tag{2.166}$$

Typically, PDF analysis is carried out on powder samples placed in a cylinder shaped specimen holder, which requires a separate scan of the latter to be performed using the same scan settings as those previously employed for the sample [350]. Subtraction of the signal characteristic of the empty specimen holder is performed using the standard PDF data reduction software [347, 352, 353], such as PDFgetX3 for X-ray measurements and equivalent software for neutrons. As an example, the raw X-ray diffraction data are given in Fig. 2.194 for an oxygenated biochar, after subtraction of scattering from the specimen holder (a cylindrical polyimide capillary was used for this experiment); it is clear that the data primarily consist of diffuse scattering.

For the data given in Fig. 2.194 the sample was measured for 2 min in transmission geometry on 11-ID-B beamline at the Advanced Photon Source, Argonne National Laboratory [354]. A high-energy X-ray beam ($\sim$60 keV for these data) and short sample-to-detector distance was required to collect data over a sufficient Q-range. A large Q-range, up to $\sim$20–30 Å^{-1}, was necessary for producing a high-quality real-space PDF, since truncation effects will become prevalent for datasets with limited Q-range [355]. Hence, PDF analysis is typically performed at high-energy synchrotron sources (for X-rays) and spallation sources (for neutrons), with neutron data typically providing enhanced contrast for the lighter elements

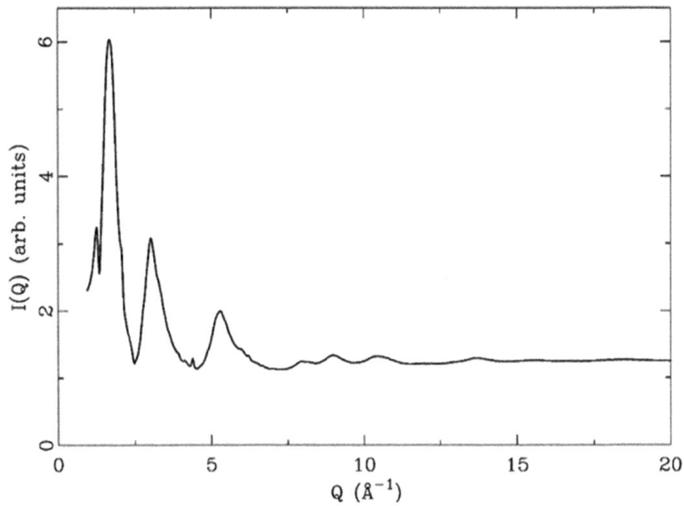

Fig. 2.194 Synchrotron-derived X-ray diffraction pattern of an oxygenated biochar

owing to the non-monotonic increases in the neutron scattering length with increasing atomic number. Laboratory-based powder XRD instruments can be used to collect limited-resolution X-ray PDFs [356], however, the Q-range of these instruments is controlled by the X-ray tube, and therefore Mo- or Ag-tubes are required since a Cu-tube can only achieve a Q_{max} value of ~ 8 Å^{-1}. Furthermore, the limited photon flux associated with laboratory XRD instruments combined with the use of point or line detectors leads to long scan times (e.g. 24 h or longer for measurement of a single sample).

- In summary, the following acquisition conditions are required:
- High energy X-rays (from a synchrotron source) or spallation neutrons;
- Wide Q-range coverage;
- Good counting statistics (achieved by a high flux radiation source and/or long measurement times);
- Additional scan of empty sample holder.

2.9.2.4 Producing the Pair Distribution Function

A standard PDF data reduction software is used to produce the normalized total scattering function $S(Q)$ and the subsequent PDF dataset ($G(r)$ function). Chemical composition of the sample is often an input parameter required by the software. Conversion from $I(Q)$ to $S(Q)$ for X-ray data involves the application of corrections including sample adsorption, polarization, Compton intensities and detector efficiency. $S(Q)$ and $G(r)$ are mathematically related as described in Eqs. (2.167) and (2.173) [350].

$$G(r)\left(\text{Å}^{-2}\right) = \frac{2}{\pi} \int_{Q_{min}}^{Q_{max}} Q[S(Q) - 1] \sin(Qr) dQ \tag{2.167}$$

Figure 2.195 shows the normalized total scattering function $S(Q)$ for an oxygenated biochar. Properly normalized scattering data oscillate around a value of 1 at high Q values. Any deviation from this normalized behavior will lead to noise/artifacts in the subsequent PDF not associated with the atomic structure of the sample. The importance of counting statistics can be observed in Fig. 2.195, where the high Q range of the data (between ~ 15 and 18 Å^{-1}) contains a certain degree of noise. Limited counting statistics will lead to exaggerated termination ripples in the PDF [357], which can swamp the sample-dependent information for the case of extremely short scan times. The choice of Q_{max} will impact the r-space resolution of individual atom-atom correlations in the PDF, with low Q_{max} values leading to significant broadening of the atom-atom correlations [355]. Hence, in general a higher Q_{max} value is beneficial, especially if subtle peak shape information is expected (such as a peak split). However, if the high-Q component of the total scattering function is particularly noisy, then a lower Q_{max} value may be advantageous to reduce the extent of noise present in the resulting PDF. In Fig. 2.196, a Q_{max} value of 18 Å^{-1} was selected and the Q_{min} was set at 0.95 Å^{-1}.

From Fig. 2.196, it is clear that the atomic structure of the oxygenated biochar is amorphous since the atom–atom correlations disappear by ~ 10 Å. A zoom of the

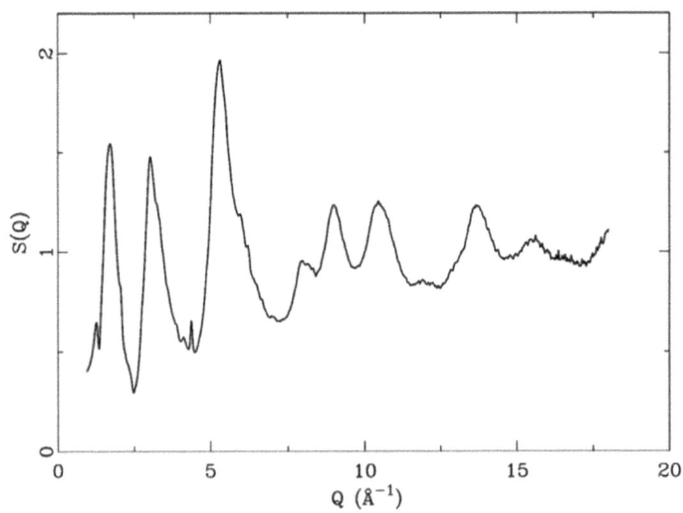

Fig. 2.195 Synchrotron-derived X-ray total scattering function of an oxygenated biochar

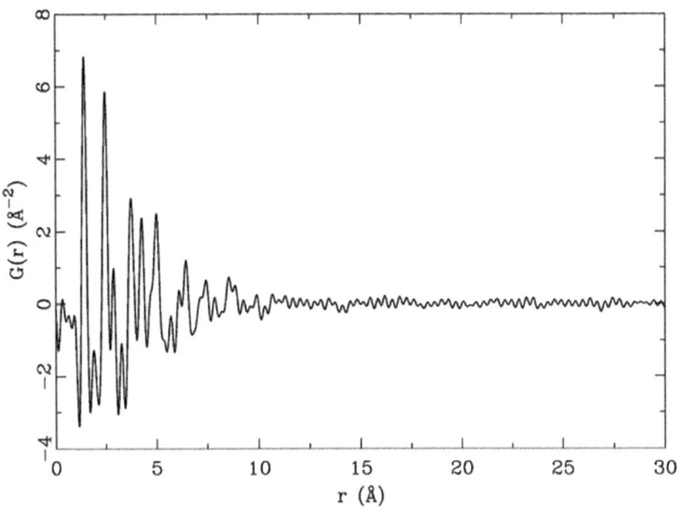

Fig. 2.196 Synchrotron-derived X-ray pair distribution function of oxygenated biochar, over an r range of $0 \leq r \leq 30$ Å

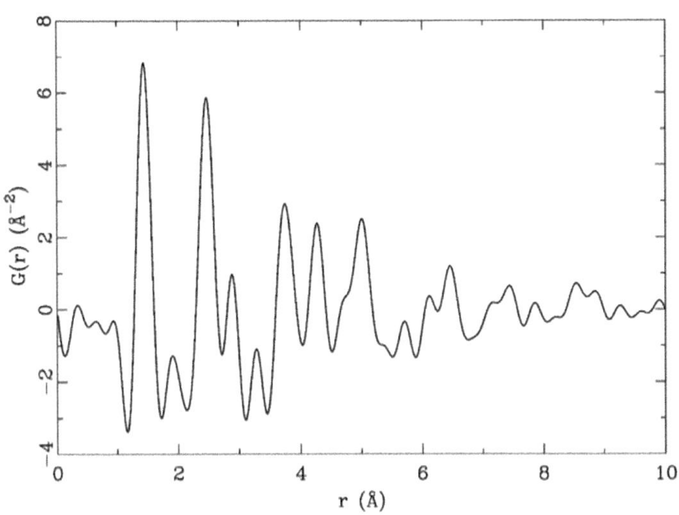

Fig. 2.197 Synchrotron-derived X-ray pair distribution function of an oxygenated biochar, over a r range of $0 \leq r \leq 10$ Å

local structure is shown in Fig. 2.197, where the nearest-neighbor atom-atom correlation is located at 1.44 Å, which corresponds to a summation of the C–C and C–O bonds in the biochar. The peak at ~2.48 Å is assigned to a combination of C–C and C–O correlations. Possible approaches for analysis of the PDF data are outlined in Chap. 15. If data analysis involves fitting of structural models to the PDF data, then the instrument parameters needs to be determined, specifically Q_{broad} and Q_{damp} (analogous to the instrument parameters needed for Rietveld refinement), and therefore a calibration standard must also be measured on the instrument, such as silicon or nickel.

2.9.3 Scanning Electron Microscopy (SEM)

2.9.3.1 Introduction

Scanning Electron Microscopy (SEM) allows the characterization of solid materials, whatever their nature (metal, polymer, ceramic, composite, etc.) and state (pulverulent or not). This technique can be used for number of industrial, commercial, and research purposes. Indeed, scanning electron microscopes are equipped with several tools for characterizing the surface and microstructural details (topography, porosity, grain size and orientation, impurities, agglomeration or aggregation state, etc.) as well as the chemical composition homogeneity of materials. Electron microscopes operate on the same basic principles as light optical microscopes (LOMs), except that they use a focused beam of electrons as a source instead of light. This leads to three major differences:

- Unlike optical microscopes the image is not "real" but formed using electromagnetic lenses, and displayed electronically;
- To allow electrons moving in straight lines (like the light), electron microscopes must operate under high vacuum;
- To focus the beam, the imaging system use electromagnetic lenses instead of glass lenses.

Because scanning electron microscopes operate with electron and lenses to direct and focus them, this type of instrument is significantly more complex than optical light microscopes. To maintain the high vacuum level in the electron microscope, the latter requires continuously-pumped high/ultra-high vacuum systems. An excessive heat is generated by high currents flowing through the electromagnetic lenses, which, accordingly, need to be continuously cooled down usually by circulating water. Furthermore this type of microscope is a large and cumbersome equipment which requires an area large enough to accommodate it, and very sensitive to vibration and external magnetic fields, hence it must stand alone in a specially designed room. For all these reasons, the acquisition and maintenance costs of microscopes are very high. Despite these disadvantages, SEM is an essential characterization tool in the field of material science.

2.9.3.2 SEM Characteristics: Resolution and Depth of Field

Resolution

The most relevant advantage of a SEM compared to a light optical microscope is its higher resolution that allows higher magnification. The microscope resolution depends on the wavelength of the electron source being used to image the specimen: long wavelengths offer poorer resolution compared to short wavelengths. The wavelength of the light (average around 500 nm for white light) limits the resolution of traditional optical microscopes to around 0.2 μm. In an electron microscope, the electron beam wavelength used is approximately one hundred thousand times smaller than that of the visible light, so that SEM can achieve resolution better than 1 nm.

Depth of Field

Another advantage is the depth-of-field (also called depth of focus), that is the extent to which features of different heights on the sample surface remain in focus, and therefore do not appear unclear to the observer. The depth of field depends mostly on the angle of the cone of beam used to image the sample. In a SEM, the aperture angle of the electron beam is much smaller than the angular aperture of the light cone captured by the objective in a LOM. This yields to a depth of field much greater. With a greater depth of field compared to a LOM (Fig. 2.198), the SEM can better image three dimensional samples.

Due to its ability to examine three dimensional details with a high resolution, from micrometer to nanometer scale, the SEM has become an indispensable tool dedicated to research in material science.

Apparatus

Figure 2.199 shows the main components of SEM equipment, which include:

- The column which contains the electron source ("Gun"), a serie of magnetic lense and apertures, and scan coils.
- The chamber which includes:

- A door (sometimes it is an air-lock) inserting and removing the sample;
- The stage that supports the specimen and can perform horizontal (X, Y) and vertical (Z) movement, the specimen tilting and rotation;
- Various signal detectors.

- The vacuum system.
- The display/data output devices.

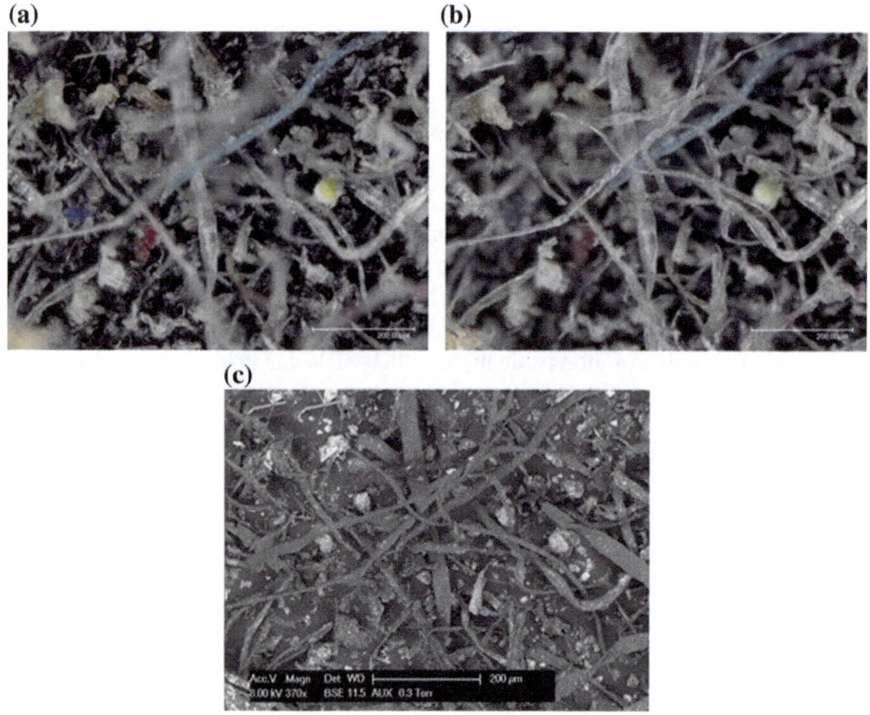

Fig. 2.198 Analysis of a solid recovered fuel using LOM and SEM. Optical images (Keyence microscope): a only particles in the background are in focus; b only particles in the foreground are in focus; c SEM image (FEI microscope) of the same area: all particles are in focus

Electron Guns

There are many types of electron guns but they all use the same principle of operation: electrons are extracted from a conducting cathode material (also named filament) and then accelerated towards an anode. Electron guns available use different cathode materials and different physical method of extraction:

- tungsten (a bent tungsten wire) and LaB_6 (sharpened rod of single crystal LaB6) guns are thermoionic gun: the extraction is achieved by heating the cathode
- in Field Emission Gun (FEG), the extraction is achieved by applying a very strong electric field, usually to a tungsten sharp point (there are different types of field emission guns).

Each of them has different costs and advantages. Thermoionic guns are robust, relatively cheap and do not require ultra-high vacuum: $\approx 10^{-5}$ torr (1.33×10^{-4} mbars) for W and 10^{-7} torrs (1.33×10^{-6} mbars) for LaB_6. Field emission guns require a very expensive vacuum system to maintain an ultra-high vacuum, i.e.

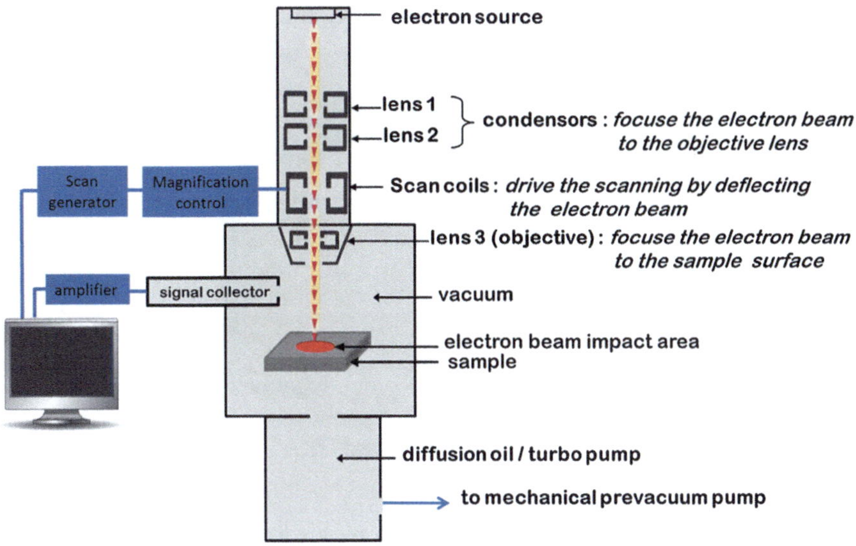

Fig. 2.199 Schematic construction of a SEM

10^{-10} torrs (1.33×10^{-9} mbars) to 10^{-11} torrs (1.33×10^{-10} mbars), but have better performances compared to that of based on the thermoionic emission.

Gun performances are evaluated against two major factors: the brightness and source size.

- Brightness (A cm^{-2} Sr^{-1}) is the current density per unit solid angle of the source. It is a very important parameter because the higher the value of brightness, the more electrons there are into the beam of a given size, and so the more information can be generated from the specimen. To obtain high quality images the source brightness should be as high as possible. Depending on the type of FEG, these electron sources have brightness 100–1000 times higher than that of tungsten filaments, and 10–100 times higher than the Lanthanum Hexaboride filament.
- Source size: since it is impossible to observe sample characteristics that are smaller than the source size, this parameter is important to obtain high resolution images. Due to its shape (sharp point), the "source size" for FEGs is of the order of nanometers while it is microns for the other emission sources. However for chemical analyses (Chap. 15) a large source size is required.

Electromagnetic Lenses

They are the magnetic equivalent to the glass lenses in an optical microscope. They consist in a coil of copper wires inside an iron pole piece. A current running through

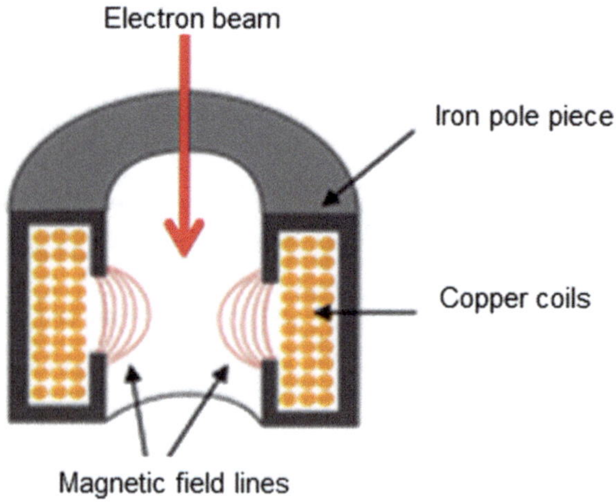

Fig. 2.200 Illustration of an electromagnetic lens

the coils creates a magnetic field in the pole pieces the role of which is to concentrate lines of magnetic force. Under the effect of the magnetic field, beam electrons are focused into one point. The focal length of the lens can be controlled by altering the current flowing through the coils (Fig. 2.200).

Apertures

Depending on the microscope, a variable number of apertures may be found in the electron column. Apertures are micron-scale holes in metal disks, centered on the optical axis. As the beam passes through the lenses in the column, apertures limit the angular width of the electron beam to reduce aberrations (Section "Electron Beam Parameters") and to improve depth-of-field in the image.

Vacuum System

Vacuum levels can be expressed by different units, the most commonly used being Torr, Pascal, and mbar (1 Torr = 133 Pa = 1.333 mbar). Various levels of vacuum being necessary for SEM analysis, different vacuum pumps are used to obtain and maintain these levels during operation. A mechanical pump (often called roughing pump) is used to bring the pressure inside the sample chamber, initially at the atmospheric pressure, down to a moderate vacuum (above 10^{-2} mbar). Once the chamber pressure is correct, a valve that isolates the column from the sample chamber is opened and high vacuum (smaller than 10^{-4} mbar) is achieved using an oil diffusion or turbomolecular pump. When the chamber is opened (beam off), the valve is closed in order to hold the column under high or ultra-high vacuum

($<10^{-9}$ mbar) depending on the source type. In modern microscopes, the vacuum system is completely automated and the vacuum levels are continuously monitored.

2.9.3.3 Imaging Principle

The electron beam emitted by the electron gun is accelerated towards the specimen using a positive electrical potential between the filament and the anode, and then focused into a thin probe (also called spot) by metal apertures and magnetic lenses. Deflector coils scan the thin probe over the sample surface, line by line, in a rectangular raster pattern (from left to right and top to bottom), and the signal generated from each point of the surface (due to interactions between the sample and the beam electrons) is collected simultaneously by a detector. This signal is then transformed in light pulses, amplified and converted into an electron current which modulates the brightness of the image displayed. Since the scan of the deflector coils is synchronized with the scan of the display device, each point on the image correspond to a point on the surface of the sample. Its brightness is proportional to the intensity of the detected signal (Fig. 2.201).

Many signals are produced (secondary electrons, back-scattered electrons, characteristic X-rays, light-cathodo-luminescence-, specimen current and transmitted electrons) but the most commonly used for imaging are secondary electrons and backscattered electrons (Sect. 2.9.3.6).

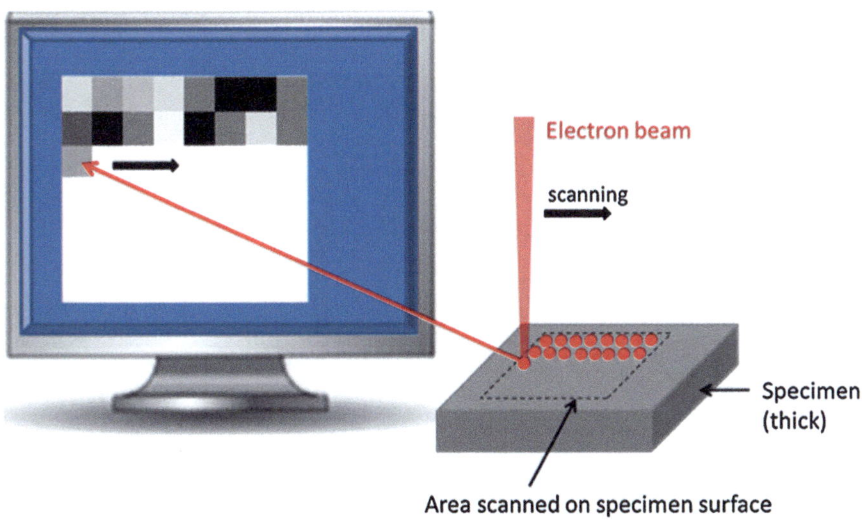

Fig. 2.201 Imaging principle

2.9.3.4 Field of View/Magnification, Resolution and Lens Aberrations in a SEM

As both light optical microscopes and scanning electron microscopes utilize the same principles, they have common notions: field of view, magnification, lens aberrations and resolution.

Field of View and Magnification

The magnification can be defined as the ratio of the surface area scanned on the monitor divided by the area scanned on the specimen (Fig. 2.202). The selection of the field of view is achieved using the XY movements of the stage. When the SEM operator adjusts the magnification via the control labelled as "magnification", the surface area to be scanned is determined. Unlike in light optical microscopes, the magnification in a SEM is not dependent on the excitation of the objective lens but is determined by the size of the beam scan, relative to the size of the display device (screen). The screen size being fixed, higher magnification is obtained while reducing the size of the scanning area on the specimen, and vice versa. When the magnification is modified, the scanning coils are excited.

Resolution

The resolution is defined as the shortest distance between two points that can be distinguished as separated entities. Resolving small specimen details requires probe diameters of similar dimensions. Consequently high resolution imaging requires the smallest possible probe diameter. In a SEM, the ability to achieve a small probe

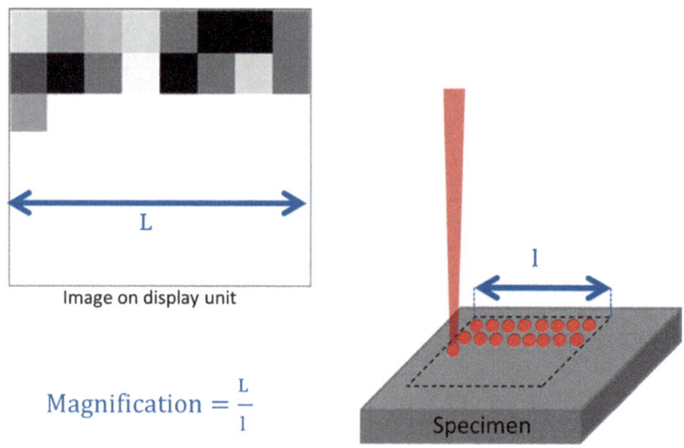

Fig. 2.202 Illustration of magnification in a SEM

diameter is directly related to the "source size". As seen previously, the "source size" for FEGs is of the order of nanometers while it is micron size for the other emission sources. So FEG SEMs are very suitable tools for high resolution imaging.

As later presented, the resolution in a SEM is controlled by beam parameters (Sect. 2.9.3.7), *and ultimately by the size of the interaction volume between the beam and the sample* (Sect. 2.9.3.5).

Lens Aberrations

As glass lenses in a light optical microscope, magnetic lenses suffer from defects and are not perfectly symmetric. Because of these aberrations, all electron paths originating from a given point of the object do not converge perfectly on the same point in the image. The defects of most importance are spherical and chromatic aberrations, and astigmatism.

- Spherical aberration arises because electrons moving in trajectories further from the optical axis are focused more strongly than those passing near or along this axis. This results in a disk rather than a point where all electrons converge. To reduce this aberration, apertures are used to eliminate the more widely diverging electrons passing through the lenses;
- Chromatic aberrations occur because the electron beam emitted by the gun is not perfectly monochromatic (there is an energy spread) and electrons with different energy coming from the same point are focused to different points. Like spherical aberrations, chromatic aberrations are strongest at the edges of the lens, so apertures can minimize their effect. But the simplest way to correct them consists in using a monochromatic source, which is accomplished by having a very stable source (with low energy spread). The cathodes used are classified by descending order of the energy spread they can generate: W > LaB_6 > FEG;
- Astigmatism exists because of manufacturing imperfections within the electromagnetic lenses. It causes an elliptical focal spot which results in a distorted image. It is corrected using a stigmator composed of a ring of electromagnets positioned around the beam, which "push" and "pull" the latter to make it more perfectly circular (Section "Astigmatism Correction" for correction).

2.9.3.5 Interaction Between Electron Beam and Sample

When the beam electrons hit the surface of the sample, they penetrate into the sample to a depth of approximately 1 μm, and interact one or more times anywhere along their paths, some of them undergoing a change in direction, as well as transferring energy to the specimen. The resulting volume over which the incident electrons interact with the sample is called "interaction volume", as illustrated in Fig. 2.203.

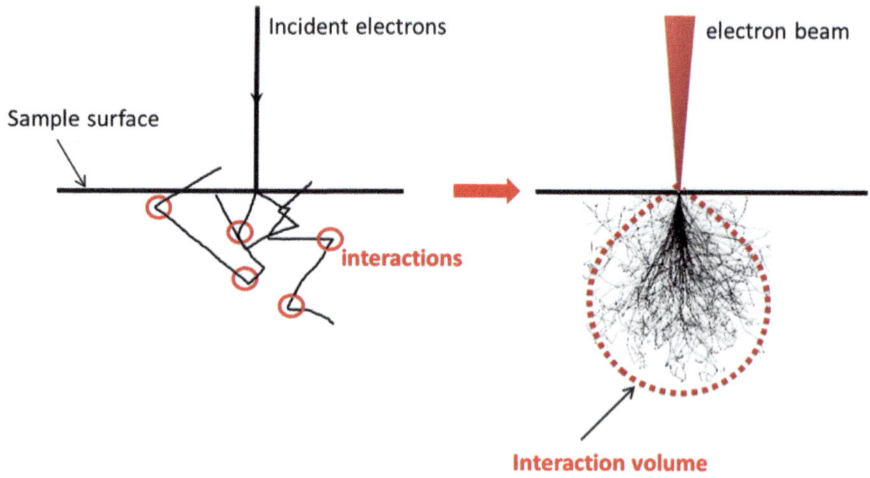

Fig. 2.203 Illustration of the interaction volume

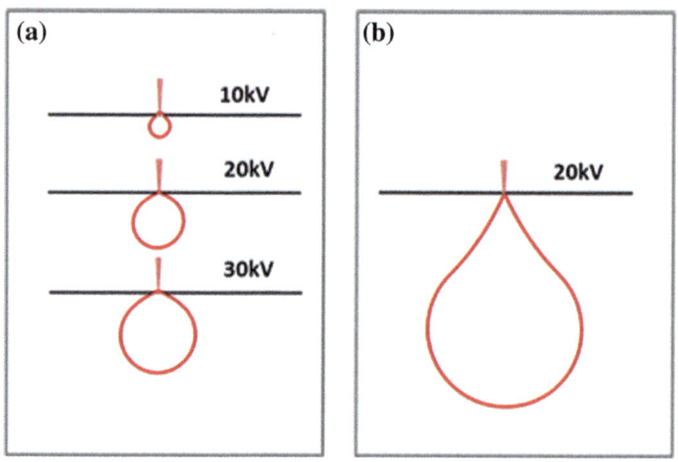

Fig. 2.204 Penetration of the electron beam depends on its incident energy (quoted in kV) as well as on sample nature. a high-atomic number; b low-atomic numbers

The size and shape of this interaction volume depend on the sample composition and electron beam energy: dense samples (or samples with high average atomic number) reduce beam penetration, whereas higher energy electrons penetrate more deeply inside the sample. As shown in Fig. 2.204, this interaction volume has pear shape for low-atomic-number specimens, rather a hemi-sphere shape for those with high-atomic numbers.

It's very important to note that in both cases, every section of the interaction volume taken at random is significantly larger than the probe diameter at the

sample surface. Consequently, the image resolution in the SEM is not determined by the size of the electron probe by itself, but rather by the size of the pear interaction volume.

2.9.3.6 Detected Signals and Corresponding Image Contrasts

If a large fraction of the incident electrons remains in the specimen, as they lose a part of their energy and are absorbed, some of them are emitted back or produce signals that can escape and be detected. Depending on which interaction with the sample is involved, these signals have different energy and can be detected separately. The most commonly used signals are secondary and backscattered electrons for imaging, and characteristic X-rays for chemical analyses. Since higher energy signals can escape from deeper within the sample than those of lower energy, the pear-shaped interaction volume can be divided into specific regions associated to each signal (Fig. 2.205).

Secondary Electrons (SE): Topographic Contrast

Generally, all SEMs are equipped with secondary electron detectors. Secondary electrons are emitted by atoms excited by the electron beam: they are ejected from the K-orbitals of the sample atoms. Due to their low energy (a few eV), they are emitted from very close to the specimen surface (1–10 nm). Consequently:

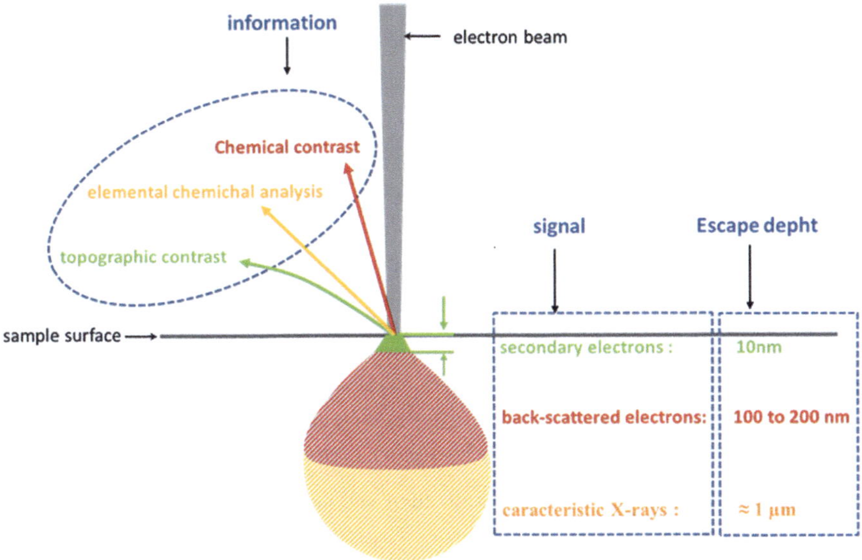

Fig. 2.205 Signals generated in the pear-shaped interaction volume

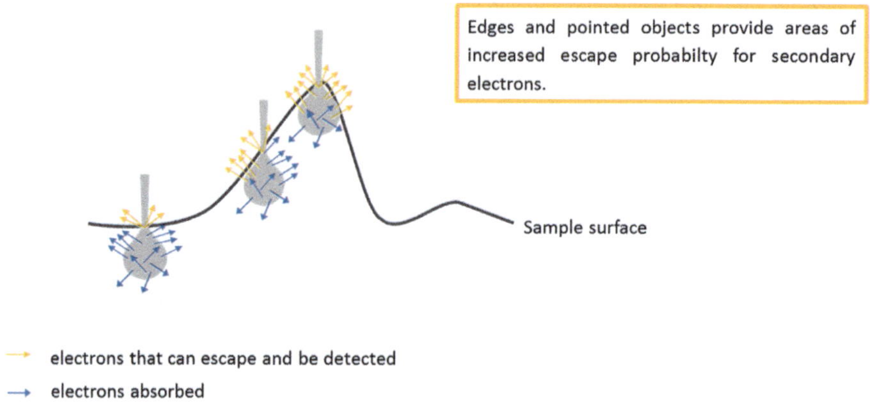

Fig. 2.206 Illustration of the detection probability for secondary electrons

The interaction volume being of the order of magnitude of the probe size, they produce very high-resolution images;

The number of electrons that can be detected depends, among other things, on the specimen topography (Fig. 2.206).

Areas at the top appear bright while valleys are dark: interpretation of SE images is very intuitive because they just look like the corresponding visual image would look. The SE detector is placed on the side of the chamber, at an angle for which more detailed surface information are provided (Fig. 2.208). In order to increase the efficiency of electron collection, a positive potential of a few hundred volts is applied to a grid which covers the detector. It's often possible to change this voltage so as to control the number of electrons to be collected. The image mode using SE detector is suitable for characterizing the size, shape and texture of three dimensional objects.

Back-Scattered Electrons (BSE): Chemical Contrast

Back-scattered electrons are primary electrons (originating in the electron incident beam) that are deflected back out of the specimen by the atoms of the sample, without a significant change in their energy. As BSE are more energetics than secondary electrons (the energy approximately the same as that of electrons of the primary beam), they are emitted from greater depth (100 to 200 nm from the specimen surface) and have the particularity of being scattered more by heavier elements (high atomic number) than by lighter elements. Consequently:

- They provide images more in depth than secondary electrons do, hence poor resolution images (Fig. 2.207a).
- Elements with a high atomic number appear brighter than those having a lower atomic number: BSE images display different contrasts between zones of the

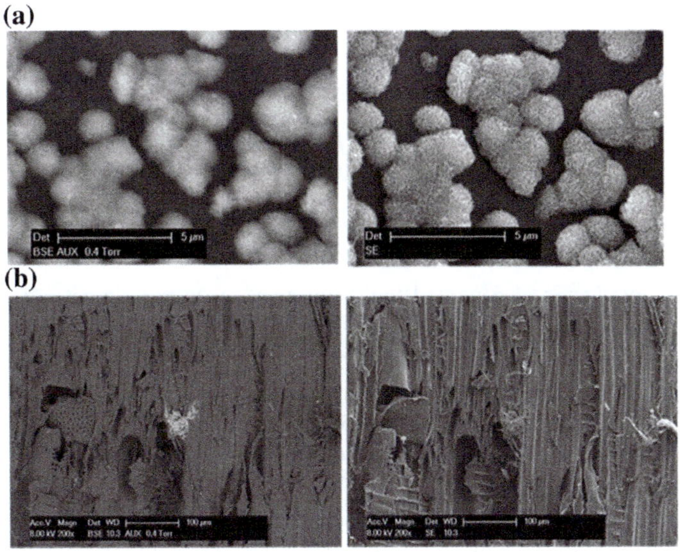

Fig. 2.207 Mineral particles on wood surface; a The texture of particles is visible using the SE detector (right) while it can't be distinguished with the BSE detector (left); b the BSE detector (left) allows highlighting the different phases (contrasting images), the latter not being distinguished at low magnification with the SE detector (right)

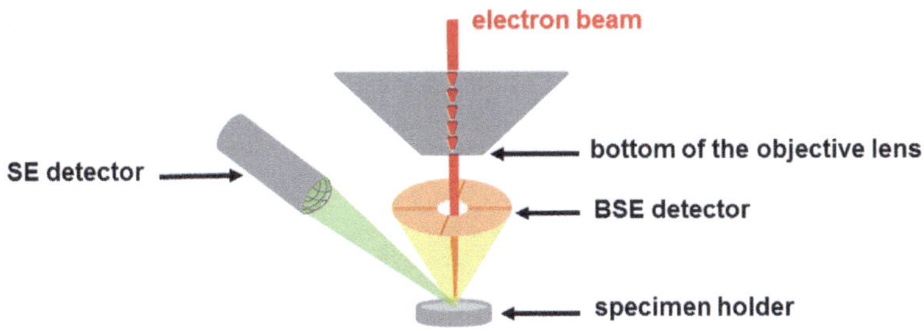

Fig. 2.208 SE and BSE detectors

sample having different average atomic numbers (Fig. 2.207b). BSE detector is suitable for highlighting the different phases contained in the sample.

As a large amount of BSE travels back along the incident beam, a dedicated backscattered electron detector is positioned above the sample in a "doughnut" arrangement, concentrically to the electron beam (Fig. 2.208).

2.9.3.7 Controllable Parameters

Electron Beam Parameters

The electron beam is defined by: (i) the probe diameter (1); (ii) the probe current (2); (iii) the accelerating voltage (3) as illustrated in Fig. 2.209 [358].

Probe Diameter

The probe diameter is also named the "spot size". It is the diameter of the beam measured at the surface of the specimen. It can be variable, in the range ≈ nm to ≈1 μm depending on the SEM equipment, by altering the current to one or two condenser lenses. This is usually a single control in microscopes, labeled "spot size" or "condenser". As previously shown, high resolution imaging requires as small spot size as possible, but there is a lowest limit on spot size. In fact, the probe diameter must provide sufficient current to obtain an acceptable image, i.e. with good a contrast and a high signal to noise ratio (Section "Frame Time/Scanning Speed (Signal-to-Noise Ratio)"). Decreasing the probe diameter decreases the associated current, and there is a limit below which the lack of current deteriorates the image quality. The best compromise between the spot size and probe current must be found during the analysis.

Notes:

- The lack of current can be compensated by increasing brightness and contrast levels (Section "Contrast/Brightness");
- Decreasing the spot size will also decrease lens aberrations.

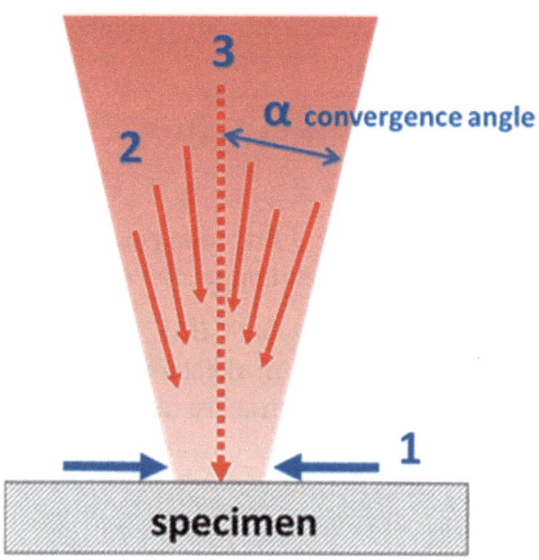

Fig. 2.209 Illustration of the electron beam parameters

Probe Current

It is the current in the final probe. In the range of pA to μA, it is controlled by the size of the final lens aperture, sometimes called "objective aperture". Depending on the microscope, the electron column may be equipped with one or more apertures. However, all SEMs have final lens apertures (located below the scanning coils) of which the size range can be selected. In the first place, this final aperture is used to eliminate the more widely diverging electrons that reach the objective lens, and thereby minimize aberrations. Decreasing the diameter of the aperture will decrease lens aberrations and thus increase resolution. It will also decrease the convergence angle of the beam. Hence the depth of field will increase (Section "Depth of Field" above). However, as a part of the electron beam is eliminated, the probe current will decrease. Finally, owing to its position at the bottom of the column, this is the final lens aperture that controls the current being carried by the final probe.

The aperture size that optimizes the tradeoff between the beam current and the spot size must be found during the analysis.

The Accelerating Voltage

Quoted in kilovolts (kV), this is the voltage used to accelerate the electrons down the column. Depending on the microscope, it often ranges from 1 to 30–40 kV. It determines the amount of energy carried by the electron beam. Considering it alone, higher accelerating voltage gives smaller probe diameter by reducing lens aberrations, and thereby better resolution. It also increases the probe current at the specimen that can be useful for sample with poor contrast. But high energy tends to reduce image resolution by enlarging the interaction volume (Sect. 2.9.3.5). The deeper the probe electrons penetrate inside the specimen, the less surface information is contained in the image. Since back-scattered electrons have higher maximum escape depth than secondary electrons, this is particularly the case for BSE imaging, and all the more for samples with low average atomic number. For this reason, for biomass which has most often low average atomic number, it is more suitable to use low voltage, below 10 kV if possible.

Notes:

- FEG sources having brightness 10–1000 times higher than the other sources, they are extremely useful for low voltage applications. It's possible to work with a voltage below 10 kV and a small probe diameter while having sufficient current to obtain a good image quality;
- Increasing voltage will potentially increase charge-up effect for non-conductive specimens and damage for specimens sensitive to electron beam (Sect. 2.9.3.8).

Finally, the most suitable accelerating voltage depends on the type of material being examined, operating conditions and signal type. The beam parameters being interrelated, a compromise must be found in order to satisfy the best conditions.

Other Parameters

Working Distance

This is the distance between the bottom face of the objective lens and the surface of the specimen. The working distance is adjustable for all SEMs using the Z movement of the sample stage. Increasing the working distance will (i) increase the depth of field; (ii) increase probe size and thus decrease resolution; and (iii) possibly decrease signal strength. The working distance that will allow a good depth of field while maintaining good resolution must be selected. When adjusting the working distance, it is imperative to ensure that the sample will not hit the pole piece or any detector. Each microscope has a safe minimum working distance (specified in the user manual) that will produce better resolution images.

Frame Time/Scanning Speed (Signal-to-Noise Ratio)

The frame time is the time needed to scan a single line multiplied by the number of lines within a frame. It is controlled by setting the scanning speed. Changing the scanning speed will change signal-to-noise ratio and subsequently the image quality. Noise is the statistically random fluctuations in the signal. It is due to the statistical nature of electrons production by the gun and electron interactions with the specimen. It is also generated by electronic devices (especially the amplifier). Signal-to-noise ratio (abbreviated SNR or S/N) is a measure that compares the level of a desired signal to the level of background noise. Good image quality will be obtained for a high SNR. To acquire an image of higher quality, slow scan mode has to be used, so that the beam spends a longer time on each pixel, and therefore produces more amount of signal.

Contrast/Brightness

Brightness and contrast are dependent on the number of grey levels in the image. To produce an image with the maximum amount of details, the number of grey levels has to be maximized. Most microscopes have an automatic brightness and contrast control. However a manual adjustment is possible. The brightness control allows making the whole image brighter or darker. The contrast control amplifies the difference between the brightest and darkest regions in the image. When adjusting contrast and brightness, the direct current level output and the gain of the amplifier (which amplifies the detected signal) are adjusted, respectively. In order to adjust contrast and brightness, an intensity profile can generally be displayed on the image (Section "Image Optimization Before Saving: Brightness/Contrast and Scan Speed").

2.9.3.8 Sample Requirements and Preparation

Sample and Vacuum Conditions

To operate the electron gun, high or ultra-high vacuum conditions are required. In conventional microscopes, a high vacuum exists in both the column and chamber. Under these conditions, samples should withstand low pressures inside the vacuum chamber. Samples which release gas molecules under high vacuum are unsuitable for examination in conventional SEMs. This is the case for wet samples such as organic materials, and for materials with strong porosity like coal or char. Wet samples must be dried and, more generally all volatile materials (including the air) must be removed from the sample so as to not degrade the vacuum quality within the SEM. However, these types of samples can be successfully examined in specialized instruments: “low vacuum” and “environmental” SEMs (Sect. 2.9.3.9).

Thermal Conductivity and Damage

Materials with low heat conductivity like biomass products or polymers are generally not resistant to heat and are easily damaged by the electron beam. For these samples it is advised:

- To use low accelerating voltage and low probe current if possible;
- To use large scanning areas, with low magnification. For a given scanning speed, the beam will spend a shorter time on each point of the specimen surface, and thereby reduce heating phenomenon;
- To adjust beforehand the imaging parameters (astigmatism, contrast and brightness, Sect. 2.9.3.9) using another field of view, and then return to the area of interest to collect the pictures as quickly as possible using short exposure time (slow scan).

Electrical Conductivity and Charging Effect

As a large fraction of the incident electrons remain in the specimen, negative charges accumulate inside this latter or at the vicinity of its surface depending on whether or not it is conductive. In all SEMs, the specimen stage is connected to ground, and when the sample is electrically conductive, these charges flow through the specimen stage. When a specimen is non-conductive, these charges cannot flow towards to ground so that the specimen becomes negatively charged, and a lot of negative charges are then accumulate around the irradiated region. This phenomenon causes a local disturbance of the trajectory of secondary electrons and/or an alteration of their collection by the detector. Some parts of the image appear extremely bright while other parts appear black, that does not correspond to the contrast generated by the features of the sample (Fig. 2.210a). If the sample surface

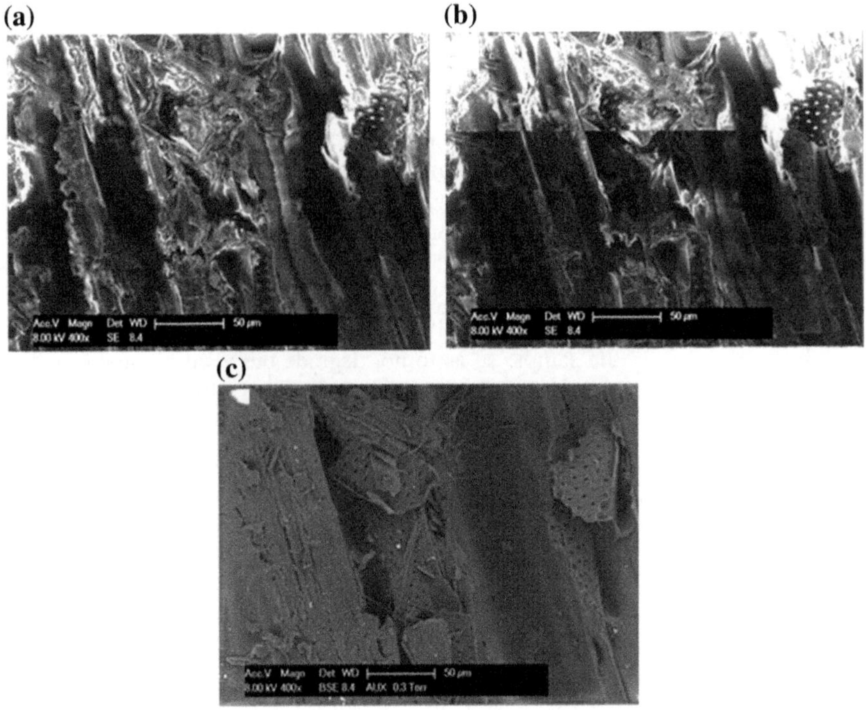

Fig. 2.210 Charging effect observed on wood surface. (a) Some parts of the image appear extremely bright while other parts appear black; (b) deflection beam that makes the image appears as if it was broken away; c the same area without charging effect (low vacuum mode, see the next section)

is strongly negative, repulsive forces deflect the probe resulting in a "break" in the image (Fig. 2.210b). All these phenomena are the so called "charging effects".

Charging effect can be minimized or even avoided by:

- Coating the sample with a conductive film (see next section);
- Reducing the spot size;
- Using lower accelerating voltage;
- Imaging with a rapid scan (using frame averaging to reduce the signal-to-noise ratio);
- Imaging with BSEs: BSEs having a larger energy than secondary electrons, they are less influenced by charge-up.

Sample Coating

In conventional microscopes, non-conductive samples can be imaged when coated with a thin layer of a conducting material (typical thicknesses are in the range 1–10 nm), commonly carbon, gold, or some other metals or alloys. Samples are attached to their holder with conductive tape or paint, and this layer provides a continuous conductive film that allows the excess electrons to be discharged towards the ground. Coating is achieved using thermal evaporation or sputter coating techniques for metals, and thermal evaporation technique for carbon. For heat sensitive samples, it is better to use sputter coating because thermal evaporation may damage them. The choice of material for coating depends on the data to be acquired:

- The coating with a light element, such as carbon, is more suitable than metal to perform elemental chemical analysis (Chap. 15), otherwise low energy X-ray radiations of the sample may be absorbed by the heavy elements of the metal, and X-ray lines of these could interfere with the X-ray lines characteristic of the sample;
- Metal coating is suitable for high resolution imaging. If possible, it is better to select a metal suitable for the selected observation magnification. Indeed, depending on metal and magnification being used, metal particles might be visible and then cover some surface details. Gold is commonly used but gold/palladium or platinum is more adapted for very high magnification observations because of their smaller grain size.

2.9.3.9 Environmental Scanning Electron Microscope (ESEM)—Low Vacuum Scanning Electron Microscope (LV-SEM)

With ESEM, wet or outgassing or insulating samples can be observed in their raw state without significant modification or preparation. Eliminating the need for sample preparation, particularly the need for conductive coatings, makes possible dynamic observations (with the appropriate sample stage): heating, cooling, hydration, dehydration, deformation. Note that with most ESEMs, it is also possible to work under low vacuum (see paragraph below) or high vacuum modes.

ESEM Principle

As seen previously, only conductive and dry samples can be directly observed in conventional microscopes. Insulating samples have to be coated and wet samples must be dried. In environmental microscopes, apertures limiting the pressure separate the chamber from the column. Different regions are pumped separately to provide a graduated vacuum from the chamber through the column to the electron gun. In some microscopes, a very low vacuum, as low as 50 torr, can be maintained in the chamber without degrading the high/ultra-high vacuum in the column and gun (10^{-5}–10^{-10} torr).

The pressure in the chamber is stabilized by the vacuum system even if the sample outgasses during the experiment. The vacuum system allows the introduction of any gas in the chamber (obviously non-toxic, non-flammable, and which may not chemically react with components of the chamber and vacuum system)

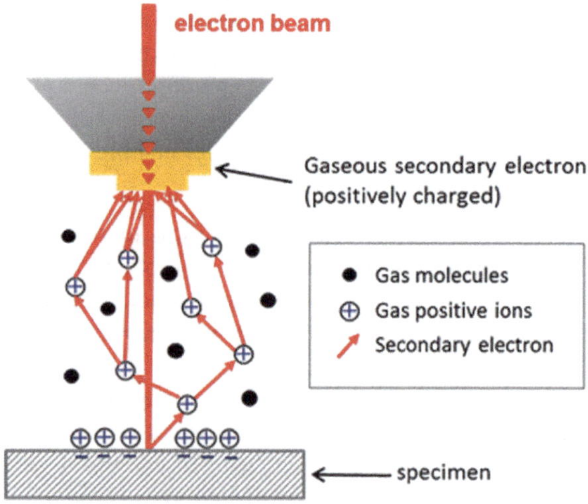

Fig. 2.211 Gaseous secondary detector: amplification of the secondary electron signal and elimination of charging effect

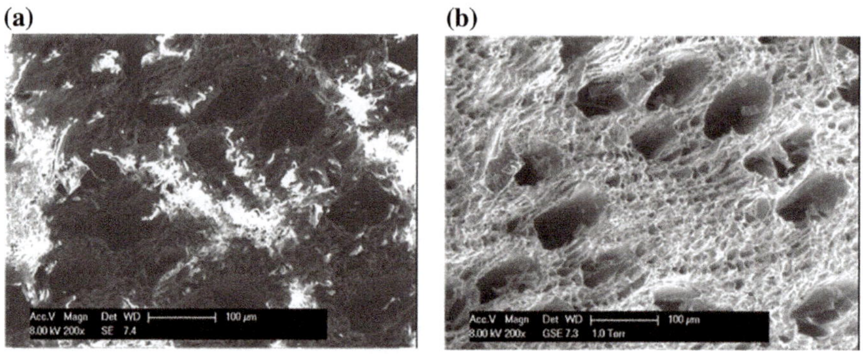

Fig. 2.212 Example of wood surface analysed by two different detectors: a high vacuum mode with secondary detector; b low vacuum mode with gaseous secondary detector

which allows to vary the sample environment. This makes possible to perform dynamic in situ experiments within the microscope chamber: heating (see example in the Section "Example of Dynamic Observation: Heating Experiment Using Heating Stage"), freezing, hydration …

Another meaning benefit of the ESEM [359] is the absence of charging effect typically observed with non-conductive samples. Conventional secondary electron detectors used in SEMs are unable to operate in the gaseous environment, that is why ESEMs have specific SE detectors using gas molecules in the sample environment to amplify the secondary electron signal, while at the same time producing positive ions which suppress charging effect. As for conventional secondary electron detectors, a positive potential is applied to the gaseous detector to attract secondary electrons emitted by the sample. As they accelerate towards the detector, they collide with gas molecules, producing gas ionization that creates additional electrons, which in turn collide with other gas molecules. This results in a cascade amplification of the original signal (Fig. 2.211). At the same time, gas positive ions produced are attracted by negative charges accumulating on non-conductive samples surface that neutralizes them and eliminates charging effect (Fig. 2.212).

Note: *water vapor is the most commonly used gas (supplied from a flask of water) because it ionizes easily, that provides high signal amplification and therefore images with excellent contrast.*

Environmental Mode and Imaging Resolution

With the low gas pressure in the chamber, some electron beams are scattered by gas molecules along their path from the final pressure limiting aperture (at the bottom of the column) to the sample surface. The resulting beam broadening, called "skirting effect" (Fig. 2.213) can affect resolution by decreasing the current in the original spot.

If a majority of the electron beam is scattered, that may occurs for very low pressures, the imaging current becomes insufficient to form an image.

To avoid the loss of resolution, the working distance and chamber pressure have to be optimized as follows:

The working distance must be reduced given that long beam gas path lengths increase loss of current, but not too much to keep enough signal amplification;

The chamber pressure must be sufficient to eliminate charging effects, but not too high because increasing pressure will increase ionization, and the latter removes electrons from the imaging current (current remaining in the original spot on the sample).

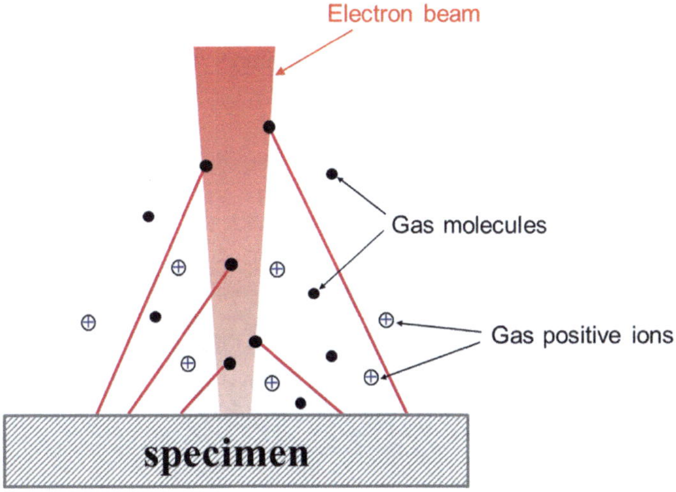

Fig. 2.213 Illustration of «skirt» formed by electrons scattered out of the beam by gas molecules

Example of Dynamic Observation: Heating Experiment Using Heating Stage

Heating stage is an optional accessory of environmental Scanning electron microscopes (ESEM) which allows continuous in situ imaging of live dynamic heating experiments.

Heating can be performed under any gas atmosphere (obviously non-toxic, non-flammable, and which may not chemically react with components of the chamber, the vacuum system and the heating system) through a range of pressures. Chamber pressure is stabilized by the vacuum system, even if the sample outgasses during the experiment.

Although there are some manufacturers of SEM heating stages, the 1000 °C heating system supplied by FEI (temperature range from 20 to 1000 °C, maximum ramping rate 300 °C/min) will be described here.

Equipment

The heating system itself is primarily composed of a heating element, thermocouple, crucible in which the sample is placed and water line tubing connected to a chiller outside the chamber (Fig. 2.214).

ESEM chamber is equipped with several accessories ports on which a feedthrough plate can be installed. Heating stage feedthrough plate provides feedthrough connectors for cooling water, heating stage power, and thermocouple input. The heating system includes:

- A stage adapter to fit the heating stage onto the sample stage;
- A programmable temperature controller equipped with a PC interface, which pilots the heating stage.

It is possible to heat the stage to a single temperature, or to set up a series of heating cycles with several consecutive temperatures with separate ramp and soak times in between.

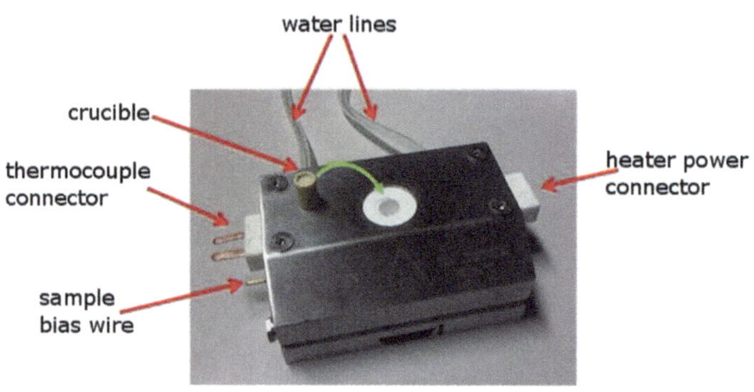

Fig. 2.214 A photo of heating stage working up to 1000 °C supplied by FEI

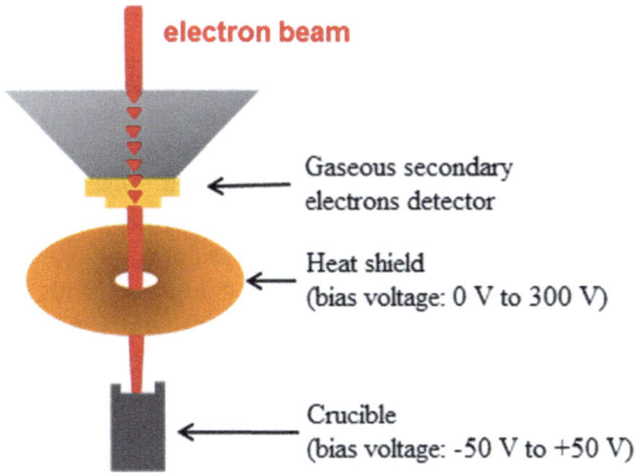

Fig. 2.215 Heating stage configuration

- A water chiller to cool the stage;
- A dedicated secondary electrons detector to image under high temperature since standard gaseous secondary detectors cannot be used for heating experiments.

Note that BSE images are not possible at high temperature since all BSE detectors will be blinded by infra-red radiations emitted from the heater. A heat shield is used which creates an "oven" effect to avoid heat loss by the sample at high temperature and keeps the temperature homogeneous throughout the sample. It is mounted to a swing arm fixed to the chamber, and placed just above the crucible during the experiment. Only 2 mm separate them (Fig. 2.215).

Sample Requirement

The sample size is limited by the size of the crucible which is about 5 mm in diameter. To obtain accurate temperature, the sample must be thin enough. Although the heating stage has been calibrated to match with the temperature of the crucible, the real sample temperature may be different. It depends on the thermal conductivity of the sample and its thickness. If the sample is non-conductive and/or thick, the temperature at the top of the sample will be lower than at the bottom.

Parameters Under Control of the Operator

Beam parameters:

Accelerating voltage, beam current and contrast and brightness are used to optimize the image (more details in the Sect. 2.9.3.7).

Heat shield bias and sample bias:

As explained in the Section “Environmental Mode and Imaging Resolution”, to avoid the loss of signal due to gas presence in the chamber when imaging in environmental mode, the working distance and the chamber pressure should be optimized. However, since the crucible must be kept close to the heat shield, the working distance can not be changed much. Furthermore, the time necessary to focus when the working distance is changed and for image stabilization while the pressure is modified, is often not compatible with dynamic observations. For these reasons, bias voltages applied to heat shield and to the crucible are used to optimize electrons collection. To increase the signal at low temperature, a negative bias applied to the sample “push” the electrons to the detector, and a positive bias applied to the shield above attracts them towards the detector. As the temperature rises to 1000 °C, the image often begins to brighten due to “thermal electrons” emission (called thermal emission). First, lowering the brightness will minimize this phenomenon. If necessary the contrast would be decreased but it will result in a loss of signal on the image. The beam current should be increased to compensate.

Example

Figure 2.216 presents an example of heating biomass modification with temperature in various atmospheres.

Low Vacuum SEM

There is another type of SEM that operates in an intermediate vacuum range between high vacuum and environmental conditions. These are low vacuum SEM (LV-SEM). As LV-SEM has maximum chamber pressures in the range of 1.5–4 torr (depending on the microscope), gas ionization may be insufficient to eliminate charging artifacts on some samples, and dynamic observations are not possible.

Basic Operation Procedure for Imaging

Sample Introduction and Field of View Selection

Sample needs to be fixed firmly to the specimen holder using an electrically conductive double-sided adhesive tape or conductive paint. Powders are dispersed on the adhesive surface, and any lose particles are removed from the surface by flowing with a small container of compressed nitrogen (or dry air) with a fine nozzle. Always wear powder-free gloves when handling samples, sample stubs, and placing samples inside the vacuum chamber. Only vent and open the chamber when the sample is ready to be loaded because when the chamber is exposed to ambient air, humidity and contaminants may enter and deposit on the walls of the sample chamber.

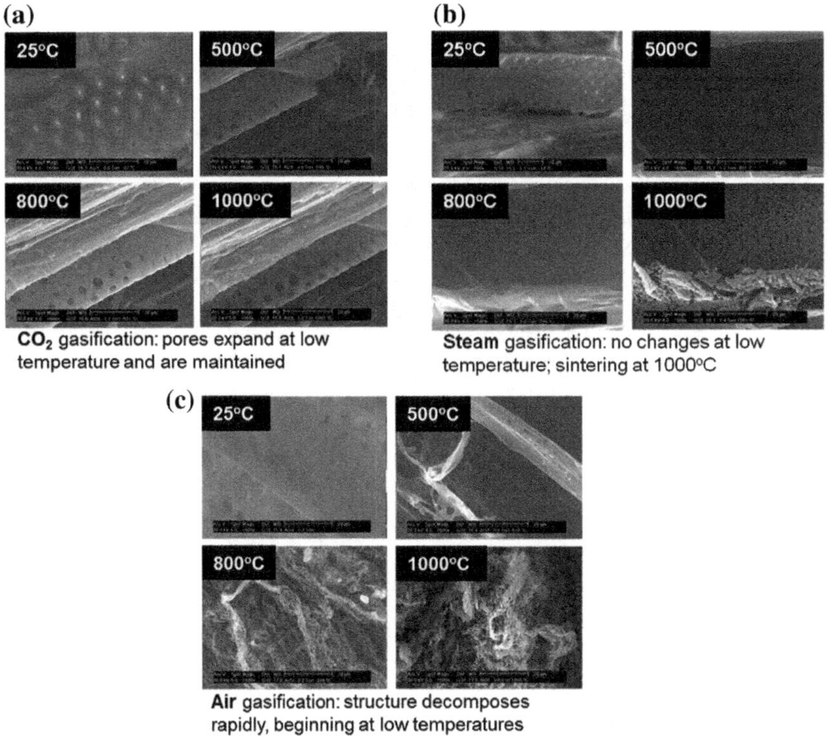

Fig. 2.216 Example of heating biomass (poplar wood) modification with temperature in various atmospheres [360]. Reprinted (adapted) with permission from [360]. Copyright 2012 American Chemical Society

- Vent the chamber, generally by pressing the "VENT" button;
- When the chamber is at atmospheric pressure, open it, insert your sample stub onto the sample holder and lock its position with the set screw;
- Slowly push the door to close the chamber while making sure that the sample will not hit the pole piece (the very end of the column) or any detector. A height tool is often provided to ensure that the sample travel will have sufficient Z range;
- While maintaining one hand on the door, pump the chamber and wait until the vacuum is correct;
- Turn on the high voltage. An image will appear, possibly out of focus and with incorrect contrast and brightness, and a message will ask you to confirm that the sample is in focus;
- Adjust the contrast and brightness and correct the focus (see following paragraphs) to obtain a clear image, and then confirm that the image is in focus. This will cause the software to store the working height so that the stage cannot damage the sample;

- Using XY movements, drive the stage to an area of interest. To achieve an easier view of stage movement and allow a faster positioning ability, use low magnification and fast scan (TV scan is ideal for navigation).

Focus Adjustment

The focus correction can be carried out manually and automatically. As soon as the magnification is increased, the focus must be optimized. The following procedure should help for this end:

- An image focused at high magnification remains in focus at lower magnification, therefore it is recommended to focus the image at magnification 2 or 3 times higher than the magnification needed for the final image. It will make the lower magnification sharper;
- To optimize the focus, after adjusting it as finer as possible, continue to move the focus control a little bit more so as to past the sharp focus, and then return to the optimum focus. This ensures that the best focus point can be observed;
- If the image moves while focusing, it is because the final aperture is not centered on the optical axis, making the electron beam not aligned through the center of the objective lens. An aperture alignment has to be done (See next paragraph)

If a stretching of the image is observed on both sides of the focus (in one direction when it is under focused and in the orthogonal direction when it is over focused), the astigmatism needs to be corrected.

Aperture Alignment

The SEM aperture alignment is performed wobbling the current to the objective lens using the "modulator" button.

- At low magnification and TV rate scanning, a rotation of the image will be observed. The rotation center should be at the center of the viewing screen. To achieve that, center an object in the middle of the screen and move the position of the aperture to align the center of rotation over the center of the screen;
- At high magnification (above ×20000) the image will move slightly in a certain direction. Minimize this movement by adjusting more precisely the aperture (Fig. 2.217).

Note: *this alignment should be done whenever the aperture is changed, or when the accelerating voltage, probe current or working distance have been changed by large amounts.*

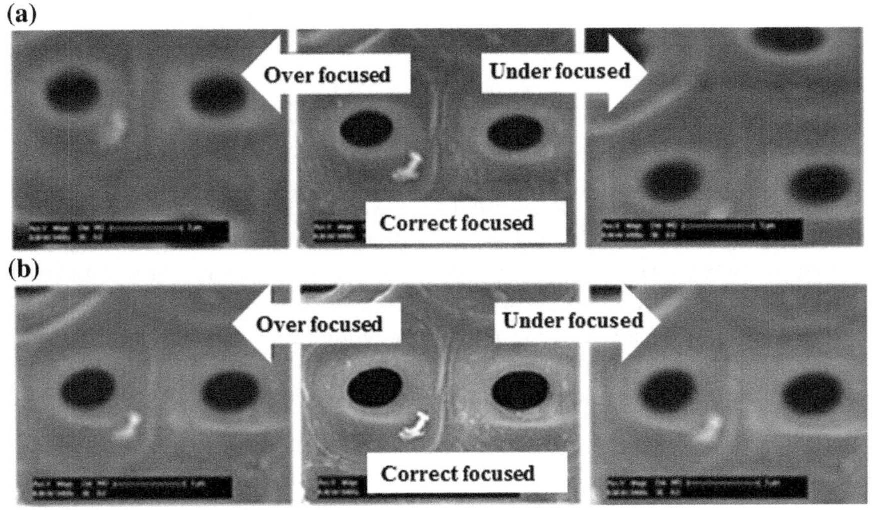

Fig. 2.217 Illustration of aperture alignment (images from pyrolysis char of virgin woods); a final aperture not centered on the optical axis; b final aperture centered

Fig. 2.218 Illustration of astigmatism correction (image of minerals on the surface of pyrolysis char from woods); a before astigmatism correction; b after astigmatism correction

Astigmatism Correction

All microscopes are generally equipped with an astigmatism corrector or "stigmator" which applies a compensating field. The correction can be performed manually or automatically. The manual astigmatism correction consists in bringing the focus condition to the mid-point between the two distortions observed, then in moving the astigmatism control until the image becomes in focus and sharpness at one point (Fig. 2.218).

Chamber Pressure Adjustment in Environmental or Low Vacuum Mode

The chamber pressure must be adjusted to a level as it will prevent the specimen from charging while maintaining a good image resolution. As the gas in the chamber scatters some electrons from the beam (see previous sections), a range of operating conditions (pressure, working distance, accelerating voltage), for which beam scattering does not degrade image resolution, must be adjusted. The relationship between working distance and chamber pressure depends largely on the nature of sample. So the analysis repetition may be necessary.

Image Optimization Before Saving: Brightness/Contrast and Scan Speed

Before saving, the image quality can be improved by adjusting brightness/contrast and scan speed.

Brightness/contrast:

In order to adjust or check the contrast and brightness, an intensity profile can generally be displayed on the image. It gives the intensity of the signal coming directly from the photomultiplier (that amplifies the detected signal) along the currently scanned horizontal line: the X-axis corresponds to the distance scanned across the specimen, and the Y-axis represents the dynamic range of the signal viewed on the monitor (Fig. 2.219).

The base level of the signal is the brightness while the spread of the signal up from this baseline is the contrast (Fig. 2.220).

To adjust correctly the brightness and contrast, the following steps can be applied:

- First reduce the contrast to zero and adjust the brightness level to the lower limit;
- Then increase the contrast so that the signal spreads over the full dynamic range.

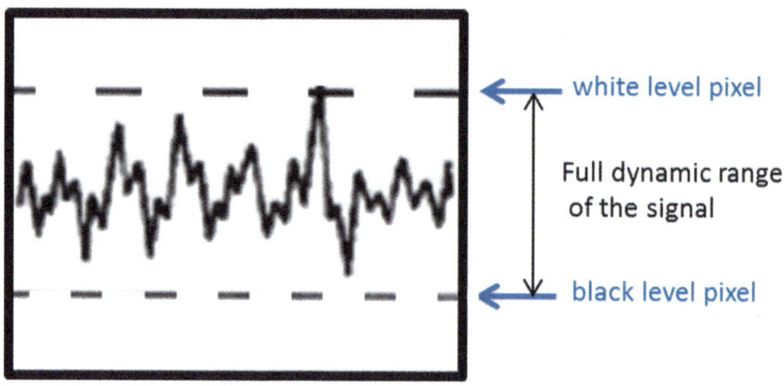

Fig. 2.219 Illustration of an intensity profile used to adjust contrast and brightness

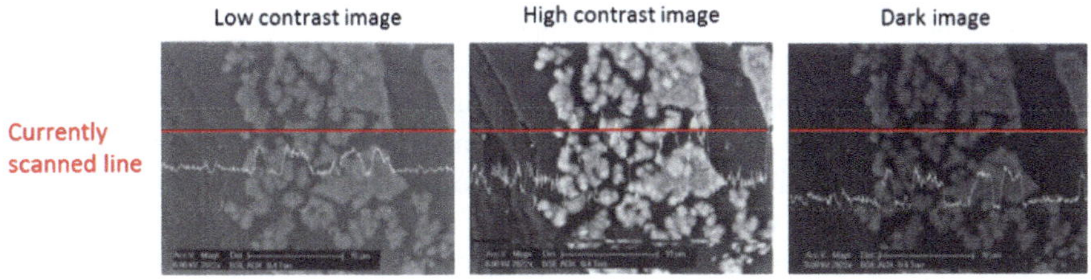

Fig. 2.220 Intensity profile of an image of minerals on a wood surface

Note: Doing this setting procedure results in images that can appear "flat" for most novice SEM operators which have the natural tendency to produce images with too much contrast. If the contrast setting is very high compared to the brightness, the image will not contain the maximum amount of scientific details and may appear noisy.

Scanning speed:

Prior to save the image, the signal-to-noise ratio can be improved using a slow scan. In this case, a particular attention must be accorded to non-conductive specimens because just like increasing voltage, a slow scan mode will potentially increase charge-up effects. Indeed, as the beam spends a longer time on each point of the surface, the amount of electrons on each point increases and the beam does not move quickly enough, so they can dissipate. For these specimens, another noise reduction method can be used, named frame averaging. A first entire frame is scanned, stored, and then averaged with a next frame for a selectable number of frames. This "frame averaging" control allows improving signal-to-noise ratio while keeping a fast scan speed.

Note: increasing the image collection time can also lead to sample damage.

Once the analysis is complete, switch off the beam, vent the chamber and slowly pull the chamber door open, being careful that the sample will not hit the pole piece or any detector.

2.9.4 Energy Dispersive Spectroscopy (EDS)

2.9.4.1 Introduction

EDS (Energy Dispersive Spectroscopy) is a technique used to characterize the elemental composition of a sample based on a specific electron-matter interaction in an electron microscope (EM). Electron-matter interactions in an EM produce backscattered (BSE) and secondary (SE) electrons which generate a grayscale image of the sample at very high magnifications (Sect. 2.9.3). As well as producing electrons, interactions also produce X-ray radiations the energies of which are characteristic of the elements contained in the sample. The equipment consists of an X-ray detector, operating inside the chamber, and suitable hardware and software outside the microscope to analyze collected data. This is a non-destructive technique that can be qualitative, semi-quantitative or quantitative depending on the sample preparation and calibration. It also provides spatial distribution of elements through mapping. Basically, all elements from atomic number 4 (Be) to 92 (U) can be detected however all instruments are not equipped for 'light' elements ($Z < 10$) detection.

2.9.4.2 Theory

The basic units of matter—are the smallest components of it consisting of a nucleus which contains neutrons and protons, and one or more electrons on the periphery bound to the nucleus by electrical attraction. Electrons surrounding the atomic nucleus are moving according to a trajectory so called atomic orbitals. Each atom can have many orbitals, called electron shells. The associated principal quantum number (n) of electron shells indirectly describes its shape and energy. When the electron beam hits the sample, it might eject inner shell electrons of the specimen atoms. If such an ionization occurs, an ejected electron leaves a vacancy (hole). When an electron from a higher-energy shell fills the hole, the difference in energy between the higher-energy shell and the lower energy shell is released in the form of an X-ray radiation (photon X). Because the energy difference between shells is well defined and specific to each chemical element, the energy of an X-ray is characteristic of the emitting atom. Therefore EDS allows the identification of chemical elements present in a sample. To ionize an atom, the electron beam must possess a minimum energy, namely the 'critical' excitation (or ionization) energy (E_c). In practice, to give reasonably high excitation efficiency, the incident electron energy (E_0) must exceed E_c by a factor of 1.5 to 3; this is called the overvoltage factor.

2.9.4.3 Spectrum

Spectrum Structure

The data obtained by EDS analysis consist of spectra which display characteristic X-ray (peaks) lines characteristic of chemical elements present in the sample (Fig. 2.221). The EDX spectrum represents the number of characteristic X-ray received and processed by the detector (X-axis), the so called "counts", as a function of their energy displayed on the X-axis. As shown in the schematic representation of a spectrum (Fig. 2.221), peaks overhang a background called continuum X-rays (or Bremsstrahlung) which corresponds to energy lost by the electron beam when its interact with the nucleus of the specimen atoms while being

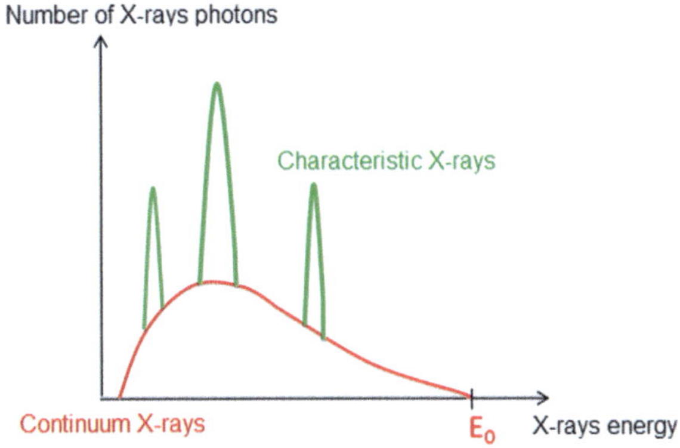

Fig. 2.221 Schematic representation of a spectrum

slow down. This background of continuum is not characteristic of the specimen atoms and covers all energies from zero to the incident electron energy (E_0 in Fig. 2.221). The peak-to-background ratio (P/B) is a very important factor that determines the detection limit of the detector since small peaks could not be detected. The peak intensity increases with the probe current, atomic number and accelerating voltage (see Section "The Accelerating Voltage" for accelerating voltage).

Characteristic X-ray Nomenclature

Most elements have multiple energy shells and may emit X-rays of several different energies. X-ray lines are identified by a capital letter indicating the shell in which the initial vacancy occurs (K, L or M), and a greek letter subscript according to the shell from which an electron drops to fill that vacancy: if the vacancy filling electron drops from the adjacent shell an α ray is emitted, if the electron drops from two shells away, the emitted ray is a β ray etc. (Fig. 2.222).

Energy and Intensities of X-ray Lines

General Rules

Some general rules applicable to the various spectral lines are presented as follows:

- Within a line series (K, L, M etc.), the energy of the emitted X-ray increases with the average atomic number. For example: oxygen K lines have higher energy than carbon K lines.
- K lines are simple and form distinct peak. L and M lines become progressively more complex, and overlapping between lines appears as Z increases.
- For a given element:

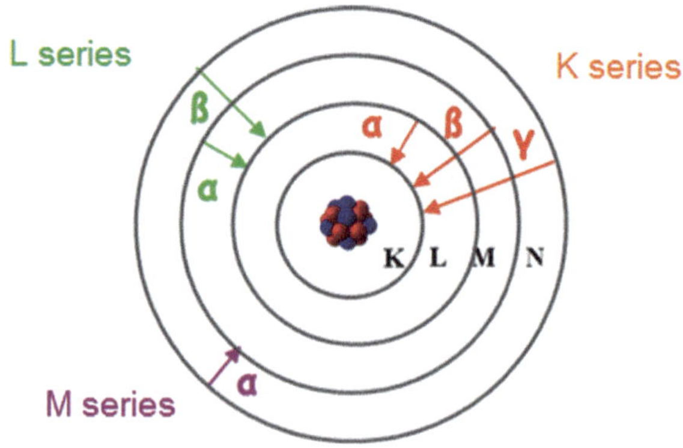

Fig. 2.222 Characteristic X-ray nomenclature

- Lower line series have higher energy: K lines have higher energy than L lines, which have higher energy than M lines etc.;
- The intensity of the lines depends on the transition probability: since L and K are adjacent energy shells, the most probable transition when a K-shell vacancy is created is the L to K transition. Therefore K_α radiation will always be more intense than K_β radiation.

Effect of Accelerating Voltage on the Line Intensities

When making identification of an element based on relative intensities, it is important to remember the influence of accelerating voltage. If the latter is under the overvoltage of that of elements, the ratio of the line intensities will dramatically be changed.

2.9.4.4 Equipment

Main Components of an EDS

The main components of an EDS system include:

- An X-ray sensor;
- A pulse processor;
- A multiple channel analyzer.

The sensor is used to convert X-ray energy into voltage signals; this information is sent to a pulse processor, which measures the signals and passes them onto an analyzer so as to display and interpret X-rays data. Two types of sensors are used for X-ray detection:

- Traditional silicon crystals drifted with lithium, so called the Si(Li) detector, cooled to cryogenic temperatures with liquid nitrogen;

- Newer silicon drift detector, "SDD", which have largely replaced the previous ones. These detectors, in addition to not needing liquid nitrogen to be cooled down, have better resolution at high count rates.

EDS detectors are characterized by their energy resolution. This latter parameter is defined as the full width of the peak at half maximum (FWHM). Typical detector resolutions, which are conventionally measured using the Mn K_α peak at 5.9 keV, are in the range 130–150 eV.

Parameters Under Control of the Operator

Time Constant and Dead Time

These two parameters are correlated. Time constants (quoted in μs), also called amplification process time or processing times, is the time during which a pulse is formed (shaped) from an incoming X-ray. There is a range of available time constants to be selected, and each of them can handle a maximum number of counts called «throughput». For a given count rate the time constant determines the dead time, corresponding to the unresponsive duration time of the system. This dead time starts when an X-ray photon reaches the detector and ends after its processing. The longer the time constant, the lower the noise and the better the resolution (it becomes easier to separate peaks that are close in energy). However, the longer the process time, the more time is spent measuring each X-ray, and the fewer events that can be measured. There is thus a compromise between resolution and throughput.

The probe current (beam parameter) must be adjusted to maximize the throughput at a given time constant. It is generally recommended to keep detector dead time below 30% of analysis time.

Note that the maximum throughput rate depends on the system, particularly on the type of detector. Typically values for Si(Li) detectors are around 20,000 counts per s, rather than 100,000 counts per s and above for SDD.

Collection Time

This is the duration of the signal acquisition. To obtain good peak-to-background ratios, the number of counts must be statistically significant. The time collection is determined in relation to the throughput (which depends on probe current and time constant) so has to have more than 100,000 counts in the spectrum.

Note: in an EDS system, the real time, called "clock time" is divided into live time and dead time. The live time, as opposed to the dead time, is the time during which the detector is able to receive an X-ray. Typically live time varies from 10 to 100 s.

Spectral Artifacts

"Sum peaks" occur when X-ray photons reach the detector at nearly the same time, and the system record the sum of their energies instead of each one individually. This may occur with a too high dead time percentage due to a too high count rates under the chosen amplification time (time constants). Lowering the beam current lowers the count rate and usually solves the problem.

"Escape peaks" occur when the silicon detecting crystal is ionized by an incoming X-ray radiation. If an ejected silicon photon escape the crystal, it carries on with it a well-defined energy, and the energy measured for the incoming X-ray radiation will be less than its actual energy by exactly that amount: 1.74 eV corresponding to the silicon $\boldsymbol{K_\alpha}$ X-ray radiation. This may occur with any element of the sample, but for major elements for which a large amount of counts accumulate, an escape peak can be expected at an energy 1.74 below that of the peak of the characteristic element.

EDS systems employ different methods for handling these X-ray artifacts, however it is necessary to look for escape peaks concerning the peaks of greatest height and for sum peaks for these same peaks when a high count rate has been used for spectrum acquisition.

Resolution

Spatial Resolution

The spatial resolution is governed by the penetration and spreading of the electron beam within the specimen, which depends on both the accelerating voltage and its average atomic number. The unit of the spatial resolution is micron. It ranges from 0.2 to 1 μm^3 for high atomic numbers, and 1–5 μm^3 for low atomic numbers. SE or BSE images and X-ray spectra are generated from largely different volumes inside the sample. Proper interpretation of microanalysis results requires that users be constantly aware of the difference between image resolution and X-ray spatial resolution. In case of inhomogeneous sample, as illustrated in Fig. 2.223, some particles/phases present inside the sample may not be detected on the image while they may be detected by X-ray analysis.

Energy Resolution

EDS detectors are characterized by their energy resolution. The energy resolution is defined as the full width of the peak at half-maximum (FWHM). Typical detector resolutions are 130–150 eV at 5.9 keV, energy which corresponds to the Mn $\boldsymbol{K_\alpha}$ peak (conventionally used for determining resolution).

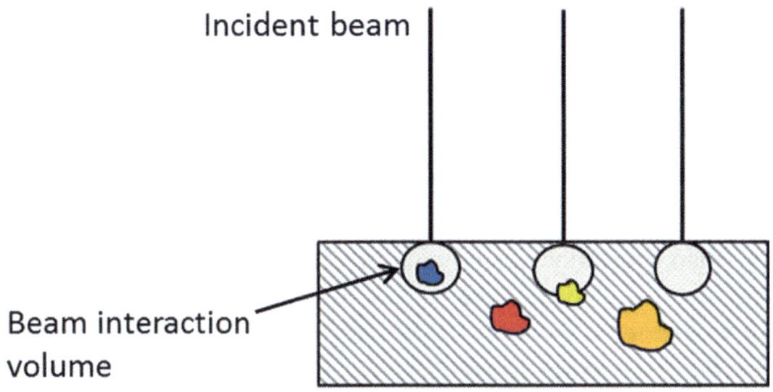

Fig. 2.223 Non homogeneous sample: depending on the location of the irradiated point, some particles may be detected while they are not visible on the image

Accuracy and Detection Limit

Accuracy

X-ray intensities are measured by counting X-ray photons. Since the emission and subsequent detection of a photon is a statistically event which has a probability of occurrence, the precision in evaluating the intensity is limited by statistical error. Another source of errors is the various corrections which need to be applied to the raw data (see further paragraph "quantitative analysis"). Precision that can be obtained is around ±1–2% (relative).

Detection Limit

The detachability of small peaks is limited by the background of the spectrum. For elements from Na to U, the detection limit is on the order of 1000 ppm (by weight). Since the size of the characteristic peaks increases linearly with acquisition time, detection limit can be reduced using long X-ray collection time.

Analysis Modes

EDS can be carried out by different modes as summarized below:

- "General analysis": generation of an X-ray spectrum from the entire scan area of the SEM: it is the most basic analysis for showing a general overview of elemental composition within the entire sample.
- Point/area analysis: the electron beam is kept stationary on a spot or is scanning a reduce area of interest and generates spectrum that will provide more localized elemental information.

- Line scan: the electron beam follows a preselected line drawn on the sample image and generates a plot of the relative proportions of previously identified elements along that line.
- Mapping: the electron beam scans a selected frame mapping and a spectrum is collected at each pixel of the image. The intensity of each element is stored at each pixel and is used to modulate pixel intensities in a single element map.

An X-ray map shows the spatial distribution and relative proportion (intensity) of previously defined elements over a particular selected area.

Unlike back-scattered electron (BSE) images, which only show average Z in a phase, elemental maps allow to identify which element are present in a phase and their relative proportions (atom% or mass%).

Qualitative Analysis

The qualitative analysis allows identifying which elements are present in the sample by comparing the energies of peaks present in the spectrum with reference tables containing all possible elements and all possible lines per element. This can be achieved manually or automatically. Since the energy resolution of EDS imposes a limit on the separation of peaks, there are several examples of peak overlapping which commonly occur. For example, the overlapping can happen for Pb M_α (2.34 keV), S K_α (2.31 keV) and Mo L_α (2.293 keV) lines; and for Zn L_α (1.009 keV) and Na K_α (1.041 keV) lines. To help the user in identifying elements when overlapping is present, "peak deconvolution" software provided with the EDS system can be used. It fits the spectrum using stored libraries of X-ray lines for each element. After performing peak deconvolution, the EDS software displays the theoretical spectrum that can be compared with the collected spectrum. For accurate peak identification, the EDS spectrometer should be calibrated periodically to ensure that the spectral peak positions are correct. An energy calibration check can be easily performed with a piece of copper attached on any sample holder because the K_β and L_α lines of the Cu, at 8.907 keV and 0.928 keV respectively, span most of the typical EDS spectral range.

Quantitative Analysis

The quantitative analysis provides weight and atomic percent data for the chemical element identification. To do that, several phases are necessary: background removal, deconvolution of overlapped peaks and calculation of concentration. It can be performed with standard compounds (samples with known concentration) or without standard (standardless analysis). In many cases, standardless analysis is the preferred method thanks to its rapid and ease use. For both methods the mass concentrations are obtained by comparing the intensity of the characteristic line emitted by each element of the unknown sample with the corresponding intensity emitted from a standard. Standards are materials in which the concentrations of all the elements are accurately known, and may be pure elements or compounds with known concentrations.

$$K_{ratio} = \frac{I_{sample}}{I_{standard}} \tag{2.168}$$

However, in order to obtain the unknown concentrations, the intensity ratios (K_{ratio}) must be corrected by a factor taking into account matrix effects (Eq. 2.168). Indeed, the detected X-ray intensity is dependent on several properties of the sample, the so called matrix effects. This correction takes into account that:

- The beam penetration, and consequently excitation efficiency, depend on elements present in the sample (average Z);
- Once created in the sample, some X-ray radiations will be absorbed before being detected;
- Some elements can be overestimated owing the fluorescence caused by characteristic X-ray radiations of other elements.

In most standardless quantitative analysis software, the calculation is carried out with a standards database made from a series of experimental standard measurements performed by the manufacturer, and subsequently adjusted for the characteristics of the local instrument. This correction assumes that measurements are performed with the standard and sample under the identical operating conditions (beam current, geometry, kV and lifetime).

2.9.4.5 Sample Preparation and Requirements for EDS Analysis

For both qualitative and quantitative analysis, samples requirements are the same than for analysis in a SEM, as already presented previously. Samples need to be able to withstand low pressures and electron irradiation. If the SEM equipment only works under high vacuum, non-conductive samples must be coated with a thin layer of conducting material. Carbon is usually used as coating material for EDS analysis.

Sample Preparation for Qualitative Analysis

Since it is possible to record an X-ray spectrum from almost any location where the beam strikes the specimen, every sample, even with complex micro-structure, may be analyzed without any preparation. However, specimen geometry effects such as shape and local surface inclination to the beam can have a strong impact on the results (Fig. 2.224). When analyzing a sample without any preparation, it is very important to keep in mind these considerations.

For qualitative mapping analysis, sample must be embedded and polished to avoid “shading effect” illustrated in Fig. 2.225:

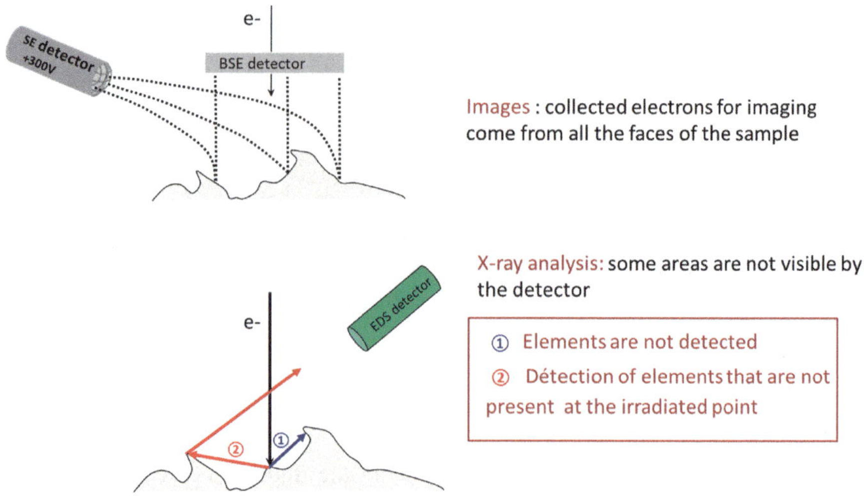

Fig. 2.224 Effect of the sample surface topography on the results

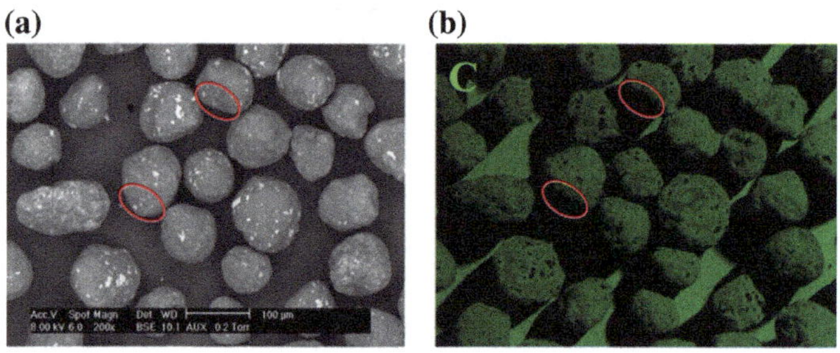

Fig. 2.225 Carbonaceous materials: a BSE image; b carbon map (X-ray image): chemical information is lost due to shading effect

Specific Sample Requirement and Preparation for Quantitative Analysis

Sample Must Be Homogeneous

Taken into account for the concentration calculation, matrix effects must be the same for specimen and standard. Thus, both specimen and standard must be homogeneous at the microscopic level. This is illustrated in Fig. 2.226.

Sample Must Be Well Polished

This avoids the effect of surface roughness that could generate artefacts on the obtained results. The absorption correction, generally the most important in concentration calculation, depends on the angle between the surface of the specimen

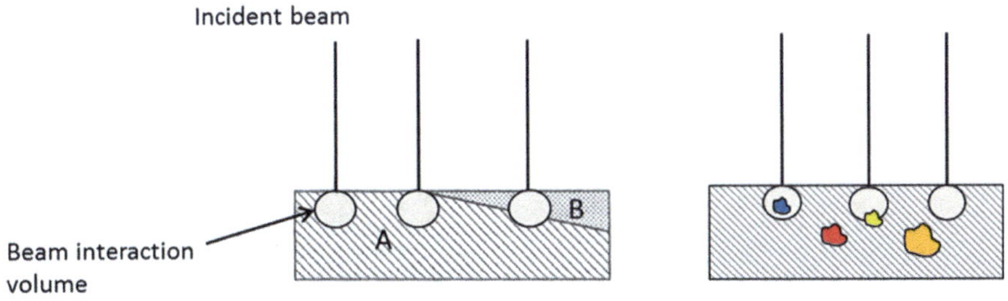

Fig. 2.226 Non homogeneous sample: depending on the location of the irradiated point, some phases or particles may be detected or not

Fig. 2.227 Illustration of the positions of the EDS detector and specimen

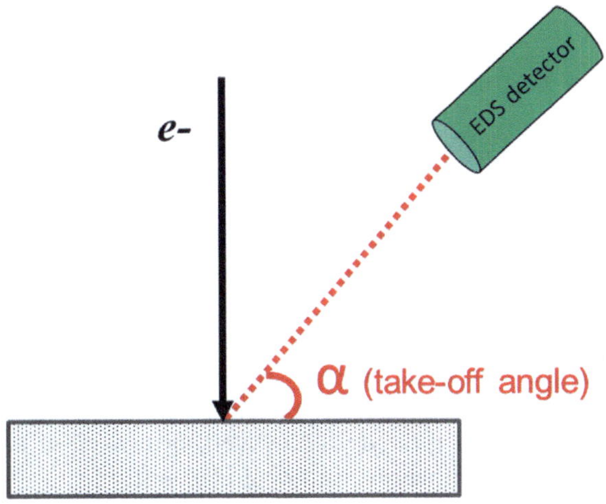

and the X-ray path to the spectrometer. Indeed, each EDS system has a specific geometry characterized by the take-off angle which is the angle between the specimen surface (at 0° tilt of specimen) and a line to the center of the detector (Fig. 2.227). This is the angle at which characteristic X-rays emitted from the specimen are received.

As shown in Fig. 2.228, if sample surface is not horizontal, the length of the X-rays escape path through the sample will vary depending on the inclination angle of the surface that will skew the results.

2.9.4.6 Basic Procedures for EDS Analysis

For Si(Li) detector, prior to run EDS, the amount of liquid nitrogen in the dewar for the detector must be enough for the planned EDS analyses.

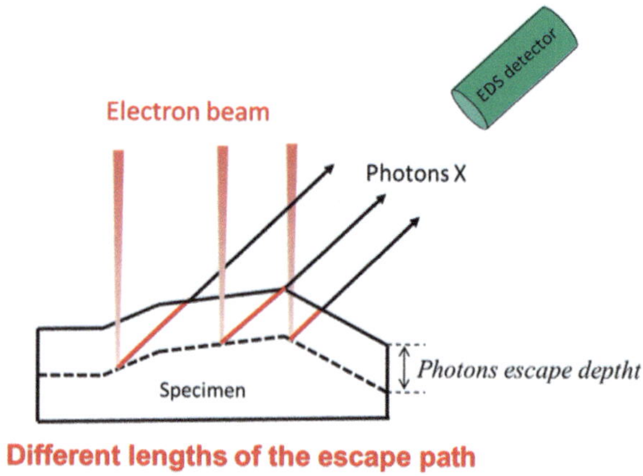

Fig. 2.228 Illustration of the influence of the sample topography on detected intensities: the variation of intensity is not only due to a variation of concentration

Check SEM Conditions

Acceleration Voltage

Select an appropriate acceleration voltage for the material to be analyzed. First, a voltage should be chosen to get an image of the sample, and second ensure that this voltage will excite all the elements of interest. For unknown sample, with a beam energy of 15 keV or more, most X-ray lines from most elements will efficiently be excited and automatically be identified in the spectrum. For small particles, a lower voltage should be chosen, since a smaller excitation volume will enhance the contribution from these features. For a sample with a low average atomic number like biomass, it is advised to work at a reduced beam energy, e.g. 5–10 kV to enhance sensitivity to the low-Z elements.

Working Distance (WD)

Ensure that the working distance, as displayed in millimeters on the microscope monitor, is the desired one for performing X-ray micro-analysis. This is specific to the geometry of the detector mounted on the SEM chamber, and specified on the user manual. To change the WD, the focus should be adjusted until the microscope monitor displays the recommended working distance. The image may now appear out of focus. Carefully adjust the sample height using the Z drive of the stage so that the image comes into focus.

Aperture and Spot Sizes

Choose the most appropriate aperture and spot sizes for imaging and EDS analysis (see Sect. 2.9.3). Typical probe currents vary greatly depending on the brightness of the gun (see Section "Electron Guns") and the acceleration voltage. More useful parameters are total count rate spectrum (cps) and dead time. With a Si(Li) detector, a count rate more than 1000 (per second) and dead time less than 30–40% will ensure that appropriate volume of X-ray radiations reach the EDS detector at the current conditions.

For mapping, energy resolution is less important than count rate. To obtain higher counts per pixel choose a higher beam current and set a shorter time constant to increase the count rate capability up to 50,000 cps over the entire spectrum. With silicon drift detectors (SDD), which have high count rate capabilities, up to 500 000 cps can be obtained.

Select the Analysis Mode: Point/area Analysis

- Set a point or frame for the area analysis on the image;
- Collect a spectrum (start EDX data collection);
- Identify the elements: most software propose both manual and automatic modes.

– Manual analysis mode: A cursor can be moved to the peak of interest. Appropriate element can be selected from a proposed list by clicking on this peak;
– Automatic analysis mode: the element is directly displayed near the peak in the spectrum.

In both cases the peak identification is achieved using a table with all possible elements and all possible lines per element. Note that for accurate peak identification, the EDS spectrometer should be periodically calibrated.

Figure 2.229 shows a point (spot) EDS analysis carried out on the surface of a solid residue recovered from the thermo-conversion of a beach wood. The EDS spectrum indicates the presence of different chemical elements at the selected analysis point (red arrow on SEM image).

Figure 2.230 shows spot EDS analysis of a CCA-wood (a wood treated with chromium, copper, and arsenic). On the left-hand-side, a metal layer (bright region) on the surface of the sample as well as a likely-virgin wood structure in the core could be observed. On the right-hand-side, at higher magnification, metals particles could be observed by contrast difference. The EDS analysis performed on one particle showed the presence of different minerals including Fe, Cr, Al, Si, Mg.

Select the Analysis Mode: Mapping or Line Scan

These analysis modes can be performed with the following steps:

- Set a mapping frame (or a bar for line scan) on the image;
- Pre-select the elements of interest;
- Select the resolution: The resolution is set by choosing the number of pixels in the X and Y dimensions of the selected frame. For a square monitor, typical options are 32×32, 64×64, 128×128, 256×256 or 512×512 pixels.

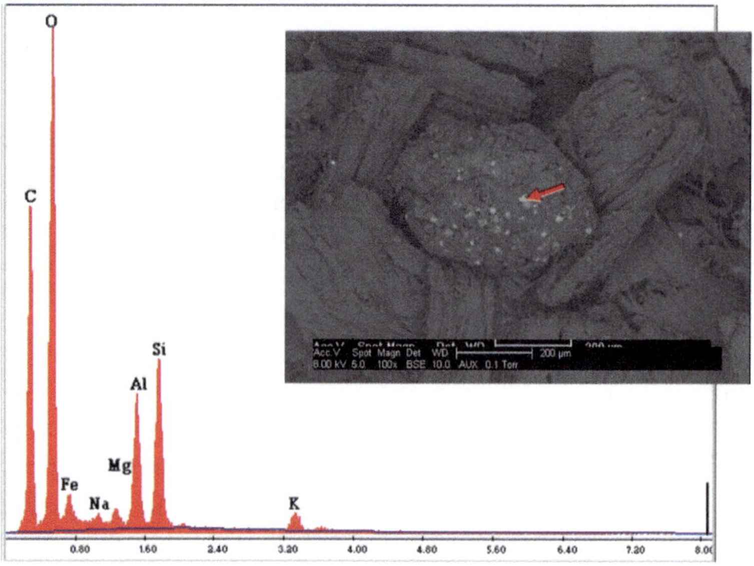

Fig. 2.229 Example of point EDS analysis on the surface of a solid residue recovered from the thermo-conversion of a beach wood

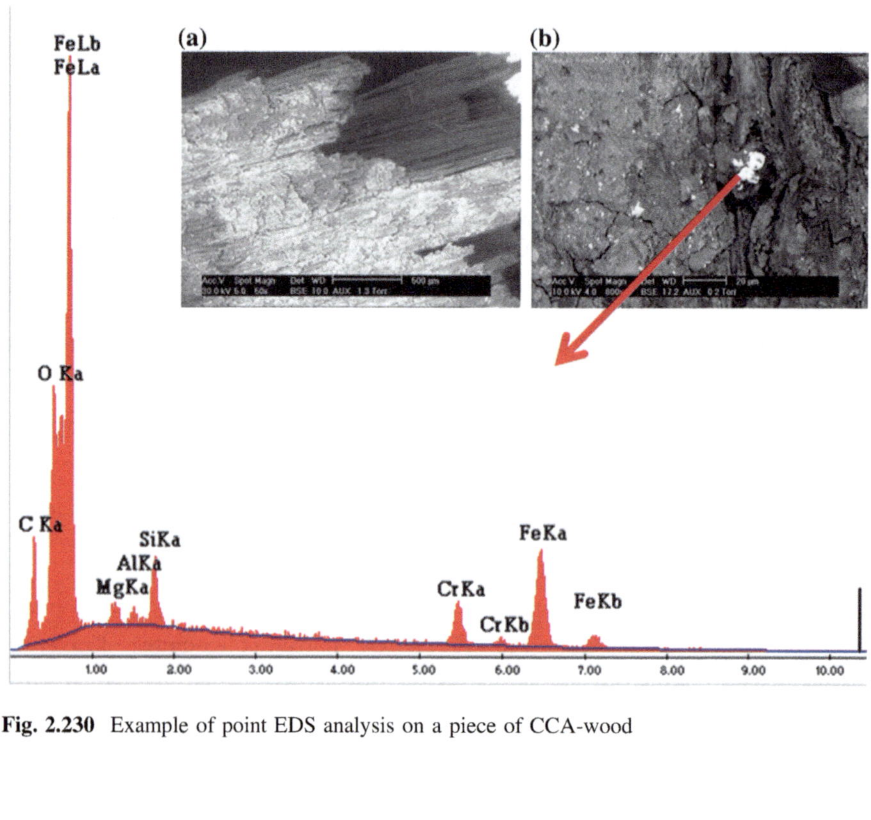

Fig. 2.230 Example of point EDS analysis on a piece of CCA-wood

When choosing the resolution, ensure that the pixel size at the current magnification approximates 1 μm (order of magnitude of the X-ray spatial resolution, which depends on both the primary beam energy and the atomic number of the specimen). These parameters are connected by the following equation:

$$M = \frac{L}{R \cdot N} \tag{2.169}$$

where M: magnification; L: width of the screen image; R: X-ray resolution; N: number of pixels in a line of the image.

Note: for samples with low average atomic number like biomass, since the critical excitation energies are low, the map collection would be performed at low accelerating voltage (8–10 kV) to improve spatial resolution (Fig. 2.231).

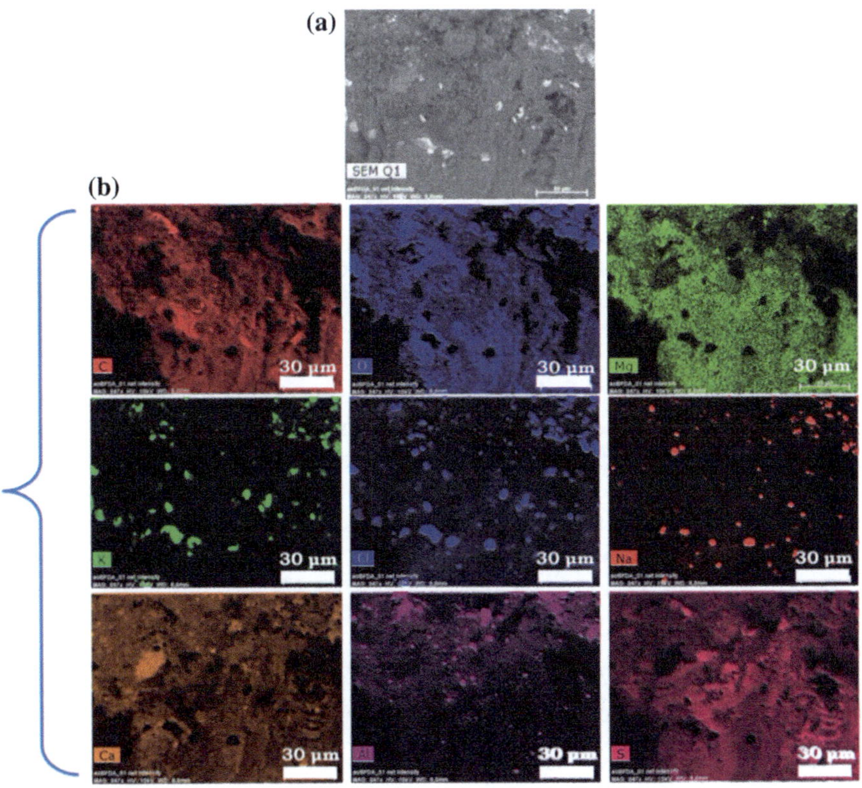

Fig. 2.231 Mapping EDS analysis of char from wood pyrolysis; a backscattered image, voltage: 10 kV; b X-ray image, voltage: 10 kV, resolution: 300 × 300 pixels (area of 150 × 150 μm)

- Select dwell time: Dwell time is the time spent by the beam on each pixel, in other words it is the collection time for each pixel. To obtain high counts per pixel, do choose a long dwell time per pixel, ~100 ms or longer.

When choosing parameters you must keep in mind that the time taken to generate a map will mostly depend on the resolution selected and the dwell time. Silicon drift detectors with their high count rate capabilities can produce X-ray maps in minutes, while hours are necessary for conventional EDS.

Silicon drift detectors with their high count rate capabilities can produce X-ray maps in minutes, while hours are necessary for conventional EDS.

2.9.5 Micro-tomography

2.9.5.1 Introduction

X-ray micro-tomography is a non-destructive characterization technique that allows identifying the internal microstructure of a material by 3D reconstruction of a representative volume, from a series of measurements carried out by slice. As a whole, the process is similar to that of the medical imaging, only the photon energies used are different. The term "tomography" defines any process of characterization photographing at regular intervals successive slices of the object to be visualized. This section can be virtual, as is the case in X-ray micro-tomography, or real, in which case the characterization method becomes invasive. The term "micro" simply points out at the observation scale to which the technique is achieved. Nowadays, the best X-ray micro-tomographs achieve spatial resolutions of the order of 0.05 μm per voxel. Several techniques dedicated to the characterization of the internal structure of a material using X-rays exist. The most common consist in analyzing transmitted and diffracted X-rays using transmission and phase contrast micro-tomographies, respectively. These two techniques are complementary: the transmission makes it possible to visualize the internal texture thanks to the X-ray absorption difference between the constituents of the material, while the phase contrast highlights the boundaries between these constituents.

2.9.5.2 Theory

The operating principle of transmission micro-tomography uses two laws:

- The Beer-Lambert law, connecting the transmission of X-ray radiations to the properties and the thickness of the matter through which it passes;
- The inverse Radon transform, a mathematical principle used to reconstruct a 3D geometry from different 2D projections recorded.

Beer-Lambert Law

The diagram in Fig. 2.232 illustrates the transmission of X-rays through the matter for a fixed wavelength, or for a constant photon energy [361].

The Beer-Lambert law is expressed by Eq. (2.170):

$$I = I_0 \times e^{\int \mu(x,y) dx} \tag{2.170}$$

where: I_0: Intensity of the emitted beam (W m^{-2}), I: Intensity of transmitted beam (W m^{-2}), μ: Linear mitigation coefficient (m^{-1}) and dx: Integration variable (m). The linear attenuation coefficient (μ) depends on the density and the average atomic number of the phase traversed by the X-ray beam and do not have the same expression according to the X-ray photon energies used.

- For E < 50 keV, the photoelectric effect predominates when X-ray photons interact with the matter, so the attenuation coefficient is simply a function of the average atomic number and the density of it. Therefore, slight variations of Z significantly modify the attenuation coefficient, especially at low energies. When the photon energy is definitely higher the energy of electrons on the K shell, it is impossible for the X-ray photon to transfer all its energy to one electron, so the Compton effect is neglected at high energies;
- For E > 100 keV, the Compton effect becomes the predominant interchange process and the attenuation coefficient (μ) essentially depends on the density of the matter and has a poor energy dependence. The atomic number of the elements does not intervene. However, at such energies, μ becomes weak and the difference between the intensity of the emitted and transmitted beam is lower.

$$\frac{\mu}{\rho} = K \times \frac{Z^4}{E^3} \tag{2.171}$$

where: ρ: Density of the medium; K: Constant; Z: Atomic number of the element crossed; E: Photon energy and $E < 200$ keV.

For an X-ray tomography performed in a laboratory and equipped with a tungsten filament, the peak of emission energies is centered on the value of around 60 keV. In this case, the attenuation coefficients for different materials having close

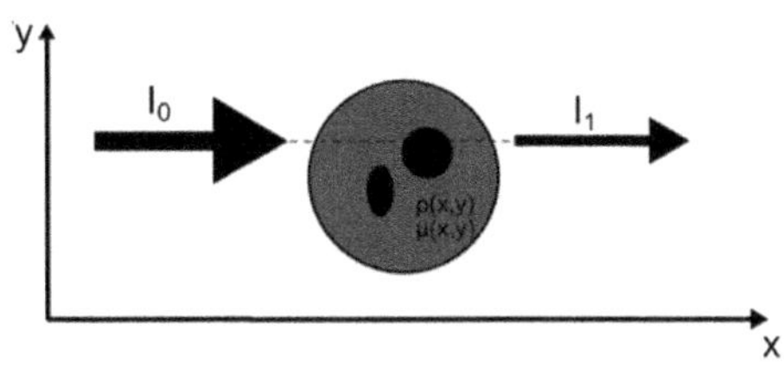

Fig. 2.232 X-ray transmission in matter. Reproduced with permission from Hauss [361]

values of Z are similar, as indicated by Eq. (2.171). As an example, alumina (Al_2O_3) and silicon nitride (Si_3N_4) have attenuation coefficients (μ) of 0.94 and 0.90 cm^{-1} respectively for an X-ray radiation energy of 60 keV.

Inverse Radon Transform

The Radon transformation is the mathematical tool behind X-ray tomography. Concretely, the Radon transform of a 2D function is all of its projections (Eq. 2.172) on a rotating plane (Fig. 2.233a).

In X-ray tomography, the 2D function corresponds to the values of the attenuation coefficient of slices of the object and its Radon transformation, to all the values recorded on the width of the detector for θ ranging from 0 to 2π. From a sufficient number of projections, it is possible to determine all the values of the attenuation coefficients of the object by inverse Radon transformation. The graphical representation of the Radon transformation is given by the sinogram in Fig. 2.233 for which the gray values are displayed for each slice according to a (U, θ) diagram, U being the axis of the detector width and θ the projection angle. In such a space, the image of a point of the object is a sinusoid while the image of a point of the detector is a straight line. This representation is particularly used for the detection of the dead or saturated pixels of the detector, responsible for ring artifact defects.

$$P_\theta(U) = \iint_{-\infty}^{+\infty} f(x,y)x\delta(x \times \cos(\theta) + y \times \sin(\theta) - U)dxdy \qquad (2.172)$$

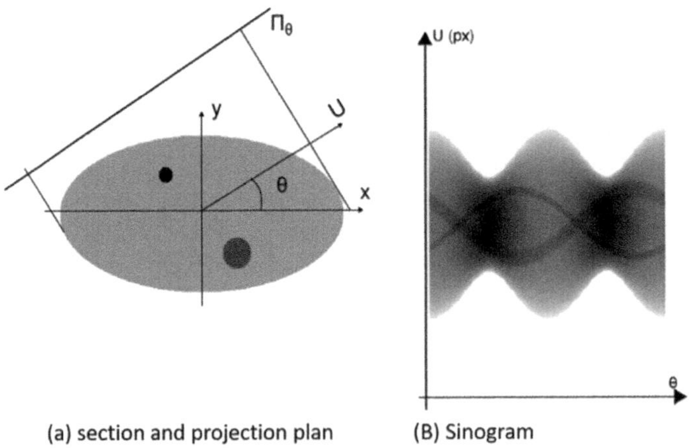

Fig. 2.233 Section reconstructed and its Radon transformation presented as a sonogram. Reproduced with permission from Grégory HAUSS [361]

where: P_θ: Projection on a detector line of a sample section; f: $2D$ function of the attenuation coefficients in a slice of the sample; δ: Distribution of Dirac.

2.9.5.3 Device of X-Ray Microtomography

Conventional Experimental Device

Different types of devices can be used to perform X-ray tomography experiments. They all include a source (synchrotron in our case), a monochromator, a turntable and an X-ray detector (Fig. 2.234).

The three elements of this chain determine the final resolution of the image. The peculiarities of the source guarantee this resolution. The mechanics of the rotation plate must be precise enough to not degrade this resolution, so the precession movement (parasitic translations less than a micrometer during the whole rotation) has to be corrected to the maximum. The X-ray detector, finally, is a particular element for which the technology has progressed a lot recently. The technical specificities of monochromators and detectors used in synchrotron X-ray tomography will be further detailed.

Monochromator

The high flux inside the synchrotron emission cone allows the selection of a precise spectral band while keeping short exposure times. Monochromaticity makes it possible to avoid artifacts in tomographic reconstruction, but also to make the latter finer since it is possible to reconstruct quantitatively the linear attenuation coefficient μ determined by the local composition of the sample. For example, this possibility has made it possible to characterize the degree of mineralization of the

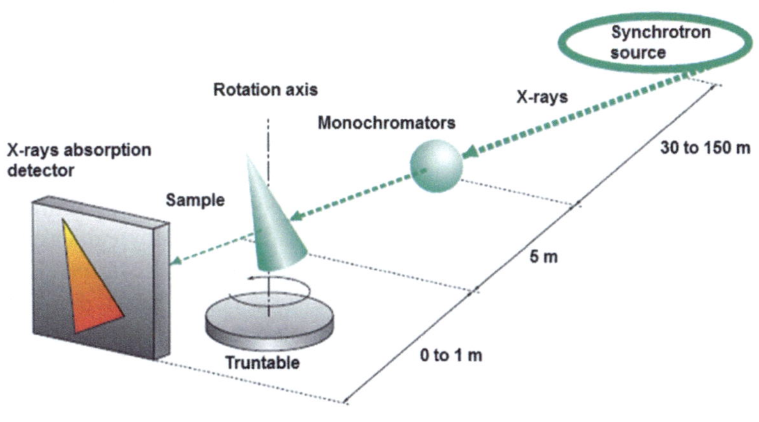

Fig. 2.234 Schematic of the classical experimental device in a synchrotron X-ray tomography experiment. Reproduced with permission from Techniques de l'Ingénieur [362]

human skeleton and dentition at high spatial resolution. The value of the energy can be chosen according to the sample for example to optimize the signal-to-noise ratio. It is possible to reveal a specific atomic element by selecting an energy close to an absorption threshold linked to an excitation of the atoms of this element. Two types of monochromators are used for X-ray tomography. The devices based on the Bragg diffraction by almost perfect silicon monocrystals give a very fine spectral width ($\frac{\Delta E}{E} \sim 10^{-4}$). It is now possible to grow periodic multilayer structures which diffractably select a controlled spectral band wider than that of given by natural crystals ($\frac{\Delta E}{E} \sim 10^{-2}$). The resulting flux gain can be crucial for very high spatial resolution tomography. Two aspects are to be addressed on monochromatic devices: an efficient cooling in order to support the thermal load of synchrotron radiation [363] and the conservation of the properties of the beam in terms of homogeneity and coherence [364].

Detector

Using a parallel X-ray beam, the image is only magnified by the detector. It is therefore understandable that the performances of the latter are crucial in the definition of the spatial resolution of the system. Since a very large number of radioscopies have to be acquired, of the order of a thousand, for the reconstruction of a single 3D image, the recording time being a second key parameter. The fabrication of X-ray lenses is constrained by the small difference in refractive index compared to 1. So, there is still some progress to be achieved in this area [365]. The use is not yet common. The principle of two-dimensional high resolution detectors consists in converting X-ray radiations into visible light by a luminescent material. The image formed is enlarged and projected by an optical system on a charge coupled device (CCD) camera. The radiation transformation to visible light enables the use of different types of high quality lenses that, in turn, limits the spatial resolution to 500 nm at best due to the diffraction phenomenon. The fluorescent screen is a thin layer deposited by liquid phase epitaxy on a transparent crystal that contains absorbent heavy elements along with a doping element that promotes visible light re-emission. The lenses used may be conventional microscope objectives or particularly robust mirror lenses in environments with very harsh X-rays. The CCD camera must simultaneously provide good dynamics, low noise and the shortest possible read time.

Note 1: A specific FRELON camera (fast readout low noise) fulfilling these conditions has been developed at the ESRF [366]. It is cooled by Peltier effect and has well-adapted characteristics: 1024 × 1024 or 2048 × 2048 pixels, 14-bit dynamic, 65 ms reading time, 3 electron/s electronic noise. This type of detector makes it possible to routinely obtain a spatial resolution of the order of one micron [367]. The ESRF also has a camera (DALSA) with a lower dynamic range (12 bits). It is characterized by a reading time of 14.7 ms and currently allows faster projections.

The massive development of complementary metal oxide semiconductor (CMOS) cameras is a very interesting alternative to CCD cameras when acquisition time becomes paramount. Indeed, for these cameras the reading time is reduced by 2 orders of magnitude, the dynamics (signal-to-noise ratio) being strongly degraded in return. However, only the use of a CMOS camera makes it possible to perform the acquisition of a 3D image in less than a few seconds. The evolution of acquisition times and resolutions in synchrotron X-ray tomography as a function of the resolution could be found in the literature [368]. Fast acquisitions (<10 s) are usually obtained with CMOS cameras.

Architecture of an Industrial X-ray Tomograph (TIX)

The tomographic analysis is done in two steps (Fig. 2.235):

- The acquisition during which the object is X-rayed under a multitude of angles of view. The number of X-rays varies between 1000 and 3000.
- The reconstruction where X-rays are processed via a reconstruction algorithm based on the inverse Radon transformation.

The reconstructed volume is a three-dimensional cartography of the attenuation coefficients of the studied object.

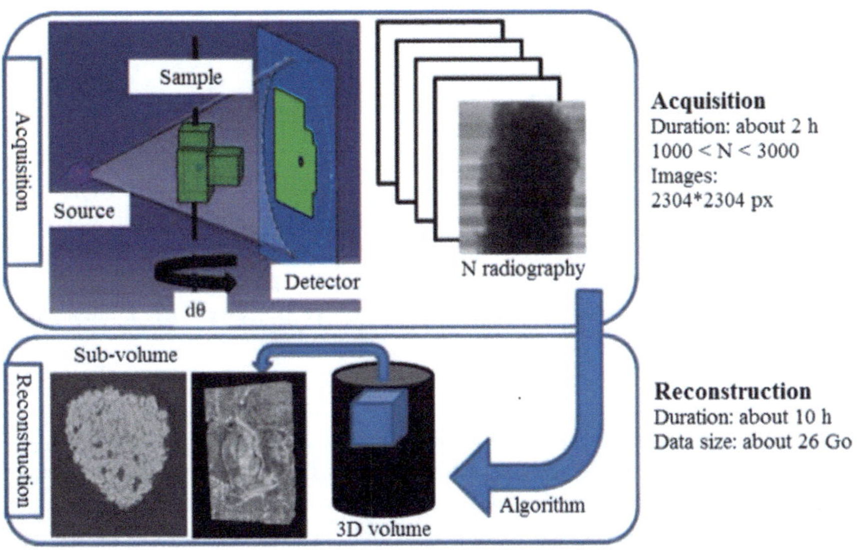

Fig. 2.235 Steps of the tomography test: acquisition and reconstruction of image. Reproduced with permission from Hauss [361]

Acquisition System

The role of the acquisition system is to capture and pre-process the attenuation measurements required to rebuild the section. It is composed of:

- The X-ray generator;
- The mechanical bench for positioning and scanning the object;
- The detection assembly associated with the scanning electronics.

A management computer allows to execute the animation as well as to control and dialogue programs with the user. The mechanical bench ensures the scanning of the workpiece in the measuring beam and the z-positioning of the cutting plane. The positioning of the support of the object requires great precision along the axis of translation where the measurement takes place. In particular, the center of rotation must be perfectly known. Its adjustment goes through a careful alignment of the X-ray beam. It is the mechanical bench that appears as the most variable element of a tomograph. Indeed, it has to endure the specific constraints of the control to carry out. Its parameters (shelf diameter, stroke height, weight) depend directly on the type and nature of the parts to be examined.

Image Reconstruction and Analysis System

The image reconstruction and analysis system implements the reconstruction algorithm with a sufficient speed and precision that is to say, in real time (display on the screen in a few seconds with a minimum resolution of 512 × 512) up to 2048 × 2048 in the medical field. It is constituted of:

- A medium power computer;
- Standard peripherals (console, optical disk, tape magnetic etc.);
- More specific unit dedicated to the reconstruction of images;
- An image display unit and/or image analysis unit equipped with a color terminal.

The use of color on the terminals is possible, but can be a limitation for the interpretation, in the case of the appreciation of discrete phenomena such as defects. Indeed, it should be pointed out that the human eye perceives better contrasts of black and white pallets.

Note 2: The software solutions tend to replace computer associations—specialized maps of yesteryear to perform these reconstructions. They allow access to completely acceptable reconstruction times.

Note 3: The extension of the use of X-ray tomography in the industrial sector, especially with the arrival on the market of new machines, and the very wide range of radiation/matter interactions, require the implementation of a reflection of specialists and users of the technique (working group created in 1996 at the French Confederation for Non Destructive Testing COFREND) in order to define its performance in terms of resolution. This work led to the creation of a CEN group (CEN/TC138/WG1) on a European scale under the direction of the AFNOR, and resulted in 2008 in the filing of four standards in X-ray tomography in which the methods for measuring the resolution initiated are validated by the COFREND group:

- EN 16016-1:2011: Non destructive testing—Radiation methods—Computed tomography—Part 1: Terminology [369];
- EN 16016-2:2011: Part 2: Principles, equipment and samples [370];
- EN 16016-3:2011: Part 3: Operation and interpretation [371];
- EN 16016-4:2011: Part 4: Qualification [372].

As example, Figure 2.236 shows images from the 3D reconstruction of ceramic composites, and highlights the "air phase" of the samples [160]. This "air phase" includes the porosity and micro-cracks that could be due to the use of materials.

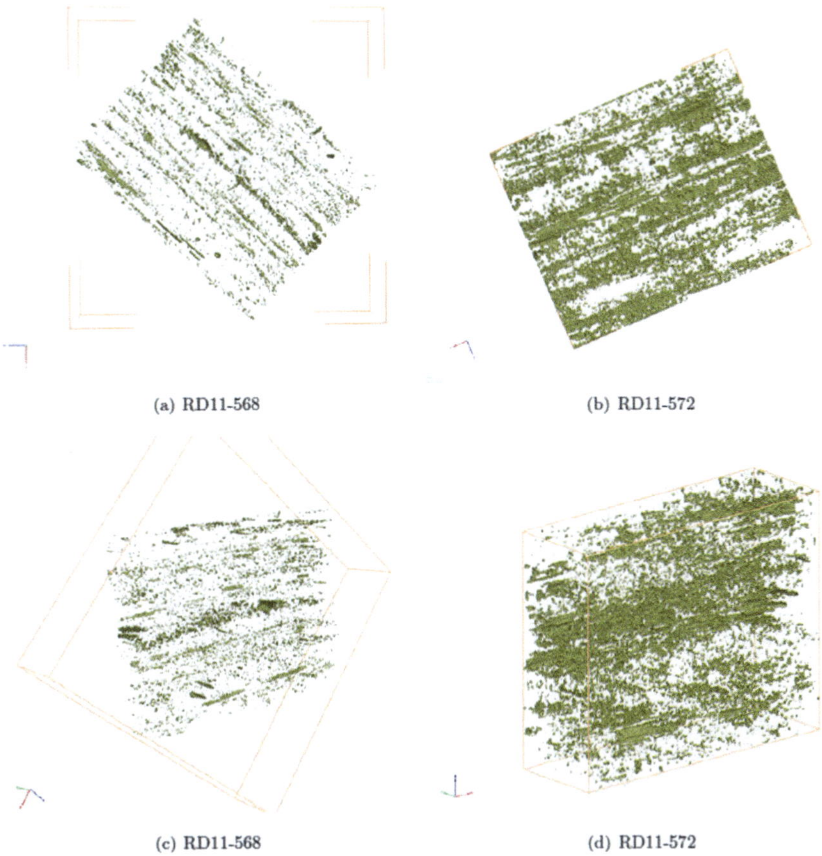

Fig. 2.236 Example of image reconstruction: Observation of 3D reconstruction images obtained by X-ray tomography of ceramic composites: RD11-568 and RD11-572. Reproduced with permission from Farrugia [160]

The calculation of the aspect ratio is used to classify the items in one or other of these categories. It is calculated from the total area and the total volume of each item. If it is less than or equal to 10, the elements are classified as "pores". If the ratio is greater than 10, the elements are considered as "cracks".

2.9.5.4 Applications

Thermal Test

In order to follow structural transformations such as sintering or wetting at grain boundaries, ovens compatible with X-ray tomography have been developed. Different constraints specific to X-ray micro-tomography have to be taken into account. A low and homogeneous absorption of the X-ray beam during all the 180° rotation can be obtained with a cylindrical chamber made of pure quartz with fine walls. The vibration level must be as low as possible in order to reach resolution on the micrometer level. Finally, the device must be compact so as to be compatible with the X-ray micro-tomography instrument. Such an oven has been built at the ESRF, which covers a temperature range of 300–1300 K and can operate under vacuum or controlled atmosphere [373]. Another solution exists, where the sample rotates inside a fixed oven [374].

Mechanical Test

Applying the same type of concepts in terms of mechanical tests is not trivial because the amount of biasing machines necessarily masks the sample over a certain angular range. X-ray radiations are no longer available in the whole range of angles which disrupts the reconstruction by filtered back projection. To overcome this drawback, in situ mechanical stressing machines developed and used in X-ray tomography also include a cylindrical tube but, this time, in polymer, which transmits the load of the upper jaw to the bottom jaw of the machine. This tube absorbs little X-ray radiations and above all it absorbs them in the same way over the full angular range. This principle has been applied to design tensile and/or compression machines and more recently fatigue machines [362].

2.9.5.5 Illustration

Comparison Between Observation Using SEM and X-Ray Microtomography

A comparison between SEM observation and X-ray microtomography was reported in [368]. 2D and 3D images were obtained with of 2124 Al/25% SiC composite particles and Ti64/5% TiB wire metal matrix composite. Compared to back scattering electron images collected in the scanning electron microscope (SEM), nano-tomography images were of low resolution (100 nm) despite being at the current limits of X-ray tomography, however the latter allows the 3D analysis of the spatial relationships and connection between the Fe-Cu intermetallics (orange solids) and the SiC reinforcement (grey solids) in the 2124 Al/25% SiC composite. Concerning the Ti64/5% TiB composite, the 3D reconstruction shows the distribution of TiB needles (green solids) and irregularly shaped Ti grains (yellow solids). All these abovementioned details observable in 3D are not so apparent on the 2D images recorded by SEM.

Observation of Pore Size of Materials

In order to obtain a three-dimensional representation, a series of radiographs is acquired from different angles by rotating the sample and each section is reconstructed using an algorithm. The three-dimensional description is obtained by stacking the slices. In this part is presented an example of observation of a ceramic material with and without the incorporation of RDF waste (Fig. 2.237). The objective of this X-ray tomography analysis is to compare the size of the pores created during the sintering in these two materials. The experimental parameters that were used for the analysis are gathered in Table 2.91.

The tomographic analysis of the samples showed that:

- For the ceramic sample: a pore class between 0.01 and 1 μm was observed. This is related to the decarbonation of the calcite between 700 and 800 °C during the firing of ceramic sample.

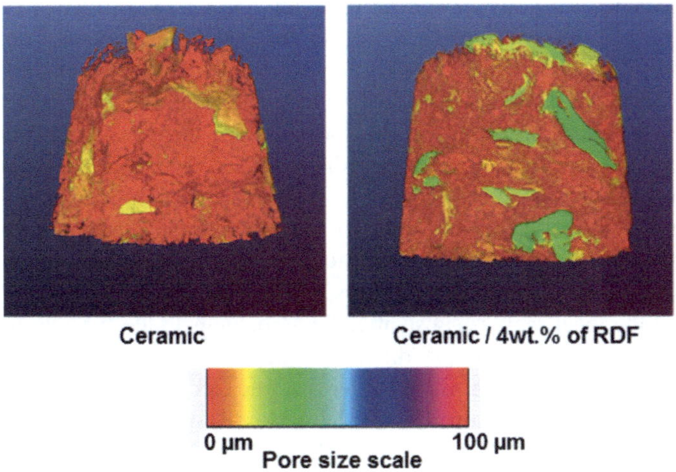

Fig. 2.237 Observation 3D of pore size of ceramic materials

Table 2.91 Experimental parameters of the X-Ray microtomography analysis

Experimental parameters	
X-ray source	ESRF ID19
Energy	35 keV
Number of projections	3500
Laying time	0.06 s
Volume of sample	1.6 × 1.6 × 1.3 mm^3

- For the sample "ceramic/4wt% RDF": in addition to the pore class of 0.01–1 μm related to the decarbonation of the material, a new class of pores has been observed, which is between 1 and 10 μm. This is associated with the decomposition of organic matter initially present in the RDF between 200 and 500 °C during the firing step.

2.9.6 Transmission Electron Microscopy

2.9.6.1 Introduction and Principle

The term "microscopy" is derived from Greek, "mikros" for "small" and "skopein" for "examine", while "transmission" refers to electrons of the incident beam to be "sorted" (through various detectors) and analyzed, which pass through a thin sample after interacting with it. Transmission electron microscopy (TEM) is increasingly used in materials science in order to observe and analyze the matter at very small dimensions, from the nanometric (<100 nm) to the atomic (<1 nm) scale. Indeed, more and more articles (about one out of two) specialized in materials science contain at least one piece of information on the (nano)structure or chemical composition of a material extracted from TEM, which demonstrates the growing interest in using such a characterization tool to locally explore the matter. Although the implementation of the various techniques (bright- and dark-field conventional imaging, selected area electron diffraction (SAED), high-resolution transmission electron microscopy (HRTEM), energy dispersive X-ray in scanning transmission electron microscopy mode (STEM-EDX), high-angle annular dark field (HAADF), electron energy loss spectroscopy (EELS)) associated with TEM is time-consuming (elapsed time from sample preparation to implementation of the analysis procedure), compared to other analytical tools, the number of information that can be extracted at the local scale is substantial. TEM belongs to the far-field microscopy since the distances between the different functions (electron source—sample—detector) of the apparatus are much longer than the electron wavelengths used. The principle of TEM was developed in 1931 by Max Knoll and Ernst Ruska, both of them having received the Nobel Prize in Physics in 1986 for this invention. TEM uses a high-energy electron beam that passes through the specimen to be characterized, under the condition that it is thin enough (<100 nm) while interacting with it. The different phenomena that occur when an incident electron beam interacts with an ultrathin specimen are schematically displayed in Fig. 2.238 [375–378].

The characteristic wavelength of an electron accelerated in the column of a TEM being much lower than an Ångstrøm, the final resolving power of an electron microscope (e.g. 1.25 pm at 200 kV and 2.5 pm at 60 kV) is such that it would enable to observe "theoretically" the matter at dimensions (typically half the wavelength of the illuminance wave according to the Abbe's law stated in 1893) even smaller than that of the atom (from 1 to 3 pm). Under these conditions, the limit magnification of a TEM would be one hundred million times. In practice, the magnification in a TEM cannot exceed one million because many defects (spherical chromatic and diffraction aberrations, astigmatism, etc.) mainly limit objective lens performances, limiting the microscope resolution and therefore the formation of images. Hence the resolution limit is not as small as the value of the reduced wavelength could make it expected. In other words, the resolution will always be inferior to the resolving power and the value of the first will depend on the characteristics of the objective lens and experimental conditions of observation.

Electron scattering is a fundamental physical phenomenon which is involved in the formation of the contrast (=image); in other words, it would be impossible to record images of a sample if no electron scattering would have occurred while incident

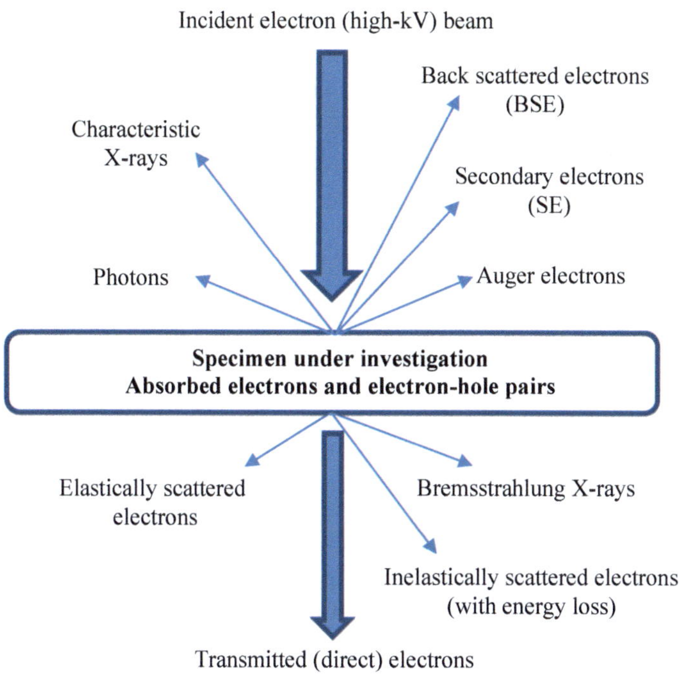

Fig. 2.238 Schematic representation of the different types of signals emerging subsequent to the interaction of an incident electron beam with the specimen under investigation

electrons interact with it. The electrons that hit the sample are named "incident electron beam" whereas those deviated by the sample are so-called "scattered or diffracted electron beam". The electrons that pass through the thin sample without deflection, called "direct electron beam" or "transmitted beam", are very useful as they participate in the TEM imaging process. Obviously, the fraction of deviated or undeflected electrons passing through the specimen will depend on its thickness, its density and the atomic number of the various elements of which it is composed.

The TEM performances mainly depend on the quality of the electron source (coherence, stability, energy dispersion, etc.), and field emission gun technology is preferred (cold-FEG, or Schottky). Performances also depend on the quality of the objective lens; its focal distance is constant and the magnification can be modulated only by playing on the focal distance of the projector lenses. The four main parts (Fig. 2.239) of a TEM instrument are:

1. An illumination system constituted of an electron gun along with condenser lenses that focus the incident electrons on the sample;
2. The sample stage located in between this first and the imaging system accommodating the specimen;
3. The imaging system combining lenses which provides a magnified image. It comprises: (i) an objective lens that focalizes the electron beam, after it has passed through the specimen, to form an intermediate image of the latter, and (ii) one or more projector lenses that magnify a portion of the intermediate image to form the final image;
4. The recording system that converts the radiation in a permanent image on photographic (fluorescent screen) or digital (cameras) supports.

A schematic set-up of a cross-section of the anatomy of a TEM is depicted in Fig. 2.239, in which an incident electron beam of high energy (100–400 keV) and intensity passes through a condenser lens leading to a parallel electron radiation that hits the sample under investigation.

In TEM, only the electrons transmitted by the sample are analyzed along with the emitted X-rays whose energies are characteristic of the ionized elements constituting it. Indeed, with respect to this final point, the measurement of the frequency distribution of the radiation (spectroscopies) after interaction with the sample makes it possible to determine the specific energy levels of the latter and consequently to identify its chemical composition or the electronic properties. The variety of signals obtained by the "electron-matter" interaction confers on electron microscopes the possibilities of a real "nano-laboratory" making it an incomparable instrument for the accurate study of materials. The interaction effects between the incident electron beam and the specimen are further exploited to provide the relevant information (crystallite and domain size, internal organization and structure of the matter, elementary composition, crystallographic structure, study of chemical bonds and electronic properties in relation with the structure, visualization of electric and magnetic fields etc.) sought by the experimenter. More detailed information about the principles of TEM techniques can be found elsewhere [376, 378–380].

Fig. 2.239 Schematic diagram of a transmission electron microscope (TEM)

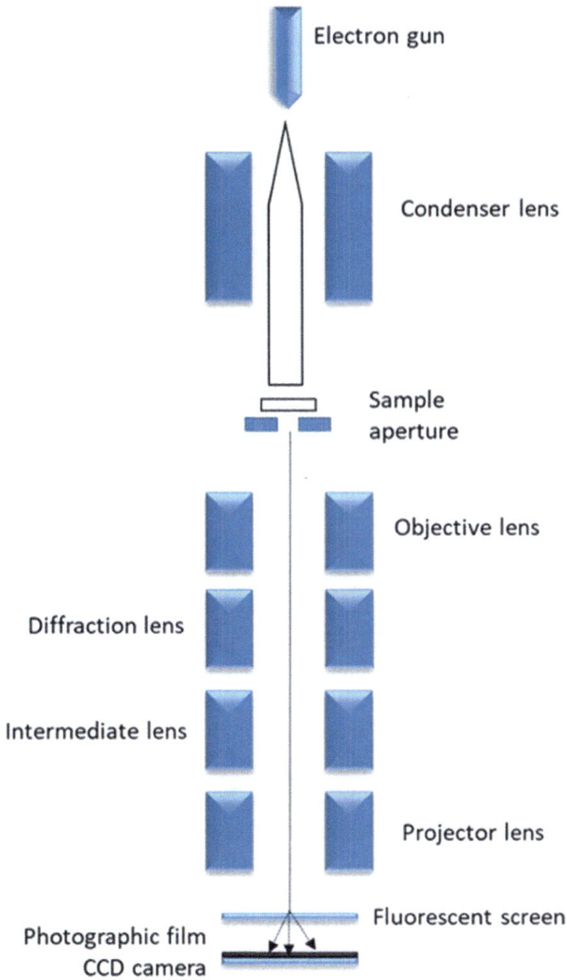

Transmission electron microscopes from different commercial suppliers are depicted in Fig. 2.240 [379–381].

2.9.6.2 Preparation of the Sample

General Points

The sample to be observed and analyzed by TEM must meet a certain number of criteria [383, 384]. On the one hand, the specimen must be as thin as possible (ca. <100 nm) so that the electrons can pass through it (electron transparency) and be

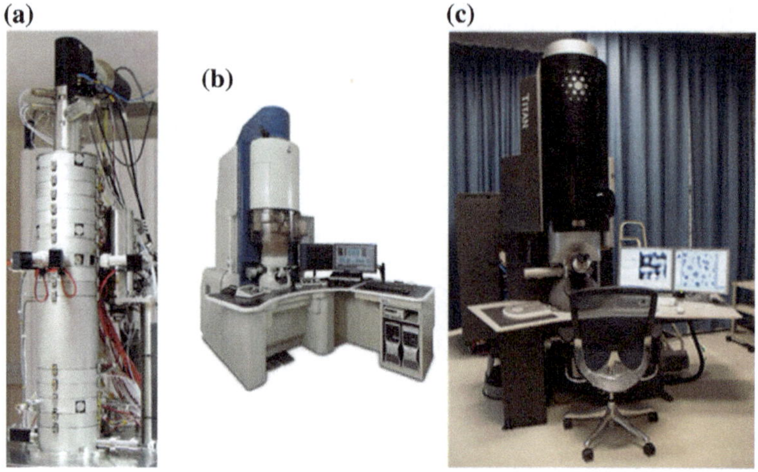

Fig. 2.240 Some of the commercial TEMs in the market: a Nion UltraSTEM™ 100, a high-performance dedicated scanning-transmission electron microscope (STEM) with many unique features; reproduced with permission from NionCompany [381]. b JEOL 200 keV TEM/STEM; reproduced with permission from JEOL [382]. c FEI Titan 80-300 Analytical TEM/STEM (Credit: A. Myers/NIST) [379]

collected. The interaction effects between the electrons and the sample can give rise to an image (2D projection of the specimen), depending on the mode used, the contrast of which will depend on its composition, density and thickness. Further thinning of the specimen might be necessary, depending on its mean atomic number of the specimen: the higher the atomic number, the higher the electron scattering factor, therefore the specimen needs to be thinner. After polishing of the specimen, its final thinning can be carried out using chemical or mechanical techniques to provide a better electron transparency. The preparation of such thin samples free of artifacts in which there is no alteration of the internal structure is not easy, making it a crucial step in the TEM analytical technique. On the other hand, the sample should not be sensitive to electron irradiation and be stable under high vacuum to not be degraded during the observation and analyses. According to the nature of the sample, either organic or inorganic, the preparation methods differ but in any case, the size of the sample will be limited in surface (1×1 mm^2). If the sample is in the form of powder, it can be directly deposited on a mesh grid covered with film (plain or holey carbon, silicon nitride, polymer). Hard materials can be cut using a diamond knife or thinned down under ionic bombardment depending on their rigidity and hardness. Rubber-like materials can be cold cut in cryo-ultramicrotomy (−15 to −150 °C). For soft materials, the preparation methods are close to that used for biological samples; otherwise, they can be cut and observed under cold conditions. The biological samples must be embedded in a resin (epoxy, methacrylate, lowicryl, etc.) before slicing; for such samples, high-Z number contrasting agents (uranyl acetate, phosphotungstic acid, lanthanides, etc.) are used.

Preparation of Biomass and Biowaste Sample for TEM Analysis

An example is given for sample preparation to image the xylem and sclerenchyma cells of pretreated (dilute acid) corn stover stalks [385]. Corn stover tissue is softened using a microwave processing technique for electron microscopy [386]. Being soft materials, the preparation method get closer to that used for biological samples, and it comprises different steps among which initial and post-fixation, dehydration, embedding and resin polymerization, cutting and observation, detailed below:

- Initial (chemical) fixation: samples are fixed two times 6 min in a 0.1 M sodium cacodylate buffer (EMS, Hatfield, PS) containing 2.5% gluteraldehyde at pH 7.3 and under vacuum;
- Dehydration: the samples are then dehydrated using a series of graded ethanol for 1 min each (15, 30, 60, 90%, two times in 100% ethanol);
- Embedding: dehydrated samples are impregnated with a Ladd Research (LR) white resin in the microwave under vacuum and left for incubation overnight at ambient temperature in increasing concentrations of resin (15, 30, 60, 90%, three times in 100% resin, diluted in ethanol);
- Resin polymerization: the samples are then transferred to gelatin capsules and the resin polymerization was achieved by heating to 60 °C overnight;
- The LR white resin embedded specimens are cut to 60 nm size with a Diatome diamond knife using an EM UTC ultramicrotome from LEICA (Wetzlar, Germany).

The cut samples from the previous step are transferred on Formvar coated copper or nickel slot grids (SPI Supplies, West Chester, PA). These grids are then post-stained for 4 min with 2% aqueous uranyl acetate, followed by a staining with Reynolds lead citrate for 2 min. For selective staining of lignin fraction from corn stover stalks, grids are treated with 1% $KMnO_4$ for 10 min.

Preparation of Sample for TEM from Char, Post-treated or not, Produced by Pyrolysis

Samples for TEM observation and analyses are prepared by ultramicrotomy using a LEICA Ultracut R Ultramicrotome. Char fragments of approximately 2–3 mm^3 were inserted into standard Beem capsules (Agar Scientific), having the shape of a truncated pyramid at the top, and embedded in an epoxy resin in order to produce blocks well-suited for cutting operation at the pyramidal top. The resin polymerization was carried out at 55 °C for 4 h. Such a resin is well-known to provide appropriate results in terms of cutting quality for graphenic (sp^2 carbon) materials. After being removed from the mould, a fraction of the surface of the sample is found being tangent to the truncated vertex of the pyramid, of square section (1 mm × 1 mm). The block top was perfectly flatten using a diamond trimming tool 45° (from DiATOME and provided by LFG Distribution). Thin slices of 50–100 nm thick were then cut using a 35° ultra diamond knife (from DiATOME and provided by LFG Distribution) with a cutting speed of 1 mm/s and a clearance angle of 6°. The undamaged rectangular sections floating on the surface of the water bath were gently collected with a perfect loop and transferred onto 200 square mesh cupper/palladium grids (from Agar Scientific). The ultrathin slices have a size of about 700 μm × 500 μm × 0.1 μm. The Cu/Pd grid is placed in the cavity of a double tilt sample holder allowing rotations around the X-Y axes.

2.9.6.3 Preliminary Procedure for TEM Characterization

After sample preparation, the procedure to obtain bright-field (BF) and dark-field (DF) images is schemed in Fig. 2.241.

Introduction of the Sample Holder Into the Specimen Stage

First of all, the sample is placed in the cavity of the sample holder devoted to this purpose (grid TEM + O-ring) with the help of a light optical microscope (LOM) to do it right. It is checked that the sample holder has no damages or foreign substances on its surface that could degrade the quality of the vacuum in the column of the microscope when it is inserted into the specimen stage. Available sample holders are single- or double-tilt. Do not forget that magnetic sample should not be analyzed with TEM equipment using single-tilt sample holder. The specimen moving along X, Y, Z, and α is supported by the goniometer. A double-tilt sample holder will additionally control the rotation of the sample around the β angle (Fig. 2.242). After controlling that the stage is at the starting position, the sample holder is inserted in the specimen stage; a particular attention should be paid to maintain horizontally the sample holder and go straight while introducing or removing it from the specimen stage in order to avoid suddenly degrading the vacuum in the column. When the sample holder is properly inserted, its type is automatically recognized as new hardware and appears in the pre-recorded list of sample holders as soon as it is connected to the microscope with a cable. Indeed, in most of the case, TEM instruments are equipped with "plug-and-play" accessories that are automatically recognized by the microscope control software. The pumping system could now be started for about 5 min before opening the electromagnetic valve that separates the electron gun from the column.

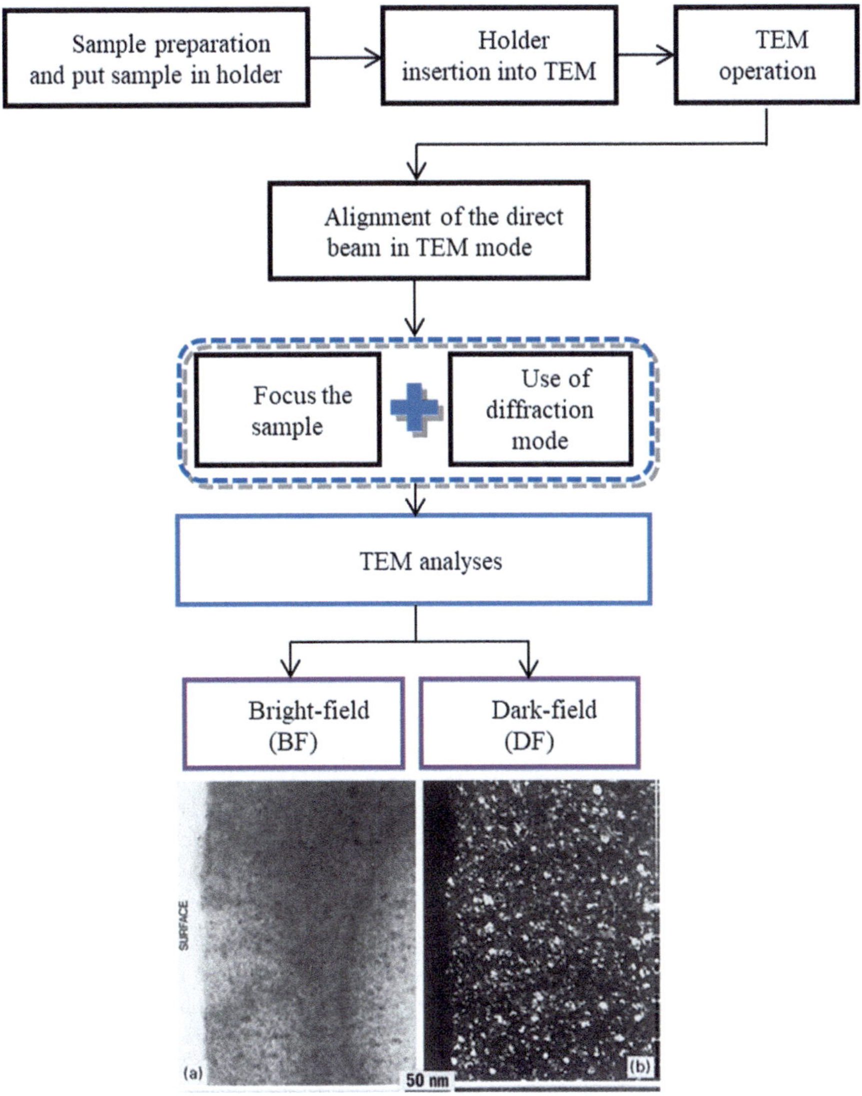

Fig. 2.241 TEM imaging procedures. The images show the same area of a portion of the surface and sub-surface of an experimental ceramic fiber seen in cross-section. Such a fiber was studied for potential use in composite materials for the European Space Shuttle Hermès in the nineties. a Contrasted BF image, b DF image mostly built from beams diffracted by nanosized SiC crystals. *Photo credit* F. Laanani, M. Monthioux (CEMES—CNRS, Toulouse)

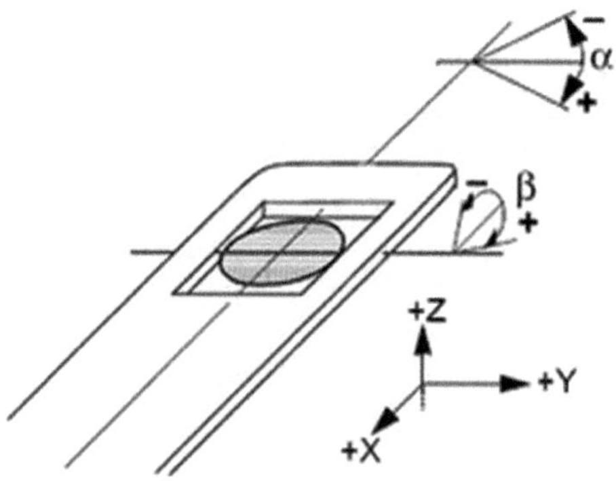

Fig. 2.242 Scheme of a double-tilt sample holder and coordinates of the goniometer (Image courtesy of Thermo Fisher Scientific) [387]

- TEM operation

First parameters should be checked: acceleration voltage, FEG current and the vacuum level in the column. The column valve could be then opened and the beam should appear at the viscinity of the middle of the screen.

- TEM alignment

The general steps to look for in the TEM alignment procedure are the condenser lens (aperture and astigmatism), and the beam alignment. To get a proper alignment (with an appropriate intensity level), the electron gun should be adjusted by checking its tilt and shift, the eucentric focus, the rotation centers and the astigmatism of the objective lens. For speeding up the procedure, the TEM alignment could directly be done from a file by loading a previous set of optical conditions preregistered.

Focus the Sample

The image is "in-focus" with a high-intensity electron potential projection (objective lens = 91.8%) once the current in the objective lens (which makes the focal distance vary) is adjusted so that the image of the specimen is formed on the image plane (Gauss plane) of the objective lens. Defocus is measured in units of length and is usually of the order of a few dozen nanometers as illustrated elsewhere [388]. There are specific focus values depending on the operating mode. For conventional imaging, the image should be registered with a slight under-focus (objective lens < 91.8%). For high resolution imaging, the spatial frequencies depend on the specific focus conditions. At Scherzer defocus, the number of spatial frequencies will be maximized.

TEM Analysis

Depending on the information to be extracted from the TEM investigation, different operating modes can be selected, and a specific procedure should be applied. In the next section, some of the TEM modes are described.

2.9.6.4 TEM Operating Modes

Since Transmission Electron Microscopy (TEM) allows observing the local structure of materials, at a scale not accessible by other characterization tools, such as photonic microscopy (LOM) or scanning electron microscopy (SEM), it is a suitable technique able to highlight the modification of carbon structures subsequent to chemical treatment from the nanoscale to the atomic scale. With regard to this multitude of signals (radiative or not) emitted by the thin specimen, following the electron-matter interaction, as well as the anatomical structure of the microscope, consisting of the combination of lenses and diaphragms, different operating modes are possible depending on the configuration chosen and the type of radiation analyzed. These modes are detailed below. As an example, Fig. 2.243 indicates the type of electron beam that participates to the formation of image contrast for some imaging modes. As the wavelength of accelerated electrons is about one hundred times smaller than the spacing between atoms in a solid, the atoms act as a diffracting lattice for the electrons which are scattered. Some fraction of these electrons will be scattered to particular angles, imposed by the crystal structure of the sample, while the others pass through the sample without deflection. As a result, a series of spots is formed in the back focal plane (Abbe plane) of the objective lens which is referred to as the "selected area diffraction pattern" (see Chap. 15 for detailed information), and, each spot corresponds to a specific diffraction condition. Switching from the imaging mode to the diffraction mode only requires adjusting the bottom lens currents (diffraction, intermediate, and projector lenses, see Fig. 2.239) so that either the image plane or the back focal plane of the objective lens is put on the TEM CCD camera. Thanks to the use of pre-set values, this only requires one second. These modes are detailed in Fig. 2.243.

Table 2.92 indicates specific experimental manipulations that should be done after TEM alignment to get proper images

Conventional Imaging (Bright Field and Dark Field)

The transmitted electron beam that is attenuated following the electron-matter interaction forms a two-dimensional electron projection (=image) which is subsequently magnified onto the fluorescent display screen to get an image of the specimen [378]. The spatial contrast will result from differences in thickness, density and chemical composition of the specimen. Since the electrons are highly accelerated, the image resolution is enhanced compared to that of given by SEM.

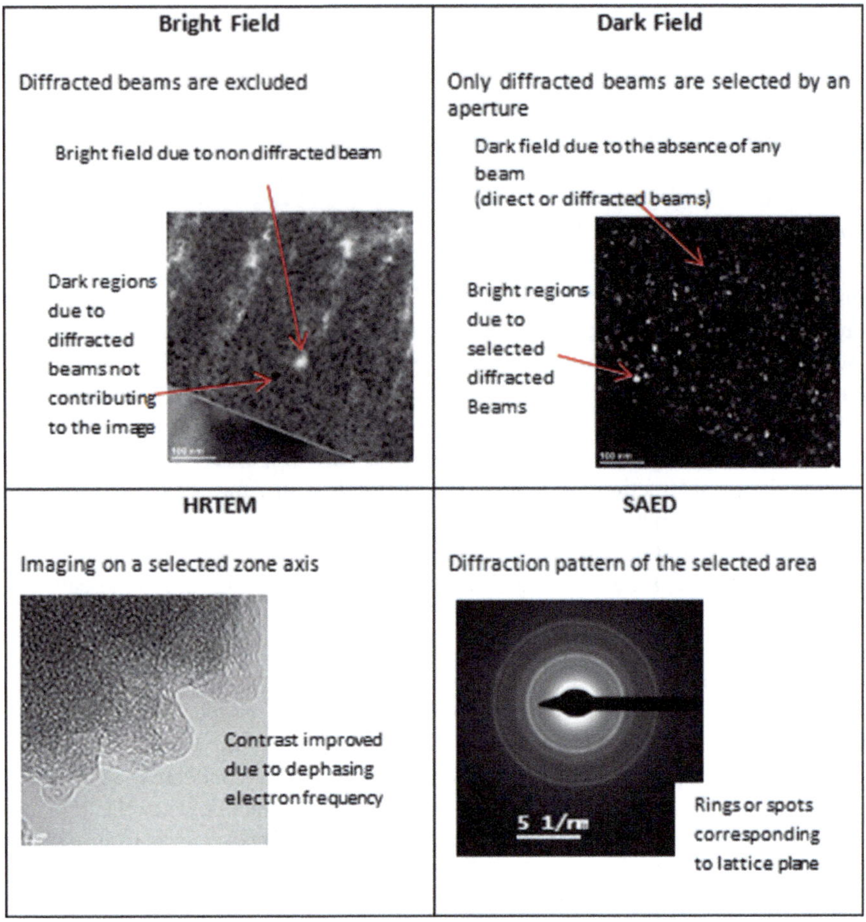

Fig. 2.243 Examples of images obtained for four TEM modes (solid residues are various biochars). BF and DF photo credit: A. Pugliara (Centre de Microcaractérisation Raimond Castaing - UMS3623, Toulouse).

Table 2.92 Specific steps of the procedure for each of the 2 (bright- and dark-field) TEM imaging modes

Bright-field TEM	Dark-field TEM
Focus sample	Focus sample and select zone axis
Select diffraction mode	Select diffraction mode and dark field mode
Center the scattered spot	Center the desired diffracted spot
Insert small objective aperture	Insert small objective aperture
Center aperture	Adjust aperture around diffraction spot
Go back to imaging mode	Go back to imaging mode
Record image with CCD camera	Record image with CCD camera

Adapted from [388]

In case the objective aperture placed in its focal plane is centered on the incident electron beam, only the non-diffracted electrons will form a contrasted bright field (BF) image on the screen. Indeed, the areas of the sample that strongly diffract the incident beam will then appear darker; in the areas where there is no sample, 100% of the beam is transmitted and the image appears clear, hence the name "bright field" or "BF". This mode also exists without using a diaphragm, so all the diffracted and transmitted electrons participate to the formation of the image. The image contrast then depends on the atomic number of the different elements constituting the sample; the heavy elements appear darker while the light elements appear brighter. At the same time, the thickest areas of the sample appear darker than the thinnest ones. The BF-TEM imaging mode is often used to observe particle morphology and to evaluate the grain or particle size distribution of a sample, depending on its agglomeration state, provided that this statistical analysis is performed on a series of randomly collected images. A dark-field (DF) image can be also obtained by centering, always in the focal plane, the objective diaphragm on an electron beam diffracted by a specific series of (hkl) planes of the sample, while the direct transmitted electron beam is blocked; thus, only parts of the sample whose diffracted electrons are selected appear bright on a dark-field image. In the absence of sample, no diffraction event occurs and then the image is dark, hence the name "dark field" or "DF". The DF-TEM imaging mode is useful to a variety of purposes, e.g. observing 1D and 2D (such as dislocations) crystal defects in crystalline samples, performing orientation mapping of crystals, spatially locating a phase with respect to another, identifying amorphous with respect to crystalline structural state, etc. Bright- and dark-field images are informative and in fact complementary to each other. The procedure to obtain bright-field (BF) and dark-field (DF) images is schemed in Fig. 2.241 and Table 2.92 indicates specific experimental manipulations that should be done after TEM alignment to get proper images.

Energy Dispersive Spectroscopy (EDS)

When the incident electron beam interacts with the sample (Fig. 2.238), the atoms constituting it are ionized and then emit X-rays radiations upon deexcitation; the energy of the emitted radiation is characteristic of the chemical nature of the sample under investigation and therefore can be used to determine its elementary (chemical) composition. Experimentally, the measurement of the frequency distribution of the radiation after interaction with the sample allows determining the specific energy levels of the latter and consequently to identify its chemical composition or electronic properties. This analytical technique associated with TEM is referred to as energy-dispersive X-ray analysis (EDX) or X-ray energy-dispersive spectroscopy (XEDS or EDS) [389]. The energy of the characteristic X-ray radiations emitted enables the constituent elements to be identified (qualitative analysis), while their intensity is proportional to the concentration (quantitative analysis), under the condition that the thickness of the slice is homogeneous. The spatial resolution, determined by the size of the electron probe and the thickness of the sample, is of

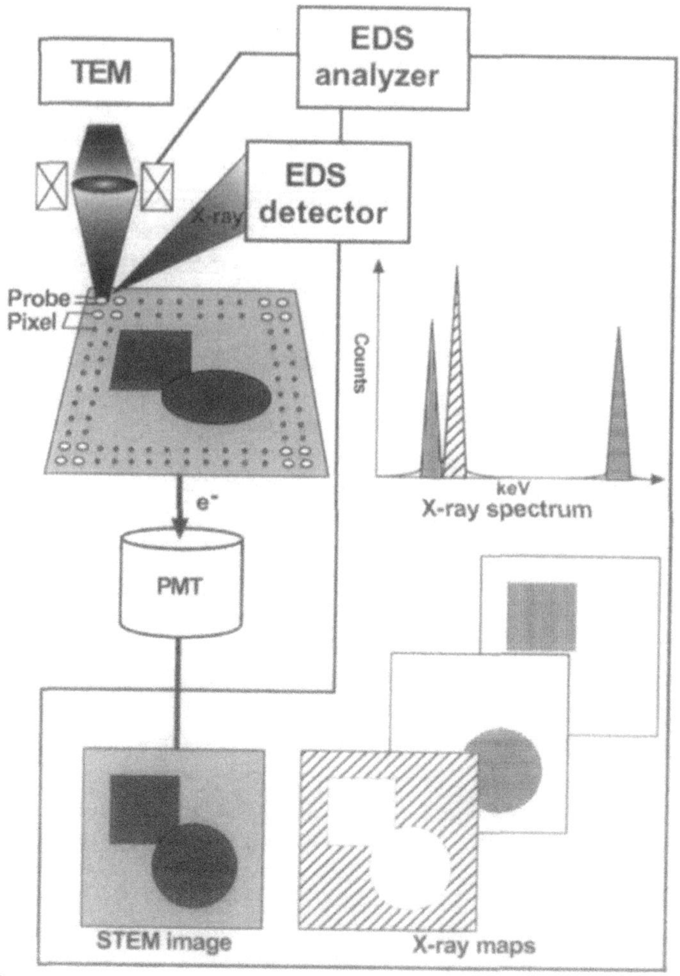

Fig. 2.244 Schematic diagram of digital EDS elemental mapping. After the energy windows of desired elements are set, the EDS computer controls the probe scan digitally. Meanwhile, a STEM image signal from a photomultiplier (PMT) and X-ray intensities of the desired elements from an X-ray analyser are monitored in accordance with the scan position. Reprinted with permission from Oxford University Press [390]

the order of few nanometers which makes composition mapping possible, provided microscope is equipped with a STEM stage, which allows scanning the beam across a selected area of the specimen. The schematic diagram of the digital EDX elemental mapping is presented in Fig. 2.244 [390]. The EDX analysis enables to detect elements with atomic numbers ranging from that of beryllium to uranium. The detection limit varies from 0.1 to a few atom percent, depending on the constitutive element and the sample matrix. In some cases, the analysis of EDX spectrum becomes difficult when peaks overlap with each other because of the presence of elements whose atomic numbers are very close in the periodic table; therefore the data treatment should be led carefully in order to avoid misinterpretation. Detailed and in-depth description of EDS analysis in the (transmission) electron microscope can be found elsewhere [390, 391], and also evolved in Sect. 2.9.4 of this chapter.

Selected Area Electron Diffraction (SAED)

As a diffraction technique, SAED is mostly applied to identify crystal structures and evaluate lattice parameters. There is no specific setting related to this mode as the diffraction pattern of the specimen automatically forms in the back focal plane (Abbe plane) of the objective lens. The operator has only to decide to image the Abbe plane on the screen, instead of the image plane (Gauss plane). This mode is essential for setting up dark-field imaging conditions as the crystalline orientation to be mapped is selected in this mode. In addition SAED is implemented in order to examine crystal defects and assess the crystallinity of samples by selecting with a diaphragm an extended area of itself to be diffracted. More detailed information and examples of SAED TEM mode are provided in Chap. 15.

High Resolution Transmission Electron Microscopy (HRTEM)

The high resolution TEM (HRTEM) mode is used to observe the matter at the atomic scale by enabling the transmitted beam and several diffracted beams to pass and form the image by interacting with each other (interference pattern). In case the microscope objective lens is aberration-corrected, the corrector is adjusted so that there is no dephasing between the diffracted beams over the largest as possible 2θ range. If it is not, a wide aperture is used and the focusing value of the objective lens is adjusted so that to limit dephasing between the diffracted beams and eliminate the other beams while enhancing the dephasing between the transmitted beam and the diffracted ones. In case atomic resolution is targeted, the image contrast is directly correlated to the projected potential of the sample. The interpretation of such a high resolution image is not direct insofar as dark or white parts can not be directly attributed to a column of atoms. Indeed, the image contrast drastically changes according to the defocalisation (which acts on the dephasing between the direct and the diffracted beams) and the sample thickness. So, the interpretation should be supported by simulation of the interference pattern in order to localize the position of atoms. HRTEM has been extensively and successfully used for characterizing crystal structures and lattice imperfections in various kinds of advanced materials at the atomic scale. This visualization mode enables crystalline organization as well as defects (grain boundaries, dislocations, etc.) to be extracted.

Both HRTEM and EDX techniques enable to image the internal structure (for e.g. interior of the cells) and analyze the chemical composition of the pretreated biomass and their derived products after pyrolysis treatment, as biochar, and consequently various elements present in the specimen can be identified. More detailed information and some examples are displayed in Chap. 15. The heterogeneous and fragile nature inherent in biomass along with the multi-scale complexity represent a challenge to select and implement the most appropriate sample preparation technique that will not alter its internal (nano)structure and chemical composition.

Scanning Transmission Electron Microscopy (STEM)

Scanning transmission electron microscope (STEM) essentially combines the techniques of both SEM and TEM to provide topographic images of the specimen surface with a resolution (<1 nm) much higher than that of the images obtained from scanning electron microscopy (SEM). In addition, the elementary composition of a specimen can be characterized using EDS or EELS. A probe aberration corrected STEM equipped with an electron energy loss spectrometer (EELS mapping) analyzes the electron having lost energy (inelastically scattered electrons in Fig. 2.238) after interaction with the specimen. An example of STEM application is provided in Chap. 15.

Electron Nanotomography

Otherwise, the external 3D nanostructure of a sample, associated with a 3D mapping of its elementary composition at the nanometric scale, can be reconstructed by electron tomography [378, 392], from successive 2D slices recorded by stepwise rotation of it placed in a dedicated rotating sample holder. TEM and associated techniques are valuable tools for analyzing the 2D and 3D cellular (nano)structure of samples obtained before and after chemical (pre-)treatment of the biomass [393]. Further, the biochar produced from pyrolysis and gasification process of the biomass could also be visualized to extract structural and morphological information.

2.9.6.5 Examples of Sample Analysis

TEM analysis is usually performed on dry samples and under vacuum (environmental TEM is an exemption), therefore samples must be dehydrated carefully before being introduced into the sample stage of the microscope in order to avoid the problems arising from the moisture contained in the biomass.

- The grid containing the as prepared specimen is introduced into the sample stage of the TEM [392];
- TEM images are acquired with a Gatan UltraScan 1000 camera (Gatan, Pleasanton, CA) on a FEI Tecnai G2 20 Twin operating at 200 kV TEM

(Hillsboro, OR) and equipped with a LaB_6 filament; the operation is driven using the Digital Micrograph standard imaging process;
- Image processing and analysis are performed using a range of software and analysis tools such as ImageJ, Adobe Photoshop, IMOD, Digital Micrograph and Chimera.

Representative TEM images recorded on raw and pretreated corn stover stalk samples are displayed in Fig. 2.245.

EDS analyses are performed on regions of interest to confirm the presence or not of specific elements under investigation. Representative TEM images with EDS analysis of reference lignin and corn stover stalk samples are shown in Fig. 2.246.

Analysis of TEM images must be performed carefully to avoid any possible misinterpretation [391]. The right sample controls and having enough experience on using TEM technique (well trained eye) are essential for avoiding misapprehension and would enable for example, to distinguish possible disruption caused to the cell walls by the mechanical milling of samples or from the chemical disruption caused by pretreatment and digestion, inter- versus intra-sample variability. Further, it is important to note that the direct observation and analysis are not always adequate for statistical analysis and often requires considerable working experience and expertise of the user.

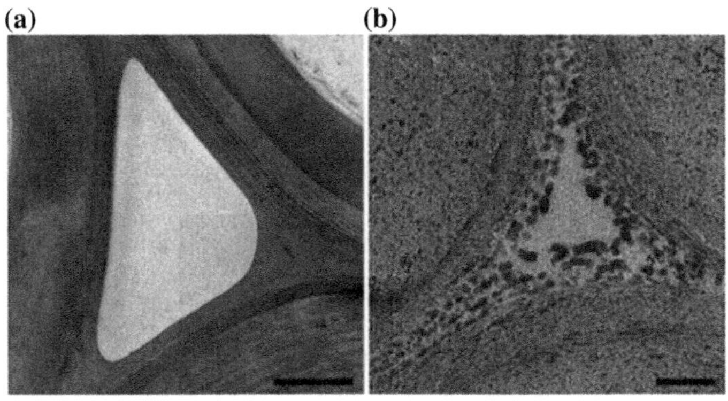

Fig. 2.245 TEM images of unpretreated (a) and dilute acid pretreated (b) corn stover stalk samples. The specimen is stained with $KMnO_4$ to emphasize the location of lignin. In the native sample a, lignin appears relatively evenly dispersed with typical lamellar concentrations. After the pretreatment process at 150 °C for 20 min, lignin can be seen as the small dark droplet-like pattern within the primary and secondary cell walls (b). Scale bars: 1 mm. Reproduced with permission from John Wiley and Sons [385]

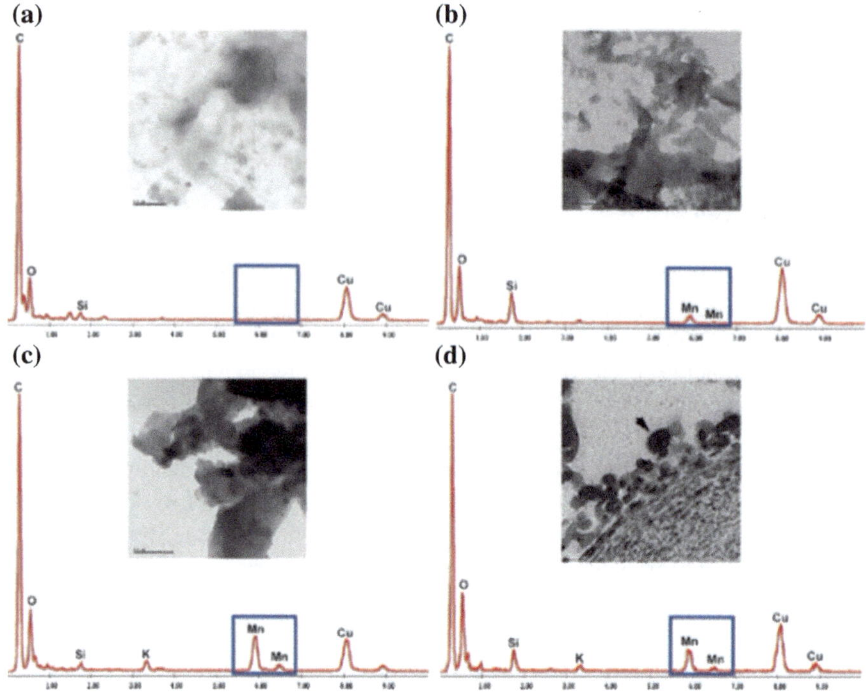

Fig. 2.246 TEM and EDS analysis of reference lignin samples and droplets on corn stover stained with: a organosolv and acid chlorite treated solids with little or no lignin content shows no $KMnO_4$ staining and no detectable Mn EDS peak, b organosolv solids with minimal lignin content show light staining by $KMnO_4$ and minimal Mn signal from EDS, c organosolv lignin shows dense $KMnO_4$ staining and a strong Mn peak, and d pretreated corn stover cell corner with dense staining droplets (arrow) and a moderate Mn peak by EDS. Reproduced with permission from John Wiley and Sons [385]

2.10 Mechanical Characterization

2.10.1 Thermomechanical Analysis

2.10.1.1 Introduction

Thermomechanical Analysis (TMA) is used to measure variations in the size of solids or powders and to characterise various materials (softening, glass transition, degradation, cross-linking) under different constraints. TMA can help with the study of films, fibres, sintering of ceramics, thermal behaviour of alloys and more generally the measurement of expansion coefficients.

Thermomechanical analysis involves monitoring the changes in dimensions of a sample over time following a defined temperature program under a fixed load and in a controlled atmosphere. Changes in dimensions are sent to a movement sensor by means of a moveable stem or probe (Fig. 2.247).

The term dilatometry is used if the applied force cannot deform the sample.

Thermomechanical analysis can monitor the following physico-chemical transitions and transformations:

TMA as the most appropriate technique:

- Sintering;
- Coefficient of expansion;
- Melting and crystallisation;
- Glass transition;
- Polymorphism (solid-solid transitions).

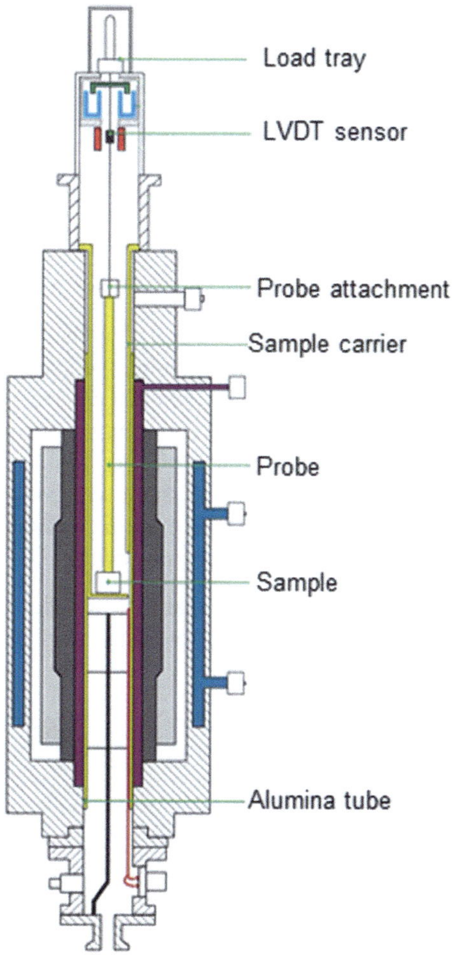

Fig. 2.247 Example of scheme of TMA measurement. Reproduced with permission from SETARAM

TMA as an alternative solution:

- Kinetics, enthalpy of reaction;
- Crosslinking, vulcanization (process parameters).

2.10.1.2 Principle of the Measurement and Apparatus

The general description of an experimental TMA apparatus includes the following components:

- An electric oven. This oven can heat the sample at a rate of between 2 and 50 °C/min over a temperature range from 20 to 1200 °C. However, these limits may vary between models and suppliers;
- A thermocouple placed near the sample;
- A Linear Variable Differential Transformer that very precisely measures the dimensional variation in the sample if its length does not exceed 20 mm. The freedom of movement that can be measured is between ±2 and ±5 mm;
- A system controlling the application of the force transmitted to the sample by the moveable measurement probe.

Figure 2.247 illustrates an example of TMA measurement, designed by SETARAM.

Figure 2.248 shows the different possible modes of TMA.

- Expansion: the probe is flat and has a large surface area and is used without load on the sample to determine the coefficient of dilatation;
- Penetration: a heavy strain is created by a small-diameter probe applying a high load, increasing the contribution of penetration with respect to expansion;
- Traction: the sample is fixed at either end for the study of films or fibres;
- Three-point flexion: system with supports in the form of knife blades;
- Expansion in volume: system consisting of a crucible and a flat-bottomed probe to study the dilatation of powders, for example;

Before starting the analysis, the user must place the probe in contact with the sample, being careful to ensure that the movement sensor is well within its

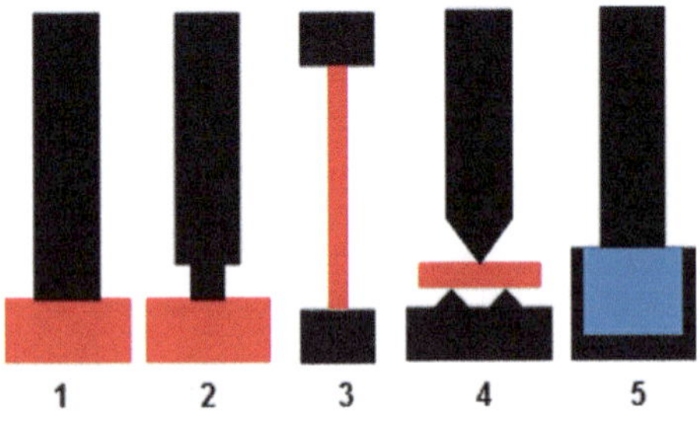

Fig. 2.248 Different probes and measurement modes for TMA; 1: Expansion; 2: Penetration; 3: Traction; 4: Three-point flexion; 5: Volume expansion

measurement range which can be between ±2 and ±5 mm depending on the apparatus (Fig. 2.247).

The sample, of initial length L_0 at initial temperature T_0, is found to have length L_T at temperature T. TMA measures the absolute expansion:

$$\Delta L = L_T - L_o \tag{2.173}$$

The variation in length (and thus in volume) depending on the variation in temperature can thus be calculated for all homogenous isotropic materials:

$$\Delta L = \alpha L_o \Delta T \tag{2.174}$$

where: $\Delta L = L_T - L_o$: variation in length (m); L_T: length at temperature T (m); L_o: initial length (m); α: coefficient of linear expansion (K^{-1}); $\Delta T = T - T_o$: variation in temperature (°K or °C).

The mean expansion coefficient α_{mean} between temperatures T_1 and T_2 is calculated as follows:

$$\alpha_{mean} = \frac{L_2 - L_1}{L_1 \times (T_2 - T_1)} \tag{2.175}$$

where: L_1: length (m) at temperature T_1; L_2: length (m) at temperature T_2.

2.10.1.3 Experimental Conditions

Atmosphere

Analyses can be performed under an oxidising (i.e. air) or inert (i.e. He, N_2, Ar) atmosphere. Helium enables the temperature to be made uniform because of its high thermal conductivity. The gas flow rate can be between 3 and 6 L/h.

Crucibles

If the sample is sufficiently solid, a crucible is unnecessary, but it can help to correctly orientate the probe. The crucibles used are often made of alumina because it is a refractory material that can withstand very high temperatures. However, its thermal expansion coefficient is non-negligible (cf. Table 2.93).

Table 2.93 Different probes and measuring methods for TMA

Material	
Silica probe	This has the lowest expansion coefficient, of the order of $0.4 \cdot 10^{-6}$/°C, i.e. 0.4 μm of expansion for 1 m of silica for a 1 °C temperature difference ($=0.4 \frac{\mu m}{m \cdot °C}$) This type of probe must be used for the determination of small variations for temperatures below 1000 °C
Alumina probe	This has an expansion coefficient of the order of 9×10^{-6}/°C which is thus non-negligible. This can be corrected by performing an analysis without sample, which is considered as blank test
Graphite probe	This has an expansion coefficient of the order of 5×10^{-6}/°C. There is great variation in the value of the expansion coefficient from one graphite probe to another, which means that each one must be individually verified
Shape (Fig. 2.248)	
Hemispherical end	Only used for analysing hard, flat samples, because it can slightly penetrate or slide on an angular shape
Flat end	Used for analysing samples with a rough surface or where the pressure load can be applied over a bigger area, which avoids penetration of the sample
Knife-blade probe	Mainly used for three-points analysis, but also to measure expansion of lengths of fibres, wire, etc., which avoids releasing the strain
Powder probe	In addition to being used to analyse powders, it is also used in the case of soft samples for better distribution of the pressure (for example, analysis of a mixture between a polymer and a metallic powder)

Fig. 2.249 Probes and crucible from the SETARAM SETSYS TMA

Probes

In conjunction with Fig. 2.248, Table 2.93 presents the different types of probe. Figure 2.249 shows a photograph of the probes and crucible from the SETARAM SETSYS TMA.

Charges

For a laboratory-scale analysis, and depending on the nature, shape and fragility of the sample, it is possible to apply a pre-load of between 20 and 50 g during the descent of the probe. This enables the load to be well-distributed on the sample. During the analysis, the load is usually set to between 5 and 20 g.

Samples

Sampling and preparation are important steps. They must be carried out using a sampling procedure well-adapted to the material being analysed.

The shape of the sample must be optimised, with two parallel faces and one surface as smooth as possible, in order to provide the probe with good stability. Silica spacing discs enable the applied load to be evenly distributed over the whole surface of the sample. They are also used to separate sticky samples and samples that break down in contact with the probe or the support. The smaller the granulometry of the sample the more it will tend to stick to the probe at high temperature.

The initial length L_o must be known (less than 1% uncertainty) and recorded at the same level of precision as the variation to be measured. It should preferably be less than or equal to 10 mm.

Measurements of the coefficient of thermal expansion (CTE) should be made under the smallest load possible in order to avoid any compression of the sample. The sample should be about 5 mm long in order to accurately measure the CTE.

Temperature

In order that the various signals can stabilise and that the oven has the time to flush itself with air or oxygen, it is important to start an analysis with an isothermal stage of 10 min using air or 20 min if another gas is used, at a temperature controlled by the apparatus (often equal to or greater than 30 °C). Standard heating rates are between 2 and 20 °C/min.

TMA Correction Without a Sample

The TMA correction graph enables the operator to avoid certain of the instrument's parameters affecting the results (expansion of oven or probe, crucible …). Since the instrument is not symmetrical, an analysis using an empty crucible or reference benchmark may be necessary in order to ensure that the actual movement of the sample is measured.

2.10.1.4 Instrument

There are several manufacturers of TMA equipment on the market including SETARAM, TA INSTRUMENTS and NETZSCH. Their models are, respectively, SETSYS Evolution TMA, Q400EM and Hyperion TMA 402.

2.10.1.5 Experimental Protocols

Below are two protocols frequently used in the laboratory with the SETARAM SETSYS TMA instrument on monolith and powder samples. Of course, other experimental protocols can be used as a function of the sample nature and the analysis purpose.

Monolithic samples (Fig. 2.250, top):

- The crucible used is in alumina;
- The load applied is 50 g;
- Isotherm for 10 min at 30 °C;
- Ramp from 30 to 1000 °C at 2 °C/min;
- Isotherm for 30 min at 1000 °C;
- Cooling from 1000 to 30 °C at 2 °C/min.

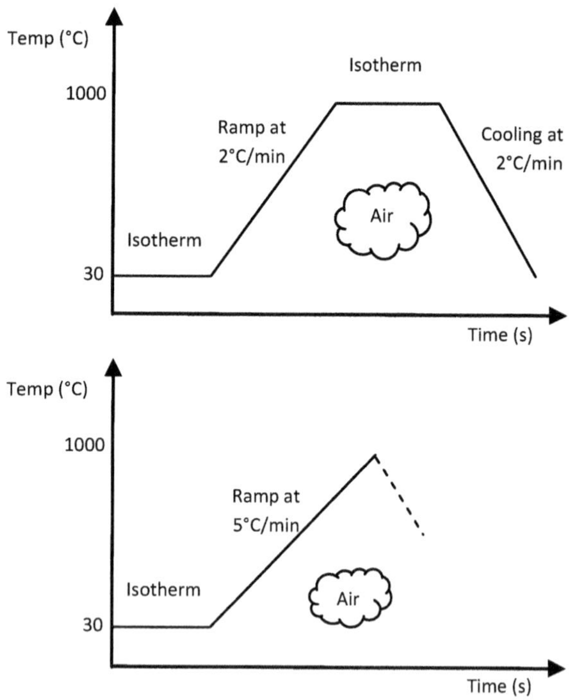

Fig. 2.250 Illustration of experimental protocols

Power samples (Fig. 2.250, bottom):

- The crucible used is in alumina;
- The load applied to the TMA is 10 g;
- Isotherm for 10 min at 30 °C;
- Ramp from 30 to 1000 °C at 5 °C/min;
- Cooling down to 30 °C.

2.10.1.6 Examples and Exploitation of the Results

TMA Analysis of a Refused Derivative Fuels (RDF)

As previously discussed, refused derivative fuels from municipal solid wastes contain mainly organic matters (cardboard, plastics, papers …) with some inorganic fractions. They can be valorized in different processes such as porous agents for producing fired clay brick with low thermal conductivity (high thermally-insulant character). This example shows TMA analyses of a RDF sample under fine particles form, at different heating rates of 2, 5 and 10 °C per min. The operating conditions are as follows:

- The crucible used is alumina.
- The load applied to the TMA is 50 g.
- The air flow is fixed at 3 L/h.

The thermal cycle applied is as follows:

- Isotherm 10 min at 30 °C.
- Ramp from 30 °C to 1000 °C at 2, 5 and 10 °C/min.
- Cooling up to 30 °C.

Figures 2.251 and 2.252 present the linear deformation and their derivatives at 2, 5 and 10 °C/min of the sample RDF in function of the temperature. Three zones of deformation were observed as a function of the heating temperature:

- A shrinkage is observed after the dehydration of the water presents in RDF. For example, in the curve of 2 °C/min (Fig. 2.251), the shrinkage related to the dehydration of RDF was 8%.
- Then, during the thermal decomposition of the organic matter and calcite present in the RDF, a very small shrinkage is recorded. Increasing the heating rate tends to shift the RDF deformations to the higher temperatures. This is due to heat transfer limitations resulting from the temperature difference between the sample and the oven. As an example for the heating rate of 10 and 5 °C/min, these physico-chemical transformations took place in the zone of 400–800 °C and 450–700 °C, respectively.
- Beyond the decarbonation of the RDF, a large shrinkage is recorded. This is associated with the densification of the material.

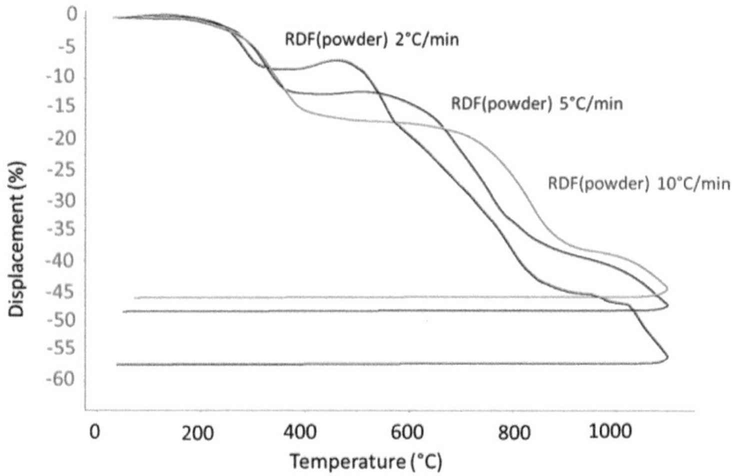

Fig. 2.251 Evolutions of deformations of RDF at 2, 5 and 10 °C/min between 30 and 1100 °C

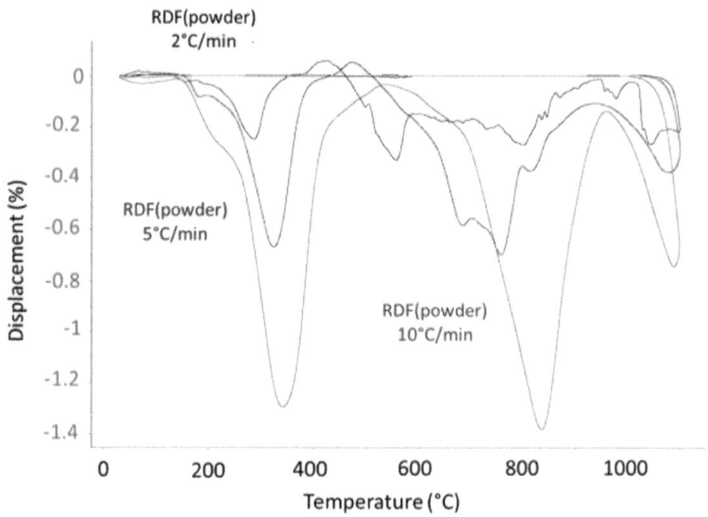

Fig. 2.252 Derivatives of deformations of RDF at 2, 5 and 10 °C/min between 30 and 1100 °C

<u>Note</u>: These dimensional variation curves can be very useful for kinetic studies of the sintering of ceramic materials or by-products [394–396].

TMA Analysis of a Monolithic Ceramic Prepared from Clay and RDF

This example describe the TMA analysis of a monolithic sample composed of clay with the incorporation of 4 wt% of a RDF. The operating conditions are as follows:

- The crucible used is alumina;
- The applied load is 50 g;
- The air flow is fixed at 3 L/h.
- The thermal cycle applied is as follows:
- Isotherm 10 min at 30 °C;
- Ramp from 30 to 1000 °C at 2 °C/min;
- Isotherm 30 min at 1000 °C;
- Cooling from 1000 to 30 °C at 2 °C/min.

Figure 2.253 shows the linear deformation of this sample (Displacement, %) as a function of the heating temperature, as well as the evolution of α_{mean}. We distinguish five zones of dimensional variations:

- Between 30 and 200 °C: a first expansion is observed, it is related to evaporation of the residual water after drying of the clay mixture and hygroscopic water. A peak is recorded in the curve of α_{mean}.
- Between 200 and 500 °C: in this zone the clay mixture undergoes microstructural rearrangements. This leads to low expansion.
- In the zone of 500–700 °C: a significant expansion is recorded in the deformation curve accompanied by a peak in the α_{mean} curve. This expansion is associated with the allotropic transformation of quartz-α to quartz-β at about 573 °C and the dehydroxylation of clay minerals (kaolinite and illite).

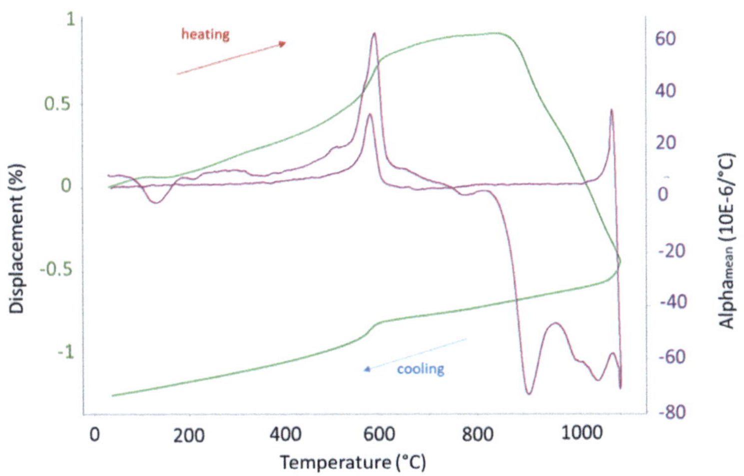

Fig. 2.253 Evolution of linear deformation and α_{mean} of the RDF-containing clay monolith during heat treatment

- Between 700 and 800 °C: the decarbonation of calcite leads to a weak expansion of the material.
- Above 800 °C: a shrinkage of 1.35% is recorded, which is associated with the densification of the material.

During cooling: the material undergoes a contraction from the passage through the quartz point to 573 °C (quartz-β to quartz-α).

2.10.2 Compressive and Tensile Strength Analysis

2.10.2.1 Introduction

This section concerns the measurement of the compressive and tensile strength.

When a material is loaded in such a way that it extends, it is said to be in "tension". On the other hand, if the material compresses and shortens it is said to be in "compression". Compressive strength is the capacity of a material to withstand compression loads whereas the tensile strength represents the maximum amount of tensile stress that it can take before failure.

Compressive and tensile strength are largely used to determine the mechanical resistance of large range of materials as construction materials (concrete, bricks …), advanced ceramics, rocks, pellets (from biomass and biowastes etc.), tablets etc. They are useful indicators in the materials development area, design and data generation and quality control purposes. They are usually used by engineers to make sure their materials can withstand a sufficient amount of force for the product they are designing.

Compressive strength is obtained experimentally by the compression test. The diametral compression test was originally developed to study the mechanical strength of brittle construction materials as concrete, rocks and hydraulically bound mixtures. Nevertheless, it has been adapted to the characterization of pharmaceutical tablets and more recently, biomass and biowaste pellets. This test allows the evaluation of the tensile strength in an indirect manner [397] as the molecules of the material are pushed apart in opposite directions, similar to what happens to molecules in a direct tensile strength test.

Focusing to pellets produced from biomass and biowastes, an appropriate characterization is important in order to determine and adapt the required conditions to produce high quality densified products that fulfill the market standards [398, 399]. In particular, the mechanical properties are considered as a quality measure, since pellets must support compression loads during their storage and handling. Accordingly, the evaluation of the mechanical properties of pellets is usually performed under compression force (e.g. diametral compression test). Proper mechanical durability and high tensile strength are desirable properties that depend on many factors including the feedstock composition, particle size, compression temperature and force, and binding agents [400–403]. Therefore, the mechanical characterization of pellets can help adjusting the production process to obtain the desired properties according to their applications.

For both compression and tensile tests, the stress-stain curve is plotted and shows the maximum force to apply to the sample before his failure.

2.10.2.2 Apparatus Description

The compressive and tensile strength can be measured using the same machine.

The typical machine consists of a load frame, compression platens, load cell, control panel and data storage unit.

A load cell is attached on the moving structure that can move up and down. Two compression platens are used; one is attached on the load cell (upper platen) and the second is based on the fixed lower bedplate (lower platen). The load is applied to the sample placed between these two platens by the top.

A control panel controls the displacement of the upper platen. It shall enable the machine to be verified and to allow force to be applied smoothly and without shock.

The computer with adequate software stores the data during the test, plots the stress-stain curve and calculates the compressive or the tensile strength.

Note: Machine and auxiliary platens shall be made of a material that shall not deform irreversibly during the test. They shall have a hardness value of at least 550 HV 30 (HRC 53) when tested in accordance with ISO 6507-1.

Concerning the platens, they shall be flat. The flatness tolerance reported in the standard (NF EN 12390-1 [404]) is 0.03 mm for the area in contact with the sample. In addition, the area of platens in contact with the sample shall be at least as great as the area of the sample being tested.

Figure 2.254 is a schematic illustration of the compression and the diametric compression tests. As shown, the sample is differently positioned on the platen, vertically for the compression (a) and horizontally for the diametric-compression test (b).

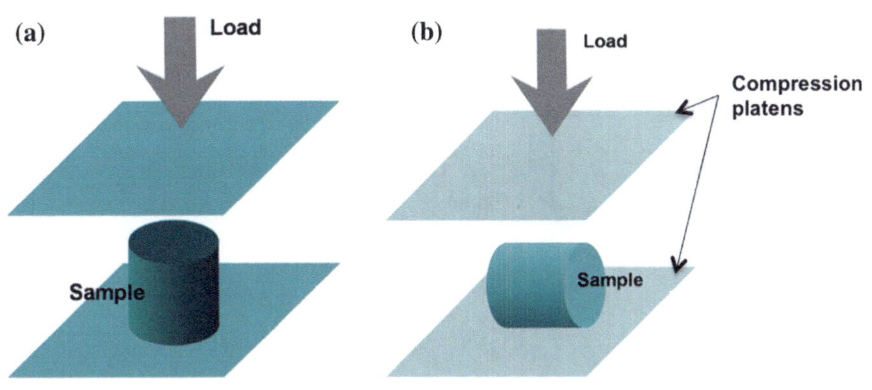

Fig. 2.254 Schematic illustration of the compression test (a) and the diametric compression test (b)

Note that in both cases, the sample shall be strictly centered between the two platens in order to have good repartition of the load on the sample.

2.10.2.3 Sample Preparation

The test samples could be cylinder, cube and prism [404]. It was demonstrated that samples with a designed shape and non-shape sample as rocks can be used to measure with high accuracy the compressive and tensile strength. For diametric compression tests, the samples are generally cylinders [405]. Tables 2.94 and 2.95

Table 2.94 Examples of samples shapes and dimensions for axial compression test

Standards/ references	Shape	Dimensions d (mm)	H (mm)	L (mm)	Materials
EN 12390-1 [404]	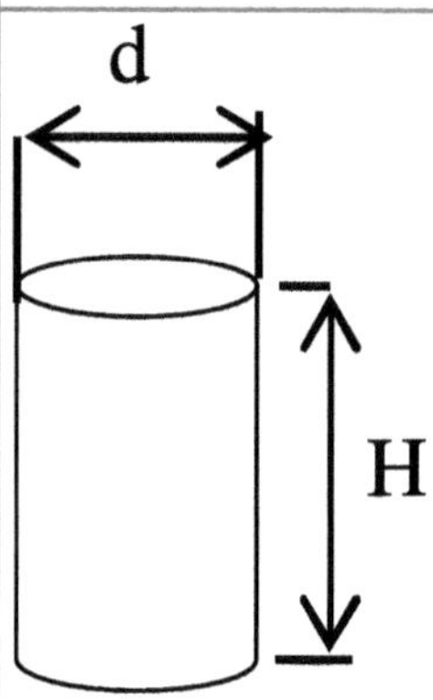		(2.d)	–	Hardened concrete
		100	200		
		113	226		
		150	300		
		200	400		
		250	500		
		300	600		
Wang et al. [407]		6	11	–	Magnesium potassium phosphate chemically bonded ceramics
Suvorova et al. [408]		20	20	–	Construction ceramics
EN 12390-1 [404]	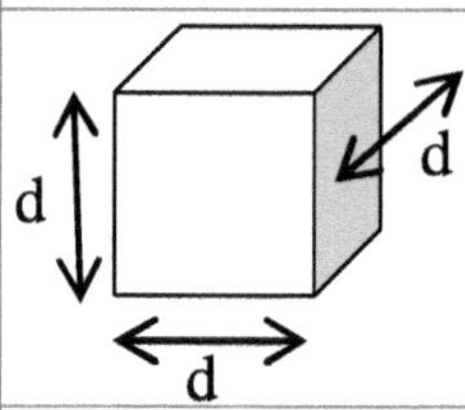	100 150 200 250 300	–	–	Hardened concrete
EN 12390-1 [404]	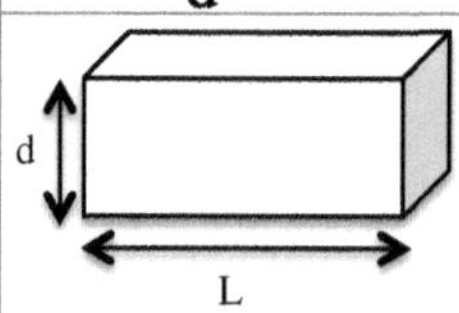	100 150 200 250 300	–	≥ 3.5d 350 525 700 875 1050	Hardened concrete
EN 196-1 [409]		40		160	Cement
ISO 13061-17 [410]		20		60	Wood

Table 2.95 Diametric compression test: examples of samples shapes and dimensions

Standards	Shape	Dimensions			Materials
		d (mm)	H (mm)	H/d (−)	
EN 12390-6 [405]	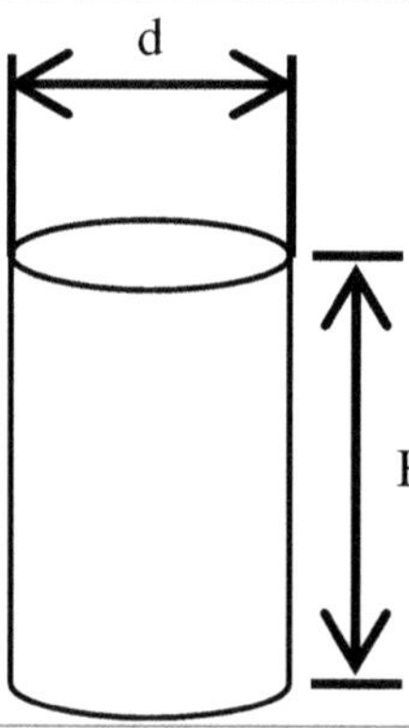	150	300	2	Hardened concrete
NF P94-422 [411]		40–150		0.8–1.5	Rocks
EN 12697-23 [412]		80–160	35–75	–	Bituminous mixtures
ASTM D6931-17 [413]		–	–	<0.5	Asphalt mixtures
EN 13286-42 [414]		–	–	0.8–2	Hydraulically bound mixtures

show some examples of dimensions for axial and diametric compression tests. Even if the characteristics and applications of biomass and biowaste pellets are not related with these materials, these standards give some insights about the test procedure for pellets characterization.

For cylinder samples, it is recommended to keep the testing samples with a ratio of the length to the sample diameter (H/d) less than 3 [406].

Samples preparation process can include compression (to produce pellets, tablets, etc.), moulding (concrete), cutting and polishing (for ceramics materials available in formats well defined by the production), etc.

Note: Whatever the preparation method used, it is recommended to verify the sample parallelism. If it is not respected, the dimensions must be adjusted or the sample should be removed.

For the specific case of pellets or tablets manufacturing, it is important to note in a laboratory notebook the pressure applied to the raw materials; it could be helpful in the field of fracture interpretation and analysis.

The dimensions measurements shall be carried out using calipers or any other standardized device. For each sample, at least three measurements should be made; an average is made to obtain the exact dimension (diameter, length and width).

For biomass pellets, from the production process, the most variable characteristics are probably the pellets dimensions. According to the production materials and targeted applications, the diameter d and length H of pellets can vary. Consequently, for characterization and comparison purposes it is recommended to keep the testing specimens with a slenderness ratio (H/d) between 0.8 and 2. This is in accordance with the existing standards using diametral compression test for materials characterization. It is also important to note that due to the great diversity of biowastes as well as the different parameters that can be adjusted and modified during the production of pellets, a large variety of characteristics and properties can be obtained for the manufactured materials. In this regard, the characterization and subsequent comparison of pellets properties should be done carefully. Similar testing conditions should be guaranteed in order to have a solid comparison basis related to pellet characteristics and quality.

2.10.2.4 Test Procedure

The procedure used to measure compressive strength and indirect tensile strength is the same and follows several steps:

- Start the test machine and the computer.

Note: If it is necessary, wipe all testing machine bearing surfaces clean and remove any loose grit or other extraneous material.

- Install the load cell and the compression platens
- It is important to note that the maximum force applied to the sample shall not exceed the cell load capacity, so it should be chosen accordingly.
- Carefully centered the sample (cylinder, cube etc.) with respect to the lower platen to accuracy of 1% of the designated diameter of cylindrical specimens or designated size of cubic if the test is compression.
- For diametric compression test, the sample horizontally placed on the lower platen must be also carefully centered.
- Enter in the software the samples references and dimensions and chose the data storage folder
- Apply an initial load to the sample without shock. Be careful, this load must not be a high value.
- Ensure that the upper platen is parallel with the lower platen, during preloading.
- Start the test by pressing the button start on the control panel or on the computer.
- When start button is pressed, the load is applied to the sample without shock and increase continuously, at the selected constant rate until the sample failure. It is generally recommended to apply low deformation speeds from 1 to 30 mm/min (for diametral and axial compression).

Note: For security reasons, it is highly recommended to put crumbled tissues or other appropriate material around sample in order to prevent pieces from flying out upon fracture.

- Stop the test and data recording after the failure of the sample by pressing the button stop on the control panel.
- Retrieve and identify the fracture fragments for later examination if it is necessary.
- Repeat the procedure for each test sample; at least 5–10 for each composition.

When reporting the test results and in order to provide valuable data for comparison purposes, it is recommended to include at least the following information:

- Sample composition
- Sample dimensions (D, H or L, d)
- Sample preparation conditions (pressure, temperature etc.)
- Deformation speed or stress rate
- Compression strength or indirect tensile strength
- Stress-Strain or Load-Distance plot.

2.10.2.5 Calculation

The results obtained are influenced by a large number of factors principally associated with the microstructure of the material, the shape and size of the sample. Consequently, different results could be obtained from samples with an identical test composition. So, several measurements (at least 3) should be performed and the results should be described in terms of a mean value and a standard deviation.

The compressive strength is given by the Eq. (2.176) below [408, 415]:

$$\sigma_c = \frac{F_{max}}{A_c} \tag{2.176}$$

where: σ_c = Compressive strength (MPa); F_{max} = Maximum load at failure (N); A_c = Cross-sectional area of the sample on which the compressive force acts, calculated from the designated shape and size of the sample (mm^2).

The tensile strength is calculated by the following formula [416, 417]:

$$\sigma_{cT} = \frac{2F_{max}}{\pi \cdot H \cdot d} \tag{2.177}$$

where: σ_{cT} = Tensile strength (MPa); F_{max} = Maximum load at failure (N); H = Length of the line of contact of the specimen (mm); d = Sample diameter (mm).

2.10.2.6 Example of Tensile Strength Measurement

This example is based on the work of Mateusz Stasiak et al. [417]. The objective of this work was to determine the mechanical and combustion properties of pellets made of pine sawdust mixed with different proportions of wheat and rapeseed straw. The mixtures were compressed to perform pellets. The strength of the pellets was determined by diametric compression between circular flat plates with a constant deformation rate of 0.033 mm/s.

Sample Preparation

The biomass was compacted into the cylindrical die 10 mm in diameter and 25 mm high. Figure 2.255 shows the arrangement of the experimental pelletizer. The sample fields the entire volume of the cylinder. After filling, a piston was placed on the material surface and conducted compaction using high-pressure actuator and yoke. The compaction pressure applied was 60 and 120 MPa. After compaction, the pellets were obtained.

Measurement Procedure

The test procedure included the following steps:

(1) Test machine and the computer were turned on;
(2) The load cell with the compression trays was fixed;
(3) The sample was carefully place horizontally on the lower tray and centered;
(4) Dimensions and references of the sample were inserted in the software;
(5) The appropriate preload was applied to the sample;
(6) The start button was pressed to begin the test;
(7) The stop button was pressed to stop data recording after the failure of sample.

Data storage unit store applied force, displacement and automatically made the calculation to give tensile strength by Eq. (2.177).

Results

The tensile strength of the produced pellets is shown in Fig. 2.256 as a function of the percentage of wheat straw (0–100%). For each composition, at least three measurements were performed in order to obtain a mean value of the strength with a standard deviation. The results reveal that the mechanical strength of pellets decreases with an increase in the proportion of straw in the blend. Also, it was seen

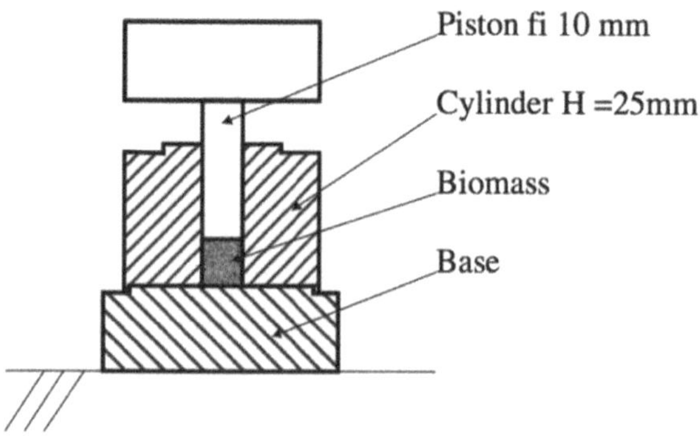

Fig. 2.255 Test pelletizer used for biomass compression. Reproduced with permission from Elsevier [417]

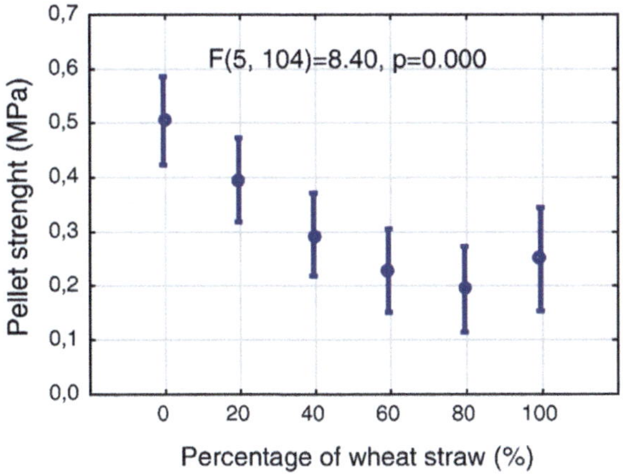

Fig. 2.256 Mean values of strength of the pellets as dependent on percentage of wheat straw. Reproduced with permission from Elsevier [417]

that the compaction pressure and the moisture in the biomass have a strong influence in the pellets strength.

2.10.2.7 Example of Compressive Strength Measurement

This example is based on the work of Suvorova et al. [408]. This research aims to explore the possibility of involving the concentration of inorganic wastes (copper-nickel, apatite-nepheline and ferruginous quartzite ores) in the production of ceramic construction materials using the compression molding technique and to study their physical and mechanical properties.

Sample Preparation

The compressive strength was obtained from cylinder samples with identical diameters and heights 20 mm. To do, the starting materials were ground in a porcelain drum until the mesh 0.1 mm. The mix was carefully homogenized, wetted until the optimal moisture content and molded at a specific pressure of 20–150 MPa. The temporary binding agent imparting strength to the raw brick was a 0.5 wt% sulphite waste liquor. The samples were finally fired at different temperatures before the mechanical tests.

Measurement Procedure

The procedure is the same as presented in the previous example. The only difference is that the sample is positioned vertically on the lower tray and centered between the two compression trays.

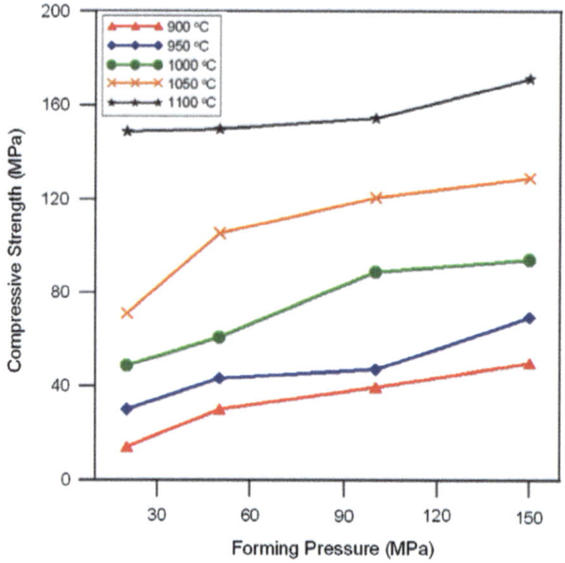

Fig. 2.257 Relationship between the compressive strengths of the ceramic materials and forming pressure at different sintering temperature. Reproduced with permission from Elsevier [408]

Results

Figure 2.257 shows the compressive strength of the optimal composition of inorganic wastes as function of the samples forming pressure and the firing temperature. This study revealed on the one hand that increasing the firing temperature and the forming pressure promote increasing of the strength of samples. On the other hand, the samples sintered at high temperature and moulded at 20–150 MPa can be referred to clinker ceramics and be attractive as high quality full and hollow bricks.

2.10.3 Flexural Resistance Analysis

2.10.3.1 Introduction

This section is focused on the measurement of the monolithic materials flexural strength. Flexural strength, also known as bend strength, or transverse rupture strength is a material property, defined as the stress in a material just before it breaks in a flexure test [406]. It represents maximum nominal stress supported by the monolithic materials at the instant of failure when loaded in linear elastic bending. It is an important indicator in the ceramics development area.

This property is intended to be used for material development, quality control, characterization and design data acquisition purposes [418, 419].

The strength level determined by the test is calculated on the basis of linear elastic bending of a thin beam on the assumption that the material being tested is elastically homogeneous and isotropic, and shows linear stress-strain behaviour.

The values needed to calculate flexural strength are measured by experimentation, with rectangular samples of the material placed under load in a three-point or four point flexural testing setup. These methods determine the flexural strength at ambient temperature and environmental conditions.

Three-point flexure means of the test piece is supported on bearings near its ends and a central force is applied and four point flexure means the test piece is supported on bearings near its ends and is loaded equally at two positions symmetrically disposed about the center of the supported span. The first part of this section concerns the flexure test, the second part will discussed the sample preparation, test procedure and calculations and finally an example will be presented.

2.10.3.2 Apparatus Description

A conventional machine used for flexural strength measurement typically consists of a load frame, flexure fixture (three or four points), load cell, control panel and data storage unit. An example of suitable machine is Instron model 5800R.

The load frame has a moving structure (crosshead) where the load cell is attached; and fixed lower bedplate on which the flexure fixture is based. The crosshead slides up and down with a constant displacement rate in order to apply a constant force to the test piece. The displacement is controlled by the control panel attached to one of the column covers as well as the start or stop of a measurement. The load frame has also upper and lower limit stop that are adjustable to restrict crosshead travel to within safe boundaries and emergency stop button.

Flexure Fixture

Schematic arrangements of the flexure fixture with the standard spacemen between rollers are shown in Fig. 2.258 for three-point flexure, and Fig. 2.259 for four-point flexure.

- Three-point flexure: a third roller is located at the mid-point between and parallel to the two support rollers (Fig. 2.258). This roller shall have the same diameter as the support rollers and apply a central force F on the sample.

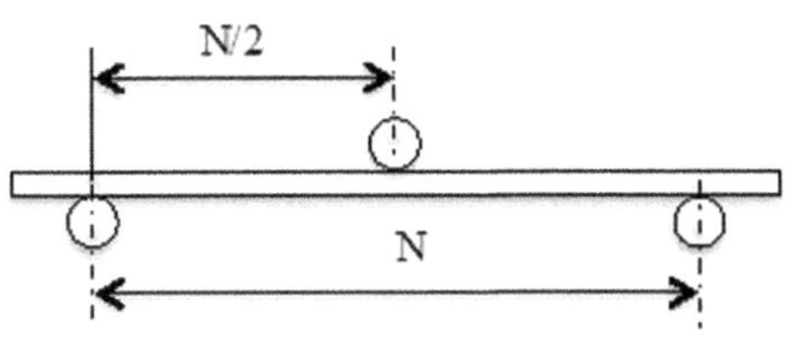

Fig. 2.258 Illustratione of three-point flexure configuration

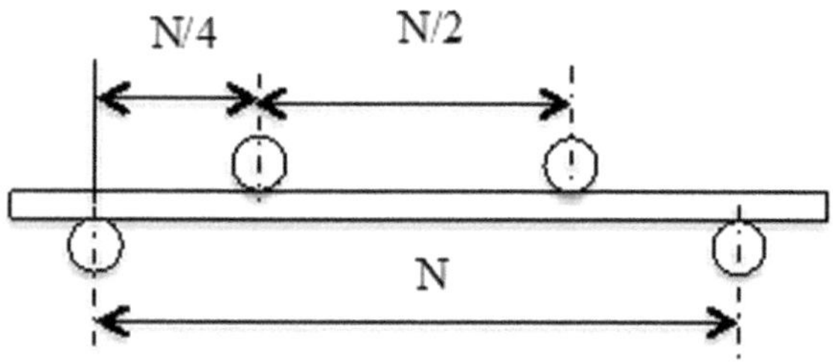

Fig. 2.259 Illustration of four-point flexure configuration

- For four-point flexure, two loading rollers are located at the quarter points (Fig. 2.259) and shall be free to roll inwards. The loading rollers shall be symmetrically positioned. The arrangement for loading shall ensure that equal forces are applied to the two loading rollers. To do, the distances between the center of the support rollers and adjacent loading rollers shall be measured along the length of the test piece perpendicular to the direction of loading, using a suitable device and shall respect the arrangement presented in Fig. 2.259.

The three-point test configuration exposes only a very small portion of the specimen to the maximum stress. Therefore, three-point flexural strength is likely to be much greater than four-points flexural strength. Three-point flexure has some advantages. It uses simpler test fixtures, it is easier to adapt to high temperature and fracture toughness testing.

Rollers/Three and Four Point Flexure

Cylindrical bearing edges are used for the support of the sample and for the application of load. The rollers diameters shall be approximately 1.5 times the beam depth of the test specimen size. In the same time, they must have a length at least three times the specimen width [418, 419]. In some cases, they shall be at least 10 mm longer than the width of the test specimen [420].

Normally, the rollers shall be made of hardened steel which has hardness of at least HRC 40 or which has an yield strength of at least 1240 MPa. Alternatively, the cylinders may be made of a ceramic with an elastic modulus between 2.0 and 4.0×10^5 MPa and a flexure strength of at least 275 MPa [419]. Also, the portions of the test fixture that support the rollers may need to be hardened to prevent permanent deformation and then impact the test results.

Note: During the installation, the rollers shall be carefully positioned such that the spans are accurate within ±0.10 mm. Also it is possible to use rubber fasteners to attach them. The load application rollers for three-point configurations shall be positioned midway between the supports bearing within ±0.10 mm if it is possible. The load application (inner) rollers for the four-point configurations shall be centered with respect to the support rollers within ±0.10 mm.

Table 2.96 Flexure fixture spans and bearing diameters

Standard	Configuration	Fixture spans	Rollers	Materials
		Support span (L)	Diameter, mm	
ASTM C1161-13 [418]	A	20	2.0–2.5	Ceramics
	B	40	4.5	
	C	80	9.0	
EN 12390-5 [420]	–	3b (b = width of the sample)	20–40	Hardened concrete
EN 310 [421]	–	20 h (h = thickness of the sample)	15–30	Wood

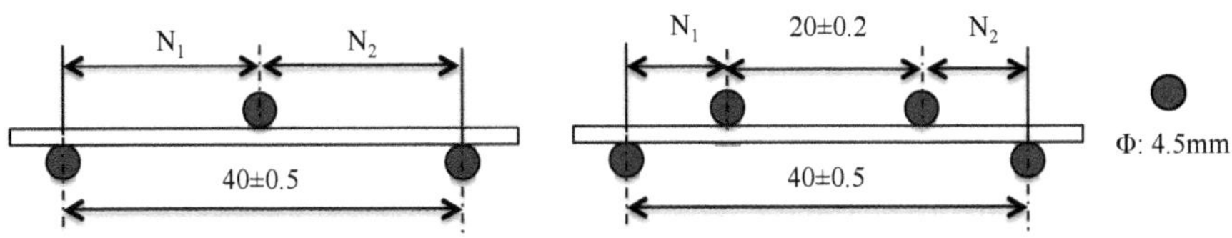

Fig. 2.260 Illustration of the "Configuration B" of test span dimensions and tolerances

Test Span Dimensions

Table 2.96 shows fixture spans and nominal rollers diameters for different configurations. An example of test span dimensions for three and four point flexure tests are shown in Fig. 2.260 for one configuration (configuration B). Configurations A and C follow the same principle.

The separation of the centers of the rollers in their starting positions shall be measured using a suitable device as a caliper.

2.10.3.3 Sample Preparation

Normally, samples must respect precise dimensions given for example in Table 2.97 for the specific cases of the configurations A, B and C showed before. Specifications of configurations A to C are available in the Table 2.96.

Generally, ceramic materials must follow different preparation procedures in order to respect these dimensions. This procedure can include cutting and polishing. In fact, generally, ceramic materials are available only in formats well defined by the production. So depending on the production conditions, the monolithic materials may have any shapes and dimensions that are not immediately adapted for flexural strength measurement. The process below is an example of sample preparation method that could be used typically in the case of building materials as

Table 2.97 Sample dimensions for flexural strength measurement

Standard	Shape	Configuration	Width (b), mm	Thickness (h), mm	Length minimal (L), mm	Materials
ASTM C1161-13 [418]	Parallelepiped	A	2.0	1.5	25	Ceramics
		B	4.0	3.0	45	
		C	8.0	6.0	90	
EN 12390-5 [420]		–	b	b	3.5b	Hardened concrete
EN 310 [421]		–	50	h	20h + 50	Wood

fire clay ceramics or biomass, industrial by-products containing monoliths. Other samples preparation technics exist and could be found consulting these references [418, 419, 422–424].

Cutting

This operation calls for the use of rotatable circular saw (diamond saw blade) or other suitable device. The objective is to have pieces with the least damage as possible.

Note: *It is important for security reason to be very attentive during this step and wear appropriate protective devices (gloves, eye and ear protections).*

Polishing

The test face and the adjacent of the test pieces shall be polished using a SIC abrasive paper with a suitable grit size. The polished test pieces shall be free from scratches observed and shall have parallel faces. For the special case of clay-based ceramics, the abrasive paper with 82 μm grit size is recommended. Examples of abrasive papers brands are BUEHLER, BY PRESI and CarbiMet available with different grit size between 10 to more than 1000 μm. During this step, protection devices are needed.

2.10.3.4 Dimensions Measurement

Due to the desired accuracy, the measurements are carried out using calipers or any other standardized device. For each sample, three measurements are made in the center and at both ends on the left and right. An average is made to obtain the exact dimension (thickness and width). This method could be used to check the samples parallelism and then make adjustments if necessary.

Number of Test Samples

For material development, characterization or quality control, the minimum number of test pieces shall be 10. For statistical evaluation of strength data (e.g. Weibull parameters) in accordance with EN 843-5 [422] and ASTM C1239-13 [424], the minimum number shall be 30. The number of specimens required by this test method has been established with the intent of determining not only reasonable confidence limits on strength distribution parameters, but also to help discern multiple-flaw population distributions

The prepared samples shall be handled with care to avoid the introduction of damage subsequent to the machining process. If it is possible the test pieces shall be kept separate at all times and shall be individually wrapped for transport.

Test procedure

The procedure used to measure flexural strength follows several steps:

(1) Three or four point flexure fixation on the test machine with appropriate configuration such as fixture spans, bearing diameters etc.;

(2) Methodically identify and take the sample dimensions if it's not done;
It is strongly advised to identify the samples in a precise order, example: initials of the manipulator/date/number of the sample;

(3) Carefully place the sample into the test fixture to preclude possible damage and ensure alignment of the sample in the fixture;
Note: *It is strongly recommended that if one of the wide specimen surfaces has been marked for extraneous damage, place the damaged surface so that it is loaded in compression. Also, if a side surface is marked as damaged, then the specimen may be tested, but shall be inspected after the test to confirm that the scratch or damage did not cause fracture. If that is the case, this result is not usable.*

(4) Slowly apply the load at right angles to the fixture in order to apply a preload to the sample. It is recommended to inspect the point(s) of contact between the roller(s) and specimen to ensure even line loading and that no dirt or contamination is present;

(5) Reset the displacement on the control panel because of sample preloading;

(6) Data storage folder selection on the computer;

(7) Identify the point of load application and also the tensile and compression faces;
Note: This step is not necessary but these identifications could be helpful in the field of fracture interpretation and analysis.

(8) Start the test by pressing the button start on the control panel. For security reasons, it is highly recommended to put crumbled tissues or other appropriate material around sample in order to prevent pieces from flying out upon

fracture. It is also important before to check the crosshead speed. Table 2.98 shows examples of crosshead speeds for different testing configurations [418];

(9) Stop data recording after the failure of the sample;

(10) Retrieve and identify the fracture fragments for later examination if it is necessary;

(11) Repeat the procedure for each test piece.

Note that during the test, the applied load must not exceed the permissible load by the load cell. If this is the case, stop immediately the test.

2.10.3.5 Calculations

The result obtained from a strength test is determined by a large number of factors principally associated with the microstructure of the material, the sample preparation method, the size and shape of the test piece, etc. As a consequence of the brittle nature of ceramics, there is usually a considerable range of results obtained from a number of nominally identical test pieces. For many purposes, the results of strength tests may be described in terms of a mean value and a standard deviation.

For a rectangular sample, the resulting stress under an axial force is given by the following equations [418, 419, 425–427]:

For three-point flexure:

$$\sigma_f = \frac{3FL}{2bh^2} \tag{2.178}$$

For four-point flexure:

$$\sigma_f = \frac{3F(L-l)}{2bh^2} \tag{2.179}$$

where σ_f: the fracture strength expressed as N mm^{-2}, equivalent to MPa; F: the peak force at fracture, expressed as newtons (N); b: the sample width, expressed in millimeters (mm); h: the sample thickness, expressed in millimeters (mm); L: the distance between centers of the outer support rollers (three-point flexure), expressed in mm; l: the distance between centers of the inner support rollers (four-point flexure), expressed in millimeters (mm).

Table 2.98 Examples of crosshead speeds for different testing configurations

Configuration	Crosshead speeds, mm/min
A	0.2
B	0.5
C	1.0

The data is usually saved in a .txt file in terms of applied force (kN), displacement (mm) and stress (MPa). The maximum stress during the failure of the specimen is obtained using the maximal force (F_{max}) recorded.

2.10.3.6 Example of Flexural Strength Measurement

This example is based on the work of Korah et al. [427]. This research investigated the impact of the particle shape of organic additives on the anisotropy of a clay-based ceramic and its thermal and mechanical properties. Organic additives used in this study concern Olive Stone Flour (OSF) in the form of powder and Wheat Straw (WS) in the form of fibers. The mechanical properties of the ceramics were measured by three-point bending after a firing process using an Instron 5800R machine with a 40 mm loading span and a 500 N load cell at a constant displacement of 1 mm/min.

Sample Preparation

The samples of 60 × 30 × 5 mm^3 were used for flexural strength tests. They were prepared from the dried ceramics (with initial dimensions of 180 × 80 × 18 mm^3) by cutting with circular saw and polishing with P80, P120, P180 and P280 SiC abrasive papers (CarbiMet, Buehler).

Measurement Procedure

The experimental procedure included the following steps:

(1) Three point flexure was fixed on the Instron machine with appropriate configuration (fixture span, bearing diameters);
(2) The sample was placed into the test fixture;
(3) The sample dimensions (width and thickness) were measured and introduced in the computer;
(4) The appropriate preload was applied to the sample;
(5) The displacement has been reset;
(6) The start button on the control panel was pressed to begin the test;
(7) The stop button was pressed to stop data recording after the failure of sample.

Data storage unit stores applied force, displacement and automatically made the calculation to give flexural stress by Eqs. (2.178) and (2.179).

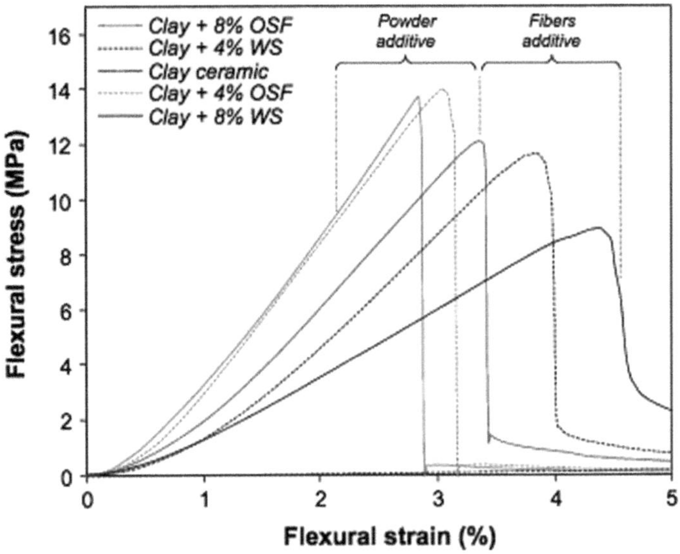

Fig. 2.261 Flexural stress-strain curves of the clay ceramic and of the clay ceramic with a 4 and an 8 wt% addition of OSF and WS after a firing process at 900 °C. Reproduced with permission from Elsevier [427]

Results

Figure 2.261 shows the flexural stress-strain curves obtained after different tests with ceramics made from clay/sand mixture and clay/sand/organic additives (4 and 8 wt% under powder form) mixtures. The addition of organic additives led to the improvement of the flexural strength of at least 13%.

2.10.4 *Young Modulus Determination*

2.10.4.1 Introduction

This section is focused on the experimental determination of monolithic materials Young's modulus (room to 1000 °C).

Young's modulus (E) is a mechanical property that measures the stiffness or the elasticity of a monolithic material. It defines the relationship between stress (force per unit area) and strain (proportional deformation) in a material [406].

In fact, a monolithic material will undergo deformation when a small load is applied to it. If it returns to its original shape after the load is removed, this is called elastic deformation. In the range where the ratio between load and deformation remains constant, the stress–strain curve is linear. A stiff material needs more force to be deformed compared to a soft material. An infinite force would be needed to deform a perfectly rigid material, implying that it would have an infinite Young's modulus. Although such a material cannot exist, a material with a very high Young's modulus can be considered as rigid.

Young's modulus predicts how much a material sample extends under tension or shortens under compression. It is a useful indicator in the materials development area, design and data generation and quality control purposes.

There are several experimental techniques for the Young's modulus determination. These techniques can be classified into two groups: destructive and non-destructive methods [428]. It is also possible to classify them as: Static and dynamic methods [429]. Destructive (and static) methods measure directly the strain and the stress during the test. These methods include tensile, compressive and flexural tests presented in the Sects. 2.10.2 and 2.10.3. With these techniques, Young's modulus is calculated at ambient temperature.

Non-destructive (and dynamic) methods are based on knowing the dimensions and densities of the samples. They are therefore called indirect methods. These techniques provide an advantage over static techniques because of the easy specimen preparation, wide variety of specimen shapes and sizes, good accuracy, applicability to porous and brittle materials due to small strains and measurement over a wide temperature range (up to 1700 °C) [429, 430].

Impulse excitation technique is one of these methods. It is used to determine the elastic properties and internal friction of a material of interest. This technique allows to calculate Young's modulus at room temperature as well as at high temperature with suitable equipment modifications and appropriate modifications to the calculations [431–433].

The second dynamic method is the Sonic resonance technique. Details about this technique can be found elsewhere [434].

The first part of this section concerns the Impulse excitation method. The second part will present the samples preparation, test procedure and calculation. Finally examples of measurement will be presented.

2.10.4.2 Impulse Excitation Technique

This method is based on the measurement of the resonant frequencies in order to calculate the Young's modulus of predefined shaped materials [433].

Apparatus suitable for accurately detecting, analyzing and measuring the fundamental resonant frequency consists of an impulser, a suitable pickup transducer, a support system, a signal conditioner/amplifier, a signal analyser and data processing system (Ex: RFDA basic apparatus model commercialized by IMCE).

The measurement principle is based on tapping the sample with an impulse tool (automated tapping device with a small hammer). A non-contact transducer (microphone) detects the resulting mechanical vibrations of the sample and transforms them into electric signals.

Note: *The frequency range of the transducer shall be sufficient to measure the expected frequencies of the sample of interest. A suitable range would be from* 20 Hz *to* 50 kHz *(for most ceramic test samples)* [432].

The signals are analyzed, and the fundamental resonant frequency is isolated and measured by the electronic system. It consists of a signal conditioner/amplifier and signal analyser. The system should have accuracy and precision sufficient to measure the frequencies of interest to an accuracy of 0.1%. Fast Fourier transformation analysis is used for analysing and identifying the fundamental resonant frequency.

The appropriate fundamental resonant frequencies, dimensions and mass of the sample are used to calculate the elastic properties.

For Young's modulus, the flexural configuration (as support system) is used as shows in Fig. 2.262. The support shall serve to isolate the sample from extraneous vibration without restricting the desired mode of sample vibration.

Appropriate materials should be able to work at the test temperatures. Wire suspension can be used to support samples (ex: RFDA basic apparatus commercialized by IMCE). In this case, samples are supported along nodes lines appropriate for the desired vibration in the location.

Note: *All equipment used shall withstand the working temperature.*

Figure 2.263 shows suitable configuration for the Young's modulus measurement at high temperature in air atmosphere or in an inert gas flow. The same principle as at room temperature is used and measurements can be performed in

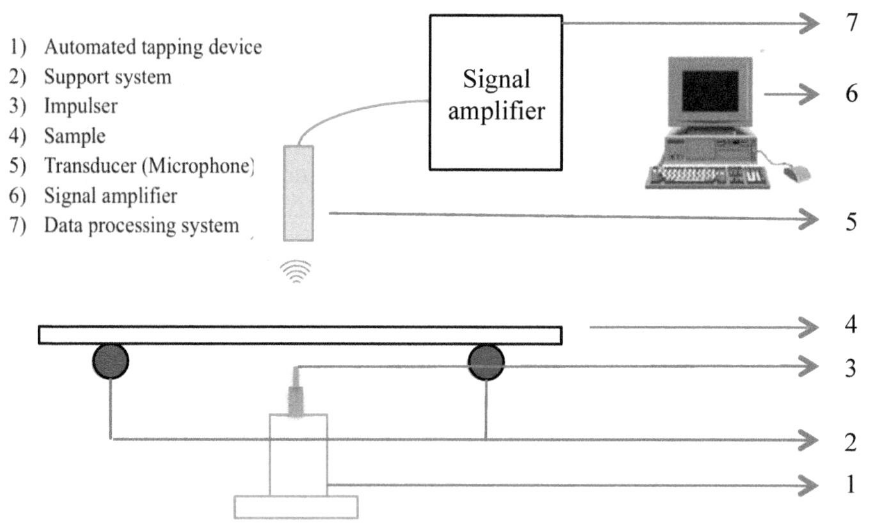

Fig. 2.262 Principle of the resonance frequency method

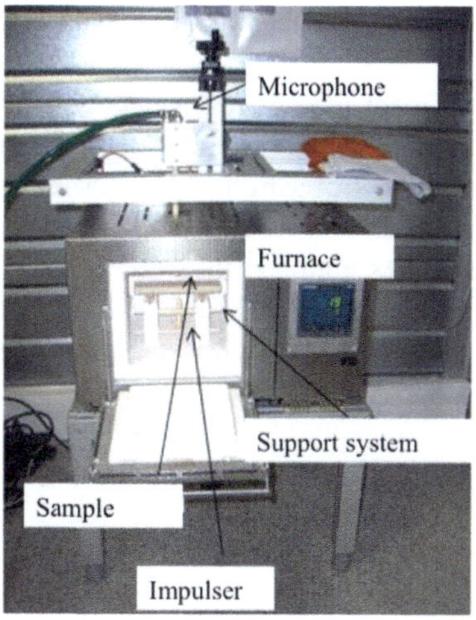

Fig. 2.263 RFDA HT650 with suitable equipment modifications (high temperature). Reproduced with permission from Aurock Company [435]

predefined intervals during heating and cooling. The microphone is fixed in the furnace by the top and automated tapping device at the bottom (impulse stay in the furnace). Support system with sample is also inserted in the furnace.

2.10.4.3 Sample Preparation

The samples shall be prepared so that they are either rectangular or circular in cross section [432, 433].

Dimensions should be selected regarding the relationship between the sample size and its mass because of resonant frequencies are function of these two parameters.

Note: *The selection of the samples size shall be made so that, for an estimated Young's modulus, the resonant frequencies measured will fall within the range of frequency response of the transducers and electronics used.*

It is recommended that the ratio of length to minimum cross-sectional dimension have a value of at least 5. A ratio of approximately 20 $\approx$ 25 is preferred for easing the calculation [432].

The samples preparation procedures are the same as presented in the Sect. 2.10.3 (flexural strength measurement). In order to have specific dimensions, solid materials could be cut and polished as shown in the previous section.

Note: *For specimens with as-fabricated and rough or uneven surfaces or moulded, variations in dimension can have a significant effect in the calculation. Uniform sample dimensions and precise measurements are essential for accurate results.*

Dimensions and Mass Measurements

To meet the desired accuracy, the measurements are carried out using calipers or any other standardized device. Special care must be taken when measuring the thickness of samples. Concerning the mass, a balance with high precision must be used. Example of suitable balance is BALCO/AT400 (accuracy of 0.1 mg) for samples with weight <405 g.

Note: *For each sample, at least three measurements of dimensions (length, width and thickness) and weight shall be done in order to estimate to obtain the mean values.*

Recommendations

All the surfaces on the rectangular specimen shall be flat. Opposite surfaces across the length, thickness and width shall be parallel. For cylindrical samples, they shall be round and constant in diameter to within 0.1%. Concerning the mass, it shall be determined to within 0.1%. Table 2.99 illustrates the impact of measured parameters on the modulus calculation.

The prepared samples shall be handled with care to avoid the introduction of damage subsequent to the machining process. If it is possible, the test pieces shall be kept separate at all times and shall be individually wrapped for transport.

2.10.4.4 Test Procedure

The procedure used to measure Young's modulus at ambient temperature and high temperature follows several steps:

- Activate all electrical equipment;
- Verify the equipment response and accuracy;

Note: *The device shall be systematically checked at the beginning with standard sample.*

Table 2.99 Impact of measured parameters on the modulus calculation [432]

Variable	Measurement error (%)	Young's modulus calculation error (%)
Length	0.1	0.3
Width		0.1
Thickness		0.3
Mass		0.1
Diameter		0.4

- After the sample dimensions and mass measurements as shown before, place it carefully on the supports (wire suspensions) located at the fundamental nodal points;

Note: *Nodal points are located at 0.224L from each end; L represents length of the sample* [432].

Note: *In the case of high temperature measurement, sample and steel wire suspensions are positioned in the oven.*

- Enter in the software all information about sample;
- Place the microphone over an antinode point and close enough to the test sample to record the desired vibration;

Note: *Verify that the microphone is not so close as to interfere with the free vibration.*

- If it is possible, strike the sample lightly at the center or at the opposite end from the detecting microphone;

Note: *If the microphone detects no signal, be sure that it is connected with the signal amplifier/analyser;*

- Close the oven and enter the heating and cooling cycles;
- Start the test;

Note: *Tests and adjustments can be realised before to start the test (specially in the oven) in order to obtain a non-noisy, repeatable signal on which the identification of the resonance frequency is unambiguous. The temperature test is started when these adjustments are completed.*

For comparison purposes (between ceramics made with different biomass and biowastes), similar testing conditions should be guaranteed in order to have solid comparison basis and logic interpretation.

2.10.4.5 Calculations

For the fundamental flexure frequency of a rectangular sample with elastic and isotropic behaviour, the relation below (Eq. (2.180)) is used to calculate Young's modulus [431, 433].

$$E = 0.9465\left(\frac{mf_f^2}{w}\right)\left(\frac{L}{t}\right)^3\left(T_{1a} - \frac{T_{1b}}{T_{1c}}\right) \tag{2.180}$$

where E: Young's modulus (Pa); m: mass of the sample (g); f_f: Fundamental resonant frequency of the sample in flexure (Hz); w: width of the sample (mm); L: length of the sample (mm); t: thickness of the sample (mm); T_{1a}, T_{1b}, T_{1c}: correction factor for fundamental flexural mode to account for finite thickness of sample.

$$T_{1a} = 1.0 + 6.585\left(1.0 + 0.0752\upsilon + 0.8109\upsilon^2\right)\left(\frac{t}{L}\right)^2 - 0.868\left(\frac{t}{L}\right)^4 \quad (2.181)$$

$$T_{1b} = 8.34\left(1.0 + 0.2023\upsilon + 2.173\upsilon^2\right)\left(\frac{t}{L}\right)^4 \quad (2.182)$$

$$T_{1c} = 1.0 + 6.338\left(1.0 + 0.1408\upsilon + 1.536\upsilon^2\right)\left(\frac{t}{L}\right)^2 \quad (2.183)$$

where υ: Poisson's ratio. The value of $\upsilon = 0.275$ can be commonly used for the calculation when it is unknown.

For a circular cross section bar, the Young's modulus is calculated as follows [432]:

$$E = 1.6067\left(\frac{L^3}{D^4}\right)\left(m \cdot f_f^2\right)\left(T'_{1a} - \frac{T'_{1b}}{T'_{1c}}\right) \quad (2.184)$$

where D: diameter of the sample (mm); T'_{1a}, T'_{1b}, T'_{1c}: correction factor (–)
and

$$T'_{1a} = 1.0 + 4.939\left(1.0 + 0.0752\upsilon + 0.8109\upsilon^2\right)\left(\frac{D}{L}\right)^2 - 0.4883\left(\frac{D}{L}\right)^4 \quad (2.185)$$

$$T'_{1b} = 4.691\left(1.0 + 0.2023\upsilon + 2.173\upsilon^2\right)\left(\frac{D}{L}\right)^4 \quad (2.186)$$

$$T'_{1c} = 1.4.754\left(1.0 + 0.1408\upsilon + 1.536\upsilon^2\right)\left(\frac{D}{L}\right)^2 \quad (2.187)$$

Note: Adequate units for all parameters must be used throughout the calculations.

2.10.4.6 Damping

It is possible to estimate the signal damping by the relation below (Eq. 2.188). In fact, the signal damping could be used to quantify the internal friction of the sample [436]. Generally, for most of materials, increasing temperature lead to the damping signal increased. In other words, this parameter can be useful to verify or follow the microstructural modifications of the sample (phase change, damage …).

Note: For a better understanding of the sample microstructural modifications, it could be necessary to couple the damping with others experimental technics and/or literature.

$$Damping = \frac{Dissipated\ Energy}{Total\ Energy} \tag{2.188}$$

The obtained data from the Impulse excitation test must be treated with care, and the validity of the test verified, even for comparison purposes. The correct application of this characterization technique gives valuable information to: (1) develop and optimize the sustainable clay bricks production processes using biomass and biowastes; (2) and to understand the impact of raw materials and process parameters as the temperature in the quality and performance of monoliths.

2.10.4.7 Example 1 on Young Modulus Measurement

For this example, the impulse excitation technique is used to measure the dynamic Young's modulus of a monolith material obtained from a mixture of clay and 4 wt % of RDF (Refuse-Derived Fuel). This kind of material has been chosen because it appears in recent years as a relevant answer in the challenge offered to the waste and biomass valorization in the civil engineering field [437–439] by producing sustainable-fired clay bricks. In fact, the biomass addition in a clay brick leads during the firing process to the microstructural transformations (porosity with different shape, etc.) that can affect the ceramics resistance [158, 440]. That is why it is important to follow its evolution with the temperature by measuring the Young's modulus.

Sample Preparation

The clay/RDF monolith characterized in this example was obtained by extrusion and has the initial dimensions of 180 × 80 × 18 mm^3. A preparation method including cutting and polishing is used to manufacture the tests samples with the specific dimensions shown in Table 2.100. This table shows also the measurement uncertainties and the criteria for compliance with the standard ASTM E1876-01 [441]. For this sample, the measurement uncertainties are very low and the criteria for compliance with the standard are respected.

Measurement Procedure

The experimental procedure included the following steps:

- Support system with the steel wire suspensions (located at the fundamental nodal points) were installed in the furnace;
- All the electronics equipment were activated;
- The dimensions of the sample were measured as well as the mass. These data were inserted in the software;

Table 2.100 Sample dimensions and bulk density, criteria for compliance with standard and temperature program

Sample	Clay + 4%.wt of RDF ceramic
Dimensions and bulk density	
Mass m (g)	20.807
Length L (mm)	59.427
Width b (mm)	31.890
Thickness t (mm)	5.883
Bulk density (kg/m^3)	1866.200
Poisson's ratio	0.275
Measurement uncertainties	
Mass m (g)	0.001
Length L (mm)	0.010
Width b (mm)	0.001
Thickness t (mm)	0.001
Temperature T (°C)	0.100
Criteria for compliance with ASTM E1876-01	
C1: Smooth appearance sample	Yes
C2: m < 0.1%	0.0048%
C3: L < 0.1%	0.0168%
C4: b < 0.1%	0.0031%
C5: t < 0.1%	0.0170%
Temperature program	
Temperature range (°C)	20–1050–20
Heating rate (°C/min)	5

- The microphone and impulser position were checked;
- The furnace has been closed and the temperature program was entered;
- Interval time of 40 s was selected for the automatic measurement;
- The start button was pressed to begin the test.

The HT650 has dedicated software that automatically makes the necessary calculations (Eqs. 2.180–2.183) to produce the Young's modulus with the temperature evolution.

Results

Figure 2.264 shows the Young's modulus evolution as function of the temperature. Each point in this figure represents the Young's modulus value at a given temperature during the heating and the cooling. The apparatus automatically performed the measurement every 40 s during the heating/cooling process and the Young's

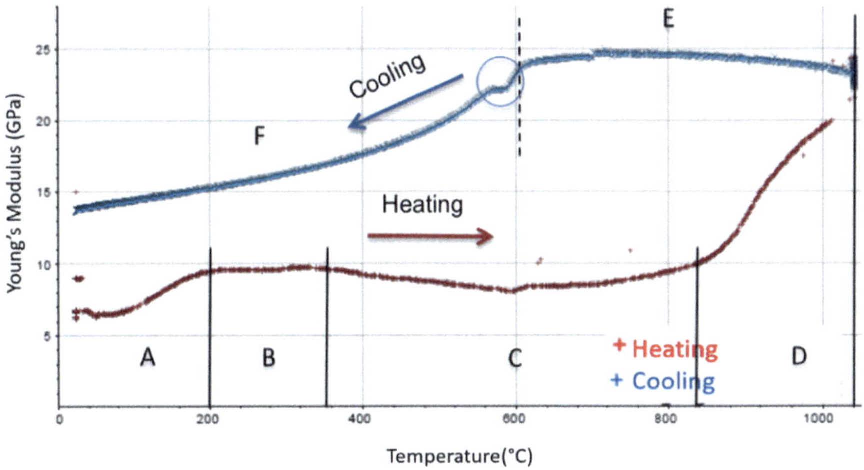

Fig. 2.264 Young's modulus evolution with the temperature [442]

modulus is calculated by Eq. 2.142. The curve is obtained by extracting the data and plotting using adapted software such as excel.

For the material characterized in this case, Fig. 2.264 shows four domains (A to D) during the heating:

- Domain A, room temperature to 200 °C: the Young's modulus increased from 6.8 to 9.5 GPa. In this range of temperature, this behavior is generally related to the material dehydration
- Domain B: 200–350 °C: the mechanical resistance is stable; there is no modification of the Young's modulus that has the mean value of 9.8 GPa
- Domain C: 350–850 °C: the results reveal firstly a slight decrease of the mechanical resistance before to increase after a temperature of 600 °C. The lower value measured is 7.5 GPa at 600 °C. This is usually related to the decomposition of the biomass used in the mixture which takes place between 350–600 °C.
- Domain D: 850–1050 °C: the results show an exponential increase of the Young's modulus from 10 to 23 GPa due to the material densification accompanied with microstructural modifications.

During the cooling, two domains can be identified:

- Domain E: The Young's modulus is stable between 1050 and 600 °C with a mean value of 24.5 GPa meaning that the sample keeps the same mechanical resistance
- Domain F: 600 °C to ambient temperature, the mechanical property slowly decreases from 24 to 16 GPa at ambient temperature. This indicates that this sample fired at 1050 °C has a Young' modulus of 16 GPa at ambient temperature. The peak identified at around 600 °C represented by the circle can be associated to the allotropic transformation of the quartz present in the clay.

The behavior presented in this study suggested that the mechanical resistance of this sample is high at high firing temperature, and during the firing process, the decomposition of the incorporated biomass leads to a weakening of the material (only during this step).

2.10.4.8 Example 2 on Young Modulus Measurement

This example is based on the work of Pierre-Marie NIGAY [158]. This work was focused on the study of microstructural modifications of clay/biomass mixtures during the firing process and the relationship with the thermal and mechanical properties.

Organic additives used in this study concern olive stone flour (FNO), wheat flour (FBL) in the form of powder and wheat straw (PB1) in the form of fibers. The Young's modulus evolution with the temperature of the ceramics were measured by impulse excitation method using a RFDA HT650 machine. The temperature program is 30–1050 °C with heating rate of 5 °C/min. This technique was also used to compare ceramics by keeping the same testing conditions.

Samples preparation and measurements procedure are similar than presented before (Sections "Sample Preparation" and "Measurement Procedure").

Figure 2.265 shows the evolution of the Young's modulus with the temperature between the room temperature to 1050 °C for clay ceramics with and without

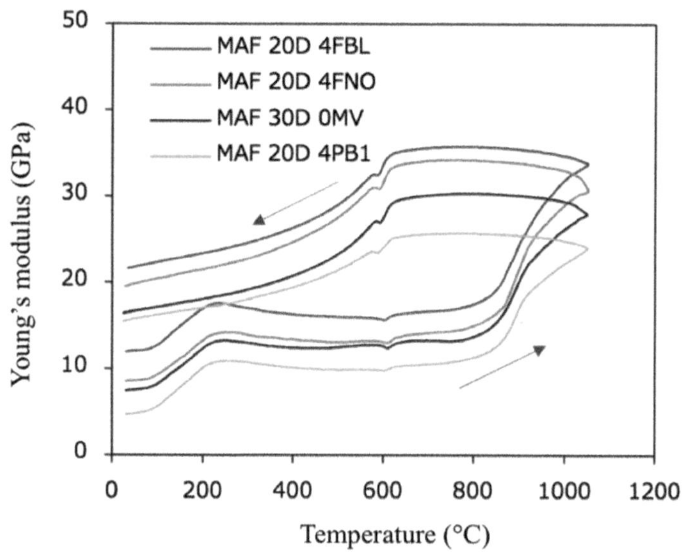

Fig. 2.265 Evolution of the Young's modulus with the temperature. Reproduced with permission from Nigay [158]

biomass mixture. The results reveal the strong dependence between the microstructure modifications during the firing process and the mechanical properties of the ceramics. Also, it was seen that the biomass addition in the conventional clay ceramics increases the mechanical properties (Young's modulus). There is a relationship between the shape, the nature of the biomass added to the clay ceramics and the mechanical resistance.

2.11 General Conclusions

The valorization of biomass, biowaste and by-product streams into energy and useful materials appears as a highly-efficient way to reduce carbon footprint for the society and generate wealth. However, they are complex and have heterogeneous structures composed with organic and inorganic components. The choice of appropriate analysis and characterization methods and tools with revelant accuracy are crucial for any work related to these resources.

This chapter is devoted to the generic and advanced characterization methods which are available in the literature. Recently-developed characterization techniques for biomass, biowaste and related by-products are presented. These characterizations are gathered into different families going from sampling and storage to proximate and ultimate analysis, thermal analysis, physico-chemical, thermal and mechanical characterization. For each method, the parameter studied was defined, followed by measurement principle, apparatus and protocol description, and usually finished by some examples of application.

With this chapter, readers can find most of the characterization methods, which enable to correctly characterize complex mediums which are biomass, biowaste and related by-products.

References

1. https://www.bp.com/en/global/corporate/energy-economics/statistical-review-of-world-energy/primary-energy/energy-and-the-environment.html. Accessed 06 August 2019
2. EN 14899: Characterization of waste—sampling of waste materials—framework for the preparation and application of a sampling plan (2006)
3. FD CEN/TR 15310-1: Characterization of waste—sampling of waste materials—Part 1: Guidance on selection and application of criteria for sampling under various conditions, Mar 2007
4. FD CEN/TR 15310-2: Characterization of waste—sampling of waste materials—Part 2: Guidance on sampling techniques
5. FD CEN/TR 15310-3: Characterization of waste—sampling of waste materials—Part 3: Guidance on procedures for sub-sampling in the field (2007)
6. FD CEN/TR 15310-4: Characterization of waste—sampling of waste materials—Part 4: Guidance on procedures for sample packaging, storage, preservation, transport and delivery (2007)

7. FD CEN/TR 15310-5: Characterization of waste—sampling of waste materials—Part 5: Guidance on the process of defining the sampling plan (2007)
8. ISO 13833: Stationary source emissions—determination of the ratio of biomass (biogenic) and fossil-derived carbon dioxide—radiocarbon sampling and determination (2013)
9. NF X 30-413: Déchets ménagers et assimilés—Constitution d'un échantillon de déchets ménagers et assimilés contenus dans une benne à ordures ménagères (2006)
10. NF X 30-437: Déchets ménagers et assimilés—Constitution et caractérisation, en entrée de centres de tri, d'un échantillon sur un lot de déchets ménagers et assimilés collectés sélectivement, Mars 2009
11. EN 15442: Solid recovered fuels—methods for sampling (2011)
12. ASTM D7204-15: Standard practice for sampling waste streams on conveyors. ASTM international, West Conshohocken, PA (2015). https://www.astm.org/
13. EN 12255-8: Wastewater treatment plants—Part 8: Sludge treatment and storage (2001)
14. EN 15413: Solid recovered fuels—methods for the preparation of the test sample from the laboratory sample (2011)
15. ISO 6322-1: Storage of cereals and pulses—Part 1: General considerations for the keeping of cereals (1996)
16. ISO 20023: Solid biofuels—safety of solid biofuels pellets—safe handling and storage of wood pellets in residential and other small-scale applications (2018)
17. EN 15002: Characterization of waste—preparation of test portions from the laboratory sample (2015)
18. EN 15443: Solid recovered fuels—methods for the preparation of the laboratory sample (2011)
19. EN 16179: Sludge, treated biowaste and soil—guidance for sample pretreatment (2012)
20. ASTM D5231-92: Standard test method for determination of the composition of unprocessed municipal solid waste. ASTM International, West Conshohocken, PA (2016). https://www.astm.org/
21. Nazaroff, W.W., Alvarez-Cohen Lisa.: Environmental Engineering Science. Wiley, New York, p 22 (2001)
22. Strezov, V., Evans, T.J.: Biomass Processing Technologies. CRC Press, Taylor and Francis Group, Boca Raton (2015)
23. ISO 18134: Solid biofuels—determination of moisture content—oven dry method (2015)
24. ASTM E790-15: Standard test method for residual moisture in refuse-derived fuel analysis samples. ASTM International, West Conshohocken (2015). https://www.astm.org/
25. ASTM D3173/D3173M-17a, Standard test method for moisture in the analysis sample of coal and coke. ASTM International, West Conshohocken (2017). https://www.astm.org/
26. ASTM D4442-16: Standard test methods for direct moisture content measurement of wood and wood-based materials. ASTM International, West Conshohocken, PA (2016). https://www.astm.org/
27. ASTM E1358-97: Standard test method for determination of moisture content of particulate wood fuels using a microwave oven, ASTM International, West Conshohocken (2013). https://www.astm.org/
28. ISO 11722: Solid mineral fuels—hard coal—determination of moisture in the general analysis test sample by drying in nitrogen (2013)
29. Samuelsson, R., Burvall, J., Jirjis, R.: Comparison of different methods for the determination of moisture content in biomass. Biomass Bioenergy **30**, 929–934 (2006)
30. Muhr, H.-J., Rohner, R.: Good titration practice in Karl Fischer titration. Mettler Toledo, Printed in Switzerland (2011)
31. Bruttel, P., Schlink, R.: Water determination by Karl Fischer titration. Metrohm, Printed in Switzerland (2006)
32. ISO 18123: Solid biofuels—determination of the content of volatile matter (2015)
33. ASTM D1762-84: Standard test method for chemical analysis of wood charcoal. ASTM International, West Conshohocken (2013). https://www.astm.org/

34. ASTM D3175-17: Standard test method for volatile matter in the analysis sample of coal and coke. ASTM International, West Conshohocken (2017). https://www.astm.org/
35. Enders, A., Lehmann, J.: Proximate analyses for characterising biochar. In: Singh, B., Camps-Aberstain, M., Lehmann, J. (eds.) Biochar: A Guide to Analytical Methods, pp. 9–22. CRC Press, Taylor and Francis Group, Boca Raton (2017)
36. ISO 18122: Solid biofuels—determination of ash content (2015)
37. ASTM D3174-12: Standard test method for ash in the analysis sample of coal and coke from coal. ASTM International, West Conshohocken (2012). https://www.astm.org/
38. ASTM D1102-84: Standard test method for ash in wood. ASTM International, West Conshohocken (2013). https://www.astm.org/
39. ASTM E1755-01: Standard test method for ash in biomass. ASTM International, West Conshohocken (2015). https://www.astm.org/
40. ASTM E887-15: Standard test method for silica in refuse-derived fuel (RDF) and RDF ash. ASTM International, West Conshohocken (2015). https://www.astm.org/
41. Antal, M.J., Grønli, M.: The art, science, and technology of charcoal production. Ind. Eng. Chem. Res. **42**, 1619–1640 (2003)
42. Phyllis database: https://www.ecn.nl/phyllis2/. Accessed 06 Aug 2019
43. Tanger, P., Field, J.L., Jahn, C.E., DeFoort, M.W., Leach, J.E.: Biomass for thermochemical conversion: targets and challenges. Front. Plant Sci. 4, article 218 (2013)
44. de Wild, P.J., den Uil, H., Reith, J.H., Kiel, J.H.A., Heeres, H.J.: Biomass valorisation by staged degasification: a new pyrolysis-based thermochemical conversion option to produce value-added chemicals from lignocellulosic biomass. J. Anal. Appl. Pyrol. **85**, 124–133 (2009)
45. Kiely, G.: Environmental Engineering. McGraw-Hill, New York (1997)
46. He, M., Xiao, B., Liu, S., Guo, X., Luo, S., Xu, Z., Feng, Y., Hu, Z.: Hydrogen-rich gas from catalytic steam gasification of municipal solid waste (MSW): influence of steam to MSW ratios and weight hourly space velocity on gas production and composition. Inter. J. Hydr. Ener. **34**, 2174–2183 (2009)
47. Cen, K., Chi, Y., Wang, F.: Challenges of power engineering and environment. Zhejiang University Press, Springer, Hangzhou (2007)
48. Xia, G., Ni, M., Chi, Y., Zhang, J., Mao, Q., Cen, K.: Research on low emission MSW gasification and melting system. Front. Env. Sci. Eng. China **1**, 498–503 (2007)
49. Onwudili, J.A., Williams, P.T.: Hydrothermal catalytic gasification of municipal solid waste. Ener. Fuels **21**, 3676–3683 (2007)
50. Suliman, W., Harsh, J.B., Abu-Lail, N.I., Fortuna, A.-M., Dallmeyer, I., Garcia-Perez, M.: Influence of feedstock source and pyrolysis temperature on biochar bulk and surface properties. Biomass Bioener. **84**, 37–48 (2016)
51. ASTM D5291-16: Standard test methods for instrumental determination of carbon, hydrogen, and nitrogen in petroleum products and lubricants. ASTM International, West Conshohocken (2016). https://www.astm.org/
52. Committee TRS of C-AM: AMC technical briefs—CHNS elemental analysers. AMCTB (2008)
53. Krotz, L., Leone, F., Galotta, W., Giazzi, G.: High automation of thermo scientific flash smart CHNS/O analyzer using the MultiValve Control (MVC) module—TN42256. Thermoscientific Tech Note
54. ISO 16948: Solid biofuels—determination of content of carbon, hydrogen and nitrogen (2015)
55. Zhu, Q.: Coal sampling and analysis standards. IEA Clean Coal Centre, London (2014)
56. ASTM D5373-16: Standard test methods for determination of carbon, hydrogen and nitrogen in analysis samples of coal and carbon in analysis samples of coal and coke. ASTM International, West Conshohocken (2016). https://www.astm.org/
57. Romero Millán, L.M., Sierra Vargas, F.E., Nzihou, A.: Kinetic analysis of tropical lignocellulosic agrowaste pyrolysis. Bioenergy Res. **10**, 832–845 (2017)

58. Arena, U., Ardolino, F., Di Gregorio, F.: A life cycle assessment of environmental performances of two combustion-and gasification-based waste-to-energy technologies. Waste Manage. **41**, 60–74 (2015)
59. Dean, J.R.: Practical Inductively Coupled Plasma Spectroscopy. Wiley, New York (2005)
60. Thompson, M., Walsh, J.N., Walton, S.J., Hall, G.E.M.: Handbook of Inductively Coupled Plasma Spectrometry. Blackie & Son Ltd. (1989)
61. Mester, Z., Sturgeon, R.E.: Sample Preparation for Trace Element Analysis. Elsevier Science (2003)
62. EN 16170: Sludge, treated biowaste and soil—determination of elements using inductively coupled plasma optical emission spectrometry (ICP-OES) (2016)
63. ISO 16967: Solid biofuels—determination of major elements—Al, Ca, Fe, Mg, P, K, Si, Na and Ti (2015)
64. Van der Sloot, H., Heasman, L., Quevauviller, P.: Harmonization of Leaching/Extraction Tests. Elsevier Science & Technology Books (1997)
65. Sun, Y., Chi, P., Shiue, M.: Comparison of different digestion methods for total decomposition of siliceous and organic environmental samples. Anal. Sci. **17**, 1395–1399 (2001)
66. Dreblow, K.: Safety-relevant aspects of microwave digestion, technical memorandum. Berghof products and instruments Gmbh. https://www.berghof-instruments.com/en/insights/article/safe-microwave-digestion/. Accessed 18 Nov 2019. (2017)
67. EN 15410: Methods for the determination of the content of major elements (Al, Ca, Fe, K, Mg, Na, P, Si, Ti) (2011)
68. EN 13656: Characterization of waste—microwave assisted digestion with hydrofluoric (HF), nitric (HNO_3) and hydrochloric (HCl) acid mixture for subsequent determination of elements (2002)
69. Low, F., Zhang, L.: Microwave digestion for the quantification of inorganic elements in coal and coal ash using ICP-OES. Talanta **101**, 346–352 (2012)
70. Coedo, A.G., Dorado, M.T., Padilla, I., Alguacil, F.J.: Use of boric acid to improve the microwave-assisted dissolution process to determine fluoride forming elements in steels by flow injection inductively coupled plasma mass spectrometry. J. Anal. At. Spectrom. **13**, 1193–1197 (1998)
71. De Moraes Flores, E.M.: Microwave-assisted sample preparation for trace element determination. Elsevier B.V. (2014)
72. https://www.bruker.com. Accessed 06 Aug 2019
73. https://www.thermofisher.com. Accessed 06 Aug 2019
74. https://www.malvernpanalytical.com. Accessed 06 Aug 2019
75. Goldstein, J., Newbury, D.E., Joy, D.C., Lyman, C.E., Echlin, P., Lifshin, E., Sawyer, L., Michael, J.R.: Scanning electron microscopy and X-ray analysis. Plenum Press, Springer, p. 341 (2003)
76. Morgan, T.J., George, A., Boulamanti, A.K., Álvarez, P., Adanouj, I., Dean, C., Vassilev, S.V., Baxter, D., Andersen, L.K.: Quantitative X-ray fluorescence analysis of biomass (switchgrass, corn stover, eucalyptus, beech, and pine wood) with a typical commercial multi-element method on a WD-XRF spectrometer. Energy Fuels **29**, 1669–1685 (2015)
77. Andersen, L.K., Morgan, T.J., Boulamanti, A.K., Álvarez, P., Vassilev, S.V., Baxter, D.: Quantitative X-ray fluorescence analysis of biomass: objective evaluation of a typical commercial multi-element method on a WD-XRF spectrometer. Energy Fuels **27**, 7439–7454 (2013)
78. Xing, P., Mason, P.E., Chilton, S., Lloyd, S., Jones, J.M., Williams, A., Nimmo, W., Pourkashanian, M.: A comparative assessment of biomass ash preparation methods using X-ray fluorescence and wet chemical analysis. Fuel **182**, 161–165 (2016)
79. Thyrel, M., Samuelsson, R., Finell, M., Lestander, T.A.: Critical ash elements in biorefinery feedstock determined by X-ray spectroscopy. Appl. Ener. **102**, 1288–1294 (2013)
80. EN 15408: Solid recovered fuels—methods for the determination of sulphur (S), chlorine (Cl), fluorine (F) and bromine (Br) content (2011)

81. Hitachi High-Tech Science corporation: Measuring controlled substances in halogen using fluorescent X-ray analysis—SEA No. 35 (2008)
82. ISO 16994: Solid biofuels—determination of total content of sulfur and chlorine (2015)
83. https://www.shimadzu.com/an/toc/lab/index.html. Accessed 06 Aug 2019
84. https://www.mt.com/ca/en/home/products/Process-Analytics/Total-Organic-Carbon-TOC-analyzer/Thornton-TOC-sensor-5000i.html. Accessed 06 Aug 2019
85. https://www.geinstruments.com/products/sievers-toc-analyzers-and-sensors-overview. Accessed 06 Aug 2019
86. Pham Minh, D.: Oxydation par voie humide catalytique des eaux usées de la production d'huile d'olives: Catalyseurs métaux nobles supportés. Ph.D. thesis, order number 2006-116, University Claude Bernard Lyon 1 (2006)
87. ISO 10304-1: Determination of dissolved anions by liquid chromatography of ions—Part 1: Determination of bromide, chloride, fluoride, nitrate, nitrite, phosphate and sulfate (2009)
88. ISO 17294-2: Water quality—application of inductively coupled plasma mass spectrometry (ICP-MS)—Part 2: Determination of selected elements including uranium isotopes (2016)
89. ISO 9297: Water quality—determination of chloride—silver nitrate titration with chromate indicator (Mohr's method) (2000)
90. ISO 10359-1: Water quality—determination of fluoride—Part 1: Electrochemical probe method for potable and lightly polluted water (1992)
91. EN 14582: Characterization of waste—halogen and sulfur content—oxygen combustion in closed systems and determination methods (2016)
92. EN 15408: Methods for the determination of sulphur (S),chlorine (Cl), fluorine (F) and bromine (Br) content (2011)
93. ISO 16994: Solid biofuels—determination of total content of sulfur and chlorine (2016)
94. EN 15414-3: Solid recovered fuels. Determination of moisture content using the oven dry method. Moisture in general analysis sample (2011)
95. Wunderlich, B.: Thermal analysis, Academic Press (1990)
96. Gabbott, P.: Principles and Applications of Thermal Analysis. Blackwell publishing (2008)
97. Lever, T., Haines, P., Rouquerol, J., Charsley, E.L., Van Eckeren, P., Burlett, D.J.: ICTAC nomenclature of thermal analysis (IUPAC Recommendations 2014). Pure Appl. Chem. **86**, 545–553 (2014)
98. Brown, M.E.: Introduction to Thermal Analysis: Techniques and Applications. Kluwer Academic Publishers, New York (2004)
99. Wirth, E., Guitteny, F., Mathonat, C.: Thermogravimétrie - Techniques de l'ingénieur P1260 v3 (Publication date: 10 June 2014). http://www.techniques-ingenieur.fr
100. https://www.nist.gov/. Accessed 06 Aug 2019
101. https://www.lgcstandards.com/FR/fr/. Accessed 06 Aug 2019
102. Nijman, M., Fedelich, N., Hammer, A., Hempel, E., Jing, N., Riesen, R., Schawe, J., Schubnell, M., Wrana, C.: Thermal Analysis in Practice, Tips and Hints. Mettler Toledo (2016)
103. Yang, H., Yan, R., Chen, H., Lee, D.H., Zheng, C.: Characteristics of hemicellulose, cellulose and lignin pyrolysis. Fuel **86**, 1781–1788 (2007)
104. Vyazovkin, S., Burnham, A.K., Criado, J.M., Pérez-Maqueda, L.A., Popescu, C., Sbirrazzuoli, N.: ICTAC Kinetics Committee recommendations for performing kinetic computations on thermal analysis data. Thermochim. Acta **520**, 1–19 (2011)
105. Burnham, A.K., Introduction to chemical kinetics. In: Burnham, A.K. (eds.) Global Chemical Kinetics of Fossil Fuels: How to Model Maturation and pyrolysis, pp. 25–74. Springer International Publishing (2017)
106. Brown, M. E.: Principles and practice. In: Brown, M.E., Gallagher, P.K. (eds.) Handbook of Thermal Analysis and Calorimetry, pp. 1–691. Elsevier (1998)
107. Šimon, P.: Isoconversional methods fundamentals, meaning and application. J. Therm. Anal. Calorim. **76**, 123–132 (2004)

108. Friedman, H.L.: Kinetics of thermal degradation of char-forming plastics from thermogravimetry. Application to a phenolic plastic. J. Polym. Sci. Part C Polym. Symp. 6, 183–195 (1964)
109. Akahira, T., Sunose, T.: Method of determining activation deterioration constant of electrical insulating materials. Res. Rep. Chiba Inst. Technol. **16**, 22–31 (1971)
110. Flynn, J.H., Wall, L.A.: General treatment of the thermogravimetry of polymers. J. Res. Natl. Bur. Stand. Phys. Chem. 70A, 487–523 (1966)
111. Starink, M.: The determination of activation energy from linear heating rate experiments: a comparison of the accuracy of isoconversion methods. Thermochim. Acta **404**, 163–176 (2003)
112. Gotor, F.J., Criado, J.M., Malek, J., Koga, N.: Kinetic analysis of solid-state reactions: the universality of master plots for analyzing isothermal and nonisothermal experiments. J. Phys. Chem. A **104**, 10777–10782 (2000)
113. Ozawa, T.: Non-isothermal Kinetics and Generalized Time, vol. 100, pp. 109–118. Elsevier Sci. Publ. B.V (1986)
114. Sanchez-Jimenez, P.E., Pérez-Maqueda, L.A., Perejon, A., Criado, J.M.: Generalized kinetic master plots for the thermal degradation of polymers following a random scission mechanism. J. Phys. Chem. A **114**, 7868–7876 (2010)
115. Vyazovkin, S.: Isoconversional kinetics of thermally stimulated processes. Springer International Publishing, Switzerland (2015)
116. Vyazovkin, S.: Alternative description of process kinetics. Thermochim. Acta **211**, 181–187 (1992)
117. Basu, P.: Biomass Gasification and Pyrolysis: Practical Design and Theory. Elsevier (2010)
118. Sánchez-Jiménez, P.E., Pérez-Maqueda, L.A., Perejón, A., Criado, J.M.: Generalized master plots as a straightforward approach for determining the kinetic model: the case of cellulose pyrolysis. Thermochim. Acta **552**, 54–59 (2013)
119. Samuelsson, L.N., Moriana, R., Babler, M.U., Ek, M., Engvall, K.: Model-free rate expression for thermal decomposition processes: the case of microcrystalline cellulose pyrolysis. Fuel **143**, 438–447 (2015)
120. Burnham, A.K., Zhou, X., Broadbelt, L.J.: Critical review of the global chemical kinetics of cellulose thermal decomposition. Energy Fuels **29**, 2906–2918 (2015)
121. Capart, R., Khezami, L., Burnham, A.K.: Assessment of various kinetic models for the pyrolysis of a microgranular cellulose. Thermochim. Acta **417**, 79–89 (2004)
122. Morin, M., Pécate, S., Masi, E., Hémati, M.: Kinetic study and modelling of char combustion in TGA in isothermal conditions. Fuel **203**, 522–536 (2017)
123. Dupont, C., Jacob, S., Marrakchy, K.O., Hognon, C., Grateau, M., Labalette, F., Perez, D.D. S.: How inorganic elements of biomass influence char steam gasification kinetics. Energy **109**, 430–435 (2016)
124. Marquez-Montesinos, F., Cordero, T., Rodriguez-Mirasol, J., Rodriguez, J.: CO_2 and steam gasification of a grapefruit skin char. Fuel **81**, 423–429 (2002)
125. Brown, M.E., Maciejewski, M., Vyazovkin, S., Nomen, R., Sempere, J., Burnham, A., Opfermann, J., Strey, R., Anderson, H.L., Kemmler, A., Keuleers, R., Janssens, J., Desseyn, H.O., Li, C.-R., Tang, T.B., Roduit, B., Malek, J., Mitsuhashi, T.: Computational aspects of kinetic analysis. Part A: The ICTAC kinetics project: data, methods, and results. Thermochim. Acta 355, 125–143 (2000)
126. Agarwal, A.K., Brisk, M.L.: Sequential experimental design for precise parameter estimation 1. Use of reparameterization. Ind. Eng. Chem. Process Des. Dev. 24, 203–207 (1985)
127. David, C., Salvador, S., Dirion, J.L., Quintard, M.: Determination of a reaction scheme for cardboard thermal degradation using thermal gravimetric analysis. J. Anal. Appl. Pyrol. **67**, 307–323 (2003)
128. MATLAB and Optimization Toolbox Release. The MathWorks, Inc., Natick, Massachusetts, United States (2012)
129. Schmidt, A.B.: Probability distributions. In: Schmidt, A. B. (eds.) Quantitative Finance for Physicists, pp. 17–27. Academic Press Advanced Finance (2005)

130. Bhattacharya, P.K., Burman, P.: Theory and methods of statistics. Elsevier (2016)
131. Weibull, W.: A statistical distribution function of wide applicability. J. Appl. Mech. 293–297 (1951)
132. Johnson, N.L., Kotz, S., Balakrishnan, N.: Continuous Univariate Distributions, vol. 1. Wiley-Interscience Publication, Wiley, New York (1994)
133. Fraser, R.D.B., Suzuki, E.: Resolution of overlapping absorption bands by least squares procedures. Anal. Chem. **38**, 1770–1773 (1966)
134. Gravalos, I. Kateris, D., Xyradakis, P., Gialamas, T., Loutridis, S., Augousti, A., Georgiades, A., Tsiropoulos, Z.: A study on calorific energy values of biomass residue pellets for heating purposes. In: Proceedings on forest engineering: meeting the needs of the society and the environment, vol. 1114. Padova, Italy (2010)
135. Montes, C.S., da Silva, D.A., Garcia, R.A., Bolzón de Muñiz, G.I., Weber, J.C.: Calorific value of *Prosopis africana* and balanites aegyptiaca wood: relationships with tree growth, wood density and rainfall gradients in the West African Sahel. Biomass Bioener. 35, 346–353 (2011)
136. Erakhrumen, A.A.: Energy value as a factor of agroforestry wood species selectivity in Akinyele and Ido local government areas of Oyo State, Nigeria. Biomass Bioenergy **33**, 1428–1434 (2009)
137. Telmo, C., Lousada, J.: The explained variation by lignin and extractive contents on higher heating value of wood. Biomass Bioenergy **35**, 1663–1667 (2011)
138. Günther, B., Gebauer, K., Barkowski, R., Rosenthal, M., Bues, C.-T.: Calorific value of selected wood species and wood products. Eur. J. Wood Wood Prod. **70**, 755–757 (2012)
139. Patel, B.: Biomass characterization and its use as solid fuel for combustion. Iran. J. Energy Environ. **3**, 123–128 (2012)
140. Basu, P.: Biomass Gasification and Pyrolysis: Practical Design and Theory. Academic Press, Burlington (2010)
141. IKA®—WERKE: IKA—Calorimeter system, C5000 duo-control, Operating Instructions (2004)
142. DIN 51900-3: Testing of solid and liquid fuels—determination of gross calorific value by the bomb calorimeter and calculation of net calorific value—Part 3: Method of using adiabatic jacket (2005)
143. Gravalos, I., Xyradakis, P., Kateris, D., Gialamas, T., Bartzialis, D., Giannoulis, K.: An experimental determination of gross calorific value of different agroforestry species and bio-based industry residues. Nat. Resour. **7**, 57–68 (2016)
144. CEN/TS 14918: Solid biofuels—method for the determination of calorific value (2005)
145. EN 15400: Solid recovered fuels—determination of calorific value (2011)
146. EN 15414-3: Solid recovered fuels—determination of moisture content using the oven dry method (2011)
147. EN 15400: Combustibles solides de récupération - Méthodes de détermination du pouvoir calorifique (2011)
148. Legendre, B.: Détermination des capacités thermiques spécifiques en fonction de la température. Techniques de l'ingénieur, R2970
149. http://www.setaram.fr/. Accessed 06 Aug 2019
150. Bohren, G.F., Huffman, D.R.: Absorption and scattering of light by small particles. Wiley Interscience (1998)
151. Kerker, M.: The Scattering of Light and Other Electromagnetic Radiations. Academic Press (1969)
152. Gouesbet, G., Gréhan, G.: Diffusion des faisceaux laser par des particules. Techniques de l'ingénieur, AF3460
153. http://nte.mines-albi.fr/STP/co/uc_RepresentationsGraphiques.html. Accessed 06 Aug 2019
154. Barth, H.G.: Modern methods of particle size analysis. Wiley Interscience (1984)
155. Seville, J.P.K., Coury, J.R., Ghadiri, M., Clift, R.: Comparison of techniques for measuring the size of fine non-spherical particles. Part. Part. Syst. Charact. **1**(1–4), 45–52 (1984)
156. ISO 13320: Analyse granulométrique - Méthodes par diffraction laser (2009)

157. Charlou, C.: Caractérisation et modélisation de l'écoulement de boues résiduaires in un sécheur à palettes. Ph.D. thesis, IMT Mines Albi (2014)
158. Nigay, P.M.: Etude des transformations microstructurales de mélanges argile/biomasse lors de la cuisson et relations avec les propriétés mécaniques et thermiques. Ph.D. thesis, IMT Mines Albi (2015)
159. World Health Organization, Bulk density and Tapped density of powder—Document QAS/ 11.450 Final, Mar 2012
160. Farrugia, A.: Relations procédé-microstructure-comportement de composites à matrice vitrocéramique mis en œuvre par voie liquide. Ph.D. thesis, IMT Mines Albi (2013)
161. NF P94-410-3: Rock—tests for physical properties of rock—Part 3 : determination of porosity (2001)
162. Serp, P., Figueiredo, J.L.: Carbon Materials for Catalysis. Wiley, Hoboken (2009)
163. Lehmann, J., Joseph, S.: Biochar for environmental management: science, technology and implementation. Taylor and Francis Group, London (2015)
164. Liu, W.-J., Zeng, F.-X., Jiang, H., Zhang, X.-S.: Preparation of high adsorption capacity bio-chars from waste biomass. Bioresour. Technol. **102**, 8247–8252 (2011)
165. Hervy, M., Berhanu, S., Weiss-Hortala, E., Chesnaud, A., Gérente, C., Villot, A., Pham Minh, D., Thorel, A., Le Coq, L., Nzihou, A.: Multi-scale characterisation of chars mineral species for tar cracking. Fuel **189**, 88–97 (2017)
166. Barroso-Bogeat, A., Alexandre-Franco, M., Fernández-González, C., Gómez-Serrano, V.: Preparation of activated carbon–metal oxide hybrid catalysts: textural characterization. Fuel Process. Technol. **126**, 95–103 (2014)
167. Qian, K., Kumar, A., Zhang, H., Bellmer, D., Huhnke, R.: Recent advances in utilization of biochar. Renew. Sust. Ener. Rev. **42**, 1055–1064 (2015)
168. Gunzler, H., Gremlich, H.U.: IR Spectroscopy, an Introduction. Wiley-VCH (2002)
169. Stuart, B.H.: Infrared Spectroscopy: Fundamentals and Applications. Wiley, New York (2004)
170. Theophanides, T.: Infrared Spectroscopy—Materials Science, Engineering and Technology. InTech (2012)
171. Larkin, P. Infrared and Raman Spectroscopy: Principles and Spectral Interpretation. Elsevier, Amsterdam (2011)
172. Lehto, J., Louhelainen, J., Huttunen, M., Alén, R.: Spectroscopic analysis of hot-water- and dilute-acid-extracted hardwood and softwood chips. Spectrochim. Acta A: Mol/Biomol. Spectrosc. **184**, 184–190 (2017)
173. Rao, C.N.R.: Ultra-violet and Visible Spectroscopy: Chemical Applications. Butterworths (1975)
174. Tkachenko, N.: Optical Spectroscopy: Methods and Instrumentations. Elsevier Science (2006)
175. Lindon, J.: Encyclopedia of Spectroscopy and Spectrometry. Academic Press (2016)
176. Chan, K.S., Tan, J.: Understanding Advanced Organic and Analytical Chemistry: the Learner's Approach. World Scientific Publishing, p. 483 (2011)
177. Vahabi, H., Shabanian, M., Aryanasab, F., Mangin, R., Laoutid, F., Saeb, M.R.: Inclusion of modified lignocellulose and nano-hydroxyapatite in development of new bio-based adjuvant flame retardant for poly(lactic acid). Thermochim. Acta **666**, 51–59 (2018)
178. Perkin-Elmer. Demountable cell diagram. http://www.interchim.fr/cat/C6-C12-Infra-Rouge.pdf. Accessed 06 Aug 2019
179. Ofudje, E.A., Rajendran, A., Adeogun, A.I., Idowu, M.A., Kareem, S.O., Pattanayak, D.K.: Synthesis of organic derived hydroxyapatite scaffold from pig bone waste for tissue engineering applications. Adv. Powder Technol. **29**, 1–8 (2018)
180. Yadav, L.D.S.: Organic spectroscopy. Springer Science (2005)
181. Lee, R.A., Bédard, C., Berberi, V., Beauchet, R., Lavoie, J.M.: UV–Vis as quantification tool for solubilized lignin following a single-shot steam process. Bioresour. Technol. **144**, 658–663 (2013)

182. Giannakopoulos, E., Isari, E., Bourikas, K., Karapanagioti, H. K., Psarras, G., Oron, G., Kalavrouziotis, I. K.: Oxidation of municipal wastewater by free radicals mechanism. A UV/ Vis spectroscopy study. J. Env. Manag. 195, 186–194 (2017)
183. Iqbal Bhanger, S.M., Niaz, A., Shah, A., Rauf, A.: Ultra-trace level determination of hydroquinone in waste photographic solutions by UV–vis spectrophotometry. Talanta **72**, 546–553 (2007)
184. Vandenabeele, P.: Practical Raman Spectroscopy—An Introduction. Wiley, New York (2013)
185. Scrader, B.: Infrared and Raman spectroscopy: methods and applications. VCH Verlagsgesellschaft, Weinheim (1995)
186. Diening, T., Hollricher, O., Toporski, J.: Confocal Raman microscopy. Springer, Berlin (2010)
187. WITec User Manual of the Raman microscope Alpha300—confocal Raman configuration for automated measurements part
188. WITEC: https://www.witec.de/. Accessed 05 May 2019
189. Xia, Y., Gates, B., Yin, Y., Lu, Y.: Monodispersed colloidal spheres: old materials with new applications. Adv. Mat. **12**, 693–713 (2000)
190. Shaw. D.J.: Introduction to Colloid & Interface Science. Elsevier Science Ltd. (1992)
191. Derjaguin, B., Landau, L.: Theory of the stability of strongly charged lyophobic sols and of the adhesion of strongly charged particles in solutions of electrolytes. Progress Surf. Sci. **43**, 30–59 (1993)
192. Verwey, E.J.W., Overbeek, J.T.G.: Theory of the stability of lyophobic colloids. Elsevier, Amsterdam (1948)
193. Bergström, L.: Hamaker constants in inorganic materials. Adv. Colloid. Interf. Sci. **70**, 125–169 (1997)
194. Butt, H.J., Graf, K., Kappl, M.: Physics and chemistry of interfaces. Wiley-VCH, Weinheim (2003)
195. Lyklema. J.: Fundamentals of Interface and Colloid Science. Elsevier, Amsterdam (1991)
196. Helmholtz. H.: Studien über electrische Grenzschichten. Ann. Phys. Chem. 337–282 (1879)
197. Mécanisme de l'électrisation de contact et solutions colloidales: Perrin. J. J. Chim. Phys. **2**, 601–651 (1904)
198. Gouy, M.: Sur la constitution de la charge électrique à la surface d'un electrolyte. J. de Phys. **9**, 457–468 (1910)
199. Chapman, D.L.: A contribution to the theory of electrcapillarity. Philos. Mag. **25**, 475–481 (1913)
200. Stern, O.: Zur theorie des electrolytischen. Z. Elektrochem. 30, 508–516 (1924)
201. Grahame, D.C.: The electrical double layer and the theory of electrocapillarity. Chem. Rev. **41**, 441–501 (1947)
202. Parson. R.: The electrical double layer at solid/liquid interfaces. J. Electroanal. Interf. Electrochem. 118, 3–18 (1981)
203. Parks, G.A.: The isoelectric points of solid oxides, solids hydroxides, and aqueous hydroxo complexes systems. Chem. Rev. **65**, 177–198 (1965)
204. Morison, S.R.: The Solid Liquid Interface, Electrochemistry at Semi-conductor and Oxidized Metal Electrodes, p. 49. Plenum Press, New York (1980)
205. Von Smoluchowski, M.: Contribution to the theory of electro-osmosis and related phenomena. Bull. Int. Acad. Sci. Cracovie **3**, 184–199 (1903)
206. Russel, W.B., Saville, D.A., Schowalter, W.R.: Colloidal dispersions. Cambridge University Press (1992)
207. Fang, H.Y., Chaney, R.C.: Introduction to Environmental Geotechnology. CRC Press (2016)
208. Masliyah, J.H., Bhattacharjee, S.: Electrokinetic and Colloid Transport Phenomena. Wiley, New Jersey (2006)
209. Wiersema, P.H., Loeb, A., Overbeek, J.Th.G.: Calculation of the electrophoretic mobility of a spherical colloid particle. J. Colloid Interf. Sci. 22, 78–99 (1966)
210. Hunter, R.J.: Zeta potential in colloid science. Academic Press, London (1981)

211. Hunter, R.J.: Foundations of colloids science. Oxford University Press, New York (2001)
212. Henry, D.C.M.A: The cataphoresis of suspended particles. Part I—The equation of cataphoresis. Proc. Roy. Soc. Lond. Ser. A 133, 106–129 (1931)
213. Ohshima, H.: Electrostatic interaction between a hard sphere with constant surface charge density and a soft sphere: polarization effect of a hard sphere. J. Colloid. Interf. Sci. **168**, 255–265 (1994)
214. O'Brien, R.W., White. L.R.: Electrophoretic mobility of a spherical colloidal particle. J. Chem. Soc. Faraday Trans. II 74, 1607–1626 (1978)
215. https://www.malvernpanalytical.com/fr/products/product-range/zetasizer-range/zetasizer-nano-range. Accessed 06 Aug 2019
216. http://www.horiba.com/fr/scientific/products/particle-characterization/technology/zeta-potential/. Accessed 06 Aug 2019
217. http://www.occhio-instruments.fr/zetametre_mesure_potentiel_zeta.html. Accessed 06 Aug 2019
218. Graber, E. R., Singh, B., Hanley, K., Lehmann, J.: Determination of cation exchange capacity in biochar. In: Singh, B., Camps-Aberstain, M., Lehmann, J. (eds.) Biochar: a guide to analytical methods, pp. 74–84. CRC Press, Taylor and Francis Group (2017)
219. NF X31-130: Soil quality—chemical methods—determination of cationic exchange capacity (CEC) and extractible cations (1999)
220. Laird, D., Fleming, P.: Analysis of layer charge, cation and anion exchange capacities, and synthesis of reduced charge clays. Methods Soil Anal. Part 5. Mineral. Methods, 485–508 (2008)
221. ISO 23470: Soil quality—determination of effective cation exchange capacity (CEC) and exchangeable cations using a hexamminecobalt trichloride solution (2011)
222. ISO 23470: Soil quality—determination of effective cation exchange capacity (CEC) and exchangeable cations using a hexamminecobalt trichloride solution (2018)
223. ISO 11260: Soil quality—determination of effective cation exchange capacity (CEC) and base saturation level using barium chloride solution (2011)
224. ISO 11260: Determination of effective cation exchange capacity (CEC) and base saturation level using barium chloride solution (2018)
225. ISO 11464: Soil quality—pretreatment of samples for physico-chemical analysis (2006)
226. Cieselski, H., Sterckeman, T., Santerne, M., Willery, J.P.: A comparison between three methods for the determination of cation exchange capacity and exchangeable cations in soils. Agronomie **17**, 9–16 (1997)
227. ISO 11732: Water quality—determination of ammonium nitrogen—method by flow analysis (CFA and FIA) and spectrometric detection (2005)
228. Denyes, M.J., Parisien, M.A., Rutter, A., Zeeb, B.A.: Physical, chemical and biological characterization of six biochars produced for the remediation of contaminated sites. J. Vis. Exp. **93**, 1–12 (2014)
229. Chapman, H.D.: Cation-exchange capacity. In: Norman, A.G. (eds.) Methods of soil analysis. Part 2. Chemical and microbiological properties, pp. 891–901. Madison, Wisconsin (1965)
230. Dohrmann, R.: Cation exchange capacity methodology I: an efficient model for the detection of incorrect cation exchange capacity and exchangeable cation results. Appl. Clay Sci. **34**, 31–37 (2006)
231. Oryschak, M., Chrapko, D., Huggins-Rawlins, N., Jones, S., Korth, C., Kryzanowski, L., Neden, J., Pauly, D., Tremblett, K., Wallace, T.: Nutrient management planning guide for Alberta (Chap. 3.4). https://www1.agric.gov.ab.ca/$department/deptdocs.nsf/all/epw11920/$FILE/3-4.pdf. Accessed 06 Aug 2019
232. Ok, Y.S., Uchimiya, S.M., Chang, S.X., Bolan, N.: Biochar: Production, Characterization, and Applications. CRC Press, Taylor and Francis Group, Boca Raton (2016)
233. Boehm, H.P.: Functional groups on the surfaces of solids. Angew. Chem. **5**, 533–544 (1966)
234. Graber, E. R., Tsechansky, L., Fidel, R. B., Thompson, M. L., Laird, D. A.: Determining acidic groups at biochar surfaces via the Boehm titration. In: Singh, B., Camps-Aberstain,

M., Lehmann, J. (eds.) Biochar: A Guide to Analytical Methods, pp. 85–94. CRC Press, Taylor and Francis Group, Boca Raton (2017)
235. Boehm, H.P.: Surface oxides of carbon. Angew. Chem. **3**, 669–677 (1964)
236. Voll, M., Boehm, H.P.: Basische Oberflächenoxide auf Kohlenstoff-IV. Chemische Reaktionen zur Identifizierung der Oberflächengruppen, Carbon **9**, 481–488 (1971)
237. Boehm, H.P.: Some aspects of the surface chemistry of carbon blacks and other carbons. Carbon **32**, 759–769 (1994)
238. Fidalgo, B., Menéndez, J.Á.: Carbon materials as catalysts for decomposition and CO_2 reforming of methane: a review. Chin. J. Catal. **32**, 207–216 (2011)
239. Suelves, I., Pinilla, J.L., Lázaro, M.J., Moliner, R.: Carbonaceous materials as catalysts for decomposition of methane. Chem. Eng. J. **140**, 432–438 (2008)
240. Muradov, N., Smith, F., Ali, T.: Catalytic activity of carbons for methane decomposition reaction. Catal. Today 102–103, 225–233 (2005)
241. Lázaro, M.J., Pinilla, J.L., Suelves, I., Moliner, R.: Study of the deactivation mechanism of carbon blacks used in methane decomposition. Int. J. Hydr. Energy **33**, 4104–4111 (2008)
242. Ducousso, M., Weiss-Hortala, E., Nzihou, A., Castaldi, M.J.: Reactivity enhancement of gasification biochars for catalytic applications. Fuel **159**, 491–499 (2015)
243. Hervy, M., Pham Minh, D., Gérente, C., Weiss-Hortala, E., Nzihou, A., Villot, A., Le Coq, L.: H_2S removal from syngas using wastes pyrolysis chars. Chem. Eng. J. **334**, 2179–2189 (2018)
244. Figueiredo, J.L., Pereira, M.F.R.: The role of surface chemistry in catalysis with carbons. Catal. Today **150**, 2–7 (2010)
245. Figueiredo, J.L., Pereira, M.F.R., Freitas, M.M.A., Orfao, J.J.M.: Modification of the surface chemistry of activated carbons. Carbon **37**, 1379–1389 (1999)
246. Goertzen, S.L., Thériault, K.D., Oickle, A.M., Tarasuk, A.C., Andreas, H.A.: Standardization of the Boehm titration. Part I. CO_2 expulsion and endpoint determination. Carbon 48, 1252–1261 (2010)
247. Oickle, A.M., Goertzen, S.L., Hopper, K.R., Abdalla, Y.O., Andreas, H.A.: Standardization of the Boehm titration. Part II. Method of agitation, effect of filtering and dilute titrant. Carbon 48, 3313–3322 (2010)
248. Fidel, R.B., Laird, D.A., Thompson, M.L.: Evaluation of modified Boehm titration methods for use with biochars. J. Env. Qual. **42**, 1771–1778 (2013)
249. Tsechansky, L., Graber, E.R.: Methodological limitations to determining acidic groups at biochar surfaces via Boehm titration. Carbon **66**, 730–733 (2014)
250. Nzihou, A., Stanmore, B.: The fate of heavy metals during combustion and gasification of contaminated biomass—a brief review. J. Hazard. Mat. **256–257**, 56–66 (2013)
251. Suliman, W., Harsh, J.B., Abu-Lail, N.I., Fortuna, A.-M., Dallmeyer, I., Garcia-Perez, M.: Influence of feedstock source and pyrolysis temperature on biochar bulk and surface properties. Biomass Bioenergy **84**, 37–48 (2016)
252. Mukherjee, A., Zimmerman, A.R., Harris, W.: Surface chemistry variations among a series of laboratory-produced biochars. Geoderma **163**, 247–255 (2011)
253. Yang, H.-H., Lee, W.-J., Chen, S.-J., Lai, S.-O.: PAH emission from various industrial stacks. J. Hazard. Mat. **60**, 159–174 (1998)
254. Ramesh, A., Archibong, A.E., Hood, D.B., Guo, Z., Loganathan, B.G.: Global environmental distribution and human health effects of polycyclic aromatic hydrocarbons. In: Loganathan, B.G., Lam, P.K.S. (eds.) Global contamination trends of persistent organic chemicals. CRC Press, Taylor & Francis Group, Boca Raton, London (2011)
255. Lerda, D.: Polycyclic aromatic hydrocarbons (PAHs) factsheet. https://ec.europa.eu/jrc/sites/jrcsh/files/Factsheet%20PAH_0.pdf. Accessed 06 Aug 2019
256. Directive 2000/76/EC of the European Parliament and of the Council Emissions from incineration plants, PAHs (Mentioned as carcinogenic compounds that might be subject to limitations in Member States' regulations)

257. Decision No 2455/2001/EC of the European Parliament and the Council, Water, PAHs (Annex: Identified as priority hazardous substance), Benzo[a]pyrene, Benzo[b]fluoranthene, Benzo[ghi]perylene, Benzo[k]fluoranthene, Indeno[1,2,3-cd]pyrene
258. Regulation (EC) No 2065/2003 of the European Parliament and of the Council, Primary smoke products, Benzo[a]pyrene, Benzo[a]anthracene
259. Directive 2004/107/EC of the European Parliament and the Council, Ambient air, Benzo[a]pyrene
260. Commission Regulation (EC) No 627/2006, Primary smoke products, 15 SCF PAHs, Benzo[a]pyrene, Benzo[a]anthracene
261. EN 16181: Soil, treated biowaste and sludge—determination of polycyclic aromatic hydrocarbons (PAH) by gas chromatography (GC) and high performance liquid chromatography (HPLC) (2018)
262. ISO 18135: Solid biofuels—sampling (2017)
263. ISO 14780: Solid biofuels—sample preparation (2017)
264. EN 15527: Characterization of waste—determination of polycyclic aromatic hydrocarbons (PAH) in waste using gas chromatography mass spectrometry (GC/MS) (2008)
265. ISO 22892: Soil quality—guidelines for the identification of target compounds by gas chromatography and mass spectrometry (2011)
266. Fabbri, D., Rombolà, A.G., Torri, C., Spokas, K.A.: Determination of polycyclic aromatic hydrocarbons in biochar and biochar amended soil. J. Anal. Appl. Pyrol. **103**, 60–67 (2013)
267. Singh, B., Camps-Arbestain, M., Lehmann, J.: Biochar: A Guide to Analytical Methods. CSIRO Publishing, Melbourne (2017)
268. Fujita, S., Watanabe, H., Katagiri, A., Yoshida, H., Arai, M.: Nitrogen and oxygen doped metal-free carbon catalysts for chemoselective transfer hydrogenation of nitrobenzene, styrene, and 3-nitrostyrene with hydrazine. J. Mol. Catal. A: Chem. 393, 257–62 (2014)
269. Laine, N.R., Vastola, F.J., Walker Jr., P.N.: The importance of active surface area in the carbon-oxygen reaction. J. Phys. Chem. **67**, 2030–2034 (1963)
270. Ehrburger, P., Louys, F., Lahaye, J.: The concept of active sites applied to the study of carbon reactivity. Carbon **27**, 389–393 (1989)
271. Burg, P., Abraham, M. H., Cagniant, D.: Methods of determining polar and non-polar sites on carbonaceous adsorbents. The contribution of the linear solvation energy relationship approach. Carbon 41, 867–879 (2003)
272. Tisserand, C., Calvet, R., Patry, S., Galet, L., Doods, J.A.: Comparison of two techniques for the surface analysis of alumina (Al_2O_3): inverse gas chromatography at finite concentration (IGC-FC) and dynamic vapor sorption (DVS). Powd. Technol. **190**, 53–58 (2009)
273. Bingol, G., Prakash, B., Pan, Z.: Dynamic vapor sorption isotherms of medium grain rice varieties. LWT Food Sci. Technol. **48**, 156–163 (2012)
274. Bisson, A., Rodier, E., Rigacci, A., Lecomte, D., Achard, P.: Study of evaporative drying of treated silica gels. J. Non-Crystalline Solids **350**, 230–237 (2004)
275. Surface Measurement System®, DVS user manual (DVS1 and DVS2) (2000)
276. https://www.surfacemeasurementsystems.com/dvs-advantage-2/. Accessed 06 Aug 2019
277. ASTM E2551-13: Standard Test Method for Humidity Calibration (or Conformation) of Humidity Generators for Use with Thermogravimetric Analyzers. ASTM International, West Conshohocken (2013). https://www.astm.org/
278. Fletcher, A.J., Yüzak, Y., Thomas, K.M.: Adsorption and desorption kinetics for hydrophilic and hydrophobic vapors on activated carbon. Carbon **44**, 989–1004 (2006)
279. Sullivan, P.D., Stone, B.R., Hashisho, Z., Rood, M.J.: Water adsorption with hysteresis effect onto microporous activated carbon fabrics. Adsorption 13, 173–189 (2007)
280. Sing, K.S.W., Everett, D.H., Haul, R.A.W., Moscou, L., Pierotti, R.A., Rouquérol, J., Siemieniewska, T.: Reporting physisorption data for gas/solid systems with special reference to the determination of surface area and porosity. Pure Appl. Chem. **57**, 603–619 (1985)
281. Burnett, D., Thielmann, F.: Organic solvent sorption using a dynamic vapour sorption instrument-an overview. Application Note 103, Surface Measurement Systems Ltd., Last Updated 24/11/06

282. Thielmann, F.: Hysteresis effects in vapour sorption. Application Note 37, Surface Measurement Systems Ltd
283. Burnett, D., Thielmann, F., Booth, J.: Determining the heat of sorption on organic and inorganic powders using DVS. Application Note 19, Surface Measurement Systems Ltd
284. Thielmann, F., Burnett, D.: Isotherm types and adsorption mechanisms of solvents on pharmaceutical excipients. Application Note 26, Surface Measurement Systems Ltd
285. Williams, D.R., Levoguer, C.L.: Measuring BET surface areas using organic probe molecules. Application note 18, Surface Measurement Systems
286. Levoguer, C.L.: Measuring the moisture sorption kinetics of cements using DVS. Application Note 09, Surface Measurement Systems Ltd
287. Burnett, D., Thielmann, F., Sokoloski, T., Wilson, R.: Calculation of diffusion constants in a pharmaceutical powder using DVS. Application Note 30, Surface Measurement Systems Ltd
288. Levoguer, C.L., Williams, D.R.: Measurement of the surface energies of pharmaceutical powders using a novel vapour adsorption method. Application Note 17, Surface Measurement Systems Ltd
289. Burnett, D., Thielmann, F.: A new gravimetric method to calculate low levels of amorphous content. Application Note 34, Surface Measurement Systems Ltd
290. Burnett, D., Thielmann, F.: Investigation of hydrate formation and loss using the DVS. Application Note 36, Surface Measurement Systems Ltd
291. Rouquerol, F., Rouquerol, J., Sing, K.: Adsorption by Powders and Porous Solids: Principles, Methodology and Applications. Academic Press, London (1998)
292. Lowell, S., Shields, J.E., Thomas, M.A., Thommes, M.: Characterization of porous solids and powders: surface area, pore size and density. Chapman & Hall, London (1991)
293. Lowell, S., Shields, J.E.: Powder surface area and porosity. Chapman & Hall, London (1991)
294. Rouquerol, F., Luciani, L., Lewellyn, P., Denoyel, R., Rouquetol, J.: Texture des matériaux pulvérulaents ou poreux. Techniques de l'ingénieur, P 1050 v3 (2003). Published on Mar 10, 2003. https://www.techniques-ingenieur.fr/base-documentaire/tian0-archives-materiaux-fonctionnels/download/p1050/3/texture-des-materiaux-pulverulents-ou-poreux.html. Accessed 06 Aug 2019
295. Thomas, M., Kaneko, K., Neimark, A. V., Olivier, J.P., Rodriguez-Reinoso, F., Rouquerol, J., Sing, K.S.W.: Physisorption of gases, with special reference to the evaluation of surface area and pore size distribution (IUPAC Technical Report). Pure Appl. Chem. **87**, 1051–1069 (Figs. 2–3) (2015)
296. Tascón, J.M.D.: Novel Carbon Adsorbents. Elsevier, Amsterdam (2012)
297. Seaton, N., Reinoso, F.R., Llewellyn, P., Kaskel, S.: Characterisation of Porous Solids VIII: Proceedings of the 8th International Symposium on the Characterisation of Porous Solids. RSC Publishing (2009)
298. ISO 9277: Determination of the specific surface area of solids by gas adsorption_BET method (2010)
299. Bottani, E., Tascón, J., Tascón, J.: Adsorption by Carbons. Elsevier Science (2008)
300. Lowell, S., Shields, J.E., Thomas, M.A., Thommes, M.: Characterization of Porous Solids and Powders: Surface Area, Pore Size and Density. Kluwer Academic Publishers, Boston (2004)
301. Brunauer, S., Deming, L.S., Deming, W.E., Teller, E.: On a theory of the Van der Waals adsorption of gases. J. Amer. Chem. Soc. **62**, 1723–1732 (1940)
302. Rouquerol, F., Rouquerol, J., Beurroies, I., Llewellyn, P., Denoyel, R.: Texture des matériaux divisés - Aire spécifique des matériaux pulvérulents ou nanoporeux, Techniques de l'ingénieur, P 1050 v4 (2017). Published on May 10, 2017. https://www.techniques-ingenieur.fr/base-documentaire/42534210-surfaces-et-structures-fonctionnelles/download/p1050/texture-des-materiaux-divises.html. Accessed 06 Aug 2019

303. www.deltalabo.fr/fr/catalogue/Elim.pdf. Accessed 06 Aug 2019
304. ISO 18757: Fine ceramics (advanced ceramics, advanced technical ceramics)—determination of specific surface area of ceramic powders by gas adsorption using the BET method (2003)
305. https://www.micromeritics.com/Repository/Files/apnote104.pdf. Accessed 11 Apr 2019
306. Llewellyn, P., Rodriguez-Reinoso, F., Rouquerol, J., Seaton, N.: Characterization of Porous Solids VII. Studies in Surface Science and Catalysis, vol. 160, pp. 49–56. Elsevier, Amsterdam (2007)
307. De Boer, J.H., Lippens, B.C., Linsen, B.G., Broekhoff, J.C.P., van den Heuvel, A., Osinga, ThJ: The *t*-curve of multimolecular N_2-adsorption. J. Colloid Interf. Sci. **21**, 405–414 (1966)
308. Pinto, M.L., Mestre, A.S., Carvalho, A.P., Pires, J.: Comparison of methods to obtain micropore size distributions of carbonaceous materials from CO_2 adsorption based on the Dubinin-Radushkevich isotherm. Ind. Eng. Chem. Res. **49**, 4726–4730 (2010)
309. Horvarth, G., Kawazoe, K.: Method for the calculation of effective pore size distribution in molecular sieve carbon. J. Chem. Eng. Japan 16, 470–475 (1983)
310. Saito, A., Foley, H.C.: Curvature and parametric sensitivity in models for adsorption in micropores. AIChE J. **3**, 429–436 (1991)
311. Cheng, L.S., Yang, R.T.: Predicting isotherms in micropores for different molecules and temperatures from a known isotherm by improved Horvath-Kawazoe equations. Adsorption **1**, 187–196 (1995)
312. Barrett, E.P., Joyner, L.G., Halenda, P.P.: The determination of pore volume and area distributions in porous substances. I. Computations from nitrogen isotherms. J. Am. Chem. Soc. 73, 373–380 (1951)
313. Joyner, L.G., Barrett, E.P., Skold,R.: The determination of pore volume and area distributions in porous substances. II. Comparison between nitrogen isotherm and mercury porosimeter methods. J. Am. Chem. Soc. 73, 3155–3158 (1951)
314. Reichenbach, C., Kalies, G., Enke, D., Klank, D.: Cavitation and pore blocking in nanoporous glasses. Langmuir **27**, 10699–10704 (2011)
315. Sing, K.S.W., Everet, D.H., Haul, R.A.W., Moscou, L., Pieroti, R.A., Rouquerol, J., Siemieniewska, T.: Reporting physisorption data for gas/solid systems with special reference to the determination of surface area and porosity. Pure Appl. Chem. 57, 603–619 (1985)
316. Sing, K.: The use of nitrogen adsorption for the characterisation of porous materials. Colloids Surf. A: Physicochem. Eng. Asp. **3**, 187–188 (2001)
317. Cazorla-Amorós, D., Alcañiz-Monge, J., Linares-Solano, A.: Characterization of activated carbon fibers by CO_2 adsorption. Langmuir **12**, 2820–2824 (1996)
318. Sigmund, G., Hüffer, T., Hofmann, T., Kah, M.: Biochar total surface area and total pore volume determined by N_2 and CO_2 physisorption are strongly influenced by degassing temperature. Sci. Total Environ. **580**, 770–775 (2017)
319. Lowell, S., Shields, J.E., Thomas, M.A., Thommes, M.: Particle Technology Series: Characterization of Porous Solids and Powders: Surface Area, Pore Size and Density. Kluwer Academic Publishers, Dordrecht (2004)
320. Thommes, M., Cychosz, K.A.: Physical adsorption characterization of nanoporous materials: progress and challenges. Adsorption **20**, 233–250 (2014)
321. Ravikovitch, P.I., Vishnyakov, A., Russo, R., Neimark, A.V.: Unified approach to pore size characterization of microporous carbonaceous materials from N_2, Ar, and CO_2 adsorption isotherms. Langmuir **16**, 2311–2320 (2000)
322. Do, D.D., Do, H.D.: Pore characterization of carbonaceous materials by DFT and GCMC simulations: a review. Adsorpt. Sci. Technol. **21**, 389–423 (2003)
323. Tarazona, P., Marini Bettolo Marconi, U., Evans, R.: Phase equilibria of fluid interfaces and confined fluids non-local versus local density functionals. Mol. Phys. 60, 573–595 (1987)
324. Landers, J., Gor, G.Y., Neimark, A.V.: Density functional theory methods for characterization of porous materials. Colloids Surf. A **437**, 3–32 (2013)

325. Balzer, C., Cimino, R.T., Gor, G.Y., Neimark, A.V., Reichenauer, G.: Deformation of microporous carbons during N_2, Ar, and CO_2 adsorption: insight from the density functional theory. Langmuir **32**, 8265–8274 (2016)
326. Frenkel, D.: Advanced Monte Carlo techniques. In: Allen, M.P., Tildesley, D.J. (eds.) Computer Simulation in Chemical Physics, pp. 93–152. Kluwer Academic Publishers, Dordrecht (1991)
327. Micromeritics Instrument Corporation: Mercury Porosemitry Theory poster
328. Webb, P.A., Orr, C.: Analytical Methods in Fine Particle Technology, pp. 165–171. Micromeritics, Norcross (1997)
329. Niemantsverdriet, J.W.: Temperature-programmed techniques. In: Niemantsverdriet, J.W. (eds.) Spectroscopy in Catalysis: An Introduction, pp. 11–38. Wiley-VCH Verlag GmbH & Co. KGaA (2007)
330. Rakić, V., Damjanović, L.: Temperature-programmed desorption (TPD) methods. In: Auroux, A. (eds.) Calorimetry and Thermal Methods in Catalysis, pp. 131–174. Springer, Berlin (2013)
331. Schwarz, J.A.: Temperature-programmed desorption and reaction: applications to supported catalysts. Catal. Rev. **25**, 141–227 (1983)
332. AutoChem 2920 Automated Catalyst Characterization System: Operator's Manual V4.00. http://www.micromeritics.com/repository/files/autochem_ii_2920_operator_manual_v4.00.pdf. Accessed 06 Aug 2019
333. http://www.quantachrome.com/chemisorption/pulsar.html. Accessed 06 Aug 2019
334. Pham Minh, D.: Oxydation par voie humide catalytique des eaux usées de la production d'huile d'olives: catalyseurs métaux nobles supportes. Ph.D. thesis, Université Lyon 1, p. 60 (2006)
335. Kanervo, J.M., Keskitalo, T.J., Slioor, R.I., Krause, A.O.I.: Temperature-programmed desorption as a tool to extract quantitative kinetic or energetic information for porous catalysts. J. Catal. **238**, 382–393 (2006)
336. Niwa, M., Katada, N., Sawa, M., Murakami, Y.: Temperature-programmed desorption of ammonia with readsorption based on the derived theoretical equation. J. Phys. Chem. **99**, 8812–8816 (1995)
337. Hahn, T.: Volume A: space-group symmetry. In: Hahn, T. (eds.) The international tables for crystallography. Springer, Berlin (2005)
338. https://www.rigaku.com. Accessed 06 Aug 2019
339. Pecharsky, V., Zavalij, P. Fundamentals of Powder Diffraction and Structural Characterization of Materials. Springer, Berlin (2009)
340. Jenkis, R., Snyder, R.: Introduction to X-ray Powder Diffraction. Wiley Publisher (1996)
341. Shih, K.: X-ray Diffraction: Structure, Principles and Applications. Nova Science Pub Inc. (2013)
342. Lee, M.: X-ray Diffraction for Materials Research: From Fundamentals to Applications. CRC Press (2016)
343. Waseda, Y., Matsubara, E., Shinoda, K.: X-Ray diffraction crystallography-introduction, examples and solved problems. Springer, Heidelberg (2011)
344. Speakman, S.A.: Basics of X-ray powder diffraction. Massachusetts, USA (2011). http://prism.mit.edu/xray/documents/1%20Basics%20of%20X-Ray%20Powder%20Diffraction.pdf. Accessed 06 Aug 2019
345. Cullity, B.D., Stock, S.R.: Elements of X-ray Diffraction. Pearson (2014)
346. Warren, B. E. (1969). X-ray Diffraction. Dover Publications, Inc., New York (1990)
347. Rietveld, H.M.: Line profiles of neutron powder-diffraction peaks for structure refinement. Acta Crystallogr. A **22**, 151–152 (1967)
348. Rietveld, H.M.: A profile refinement method for nuclear and magnetic structures. J. Appl. Crystallogr. **2**, 65–71 (1969)
349. Warren, B.E.: X-ray determination of the structure of liquids and glass. J. Appl. Phys. **8**, 645–654 (1937)

350. Egami, T., Billinge, S.J.L.: Underneath the Bragg Peaks: Structural Analysis of Complex Materials. Pergamon (2012)
351. www.ncnr.nist.gov/resources/n-lengths/. Accessed 06 Aug 2019
352. Peterson, P.F., Gutmann, M., Proffen, T., Billinge, S.J.L.: A user-friendly program to extract the total scattering structure function and the pair distribution function from neutron powder diffraction data. J. Appl. Crystallogr. **33**, 1192 (2000)
353. Qiu, X., Thompson, J.W., Billinge, S.J.L.: a GUI-driven program to obtain the pair distribution function from X-ray powder diffraction data. J. Appl. Crystallogr. **37**, 678 (2004)
354. Chupas, P.J., Chapman, K.W., Lee, P.L.: Applications of an amorphous silicon-based area detector for high-resolution, high-sensitivity and fast time-resolved pair distribution function measurements. J. Appl. Crystallogr. **40**, 463–470 (2007)
355. Proffen, T., Kim, H.: Advances in total scattering analysis. J. Mat. Chem. **19**, 5078–5088 (2009)
356. Dykhne, T., Taylor, R., Florence, A., Billinge, S.J.L.: Data requirements for the reliable use of atomic pair distribution functions in amorphous pharmaceutical fingerprinting. Pharmaceutical Res. **28**, 1041–1048 (2011)
357. Toby, B.H., Egami, T.: Accuracy of pair distribution function analysis applied to crystalline and non-crystalline materials. Acta Crystallogr. Sect. A **48**, 336–346 (1992)
358. Hafner, B.: Characterization Facility. University of Minnesota, Twin Cities, 16 Apr 2007
359. Environmental Scanning Electron Microscopy: An Introduction to ESEM. Philips Electron Optics, Eindhoven (1996)
360. Klinghoffer, N.B., Castaldi, M.J., Nzihou, A.: Catalyst properties and catalytic performance of char from biomass gasification. Ind. Eng. Chem. Res. **51**, 13113–13122 (2012)
361. Hauss G.: Microtomographie X de matériaux à comportement pseudo-fragile: Identification du réseau de fissures. Ph.D. thesis. Université Bordeaux 1, France (2012)
362. Maire, E., Salvo, L., Cloetens, P., DI Michel, M.: Tomographie à rayons X appliquée à l'étude des matériaux. Techniques de l'ingénieur, IN20 v1. Published on Sep 10, 2004. https://www.techniques-ingenieur.fr/. Accessed 22 May 2019
363. Zhang, L., Lee, W.-K., Wulff, M., Eybert, L.: The performance of a cryogenically cooled monochromator for an in-vacuum undulator beamline. J. Synchrotron Radiat. **10**, 313–319 (2003)
364. Espeso, J.I., Cloetens, P., Baruchel, J., Härtwig, J., Mairs, T., Biasci, J.C., Marot, G., Salomé-Pateyron, M., Schlenker, M.: Conserving the coherence and uniformity of third-generation synchrotron radiation beams: the case of ID19, a 'long' beamline at the ESRF. J. Synchrotron Rad. **5**, 1243–1249 (1998)
365. Schroer, C.G., Meyer, J., Kuhlmann, M., Benner, B., Günzler, T.F., Lengeler, B.: Nanotomography based on hard X-ray microscopy with refractive lenses. Appl. Phys. Lett. **81**, 1527–1529 (2002)
366. www.esrf.eu. Accessed 06 Aug 2019
367. Koch, A., Raven, C., Spanne, P., Snigirev, A.: X-ray imaging with submicrometer resolution employing transparent luminescent screens. JOSA A **15**, 1940–1951 (1998)
368. Maire, E., Withers, P.J.: Quantitative X-ray tomography. Int. Mater. Rev. **59**, 1–43 (2014)
369. EN 16016-1: Non destructive testing—radiation methods—computed tomography—Part 1: Terminology (2011)
370. EN 16016-2: Part 2: Principles, equipment and samples (2011)
371. EN 16016-3: Part 3: operation and interpretation (2011)
372. EN 16016-4: Part 4: qualification (2011)
373. Suzuki, M., Nishikawa, M., Ohsawa, H., Ogawa, K.: Clinical evaluation of TDF-500 system. In: Medical imaging II, vol. 914, pp. 1398–1406. Newport Beach, CA, United States (1988)
374. Kannan, T.S., Panda, P.K., Jaleel, V.A., Ramachandra Rao, R., Roopa, H.N., Mariappan, L., Cheluvaraju, A.: Synthesis, processing, sintering and characterisation of special ceramic materials. Met. Mater. Process. **12**, 161–180 (2000)

375. Niemantsverdriet, J.W.: Microscopy and imaging. In: Niemantsverdriet, J.W. (ed.) Spectroscopy in Catalysis, pp. 179–216. WILEY-VCH Verlag GmbH & Co. KGaA, Weinheim (2007)
376. Williams, D.B., Carter, C.B.: Transmission Electron Microscopy: A Textbook for Materials Science. Vol. 2: Diffraction. Plenum Press (1996)
377. Iourov, I.Y.: Microscopy and imaging. Spectrosc. Catal., 17–25 (2017)
378. Williams, D.B., Carter, C.B.: Transmission electron microscopy: a textbook for materials science, vol. 11. Springer, New York (2009)
379. https://www.nist.gov/laboratories/tools-instruments/nanofab-tool-fei-titan-80-300-analytical-transmission-electron. Accessed 06 Aug 2019
380. Egerton, R.F.: Physical principles of electron microscopy: an introduction to TEM, SEM, and AEM. Springer, US (2005)
381. http://www.nion.com/products.html. Accessed 06 Aug 2019
382. https://www.jeolusa.com/NEWS-EVENTS/Press-Releases/PostId/109/JEOL-Flash-Go-Speeds-Operation-of-ARM200F-Cold-FEG-Transmission-Electron-Microscope. Accessed 06 Aug 2019
383. Mukhopadhyay, S.M.: Sample preparation for microscopic and spectroscopic characterization of solid surfaces and films. In: Mitra S. (eds.) Sample Preparation Techniques in Analytical Chemistry, pp. 377–411. Wiley, New York (2003)
384. Donohoe, B.S., Ciesielski, P.N., Vinzant, T.B.: Preservation and preparation of lignocellulosic biomass samples for multi-scale microscopy analysis. In: Himmel, M.E. (ed.) Methods in Molecular Biololy, pp. 31–47. Humana Press, Totowa (2012)
385. Donohoe, B.S., Decker, S.R., Tucker, M.P., Himmel, M.E., Vinzant, T.B.: Visualizing lignin coalescence and migration through maize cell walls following thermochemical pretreatment. Biotechnol. Bioeng. **101**, 913–925 (2008)
386. Giberson, R.T., Demaree, R.S.: Microwave processing techniques for electron microscopy: a four-hour protocol. Electron. Microsc. Methods Protoc. **117**, 145–158 (1999)
387. https://www.fei.com/accessories/CompuStage-High-Visibility-Low-Background-Double-Tilt-Specimen-Holder/. Accessed 06 Aug 2019
388. Besser, A.I., Chen, Z.: FEI Tecnai F-20 Operations Manual. Center for electron microscopy and nanofabrication. Portland State University (last update 2010). https://www.pdx.edu/sites/www.pdx.edu.cemn/files/TEM_manual.pdf (Figure 2 on page 15). Accessed 06 Aug 2019
389. Bell, D.C., Garratt-Reed, A.J.: Energy dispersive X-ray analysis in the electron microscope. Bios Scientific Publishers Ltd., Oxford (2003)
390. Kawasaki, M., Oikawa, T., Ibe, K., Park, K.H., Shiojiri, M.: EDS elemental mapping of a DRAM with an FE-TEM. J. Electron Microsc. **47**, 335–343 (1998)
391. Shindo, D., Oikawa, T.: Energy dispersive X-ray spectroscopy. In: Shindo, D., Oikawa, T. (eds.) Analytical electron microscopy for materials science, pp. 81–102. Springer Japan, Tokyo (2002)
392. Hanefeld, U., Lefferts, L.: Catalysis: An Integrated Textbook for Students. Wiley-VCH (2017)
393. Valdez-Vazquez, I., Quiroz-Figueroa, F.R., Carrillo-Reyes, J., Medina-López, A.: Microscopy applied in biomass characterization. In: Vaz Jr., S. (ed.) Analytical Techniques and Methods for Biomass, pp. 173–196. Springer, Cham (2016)
394. Ptáček, P., Opravil, T., Šoukal, F., Havlica, J., Holešinský, R.: Kinetics and mechanism of formation of gehlenite, Al–Si spinel and anorthite from the mixture of kaolinite and calcite. Solid State Sci. **26**, 53–58 (2013)
395. Ptáček, P., Frajkorová, F., Šoukal, F., Opravil, T.: Kinetics and mechanism of three stages of thermal transformation of kaolinite to metakaolinite. Powder Technol. **264**, 439–445 (2014)
396. Ptáček, P., Šoukal, F., Opravil, T., Havlica, J., Brandštetr, J.: Crystallization of spinel phase from metakaoline: the nonisothermal thermodilatometric CRH study. Powder Technol. **243**, 40–45 (2013)
397. Awaji, H., Sato, S.: Diametral compressive testing method. J. Eng. Mater. Technol. **101**, 139–147 (1979)

398. EN 15234-2: Solid biofuels. Fuel quality assurance. Part 2: Wood pellets for non-industrial use (2012)
399. Obernberger, I., Thek, G.: The Pellet Handbook: The Production and Thermal Utilisation of Biomass Pellets. Earthscan from Routledge, Taylor and Francis Group (2010)
400. Kaliyan, N., Vance Morey, R.: Factors affecting strength and durability of densified biomass products. Biomass Bioenergy **33**, 337–359 (2009)
401. Tumuluru, J.S., Wright, C.T., Hess, R., Kenney, K.: A review of biomass densification systems to develop uniform feedstock commodities for bioenergy application. Biofuels Bioprod. Biorefin. 5, 683–707 (2011)
402. Peter, N., Nielsen, K., Holm, J.K., Felby, C.: Effect of fiber orientation on compression and frictional properties of sawdust particles in fuel pellet production. Energy Fuels **23**, 3211–3216 (2009)
403. Harun, N.Y., Afzal, M.T.: Effect of particle size on mechanical properties of pellets made from biomass blends. Procedia Eng. **148**, 93–99 (2016)
404. EN 12390-1: Testing hardened concrete—Part 1: Shape, dimensions and other requirements for specimens and moulds (2012)
405. EN 12390-6: Testing hardened concrete—tensile splitting strength of test specimen (2009)
406. Dorlot, J.M., Bailon, J.P., Masounave, J.: Des matériaux. Presses Polytechnique de Montréal (1993)
407. Wang, A.J., Yuan, J.L., Zhang, J., Liu, L.T., Li, J.M., Liu, Z.: Effect of raw material ratios on the compressive strength of magnesium potassium phosphate chemically bonded ceramics. Mat. Sci. Eng. C **33**, 5058–5063 (2013)
408. Suvorova, O., Kumarova, V., Nekipelov, D., Selivanova, E., Makarov, D., Masloboev, V.: Construction ceramics from ore dressing waste in Murmansk region, Russia. Construct. Build. Mat. **153**, 783–789 (2017)
409. EN 196-1: Methods of Testing Cement-Part 1: Determination of Strength (2016)
410. ISO 13061-17:2017, Physical and Mechanical Properties of Wood
411. NF P 94-422: Roches. Détermination de la résistance à la traction. Méthode indirecte - Essai brésilien (2001)
412. EN 12697-23: Bituminous mixtures. Test methods for hot mix asphalt. Part 23: Determination of the indirect tensile strength of bituminous specimens (2004)
413. ASTM D6931-17, Standard test method for indirect tensile (IDT) strength of asphalt mixtures. ASTM International, West Conshohocken (2017). https://www.astm.org/
414. EN 13286-42: Unbound and hydraulically bound mixtures. Part 42: Test method for the determination of the indirect tensile strength of hydraulically bound mixtures (2003)
415. EN 12390-3: Testing hardened concrete—Part 3: Compressive strength of test specimens (2009)
416. Fell, J.T., Newton, J.M.: Determination of tablet strength by diametral compression test. J. Pharm. Sci. **59**, 688–691 (1970)
417. Stasiak, M., Molenda, M., Bańda, M., Wiąceka, J., Parafiniuk, P., Gondek, E.: Mechanical and combustion properties of sawdust-Straw pellets blended in different proportions. Fuel Proc. Technol. **156**, 366–375 (2017)
418. ASTM C1161-13, Standard test method for flexural strength of advanced ceramics at ambient temperature. ASTM International, West Conshohocken. (2013). https://www.astm.org/
419. EN 843-1: Mechanical properties of monolithic ceramics at room temperature-determination of flexural strength (2006)
420. EN 12390-5: Testing hardened concrete-flexural strength of tests specimens (2009)
421. EN 310: Wood based panels—determination of modulus of elasticity in bending and of bending strength (1993)
422. EN 843-5: Advanced technical ceramics—mechanical properties of monolithic ceramics at room temperature—Part 5: Statistical analysis (2006)
423. technical ceramics)-test method for flexuISO 14704: Fine ceramics (advanced ceramics, advanced ral strength of monolithic ceramics at room temperature (2008)

424. ASTM C1239-13: Standard Practice for Reporting Uniaxial Strength Data and Estimating Weibull Distribution Parameters for Advanced Ceramics, ASTM International, West Conshohocken (2018). https://www.astm.org/
425. Kingery, W.D., Bowen, H.K., Uhlmann, D.R.: Introduction to ceramics. Wiley, New York (1976)
426. Yakub, I., Du, J., Soboyejo, W.O.: Mechanical properties, modeling and design of porous clay ceramics. Mat. Sci. Eng. A **558**, 21–29 (2012)
427. Korah, L.V., Nigay, P.M., Cutard, T., Nzihou, A., Thomas, S.: The impact of the particle shape of organic additives on the anisotropy of a clay ceramic and its thermal and mechanical properties. Construct. Build. Mat. **125**, 654–660 (2016)
428. Sodoke, K.F., Laperrière, L., Toubal, L., Khakestar, R.S.: Fuzzy logic response to Young's modulus characterization of a flax–epoxy natural fiber composite. Mat. Design **89**, 273–285 (2016)
429. Radovic, M., Lara-Curzio, E., Riester, L.: Comparison of different experimental techniques for determination of elastic properties of solids. Mat. Sci. Eng. A 188, 56–70 (2004)
430. Toubal, L., Karama, M., Lorrain, B.: Determination of the matrix of rigidity of a composite material by the combination of speckle interferometry and ultrasonic measurements. Appl. Mechanic. Mat. **3–4**, 155–160 (2005)
431. Raggio, L.I., Etcheverry, J., Sánchez, G., Bonadeo, N.: Error analysis of the impulse excitation of vibration measurement of acoustic velocities in steel samples. Phys. Proc. **3**, 297–303 (2010)
432. ASTM E1876-01: Standard Test Method for Dynamic Young's Modulus, Shear Modulus, and Poisson's Ratio by Impulse Excitation of Vibration. ASTM International, West Conshohocken. https://www.astm.org/
433. Etcheverry, J., Sánchez, G., Bonadeo, N., Raggio, L. I.: Analysis of the ASTM standards for impulse excitation of vibration and acoustic resonance techniques for rectangular parallelepipeds. VI Congreso Iberoamericano de Acústica—FIA (2008)
434. ASTM E1875-00e1, Standard Test Method for Dynamic Young's Modulus, Shear Modulus, and Poisson's Ratio by Sonic Resonance. ASTM International, West Conshohocken (2000). https://www.astm.org/
435. http://www.aurock.fr/. Accessed 06 Aug 2019
436. Zhang, J., Perez, R.J., Wong, C.R., Lavernia, E.J.: Effects of secondary phases on the damping behaviour of metals, alloys and metal matrix composites. Mat. Sci. Eng. **R13**, 325–390 (1994)
437. Muñoz, V.P., Morales, M.P., Letelier V.G., Mendívil, M.A.G.: Fired clay bricks made by adding wastes: Assessment of the impact on physical, mechanical and thermal properties. Constr. Build. Mater. 125, 241–252 (2016)
438. CCAC: Mitigating black carbon and other pollutants from brick production. Climate and Clean Air Coalition. http://www.ccacoalition.org/. Accessed 06 Aug 2019
439. Vo, L.T.T., Navard, P.: Treatments of plant biomass for cementitious building materials—a review. Constr. Build. Mater. **121**, 161–176 (2016)
440. Nigay, P.M., Cutard, T., Nzihou, A.: The impact of heat treatment on the microstructure of a clay ceramic and its thermal and mechanical properties. Ceram. Int. **43**, 1747–1754 (2017)
441. ASTM E1876-01: Standard test method for dynamic young's modulus, shear modulus and Poisson's ratio by impulse excitation of vibration. ASTM International, West Conshohocken (2001). https://www.astm.org/
442. Sani, R.: Étude des transformations microstructurales de mélanges argile/Combustibles Solides de Récupération (CSR) lors de la cuisson: relations entre propriétés physico-chimiques, mécaniques et thermiques. Ph.D. thesis, IMT Mines Albi (2018)

Chapter 3
Lignocellulosic Biomass

Thierry Ghislain, Xavier Duret, Papa Niokhor Diouf and Jean-Michel Lavoie

Abstract Lignocellulosic biomass is a complex mixture of biomacromolecules that varies significantly from one specie to the other and even throughout the same specie depending on many aspects such as geographical location, season, soil, etc. Cellulose-based material is the most abundant source of renewable carbon on earth and the increasing concerns toward reduction of GHG emissions around the globe will surely draw significantly more attention to it in the upcoming years. Non-edible biomass (also called second-generation biomass) is often divided in three categories, which are straws, softwood and hardwood. Each of these categories involves the same major macromolecules: cellulose, hemicelluloses and lignin. To a lesser extent, this biomass may also contain proteins and lipids (mostly for agricultural biomass belonging to the straw category). Finally, all contain extractives (or secondary metabolites) which amongst all constituents are the most unpredictable. Finally, the inorganic content of lignocellulosic biomass has also a major impact on its application, especially in the field of bioenergy. In this chapter, the focus will be given on providing different levels of quantification for each of these macromolecules. When, in some cases, only a general composition is sufficient, simple analysis based on gravimetric methods will be provided. However, some applications of biomass will require a deeper investigation of the different macromolecules and in that sense, more comprehensive protocols, involving state of the art analytics such as spectroscopy and chromatography will be provided.

T. Ghislain (✉) · X. Duret · J.-M. Lavoie
Department of Chemical Engineering and Biotechnological Engineering,
Université de Sherbrooke, 2500 Boulevard de l'Université, J1K 2R1 Sherbrooke, Canada
e-mail: Thierry.Ghislain@USherbrooke.ca

P. N. Diouf
SEREX, Rimouski College Technology Transfer Center, 25-5 Rue Armand Sinclair,
G5J 1K3 Amqui, Canada

A. Nzihou (ed.), *Handbook on Characterization of Biomass, Biowaste and Related By-products*, https://doi.org/10.1007/978-3-030-35020-8_3

3.1 Introduction

Following the COP21 conference in Paris, numerous countries around the world have committed in reducing their carbon footprint. This initiative is intended to answer to the growing concerns toward climate change, increasing sea levels and increasing desertification in certain parts of the globe. Around the world, 193 countries agreed that the situation needed to be addressed and they pledge towards specific GHG reductions. Only two countries (Syria and Nicaragua) did not commit to reduce their GHG while the US decided not to aim to meet its pledge after the agreement [1]. Amongst the 10 most important producers of GHG around the planet, China, which was the biggest producer of GHG in 2012 committed to maximize their emissions by 2030 (if not before) but agreed also to source about 20% of their energy out of low carbon sources. The European Union, which in 2012 was the third most important producer of GHG committed to at least 40% reduction below 1990 levels by 2030 [1]. India (4th most important producer) committed to a 33–35% emissions intensity reduction compared to 2005 levels by 2030 which is comparable to Brazil that pledged a 37% reduction below 2005 level by 2030. Canada, despite its small relatively small population was still the 8th most important GHG producer in 2012 [1]. Hence and following the global trend, Canada has committed in reducing its GHG emissions 30% below 2005 levels by 2030 [2]. In Quebec, one of the provinces of Canada, the target has been set to 37.5% below 1990s levels making it the most ambitious carbon footprint reduction in North America [3]. In order to reach such levels, many opportunities are considered including removing coal from thermal plants, increasing the utilization of electric/hybrid cars as well as the energy efficiency of buildings and of course, using renewable carbon as a source of energy for industry, homes and for transportation.

Biomass is a very abundant commodity in many locations around the world including but not limited to Northern Europe, North America. In Canada, although it is available readily across the country, the nature of the biomass itself depends on where it is harvested. Hence, while the eastern and western seaboards have access to abundant volumes of forest biomass, the central portion of the country has access to very large volumes of agricultural biomass. In both cases, the strong interest for renewable carbon fits perfectly with the actual challenges that both the forest and agricultural industry are facing. As an example, the forest industry has been integrated for decades in North America and from harvesting down to lumbering, different industries were using resources without competing for it. The pulp and paper industry were a major actor of this integrated system, using the wood chips produced from the lumbering industry to produce newspaper and other commodity papers. With the decline of this industry, a gap was created in the forest value chain and hence, resides an opportunity to solve two challenges with a unique opportunity.

Lignocellulosic biomass can be converted to energy through solid, liquid or gas fuels (as well as value-added products) and such conversion is often associated either to the "bio" or "thermo" approach. The latter essentially focuses on the calorific value of biomass and includes processes such as combustion, gasification, pyrolysis and torrefaction. While combustion aims at a total oxidation of the carbon and hydrogen content of biomass to produce energy, gasification aims at a partial oxidation of the feedstock providing a synthesis gas (syngas) that can downstream be used as a gas fuel for turbines or boilers. Syngas can also be used as a building block to produce liquid fuels such as methanol [4], ethanol [5] and alkanes through the Fischer-Tropsch synthesis [6]. The gas can also be used as building block for other value-added molecules such as ethylene. Pyrolysis and torrefaction both generally operated without any oxidizing agent and lead to a liquid and a solid product respectively. Pyrolysis of biomass leads to a mixture called bio-oil that can be used with minor upgrading as a raw heating fuel. It can as well be upgraded to fuel through processes comparable to petroleum cracking [7] or even used as additive or as a source of value-added molecules [8, 9]. Torrefaction (or slow pyrolysis) leads to a solid (char or biochar) commonly used as a solid biofuel for combustion in Combined Heat and Power (CHP) units. However, it can be as well used as a feed for other processes such as gasification [10] or pyrolysis [11]. Overall, when thermochemical (or "thermo") processes are considered, one of the most important aspects of biomass that needs to be addressed is the elemental analysis, water and ash content, Low Heating Value (LHV), High Heating Value (HHV) and thermogravimetric analysis.

Biological (or "bio") processes generally focus on the utilization of micro-organisms at one point during conversion of biomass into fuel/commodities. Although there is a very wide array of approaches being considered on this level, a larger majority of the work being done on the subject aims at using cellulose as a source of glucose. The latter could be used as a building block for many purposes but could mainly be fed into actual first-generation refineries to produce cellulosic ethanol [12]. However, targeting cellulose in lignocellulosic biomass requires as well taking into consideration the other macromolecules occurring in the feedstock which are usually lignin, hemicelluloses, lipids, extractives and proteins. While thermochemical processes aim at transforming a generally heterogeneous substrate into a homogeneous intermediary, bioprocesses often try to take advantage of the already existing complex structure of lignocellulosic substrates. Hence, although some conversion is done, there is often interest in maintaining the monomeric or the oligomeric integrity of some of the macromolecules involved [13]. This reality significantly increases the complexity of the analysis that are related to this type of biomass. Therefore, "bio" processes often require different information when compared to their thermo counterparts. Information such as ash and water content remain important in both cases, but in the case of "bio" approaches, data such as carbohydrate content (both C5 and C6), lignin (or fractions) macromolecular size, and inhibitor content, amino acids, free fatty ac-ids are often taken in consideration as well. Eventually, spectroscopic technologies such as NIR would be very beneficial for "bio" approaches, allowing a rapid quantification of the biomass' macromolecular content prior to any transformation process.

Hence, the growing interest towards lignocellulosic biomass as an alternative to fossil fuel in the next decade will require efficient analytical tools. The latter will have to be defined according to the targeted products in order to keep in consideration both the feedstock as well as the conversion process. In light of this, a company that will decide to produce pyrolytic oil and crack it down to diesel or jet fuel will have far different requirements when it comes to analytical support as compared to a group that would consider isolating cellulose from a lignocellulosic feedstock in order to produce second-generation ethanol. In this chapter, will be reviewed the different techniques that could be used to quantify the different properties of biomass accordingly to the type of end product targeted. Hence, by combining different tools it will be possible to adapt the different quantification methods to fit with hopefully most of the new technology that will reach market in the upcoming years. Since the second-generation biofuel industry is still on its infancy, this chapter aims at providing simple and cheap techniques as well as more expensive but fast approaches that will try as much as possible to fit with the numerous industrial context that will be using biomass as source of carbon-neutral molecules in the upcoming years.

3.2 Cellulose

3.2.1 General Description

Cellulose is the most important macromolecule present in the lignocellulosic biomass and the most important source of glucose in the world. The historical application of cellulose in the Industry was dedicated to the production of paper. Whereas, in the context of depletion of fossil resources, an increasing interest is now dedicated towards the production of biofuels (such as ethanol) from cellulose. Cellulose is the major structural component of plant cell walls explaining its abundance. It can also be found in bacteria, fungi, protozoans as well as animal's tunicate [14].

The amount of cellulose in lignocellulosic biomass depends mainly on the feedstock. As an example, the typical amount of cellulose in wood averages from 40 to 50% whereas the amount varies between 16 and 41% in bark [15, 16]. In wheat, barley and triticale straw, the cellulose content is in the 30–45% range [17]. Algae generally contains less cellulose in the cell wall as compared to higher plants while filamentous green algae have a cellulose content of 20–30% [18], and 2–20% in brown algae [19, 20].

Cellulose is a linear homopolymer of cellobiose unit composed of two glucose molecules linked by a β-(1 $\rightarrow$ 4) glycosidic bond (Fig. 3.1). Cellulose is generally constituted of a crystalline and an amorphous polymer portion fibre in a matrix of lignin and hemicelluloses. This explains why lignocellulosic biomass such as wood is often considered as a natural composite [21].

The crystalline structure of cellulose comes from the organized arrangement of polymer chains linked by intra and intermolecular hydrogen bonds between hydroxyl groups which increases the density of cellulose while decreasing its chemical availability (Figs. 3.2 and 3.3) [15, 22].

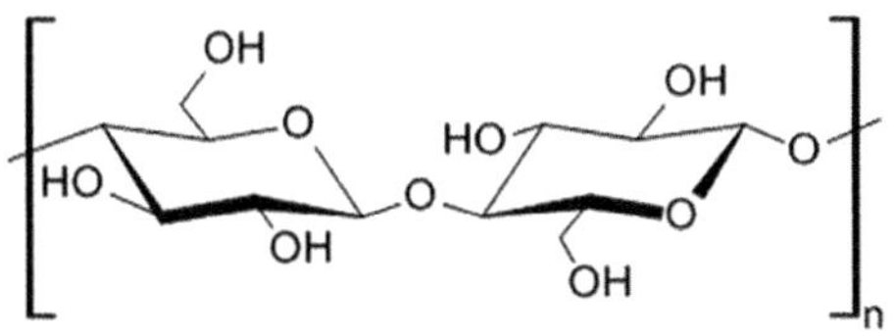

Fig. 3.1 Structure of the cellobiose unit constituting cellulose

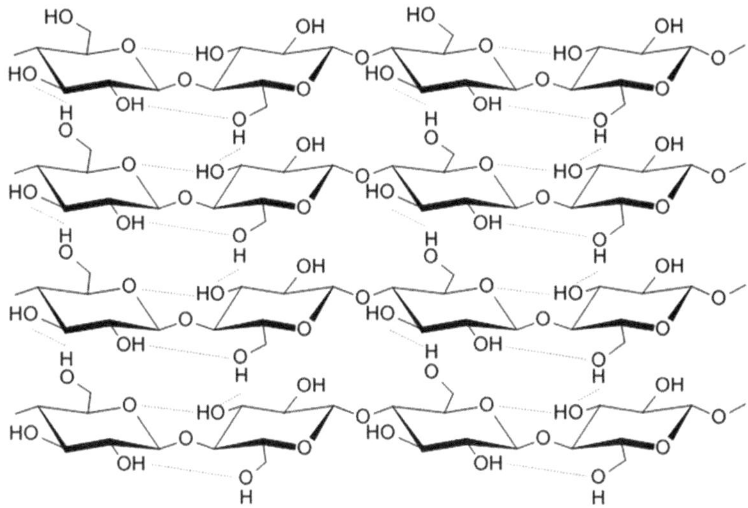

Fig. 3.2 Arrangement of cellulose chains linked by hydrogen bonds

The combination of the physical barrier formed by lignin and crystalline properties of cellulose explains its recalcitrance to biological and chemical degradation such as acid or enzymatic hydrolysis [23]. In open literature, different kind of celluloses were reported such as native cellulose, cellulose I, cellulose II, cellulose III, cellulose IV, Cross and Bevan cellulose or Kurschner cellulose [15]. Native cellulose refers to the cellulose generally found in lignocellulosic biomass (without modification), and could be identical to cellulose I. Celluloses II, III and IV are chemically modified celluloses whereas Cross and Bevan cellulose, Kurschner cellulose or alpha-cellulose refers to the method used to extract the cellulose from biomass. Cellulose II is produced by mercerization and regeneration of cellulose I. The latter involves a treatment of cellulose using concentrated sodium hydroxide in order to swell the cellulose. This process is used in the textile industry to increase the luster, strength, affinity to dye, and resistance to mildew of the cellulose-based fibers such as cotton, which is a pure cellulose. Cellulose III is produced by a treatment of cellulose I with liquid ammonia while cellulose IV is formed from the heating of cellulose III in glycerol [24]. Cross and Bevan cellulose is obtained by a

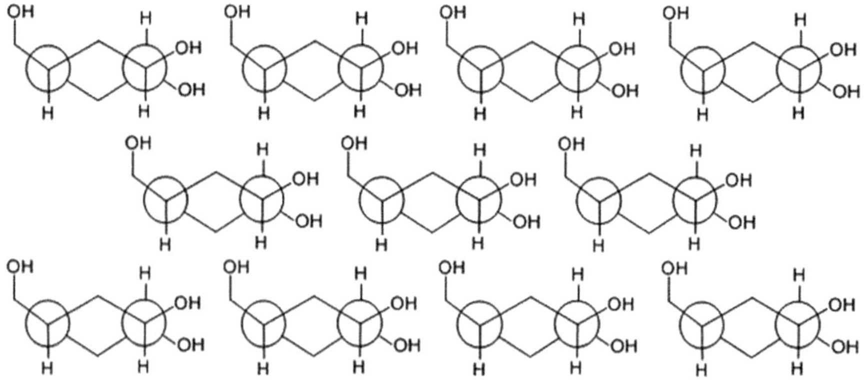

Fig. 3.3 Arrangement of cellulose chains in cross sectional view (Newman projection of glucose unit)

chlorination method while the Kurschner cellulose is obtained by a treatment with nitric acid and ethanol [25]. Alpha-cellulose is defined as a non-soluble cellulose in concentrated alkali such as a 17.5% sodium hydroxide solution and it is often obtained from holocellulose after delignification using a sodium chlorite treatment [26].

The number of glucose unit constituting the cellulose molecules is defined as the degree of polymerization (DP). The DP of cellulose depends on the biomass sources and could vary from 1000 to 11000 [27–31]. Typically, cotton has the highest degree of polymerization while hardwood and softwood celluloses have a degree of polymerization varying from 3000 to 5000. Bagasse and wheat straw contain cellulose at a degree of polymerization of 1000–1500 [32].

3.2.2 *Compositional Analysis Methods*

Cellulose content is generally determined on extractive free biomass. A Soxhlet apparatus is usually employed for successive extractions using various solvents such as ether, hexane, a mixture of ethanol and toluene, ethanol, acetone and water depending on the applied procedure. One of the most common procedure refers to the ASTM standard test method for preparation of extractives free wood [33]. Biomass is first grinded and sieved. The 40–60 mesh fraction is kept for the preparation of extractive free wood. The grinding process can have a non-negligible impact on the chemical composition since plant tissues and components have not the same grinding behavior, indeed biomass composition can vary according to particle sizes. In the case of wheat straw, for example, the hemicelluloses content can represent up to 10% of dry biomass depending on the particle size fraction [34]. Thus, in order to get a relevant compositional analysis and to limit this bias, a good practice is to perform a successive grinding step where each fraction above 60 mesh is grinded until only fractions under 60 mesh are obtained. The solvent to use depends on both the biomass and the nature of the extractives and a procedure is given in the Sect. 3.6: Extractives. The delignification of biomass is often obtained by a delignification using sodium chlorite [15]. The procedure is applied on 2.5 g of extractive free and moisture free samples. 80 mL of hot distilled water, 0.5 mL of acetic acid and 1 g of sodium chlorite are added to the sample in 250 mL Erlenmeyer flask and placed in a water bath at 70 °C. An additional 0.5 mL of acetic acid and 1 g of sodium chlorite are added to the sample hourly until biomass is delignified. The sample is then left at 70 °C overnight without addition of any other reactant. At the end of the reaction, the holocellulose is filtered on a tarred sintered glass filter and washed with acetone until the holocellulose become white and the odor of chlorine dioxide is removed. The sintered glass filter is then placed in the oven at 103 °C for 24 h, cooled in a desiccator for an hour and weighted. The Klason lignin content is determined on the holocellulose to quantify any residual lignin (procedure is provided in the Sect. 3.4: Lignin). Depending on the type of biomass, the delignification procedure may be longer (such as in the case of high lignin content). Overall, it is better to determine the residual lignin content on holocellulose rather than performing a total lignin removal in order to avoid extracting hemicelluloses during the procedure.

Since the holocellulose is a mix of cellulose and hemicelluloses, quantifying the cellulose content, requires extracting hemicelluloses, which is usually performed by an alkaline extraction. 2 g of holocellulose are added in a covered beaker with 10 mL of a 17.5% sodium hydroxide solution (at 20 °C) and mixed slightly with a glass rod until holocellulose is soaked. Then, an additional 5 mL of the 17.5% sodium hydroxide solution is added each 5 min (9 times) while stirring between each addition. The total sodium hydroxide treatment takes about 45 min. 33 mL of water (20 °C) is added for 1 h then the solution is filtered on a sintered glass filter under vacuum. The vacuum is removed after which 100 mL of the 8.3% sodium hydroxide solution is added to the holocellulose that are then filtered again. Then, holocellulose is washed twice with water by releasing vacuum. 15 mL of a 10% acetic acid solution are then added to the crucible and the vacuum is removed when the holocellulose are soaked but still covered by the acid solution. After 3 min, the vacuum is applied again to remove all the acetic acid solution and then washed with water until reaching a neutral pH. Cellulose is washed one last time using 250 mL of water and placed in an oven at 103 °C for 24 h. The cellulose is weighted, and the results are expressed as the percentage of cellulose by weight based on the initial weight of biomass with extractives.

This method has an impact on cellulose properties such as its degree of polymerization. Since if it is indeed partially degraded by acid or oxidative cleavage during the reaction, the degree of polymerization decreased especially, if complete delignification is done [35].

3.2.3 *Structural Analysis*

3.2.3.1 Crystallinity Index of Cellulose

The most common method to determine the crystallinity index of cellulose is through X-ray diffraction (XRD) technique. The XRD data permit different calculation method linking to a crystallinity index of cellulose. Among such calculation method, the XRD peak height method is the most applied since it is the easiest method to use. This method is useful to compare the relative difference between samples but does not allow determining the amount of crystalline and amorphous cellulose in cellulosic sample. Crystallinity is calculated using Eq. 3.1 considering the height of the I_{200} peak at 22.7° and the minimum between the I_{200} and I_{110} ($2\theta = 18°$) peaks called I_{AM} (Fig. 3.4). This method is also referred as the Segal method [36] and often leads to an overestimation of the crystallinity [37].

$$\text{Crystallinity index} = \frac{I_{200} - I_{AM}}{I_{200}} \tag{3.1}$$

Other methods could be used to determine the crystallinity index, such as the convolution method using a software or the amorphous standard subtraction [37, 38].

3.2.3.2 Degree of Polymerization

The degree of polymerization of cellulose can be determined by either a viscosimeric method or Gel Permeation Chromatography (GPC) method both of which

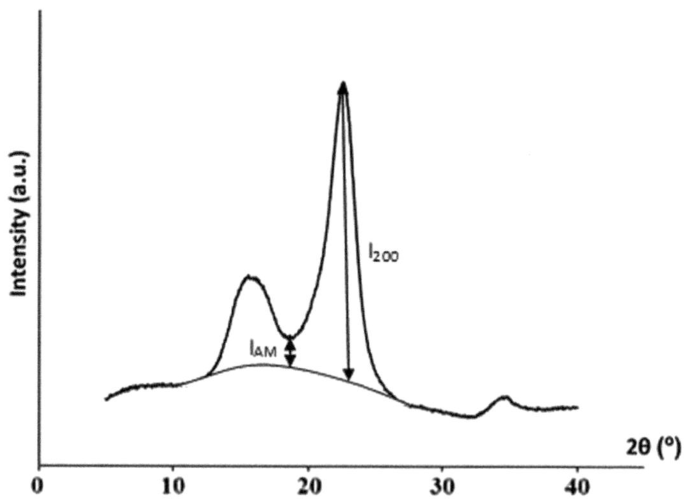

Fig. 3.4 Peak height method for the determination of the crystallinity index

are among the most used methods. In both cases, this method requires solubilization of cellulose in an appropriate solvent. Since it is not soluble in most solvents. The two most common methods to solubilize cellulose are the solubilization in a metal complex solution or the derivatization of cellulose in order to increase its solubility. The standard test method TAPPI T230 Ω-08 requires the dissolution of cellulose in cupryethylenediamine to determine the degree of polymerization using capillary viscometry. 0.25 g of dried cellulose is mixed with 25 mL of distilled water and 25 mL of a cupryethylenediamine solution. A continuous flow of nitrogen is applied during the dissolution of cellulose and the mixture is stirred (and eventually heated) in order to dissolve cellulose. When a homogeneous solution is obtained the DP is determined by viscosimetry [32].

GPC has been commonly reported for the determination of the cellulose DP. One of the most common methods is the derivatization of cellulose with phenylisocyanate, leading to the production of cellulose tricarbanilates. This cellulose derivative is soluble in THF and it can be easily analyzed by GPC [39].

3.3 Hemicelluloses

3.3.1 General Aspect

Hemicelluloses are amongst the three most important macromolecules present in lignocellulosic biomass and the second most important carbohydrate on the planet after cellulose. The term “hemicelluloses” was first reported by E. Schulze in 1891 and was considered at the time as plant polysaccharides that could be isolated by alkaline treatment, providing various sugars from a subsequent dilute acid hydrolysis [40]. This definition, based on a simple extraction method, did not take in account the structural properties of hemicelluloses. For instance, pectins are cell wall polysaccharides extracted by hot water, dilute acid or chelating agent and are mainly composed of uronic acids. Nevertheless, some pectins can be extracted only by an alkaline treatment [41] whereas some hemicelluloses can be water-soluble depending of the water temperature [42, 43]. Thus, this definition based on the extraction method is not suitable to describe, precisely, hemicelluloses and to distinguish them from other cell wall polysaccharides such as pectins.

Hemicelluloses are a relatively low mass branched heteropolymer composed of a C6 and/or C5 sugar backbone linked by β-(1 $\rightarrow$ 4) glycosidic bonds. Sugars, at the basis of hemicelluloses structure include xylose, mannose, glucose, galactose, arabinose, rhamnose, fucose, galacturonic acid and glucuronic acid. Contrary to cellulose, hemicelluloses structure varies significantly depending on its localization in the cell wall, cell types as well as species. Hemicelluloses can be classified in 5 groups: xylans, mannans, xyloglucans, glucans and arabinogalactans as presented in Table 3.1 [44].

Table 3.1 Overview of hemicelluloses in lignocellulosic biomass [15, 44]

Classes	Hemicelluloses	Biomass	Amount range
Xylans	Glucuronoxylans	Harwood	15–30 %
	Arabinoglucuronoxylans	Softwood Grasses Cereals	5–10 %
	Glucuronoarabinoxylans	Grasses Cereals	10–30 %
	Arabinoxylans	Cereals	10–30 %
	Complex heteroxylans	Seed Gum exudate Mucilage	– – –
	Homoxylans	algae	–
Mannans	Glucomannans	Softwood Hardwood	2–5 %
	Galactoglucomannans	Softwood Harwood	10–25 % 3–5 %
	Galactomannans	Various angiosperm seeds	–
Xyloglucans	-	Hardwood grasses	2–25 %
Glucans	-	Cereals (barley, oat)	3–7 %
Arabinogalactans	Arabinogalactans	Softwood Larch wood	1–3 % 10–20 %

3.3.1.1 Xylans

Xylans are constituted of a xylose-based backbone and this group includes glucronoxylans, arabinoglucuronoxylans, glucuronoarabinoxylans, arabinoxylans, complex heteroxylans and homoxylans. Glucuronoxylans are the main hardwood hemicelluloses (Fig. 3.5) and are composed of a D-glucuronic acid or 4-*O*-methyl-D-glucuronic acid linked in position 2 of the xylose backbone. They can be found in secondary cell wall of dycotiledon plant such as hardwood and in softwood at a lesser extent. Arabinoglucuronoxylans have the same structure than glucronoxylans except for the addition of arabinose, linked at position 3 of the xylose units. Glucuronoarabinoxylans have an arabinose: glucuronic acid ratio of ten, which is higher than arabinoglucuronoxylans. In addition, disaccharides sidechains composed of xylose and arabinose units and both hemicelluloses are found in softwood, grasses and cereals. Arabinoxylans are composed of a xylose backbone with a higher level of arabinose substitutions and potentially hydroxycinnamic, ferulic or p-coumaric acids. They are mainly found in cereals such as wheat, rye, barley, oat, rice, sorghum and also in flax seed, pangola grass, bamboo shoots, and rye grass [45].

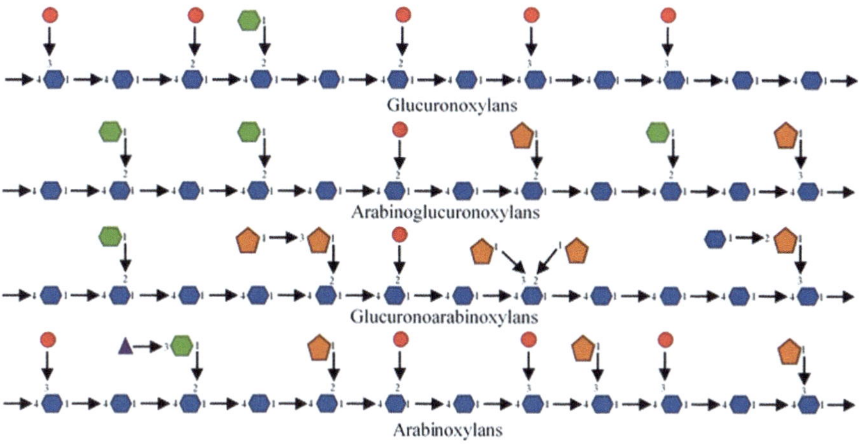

Fig. 3.5 Schematic representation of typical xylan hemicelluloses (=xylopyranose, =glucuronic acid, =acetyl, =arabinofuranose, ▲= hydroxycinnamic, ferulic or p-coumaric acids)

3.3.1.2 Mannans

The mannan group includes galactoglucomannans and galactomannans, glucomannans. The latter involves a mix of mannopyranose and glucopyranose backbone with a low amount of galactopyranose linked by α-(1 → 6) bonds. Galactoglucomannans differ from glucomannans by its higher level of galactopyranose substitutions. Galactomannan backbones are only constituted of mannopyranose units with a low level of galactopyranose substitution. Glucomannans and galactoglucomannans also have always acetyl groups linked to the backbones (Fig. 3.6).

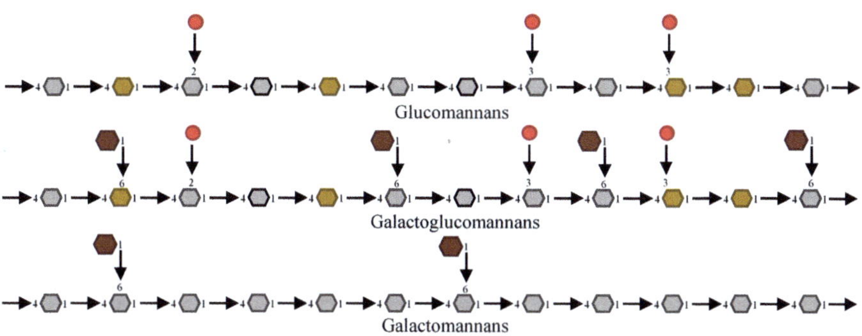

Fig. 3.6 Schematic representation of typical mannan hemicelluloses (=mannopyranose, =glucopyranose, =acetyl, =Galactopryranose)

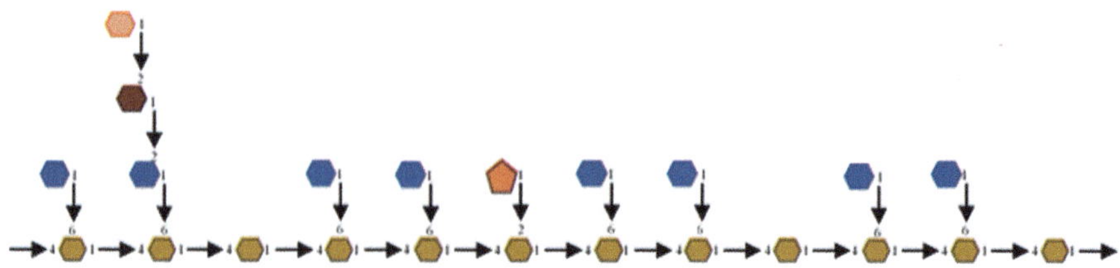

Fig. 3.7 Schematic representation of typical mannans hemicelluloses (●=Xylopyranose, ●=glucopyranose, ●=Galactopryranose, ●=fucopyranose, ●=arabinofuranose)

3.3.1.3 Xyloglucans

Xyloglucans are composed of a glucopyranose backbone linked by β-(1 → 4) bonds. Up to 75% of the glucose units are substituted by xylopyranose units at position 6. However, some glucopyranose units may be substituted by arabinofuranose units. Xyloglucans have also disaccharides and trisaccharides sidechains composed of arabinofuranose, fucopyranose and galactopyranose attached to the C2 of the xylopyranose residues (Fig. 3.7).

3.3.1.4 Glucans

B-glucans, which are generally found in cereals, are unbranched homopolymers of glucopyranose linked by β-(1 → 4) and β-(1 → 3) bonds. Generally, a block of three or four glucopyranose units are linked by β-(1 → 4) and separated by a β-(1 → 3) bond. The β-(1 → 3) over β-(1 → 4) linkage ratio may vary depending on the species.

3.3.1.5 Arabinogalactans

Arabinogalactans are present in several plants such as larch and it is a major component of gum. In addition, they can be found in high quantity in larch wood (up to 35%) [46]. These hemicelluloses are a polymer of galactopyranose linked by β-(1 → 3) bonds. Almost every galactopyranose from the backbone is substituted by another galactopyranose in position 6 or, in some cases, arabinopyranose at position 3 [15]. Arabinogalactans are often classified as pectin substances and as hemicelluloses in wood chemistry depending of the authors. In addition, these hemicelluloses are not localized in the cell wall but in lumen contrary to other hemicelluloses.

3.3.2 *Compositional Analysis*

3.3.2.1 Hemicelluloses Content

After producing extractive free biomass, hemicelluloses can be extracted either from delignified or non-delignified biomass. When aiming quantifying hemicelluloses in biomass, the best way is to extract them from delignified biomass in order to avoid a possible overestimation due to a lignin contamination. However, the delignification method is generally performed using a sodium chlorite procedure that leads to a partial degradation of hemicelluloses [47]. Thus, when aiming determining the structure of hemicelluloses present in biomass, it is better to extract hemicelluloses from a non-delignified biomass involving a further purification to remove the lignin components [48].

The following method to determine hemicelluloses content are based on the delignification method by sodium chlorite presented in the cellulose section of this chapter. Hemicelluloses are extracted from holocellulose by alkali treatment. The hemicelluloses content is expressed using the mass of biomass with extractives following Eq. 3.2.

$$\%\,\text{hemicelluloses} = \frac{m_{\text{holocellulose}} - m_{\text{cellulose}}}{m_{\text{biomass}}} \tag{3.2}$$

where m_{biomass} is the mass of biomass with extractives content.

After extraction, the hemicellulosic fraction is precipitated by neutralization of sodium hydroxide solution containing hemicelluloses with acetic acid at pH 5.5 and a further sequential precipitation by ethanol addition. Precipitated hemicelluloses are recovered by centrifugation.

In the case of biomass containing xylans and mannans such as softwood, these hemicelluloses are separated by addition of barium hydroxide.

3.3.2.2 Pentosan Content

The pentosan content is based on the quantification of all C5 sugars. This determination could be made from chromatographic analysis or by a colorimetry. The colorimetric method is based on the formation of furfural from pentoses in HCl solution with a subsequent color development using orcinol reagent. The following procedure is based on the TAPPI procedure T223 cm-01 [49].

The sample quantity to use depends on the average pentosan content. For example, for hardwood between 0.2 and 0.3 g is needed whereas for bleached pulp from softwood 3–5 g are used. Thus, 0.2–5 g of biomass sample are placed in a boiling flask with 20 g of NaCl and 100 mL of 3.85 N of HCl and boiling stone. The acid level is marked on the flask. The flask is connected to a distillation apparatus with a separation funnel containing 250 mL of 3.85 N of HCl. The flask is heated in order to distillate the acid at a rate of 2.5 mL/min. The distillate is collected in a 250 mL volumetric flask immersed in ice bath. Acid is added to the flask to maintain a volume of 100 mL of acid during 90 min. At the end of the distillation, a volume of 225 mL should be collected in the volumetric flask that is cooled to 20 °C. 3.85 N of HCl solution is added and mixed to get a total of 250 mL of solution. After, 5 mL of the resulting solution is placed in 50 mL volumetric flask and 25 mL of orcinol reagent are added, mixed and placed in a water bath at 25 °C for 60 min. After, the volumetric flask is completed by addition of ethanol up to the 50 mL mark, mixed and placed again in the water bath at 25 °C for one more 60 min and the absorbance is measured at 630 nm. A calibration curve is made from several portions of 10, 20, 40, 60, 80 and 100 mg of dried xylose. Distillation and development of the color are performed exactly as the other samples. The absorbance is also read at 630 nm using a blank of 3.85 N HCl solution as a reference. The corresponding xylan weight is calculated according to the Eq. 3.3.

$$m_{\text{xylan}}(\text{mg}) = m_{\text{xylose}}(\text{mg}) \times 0.88 \tag{3.3}$$

3.3.3 *Structural Characterization*

3.3.3.1 Sugar Composition

In order to determine the free sugars content in the hemicelluloses, an hydrolysis is required to release the free sugars. NREL provides the determination of total carbohydrate content by a two-step hydrolysis of the whole biomass in concentrated sulfuric acid. However, in the case of hemicelluloses this procedure may not be fully efficient since this procedure does not differentiate glucose from cellulose and hemicelluloses. For xyloglucan or glucan rich biomass, this could introduce an important bias. Moreover, this procedure often leads to an underestimation of the hemicelluloses content in biomass caused by sugar degradation in addition to glucose not being taken into account, the structural analysis of hemicelluloses is not possible in this case and requires hemicelluloses to be isolated. To determine the sugars composition of hemicelluloses, the procedure involves hemicelluloses hydrolysis using a 4% sulfuric acid solution at 120 °C for 1 h, followed by free sugars concentration determined by HPLC, ion chromatography or gas chromatography.

3.3.3.2 Glucosidic Linkage

In order to determine the sugar sequence constituting hemicelluloses, several techniques are available such as methylation, periodate oxidation, partial hydrolysis, acetolysis, degradation by β-elimination, and analysis by nuclear magnetic resonance [48].

The methylation method is often used for the elucidation of the polysaccharides structure. The principle of the methylation method is based on the total methylation of hemicelluloses samples followed by the hydrolysis of methylated hemicelluloses. Then, the partially methylated sugars are reduced and acetylated to produce partially methylated alditol acetates, which are then analyzed by GC-MS. The non-methylated hydroxyl group are those involved in glycosidic linkages. Due to the difficulties to find standards for each sugar derivatives, this method is more qualitative than quantitative [50]. The two most widely used methods are the Hakomori method [51] in which the methylation is performed in a DMSO solution containing a methylsulfinyl carbanion and the method of Ciucanu and Kerek that uses sodium hydroxide in DMSO [52]. The latter is preferred for several reasons. First, the Hakomori method produces non-sugars which leads to complex chromatograms in addition to the produced a methylsulfinyl carbanion which is instable and potentially explosive. The following protocol for hemicelluloses methylation was proposed by Pettolino et al. [50] and is based on the Ciucanu and Kerek method [53].

This method can be used to determine neutral and acidic monosaccharides linkage and only if a reduction of the carbonyl is performed, without that, only neutral monosaccharides are detectable.

Place 5–50 μg of freeze dried hemicellulosic sample in a 7 mL capped glass tube. It is important to have a dried sample (without salt) in order to get a total methylation. Add 10–20 μl of methanol and dry again to dehydrate the sample. Add 50 μl of DMSO, cap and sonicate the glass tube for 20–30 min. For DMSO-insoluble sample, leave overnight under argon. Grind approximatively 120 mg of NaOH pellet in 1 mL of DMSO. It is important to act quickly by avoiding to weight NaOH due to the fast water absorption of the NaOH pellet. One NaOH pellet might be weighed (but not used) to determine how many pellets are required per mL of DMSO. Add 50 μl of the DMSO/NaOH slurry into the glass tube. Cap and sonicate for 20–50 min at a temperature under 40 °C. Add a first 10 μl portion of methyl iodide and sonicate for 10 min, then add another 10 μl of methyl iodide and sonicate again for 10 min. Finally, add 20 μl of methyl iodide and sonicate for 20 min. After that, 1 mL of water and 500 μl of dichloromethane are added after which the solution is vortexed for more than 40 s. Briefly centrifuge to separate the phases and if non-soluble particles appear at water/dichloromethane interface, an additional methylation could be required. Remove the water phase and wash the dichloromethane fraction three times with 1 mL of water, remove again the water and dry the dichloromethane phase using nitrogen flow.

Following the methylation step, a hydrolysis of the sample is required in order to produce partially methylated monosaccharides. The hydrolysis is performed by adding 100 μl of a 2 M trifluoroacetic acid solution to the methylated sample at 120 °C for 90 min. Then, the sample is cooled, and myo-inositol is added as internal standard (1/20 of total CHO). The tube is placed in a water bath at 30 °C and dried with a nitrogen stream.

The reducing step is performed by dissolving the previously obtained sample in 50 μl of 2 M hydroxide ammonium solution and 50 μl of freshly prepared 1 M sodium borodeuteride in a 2 M hydroxide ammonium. The sample is then sonicated and placed at room temperature for 2.5 h. The excess reductant is neutralized by addition of 20 µl of acetic acid (bubbles should be seen in the sample, if not, the reduction was incomplete or $NaBD_4$ was inactive). The tube is placed in a water bath at 30 °C and the sample is dried under a nitrogen flow. Finally, the sample is washed twice with 250 µl of 5% (v/v) acetic acid in methanol then dried and washed again with 250 µl of methanol to remove boric in order to avoid borate complexes of sugars and to get quantitative recoveries. Store the sample in methanol overnight and dry it one more time.

The next step is the acetylation of sample by addition of the 250 µl of acetic anhydride followed by a sonication for 5 min and finally an incubation at 100 °C for 2.5 h. The excess of acetic anhydride is destroyed by adding 2 mL of water for 10 min after mixing. The partially acetylated alditols are extracted using 1 mL of dichloromethane, the sample is vortexed and centrifuged at 2500 rpm for 2 min allowing phase separation. The dichloromethane phase is washed twice with 2 mL of water. The aqueous phase is removed. The dichloromethane phase is transferred in a small glass vial, 5 µl of the sample are redissolved in 20 µl of dichloromethane while 1 µl of the dissolved sample is used for GC-MS analysis on the high-polarity such as the BPX70 column.

3.3.3.3 Uronic Acid

Uronic acid is present in xylan hemicelluloses such as glucuronoxylans and arabinoglucuronoxylans. The quantification of total uronic acid can be performed either by a decarboxylation method with strong mineral acid or using a colorimetric method. However, the decarboxylation method gives an overestimated yield of uronic acid when compared to colorimetric methods, due to the production of CO_2 from non-uronide extractives and polysaccharides [54]. The following procedure describes the colorimetric method developed by Scott [54]. This colorimetric method is based on the formation of 5-formyl-2-furancarboxylic acid from uronic acid followed by a color development with 3,5-dimethylphenol. Free extractives biomass is grinded at 40–80 mesh and moistened with ethanol. Biomass sample containing 25–50 mg of water-insoluble polysaccharides are dissolved in 2–4 mL of a 72% H_2SO_4 solution and heated at 50 °C for 10 min. Depending of the sample, trituration and longer heating period may be required. The sample is then diluted to reach 3% or 4% H_2SO_4 and heated at 120 °C for 1 h in an autoclave. 125 µl of sample containing 20–80 µg/mL of uronic acid are then mixed with 125 µl of 2% $NaCl–H_3BO_3$ solution. Then, 2 mL of concentrated H_2SO_4 solution are added and mixed vigorously and the sample is placed at 70 °C for 40 min. At the end of the reaction, the sample is cooled and 100 µl of 3,5-dimethylphenol are added to the mixture and it is left at room temperature for 10–15 min. Then, the absorbance is read at 400 and 450 nm against water as a reference. Uronic anhydride concentration is calculated from the difference of absorbance at 450 and 400 nm. The same analysis is performed on blank solution, i.e. not containing the sample. A commercial galacturonic acid is a convenient standard.

The quantification of individual uronic acid requires a depolymerization of the hemicelluloses containing uronic acid. This depolymerization can be performed by acid hydrolysis, enzymatic hydrolysis or methanolyzis. However, the linkage between uronic acid and xylose units is very resistant to acid hydrolysis. Thus, the quantification of uronic acid by acid hydrolyzis is underestimated [48, 55] while the methanolyzis is a more powerful technique but requires the derivatization of the sample for GC analysis. The following procedure has been described by Willför et al. [55]. It also allows determining all the non-cellulosic polysaccharides content in samples.

10 mg of dried biomass are depolymerized using 2 mL of 2 M HCl in anhydrous methanol. The sample is placed at 100 °C for 5 h. At the end of the reaction, the sample is cooled to room temperature and neutralized by addition of pyridine. The calibration solution is a mix of monosaccharides and uronic acid (at an equal amount) that were treated by acid methanolysis under similar conditions. 4 mL of the methanol solution containing 0.1 mg/mL of sorbitol were added to the samples as internal standard. Methanolysates and calibration samples were dried under nitrogen and placed in a vacuum desiccator at 40 °C for 20 min. The silylation reaction is performed overnight by addition of 150 µl of hexamethyl disilazane, 70 µl of trimethylchlorosilane and 150 µl of pyridine. Samples are analyzed by GC-MS. Ultimately, 1 µl of sample is injected via a split injector (260 °C, split ratio 1:50) into a HP-5 column from Agilent Technologies. The column temperature is maintained at 100 °C for 2 min then increase at a 4 °C/min rate up to 220 °C, then increases again at 15 °C/min rate up to 300 °C. Helium is used as carrier gas at a flow rate of 1 mL/min. The detector conditions were 70 eV and 40–600 amu and quantification can be achieved using relative molar response factors for each sugar.

3.3.3.4 Acetyl Group Content

The easiest way to determine the acetyl content in hemicelluloses is to combine the simple sugars from hemicelluloses or biomass and acetic acid content by liquid chromatography. This is a standard procedure from NREL [56] which provides as well a conversion factor to convert acetic acid content into acetate content to express the results. However, in order to determine the mass ratio of acetyl group in the samples, it is more adequate to express the results as acetyl group percentage based on initial anhydrous sample mass. This sample could be the biomass or the hemicellulosic fraction, depending if the hydrolysis is performed on the whole biomass or on the reported extracted and purified hemicelluloses. The acetic acid concentration determination by liquid chromatography is described in the inhibitors analysis section.

$$\%\,\text{acetyl} = \frac{m_{\text{acetic acid}} * 0.7169}{m_{\text{sample}}} \tag{3.4}$$

where m_{acetic} acid is the mass of acetic acid found in m_{sample} mass of biomass or hemicelluloses fraction.

Another approach to determine acetyl content is to perform a saponification with hot alkali solution splitting the acetyl groups and a further acidification to form acetic acid, the acetic acid concentration is determined by gas chromatography [15].

3.3.3.5 Degree of Polymerization

The determination of the molecular weight of hemicelluloses can be performed by several techniques such as end group analysis, viscometry, isothermal distillation, size exclusion chromatography [48].

Nowadays, the most used method for molecular weight determination of hemicelluloses is gel permeation chromatography or size exclusion chromatography depending of the hemicelluloses solubility. However, this method requires standards for the calibration and in the case of hemicelluloses, which is a branch heteropolymer, it is impossible to find standards with the same hydrodynamic volume leading to the introduction of an important bias depending of the chosen standard.

3.4 Lignin

3.4.1 General Aspect

Lignin is an amorphous, complex, nonlinear polymer consisting of an aromatic and phenolic system. It would be rather more accurate to talk about lignins, in the plural, since their structure is extremely variable [57]. However, by convention, the term will be kept in its singular form. While cellulose is present in all plants, lignin is a characteristic constituent of vascular plants. It is the second most abundant organic polymer in the plant world after cellulose. The lignification mechanism is necessary for all tissues that serve to conduct the sap. Lignin is a highly resistant polymer that, once inserted into the cell walls, ensure the impermeability of the sap-conducting vessels due to its hydrophobic nature. Lignin also contributes to the rigidity of the cell walls and thus ensures the growing of the plants. Lignin is biosynthesized by the enzymatic dehydrogenative polymerization of coniferyl alcohol (common in softwoods), syringyl alcohol (common in hardwoods) and coumaryl alcohol (common in grasses) (Fig. 3.8).

Usually, their contents vary from 25 to 35% in softwood [15], from 18 to 25% in hardwood and from 2 to 28% in herbaceous plants [58, 59]. Table 3.2 presents the lignin contents of some lignocellulosic materials.

Lignin is composed of highly branched phenylpropenyl (C9) units, designated as H-p (p-hydroxyphenylpropane), G-p (guaiacylpropane) and S-p (syringylpropane) (Fig. 3.8). Position 3 in the aromatic nuclei of G-p and positions 3 and 5 for S-p is occupied by a methoxyl group whereas H-p do not contain any methoxyl group.

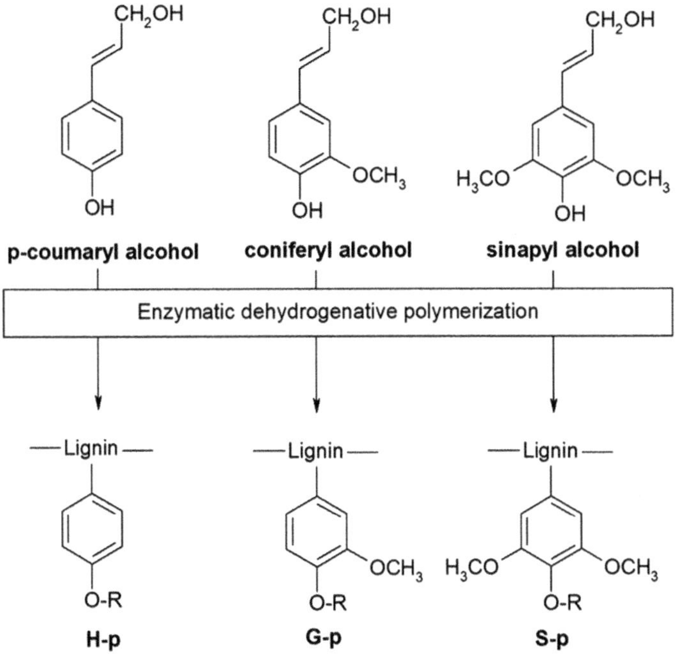

Fig. 3.8 Chemical structures of lignin precursors: (1) p-coumaryl alcohol, (2) coniferyl alcohol, and (3) sinapyl alcohol

Table 3.2 Lignin contents by weight (dry basis) of different families of lignocellulosic materials

Family	% Lignin	Reference
Harwood	18–25	[57]
Softwood	25–35	[57]
Wood bark	20–40	[60]
Gramineae	17–22	[58]
Leguminosae	9–28	[58]
Malvaceae	23–28	[58]
Linaceae	2–20	[58]
Tiliaceae	19–24	[58]
Compositeae	11–27	[58]

C–C (condensed units) or C–O–C (diarylether) bonds are possible on G-p position 5, and H-p positions 3 and 5. The main inter-unit linkages within the lignin monomer (H, G and S) are β-*O*-4 ether linkages which account from 40 to 50% of total linkages, followed by other type of ether and C–C linkages such as α-*O*-4 β-β, β-5, β-1 and 5-5 arrangements (Fig. 3.9) [57]. Although, few phenylpropane units are substituted on other positions. Lignin is often designed by the letter H, G or S based on the majority participation of p-hydroxyphenylpropane (H), guaiacyle

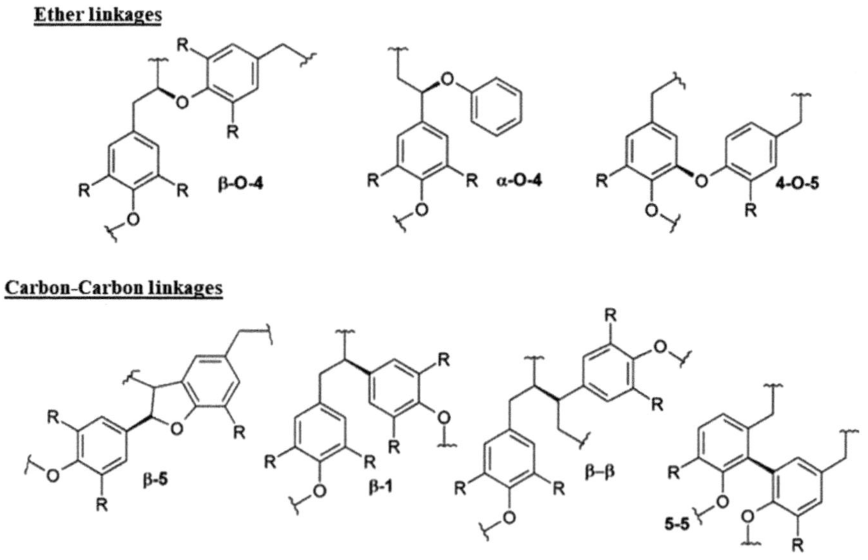

Fig. 3.9 Common phenylpropane linkages in lignin

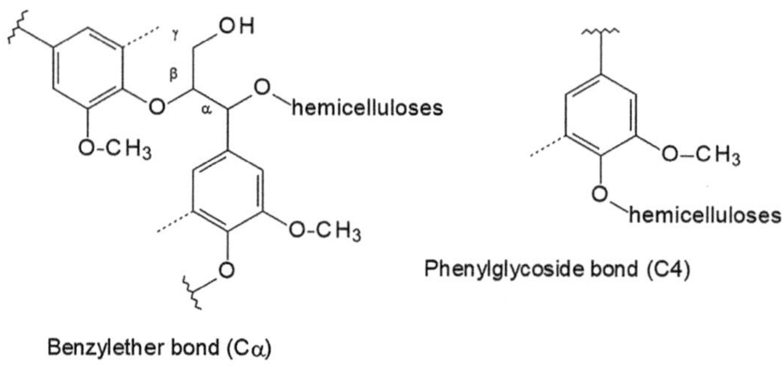

Fig. 3.10 Main lignin-hemicellulose bonds

(G) or syringyl (S) units in the lignin structure. The lignin from lignocellulosic materials differ mainly in methoxyl group substitution at position 3 and/or 5 of the phenyl groups. According to this nomenclature, the lignin of the gymnosperm (softwood) is G type, the dicotyledonous angiosperm (hardwood) one is GS type whereas the dicotyledonous angiosperm one (grasses and annual plants) are HGS type lignin.

Lignin is always associated with carbohydrates (in particular with hemicelluloses) via covalent bonds at two sites: α-carbon and C-4 in the benzene ring, and this association is called lignin carbohydrate complexes (LCC) (Fig. 3.10) [60–62].

3.4.2 Structural Characterization

A set of quantification methods have been developed to determine the lignin content within a wide range of lignocellulosic materials, which can be divided into two categories. The first category removes the lignin out of the cell wall samples based on the decomposition of lignin into soluble degradation products while quantification is done by measuring their absorbance by UV spectrophotometry [61]. These include the acetyl bromide method and the thioglycolysis method [63]. The second category removes all the cell wall constituents except for lignin. It generally consists in using strong mineral acids to solubilize and hydrolyze cell wall carbohydrates, leaving an acid-insoluble lignin residue that can then be quantified by gravimetric method. These included the hydrochloric acid method (Willstätter lignin) [64], the concentrated sulphuric acid method (Klason lignin) [65] and the acid detergent fiber method (ADF) [66].

Among all these methods, the two-stage sulfuric acid hydrolysis combined with spectrophotometric analysis of acid soluble lignin is the most common procedure to determine the lignin content within the lignocellulosic materials. This procedure has also been adapted and downscaled to a sample size of 50 mg using glass microfiber filters [67].

The following procedure describes the two-stage sulfuric acid method, which is based on ASTM, SCAN or NREL procedures with slight modifications. First, 25 g of sample are air-dried to a moisture content below 10%, then ground with a milling machine and sieved to retain 40 to 60 mesh particles (250–425 μm in size). 300 ± 10 mg (at the nearest 0.1 mg) of dried extractive-free raw biomass is weighed in glass test tubes and treated with 3 mL of 72% H_2SO_4. The procedure to obtain extractive-free raw biomass will be described in the Extractive section. The sample is maintained at 30 °C for 1 h, carefully stirred at 10 min intervals to allow decrystallization and partial depolymerization of the cellulose. After the primary stage, the acid is diluted to 4% H_2SO_4 by adding 84 mL of distilled water. The secondary stage is made in an autoclave for 1 h at 120 ± 5 °C (pressure cooking) or for 4 h at reflux boiling for complete total carbohydrate hydrolysis. The slurry is then cooled at room temperature and hydrolyzates are vacuum filtered using a medium porosity filtering crucible. The remained residue, considered as acid insoluble lignin, is washed with hot water until neutral pH is reached. The acid insoluble lignin is determined by drying the residues at 105 °C while the remaining ash content is obtained after incinerating the hydrolyzed samples at 575 °C in a muffle furnace. The acid-soluble lignin content is determined by measuring the absorbance of the acid-hydrolyzed filtrate. First, the filtrate is diluted with water until an absorbance of 0.2–0.7 is reached and the acid-soluble lignin content is quantified using UV absorbance intensity at 205 nm using an absorptivity of 110 L g^{-1} cm^{-1} (according to TAPPI UM 250). Sulfuric acid of the same concentration with the acid-soluble lignin sample is used as a blank. The total lignin (acid insoluble lignin + acid-soluble lignin) content is then calculated and determined as the mean of triplicate. For non-woody samples (such as grasses), protein contaminants may be present in the acid-insoluble residue thus leading to an overestimation of the lignin content. Protein contaminants can be estimated as described in the proteins section. Biomass containing high amounts of extractives such as barks (specifically suberin and tannins) will input on the course of lignin quantification. Thus, a prior step of mild alkaline extraction with 1% NaOH at 70 ° C for 1 h is required before the two-stage sulfuric acid procedure in order to remove the tannins and often potentially interfering molecules [68].

3.4.3 Compositional Analysis Methods

Many structures have been proposed for the lignin macromolecule. The latter exhibits a great variability of functional groups including aliphatic hydroxyl groups, phenolic groups, carbonyl and carboxylic groups (Fig. 3.11) [69].

Numerous chemical and physical techniques are available to characterize lignin's structure. However, the most informative methods to investigate lignin structures and their functional groups are often based on spectroscopic techniques such as infrared spectroscopy, UV-visible spectrophotometry (UV), nuclear magnetic resonance spectroscopy, Raman spectroscopy, etc. Recently, Lupoi et al. [70] have well reviewed the analytical methods for the qualitative and quantitative assessment of lignin. Fourier transform infrared spectroscopy (FTIR) is a technique that could provide a lot of information about the origin of the lignin as well as its structure. FTIR is by far the most common tool used for the classification of lignin [71] estimating G, GS and HGS type lignin. This analytical tool can be used to estimate the OH phenolic/OH aliphatic ratio reading 1765/1743 cm^{-1} after an acetylation step of the lignin prior to analysis. It can also estimate an index of lignin condensation by the following equation proposed by Faix [71–73] and illustrated in (Figs. 3.11 and 3.12):

$$\text{Condensation Index (CI)} = \frac{\text{Sum of all minima between 1500 and 1050 cm}^1}{\text{Sum of all maxima between 1600 and 1030 cm}^1} \tag{3.5}$$

Nuclear magnetic resonance (NMR), another spectroscopic tool has also been widely used for lignin characterization, particularly to understand or reveal its structure. These include 13C NMR and 31P NMR. For the latter, a step of phosphorylation is necessary prior to analysis. These two NMR techniques can be combined to quantify aliphatic hydroxyl, phenolic hydroxyl in guaiacyl and syringyl units as well as carboxylic groups. A UV spectrophotometry method has also been developed for the quantification of the phenolic hydroxyl content, based on the difference in absorption at 300 and 350 nm between the free phenolic units in neutral and alkaline solutions [74, 75]. Other spectroscopic tools like X-ray, XPS, SEM-EDX and TEM have also been used for surface analysis [61, 72].

Chromatographic techniques have also been reported for the characterization of lignin such as gel permeation chromatography (GPC), generally used for molecular

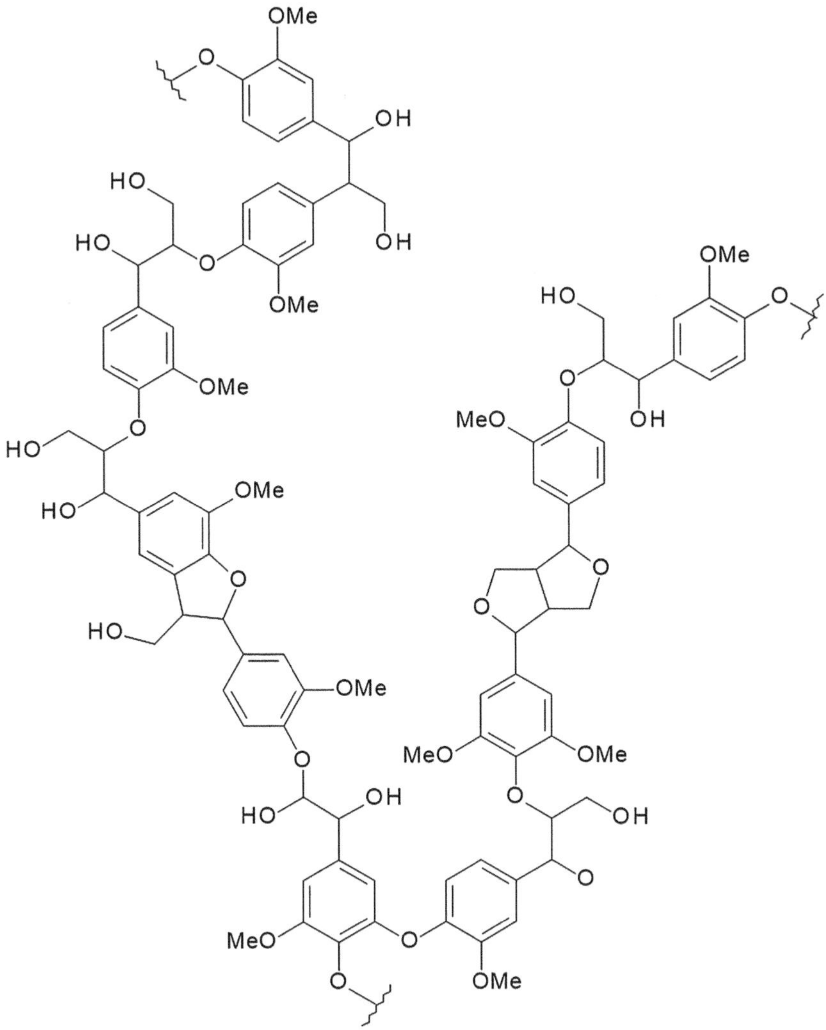

Fig. 3.11 Schematic structural representation of hardwood lignin fragment

weight distribution analysis. The monomeric content of lignin polymer such as H/G/S ratios and some other structural features can be analyzed using different chemical degradation methods such as nitrobenzene oxidation, derivatization followed by reductive cleavage (DFRC) and thioacidolysis. Analytical pyrolysis (namely Py-GC-MS/FID) is also currently used to determine the monomeric content. Meier and Faix 76 showed that pyrolysis GC-MS is an interesting technique offering the possibility to calculate the ratio HGS which can help determine the type of lignin [76].

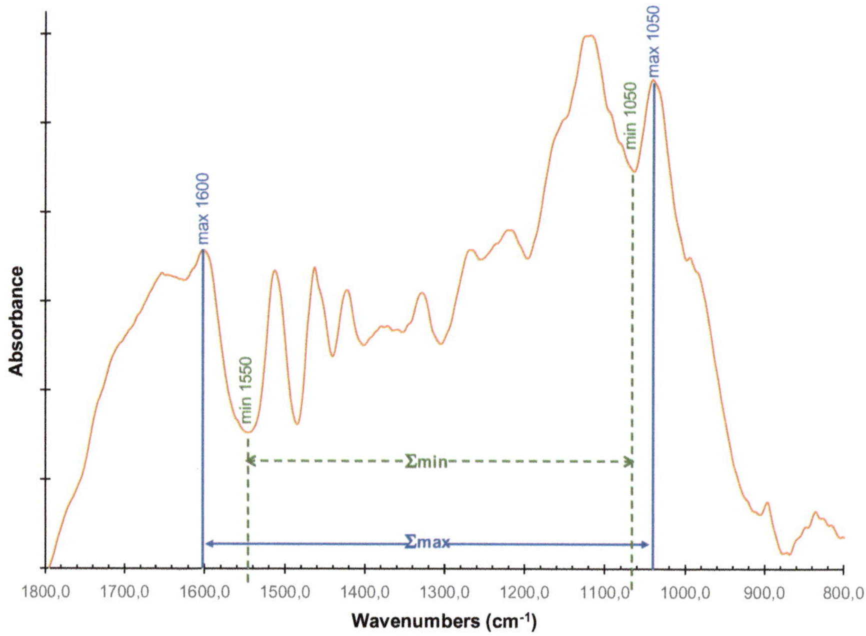

Fig. 3.12 Schematic representation of the determination of lignin condensation index [73]

To elucidate its structural characteristics, various methods of isolation have been proposed. However, since lignin undergoes changes in its chemical and physical composition during such isolation procedures, care must be taken to choose the right procedure to extract its most representative form without altering its native structure. Among the methods, the Milled Wood Lignin (MWL) and the Dioxane Lignin (DL) methods are frequently used for structural investigations. Both procedures allow high purity lignin extraction with minimal structural changes. In addition, a mild-alkaline extraction step prior to lignin isolation greatly improve its purity.

The following procedure describes the well-established MWL method [77] considered as the best preparation for lignin structural studies with slight modifications. 50 g of sample are first air-dried to a moisture content below 10%, then ground with a milling machine and sieved to retain the 40 to 60 mesh particles (250–425 μm in size). The wood meal is then refluxed in an inert atmosphere (N2) with 1 L of 0.3% NaOH solution during 1 h, vacuum-filtered using a filtering crucible of coarse porosity and washed with distilled water until neutral pH. After vacuum-drying on phosphorus pentoxide at room temperature, the wood meal is grinded to micron scale with a planetary ball milling during 100 h and extracted with 250 mL of a dioxane/water solution (v/v, 96/4), while stirring at 250 rpm at ambient temperature and under an inert atmosphere. After 24 h, the mixture is centrifuged at 1000 g during 20 min. The supernatant, containing the MWL is preserved. A fresh solution of dioxane/water is added to the residue and the previous procedure is repeated. The two supernatants are then combined, after which the volume is reduced to 100 mL by vacuum evaporation at 40 °C. Distilled water is then added to the solution which is then freeze-dried to obtain crude MWL. In order to get high purity MWL, a purification (i.e. removing cell wall carbohydrate residues) is required. The crude MWL is dissolved in 40 mL of acetic acid/water solution (v/v, 9/1) then precipitated in 800 mL deionized water. The solid residue (MWL) is vacuum-filtered using a fine porosity filtering crucible and dissolved again in 20 mL of a 1,2-dichloroethane/ethanol solution (v/v, 2/1), then precipitated first in ether and finally in petroleum ether. The purified MWL is vacuum-dried during 24 h and stored at −20 °C before further analyses.

3.5 Proteins and Lipids

Proteins and lipids are rarely found in lignocellulosic biomass such as in hardwood and softwood. However, they can be recovered in straws and in agricultural biomass at varying concentrations. Thus, in order to reach a complete chemical characterization of biomass it is important to determine their content.

3.5.1 Proteins

Proteins are polymers composed of amino acids linked by peptide bonds and their concentration and composition vary significantly from one specie to the other. Proteins originated from biomass have regained interest as they can possibly be used as food for animals and potentially humans as well [78]. It can represent up to 30% in some leaves [79]. In addition, it has been reported that for some biomass conversion reactions such as hydrothermal gasification and liquefaction, proteins may have a negative effect on the yield [80]. Thus, their quantification in biomass represents an important step in biomass characterization.

Protein content quantification in biomass is usually difficult and time-consuming to estimate by absolute quantification. Thus, in order to determine the protein content, the main procedure relies on an approximate determination based on the nitrogen-to-protein content. To this purpose, the method consists in determining the nitrogen content mainly by acidic digestion followed by the application of a conversion factor (usually 6.25) to the nitrogen content allowing deducing the protein content. The nitrogen content is generally estimated using two methods: the [81] and the [82] methods.

The most common method is the Kjeldahl method and this method is based on a three-step procedure. First, the mineralization step in which the nitrogen compounds (amines, amides, peptides etc.) are converted into ammonia using catalysts. In the second step, ammonia is distillated into a strong acid and finally, a back titration of the solution is performed. Even though, today, automatic instruments which follow the AOAC 976.05 procedure are available on the market [83], the procedure can also be done manually in a laboratory. The usual procedure follows the AOAC 984.13 methodology [84] even though, many procedures have been described based on the Kjeldahl method depending on the amount of sample, its physical state, etc. However, the procedure described below is based on the AOAC 984.13 with some modifications [85].

Considering that, protein determination can be done at the same time with the usual chemical composition analysis for biomass, it is recommended to use the same grinding approach as previously suggested. 0.2 g of oven-dried sample (55 °C) are added into a Kjeldahl flask with 15 g of K_2SO_4, 0.04 g of anhydrous $CuSO_4$ and 0.5–1 g of alundum granules and 20 mL of concentrated H_2SO_4 73% (an internal amino acid standard can be added to verify the digestion process). The mixture is brought from 25 °C to a roiling boil (up to 350 °C) in 5 min until the dense fumes disappear, then the mixture is heated for another 90 min under stirring. After cooling, 250 mL of water is added to the mixture. Meanwhile, the titration solution can be prepared by accurately add an acid standard solution to a sufficient amount of water to immerse the condenser tip then add 3–4 drops of red methyl indicator. To reduce the foaming in the mixture containing the sample, 2–3 drops of tributyl citrate can be added. Then, a NaOH solution is poured to bring the mixture in strong alkali conditions. The mixture can be then slowly distillated up the boiling point until at least 150 mL of condensate are recovered in the titration solution flask. The excess of acid in the condensate is then titrated with a NaOH standard solution. The nitrogen content can be easily calculated, and the protein content is then determined as followed:

$$\text{protein} = \text{conversion factor (usually 6.25)} \times \%\,\text{nitrogen} \tag{3.6}$$

Several variations of the Kjeldahl method exist including the use of different conditions and/or reagents (AOAC 988.05). It can be noted that the Kjeldahl is accepted by many international scientific organizations (ISO, IUPAC, AOAC, ASTM, …).

The Dumas method is simpler and faster by comparison. The latter is based on a complete combustion of a sample under an oxygen atmosphere using catalysts (AOAC 968.06, ASTM E258-07). The principle is similar to elemental analysis and can be done as such. The main advantage of the method is its fast determination of nitrogen content. The main inconvenient resides in the need for an expensive instrument such as an elemental analyzer. In addition, since the method is based on the combustion of all nitrogen containing compounds, the application of a conversion factor leads to an overestimation of the protein content when the usual conversion factor is used in case of a high inorganic nitrogen content for instance.

However, both methods suffer limitations such as the determined nitrogen content is not selective to proteins and some nitrogen-containing compounds may lead to an overestimation of the protein content. In the described methods,

Table 3.3 Examples of nitrogen-to-protein conversion factor

Sample	Conversion factor
Millets (grain)	5.83
Sorghum (grain)	6.25
Pine nuts	5.30

nitrogen-containing derivates such as nitro, azo, etc. are not converted in ammonia and other methodologies are suggested for a complete nitrogen determination [86]). In addition, considering that most proteins usually contain around 16% of nitrogen, a correction factor of 6.25 is commonly used [87]. However, some studies have noticed that the correction factor of 6.25 may not be suitable for biomass crop residues [87, 88]. Other possible conversion factors are presented in Table 3.3.

Different procedures have been suggested to determine more accurately the correction factor (described in Eq. 3.5) based on a specific sample such as the NREL [89]. First, the nitrogen content is determined by either the Dumas method (following the AOAC 990.03 protocol) or the Kjeldahl method (following the AOAC 984.13). Then the complete amino acid profiles can be determined by the method described by Mossé [90] which provides the exact protein content by HPLC analysis. Furthermore, the authors have proposed that two conversion factors can be determined, k_A: being the ratio of the content in g of amino acids per 100 g of sample over the content in g of nitrogen in amino acids per 100 g of sample and k_P: being the ratio of the content in g of amino acids per 100 g over the content in g of nitrogen per 100 g. They explained that in fact the most reliable conversion factor is an average of both. Others have indicated that k_P is more suitable for biomass samples such as wheat straw, rice straw, etc. [88].

Other techniques exist but are less common such as the colorimetric method using dye-protein complex [51] or copper ions-proteins complex. Unfortunately, none of those techniques allows an absolute quantification since other nitrogen containing compounds may react and combine the difficulty of recovering all amino acids when hydrolyzing proteins. The most reliable methods are the Kejdahl method and the Dumas method since many scientific papers can be found on the subject, giving background experiences and for some biomass, a more accurate conversion factor.

3.5.2 *Lipids*

From the first classification of the lipids presented by Bloor in 1920, there is still not a well-defined and well accepted exhaustive classification of the lipids (or lipoids) [91] that can be found from official standard organizations. Bloor proposed a classification in three groups: the simple lipids (including fats and waxes), the compound lipoids (including phospholipids and glycolipids) and the derivative lipids (including fatty acids and sterols). However, the LIPID Metabolites And

Pathways Strategy (LIPID MAPS®), an initiative which regroups a team of international researchers have proposed a 8 group classification: Fatty acyls (FA), Glycerolipids (GL), Glycerophospholipids, (GP), Sphingolipids (SP), Sterol Lipids (ST), Phenol Lipids (PR), Saccharolipids (SL), Polyketides (PK) [28]. Although this classification appears exhaustive, the only agreement may reside in a definition such as the one proposed by Christie: "Lipids are fatty acids and their derivatives, and substances related biosynthetically or functionally to these compounds" [92]. On the other hand, others have proposed to keep the definition for lipids based on their physical properties and especially their solubility. Thus, another definition could be that lipids are defined by their solubility in non-aqueous solvent such as chloroform, alkanes or alcohols [93].

In this section, different procedures will be described based on the literature and considering that in plant tissues, it is established that the main lipid group is the fatty acid derivates [92] hence based on the solubility definition and/or on the Christie's proposed definition, three methodologies will be described in this section for lipid quantification.

The first methodology resides on the solubility definition which is easy and fast allowing for the determination of the crude fat content. The most common methodology to quantify crude fat (such as waxes) following this definition is by gravimetric method using solvent extraction by hexane and diethyl ether in a Soxhlet apparatus. An official AOAC 2003.05 procedure exists using semi-automated Soxtec [94]. The procedure can also be adapted to using an accelerated solvent extractor (ASE). Typically, the biomass sample should be grinded the same way as it is for biomass chemical analysis (between 425 and 650 μm). The fraction is then dried at 105 °C overnight and extracted by hexanes and/or diethyl ether using a Soxhlet apparatus during at least 6 h at rate of 4 cycle/h minimum. The solvent fraction is then dried under a nitrogen flow or at ambient and the residue is then weighted providing the crude fat amount. It is also possible to follow the ASTM D1105-96 to perform a similar experiment.

The second methodology is based on Christie's definition considering that lipids are composed of fatty acids esterified to glycerol and that the nature of the fatty acids varies depending on the species. Fatty acids from biomass are generally composed of an even number of carbon atoms, a carboxyl group and eventually double bonds [92]. The most common saturated fatty acids are 14:0, 16:0 and 18:0 also known as myristic acid (tetradecanoic acid), palmitic acid (hexadecanoic acid) and stearic acid (octadecanoic acid) respectively (14, 16 and 18 indicate to number of carbon atoms whereas ":0" defines the number of insaturation(s)). Indications of isomers configuration can also be added to the nomenclature.

In literature, the analysis of the lipid content based on fatty acids mostly concern biodiesel production [95] but the from determination from raw biomass mostly concerns algal biomass [96]. However, a procedure, (i.e. acidic hydrolysis) could be suitable for lignocellulosic biomass. This method is also used by NREL [97] for algal biomass and consists in expressing the total lipids as the fatty acid methyl esters content (FAME). In addition, Lauren et al. [98] have compared the AOAC 991.39, AOAC 922.06 and AOAC 985.05 to their one-step in situ-acidic hydrolysis, based on the work of Lepage and have demonstrated its efficiency for algal biomass [99]. The main advantage of this procedure is that it can be used on small quantities on the entire biomass and has shown results comparable to the AOAC 991.39 [84]. Since the method was first developed for microalgal biomass, there might be some challenges in adapting it to lignocellulosic biomass.

This procedure is based on an in situ acid-hydrolysis and adapted from Laurens et al. [99]. Between 5 and 11 mg of lyophilized biomass (previously dried overnight at 40 °C under vacuum) are mixed with 200 μL of 2/1 chloroform/methanol (v/v) mixture in order to extract the lipids. 300 μL of HCl 5% in methanol are then added to in situ transesterify the lipids at 85 °C for an hour with 20 μg of a 13:0 FAME standard serving as internal standard. FAMEs are ultimately extracted using 1 mL of hexanes and quantified using GC-FID or GC-MS. An internal standard need to be used since the method does not recover 100% of the FAMEs.

Finally, the third methodology is also based on the solubility but using different solvents and considering that lipids are fatty acids. This methodology is older and reports that when lignocellulosic materials is used, it is necessary to first deactivate the enzymes (that are responsible for lipid degradation) by an isopropanol extraction prior to the procedure to avoid lipid degradation [92, 100]. This procedure is based on a modified Nichols procedure, which was originally applied to lettuce. The preparation of the sample follows the same steps than the previous methodology. A first extraction using a Soxhlet apparatus using isopropanol is followed by a second extraction using a mixture of chloroform/isopropanol 1/1 (v/v). Both filtrates are then extracted by a mixture of chloroform/methanol 2/1 (v/v). Consequently, the lipids should be recovered in the chloroform fraction [101]. The general concept behind the procedure for lipid content quantification, described below is based on the definition of lipids being non soluble in water but in organic solvent [92] and especially a mixture of chloroform and methanol. The extraction procedure is based on works published by Bligh and Dyer [102] and Floch, Lees and Stanley [103] and both compared [104, 105] leading to the previous protocol. Previously, the quantification of the lipid content by this procedure was determined by gravimetric method and thus additional steps were needed. However, using gas-phase chromatography, those steps are irrelevant, and often replaced by a dilution of the extract in hexane followed by an analysis using gas chromatography.

3.6 Extractives

3.6.1 General Aspects

Extractives include non-structural components with low molecular weight that are non-chemically bound components of biomass and found in the porous structure of lignocellulosic biomass. They are not related to the structure of the cell walls by covalent bonds and are therefore accessible to be easily extracted with organic solvents (or water). Most are secondary metabolites i.e. plant metabolism products that are not essential to its growth, as opposed to primary metabolites such as proteins and carbohydrates. The nature and content of these extractable substances give the specific properties of a biomass: color, odor, biological durability, color durability, etc.

The term "extractives" covers a large number of compounds which can be extracted as well with polar solvents (acetone, ethanol, water, …) than with apolar solvents (petroleum ether, hexane, …). Their content and composition vary with the species and even within the species. Variations are also observed depending on the geographical location and the season. The classification of extractives can be a difficult task due to their great variety of which some are still unidentified. However, their chemical nature can be briefly summarized in the following manner (Haluk 1994 in [106]):

- The aliphatic acids are mainly represented by saturated and unsaturated fatty acids with long carbon chain esterified either with glycerol (oils, fats) or with heavy alcohols (waxes). Acetic acid may be linked to hemicelluloses in the form of an ester group. Diethyl-hydroxy-carboxylic acids can be found as calcium salts. Aliphatic alcohols exist as esters, while aromatic sterols (steroids) are found a glycosides form;
- The terpenes are a very large group of natural substances. These are oligomeric derivatives of isoprene. Monoterpenes are composed of two isoprene units, sesquiterpenes of three isoprene units, diterpenes of four isoprene units, sesterterpenes of five units and triterpenes of six isoprenes units …
- The aromatic and phenolic compounds are mainly represented by tannins. Hydrolyzable tannins and condensed tannins are distinguished. Other polyphenolic substances such as flavonoids, lignans and stilbenes can also be encountered. Any identified simple phenols are likely derived from lignin metabolism [64];
- Carbohydrates occurs mainly as by starch, sucrose, glucose and fructose. These components are particularly found in herbaceous plants and grains. Sucrose is common in sap of trees such as maple (Acer sp.) whereas glucose and fructose are the main carbohydrates in birch (Betula sp.) [107]. Hydrocolloid gums which are complex mixture of polysaccharides, mainly constituted of arabinose, galactose, mannose and glucuronic acid, are produced by exudation from tree stem [108]. They are also found in condensed tannin extract (up to 8%) [109].

Since biomass extractives are highly heterogeneous (a plant extract may contain more than a hundred compounds from the different categories described previously), this section will not address the topic of their structural characterization.

3.6.2 Compositional Analysis Methods

For woody plants, sapwood generally contains low amounts of extractives mainly composed of lipophilic components such as waxes, fats, resins and terpenes [110]. At the opposite, heartwood and forest residues such as bark contain higher amounts of extractives like phenolic compounds and suberin [110]. Non-woody plants generally contain high amounts of water-soluble components such as non-structural carbohydrates [111]. For quantitative determination of cell wall component, the extractives are removed because they can potentially interfere with downstream analysis of biomass sample. However, due to their complexity (mixture of lipophilic and hydrophilic components), it is quite impossible to use only one extraction solvent to remove all the extractives out of the cell wall, specifically for extractive-rich lignocellulosic feedstock.

Usually, non-polar solvents such as hexane and low polarity solvents such as dichloromethane are currently used to extract/remove lipophilic components whereas polar solvents such as alcohols, water or mixtures are to recover for hydrophilic components. A set of standard methods have been developed to determine the extractive contents within a wide range of lignocellulosic materials. According to ASTM D1105 standard, sequential extraction with ethanol/toluene (waxes, fats, and some resins) for 4 h, 95% ethanol (terpenoids and some phenolic compounds) for 4 h and hot water (tannins, gums and starch) for 1 h is used for woody plants to completely remove all the extractives. However, 190 proof ethanol extraction in a Soxhlet-type apparatus for 24 h (according to ASTM E1690), may be sufficient to remove interfering non-structural components (waxes, fats, resins, tannins, gums, sugars, starches, and pigments) prior to cell wall component determination. However, this method may not be suitable if the extractive content is above 10% such as for bark. For suberinrich biomass such as bark, a sequential extraction method with ethanol/toluene and 1% NaOH [68] is required to remove all the extractives. For herbaceous plants, a two-step extraction procedure is adopted as recommended by NREL, which consist first in removing water-soluble and then removing ethanol-soluble extractable components. Recently, this procedure has been adapted and downscaled to a sample size of 250 mg [112]. Note that herbaceous materials generally contain more water-soluble components, specifically non-structural carbohydrates, than woody materials.

The following procedure describes the extraction protocol adapted from NREL procedure for total extractive determination (or removal) in biomass with slight modifications. 10 g of 40–60 mesh air-dried raw biomass is loaded into the cellulose thimble. With the Soxhlet apparatus set up, ground biomass is first extracted with 190 ± 5 mL of water at approximately 5 cycles/h for 24 h. The water in the round-bottomed flask is vacuum-evaporated to 19 ± 5 mL at 50 °C and then dried in the oven at 105 °C for 24 h to determine the weight of extractives present in the sample flasks. After water extraction, the biomass residue is air-dried at room temperature to a moisture content below 10% before being subjected to 190 proof ethanol extraction by the same method describes previously with water. The extractives are subsequently weighed to determine the amount of ethanol-soluble extractives. Additional step consisting in mild alkali extraction step is required with suberinrich biomass such as bark. The air-dried biomass ($\leq 10\%$ moisture content), free of water and ethanol soluble extractives, is refluxed at boiling temperature with 500 mL of 1% NaOH solution for 1 h. Then the slurry is filtered through vacuum using a filtering crucible of coarse porosity, washed first with 500 mL of hot distilled water, then with 250 mL of 10% acetic acid solution and finally with hot water until neutral pH was reached. The alkali extractive-free biomass is vacuum-dried on phosphorus pentoxide or air-dried at room temperature before cell wall component determination. The alkali-extracted biomass is subsequently weighed to determine the amount of alkali-soluble extractives.

References

1. Paris, 2015: Tracking country climate pledges. In: Carbon Brief. https://www.carbonbrief.org/paris-2015-tracking-country-climate-pledges (2015). Accessed 20 January 2019
2. Government of Canada PW and GSC Pan-Canadian framework on clean growth and climate change: En4-294/2016E-PDF—Government of Canada Publications (2002). http://publications.gc.ca/site/eng/9.828774/publication.html. Accessed 7 Mar 2018
3. Québec (Province), Ministère du développement durable environnement et parcs (2013) Le Québec en action vert 2020: plan d'action 2013–2020 sur les changements climatiques : Phase 1. Ministère du développement durable, environnement et parcs, Québec, Que (2013)
4. Yung, M.M., Jablonski, W.S., MagriniBair, K.A.: Review of catalytic conditioning of biomass-derived syngas. Energy Fuels **23**, 1874–1887 (2009)
5. Spivey, J.J., Egbebi, A.: Heterogeneous catalytic synthesis of ethanol from biomass-derived syngas. Chem. Soc. Rev. **36**, 1514–1528 (2007)
6. Tijmensen, M.J.A., Faaij, A.P.C., Hamelinck, C.N., Hardeveld, M.R.M.: Exploration of the possibilities for production of Fischer Tropsch liquids and power via biomass gasification. Biomass Bioenergy **23**, 129–152 (2002)
7. Qi, Z., Jie, C., TieJun, W., Ying, X.: Review of biomass pyrolysis oil properties and upgrading research. Energy Convers. Manag. **48**, 87–92 (2007)
8. Mohan, D., Pittman, C.U., Steele, P.H.: Pyrolysis of wood/biomass for bio-oil: a critical review. Energy Fuels **20**, 848–889 (2006)
9. Czernik, S., Bridgwater, A.V.: Overview of applications of biomass fast pyrolysis oil. Energy Fuels **18**, 590–598 (2004)
10. Brewer, C.E., Schmidt-Rohr, K., Satrio, J.A., Brown, R.C.: Characterization of biochar from fast pyrolysis and gasification systems. Environ. Prog. Sustain. Energy **28**, 386–396 (2009)
11. Manyà, J.J.: Pyrolysis for biochar purposes: a review to establish current knowledge gaps and research needs. Environ. Sci. Technol. **46**, 7939–7954 (2012)
12. Badger, P.C.: Ethanol from cellulose: a general review. Trends New Crops New Uses Proc. Fifth Natl. Symp. Atlanta. Ga. USA 10–13 November 2001, 17–21 (2002)
13. Holladay, J.E., White, J.F., Bozell, J.J., Johnson, D.: Top value-added chemicals from biomass-volume II—results of screening for potential candidates from biorefinery lignin. Pacific Northwest National Laboratory PNNL16983 (2007)
14. Wertz, J.L.: Cellulose Science and Technology, EPFL Press (2010)
15. Rowell, R.: Handbook of Wood Chemistry and Wood Composites, 2nd Edn. CRC Press (2012).

16. Klemm, D., Philipp, B., Heinze, T., Heinze, U., Wagenknecht, W.: Volume 1: fundamentals and analytical methods. In: Comprehensive Cellulose Chemistry, pp 260. Wiley, Weinheim, Germany (1998)
17. Plazonić, I., Barbarić-Mikočević, Ž., Džimbeg-Malčić, V.: V Chemical composition of triticale straw as a paper fiber source. WPP PA Wood, Pulp and Paper Polig. Acad. 292 (2014)
18. Mihranyan, A.: Cellulose from cladophorales green algae: From environmental problem to high-tech composite materials. J. Appl. Polym. Sci. **119**(4), 2449–2460 (2010)
19. Davis, T.A., Volesky, B., Mucci, A.: A review of the biochemistry of heavy metal biosorption by brown algae. Water Res. **37**, 4311–4330 (2003)
20. Siddhanta, A.K., Prasad, K., Meena, R., Prasad, G., Mehta, G.K., Chhatbar, M.U., Oza, M. D., Kumar, S., Sanandiya, N.D.: Profiling of cellulose content in Indian seaweed species. Bioresour. Technol. **100**, 6669–6673 (2009)
21. Saka, S.: Structure and chemical composition of wood as a natural composite material. In: Shiraishi, N., Kajita, H., Norimoto, M. (eds.) Recent Research on Wood and Wood-Based Materials, pp. 1–20. Elsevier (1993)
22. Zhang, N., Li, S., Xiong, L., Hong, Y., Chen, Y.: Cellulose-hemicellulose interaction in wood secondary cell-wall. Model Simul. Mater. Sci. Eng. 23(8) (2015)
23. Zhang, Y.-H.P.: Reviving the carbohydrate economy via multi-product lignocellulose biorefineries. J. Ind. Microbiol. Biotechnol. **35**, 367–375 (2008)
24. Lavoine, N., Desloges, I., Dufresne, A., Bras, J.: Microfibrillated cellulose—its barrier properties and applications in cellulosic materials: a review. Carbohydr. Polym. **90**, 735–764 (2012)
25. Cross, C.F., Dorée, C., Bevan, E.J.: Researches on cellulose. Longmans, Green, London, New York (1907)
26. Ritter, G.J.: Determination of alpha-cellulose. Ind. Eng. Chem. Anal. Ed. **1**, 52–54 (1929)
27. Sun, X.-F., Fowler, P., Baird, M.S.: Extraction and characterization of original lignin and hemicelluloses from wheat straw. J. Agric. Food Chem. **53**, 860–870 (2005)
28. Fahy, E., Subramaniam, S., Murphy, R.C., Nishijima, M., Raetz, C.R.H., Shimizu, T., Spener, F., van Meer, G., Wakelam, M.J.O., Denni, E.A.: Update of the LIPID MAPS comprehensive classification system for lipids. J. Lipid Res. **50**, S9–S14 (2009)
29. Timpa, J.D.: Application of universal calibration in gel permeation chromatography for molecular weight determinations of plant cell wall polymers: cotton fiber. J. Agric. Food Chem. **39**, 270–275 (1991)
30. Hon, D.N.S., Srinivasan, K.S.V.: Mechanochemical process in cotton cellulose fiber. J. Appl. Polym. Sci. 28, 1–10 (1983)
31. Bali, G., Khunsupat, R., Akinosho, H., Payyavula, R.S., Samuel, R., Tuskan, G.A., Kalluri, U.C., Ragauskas, A.J.: Characterization of cellulose structure of populus plants modified in candidate cellulose biosynthesis genes. Biomass Bioenergy **94**, 146–154 (2016)
32. Hallac, B.B., Ragauskas, A.J.: Analyzing cellulose degree of polymerization and its relevancy to cellulosic ethanol. Biofuels Bioprod. Biorefining **5**, 215–225 (2011)
33. ASTM D 1105—Standard test method for preparation of extractive-free wood/note. https://www.boutique.afnor.org/norme/astm-d-1105/-/article/765498/us003100 (2007). Accessed 11 May 2018
34. Silva, G.G.D., Rouau, S.G.X.: Successive centrifugal grinding and sieving of wheat straw. Powder Technol. **208**, 266–270 (2011).
35. Hubbell, C.A., Ragauskas, A.J.: Effect of acid-chlorite delignification on cellulose degree of polymerization. Bioresour. Technol. **101**, 7410–7415 (2010)
36. Segal, L., Creely, J.J., Martin, A.E., Conrad, C.M.: An empirical method for estimating the degree of crystallinity of native cellulose using the X-ray diffractometer. Text. Res. J. **29**, 786–794 (1959)
37. Park, S., Baker, J.O., Himmel, M.E., Parilla, P.A., Johnso, D.K.: Cellulose crystallinity index: measurement techniques and their impact on interpreting cellulase performance. Biotechnol. Biofuels **3**, 10 (2010)

38. Thygesen, A., Oddershede, J., Lilholt, H., Thomsen, A.B., Ståhl, K.: On the determination of crystallinity and cellulose content in plant fibres. Cellulose **12**, 563 (2005)
39. Potthast, A., Radosta, S., Saake, B., Lebioda, S., Heinze, T., Henniges, U., Isogai, A., Koschella, A., et al.: Comparison testing of methods for gel permeation chromatography of cellulose: coming closer to a standard protocol. Cellulose **22**, 1591–1613 (2015)
40. Schulze, F.: Chemical composition of vegetable cell membranes. J. Soc. Chem. Ind. **11**, 2277–2287 (1891)
41. Scheller, H.V., Ulvskov, P.: Hemicelluloses. Annu. Rev. Plant Biol. **61**, 263–289 (2010).
42. Li, Z., Qin, M., Xu, C., Chen, X.: Hot water extraction of hemicelluloses from aspen wood chips of different sizes. BioResources **8**, 5690–5700 (2013).
43. Rissanen, J.V., Murzin, D.Y., Salmi, T., Grénman, H.: Aqueous extraction of hemicelluloses from spruce from hot to warm. Bioresour. Technol. **199**, 279–282 (2016).
44. Ebringerová, A., Hromádková, Z., Heinze, T.: Hemicellulose. In: Heinze, T. (ed.) Polysaccharides I, pp. 1–67. Springer, Berlin, Heidelberg (2005).
45. Izydorczyk, M.S., Biliaderis, C.G.: Cereal arabinoxylans: advances in structure and physicochemical properties. Carbohydr. Polym. **28**, 33–48 (1995).
46. Nikolaevna, N., Nikolaevna, E., Viktorovna, N., Alekseevich, Y., Anatolievich, V.: Polysaccharides from larch biomass. In: Karunaratne, D.N. (ed.) The Complex World of Polysaccharides. Tech, Rijeka, Croatia (2012)
47. Ahlgren, P.A., Goring, D.A.I.: Removal of wood components during chlorite delignification of black spruce. Can. J. Chem. **49**, 1272–1275 (1971)
48. Shiraishi, N., Hon, D.N.S.: Wood and cellulosic chemistry, 2nd Edn. CRC Press (2000)
49. Pentosans in wood and pulp, Test Method T 223 cm^{-10}. https://imisrise.tappi.org/TAPPI/Products/01/T/0104T223.aspx. Accessed 11 May 2018
50. Pettolino, F.A., Walsh, C., Fincher, G.B., Bacic, A.: Determining the polysaccharide composition of plant cell walls. Nat. Protoc. **7**, 1590–1607 (2012).
51. Bradford, M.M.: A rapid and sensitive method for the quantitation of microgram quantities of protein utilizing the principle of protein-dye binding. Anal. Biochem. **72**, 248–254 (1976)
52. Ciucanu, I., Kerek, F.: A simple and rapid method for the permethylation of carbohydrates. Carbohydr. Res. **131**, 209–217 (1984). (Accessed May 9, 2018)
53. Scott, R.W., Libkie, K.A., Springer, E.L.: Comparison of a gravimetric CO_2 method for uronic anhydride with a colorimetric method. J. Wood Chem. Technol. **4**, 497–504 (1984).
54. Scott, R.W.: Colorimetric determination of hexuronic acids in plant materials. Anal. Chem. **51**, 936–941 (1979).
55. Willför, S., Pranovich, A., Tamminen, T., Puls, J., Laine, C., Suurnäkki, A., Saake, B., Uotila, K., Simolin, H., Hemming, J., Holmbom, B.: Carbohydrate analysis of plant materials with uronic acid-containing polysaccharides–a comparison between different hydrolysis and subsequent chromatographic analytical techniques. Ind. Crops Prod. **29**, 571–580 (2009).
56. Sluiter, A., Hames, B., Ruiz, R., Scarlata, C., Sluiter, J., Templeton, D., Crocker, D.: Determination of structural carbohydrates and lignin in biomass: laboratory analytical procedure (LAP) (2008).
57. Stevanović, T., Perrin, D.: Chimie du bois. Presses polytechniques et universitaires romandes (2009)
58. Dalimova, G.N., Abduazimov, K.A.: Lignins of herbaceous plants. Chem. Nat. Compd. **30**, 146–159 (1994)
59. Lupoi, J.S., Smith, E.A.: Characterization of woody and herbaceous biomasses lignin composition with 1064 nm dispersive multichannel raman spectroscopy. Appl. Spectrosc. **66**, 903–910 (2012)
60. Oe, E., Lindgren, B.O.: About the linkage between lignin and hemicelluloses in wood. Sven Papperstidning (1977)
61. Ghaffar, S.H., Fan, M.: Structural analysis for lignin characteristics in biomass straw. Biomass Bioenergy **57**, 264–279 (2013)

62. Buranov, A.U., Mazza, G.: Lignin in straw of herbaceous crops. Ind. Crops Prod. **28**, 237–259 (2008)
63. Hatfield, R., Fukushima, R.S.: Can lignin be accurately measured? Crop Sci. **45**, 832 (2005)
64. Fengel, D., Wegener, G.: Hydrolysis of polysaccharides with trifluoroacetic acid and its application to rapid wood and pulp analysis. In: Hydrolysis of Cellulose: Mechanisms of Enzymatic and Acid Catalysis. American Chemical Society, pp. 145–158 (1979)
65. Schwanninger, M., Hinterstoisser, B.: Klason Lignin: Modifications to improve the precision of the standardized determination. Holzforschung 56, 161–166 (2002)
66. Van Soest, J.P.: Use of detergents in the analysis of fibrous feeds. 2. A rapid method for the determination of fiber and lignin. J. Assoc. Off Agric. Chem. **46**, 829–835 (1963)
67. Ibáñez, A.B., Bauer, S.: Downscaled method using glass microfiber filters for the determination of Klason lignin and structural carbohydrates. Biomass Bioenergy **68**, 75–81 (2014)
68. Harun, J., Labosky, P.: Chemical constituents of five Northeastern barks. Wood Fiber Sci. **17**, 274–280 (2007)
69. Monteil-Rivera, F., Phuong, M., Mengwei, Y., Halasz, A., Hawariet, J.: Isolation and characterization of herbaceous lignins for applications in biomaterials. Ind. Crops Prod. **41**, 356–364 (2012)
70. Lupoi, J.S., Singh, S., Parthasarathi, R., Simmons, B.A., Henry, R.J.: Recent innovations in analytical methods for the qualitative and quantitative assessment of lignin. Renew. Sustain. Energy Rev. **49**, 871–906 (2015)
71. Faix, O.: Fourier transform infrared spectroscopy. Methods in lignin chemistry, pp. 223–241. Springer, Berlin, Heidelberg (1992)
72. Schorr, D., Diouf, N., Stevanovic, T.: Comparison of physicochemical and thermal properties of esterified and non-esterified Kraft lignins for biocomposite application, pp. 309–328 (2014)
73. Faix, O.: Condensation indices of lignins determined by FTIR-spectroscopy. Holz Als Roh-Werkst. **49**, 356–356 (1991)
74. Goldschmid, O.: Determination of phenolic hydroxyl content of lignin preparations by ultraviolet spectrophotometry. Anal. Chem. **26**, 1421–1423 (1954)
75. Liitia, T., Tamminen, T.: Direct method for the determination of phenolic hydroxyl groups in pulp. Holzforschung **61**, 623–627 (2007)
76. Meier, D., Faix, O.: Chromatography-mass spectrometry. Methods in lignin chemistry, pp. 177–199. Springer, Berlin, Heidelberg (1992)
77. Bjorkman, A.: Studies on finely divided wood part 1. Extraction of lignin with neutral solvents. Sven Papperstidn **59**, 477–485 (1956)
78. Chiesa, S., Gnansounou, E.: Protein extraction from biomass in a bioethanol refinery—possible dietary applications: use as animal feed and potential extension to human consumption. Bioresour. Technol. **102**, 427–436 (2011)
79. Nagy, S., Telek, L., Hall, N.T., Berry, R.E.: Potential food uses for protein from tropical and subtropical plant leaves. J. Agric. Food Chem. **26**, 1016–1028 (1978)
80. Kruse, A., Maniam, P., Spieler, F.: Influence of proteins on the hydrothermal gasification and liquefaction of biomass. 2. Model compounds. Ind. Eng. Chem. Res. **46**, 87–96 (2007)
81. Kjeldahl: Neue methode zur bestimmung des stickstoffs in organischen Körpern. Fresenius, Zeitschrift f. Anal. Chemie. **22**, 366–382 (1883)
82. Dumas, A: Annales de Chimie. **33**, 342 (1826)
83. Sáez-Plaza, P., Navas, M.J., Wybraniec, S., Michalowski, T., Garcia Asuero, A.: An overview of the kjeldahl method of nitrogen determination. Part II sample preparation, working scale, instrumental finish, and quality control. Crit. Rev. Anal. Chem. **43**, 224–272 (2013)
84. AOAC International AOAC: Official Methods of Analysis (Volume 1) (1990)
85. Carter, M.R., Gregorich, E.G.: Soil sampling and methods of analysis. CRC Press, Canadian Society of Soil Science (2007)

86. Prusisz, B., Jaśkiewicz, L., Pohl, P.: High-performance ion chromatography assessment of inorganic and organic nitro-gen fractions in potatoes. Microchim. Acta **156**, 219–223 (2006)
87. Merrill, A.L.: Energy value of foods: basis and derivation, Washington: human nutrition research branch, agricultural research service. U. S. Department of Agriculture (1955)
88. Chen, X., Zhao, G., Zhang, Y., Han, L., Xiao, W.: Nitrogen-to-protein conversion factors for crop residues and animal manure common in China. J. Agric. Food Chem. **65**, 9186–9190 (2017)
89. Hames, B., Scarlata, C., Sluiter, A.: Determination of protein content in biomass. Natl Renew Energy Lab 1–5 (2008)
90. Mosse, J.: Nitrogen-to-protein conversion factor for ten cereals and six legumes or oilseeds. A reappraisal of its definition and determination. Variation according to species and to seed protein content. J. Agric. Food Chem. **38**, 18–24 (1990)
91. Bloor, W.R.: Outline of a classification of the lipoids. Exp Biol Med **17**, 138–140 (1920)
92. Lipid Analysis—4th Edn. https://www.elsevier.com/books/lipid-analysis/christie/978-0-9552512-4-5. Accessed 29 Apr 2018
93. Kochhar, P.: Lipid biochemistry. An introduction (5th Edn.). Int. J. Food Sci. Technol. 37, 899–900 (2002)
94. Thiex, N.J., Anderseon, S., Gildemeister, B.: Crude fat, diethyl ether extraction, in feed, cereal grain, and forage (Ran-dall/Soxtec/Submersion method): Collaborative Study, 11 (2002)
95. Ma, F., Hanna, M.A.: Biodiesel production: a review 1. J. Ser. #12109. Agricultural research division, institute of agriculture and natural resources, university of Nebraska–Lincoln 1. Bioresour. Technol. 70, 1–15 (1999)
96. Chisti, Y.: Biodiesel from microalgae. Biotechnol. Adv. **25**, 294–306 (2007)
97. Van Wychen, S., Ramirez, K., Laurens, L.M.L.: Determination of total lipids as fatty acid methyl esters (FAME) by in situ transesterification: laboratory analytical procedure (LAP) (2016)
98. Lepage, G., Roy, C.C.: Direct transesterification of all classes of lipids in a one-step reaction. J. Lipid Res. **27**, 114–120 (1986)
99. Laurens, L.M.L., Quinn, M., Van Wychen, S., Templeton, D.W., Wolfrum, E.J.: Accurate and reliable quantification of total microalgal fuel potential as fatty acid methyl esters by in situ transesterification. Anal. Bioanal. Chem. **403**, 167–178 (2012)
100. Dutta, P.C., Appelqvist, L.-A.: The effects of different cultural conditions on the accumulation of depot lipids noly petroselinic acid during somatic embryogenesis in Daucus carota L. Plant Sci. **64**, 167–177 (1989)
101. Nichols, B.W.: Separation of the lipids of photosynthetic tissues: improvements in analysis by thin-layer chromatography. Biochim. Biophys. Acta. BBA—Spec. Sect. Lipids Relat. Subj. **70**, 417–422 (1963)
102. Bligh, E.G., Dyer, W.J.: A rapid method of total lipid extraction and purification. Can. J. Biochem. Physiol. **37**, 911–917 (1959)
103. Folch, J., Lees, M., Stanley, G.H.S.: A simple method for the isolation and purification of total lipides from animal tissues. J. Biol. Chem. **226**, 497–509 (1957)
104. Iverson, S.J., Lang, S.L., Cooper, M.H.: Comparison of the Bligh and Dyer and Folch methods for total lipid determination in a broad range of marine tissue. Lipids **36**, 1283–1287 (2001)
105. Sheng, J., Vannela, R., Rittmann, B.E.: Evaluation of methods to extract and quantify lipids from Synechocystis PCC 6803. Bioresour. Technol. **102**, 1697–1703 (2011)
106. Jodin, P.: Association pour la recherche sur le bois en Lorraine Le Bois: matériau d'ingénierie. Arbolor, Nancy (1994)
107. Kūka, M., Čakste, I., Geršebeka, E.: Determination of bio-active compounds and mineral substances in latvian birch and maple saps. Proc. Latv. Acad. Sci. Sect. B Nat. Exact. Appl. Sci. 67 (2013)
108. Coppen, J.J.W.: Gums, resins and latexes of plant origin. Food and Agriculture Organization of the United Nations (1995)

109. Pizzi, A.: Wood adhesives. Taylor and Francis (1983)
110. Argyropoulos, D.S.: Wood and cellulosic chemistry. 2nd Edn. J. Am. Chem. Soc. 123, 8880–8881 (2001)
111. Thammasouk, K., Tandjo, D., Penner, M.H.: Influence of extractives on the analysis of herbaceous biomass. J. Agric. Food Chem. **45**, 437–443 (1997)
112. Kuchelmeister, C., Bauer, S.: Rapid small-scale determination of extractives in biomass. BioEnergy Res. **8**, 68–76 (2015)

Chapter 4
Microalgal Biomass of Industrial Interest: Methods of Characterization

Catherine Dupré, Hugh D. Burrows, Maria G. Campos, Cédric Delattre, Telma Encarnação, Marilyne Fauchon, Clément Gaignard, Claire Hellio, Junko Ito, Céline Laroche, Jack Legrand, Philippe Michaud, Alberto A. C. C. Pais, Guillaume Pierre, Benoît Serive and Makoto M. Watanabe

Abstract Microalgae represent a new source of biomass for many applications. The advantage of microalgae over higher plants is their high productivities. The photoautotrophic microalgae include all photosynthetic microorganisms, i.e. Cyanobacteria (prokaryotes) or microalgae (eukaryotes). These microorganisms are characterized by a large biodiversity and chimiodiversity. Then, the analysis of microalgal and cyanobacterial biomass often needs specific adaptations of the classical protocols for extraction as well as for quantification of their contents. This chapter reviewed the main analytical methods used for the analysis of microalgae biomass and its main vaporizable compounds: proteins, polysaccharides, lipids, pigments and secondary metabolites.

C. Dupré (✉) · J. Legrand
GEPEA UMR 6144 CNRS, 37 Bd de l'Université BP 406,
Saint-Nazaire Cedex 44602, France
e-mail: catherine.dupre@univ-nantes.fr

H. D. Burrows · M. G. Campos · T. Encarnação · A. A. C. C. Pais
CQC, Department of Chemistry, University of Coimbra, Coimbra, Portugal

C. Delattre · C. Gaignard · C. Laroche · P. Michaud · G. Pierre
Université Clermont Auvergne, CNRS, Institut Pascal, Polytech Clermont-Ferrand,
2 avenue Blaise Pascal, Aubière Cedex 63178, France

M. Fauchon · C. Hellio
Biodimar, LEMAR UMR 6539, Institut Européen de la Mer, Université de Bretagne Occidentale, Brest 29200, France

J. Ito · M. M. Watanabe
Algae Biomass and Energy System R&D Center, University of Tsukuba, Tsukuba, Japan

B. Serive
USR 3151, CNRS Sorbonne Université, Station Biologique de Roscoff, Place Georges Teissier, Roscoff Cedex 29680, France

A. Nzihou (ed.), *Handbook on Characterization of Biomass, Biowaste and Related By-products*, https://doi.org/10.1007/978-3-030-35020-8_4

4.1 Introduction

Microalgae represent a unique source of biomass and biodiversity in the living world. There are several hundred thousand species colonizing almost the entire globe. They represent the first link in the trophic chain, the main supplier of oxygen and the first consumer of CO_2, some of whose storage is in the form of calcareous sediments. The advantage of microalgae over higher plants is their rapid growth. Their continuous cultivation makes it possible to direct their metabolism, thus opening the way to a renewable chemistry which does not encroach on the grounds reserved for agriculture and the forest.

Photosynthetic microorganisms include all unicellular organisms, both eukaryotic and prokaryotic able to convert light radiant energy into chemical energy for the synthesis of their cellular constituents. Some are defined photoautotrophic, as they use inorganic carbon for growth. Depending on the level of complexity of their intracellular structure, these organisms are either classified as Cyanobacteria (prokaryotes) or microalgae (eukaryotes). Microalgae show huge diversity in terms of intracellular structures, morphologies and life styles which can lead to a wide range of biotechnological applications. Most studied groups of microalgae include green algae (Chlorophyta and Charophyta), red algae (Rhodophyta), diatoms (Bacillariophyceae), (Haptophyta) and Euglenozoa.

With only a few dozen species of cultivated microalgae, annual world production is in the order of a few tens of thousands of tons. The last decades have shown the presence of a great diversity of molecules potentially recoverable from the biomass of microalgae and Cyanobacteria. The algosourced molecules currently used are essentially water-soluble (phycobiliproteins) or liposoluble (carotenoids) pigments or lipids such as terpenes and terpenoids, lipids enriched with polyunsaturated fatty acids from the omega-3 series, eicosapentaenoic acids (EPA) and docosahexaenoic acids (DHA), which play an important role in preventing cerebral and cardiovascular diseases. Other niche markets are associated with cosmetic or nutraceutical applications, directly using biomass or crude extracts. Microalgae valorisation must examine all cellular constituents according to the biorefinery principle. The objective of biorefinery or algal refining is to develop a process for exploiting microalgae, based on a series of dedicated unit operations allowing the selective extraction of metabolites of interest by integrating the recycling of waste and by-products, paving the way for exploitation on an industrial scale of microalgae.

The wealth of microalgae is not limited to molecules with high added value, and one of the major challenges of microalgal research is to enable developments on a very large scale and at competitive costs. Microalgae can provide molecules of varied interest in the field of Chemistry:

- lipids compounds, including triacylglycerols or hydrocarbons for use as biofuels and various organic compounds (aromatics, esters, etc.). For example, isomerization/methoxycarbonylation of unsaturated fatty acids leads to diesters and then polycondensation to polyesters,

- polysaccharides (gelling agents, thickeners),
- proteins (for human food or animal feed).

The aim of the present chapter is to give a critical review of the analytical methods used for the different classes of molecules of the algal biomass: global biomass, proteins, polysaccharides, lipids, pigments and secondary metabolites.

4.2 Methods for Biomass Global Characterization

4.2.1 Introduction

Any analysis of biochemical composition as well as of process efficiency in regard to a valorization of microalgae biomass should be based on accurate and reproducible methods for determining its proximate composition. Indeed, the cellular components are usually expressed as a percentage on a weight basis or as pg per cell. These data are used for mass balance calculations of the bioprocess to understand the relations between the biochemical species produced and the substrates consumed. Then proximate analysis of microalgae biomass includes the determination of the dry biomass, ash content but also elemental composition. Some of characteristics could be taken into account in assessing the effectiveness of biorefinery process. Relatively recent documents detail Standard Operating Procedure in order to minimize potential errors related to such routine analysis [1, 2].

According to the way the microalgae are produced, two preliminary studies should be necessary. A quality control (QC) could be required in case of non-application of good manufacturing practices and then those cultures could be suspected to be contaminated [3]. The method of microalgae identification by microscopical observations is based on the morphology [4]. Some database such as AlgaeBase [5] or some microalgae identification guides are available [6]. However, this cytological approach, as it can be misleading, especially in the case of species with few morphological features (e.g. coccoid cells), has been associated with biochemical traits, mainly the carotenoid and/or the fatty acid profiles [7, 8]. Some authors have proposed the FTIR spectra of whole cells for identification of the microalgae species [9]. This analysis requires few cells, which have to be washed, as some components of the culture medium such as nitrate could interfere with the infrared spectra. However, it was shown that these spectra are changing as a function of metabolic activities [10]. The application of FTIR analysis as QC is only suitable to continuous cultures at steady state, to detect any deviation of the cell suspension. For industrial production, DNA barcoding could be used for species identification, based on DNA sequence similarity against a sequence database of a defined species [11]. The Consortium for the Barcode of Life (CBOL) has recommended for microalgae DNA barcodes the use of a two-step approach with the use of a universal pre-barcode marker followed by the use of a more specific second marker [12]. DNA-based identification should be particularly useful for identifying species with no or few structural characters [13] DNA metabarcoding could be used to trace microalgae of industrial interest, as already done for some plants of industrial interest [14].

The other preliminary study deals with the sample collection methods, which should be adapted to the behavior of the microalgae culture. By contrast with cell suspensions cultivated in well-stirred photobioreactors, some species grow as colonies, such as *Botryococcus braunii*, as cell aggregates, as a result of natural palmelloid stage or flocculation. The sampling protocol has to take into account the increase in the settling rate of such heterogeneous suspension to limit the variability of the results. In some processes, microalgae are cultivated as biofilm [15, 16]. The sampling is done by scrapping biomass from a known area of the support, which size depends on the regularity of the biofilm thickness on the support [17, 18].

4.2.2 Dry Weight Method

4.2.2.1 Centrifugation or Filtration

The gravimetric measurement of dry weight (DW) is the reference method used for each species to calibrate the other indirect methods. The Standard Operating Procedures (SOPs) corresponding to this method are well described elsewhere [2]. Accurate and precise values of microalgal dry weight are obtained thanks to the use of appropriate protocols [19]. A precise volume (v) of the culture (suspension cultures) or of the buffer with the homogenized biofilm sample has to be filtered or centrifuged. The filtration method on glass filters, usually GF/C or GF/F, is faster than the centrifuge method. It has to be used for buoyant microalgae, either oleaginous or gas-vacuolated species, a part of the biomass being lost in the supernatants. These filters retain particles down to 0.7 μm, just enough for collecting cells of *Ostreococcus tauri* (0.8 μm). For picoplankton cells (0.2–2 μm) and some Cyanobacteria (size down to 0.5 μm) the use of membrane filters with a 0.2 or 0.1 μm pore size is necessary. The filters inside an aluminum cup and the centrifuge tubes should be pre-dried at 103–105 °C for 1 h followed by incubation for 15–30 min in a desiccator to prevent moisture absorption. Then the aluminum cup + the filter or the centrifuge tube are pre-weighted (mass m_1).

4.2.2.2 Washing the Biomass

Biomass of marine species as well as from (photo)heterotrophic strains grown in hypertonic culture medium as the result of high salt and/or organic substrates concentrations have to be washed before drying [20]. The aim is to eliminate the components of the culture medium to avoid overestimation of the biomass concentration. However, rinsing with distilled water could induce some cell lysis, as the result of the hypotonic shock, and then an underestimation of the DW. The biomass should be rinsed with an isotonic solution. Fast, accurate and precise values of the actual osmotic strength of the culture medium are obtained by using freezing-point micro-osmometers; it needs very small samples (<100 μL). Isotonic solutions of either ammonium formate or ammonium bicarbonate are generally used. Both salts decompose to volatile compounds during the drying, the first one at 105 °C and the second at 60 °C. If the biomass is presumed to contain volatile components, ammonium bicarbonate should be used. Moreover, the dewatering could be done in a vacuum oven at 40 °C. When the pH of the cell cultures is higher than pH 9, some protocols recommend a washing with acidified solutions (5–10 mM HCl) to remove salts, mainly carbonates, adsorbed on the cell walls. It works with distilled water for freshwater species but also with ammonium formate solutions (adjusted to pH 5.5). However, such pH adjustments may not be suitable for all the microalgae; the coccolithes calcified microalgae could be affected by pH decrease, with an impact on the microalgae dry weight but not on the ash-free dry weight (AFDW).

4.2.2.3 Biomass Dewatering

The filters removed from the holder of the vacuum filter system and back in the aluminum cup or the centrifuge tubes are placed in an air oven at 104 °C (ammonium formate) or 60 °C (ammonium bicarbonate). An alternative is the use of a vacuum oven, for an efficient temperature of 40 °C to prevent volatile organic compounds (VOCs) to escape, but also for a shorter period to reach constant weight, indicative of the end of the drying kinetic. Fast drying is obtained with halogen moisture balances, provided that, for reliable DW measurements, the drying temperature is set to 105 °C. Indeed, too high surface temperatures (>400 °C) should be avoided [21].

The dried filters or tubes are quickly transferred in a desiccator for 30 min minimum before the last weighing m_2. The biomass concentration (g L^{-1}) is deduced from the ratio of the dry biomass to the volume of the sample $(m_2–m_1)/v$. The biomass concentration X in attached cultures is expressed as biofilm areal density (X, g m^{-2}).

4.2.3 Ash and Ash-Free Dry Weight Method

The dried biomass contains both minerals and organic compounds. In diatoms, ash contents can reach up to 59% of the DW, about two to five times the values measured in non-silicified or non-calcified microalgae [22]. For accurate mass balances of biological reaction within microalgae cultures, this mineral fraction should be taken into account. Then ash free dry weight (AFDW) calculated as the difference of the weights obtained after drying at 105 °C and after ashing at 575 °C should be preferred over total dry weight measurement when the presence of mineral ash would induce a significant error in the estimation of organic content dry weight. AFDW is an estimation of the organic weight. Data are expressed as mg/L AFDW and ash content as a % of total dry weight.

4.2.3.1 Classical Gravimetric Method

The classical gravimetric method corresponds to the determination of the remaining ash obtained after further combustion of the dry biomass in a furnace at high temperatures (>550 °C). The difference gives the ash free dry weight (AFDW), corresponding to the global organic fraction of the microalgae biomass. It is also named as the volatile suspended solids (VSS) fraction. The values of biochemical fractions (lipids, proteins, sugars …) are higher when calculated on an AFDW basis. Moreover, the AFDW allows avoiding the biomass rinsing problems. The samples are placed within porcelain, silica or platinum crucibles to prevent loss; they are pre-weighted. Then, they are placed in a combustion furnace to burn up to 700 °C with a recommended value of 575 °C and for a minimum of four hours. The duration corresponds to the time needed to reach a constant weight, defined as that obtained when mass variation is less than 0.3 mg after one hour of re-heating the crucible.

This time is a function of the initial mass and the structure of the biomass to be treated. After its total oxidation in the muffle furnace, the samples within their crucibles are placed in a desiccator to cool to ambient temperature (1 h recommended) before the final weighing.

The ash content and composition of the biomass is important in furnace operations during combustion processes, but also during bio-refinery processes [23]. When necessary, the ash composition (Si, Ca, Mg, K, Na, P, …) could be easily determined using inductively coupled plasma mass spectroscopy (ICP-MS) or X-ray fluorescence (XRF).

4.2.3.2 Thermogravimetric Analysis

Thermogravimetric analysis is a technique in which the mass of a substance is measured as a function of temperature or time as the sample is subjected to a controlled temperature program in a controlled atmosphere within a furnace. The interest of thermogravimetric (TG) techniques has been proven for the study of thermal characteristics of microalgae biomass during combustion and pyrolysis [24]. Indeed, the combustion behavior of algae is still far to be well understood [25].

Microalgal ash content could be determined using an usual protocol that includes a temperature gradient with a heating rate of 20 °C/min from ambient temperature to 600 °C followed by an isothermal step a temperature of 600 °C for 30 min under an air atmosphere for TG analysis. Indeed, at higher temperatures, the ash content could be reduced by the loss of some volatile minerals. The analysis of the peaks obtained by derivative thermogravimetric (DTG) graph has been used for qualitative or semi-quantitative analysis of the lipid, carbohydrate and protein fractions in microalgae by fitting peaks for carbohydrates, protein and lipids [26]. The ash content can be quickly (about one hour) and easily performed using only 5–10 mg of sample.

4.2.4 Cell Counts Methods

Different methods are used for measuring the cell populations, including microscopic numeration with hemocytometers, impedance measurements or flow cytometer.

4.2.4.1 Hemocytometry

The cell count method by hemocytometry is commonly used as it is easy to implement and not expensive, although time consuming [27]. Some errors could be introduced by improper use of the hemocytometer. However, this microscopy analysis can help to detect any morphological changes or microbial contamination in the cultures in contrast to gravimetric or optical methods. The processing time is accelerated when the method associates an image processing program (ImageJ with the plugin cell counter.jar). For later analysis, the microalgae samples could be fixed with glutaraldehyde, lugol or formaldehyde and stored at 4/6 °C [28]. The relations between dry weight and cell population data depend not only of the species, but also of the eco-physiological conditions [29].

The measured counts are total cells since it is impossible to differentiate between live and dead cells.

Taking into account the autofluorescence of chlorophylls or phycobiliproteins, epifluorescence microscopy is advantageously used for numeration of photosynthetic microalgae. In complex culture media such property is used to distinguish microalgae from inert particles. Moreover, some protocols include fluorescence staining, such as Sytox green, for rapid dual fluorescence assays to distinguish dead from live microalgae [30]. At least 300 cells in random fields have to be counted in numeration and viability calculations for each assay.

4.2.4.2 Flow Cytometry

The flow cytometric method (FCM) is considered as an efficient, although expensive, substitute for the time-consuming direct methods (e.g. microscopic or gravimetric methods). FCM represents an automated system, allowing precise and fast determination of unicellular microalgae number in a culture [28–31]. Their auto-fluorescent properties, due to the presence of chlorophylls, allow discrimination from other non-photosynthetic microorganisms and non-living matter (excitation with 488 nm using the argon laser). Fresh unfixed samples are preferably used to preserve their fluorescent properties. Some specific protocols have to be used for colonial microalgae or filamentous Cyanobacteria, aiming to disrupt the filaments without cell lysis, such as through sonication with Triton-X100 treatment [32]. In addition, FCM offers the possibility to distinguish the smallest microalgae (picoplankton < 2 μm) from bacteria which is relatively difficult by optical microscopy.

Imaging flow cytometry is a hybrid technology, integrating the capabilities of flow cytometry with imaging features of microscopy. The combination of fluorescence microscopy with flow cytometry (1000 cells s^{-1}) results in both quantitative and qualitative data. Thanks to on-line image processing by visualization software, each enumerated particle is linked to an image, with access to size distribution, detection of any morphological changes, viability or metabolic activity [33].

4.2.4.3 Image-Based Cytometry

Solid-phase cytometry (SPC) is an image-based technique for quick and accurate enumeration of microorganisms such as microalgae with similar precision to flow cytometry [34, 35]. Microalgae are collected on a membrane filtration. The scanning involves using either chlorophyll auto-fluorescence or labelling by a fluorescent dye. The entire membrane filter surface is scanned and the fluorescent light emitted by labeled cells is automatically detected and counted. The software allows differentiating the target signal from electronic noise or non-target particles [36]. The technique allows an accurate enumeration down to a detection limit of one cell per filter.

Other automated cell counters are adapted to disposable cell counting chambers (20 μL sample size) or even to standard hemocytometers. They also provide cell sample analysis results, including cell count, concentration, diameter, images, and viability, in less than 30 s.

After scanning, cells can be examined by epifluorescence microscopy. The dry weight concentration could be estimated on the basis the single cell density and the cell size distribution.

4.2.5 Elemental Analysis

The mass balance of a microalgae culture is now recognized as a valuable tool not only for overall stoichiometry equations used in process control, e.g. identification of nutrient limitations, but also for analytical data validation, e.g. detection of measurement errors and/or unnoticed products [37]. For instance, the determination of the main elemental components of a microalgal biomass, carbon, nitrogen, hydrogen, oxygen, phosphorus and sulfur, could highlight the presence of an S-enriched fraction, such as some sulphated polysaccharides. Different instruments have been developed for measuring the elemental composition. Most of them employ catalytic combustion with pure oxygen to decompose the sample to nitrogen, water, carbon dioxide and sulphur dioxide, which are then quantitatively determined by chromatography with flame ionization or thermal conductivity detectors. Oxygen is determined by catalytic conversion to carbon monoxide.

One advantage of elemental analysis is the relatively small sample weights required. The main constraint is that samples should be dry and free of foreign substances. Indeed, total mass of every main element (C, N, O, H, P and S) is determined and reported to an exact global dry weight. The good practices previously described for sample preparation for the dry weight determination have to be scrupulously applied. Other elements which occur in smaller, but significant quantities are sodium, potassium, calcium, magnesium and silicon. Some of them have to be considered for characterizing biomass from some bio-mineralized microalgae, such as diatoms or calcareous species [38].

The large biodiversity and chemodiversity of microalgae are of great interest in tapping in never exploited natural molecules, thanks to recent and ongoing screening studies. However, such diversity implies that the standard analytical protocols for biomass characterization have to be critically examined for potential interference.

4.3 Methods for Protein Determination in Microalgae

4.3.1 Introduction

Microbial biotechnology focuses more and more on the potential of microalgae as a source of biomass, biofuels, carbohydrates, lipids, fishmeal replacement and food, and high value molecules [39, 41]. The chemical compounds synthesized by microalgae have several applications. The high-protein content of some algae species is one of the main reasons to consider them as a non-conventional source of proteins [39]. For example, *Chlorella* was amongst the first microorganism species to be commercialized as a health food in Japan, Taiwan and Mexico [42, 43]. Microalgal protein has high potential for animal feed and human consumption, and recombinant protein technology. The use of marine strains could avoid conflicts with agriculture for freshwater supplies; thus strains such as *Nannochloropsis* are used both in fish-farm aqua-feed and large scale biofuel production [44, 45]. Moreover high-value products from microalgae are usually produced within a biorefinery model, since the composition of the microalgal cell allows for extraction of different co-products, and thus proteins are valuable by-products of biofuel production [46]. Commercialization prices are still high due to the process expenses associated to the extraction and purification of intracellular metabolites. In microalgal biotechnological processes, the downstream stage can account for 50–80% of total production costs, depending on the biochemical characteristics of the compound and the purity ratio that needs to be achieved [47].

Proteins are large and complex naturally-produced molecules composed of one or more long chains of amino acids, in which the amino acid groups are held together by peptide bonds. They make up a large fraction of the biomass of actively growing microalgae [48]. Depending on the microalgae strains, protein extraction may be more or less easy. The composition of the cells walls has a significant impact on the protein extraction [40].

Table 4.1 Example of protein content (% dry weight basis)

Alga	Class	Protein content (%)
Chlorella pyrenoidosa	Trebouxiophyceae	57
Chlorella vulgaris		53.3
Chlorella ellipsoidea		42.2
Chlorella ovalis		10.97
Chlorellla spaerckii		6.87
Dunaliella salina	Chlorophyceae	57
Dunaliella primolecta		12.26
Dunaliella tertiolecta		11.4
Scenesdesmus obliquus		48
Scenesdesmus almeriensis		41.8
Tetraselmis chui	Chlorodendrophyceae	46.5
Porphyridium cruentum	Porphyridiophyceae	35
Porphyridium aerugineum		31.6

Most microalgae exhibit a high protein content (Table 4.1) with the highest content being found in *Chlorella pyrenoidosa* and *Chlorella vulgaris* (Trebouxiophyceae) where protein content represents more than 50% of dry weight. It is important to note that protein content amongst the strains of a species can vary significantly; for example, for *Chlorella* species, there are significant differences within protein content, from only 6.87% in *Chlorellla spaerckii* and 10.97% in *Chlorella ovalis* to 57% in *C. pyrenoidosa.* [49].

4.3.2 *Applications of Algal Proteins*

4.3.2.1 Human Nutrition

The nutritional quality of a protein is determined by the content, proportion and availability of its amino acids [50]. Four indices can be calculated to characterize the nutritional value of microalgae (Table 4.2):

Table 4.2 Comparative data on biological value (BV), digestibility coefficient (DC), net protein utilisation (NPU) and protein efficiency ratio (PER) of selected processed microalgae

Source	Processing	BV	DC	NPU	PER
Scenedesmus obliquus	Drum dried	75	88	67.3	1.99
	Sun dried	72.1	72.5	52.0	1.14
Chlorella sp.	Air dried	52.9	59.4	31.4	0.84
	Drum dried	76.6	89.0	68.0	2.10

Table 4.3 Essential Amino acid profile of different microalgae as compared with conventional protein sources (g 100 g^{-1} protein)

Source	Ile	Leu	Val	Lys	Met	Cys	Thr	His
Egg	6.6	8.8	7.2	5.3	3.2	2.3	5.0	2.4
Soybean	5.3	7.7	5.3	6.4	1.4	1.9	4.0	2.6
Chlorella sp.	4.4	9.2	6.1	8.9	2.2	0.4	4.7	2.4
Dunaliella sp.	4.5	9.3	6.0	6.2	2.5	4.0	5.0	2.5
Scenedesmus sp.	4.7	9.4	6.0	6.8	2.4	0.1	4.9	2.6

- PER: Protein efficiency ratio, expressed in terms of weight gain per unit of protein consumed by the test animal in short-term feeding trials.
- BV: Biological value, a measure of nitrogen retained for growth and maintenance.
- DC: Digestibility coefficient
- NPU: Net protein utilisation (BV $\times$ DC), a measure of both the digestibility of the protein and the biological value of the amino acids absorbed from the food.

Cell walls of microalgae consist of a polysaccharide and glycoprotein matrix providing the cells with a formidable defense against its environment. It represents about 10% of the algal dry matter and as it is non-digestible for humans and non-ruminants animals, a post-harvesting treatment of the microalgal cells is necessary to make the proteins accessible for digestive enzymes and can affect the various parameters (BV, DC, NPU and PER) (Table 4.2) [51, 52].

One of the quality criteria for a protein is its amino acid composition (specifically the essential amino acids which are important for human feed application). Some microalgal protein can compare with proteins derived from conventional sources such as egg and soybean (Table 4.3) and thus can be used as substitute for eggs or animal proteins [49].

Microalgae are usually consumed as a dietary supplement in the form of powder, pills or tablet [53]. Some microalgae can also be incorporated in the composition of industrial food, such as noodles, bread, biscuits, drinks, sweets and beer [54]. *Chlorella* is one of the most consumed microalgae.

One of the quality criteria for a protein is its amino acid composition (specifically the essential amino acids which are important for human feed application). Some microalgal protein can compare with proteins derived from conventional sources such as egg and soybean (Table 4.3) and thus can be used as substitute for eggs or animal proteins.

4.3.2.2 Industrial Application

High Value Metabolites: Phycobiliproteins

Among functional ingredients from microalgae, natural pigments exhibit beneficial biological activities such as antioxidant, anti-carcinogenic, anti-inflammatory, anti-obesity and neuroprotective agents [39, 55]. Phycobiliproteins are brilliant-coloured and water-soluble antennae-protein pigments which originate mainly from covalently bound prosthetic groups that are open-chain tetrapyrrole chromophores bearing A, B, C and D rings named phycobilins [56]. Phycobiliproteins absorb energy in portions of the visible spectrum (450–650 nm) and function as accessory pigments for photosynthetic light collection [57] In microalgae, three main classes of phycobiliproteins are produced: allophycocyanin (APC, bluish green), phycocyanin (PC, blue) and phycoerythrin (PE, purple) [56]. Due to its physical properties, Phycoerythrin can be used in clinical research and molecular biology [39]. Phycoerythrin can be used as label for biological molecules, as a reagent in fluorescence immunoassays, flow cytometry, fluorescence microscopy and diagnostics [58].

Phycocyanin is found as a complex mix of monomers, trimers, hexamers and various oligomers, and thus its molecular weight ranges from 44 to 260 kDa [59]. Phycocyanin is a colorant commonly used in food and cosmetics. However its use is limited because phycocyanin is sensitive to heat treatment, which results in precipitation and fading of the blue colour. Phycocyanin can as well be used as a probe for immunodiagnostics thanks to its fluorescence properties [60]; it is an efficient scavenger of oxygen free radicals and therefore its therapeutic use is promising since many diseases are related to an excessive formation of reactive oxygen species (ROS) [61].

Functional Food-Ingredients

Protein quality is determined by the essential amino acid content and bioavailability, which is defined as the proportion of a nutrient that can be absorbed and used. Indeed, during digestion, some proteins are not completely decomposed into free amino acids, but rather into peptides that can be absorbed and thus influence physiological processes. Thus, microalgae have been identified as valuable and sustainable sources of protein for the industrial production of peptide-based functional foods to prevent or treat cardiovascular disease [62, 63]. Peptides that have antihypertensive (by inhibiting angiotensin converting enzyme (ACE)), anti-obesity (by stimulating the hormones that regulate satiety) and antioxidant properties have been successfully produced from microalgae (Table 4.4) [62]. Such peptides are key for pharmaceutical industry as worldwide mortality from cardiovascular disease was estimated at 37% in [64]; 50% of deaths from stroke and coronary artery disease are caused by hypertension. Moreover, oxidative stress has been identified as one of the key factors in the development of hypertension [65] because high levels of ROS can cause the oxidation of biological macromolecules, ultimately leading to pathological conditions. However, even if peptides give promising results in lab-assays, in most cases, the link with a beneficial health outcome in humans is currently theoretical [66].

Navicula incerta and *Chlorella ellipsoidea* are promising sources of protein for the production of bioactive peptides [62]. The anti-hypertensive properties of

Table 4.4 Examples of bioactive peptides from microalgae [62]

Source of peptide	Type of treatment	Amino acid sequence	Activity
Navicula incerta	Hydrolysis with papain	Pro-Gly-Trp-Asn-Gln-Trp-Phe-Leu-Val-Glu-Val-Leu-Pro-Pro-Ala-Glu-Leu	Antioxidant
Chlorella vulgaris	Hydrolysis with pepsin	Val-Glu-Cys-Tyr-Gly-Pro-Asn-Arg-Pro-Glu-Phe	Antioxidant
Chlorella vulgaris		Phe-Ala-Leu	ACE inhibition
Chlorella ellipsoidea		Leu-Asn-Gly-Asp-Val-Trp	Antioxidant
Chlorella ellipsoidea	Hydrolysis with alcalase	Val-Glu-Gly-Tyr	ACE inhibition

Adapted with permission from Elsevier

peptides derived from microalgae have been widely studied and it was demonstrated that the short chain and low molecular weight peptides are significantly more efficient than the larger peptides. This is partly due to their bioavailability and their accessibility to the active sites [62].

Animal Feed

The increase in the global population combined with the increase in income may double the global demand for animal products by 2050, with a particular impact on the most consumed meats in the world. This increased demand for meat will be particularly dramatic for farmed agriculture, as maize and soybean food crops are the two main conventional feeds. Microalgal biomass can provide natural ingredients and supplements in animal diets to meet the growing demand for protein and energy and replace synthetic additives in foods, as it has been claimed that the average quality of most fractions of microalgae proteins is equal to or even higher than that of conventional vegetable protein fractions [50]. Dry *Chlorella* has maximal protein concentrations of between 50 and 60% of dry weight, similar in quality to yeast, soy flour and milk protein [67].

Agriculture
Supplementing poultry feed with microalgae as a protein source can improve their health, productivity, and value. This has been demonstrated using a variety of species, including *Chlorella* sp., *Arthrospira* sp., *Porphyridium* sp., and *Haematococcus* sp. [68–70]. The replacement of up to 33% of soy proteins with proteins from *Chlorella vulgaris* in pig feed has been reported as being suitable without any adverse effects [71]. The cell wall of microalgae is largely indigestible by monogastric animals. Therefore, it becomes imperative for the feed industry to develop appropriate technologies to improve the bioavailability of microalgae nutrients in animals [72]. Of all the animals evaluated for algae supplementation,

ruminants are the most promising in terms of digesting the high fibre content for the greatest extraction efficiency of algal proteins. This is in contrast to monogastric animals, for which it has been suggested that some form of prior processing may be required in order for animals (and humans) to utilise algal proteins more efficiently.

Aquaculture
Microalgae are the base of the aquatic food chain and thus are essential for the artificial reproduction of many aquaculture species, such as molluscs, shrimps and rotifers [50]. Filtering molluscs are the greatest consumer of microalgae [73]. Due to the rapid global expansion of the aquaculture industry, access to major foods (fishmeal and fish oil) is increasingly difficult due to limited wild fish resources [74]. Fisheries are the most important sources of feedstock for fishmeal. Only a small percentage of global fish production is indeed channeled to human consumption, with the remainder being used for fish and animal feed. Fishmeal is a protein-rich food and sets the basis for any balanced formulation used in commercial aquaculture. Nevertheless, microalgae can be considered as a promising alternative that can replace fishmeal and fish oil as the chemical composition of some commercially important microalgal species is comparable to the available food ingredients used in the aquaculture feed industry. The chemical composition of some commercially important microalgal species is comparable to the available food ingredients used in the aquaculture feed industry. The nutritional value of microalgae is determined primarily by the protein content and, secondly, by the polyunsaturated fatty acid content [eicosapentaenoic acid (EPA), arachidonic acid (ARA) and docosahexaenoic acid (DHA)]. Microalgae are a more reliable and less volatile source of protein, and their availability is not dependent on fish captures. This provides industry with a better control of their costs and supports a potential for future investment due to the reduction of risk in aquaculture farming operations [75]. Studies highlighted that biomass of *Chlorella* sp., *Scenedesmus* sp., *Nanofrustulum* sp. and *Tetraselmis suecica* can be used as additional protein sources or partial substitutes for fish meal proteins in the diet of various species of omnivorous and carnivorous fish [76].

Recombinant Proteins

Microalgae show great potential as bioreactors for large-scale production of recombinant proteins. Microalgae combine high growth rates like prokaryotic cells with all advantages of eukaryotic expression systems, i.e. post-transcriptional and post-translational modifications and the assembly of multimeric protein complexes [77]. Monoclonal antibodies are important tools in medical therapy, diagnostics and research and are mainly produced in mammalian cell lines, since the establishment of hybridoma technology in 1975. As cultivation of mammalian cells is expensive, alternative expression systems are studied. A full-length IgG antibody was synthesized in the chloroplast of *Chlamydomonas reinhardtii* demonstrating that antibody expression in an algal system is feasible [78]. Following this study, Hempel et al. [77] published the synthesis of a fully-assembled and functional antibody (CL.4 mA) against the Hepatitis B Virus surface protein in *P. tricornutum* and manage to obtain an accumulation of 8.7% of total soluble protein in two days of induction. When expressed in *Nicotiana tabacum* the same antibody reached much lower expression levels of only 0.2–0.6% in several weeks [79]. Another frequently reported problem with antibody expression in plants is rapid protein degradation resulting in low antibody levels and many fragmented products [80]. However in *P. tricornutum* no degradation products were detected [77]. Antibody purification was carried out with protein A-Sepharose and ELISA assays demonstrated that the algal-produced antibody is functional and binds the respective target antigen (HBsAg) very efficiently in vitro [77]. The Hepatitis B surface protein was functional and recognized by algae-produced and commercial antibodies.

By-Product of Biofuel Consumption

Microalgae are considered as one of the most promising biobased feedstocks, as their productivity in converting CO_2 into lipids exceeds that of oilseed crops [81]. Applying biorefinery, not only for oil, but as well for protein, may lead to the sustainable and economical microalgae-based fuels. With this approaches, the biomass treatment for protein extraction, alongside other compounds of interests (lipids, carbohydrates), is different to when only proteins are extracted and may affect the yield [82].

4.3.3 Protein Extraction and Quantification

4.3.3.1 Pre-treatment Prior to Extraction

Prior to protein extraction, most microalgae have to be pre-treated in order to break or fragilise the cell wall to enhance the protein extraction yields. For this, various treatment are available such as: hydrolytic enzymes or sodium dodecyl sulphate (for chemical treatments), or milling a cell slurry in presence of glass beads or fine ceramic particles can be efficient as well for the hardiest cells. Ultrasonication is another effective method for disrupting the cells but has the inconvenient to sometimes damage the proteins. In a study by Gonzalez Lopez et al. [48] several pre-treatments methods, prior to proteins analysis of *Porphyridium cruentum* and *Muriellopsis* sp., were compared. The following pre-treatments were tested: suspension in lysis buffer, ultrasonication at high power for 10 min in lysis buffer, or milling for 5 min with a pestle and mortar. They concluded that depending on the species studied, different pre-treatment must be used for optimisation of cell lysis. Moreover, the growth phase has also an impact on the ease of cell disruption, with rapidly growing cells likely to be less robust than slow growing or stationary-phase cells [83], therefore milling with ceramic particles is expected to be critically important to maximise the intracellular proteins released.

Extensive studies on cell disruption techniques have been conducted on *Chlorella vulgaris* because it has a resistant cell wall, which limits the digestibility and extraction of cellular contents. Multiple techniques for breaking the wall were tested on *C. vulgaris* and are summarised in Table 4.5. Enzymatic treatment is a promising technique that requires a thorough knowledge of ultrastructure and cell wall composition to select the appropriate enzyme and lysis conditions. According to Lee et al. [84] and Zheng et al. [85], the best cell disruption techniques of *C. vulgaris* are autoclaving, microwave treatment, enzymatic lysis and liquid nitrogen grinding. The success of cell disruption techniques is generally evaluated by performing microscopic observations or comparing the extraction yield of a compound before and after the cell disruption treatment [86].

Table 4.5 Different cell disruption techniques carried out on *C. vulgaris*

Cell disruption	Time	Experimental set-up
Acid treatment	25 min	Hot Ac2O[a] + H_2SO_4 (9:1, v:v)
Alkaline treatment	60 min	2 N NaOH
Autoclaving	5 min	125 °C + 1.5 MPa
Bead milling	20 min 5 min 2 min	Beads: 0.4–0.6 mm Rotational speed 1500 rpm Beads: 0.1 mm Rotational speed 2800 rpm Beads: 1 mm
Electroporation	Not available	Electric field: 3 kV/cm Electrode 2 cm
Enzymatic lysis	60 min 10 h Not available 24 h	Snailase (5 mg L^{-1}), 37 °C Cellulase or Lysozyme (5 mg L^{-1}), 55 °C 4% Cellulase + 1% others (w/v) 25 mM Sodium Phosphate buffer pH 7.0. 0.5 M Mannitol Cellulase 0.5 mg L^{-1} 0.5 M Mannitol
French press	Not available	138 MPa
Manual grinding	1–10 min not available	With liquid nitrogen or quartz With dry ice
Microwaves	5 min	100 °C, 2450 MHz 40–50 °C, 2450 MHz
Osmotic shock	48 h 60 min	10% NaCl 2 N NaOH
Ultra-sonication	6 min 20 min 5 min	10 W 600 W 10 kHz

[a]acetic anhydride

4.3.3.2 Protein Extraction

Ion-exchange chromatography, gel filtration chromatography, gel electrophoresis and ultrafiltration have been employed for the fractionation, purification, characterization and structural elucidation of proteins, peptides and amino acids from microalgae [87]. The protein extraction procedure is technically the same for all microalgae and is mainly carried out by solubilisation of proteins in alkaline solution (pH 10–12) in the presence of NaOH [88–90]. Further purification can be achieved by precipitating the solubilized proteins with trichloroacetic acid (25% TCA) or hydrochloric acid (0.1 N HCl) [91, 92].

4.3.3.3 Protein Content Determination

For a given algal species, the intracellular concentration of inorganic and organic nitrogen depends on the growth phase [93]. Various methodologies have been developed for proteins quantification and analysis. Main methods are based on colorimetry (Coomassie, Lowry, BCA, Amido Black, Biuret), fluorescence (fluorescamin, Nano-orange, CBQCA, OPA), immunology (Elisa) and spectroscopy (amino acids analysis, UV, mass spectrometry) protocols [94].

For microalgae, commonly proteins are quantified by elemental analysis, Kjeldahl, Lowry assay, Bradford assay [95] or dye binding method [91]. However, the first two analyses take into consideration total nitrogen present in the microalga, multiplying it by the standard nitrogen to protein conversion factor (NTP) 6.25 which may lead to overestimation or underestimation of the true protein quantity because microalgae contains high concentration of non-protein nitrogenous substances such as pigments (chlorophyll or phycoerythrin for example), nucleic acids, free amino acids and inorganic nitrogen (nitrate, nitrite, ammonia) [93] whose presence makes the factor 6.25 unsuitable since it overestimates the real protein content [96]. Several authors recommended to use a value of NTP lower than the standard 6.25 [48, 97]. Moreover, a study by Safi et al. [89] highlighted that no universal conversion factor could be recommended due to the fact that multiple factors, such as cell wall rigidity, growth conditions, growth media and environmental uncertainty, can affect the NTP value. Gonzales Lopez et al. [48] determined the NTP using a technique that correlates protein content (Lowry assay) to total nitrogen content (Kjeldahl and elemental analysis). In addition, Servaites et al. [98] quantified proteins of microalgae by staining the protein isolate with Coomassie brilliant blue R-250 (CBB) on a paper and then eluting the remaining stained proteins in 1% sodium dodecyl sulphate (SDS) followed by measuring the absorbance at 600 nm. On the other hand, the colorimetric method of Lowry [99] was also considered as one of the most accurate methods to quantify proteins, but with time this method showed to only quantify hydro-soluble proteins [89], which represents the major part of proteins. Lowry assay is considered as more acceptable then Bradford assay because the latter does not react with all amino acids present in the extract and thus giving lower protein concentrations [91].

General methodologies can be used but it is important to state that some protein may lead to challenges depending on its structure and mode of action. So far, no single method of quantification can be used for the determination of the true protein concentration for all proteins in all kinds of buffers to serve as a "gold standard" [94]. This can be explained by the fact that proteins can have a very large variety of structures and physicochemical properties. Thus, the choice of the right quantification method is crucial. The different assays available for protein quantification have various characteristics that must be considered before choosing the right protocol depending on the type of experimental work planned. Many different methodologies to assay proteins have been developed; the purpose of this chapter is not to review all but describe the most commonly applied and/or appropriate assays.

Total Nitrogen Content

Techniques that can be used to measure protein content rapidly in lyophilized material include Dumas-based combustive methods of elemental analysis and Kjeldahl to measure N-content [40]. In both methods, the total nitrogen in the sample is liberated at high temperature. In the Kjeldahl method, the nitrogen is released into a strong acid and the content is measured after neutralization and titration. In the Dumas method, the nitrogen is liberated in a gaseous form and is determined with a thermal conductivity detector, after removal of carbon dioxide and water aerosols. The Kjeldahl method was chosen as an example of this analytical principle as it is still recognized as the official method for food protein determination by the AOAC International [87].

For microalgae, measurements can be done using 200 mg samples of the dry biomass. The biological material is hydrolyzed with 3 mL concentrated sulfuric acid (H_2SO_4) containing one copper catalyst tablet in a heat block (for example, Kjeltec system 2020 digestor, Tecator Inc., Herndon, VA, USA) at 420 °C for 2 h. After cooling, H_2O was added to the hydrolysates before neutralization and titration. Following the nitrogen determination, crude protein content is estimated using a conversion factor (see Sect. 4.3.3.3).

Colorimetric Methods

Colorimetric assays rely on the appearance of a chromophore as a consequence of either the binding of a dye to a protein or the protein being involved in a redox reaction. Such methods are sensitive to interferences and their accuracy depends significantly on the methods used for pre-treating the samples. Indeed, for microbial proteins to be measured accurately, the cells must be pretreated in order to fully release the intracellular proteins [48]. For better results, Barbino and Lourenço [91] suggested to start the protein extraction from 50 mg of freeze-dried material. The most common treatments involved physical or chemical disruption of the cell wall (see Sect. 4.3.3.1 pre-treatment).

Prior to protein quantification by Bradford, BCA or Lowry methods, it is necessary to proceed to a protein precipitation step. For this a commonly used procedure is the Berges et al. [100] methodology which is based on the use of trichloroacetic acid (TCA) to allow for protein precipitation. The advantage of this method is that the pellet containing the proteins from microalgae can be stored at −20 °C until further analysis. Prior to colorimetric analysis, precipitated proteins are respectively suspended in 0.5 mL 1.0 N NaOH for the Bradford assay and 2.0 mL for the Lowry assay (when starting from 50 mg freeze dried biomass before pre-treatment).

A study by Berges et al. [100] stated that even if it remains unclear which spectrophotometric assay is the most accurate (between Lowry, BCA and Coomassie), it appears that Coomassie assay is faster and simpler, and, has the advantage of being less affected by non-protein compounds found in marine phytoplankton. The reactivity of either of the assays to a particular protein will be a function of that protein's composition as well as any other compound which might oxidize the Folin phenol reagent or bind the Coomassie dye. The bicinchoninic acid (BCA) assay, similarly to the Lowry assay, measures the conversion of Cu^{2+} to Cu^{+} under alkaline conditions, and, both assays have similar sensitivity but BCA is stable under alkaline conditions.

Lowry Assay

The Lowry method has been widely used for protein determination for many decades, due to its simplicity and availability. However, besides aromatic amino acids, a wide range of other compounds react with the Folin–Ciocalteu reagent [101]. The Lowry assay detects protein through a copper-catalysed reduction of Folin phenol reagent [99] under alkaline conditions. The reactions result in a strong blue color with a maximum absorbance at 750 nm. This reaction will detect peptide bonds, but it is also highly sensitive to specific amino acids such as tyrosine and tryptophan, and to a lesser extent on cystine/cysteine and histidine contents. The method is sensitive to about 0.01 mg of protein/mL^{-1} within a range of 0.01–1.0 mg protein/mL^{-1}. However, some caution should be taken as the Lowry assay is subjected to interference from many substances including buffers, detergents, EDTA, nucleic acids and sugars [48].

Example of experimental protocol

Step 1 Pre-treatment of microalgal samples: for example, 20 mg of aliquots of microalgal freeze-dried biomass suspended for 20 min in 10 mL of lysis buffer

Step 2 An aliquot of this suspension is diluted with the lysis buffer such that the protein concentration is within the range of 0 and 1000 mg L^{-1}.

Step 3 0.1 mL SDS and 1 mL of reagent C is added to 0.1 mL of the above solution, the solution is then vortexed.

Step 4 10 min later 0.1 mL of Folin reagent is added and well mixed. The sample is then kept in the dark to avoid the degradation of the Folin reagent.

Step 5 After 30 min the absorbance of the sample is measured at 750 nm (a blank is prepared is prepared without the algal extract).

Step 6 The spectrophotometric absorbance is converted to protein concentration using a calibration curve established with bovine serum albumin (BSA) dissolved in lysis buffer. The protein content of the biomass is calculated using the formula:

$$\text{Protein}\ (\%;\ \mathrm{w} = \mathrm{w}) = (\mathrm{CVD}/\mathrm{m}) \times 100 \tag{4.1}$$

with C the protein concentration (mg L^{-1}) obtained from the calibration curve, V the volume (L) of the lysis buffer used to resuspend the biomass, D is the dilution factor and m is the amount of biomass (mg).

Coomassie (Bradford) Bradford Assay

The Bradford [95] protein assay is used to measure the concentration of total protein in a sample. The principle of this assay is that the binding of protein molecules to Coomassie dye (G-250) under acidic conditions results in a color change from brown to blue which has an absorbance maximum at 595 nm. The quantity of proteins can thus be estimated by determining the amount of dye in the blue ionic form by measuring the absorbance of the solution at 595 nm. The method is sensitive within a range 0.01–1.0 mg protein mL^{-1}. Coomassie Brilliant Blue dye is bound by protein, primarily by arginine residues, but also to a lesser degree by histidine, lysine, tyrosine, tryptophan and phenylalanine [102]. The binding of the dye is very rapid and the protein-dye complex remains soluble and stable for 1 h. The presence of SDS even at low concentrations can interfere with protein-dye binding, moreover this assay is not suitable for quantifying free amino acids or peptides smaller than 3000 Da as the dye will not bind to them [102].

Example of experimental protocol

The Quick Start™ Bradford protein assay is a simple and accurate procedure for determining the concentration of protein in solution. The assay supplies ready-to-use dye reagent at 1× concentration and two protein standards at seven prediluted concentrations (0.125, 0.25, 0.5, 0.75, 1.0, 1.5, and 2.0 mg mL^{-1}). Protein concentration is determined in one step, with no need to dilute standards. Quick Start Bradford protein assay kits offer either bovine serum albumin or bovine γ-globulin standard sets.

Step 1 Prepare calibration curves using either Bovin Serum Albumin (BSA) with a range of 1.5–10 μg mL^{-1} or bovine γ-globulin in the range of 1.25–20 μg mL^{-1}

Step 2 Pour 150 μL of samples/standard in a microplate

Step 3 In each well of the microplate, add 150 μL of 1X Dye Reagent and shake manually
Step 4 Incubate 5 min at room temperature
Step 5 Take absorbance readings at 595 nm against a blank.

Note: In classical protein staining protocols using Coomassie Brilliant Blue (CBB), solutions with high contents of toxic and flammable organic solvents (Methanol, Ethanol or 2-Propanol) and acetic acid are used for fixation, staining and destaining of proteins in a gel after SDS-PAGE. To speed up the procedure, heating the staining solution in the microwave oven for a short time is frequently used. This usually results in evaporation of toxic or hazardous Methanol, Ethanol or 2-Propanol and a strong smell of acetic acid in the lab which should be avoided due to safety considerations.

In a protocol published by Lawrence and Besir [103], an alternative composition of the staining solution is described in which no organic solvent or acid is used. The CBB is dissolved in bidistilled water (60–80 mg of CBB G-250 per liter) and 35 mM HCl is added as the only other compound in the staining solution. The CBB staining of the gel is done after SDS-PAGE and thorough washing of the gel in bidistilled water. By heating the gel during the washing and staining steps, the process can be finished faster and no toxic or hazardous compounds are evaporating. The staining of proteins occurs already within 1 min after heating the gel in staining solution and is fully developed after 15–30 min with a slightly blue background that is destained completely by prolonged washing of the stained gel in bidistilled water, without affecting the stained protein bands.

Bicinchoninic Acid (BCA) Microplate Assay

The bicinchoninic acid (BCA) assay measures the conversion of Cu^{2+} to Cu^{+} under alkaline conditions. The amount of Cu^{2+} reduced is a function of protein concentration that can be determined spectrophotometrically by a color change of the sample solution into purple, which absorbs at 562 nm. The absorbance is directly proportional to the amount of protein present in the solution and it can be estimated by comparison with a known protein standard, such as bovine serum albumin (BSA). This assay is also generally less affected than the Lowry assay by interfering compounds such as some detergents and denaturing agents (urea, guanidine chloride). However, the BCA assay is more sensitive to the presence of reducing sugars and high level of lipids [104]. The effects of these interferences can be eliminated or reduced through several strategies such as removing the interfering substances through dialysis, gel filtration or if the protein concentration is high enough, by diluting the sample [105]. Compared to other methods BCA assay is one of the most sensitive (it can detect proteins at concentrations as low as 5 μg mL^{-1}). It has less variability than others (i.e., Bradford assay), and it can be used to measure a wide range of protein concentration [106].

Example of experimental protocol for samples containing 0.1–1 mg protein mL^{-1}

Step 1 Prepare calibration curves with Bovin Serum Albumin (BSA) with a range of 5–2000 μg mL^{-1}.
Step 2 Prepare BCA working reagent (WR)
Step 3 Dilute the microalgal samples with distilled water (1:3, v/v).
Step 4 Pour 10 μL of distilled water with 10 μL of diluted samples/standard in a microplate.
Step 5 Add 200 μL of working reagent.
Step 6 Cover the microplate and incubate for 30 min at 37 °C in the dark.
Step 7 Keep all tubes at room temperature for 15 min before measurement.
Step 8 Take absorbance readings at 562 nm against a blank (distilled water).

Amino Acids Analysis

Amino acid analysis is one of the analytical principles for protein determination. The principle is that the proteins are broken down into their constituent amino acids by hydrolysis of the peptide bonds. The liberated amino acid residues are then determined, most often chromatographically, and protein content is calculated as the sum of individual amino acid residues after subtraction of the molecular mass of H_2O.

After sample preparation (hydrolysis of the protein with 1 mL of 6 N HCl in vacuum-sealed hydrolysis vials at 110 °C for 22 h), sodium or lithium buffers (depending on the amino acids present in the sample) are prepared for separation of the amino acids by Ion Exchange Chromatography. The eluate from the ion exchange column is passed through in a Teflon coil placed in a boiling water bath, or other heating apparatus. Before entering, the column effluent is mixed with reduced ninhydrin reagent, which is dissolved in acetate buffer. The ninhydrin reacts with amino acids forming a dye complex. The absorption is determined in a flow photometer, and registered on the chart of a recorder or a computer. The area under the peaks corresponds to the amounts of amino acids present in the sample. The evaluation can be done manually or automatically with an integrator or a computer. The circumstances of the analysis make it possible to quantitate as little as one nanomol amino acid with a high degree of accuracy.

Example of protocol (from Mæhre [107])

From microalgal crude samples: 50 mg of extracts are dissolved in 0.7 mL distilled H_2O and 0.5 mL 20 mM norleucine (internal standard).

Step 1 From protein extract samples: 500 μL extract is mixed with 50 μL 20 mM norleucine (internal standard).
Step 2 Subsequently, for all samples, concentrated hydrochloric acid (HCl, 12 M) is added, to a final concentration of 6 M.

Step 3 The sample mixture is flushed with nitrogen gas for 15 s in order to minimize oxidation, before hydrolysis at 110 °C for 24 h.

Step 4 Following hydrolysis, 100 μL aliquots of the hydrolysates are evaporated under nitrogen gas until complete dryness and re-dissolved to a suitable concentration in lithium citrate buffer at pH 2.2.

Step 5 All amino acids are analyzed chromatographically using an ion exchange column followed by ninhydrin post column derivatization on a Biochrom 30 amino acid analyzer (for example, Biochrom Co).

Step 6 Amino acid residues were identified using the A_{9906} physiological amino acids standard (Sigma).

Step 7 Protein content was calculated as the sum of individual amino acid residues (the molecular weight of each amino acid after subtraction of the molecular weight of H_2O).

Validation of Protein Quantification Methodology

Once a bioassay has been selected, it is of high importance to validate that assay for use by running risk analysis techniques such as failure mode and effect analysis (FMEA) [94]. The goal is to assess the influence that each experimental step will have on the final result of an analysis and to produce a list of parameters that must be evaluated to obtain a robust assay. Key validations parameters consist of accuracy, precision, repeatability, intermediate precision, specificity, detection limit, quantification limit, linearity and range of the assay.

4.3.3.4 Extraction and Purification of High Value Phycobiliproteins from Microalgae

Phycoerythrin

On order to meet the standards of pharmaceutical and molecular biology fields, this protein must be highly purified. Two methods are used to determine the purity:

- The absorbance ratio A_{565}/A_{280} defines the relationship between the presence of phycoerythrin and other contaminating proteins. A purity ratio $A_{565}/A_{280} > 4$ corresponds to diagnostics and pharmaceutical grade phycoerythrin [108].
- The absorbance ratio A_{615}/A_{565} determines phycoerythrin purity in relation to phycocyanin, which is its closest contaminating protein [39].

The purification protocol requires three steps: protein extraction by cell disruption, primary recovery and purification.

- Cell disruption: several methods can be used (sonication, mechanical maceration or lysozyme treatment). From the literature, the best recovery yields are obtained with sonication [108]. Jubeau et al. [109] evaluated the extraction of

B-phycoerythrin from *Porphyridium cruentum* by high-pressure cell disruption varying the parameters of pressure (25–270 MPa) and they proposed a two-step selective extraction with a first passage at 50 MPa (to eliminate the contaminating protein present in the cytoplasm) in culture medium followed by a second passage at 270 MPa (to extract the B-phycoerythrin) in distilled water, achieving a 0.79 purity ratio.

- Primary recovery: the most commonly, fast and inexpensive method used for primary recovery is selective precipitation with ammonium sulfate. Various concentration of ammonium sulfate is used depending of the microalgal strains, for example 65% for *Porphyridium cruentum* [110] or a two-step ammonium sulfate precipitation, at 20 and 70% saturation for *Phormidium* spp. [111].
- Purification: Purification is typically achieved by chromatographic methods like ion exchange chromatography, hydroxyapatite chromatography, gel filtration and expanded bed adsorption chromatography [39].

  - Parmar et al. [111] purified phycoerythrin from *Phormidium sp.* A27DM with a single-step gel permeation chromatography using a Sephadex G-150 matrix pre-equilibrated and eluted with a 10 mM Tris-HCl buffer (pH 8.1) at a flow rate of 60 mL h^{-1}. This protocol yielded a final purity ratio of 3.9.
  - Bermejo et al. [110] purified phycoerythrin from *P. cruentum* using an anionic chromatographic column of Diethylaminoethanol (DEAE) cellulose. Elution was performed as a discontinuous gradient of acetic acid-sodium acetate buffer (pH 5.5). The best results were achieved with flow rate of 100 mL h^{-1}. Later, Bermejo et al. [112] developed a method of purification of phycoerythrin by expanded bed adsorption chromatography (EBA) using a DEAE adsorbent. The authors focused on maximizing product recovery rather than purity, since the process is intended to replace low-resolution methods. The use of EBA chromatography allowed partial concentration of the product and therefore works as a preparative method with little product loss and is suitable to large scale production [113].

Phycocyanin

The purity ratio of the phycocyanin extract is determined by the A_{620}/A_{280} ratio. Absorbance ratio $\geq$ 0.7 refers to food grade pigment, while reagent and analytical grade correspond to 3.9 and $\geq$ 4.0 respectively [42].

General protocol for purification includes a first extraction in buffer solutions (phosphate buffer) with sonication or ultrasound as cell disruption pretreatment which should be as quick as possible in order to avoid proteins destabilization. Later, proteins precipitation and recovery by ultracentrifugation or filtration are carried out.

Phycocyanin is water-soluble and thus can be easily extracted as a protein–pigment complex. Sørensen et al. [114] evaluated different extraction techniques for C-Phycocyanin extraction from *Galdieria sulphuraria* and obtained contents of 25–

30 mg/g^{-1}. Ammonium sulfate concentration above 1.28 mol L^{-1} ensured only C-Phycocyanin precipitation with purity of 0.7. In case of ultrafiltration, more than 50% was lost at 100 kDa tangential flow filter, while 79% was retained by the 50 kDa filter. The authors also proposed to combine ammonium sulfate fractionation with the other methodologies tested (anion exchange chromatography, tangential flow filtration) in order to enhance purity of the recoveries (3.5–4.5).

4.3.4 *Proteomics*

Numerous scientific publications report proteomic studies on microalgae [115–118], mainly targeting on topics such as the response of microalgae to different stress sources or environmental signals and issues related to taxonomy and identification of potentially harmful species [119]. With the fast development of microalgal biofuel researches, the proteomics studies of microalgae have increased quickly [120]. Proteomic studies provide a platform for discovery of some yet unidentified genes and proteins [121]. Protocols used have significantly evolved with the development of two-dimensional electrophoresis (2-DE) which represents a powerful approach for high-throughput analysis of complex protein mixtures [121]. This technique is based on the integration of multi-dimensional chromatography and mass spectrometry (MS) with the proteomic informatics. Most experiments are broadly classified as either "gel-based proteomics" or "gel-free proteomics", depending on whether 2-DE is used for protein separation and quantification or not [121].

Nowadays, 2-DE is the most common strategy for protein separation and quantification, enabling the separation, detection and quantification of different proteins from a single extract. With this method, information on small post-translational modifications (PMTs) and highly homologous isoforms can often be obtained directly, since these tend to shift the isoelectric point of a protein without extensively changing its molecular weight. However, technical difficulties such as gel-to-gel variation, limited linear dynamic range, limited throughput, and protein co-migration have been reported. This can be overcome by the utilisation of software such as Progenesis SameSpots (Nonlinear Dynamics), PDQuest (Bio- Rad Laboratories) and DeCyder (GE Healthcare) for the analysis of 2-DE gels.

Jia et al. [121] described 2-DE as an elaborate pre-fractionation step that precedes MS analysis in a typical proteomic workflow and, as such, some researchers prefer to omit this step and apply different strategies which scale up much better, thus gel-free strategies are used more and more, since they allow higher analytical throughput and deeper proteome coverage than gel-based methods.

The common workflow (gel-based and gel-free) analysis includes: (1) sample preparation; (2) protein separation and quantification; and, (3) Protein identification and characterization, and, is described below.

4.3.4.1 Gel-Based Approach

Sample Preparation

The first step consists of protein extraction, since most analytical techniques used in proteomics require prior solubilization of proteins in an appropriate solvent (aqueous buffers, organic solvents). Commonly used aqueous extraction buffers often contain (besides buffering agents) detergents, chaotropes, reducing agents and protease inhibitors, ensuring that enzymatic activity is inhibited during extraction and that interactions between proteins are minimized, preventing aggregation. Fractionation prior to proteome can be done using chromatography, electrophoresis, differential solubility and/or centrifugation.

Protein Separation and Quantification

Detection and quantification methods for 2-DE are based on Coomassie Brilliant Blue. Recently, the development of multiplex 2-DE ("difference gel electrophoresis" or DIGE), involves tagging the protein samples with different fluorophores prior to 2-DE, leading to two main advantages as several samples can be run on a single gel and improves gel-to-gel variability, by providing a common reference channel across all gels of an experiment [122].

Protein Identification and Characterization

Most gel-based proteomic studies rely on digestion of detected proteins, followed by the analysis of the resulting pep-tides by MS for their identification and characterization. Instruments currently employed for this purpose include ESI-Ion Trap, MALDI-TOF/TOF and ESI-QTOF mass spectrometers to a lesser extent. Identification of proteins can be assessed either directly through its peptide mass fingerprint (PMF), for the case of organisms with fully sequenced genome, or by analysis of the fragmentation spectra of such peptides (PFF, peptide fragment fingerprinting or even de novo sequencing) obtained through tandem MS.

4.3.4.2 Gel Free Approach

With this technic, proteins are digested from the start and analyses (separation, quantification, characterization) are done at the peptide level. Gel free methods can be combined with fractionation methods, such as liquid-phase chromatography procedures (coupled to ESI-based mass spectrometers), in order to reduce the number of different peptides entering the mass spectrometer for maximization of the total number of distinct peptides detected over the course of a sample run.

Multidimensional chromatographic separations are commonly used, as in the case of MudPIT, where peptides are separated by charge (SCX-HPLC) and hydrophobicity (RP-HPLC) prior to MS analysis. Most gel-free workflows rely on stable isotope labeling for peptide quantification, either by metabolic incorporation of radioactive amino acids in proteins (SILAC) or by post-extraction chemical modification (ICAT, TMT, iTRAQ). With stable isotope labeling, several samples can be analyzed in parallel on the same MS run and relative abundance can be estimated.

4.3.4.3 Obtention of Peptides and Identifying Proteins from Peptides

Peptides Obtention

Peptides are short-chain protein-linked amino acid residues linked by peptide bonds. They are produced by enzymatic cleavage of proteins during food digestion, microbial fermentation, food processing or exogenous enzymatic hydrolysis [63, 64]. In order to isolate the bioactive peptides, the protein hydrolysates are subjected to several fractionation and purification techniques, in particular membrane ultrafiltration and ion exchange chromatography, affinity and gel permeation techniques. The choice of treatment techniques depends largely on the structural characteristic of the peptide of interest [61]. Peptides have shown interesting biological activities that would prevent hypertension, oxidative stress, cancer, diabetes, inflammation and immune disorders [64]. The bioactivity of peptides depends on several factors such as the molecular weight of the peptide, the amino acid composition, the molecular and surface hydrophobicity [64, 123].

For example, hydrolysis of *Chlorella vulgaris* with pepsin has been shown to produce a short-chain peptide (of 11 amino acids: Val-Glu-Cys-Tyr-Gly-Pro-Asn-Arg-Pro-Gln-Phe) with dose-inhibiting-dependent anti-hypertensive properties in vitro and in vivo [61, 125]. Another peptide (Val-Glu-Gly-Tyr) derived from the hydrolysate of *Chlorella ellipsoidea* has also been reported to have ACE inhibitory activity in vitro and an in vivo effect of blood pressure reduction [61].

Identifying Proteins from Peptides

Most proteomic studies attempt to identify proteins by analysing peptides, as large proteins constitute a challenge for MS-based methods. The classical method used ("peptide mass fingerprinting" or PMF) is based on in silico digestion of genomic/EST sequences following the pattern of a predictable endonuclease (usually trypsin) to obtain a list of peptide masses (or "mass fingerprint") for each database entry. Identification is performed by comparing experimentally obtained MS mass lists with those generated in silico, and choosing significantly similar matches.

It is possible as well to tandem MS instruments in order to obtain a peptide mass list and information on their fragmentation mass spectra, leading to a fingerprint for each peptide that directly reflects its sequence. For this, "peptide fragment fingerprinting" (or PFF) involves starting with a genomic/EST database and performing in silico digestion with a standard endonuclease. Then, for each peptide, the masses of all likely fragments are deduced from prior models. Identification of peptides is deducted by comparison of experimentally obtained MS/MS spectra against all possible fragmentation spectra in the database. Since a fragmentation spectrum (unlike mass) is usually very specific for a certain peptide sequence, identification of proteins can often be attained from a single high-quality peptide match. However, this strategy works only for species for which there is genomic/EST data available or for highly conserved peptides.

4.3.5 Challenges and Future Perspectives

To produce quality microalgal biomass as a whole-feed ingredient, cultivation techniques should aim to balance the lipid profile and the protein content. An alternative and potentially more efficient approach is a biorefinery type system where microalgal oils could be separated from the cell biomass and used as concentrated feed or food supplements [126]. Thus, oil production could be maximized, nitrogen consumption minimized, and the residual biomass used for other processes including energy production [127]. In order to achieve this, microalgae containing sufficient ω-3 fatty acids and protein are searched [126].

Microalgae biorefinery uses the overall composition of the cell, and next to lipids, including carbohydrates and proteins [81]. When proteins are extracted in a context of biorefinery, mild conditions of extractions must be used such as alkali or enzymatic extractions [82]. Regarding economical costs, it was demonstrated that protein production using enzyme extraction was less expensive than when obtained after an alkali extraction. Higher revenue can be generated when the residue after protein extraction can be sold as fuel or feed for animal.

Nowadays scientists are convinced that the biorefinery concept (i.e., a sequence of unit operations to achieve the whole fractionation and/or transformation of biomass to produce multiple products) applied to microalgae would render this sector profitable. The biorefinery scheme is the key for the utilization of microalgal biomass, but the main constraint is that in most cases final recoveries are low due to the number of steps required to achieve the purity levels specified for each compound and improving these yields represents a real challenge for the future. Moreover, conditions of cultivation of strains that produce high value compounds have been well studied, however recovery of intracellular metabolites at large scale is still a challenge since not all cell disruption, extraction or purification methods are scalable. Nevertheless, some technologies such as high-pressure homogenization can be viable for scale up [128]. Another key point to challenge is the set-up of more environmentally friendly process regarding extraction of high value compounds and finding new way to avoid toxic solvent and the use of too much energy (replacing high temperature processes for example).

4.4 Methods for Polysaccharides Determination in Microalgae

4.4.1 Introduction

Polysaccharides are highly complex macromolecules exhibiting a great variety of biochemical structures based on glycosidic linkages of up to 50 different monosaccharides including numerous isomers (hexoses and pentoses), some of them being complex sugars. Various non-sugar substituents such as sulphates, acyl or methyl groups but also amino acids may be attached on these linear or ramified backbones. Understanding this complexity needs to consider that two monosaccharides such as hexoses may be theoretically polymerized by glycosyltransferases or transglycosylases by 5 different glycosidic bonds. This wide range of polysaccharidic structures dominates their conformation, flexibility and interactions as they adopt different architectures in solution [129]. In their meeting report, published in 2007 in the famous review, Nature Chemical Biology, Jeremy E. Turnbull and Robert A. Field, wrote "the key bottlenecks in the development of glycobiology has been analysis of glycan structures, from natural sources" [130]. Ten years later, this fact is definitely true in the field of polysaccharides from animals, terrestrial plants, macroalgae and, above all, microalgae. For a long time, the structure identification of carbohydrate polymers was limited by suitable and sensitive approaches and despite the recent development of powerful analytical tools, more and more efficient, the analysis of polysaccharidic sequences is still a challenge compared to those of nucleic acids and proteins. Moreover, the sole overview of methodologies currently used in polysaccharide analysis resumes well its complexity [131]. These methodologies include monosaccharide analysis, linkage methylation analysis, chemical/enzymatical degradation procedures, fractionation and purification of degradation products, infrared spectroscopy, mass spectrometry using different ionization techniques and 1-D/2-D NMR spectroscopy using $^{1}H/^{13}C$ dual probe.

Numerous microalgae are known to produce polysaccharides with several cellular locations (Fig. 4.1). They are sometimes excreted as extracellular polysaccharides [132] or are cell wall polysaccharides, intracellular starch or chrysolaminarin. Excepting starch and chrysolaminarin, their structures are highly complex including often up to 10 different monosaccharides and non-sugars groups such as methyl, pyruvyl, acetyl, sulphate and others. Note also the lack of units of repetition for the major part of them limiting significantly the structural investigations despite the potential of these biopolymers as texturing or biological agents [132–134]. Exopolysaccharides from microalgae can remain associated to the cell

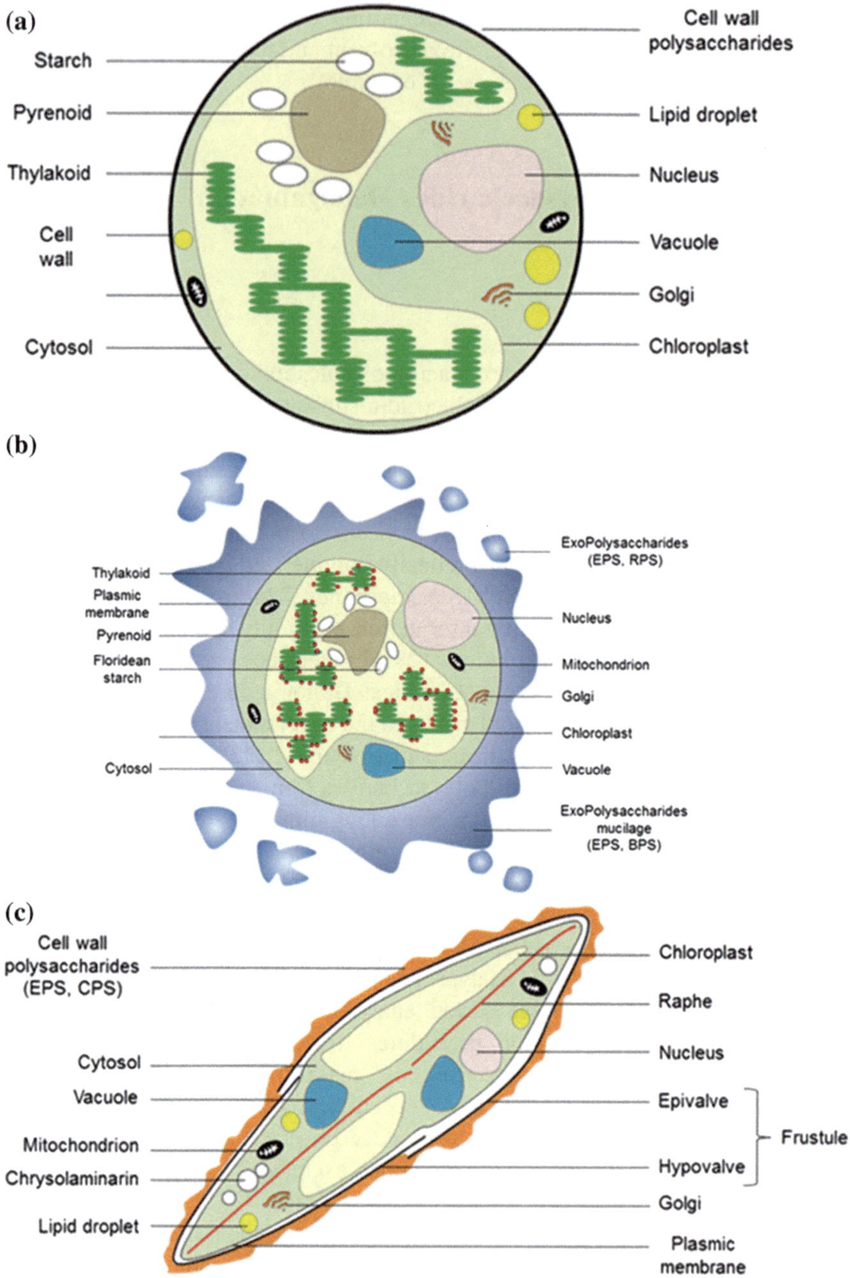

Fig. 4.1 Schematic representation of red/green microalgae (**a**, **b**) and Diatoms (**c**)

surface (cell-bound polymers) and/or be liberated into the surrounding environment as released polysaccharides [135–137]. The status of cell-bound exo-polysaccharides is often unclear. Exopolysaccharides (EPS) are sometimes called ExoPolymeric Substances (EPS), Extracellular Polysaccharides (ECPS), Released polysaccharides (RPS), Extracellular Proteoglycan (EPG), capsular polysaccharides (CPS), Polysaccharide (PS) and sulphated Polysaccharides (sPS) [132–134, 136, 138, 139]. Exopolysaccharides are the most studied compounds among polysaccharides from microalgae. This popularity is explained by their structural diversity, extractability and solubility, as these biopolymers have not covalent linkages with cell walls. Their characterization may be impacted by culture conditions for their production, but also by the extraction and purification used to collect them [140, 141]. The identification by some authors of structures differences depending on culture conditions could be in some cases artefacts related to different extraction, purification and/or structural characterization protocols. From today and at our knowledge, only few full structures from microalgae polysaccharides have been described in the literature and a significant part of them are just oligosaccharides extracted and purified from complex polysaccharides [132]. These partial characterizations are not totally representative of the native macromolecular structures.

4.4.2 Polysaccharides Sampling and Extraction Strategies

Mostly, the largest part of microalgae polysaccharides and notably EPS stay poorly understood because of the low extraction yields and the difficulty to properly separate them. In the following sections are presented the most important strategies described and developed for the extraction and purification of microalgae polysaccharides (Fig. 4.2).

4.4.2.1 Alcoholic Precipitation of Polysaccharides as Conventional Extraction Processes

Production yields of microalgae polysaccharides are closely dependent to growth and cultures conditions [142]. Then, many cost-effective and environmentally friendly parameters have to be investigated to selectively extract and purify EPS and cell-bound or intracellular polysaccharides (IPS) such as chrysolaminarin, starch, capsular or sheath polysaccharides [143, 146]. Works of Ramus were the first to recommend an experimental approach to separate and isolate the different categories of polysaccharides from *Porphyridium*. In this strategy, after a preliminary depigmentation of microalgae biomass using ethanol/acetone treatment, cell bound exopolysaccharides were extracted using hot water. A second step with cetyl pyridinium chloride selectively isolated EPS by precipitation in calcium salt prior to a final ethanol precipitation. Focusing only on EPS, Delattre et al. [132] have proposed a general process to selectively extract and purify these hydrocolloids

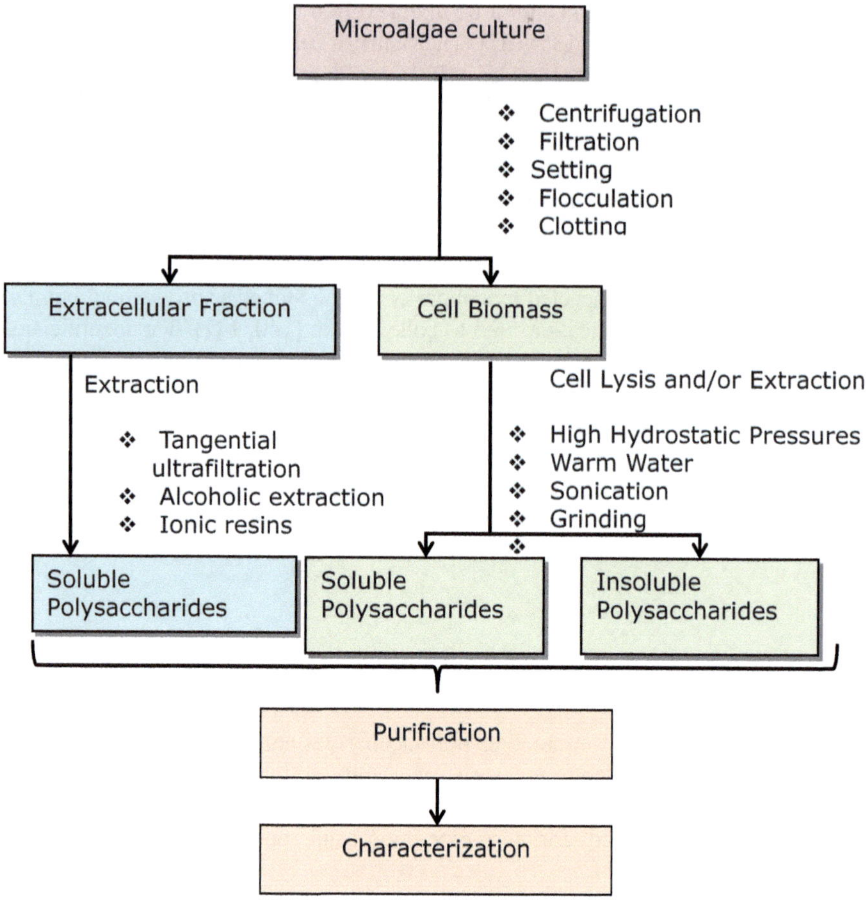

Fig. 4.2 Overview of polysaccharide extraction from microalgae

from culture media of several microalgae. Briefly, after microalgae cultivation, biomass is extracted from culture medium using centrifugation or microfiltration [147, 148]. Afterwards, soluble EPS (in supernatant or in permeate) are concentrated under vacuum (depending on the viscosity) and finally precipitated with 2–3 volumes of cold alcohol (−20 °C) such as ethanol, isopropanol, or methanol [132, 149–151]. The main advantage of this alcoholic precipitation in industrial point of view is the possibility to work with highly viscous solutions and to recycle the solvent (water, alcohol) by distillation to reduce the production cost. Nevertheless, the main drawback of this method is the co-precipitation of a part of salts coming from the culture medium, notably when the cultivated microorganisms are marine microalgae (up to 30–40 g L^{-1} of NaCl depending on culture media). This phenomenon can be partially avoided repeating the alcoholic precipitation 2 or 3 times and or using several selective alcohol concentrations. Then, tangential ultrafiltration

(TUF) in diafiltration operating mode is commonly applied to purify polysaccharide and remove salts. Patel et al. clearly showed that the desalting of EPS from *Porphyridium cruentum* was optimal using diafiltration of medium onto 300 kDa membranes comparatively to the conventional alcoholic precipitation which gave EPS fraction polluted by high salt contents [150]. Finally, note to mention that additional combined purification steps such as trichloroacetic acid (TCA) treatment, selective precipitation (alcoholic gradient) and TUF lead also to the removing of all low molecular weight impurities (phenolic compounds, pigments, proteins, salt, mineral, …) [132, 150]. Lastly, the high purity EPS is dried (freeze-drying or in ventilated oven at 40–50 °C). When polysaccharides are dried in a ventilated oven at 40–50 °C, this step is sometimes preceded by a precipitation of polysaccharide and it's washing by acetone. It is important to notice that as mentioned by Delattre et al. [132] during the purification step of polysaccharides, other high-value molecules such as proteins, pigments, and lipids could be valorized for industrial purposes [152]. The next section gives an overview of the extraction/purification optimization of polysaccharides using TUF.

4.4.2.2 Tangential Ultrafiltration Process for EPS Purification

Recently, many researchers have proposed TUF processes as an alternative to the conventional extraction of EPS using alcoholic precipitation. This methodology was largely investigated to extract biomolecules with high yields from complex medium without using additives (chemical, enzymes, solvents …) [153, 154]. Therefore, many proteins and polysaccharides purification processes were investigated on several ultrafiltration modules such as rotating, hollow fiber, tubular, spiral wound or flate plate system [153]. Generally, the clarification/desalting/concentration steps were carried out using nominal molecular weight cut-off (NMWCO) from 1 to 500 kDa [154]. To be efficient for the purification of polysaccharides, the membrane system has to be studied depending on the concentration and viscosity of media but also the trans-membrane pressure and flow rate velocity. In the works of Li et al. [147] a combined microfiltration (polypropylene membrane, 0.2 μm) and ultrafiltration (polyethersulfone membrane with NMWCO of 5 kDa) pilot-scale extraction processes was proposed to concentrate EPS from varied microalgae and Cyanobacteria including: *Nostoc sphaeroide, Nostoc commune, Chlorella pyrenoidosa, Chaetoceros muelleri, Haematococcus pluvialis* and *Spirulina platensis*. Recently, the same approach was used to efficiently isolate and purify polysaccharides from *Porphyridium sp.* using two ultrafiltration steps with two polyethersulfone membranes (NMWCOs of 10 and 300 kDa) [155]. Generally, as demonstrated by the works of Zhang and Santschi [156] on the purification of *Amphora sp.* EPS, the ultrafiltration technology was well-recommended to purify microalgae EPS from sea water culture media. However it should be noticed that to be generalized to all microalgae EPS extraction/purification processes, TUF technology system must be optimized for industrial scale-up potential. There are some drawbacks limiting TUF process such as (i) the great amount of water used for desalting/purification/concentration of EPS and, (ii) the clogging of membranes due to the high viscosity of microalgae EPS solution [147, 148, 157, 158] which reduce considerably the EPS extraction yield and increase the operating prices compared to the alcoholic precipitation. Consequently, it is essential for the development of new efficient membrane systems to study rigorously the combination of both numerical and theoretical analyses in order to upgrade both mass-transfer and hydrodynamics setting [153, 159]. For this purpose, the works of Jhaveri and Murthy [160] gave a very attractive overview on anti-fouling nanocomposite membranes development for industrial application. Indeed, authors clearly highlighted the importance for innovation in synthetic material fields in particular to enhance the membrane filtrations system performance by improving the hydrophilicity, the anti-fouling and the self-cleaning properties. In this context, the use of organic (polyacrylonitrile) and inorganic (AZT: Aluminium/ Zirconium/Titanium Oxide) membranes has been efficiently performed with the cyanobacterium *Arthrospira platensis* [161, 162].

4.4.2.3 Specific Treatments for the Extraction of Cell-Bound Exopolysaccharides

As well-established, around 50% of the total amount of EPS are cell-bounded to the microalgae cell as slime layer [135]. Numerous procedures have been described in literature to purify and isolate cell-bound exopolysaccharides. Among them, treatment of microalgae with ethylene diamine tetracetic acid (EDTA), formaldehyde (FA), sodium hydroxide (NaOH), water, sonication, heating and ionic resins have been proposed [146, 151, 163–166]. Treatment with cationic resins is known to break specific interactions between microorganisms and EPS without any cellular damage [145, 146, 151]. By this way, Pierre et al. [151, 167, 168] isolated several microalgae EPS from a diatom-dominated intertidal mudflat using the cationic exchange resin Dowex Marathon C. In other studies, researchers proposed during the extraction of EPS, the preliminary combined treatment of microalgae cell by FA and glutaraldehyde (GTA) as fixative agents in order to reduce cell lysis [143, 145, 146, 164]. Nevertheless, these chemical treatments caused structural modifications of EPS disturbing extraction [143, 164, 169]. That is why the general alternative to extract cell-bounded EPS is to use washing treatments with water at different temperatures (30–95 °C) and pH (basic of acid). According to the treatment duration (1–4 h) and biomass/water ratio, microalgae cell lysis was observed. Thus, the contamination of EPS by intracellular compounds such as starch, chrysolaminarin and proteins was described for the extraction of EPS from *Craspedostauros australis, Nitzschia epithemioides, Thalassiosira pseudonana, Anacystis nidulans, Navicula jeffreyi, Cylindrotheca fusiformis, Anabaena cylindrica, Cyanospira capsulata, Navicula phyllepta* and *Phaeodactylum tricornutum* [144, 146, 170–172].

4.4.2.4 Specific Treatments for the Extraction and Quantification of Starch

Despite growing interest for the use of starch from microalgae as feedstock for bioethanol production within a biorefinery point of view, only few studies describe a protocol for extraction and purification of starch. Extraction of starch is only performed when structural characterization is needed, especially for the understanding of mechanisms ruling the metabolic patterns depending on culture conditions. As starch is an intracellular polysaccharide, first step involves a cellular lysis. Several methods exist, that can be classified as physical, enzymatical, or chemical, but physical ones are generally preferred. Delrue et al. [173] and Deschamps et al. [174], have purified starch from microalgae using similar protocols. The cellular lysis was achieved by sonication, of a cellular suspension in phosphate buffer (pH 7.5–8.0) containing 5 mM EDTA. For Decamps et al. [174] the buffer was specifically supplemented by 1 mM of dithiotreitol. After centrifugation, the pellets (starch and cell fragments) were resuspended in 90% Percoll and centrifuged (10,000 g, 30 min) to separate high-density starch granules from cell debris of lower density. The Percoll gradient step is repeated to ensure complete removal of cell debris from the starch pellet. The starch is then washed in sterile water and freeze-dried after its recovery by centrifugation. Kobayashi et al. [175] used a 10% toluol solution in order to separate starch and cell debris. After mixing vigorously during 10 min, an emulsion layer, containing cells debris, was formed at the top of the tube. After discarding this layer, the starch was recovered and toluol step repeated, until no coloration appeared. Starch was then washed with water and ethanol before drying. Other alternative can be, after cellular lysis and the first centrifugation step, to resuspend pellets in water and to boil (30 min), in order to solubilize starch. After a new centrifugation (10,000 g, 15 min), the supernatant containing starch was precipitated with 3 volumes of ethanol, or more specifically with iodine [176]. Nevertheless, starch from microalgae is structurally different from that of higher plants regarding the ratio amylose: amylopectin. Whereas starch from higher plants contains generally around 30% of amylose, amounts of 5–25% have been reported for chlorophyceae [175] and is rather absent in starch from Rhodophyceae except in few unicellular strains [177]. Consequently the efficiency of iodine precipitation can differ from one genus to another [178].

After starch extraction and purification, classical methods can be used for its quantification and characterization. Nevertheless, most authors are interested in quantifying starch without time consuming extraction. Several methods are found in literature for quantification of starch content in microalgae. Most of them are derived from the characterization and quantification of carbohydrates from terrestrial plants. For example, starch content can be estimated after starch solubilization from ethanol-treated biomass (defatted and depigmented) by boiling. After centrifugation, supernatant containing starch is mixed with iodine solution and color is compared with standards of known starch content [179]. Drawback of this method is that, as previously mentioned, structural differences of starch from microalgae induce differences in color when mixed with iodine and the coloration is then not directly proportional to amount of starch. As an example, floridean starch gives a pink color due to low amylose content [180], whereas starch from Chlorophyceae gives a blue-dark coloration [181]. One consequence is that maximum of absorbance is shifted depending of microalgae class: 597 nm for Chlorophyceae [175], and 530–550 nm for Rhodophyceae [180–182]. For accurate quantification, the standard should then be of the same type as the analyzed sample. Before using this method as routine procedure, an extraction of the starch is required to use it as standard (as the starch type could be not commercially available), and a more accurate method of quantification is necessary to calibrate it.

Chemical treatments such as 45–52% perchloric acid, 3–15% sulphuric acid, or 90% dimethyl sulfoxide, are often used to degrade starch before quantification of released sugars by enzymatical or colorimetric methods [175, 179, 183]. Even if this method is frequently used by authors, the efficiency can be discussed as several drawbacks may be highlighted. First, an incomplete hydrolysis can lead to underestimation. Second, celluloses and hemicelluloses found in cell walls can be degraded together with the starch, and result will be in that case overestimated, especially if colorimetric assay is used to quantify sugars. If enzymatic assay is used (generally based on glucose oxidase activity), only cellulose will interfere as hemicelluloses from microalgae are often xylan-type.

In order to circumvent the problem of non-selective degradation, enzymes can be used. A mix of α-amylase and amyloglucosidase, in appropriate amounts and reaction conditions, will provide efficient degradation of starch, without contamination from cellulose degradation as β-links are not attained by these enzymes. The released sugars are further quantified as previously described. This protocol is at the origin of starch assay kits from Megazyme, Bohringer, or Sigma-Aldrich.

Recently, Ji et al. [184] have proposed a fast method to quantify starch by direct in situ measurement in individual cells by Raman spectrometry. Method requires expensive equipment, as Raman spectrometer coupled to microscope is needed. Nevertheless, results revealed a nearly linear correlation between the signal intensity at 478 cm^{-1} and the starch content of the cells. This method has been tested both on *Chlamydomonas* and *Chlorella* strains, and the specificity of the band at 478 cm^{-1} has been validated on a *Chlamydomonas* mutant (non-producing starch strain).

4.4.2.5 Specific Treatments for the Extraction of Fibrillar Polysaccharides

Some microalgae such as green microalgae have a rigid cell wall composed of fibrillar polysaccharides including chitin-like glycan or glucose-mannose polymers [185]. For fibrillar polysaccharides extraction, the method of Mc Cleary [186] can be used. This method has led to the AOAC 2009.01 method for extraction and quantification of soluble and insoluble fibers in food samples. Briefly, microalgae biomass is subjected to action of α-amylase and amyloglucosidase (16 h, 37 °C, pH 6) in order to hydrolysate starch. For better results, a mechanical cellular lysis is recommended in order to improve access of enzymes to the starch. After heating to 95 °C (for inactivation of enzymes) and pH shift to 8.2, protease is added (30 min, 60 °C) to degrade proteins. After filtration and washing of residues, they are dried and correspond to insoluble fibers (fibrillar polysaccharides in microalgae case). It is to notice that the filtrate obtained at the last step could be precipited by ethanol and should correspond to soluble fibers (matricial polysaccharides in microalgae case).

4.4.3 How to Determine the Global Composition of Polysaccharides?

Identification and quantification of carbohydrates are of primary importance when it comes to the characterization of microalgal biomass. Owing to their low absorption in UV-Visible regions and their structural complexity, the analysis of polysaccharides can be a very challenging work. Overall, a deep investigation is needed to accurately determine the structure of carbohydrates and strategy for their analyses can change depending on their sources, extractions and compositions.

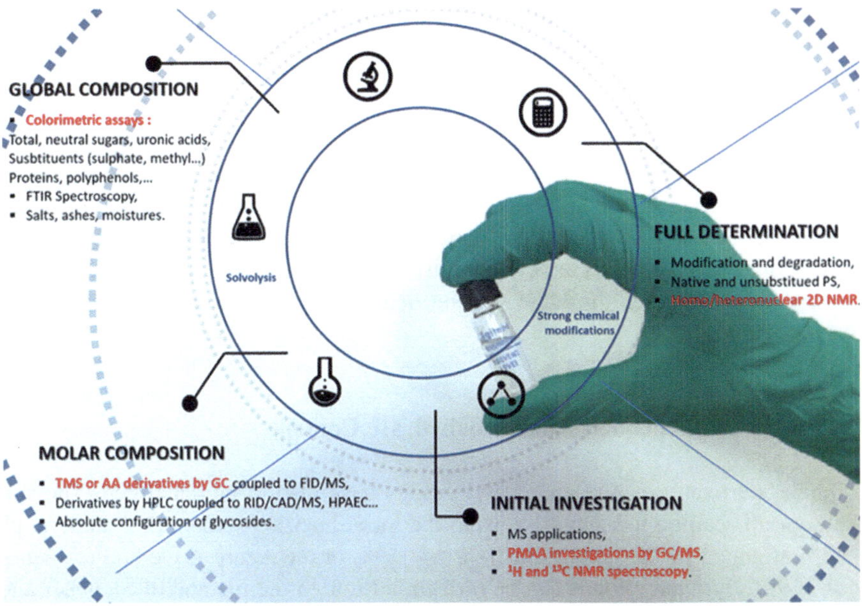

Fig. 4.3 Overview of the strategies needed for the structural analysis of polysaccharides from microalgae. HPLC: High-Performance Liquid Chromatography, RID: Refractive Index Detection, CAD: Charged Aerosol Detection, MS: Mass Spectrometry, GC: Gas Chromatography, FID: Flame-Ionization Detector, TMS: TriMethylSilyl, AA: Alditol Acetates, HPAEC: High-Performance Anion-Exchange Chromatography, NMR: Nuclear Magnetic Resonance spectroscopy, EI: Electron Impact Ionization, PMAA: PerMethylated Alditol Acetates, FTIR: Fourier Transformed InfraRed spectroscopy, PS: Polysaccharides

Figure 4.3 gives an overview of the main ways for the structural analysis of carbohydrates, from the determination of the global composition to the fine analysis of the branching patterns.

4.4.3.1 Total Carbohydrates

The phenol [187] or orcinol [188] sulfuric acid methods, as well as the recently updated anthrone assay [189, 190] can be used for determining the total amount of carbohydrates. Based on the dehydration of monosaccharides to furfural compounds in acid, the phenol-sulfuric acid is still the most used method with few alternatives (miniaturization, heating, acid concentrations) described in the literature [191, 192]. Briefly, the absorbance is measured at $\lambda = 485$ nm and D-glucose is used as standard. Attention should be given to other compounds, e.g. pigments, proteins, lipids, polyphenols and salts to limit interferences.

4.4.3.2 Uronic Acids and Neutral Sugars

Based on the formation of furfural derivatives in presence of resorcinol solubilized in acid, the method of Monsigny et al. [193] can be used to determine the neutral sugars content. The absorbance is classically measured at $\lambda = 510$ nm and D-galactose is used as standard. The meta-hydroxydiphenyl (m-HDP) assay [194] modified by Filisetti-Cozzi and Carpita [195] can be used to measure the uronic acids content. In presence of potassium sulfamate and m-HDP, uronic acids form chromogens which absorb at $\lambda = 525$ nm. In general, D-galacturonic acid is used as standard. It is noteworthy that the corrective formula from Montreuil and Spick [196] should be used to accurately quantify both neutral sugars and uronic acids in the same sample since the latter can be detected by the resorcinol-sulfuric acid assay.

4.4.3.3 Substituents and Non-carbohydrate Content

Sulphate, pyruvate, methyl and acetyl groups have also to be quantified. The sulphate groups can be measured by using the turbidimetric protocol of Craigie et al. [197], estimated by the gelatin/$BaCl_2$ assay [198] or the Azure A method of Jaques et al. [199]. Pyruvate groups can be measured through the method of Sloneker and Orentas [200]. Methoxy and acetyl groups are often observed by ^{1}H NMR since these groups have specific resonances. However, a saponification step can be applied to release $-CH_3$ and $-COCH_3$ groups, which can be directly or indirectly monitored by High-Performance Liquid Chromatography (HPLC) [201, 202]. Even if the extraction step and more specifically alcoholic precipitation (see the previous section) decreases the presence of proteins and/or glycoproteins, total proteins (which can also be covalently linked to polysaccharides) can be quantified by the

Smith et al., Bradford or Lowry et al. methods [95, 99, 203]. For determining the polyphenol content, the Folin-Ciocalteu assay can be used from [204]. Finally, salts, ashes and moisture should be also quantified.

4.4.3.4 Identifying Groups by Infrared Spectroscopy

Fourier Transformed InfraRed (FTIR) spectroscopy can be used for identifying and/or discriminating polysaccharides as well as observing changes in their composition (footprint) by providing details on a range of vibrationally active functional groups [205]. The structural analysis of carbohydrates by FTIR spectroscopy focuses on five specific regions from 3600 to below 700 cm^{-1}, which correspond to O–H and C–H stretching vibrations, local symmetry, C–O stretching vibration and fingerprint or anomeric region and skeletal region [206]. Polysaccharides usually display a strong and broad absorption peak at 3360 cm^{-1} corresponding to (O–H) stretching vibrations as well as water adsorption [207, 208]. A weak signal at 2920 cm^{-1} can be attributed to the asymmetric vibration of ($C–H_2$) groups [209, 210]. Overall, the region from 1500 to 500 cm^{-1} is defined by several modes concerning the type of polysaccharides and glycosidic linkages. As an example, furanose derivatives from pentoses can be identified at 1250 cm^{-1}. At 1250–1170 and around 1300–1250 cm^{-1}, aliphatic esters and aromatic esters can be respectively observed. Sulphate ester ($O–SO_3^-$) groups can be identified in the region from 1260 to 1210 cm^{-1} [211], carboxylic acid ester form (C=O) or carboxylate anion form (COO^-) can be observed around 1730 and 1600 cm^{-1} [212] whereas acetyl groups can be found at 1250–1220 cm^{-1}. The signal observed around 1040 cm^{-1} is in general from the presence of (C–O) of polysaccharides [213]. Anomeric (α and β) configurations can be identified by three types of bands from 950 to 750 cm^{-1} [214]. Note also that glycosidic methoxy groups give a specific symmetric $C–H_3$ stretching band around 2880–2830 cm^{-1}. FTIR spectroscopy also benefits the possibility to detect proteins, nucleic acids (NA) and other potential "contaminants" of the carbohydrate fraction through the amide I (1709–1583 cm^{-1}), amide II (1585–1481 cm^{-1}) and NA (1356–1191 cm^{-1}) bands [215].

4.4.4 How to Determine the Monosaccharides Composition of Polysaccharides?

4.4.4.1 Preliminary Solvolysis

Monosaccharides have to be released by cleaving the glycosidic bonds before their analyses by Chromatography. Two preliminary solvolysis, i.e. (i) methanolysis and (ii) hydrolysis, are well described in the literature for this purpose. The first one involves the cleavage of the linkages by methanolic HCl (around 2 M) in milder conditions than for classical hydrolysis. Less degradation of monosaccharides to furfural derivatives (or more degraded products) is often reported and it is possible to analyze both neutral sugars, uronic acids, sialic acids and acetamido sugars. Moreover, methyl groups substitute the hydroxyl groups previously engaged in the glycosidic bonds [216]. Briefly, the protonation of the glycosidic-O-atom results to the cleavage of the glycosidic linkage from a specific ring form. An anomeric carbocation will be formed and be able to react with water and give α/β anomers. Thus, ring form conversions can occur through protonation of the ring-O-atom and/or mutarotation [131]. On the other hand, hydrolysis by mineral acids (H_2SO_4, HCl, CF_3COOH) can be used to release the monosaccharides. Numerous conditions are reported in the literature and the use of TriFluoroacetic Acid (TFA) is often preferred (2–4 M, 90–240 min, 100–120 °C). Indeed, some authors reported the degradation of monosaccharides in particular when strong conditions are used to ensure the cleavage of resistant glycosidic linkages [217]. The effect of these conditions should not be neglected since it may cause underestimation of specific monosaccharide and more specifically ketoses. Here again, released monosaccharides from the polysaccharide structure give a monosaccharide-specific anomeric distribution. Finally, methyl glycosides or released monosaccharides can be analyzed by Gas Chromatography (GC) or HPLC including High-Performance Anion-Exchange Chromatography (HPAEC).

4.4.4.2 Chromatography

First, monosaccharides composition can be determined by HPLC combined with various detection apparatus such as Refractive Index Detection (RID), Charged Aerosol Detection (CAD) or Low-Wavelength Ultraviolet (UV). HPLC coupled to ElectroSpray Ionization (ESI) or High-Resolution Mass Spectrometry (HRMS) should be also considered as decent methods reported in the literature. One main advantage of LC methods is probably from the sample preparation which is strongly reduced compared to classical derivatizations of monosaccharides needed for GC analysis. On the contrary, one main disadvantage could be the low resolution especially for complex mixtures [218]. Overall, the separation is based on polarity allowing to work with normal and Reversed-Phase Chromatography (RPC). Ions-exchange, polymer, silica or metal based and amino bonded can be used with heating (30–80 °C) under various mobile phases (diluted H_2SO_4, water, acetonitrile) for the quantification of the released monosaccharides. Few papers also reported derivatization of monosaccharides prior the analysis [219]. HPAEC methodology should be also considered as a first-rate alternative for quantifying monosaccharides (using CarboPac PA-1), as reported by Templeton et al. [220]. Coupled with Pulsed Amperometric Detection (PAD), the method is resolutive and based on the pKa of monosaccharides and changes in temperature or alkalinity. Schäeffler et al. [221] reported tuning procedures to improve HPAEC analyses for carbohydrates even if Fu and O'Neill [219] highlighted the drop of electrochemical response in presence of proteins.

Secondly, GC is still used in many papers for quantifying monosaccharides after solvolysis due to higher sensitivity and resolution of spectra [222]. On the contrary to HPLC, released monosaccharides must be converted into volatile compounds via derivatization procedures including silylation and acylation. Silyl derivatives, which are the most used for GC applications, are generated by displacement of active hydrogen on –NH, –OH or –SH groups and various derivatizating agents can be used for this purpose, e.g. *N*-methyl-*N*-trimethylsilyl trifluoroacetamide (MSTFA), *N*-trimethylsilyimidazole (TMSI) or N, O-bis-(trimethylsilyl) trifluorocetamide (BSTFA). The latter is often combined to trimethylchlorosilane (TMCS) to increase the silyl donor strength. Thus, TriMethylSilyl (TMS) alditols, aldonitrile acetates, TMS methyl ester glycosides or oxime derivatives can be prepared for GC analysis. In general, Flame-Ionization Detector (FID) is preferred to MS for the quantification since the results are more reproducible. As reported by Peña et al. [223], modifying the mass tuning and the possibility to use ion suppression strongly impact MS quantification and can result in inaccurate quantification. Note that MS benefits mass spectra (associated to specific fragmentation patterns) which are essential for discriminating carbohydrates to non-carbohydrate contaminants. Finally, GC/MS experiments can be performed using MS-Chemical Ionization (CI) or Electron Impact Ionization (EI). Pyranose and furanose ring forms can be distinguished comparing the intensities of the specific ions m/z 204 and m/z 217 for TMS (methyl) glycosides [216] as well as regarding the ion m/z 205. Ions such as m/z 379, 319, 305, 217, 205, 204, 117 are characteristics of the fragmentation of TMS (methyl) glycosides by GC/MS-EI [151]. Beside to TMS derivatives, the conversion of released monosaccharides to per-O-acetylated alditol (AA) derivatives is another way to quantify monosaccharides composition [223]. Each monosaccharide gives only one peak with specific fragmentation patterns and main ions such as m/z 289, 217, 187, 157, 115, 145, 103, 43. Overall, a wide range of columns can be used for GC/MS-EI analysis, e.g. AT-1 [131], HP-1, CP-Sil 5 CB, DB-1701 [224], DB-225MS [220], Optima-1MS [225], OV-17 [226] or SP 2330 [223].

4.4.5 *How to Elucidate the Branching Patterns of Polysaccharides?*

4.4.5.1 Absolute Configuration Analysis

Usually, GC and chiral stationary phase are used for determining the absolute D or L configuration of monosaccharides. Glycosidation reactions can also be used with a chiral alcohol ((−)-2-butanol or (+)-2-octanol) to generate an extra chiral center, allowing the use of a nonchiral stationary phase for analyzing TMS (−)-2-butyl or (+)-2-octyl glycosides [131].

4.4.5.2 NMR Analysis

NMR is probably one of the best tools for establishing native glycan structures of polysaccharides [227, 228]. Two decades ago, Duus et al. [229] published an extensive review giving a deep overview of how to determine the structure of carbo-hydrates by NMR. Basically, 1D NMR (^1H and ^{13}C) spectroscopies give primary information about the structure and conformation, e.g. the type of monosaccharides, anomeric configurations, branching types, presence of substituents such as sulphate, methyl, acetyl or carboxyl groups [131]. Indeed, signals at specific resonance frequencies (shifts) are from carbon and non-equivalent hydrogen and the shape of each signal corresponds to the chemical environment of the corresponding carbon and hydrogen [230]. Considering the wide literature and numerous fingerprints spectra (^1H and ^{13}C) of polysaccharides described, authors usually compare chemical shift values for attribution. Today and especially for new and/or original structures, homo- and heteronuclear 2D techniques should be performed to assess and clarify ambiguous assignments and give accurate and deep understanding of polysaccharides linkages [231]. For NMR users, partial depolymerization of the analyzed polysaccharide (combined to a long accumulation at 60 °C), e.g. by using hydrolysis, ozonolysis, chemical degradation or hydrogenation, should be done since it greatly helps solubilizing the polymer (molecular weight under 100 kDa), increasing the quality and accuracy of NMR spectra and thus facilitating the structural elucidation. Note that various conditions should be performed since hydrolysis for example is well-known for removing some branching residues, side chains and/or terminal units [232]. Mass Spectrometry (MS) should also be used to analyse kind of derivatives.

4.4.5.3 Mass Spectrometry Analysis

MS applications can be combined with a broad range of chromatographic technologies such as MALDI-TOF (Matrix-Assisted Laser Desorption/Ionization with Time of Flight mass spectrometry), GC/MS-CI, HPLC-ES-MS, HPAEC-PAD/MS, HPLC-ESI-MS-MS, etc. [131]. MALDI-TOF MS, electrospray ionization mass spectrometry as well as GC/MS-EI are often preferred for the structural studies of glycans. The latter involves the preparation of PerMethylated Alditol Acetates (PMAA) which can be used to determine terminal units, glycosidic linkages, branching points and the ring size of monosaccharides. Before hydrolysis, permethylation of the polysaccharide is done by using CH_3I in alkaline medium, e.g. solid NaOH in C_2H_6OS [223]. As previously described, various conditions of hydrolysis can be used but 2–4 M TFA give good depolymerization results. Reduction using ($NaBH_4$ or $NaBD_4$ for isotopic labeling in NH_4OH/C_2H_6OS) then peracetylation $(CH_3CO)_2O$ in presence of co-catalyst $C_4H_6N_2$) are performed to yield PMAA. Interpreting the fragmentation patterns from mass spectra (mostly without molecular ion peaks) leads up to determine the position of glycosidic bonds initially composing the polysaccharide. A comprehensive study of these fragmentations patterns and tables can be found in Kamerling and Gerwig [131]. Further analyses are needed to complete the initial monosaccharides linkages analysis such as (i) uronic acid reduction due to their non-esterified carboxyl function and low volatility, (ii) partial hydrolysis to determine repeating unit patterns, (iii) acetolysis, (iv) uronic acid, smith and periodate degradations, (v) partial alkaline hydrolysis, (vi) removing of substituent(s) and/or (vii) enzymatic depolymerizations for determining anomeric configuration.

4.4.5.4 Conclusion

In the emerging field of glycosciences the specific detection and quantification of polysaccharides from microalgae is in progress even if the cracking of polysaccharidic structures is always a real challenge. However, the recent interest of scientific and industrial communities for the culture of microalgae (mainly for production of high value molecules) led to a better understanding of the physiology of these microorganisms and to the development of technologies for their culture. At this time the ability of scientific community to cultivate in controlled conditions numerous strains of microalgae isolated from diverse environments open the way to identification of new polysaccharides with original structures and physico-chemical properties. Many exciting applications after analysis of polysaccharides from microalgae will probably become evident with the development of essential tools for their structure analysis and modelling.

4.5 Methods for Lipids Determination in Microalgae

Sources of lipids include soybean, corn, palm, canola, and jatropha as oleaginous plants and animal fat and waste cooking oil [233–235]. It is also widely anticipated that microalgae can be used as a new lipid source [234, 236, 237]. Microalgae grow rapidly, and many species are exceedingly rich in oil. Microalgae typically double their biomass within 24–72 h, and the doubling times of the fastest-growing algae are as short as 3.5 h during the exponential growth phase. The oil content of microalgae is 20–50%, and some microalgae can produce oil content of 80% per weight of dry biomass [58, 232, 236–239]. Therefore, the oil productivity of microalgae (the mass of oil produced per unit volume of broth per day) can reach tens to hundreds of times higher than that of oil crops.

Lipids accumulate in microalgae in the forms of acylglycerols (triacylglycerol, diacylglycerol, and monoacylglycerol), phospholipids, glycolipids, lipoprotein, free fatty acids (FFA), sterols, hydrocarbons, and pigments [240]. Figure 4.4 shows the chemical structures of major representatives from microalgal lipids. This section explains qualitative and quantitative analyses of these lipids (the summary is shown in Fig. 4.5).

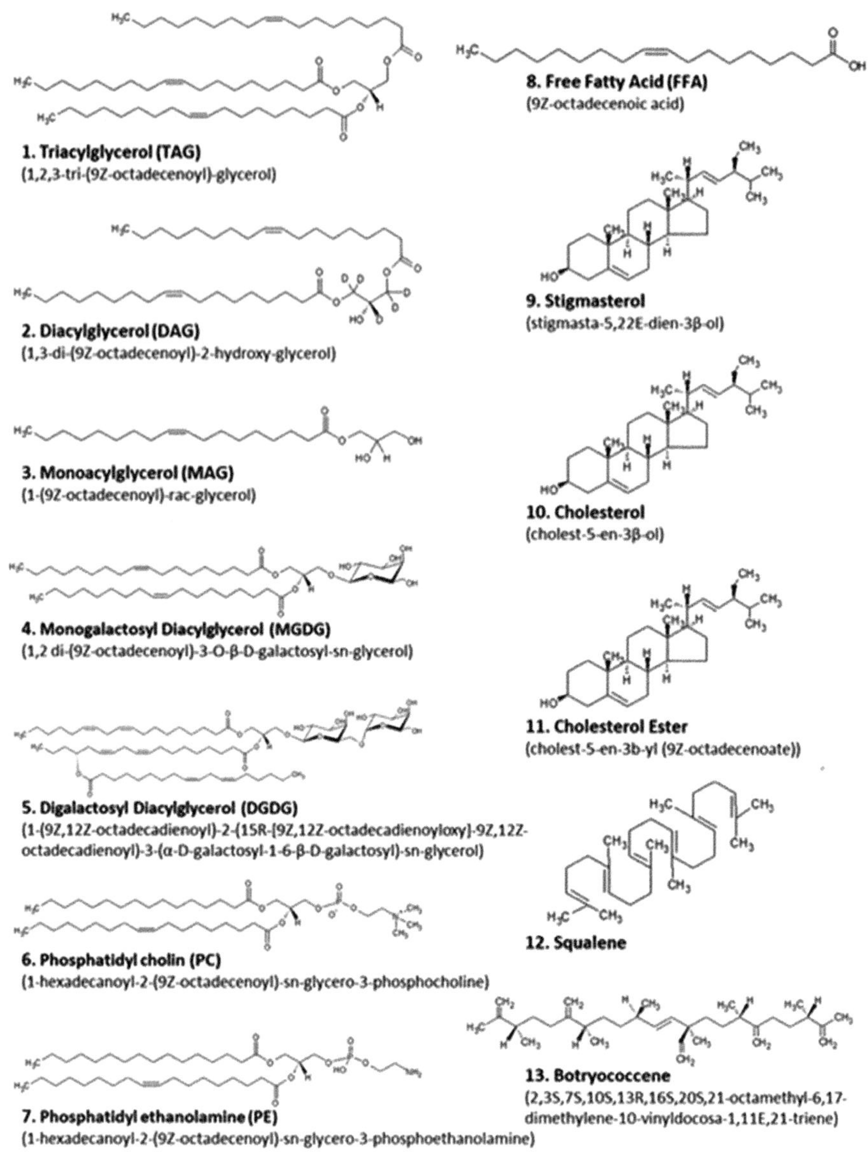

Fig. 4.4 Chemical structures of the major representatives from microalgae lipid

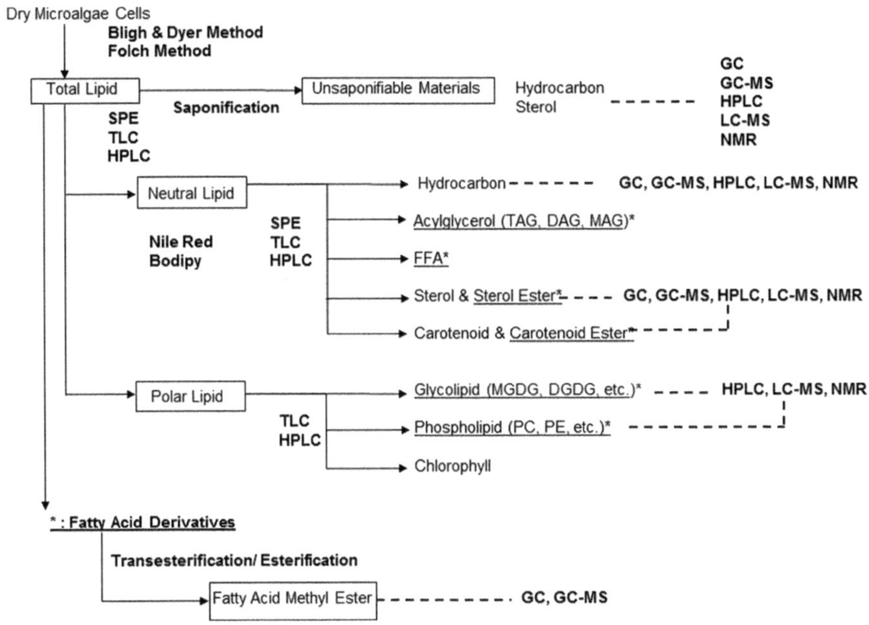

Fig. 4.5 The summary of lipid analysis steps and available analysis technics. SPE: Solid Phase Extraction, TLC: Thin-Layer Chromatography HPLC: High-Performance Liquid Chromatography, GC: Gas-Chromatography, MS: Mass Spectrometry, NMR: Nuclear Magnetic Resonance, TAG: Triacylglycerol, DAG; Diacylglycerol, MAG: Monoacylglycerol, FFA: Free Fatty AcidMGDG: MonogalactosylDiacyl-glycerol, DGDG: Digalactosyl Diacylglycerol, PC: Phosphatidylcholine, PE: Phosphatidylethanolamine

4.5.1 Preparation of Microalgal Samples

Microalgae are a highly diverse group of organisms, having a wide range of different cell walls, and various molecular components, intramolecular and intermolecular linkages, and overall structures [241, 242]. The cell wall characteristics and chemical compositions vary not only according to the microalgal species but also depending on the growth conditions and growth phase [241, 243]. These differences affect the barrier of intracellular lipid recovery and the ease of cell disruption.

The basic microalgal samples for lipid analysis are dried cells. To obtain such samples, microalgae are cultivated in medium and harvested by centrifugation. The obtained pellet is then washed with sterile water or medium, and the wet cells are obtained after centrifugation. Although the wet cells can be dried by either heating [244, 245] or lyophilization, the optimal approach is lyophilization [246]. After washing, the wet pellets are stored at −80 °C. These frozen pellets are then lyophilized in a lyophilizer. This removes water effectively without reducing the quality of the molecular components, compared with drying using heat. Lyophilization al-so increases the surface area of the sample, leading to better lipid extraction and easy rehydration. Lyophilized cells are also easily weighed, stored, and handled [246, 247]. Furthermore, lyophilized cells contain very little water, and there is no need to add lipase deactivators during lipid extraction.

Oil extraction from wet microalgae has been a subject of intense investigations [58, 236–240, 248–254]. These techniques are very important for practical and scalable processes for extracting lipids from microalgae because the drying of microalgae involves high energy costs. Some lipid analyses can use wet cells without any dehydrating, but there is a need to pay close attention to the cell disruption and the efficiency of oil extraction [236, 240].

4.5.2 Analysis of Total Lipid Content

Lipids are defined as being molecules that are not soluble in water but in nonpolar solvent instead, although the definition is relatively vague [19]. In this section, the entire oil content of microalgae is defined as total lipids. The total lipid content in microalgae is a very important value for considering the characteristics of microalgae and the possibility of practically using them as a lipid source.

The standard method of analyzing total lipid content is the method of Folch et al. [253], or that of Bligh and Dyer [254], involving the use of chloroform-methanol mixture as an extracting solvent [245, 246]. The co-solvent containing methanol as a polar solvent can extract not only neutral lipids like hydrocarbon and acylglycerols but also polar lipids like glycolipids and phospholipids. Furthermore, the polar solvent releases the lipids from their protein–lipid complexes, and the lipids subsequently dissolve in the nonpolar solvent [246, 252]. Modified versions of the methods of Bligh and Dyer [254] or Folch et al. [253] have also been proposed in many studies [19, 241, 245, 246, 248, 249, 255–260].

Ryckebosch et al. [246] explained that the method of Folch et al. [253] is the most widely used for the extraction of total lipids from microalgae. This method was originally optimized for the isolation and purification of total lipids from animal tissues. It uses chloroform/methanol at a 2:1 ratio for the extraction of lipids and water along with a small amount of NaCl to remove non-lipid substances from the extract. The method of Bligh and Dyer has also often been used for the extraction of total lipids from microalgae and was originally optimized for the extraction of phospholipids from fish muscle. It uses chloroform/methanol at a 1:2 ratio and followed by extraction with chloroform. Iverson et al. [254] compared the methods of Bligh and Dyer [254] to Folch et al. [253] for the determination of total lipids in marine fish tissue. They concluded that the unmodified method of Bligh and Dyer resulted in lower estimation of lipid content in samples containing > 2% lipid, and this underestimation increased significantly with increasing lipid content of the sample.

Ryckebosch et al. [246] showed that the total lipids could be extracted from lyophilized microalgae through them being immersed in a chloroform/methanol mixture, without any cell disruption. However, some samples need to undergo cell disruption before the oil extraction process. This is largely influenced by the cell wall characteristics, which depend on the microalgal species, the growth condition and phase, and the existence and intensity of stress factors [241]. When the level of oil recovery is low, cell disruption such as bead beating, microwave irradiation, sonication, and exposure to a homogenizer should be performed as pretreatment for oil extraction. The presence of lipids in microalgal cells can be evaluated by the fluorescence measurement of Nile red or BODIPY-stained cells [25, 244, 261]. These dyes can stain the lipids inside cells, especially neutral lipid droplets. After cells have been stained with Nile red or BODIPY, the fluorescence measurement or microscopic observation of stained microalgae is performed. In this context, one option is first to immerse the lyophilized microalgae in co-solvent and extracted oil and then to evaluate the oil-extracted cells by fluorescence analysis. When there is residual oil inside the cells, the microalgal samples should be disrupted as pre-treatment for oil extraction.

Chloroform and methanol can be used to extract oil from microalgae, but these organic solvents are associated with serious risks. For example, chloroform has been classified into Group B2 in risk assessment for carcinogenic effects, that is, probably carcinogenic to humans, by the US EPA [257]. Moreover, methanol is a flammable solvent and damages the visual system of animals [262]. These solvents should be carefully handled to avoid damaging the environment and human health. For this reason, other solvents have been investigated as less toxic substitutes for lipid extraction [25, 249, 257, 262–265]. However, it has been difficult to determine the best solvents because of different results being obtained depending on the microalgal species, growth conditions, and growth phase. The current best way of determining the total lipid amount is to follow the methods of Folch et al. [253] or Bligh and Dyer [254]. Finally, the following should be kept in mind: all extraction processes may not recover all lipids and may extract non-lipid components, such as chlorophyll, pigments, protein, and soluble carbohydrates, among the total lipids [19].

4.5.3 Separation and Analysis of Lipid Classes

Lipids are generally separated into neutral lipids (hydrocarbons, acylglycerols, and sterol esters) and polar lipids (glycolipids and phospholipids). Neutral lipids such as triacylglycerol (TAG) are present in the cytoplasm as a form of energy storage, and polar lipids such as glycolipids and phospholipids are present in chloroplast membranes and in plasma membranes [266]. It is very useful to determine which lipid classes and how much of each class are contained among lipids extracted from microalgae, in order to characterize the extracted lipids, and to evaluate a method for applying and purifying these lipids. Solid-phase extraction (SPE) [246, 266–268], thin-layer chromatography (TLC) [246, 257, 267, 269], high-performance liquid chromatography (HPLC) [270–272], and nuclear magnetic resonance (NMR) spectroscopy [272] are well-known methods for isolating lipid classes, such as neutral lipids, glycolipids, and phospholipids, from among extracted lipids.

SPE is a simple and effective method for determining lipid class composition [246, 266–268]. This method has the advantage that the amount of each lipid class can be determined gravimetrically and yielded as a fraction for further detailed analysis. Guckert et al. [266] reported an isolation method using simple silica column chromatography for lipid class separation. In this case, the total lipids were applied to a silica column, and elution with chloroform yielded the neutral lipids, acetone yielded the glycolipids, and methanol yielded the phospholipids. An easier SPE method involves the use of a commercial cartridge. The major SPE cartridge is Waters SepPakTM of silica gel or Alltech Extract-Clean column for the separation of lipid classes [246, 267–269].

TLC is also a typical method for lipid class separation [246, 257, 269]. Yao et al. [269] reported that a spot of total lipids was applied on a silica TLC plate and developed by a mixed solvent of hexane/diethyl ether/acetic acid. This was separated into TAG and FFA as neutral lipids, and polar lipids including, sterols, and chlorophyll derivatives. The separation could be visualized by spraying with staining reagents or by using ultraviolet light. The separated spots were scraped off the plate and extracted in solvent.

The major usage of TLC for lipid separation is to analyze the profile of neutral- or polar-lipid fractions [246, 257, 269]. After separation of lipid classes by the SPE method, each fraction is applied to TLC for profiling of the lipid composition in more detail. TLC for neutral lipids can separate TAG, diacylglycerol (DAG), monoacylglycerol (MAG), FFA, carotenoids, and sterols. TLC for polar lipids can separate monogalactosyl diacylglycerol (MGDG) and digalactosyl diacylglycerol (DGDG), among others, as glycolipids, and phosphatidylcholine (PC) and phosphatidylethanolamine (PE), among others, as phospholipids.

HPLC with evaporative light-scattering detectors (ELSD) can identify major lipid classes in total lipids [270–272]. The HPLC method can be fully automated and has higher separation capacity. Furthermore, each fraction of lipid classes can be easily collected for further identification and analysis of molecular species. Jones et al. [270] showed that normal-phase HPLC with ELSD can identify and quantify all major lipid classes in total lipid extracts. The used column was poly-vinyl alcohol bonded on a silica stationary phase, such as YMC Pack PVA-Sil-NP. This method covers the lipid classes of nonpolar groups (hydrocarbons, sterol esters, and wax esters), sterols, chlorophyll, acylglycerols (TAG, DAG, and MAG), FFA, glycolipids (MGDG and DGDG), and phospholipids (PC, PE, etc.) in a single chromatographic run.

NMR spectroscopy measures the interaction of nuclear spins under a powerful magnetic field and can identify the molecular structure of materials. Nuzzo et al.

[273] showed that proton NMR (1H NMR) spectroscopy using the electronic reference to access in vivo concentration (ERETIC) method could identify and quantify all lipid classes (TAG, PL, GL, and FFA). Total lipids from microalgae were measured by ^{1}H NMR spectroscopy, and these spectra were compared with those of standard compounds under the same conditions. Some peaks showed pronounced overlaps, such as aliphatic and olefinic protons, but each class of molecules showed diagnostic signals that could be selected as NMR markers of each lipid class. This method allows assessment of the total lipid content and also of the degree of saturation.

4.5.4 Analysis of Fatty Acid Composition and Content

For the commercial application of microalgae, a wide range of products are available, such as biofuels, feeds, fertilizer, chemical feedstock, nutrient ingredients, and raw materials for cosmetics and drugs [58, 239]. It is very important to determine the fatty acid composition and content of microalgae when considering their applications.

The analysis of fatty acids can be conducted on the lipids extracted from microalgae and lipids separated into lipid classes. It is very difficult to analyze fatty acids in crude extracted lipids without any pretreatment using analytical equipment directly. The standard method is first to convert lipids to fatty acid methyl esters (FAME) through transesterification and then to analyze FAME by gas-chromatography (GC) [84, 246, 257, 259–261, 263, 267, 269, 274, 275].

Carrapiso et al. [275] reported the principles of transesterification by acidic or basic catalysts. In an acidic or basic methanol solution, the fatty acid derivatives (TAG and phospholipids) are changed to FAME. Major catalysts are hydrochloric acid (HCl) [257, 260], sulfuric acid (H_2SO_4) [246, 259, 260, 268, 269], and boron trifluoride (BF_3) [260, 273] as acids, and sodium hydroxide (NaOH), potassium hydroxide (KOH), and sodium methoxide ($NaOCH_3$) as bases.

Acidic catalysts are most widely used for microalgal fatty acid analysis. They have strong esterification power, reacting with not only TAG and phospholipids but also FFA and *N*-acyl lipids (sphingolipids). Furthermore, there is less water interference than for basic catalysts. The method of Christie [267] referenced by many papers [246, 259, 260, 268, 269, 274] uses H_2SO_4 as a catalyst. H_2SO_4 has a longer life of catalytic activity and involves safer preparation than HCl. In the case of BF_3, it has higher esterifying power than H_2SO_4 and can be used to analyze a small amount of sample [261]. However, BF_3 cannot be used for water-containing samples because it is rapidly destroyed by water [275, 276].

Basic catalysts have the advantage of having a high reaction speed and relatively mild heating conditions, compared with acidic catalysts [275, 277, 278]. Furthermore, for lipids containing acid-labile fatty acids (cyclopropane rings, epoxyl groups, or conjugated unsaturated molecules), it is better for the analysis to be performed under basic than acidic conditions. However, basic catalysts cannot react with FFA and sphingolipids, and the saponification reaction occurs in the presence of water.

A method combining the use of both basic and acidic catalysts is the American Oil Chemists' Society method [279]. This is now recognized as a standard method for oil and lipids, which means that it is not specific to microalgal oil. This method involves first transesterification (or saponification) by NaOH in methanol, followed by esterification by BF_3 in methanol [261, 279, 280]. Yi et al. [281] applied KOH in methanol to fish oil, which contains high levels of docosahexaenoic acid (DHA) and eicosapentaenoic acid (EPA), prior to BF_3 in methanol. *Schizochytrium* sp. can produce higher levels of multiple polyunsaturated fatty acids such as DHA, EPA, and docosapentaenoic acid. For the analysis of fatty acids produced by *Schizochytrium* sp., KOH and BF_3 were used in methanol [282, 283].

The FAME obtained by transesterification/esterification can be analyzed by GC-flame ionization detector (FID) or GC-MS [84, 246, 257, 259–261, 263, 267, 269, 274, 275]. Before sample analysis, each fatty acid should be identified by GC or GC-MS by comparing the retention times with those of FAME standards. In the case of the quantification of fatty acids in lipids, an internal standard should be added in samples before the transesterification/esterification reaction. The FFA, methyl ester or TAG of pentadecanoic acid (C15:0), heptadecanoic acid (C17:0), and tricosanoic acid (C23:0) were used as internal standards [259–261, 268, 274, 279, 280]. The internal standard should be a saturated fatty acid that is not included among the sample lipids.

In situ (direct) transesterification is a method of applying methylation agent directly to the biomass and reducing the extraction steps [260, 261, 284–289]. To dry microalgae, catalyst solution with methanol is added, followed by incubation with heating. Excluding the lipid extraction process leads to less organic solvent and reduces the process time. Acidic catalysts such as BF_3 [260, 261, 287], HCl [260, 285, 287], and H_2SO_4 [286, 287, 289] are commonly used. Cavonius et al. [260] compared 11 methods to analyze fatty acids in three different microalgae. These were four different conventional methods of extraction prior to transesterification, two types of two-step transesterification (saponification followed by esterification), and five different in situ transesterifications. Their results showed that the conventional method resulted in underestimation of the fatty acid content compared with two-step and in situ transesterification. Regarding the catalyst, HCl and H_2SO_4 could recover the fattiest acids, whereas BF_3 recovered slightly less. VelasquezOrta et al. [288] performed a comparison between a basic catalyst (NaOH) and an acidic catalyst (H_2SO_4) for in situ transesterification of *Chlorella vulgaris* and concluded that the basic catalyst outperformed the acidic one, obtaining higher conversion at shorter reaction times. At any rate, in situ transesterification is very useful to obtain FAME for analyses of fatty acid composition and content. The obtained FAME can be analyzed by GC-FID or GC-MS.

Compared with GC methods, HPLC has the advantage in fatty acid analysis that higher-molecular-weight and non-volatile materials can be analyzed without raising the temperature to a high level. LC methods have been employed for fatty acid analysis of FFA, TAG, phospholipids, and others [290–292]. The fatty acid analysis of FFA using GC requires the derivatization of fatty acids (FAME) by thermal and chemical reactions and high-temperature operation for volatile FAME. The high temperature is associated with a risk of isomerization of saturated fatty acids [290]. On the other hand, LC can be applied to FFA analysis directly without any derivatization [292] or with a derivatization reaction at a lower temperature [290, 291]. LC-MS and LC-MS/MS have been employed for lipid analysis without additional separation and derivatization steps [293–295]. Samburova et al. [294] analyzed the quantification of algal TAG, and the characterization and identification of TAG structures using direct LC-MS and MS/MS analysis. The advantage of this method is that the lipids are injected into the LC instrument without any derivatization, and that the structure of individual TAG can be estimated. These techniques have also been used in lipidomics research.

4.5.5 Hydrocarbon Analysis

Some microalgal species can produce hydrocarbon oil. The green alga *Botryococcus braunii* is characterized by its ability to produce and accumulate a large amount of hydrocarbons, known as botryococcene [296–299]. This alga is subclassified into four chemical races (A, B, L, and S), according to the hydrocarbon structures [299, 300]. *B. braunii* Race A produces normal alkenes of an odd number of carbons in the range of C25–C31 with two to three carbon double bonds. Race B produces triterpenic hydrocarbons of C_nH_{2n-10} (n = 30–37). Race L produces tetraterpenic hydrocarbons of lycopadiene ($C_{40}H_{78}$) [301]. Race S comprises epoxy-n-alkane and saturated n-alkane chains with carbon numbers of 18 and 20 [300]. Another well-known alga producing hydrocarbons is *Aurantiochytrium* sp., which produces squalene [302–305]. Squalene is also a polyunsaturated triterpenic hydrocarbon ($C_{30}H_{60}$). These hydrocarbons can be used in the cosmetic industry as moisturizing agents [302] and are expected to be an alternative petroleum fuel and chemicals, among others.

Hydrocarbons are classified as neutral lipids. The hydrocarbons inside microalgal cells can be stained using Nile red or BODIPY. In the case of *B. braunii*, hydrocarbons are mainly accumulated in the extracellular space and stained by Nile red [296, 306]. This fluorescence measurement or microscopic observation of stained microalgae can be used to estimate the presence of hydrocarbons (Fig. 4.6).

In general, the level of hydrocarbons in lipid-extracted microalgae is low because major components in lipids are those with fatty acids, such as TAG, GL, and PL. Hydrocarbons should be analyzed using lipids after pretreatment to remove the fatty acids. There are three major methods for purifying hydrocarbons: (1) TLC, (2) column chromatography, and (3) saponification.

TLC is a typical method for the separation of lipid classes. TLC for neutral lipids can separate TAG, DAG), MAG, FFA, carotenoids, and sterols. Hydrocarbons can also be separated by TLC [307–309]. Nakazawa et al. [307] used TLC for the

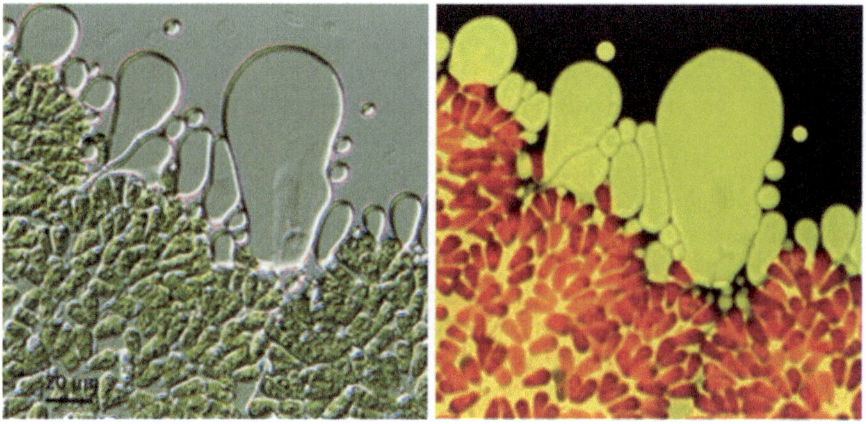

Fig. 4.6 *Botryococcus braunii*. Left: photomicrograph of colony; Right: fluorescence micrograph of Nile Red stained cell (yellow fluorescence indicates oil)

screening of thraustochytrid strains that produce squalene. The lipids extracted from lyophilized thraustochytrid cells by Folch's method [253] were subjected to TLC and hydrocarbon was separated. The existence of hydrocarbons could be estimated by the clarity and strength of the hydrocarbon spot in TLC. The hydrocarbon spot on TLC can also be picked up and used for further detailed analysis. Cavanagh et al. [309] used TLC-FID for the analysis of biodegradation hydrocarbons. The lipids were developed by TLC, and separated hydrocarbons were detected by a FID, which is well-known as a GC detector. This method involves not only qualitative analysis but also rough quantitative determination. TLC is a simple method and can be completed in a short time. When the analyses of many samples are needed in screening and there is a need to obtain rough results in a short time, TLC is a very good option.

SPE is also a simple and effective method for analyzing lipid class composition. In this approach, neutral lipids or lipids extracted using hexane are applied to a silica column, and elution with hexane yields hydrocarbons. This method can be performed gravimetrically and yields a fraction for further detailed analysis. Many studies of hydrocarbons produced by *Botryococcus* sp. and *Aurantiochytrium* sp. have been conducted using the SPE method, with analyses of the quantity, identity, and structure of separated hydrocarbons by GC, GC-MS, HPLC, LC-MS, and NMR [299, 302–305, 310–315].

Saponification can be used to make soap through the reaction of fatty acids with a base. Fatty acids in lipid extracted from microalgae are changed to soap by saponification and can be removed by washing because soap more easily dissolves in water. Meanwhile, hydrocarbons are not changed by saponification and can be recovered in organic solvents like hexane. Many studies of hydrocarbons produced by *Botryococcus* sp. and *Aurantiochytrium* sp. have been conducted using the saponification method [298, 305, 316, 317]. Saponification is useful for removing fatty acid materials and recovering unsaponifiable matter such as hydrocarbons and sterols.

The purified hydrocarbons were subjected to quantification, identification, and structural determination by GC, GC-MS, HPLC, LC-MS, and NMR. When the hydrocarbons extracted from microalgae were unknown materials, GC-MS or LC-MS was useful for the identification of each hydrocarbon. More information on unknown materials such as their structure can be obtained by NMR spectroscopy after further purification of hydrocarbon. In the case of botryococcene, there are no standard reagents on the market. Furthermore, hydrocarbons produced by *Botryococcus sp.* have many types of molecular weight and structure. Identification of these hydrocarbons has been conducted not only by analysis of purified hydrocarbons using GC-MS and NMR spectroscopy but also by GC-MS or NMR spectroscopy after ozonation of hydrocarbon [315, 318] or hydrogenation [313, 317]. Quantification of hydrocarbons can be performed using an external or internal standard method. The added internal standards were reported to be triacontane [298] or squalene [310] for botryococcene by *Botryococcus* sp., and squalene [319–321], heptadecanyl stearate [322], and 5α-cholestane [269, 323] for squalene, by GC or GC-MS analysis. The internal standard for HPLC or LC-MS analysis [324] was octadecylbenzene for squalene [316].

4.5.6 Phytosterol Analysis

Phytosterols are mainly found in the cellular membranes of microalgae and one of the most interesting compounds for plant-based ingredients [325]. Phytosterols have many health-promoting effects in humans, due to their ability to lower cholesterol levels, their antioxidant activity, and their reduction of inflammation and cancer risk [326]. As the phytosterol content in microalgae is equal to or higher than those in all plant oils extracted, microalgae have high potential to be a source of phytosterols.

There are more than 100 different types of phytosterols. Phytosterols are present in lipid-extracted microalgae. The analysis of phytosterols can be performed by the same methods as for hydrocarbons. Such analyses should be applied to lipids after their pretreatment to remove fatty acid materials. Purification of phytosterols is conducted by TLC, column chromatography, and saponification. The purified phytosterols are subjected to quantification, identification, and structural determination by GC, GC-MS, HPLC, LC-MS, and NMR spectroscopy. The GC analysis requires chemical derivatization, with trimethylsilyl ether (TMS) or acetate derivatives, prior to analysis [325, 327, 328]. Quantification of phytosterols can be performed using an external or internal standard method. Internal standards were reported to include 5α-cholestane [327], cholesterol [329], and 5β-cholestan-3α-ol [330] for GC analysis and 6-ketocholestanol [331], 1-hexacosanol [332], and d6-cholesterol [333] for LC analysis.

4.6 Methods for Pigments Determination in Microalgae

Organic pigments are widespread molecules in the whole biodiversity and particularly in microalgae diversity [334] since they play crucial roles in photosynthesis and photoprotection. Pigments from microalgae belong mostly to carotenoids (carotenes and xanthophylles), chlorophylls, and phycobiliproteins. However, the full definition of pigments includes various molecule scaffolds such as anthocyanosides, betalains, some alkaloids (e.g. scytonemin), aminoacids (e.g. mycosporines-like, eumelanin), azaphilones, polyphenols (e.g. floridorubin, quercetin), anthraquinones (e.g. hypericin), or chlorines (e.g. purpurin-18). Since phycobiliproteins have already been presented previously in this chapter, we will focus essentially on carotenoids and chlorophylls.

Since ancient times, civilizations have used pigment like Tyrian purple from gastropod molluscs or other organisms for clothes dying. Then, the link between food and health has been documented since Antiquity (i.e. Hippocrates associated the food to health and diseases) [335]. Moreover, Aztecs were one of the civilizations to use *Spirulina* as food source. Apart from the fact these Cyanobacteria are protein-rich source, it contains pigments known nowadays with health benefits. Indeed, these last decades, numerous scientific studies have shown interesting properties from pigments through notably antioxidant, anti-inflammatory, anti-AMD, antitumoral, antiangiogenic, antidiabetic, antiobesity, antiphotoaging, and neuroprotective activities. Chlorophylls are more recommended notably to stimulate liver function recovery and increase bile secretion [336]. Thus, today, pigments found applications in chemical, pharmaceutical, poultry, cosmetics, cosmeceutical, functional food, nutraceutical and aquaculture industries [337–339]. Today, the chemical synthesis market covers about 90% of the main carotenoid needs (e.g. asthaxanthin, β-caroten). However, natural sourcing benefits from a positive image since Western countries consumers are more and more sensitive to the impact of chemicals onto their health. The all-trans isomer of β-caroten is still chemically synthetized but the 9-cis form is produced only from natural sources, the natural form displaying better antioxidant capacity than the synthetized form. Although the natural pigment sources still compete with synthetic dyes, with the high development of agrifood industry, natural pigments are more and more used for coloring of food preparations and for their antioxidant properties. Beta-carotene market value has been estimated at US$261 million in 2014. With an annual growth rate of 3%, it is expected to reach US$334 million by 2018 [340]. This is favored by stringent government regulations regarding environmental and human health hazards (Global Market Insights Inc. [341]).

Although yeast, fungi, or bacteria allows the production of carotenoids [342–344] microalgae biomass remains one of the best way to produce industrial amounts of natural pigments [345]. Some phytoplankton strains are easily cultivable in large open ponds. Countries like Israel, Australia, USA, China, or India became pioneers/leaders on this market [346]. Pigment-rich microalgae can be also produced in other ways like photobioreactors systems (PBRs), chemostat, turbidostat, flat-panel (i.e.

green wall panel), disposable polymer bags, open tank or raceway, tubular glass PBR etc. (batch, continuous, or semi-continuous mode).

Whatever the mode of production, microalgae biomass collected has to be extracted and more or less purified according to the customer needs (as B2B or B2C). Pigments of interest, targeted applications, and production scale drive strongly the strategy to choose in term of extraction and purification. Meanwhile, characterization of targeted pigments has to be harnessed properly while acknowledging the limits of available detection and quantification techniques. Adapted extraction technique to the biomass is crucial since there is a strong relation between the amount detected and the capacity to extract targeted analytes. One of the first questions for choosing a technique is: Is my metabolite of interest sensitive to desiccation, heat, oxidation, or light? For pigments like carotenoids, chlorophylls and phycobiliproteins, the answer is yes most of the time. Various techniques are available from laboratory scale to industrial scale, but they are more or less efficient for large biomass and to preserve sensitive pigments. Moreover, at industrial scale, economic dimension has to be taken into account to make the whole process viable.

4.6.1 Extraction

4.6.1.1 Centrifugal Partition Extraction

Also called CPC for centrifugal partition chromatography, it is one of the various techniques constituent of countercurrent chromatography (CCC). This is a liquid-liquid partitioning chromatography that uses no solid phase as the liquid stationary phase is retained in a series of channels connected by ducts, engraved on disks, which rotate around a single axis. According to the centrifugational or centripetal direction of the mobile phase, there is a descending or ascending mode to work with this equipment. Co-current method can be applied as well. This technology has been used for the continuous extraction of numerous natural products including pigments [347]. CPC is particularly adapted to extraction and purification with level up to 99.9% (Kromaton data). Solvent choice, flow rate and rotational speed can be tuned to optimize the extraction of valuable metabolites.

4.6.1.2 Supercritical CO_2 Extraction

This technique (SC-CO_2) can be considered as a green strategy to reach microalgae metabolites since CO_2 gas stream is recyclable. However, prior drying or lyophilization step is required to implement this kind of extraction. There is a large panel of equipment available from laboratory scale (e.g. lyophylizators, centrifugal evaporators) to industrial scale (e.g. vacuum drying, spray drying, extrusion porosification technology). However, dewatering pre-treatment add a substantial cost to the entire process, which is not suitable for low valuable products. In case of pigments with high added value for nutraceutical, cosmeticeutical, or pharmaceutical application, the additional cost can be acceptable. In this particular case, SC-CO_2 stays promising in term of carotenoid extraction. Thus, numerous studies have demonstrated these last years that SC-CO_2 is quite selective to extract carotenoids like β-carotene, lutein, astaxanthin, or zeaxanthin over chlorophylls from various microalgae species [348–351]. The use of ethanol as entrainer (co-solvent) is particularly efficient. However, vegetable oils (e.g. soybean or olive oil) can be used as co-solvent in order to avoid the subsequent separation step of the co-solvent, i.e. the oil enriched with pigments can be sold straight away [352].

4.6.1.3 Milking

This technique has been developed to extract β-carotene from *Dunaliella salina* living cells in a two phase bioreactor [353, 354]. One phase is aqueous, where biomass is growing. The second phase is an organic solvent with value of Log Poctanol, which denotes hydrophobicity of the compound of interest. Log Poctanol must be >6 in order to be compatible with the species. Dodecane, tetradecane, or hexadecane are generally used for their potential to extract selectively carotenoids. This technique has the disadvantages to be not environmentally friendly and difficult to scale up.

4.6.1.4 Accelerated Solvent Extraction

Also known as pressurized liquid extraction (PLE), this technique is based on high pressure that forces the solvent into the matrix, whereas high temperature promotes high analyte solubility, decreases the viscosity and decreases the surface tension of the solvents. This technique has the advantage to use less solvent and being less time-consuming over the simple maceration. This technique can be coupled with the use of GRAS solvent (Generally Recognized As Safe) such as ethanol and limonene. A pre-treatment prior extraction like freezing-thawing can be performed to improve the extraction yield [355]. Extraction temperature, solvent composition and extraction time are parameters to watch to optimize the extraction process. However, due to the relative high temperature of this technique, it is not recommended to use for sensitive pigments.

4.6.1.5 High-Speed Homogenization

This technique causes cavitation and shear stress. Our own experience showed that this technique is not applicable for all types of microalgae since the microalgae cell wall, frustule or theca can wear blades prematurely (unpublished data). Moreover, we doubt the real efficiency of this technique to treat large biomass and the energy cost would be high to treat microalgae at large scale. If for niche application this technique has to be explored, blade design and material, speed, time, microalgae amount (or flow), growth phase and conditions, and microalgae type have to be studied. One other disadvantage of this technique is the thermal heating, which occurs quickly, and makes it not compatible with extraction of sensitive pigments.

4.6.1.6 Microwave Assisted Extraction (MAE)

This extraction technology is based on microwave irradiation and can be declined in vacuum microwaves-assisted extraction (VMAE). While the first one is simple and economical, the second one can be onerous at industrial scale. This technique has the advantage to reduce solvent consumption and extraction times. Over the fact that the heating by irradiation is homogenous, temperature of the extraction medium can be regulated. That makes this methodology interesting for pigment protection against chemical transformations. This technique has been applied successfully on the unfrustulated Chlorophyte *Dunaliella tertiolecta* and on the frustulated diatom *Cylindrotheca closterium* [356]. MAE seems particularly efficient when a mechanical resistance such as frustule of diatoms limits the solvent access into the cell. The optimization of this technique requires the control of agitation, dry weight, growth phase and conditions, microalgae type, power of microwave, solvent volume and time (intermittency ratio).

4.6.1.7 Sonication

This technique so-called ultrasound-assisted extraction (UAE) based on acoustic cavitation leading to cell rupture is not suitable for microalgae with thick wall and/or surrounded by polysaccharides like *Porphyridium purpureum* [357]. One disadvantage of this technique is that a hot spot at the top of the probe can reach a temperature estimated about 5000 °C due to the rapid adiabatic compression of gases and vapors within bubbles or cavities. Another one is that free radical reactions leading to metabolite oxidation can occur if a small amount of water is present in the extraction batch [358]. In biorefinery, this technique can be used as pretreatment to reach lipids notably but we do not consider this technique for pigment extraction regarding the whole diversity of microalgae. However, in case of a particular interest in this methodology, in order to improve its efficacy, cycle number and time, dry weight, growth phase and conditions, microalgae type, and power of ultrasound have to be tested.

4.6.1.8 Pulsed Electric Field

The goal is to apply a medium or high current field on the treatment chamber (0.3–1.5 kV/cm), which cause an electroporation allowing a selective extraction or an electro-plasmolysis for an intense extraction. Formation and propagation of streamer and arc happens followed by the formation of the vapor cavities and a shock wave [359]. Treatment time is around 1 s per batch. Thus, there is no issue in term of energy consumption. Some essays have shown that this technique is better efficient on big cells rather than little particles like bacteria and viruses. In conclusion, this technology is not universal but species dependent. It has to be tested for the species and pigments targeted. In the framework of optimization of this process for microalgae biomass, conductivity (electrolyte concentration), current, dry weight, growth phase and conditions, microalgae type, oscillation, and time should be studied. With this technique, there is a risk to heat quickly the treated biomass. In some conditions, this technology can be used as pretreatment of another extraction technique.

4.6.1.9 Enzyme Assisted Extraction

Phycobiliproteins are hydrosoluble molecules and can be extracted from microalgae by bead milling without a preliminary drying step. Due to the presence of large amounts of anionic cell-wall polysaccharides, this extraction can even be improved by enzymatic hydrolysis. While proteases are generally quite efficient to extract bioactive components from algae biomass, in the case of phycobiliprotein extraction, cellulase, xylanase or β-glucanase are preferable to improve extraction yield [360]. Inactivation of enzymes at the end of the process is crucial. Most of the time, temperature and pH are two parameters allowing inactivation of enzymes. For in-stance, in the case of alcalase, 30 min at 50 °C, pH 4, or 10 min at 85 °C, pH 8 are generally recommended (Univar data). If immobilized enzymes are used (e.g. lipases), the advantage is the possibility to reuse the enzymes up to 100 times in theory. Except for phycobiliprotein pigments, there is a lack in literature for carotenoids and chlorophylls regarding the use of enzymes as pre-treatment to improve their extraction yields from microalgae. Due to its biological specificity, mild operating conditions, low energy requirements, low capital investment, there is a great potential for biomass valorization. Moreover, this is a way to explore further since lysozyme has shown interesting results in extracting chlorophyll-*a*, chlorophyll-*b* and total carotenoids [361]. In a context of exploration of effective parameters for such a process, agitation, dry weight, enzyme concentration, enzyme type (i.e. currently, no enzyme is particularly dedicated to microalgae biomass), growth phase and conditions, microalgae type, oxygen level, type and amount of buffer, temperature, pressure, and time have to be taken into account in the design of experiment (DOE). Today, the main limitation in using enzymes in microalgae extraction may be the high cost to treat large volumes.

4.6.1.10 Bead Milling

This is one of the most promising techniques in term of pigment extraction yield both at laboratory scale and industry scale. It generates mechanical compaction and shear stress for cells. This technique has shown excellent results with a diatom and also especially with *Porphyridium purpureum* which is known to be quite resistant to disruption [357]. Laboratory scale equipment is based on agitation of a chamber or tubes filled with beads (glass, steel, ceramic, or zirconium material), while industrial scale equipment is based on a mixing chamber filled with beads and biomass to extract.

For this process, agitation disk design, speed, bead filling, size, material, microalgae amount (or concentration), feed rate, growth phase and conditions, microalgae type, time, and cooling are parameters to optimize. Our background experience has shown that it is preferable to work on a wet biomass rather than a lyophilized biomass to allow a good extraction yield. Associated with the proper solvent or buffer targeting the pigments of interest, this technique is quite suitable for most pigment projects. For microalgae biomass, a couple of studies support the idea that mechanical grinding is often the most effective technique to recover metabolites efficiently, including pigments [47, 357]. However, energy requirement is a very important parameter when large biomass is treated. It seems that mechanical disruption methods are considered highly energy inefficient [362]. Thus, bead milling should be preferred for high value compounds.

4.6.1.11 Soaking

Solvents are still widely used to extract pigments as additives for coloring food and beverages from other source than microalgae (carrot, turmeric, spinach/nettle, alphalpha) [363]. Polar solvents (e.g. acetone, ethanol, ethyl acetate) are appropriate for extraction of polar carotenoids while hexane is frequently selected to extract carotenes. Moreover, solvents like DMSO or a solvent mix (hexane/acetone/ethyl alcohol) have been used to extract chlorophylls and carotenoids like astaxanthin in order to evaluate the influence of media and environmental factor allowing their production [364–366]. Soaking in solvent can be used for microalgae with a thin wall (e.g. most of Chlorophyceae, Trebouxiophyceae, Prasinophyceae, Mamiellophyceae, and Chlorodendrophyceae). However, generally speaking, applied alone, soaking is not suitable to extract pigments efficiently since pigments can be deeply buried and bound within the ultrastructure, which is itself protected sometimes by strong walls. Most of solvents are produced from petrochemistry. Since it is not economically viable to use high purity grade solvents (e.g. HLPC/MS grades) to extract large biomass, the risk is to concentrate heavy metals with compounds of interest. One of the rules in eco-extraction is to favor the use of alternate solvents from sustainable agro-resources (bio-based), which are renewable and biodegradable. Ethanol and ethyl acetate are ones of the most green and sustainable solvents. It still used to extract active components from biomass but it is more and more avoided. Moreover, in the current framework of the European regulation (Novel Food), no more than 10 ppm of solvent is allowed in the extracts. In brief, solvent removal is crucial. Instant controlled pressure drop technique is one of the strategies available in that case.

4.6.1.12 Instant Controlled Pressure Drop

This green technology allows both drying (swell-drying) in order to obtain powder with good functional properties, and the extraction of metabolites like pigments. This process can be applied after mechanical dewatering (e.g. pressing, centrifugation, filtration). There is no particular need to reach a high level of drying. According to the targeted metabolites, organic solvent or water can be used [367]. One of the advantages of this technology is the preservation of sensitive molecules. Currently there is few literatures about pigment extraction from microalgae [339]. However, this process has been applied successfully on microalgae biomass for cosmetics industry. In order to make this process efficient, cycle number, biomass, time of treatment, number of cycles, growth phase and conditions, microalgae type, and pressure have to be watched. More information is available at www.abcar-dic.com.

4.6.1.13 Chemical Treatment

This strategy is more used as pretreatment rather than a standalone extraction technique. For instance, to improve extraction of astaxanthin from thick-walled cysts of *Haematococcus* cells, prior extraction with acetone, cells can be treated with organic and mineral acids (e.g. hydrochloric acid) at 70 °C (1 h). Treatment time, temperature, and concentration of the acid are critical factors for an optimized extractability [368]. However, acidic conditions are not suit-able if chlorophylls are targeted since they are instable at pH under 5. Alkaline hydrolysis (0.1 M NaOH) have been also tested previously [355].

4.6.1.14 Others

At laboratory scale, other techniques may be used for valuable compound extraction. Some are not desirable and/or very used anymore for carotenoids and chlorophylls. This is the case of heat reflux extraction (too degradative), liquid-liquid extraction in separating funnel. Solid phase extraction (SPE) is not very used as a standalone extraction for pigments. However, SPE can be used as purification technique or as pretreatment before analysis. QuEChERS method (Quick, Easy, Cheap, Efficient, Rugged and Safe) is derived from SPE originally dedicated to the detection of pesticides [369]. An adaptation of this technique may be an alternative at laboratory scale in the future. Then, some industrials claim that spray drying (originally dedicated for drying) may be used as a disruption method applicable to microalgae biomass. This technique may be used to extract broadly main pigments. Some authors have reported a pairing of several techniques. For instance, astaxanthin extraction from *Chlorococcum sp*. has been performed successfully (7.09 mg g^{-1} DW) coupling solvent system with methanol/dichloromethane (75:25), French pressure cell (110 MPa), and a saponification [370]. A two-stage extraction of free astaxanthin with dodecane and subsequent NaOH-added methanol (0.02 M) has been developed. This consists of two extraction units (i) a dodecane unit for astaxanthin mixture and (ii) a methanol unit for free astaxanthin [371]. This methodology closely related to milking has the disadvantage to exclude reuse of red cyst cells for continuous astaxanthin production and not be a green extraction process.

4.6.1.15 Biorefinery

Techniques presented above give some hints to overcome the extraction of valuable pigments from microalgae biomass (Fig. 4.7). However, extraction scale-up stays not trivial. A good review regarding cell disruption for microalgae biorefineries has been published by Günerken et al. [372] including a classification of the cell disruption methods. Although the techniques are not focused on pigments, this may be useful for projects seeking to valorize pigments from microalgae. Outsourcing of compounds of interest extraction can be cost-effective, especially for early stage of

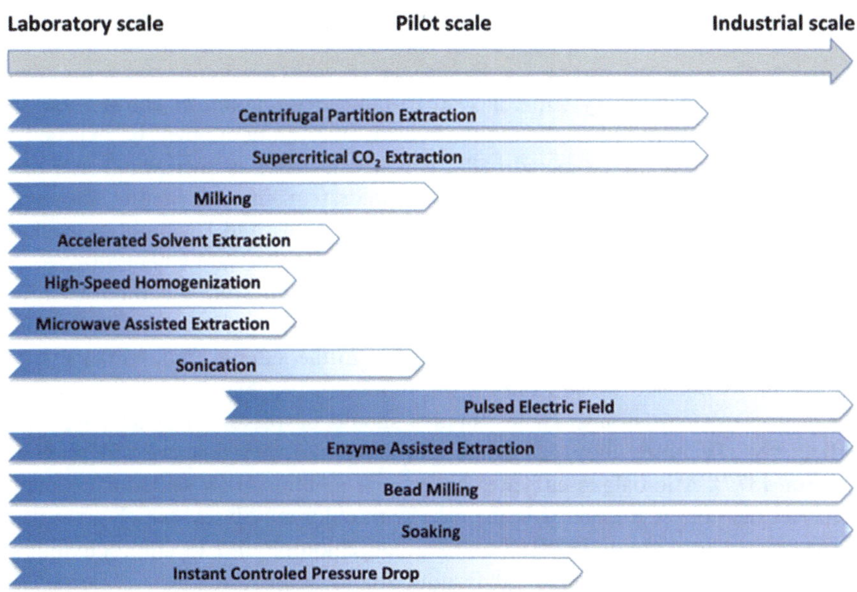

Fig. 4.7 Extraction techniques from laboratory to industrial scale

projects, since a couple of companies are dedicated to assist and co-develop flexible and qualitative solutions for drying, extraction, and separation/purification steps.

The technical and economical challenges in co-extracting marine active ingredients are still ongoing. Carotenoids represent only 0.1–0.2% of dry weight on average, and up to 14% of dry weight for β-carotene in some species like *Dunaliella salina* under stress conditions. Chlorophylls represent around 0.5–1% of dry weight, and phycobiliproteins up to 8% of dry weight [373]. These data have to be considered in the targeted applications and for the business model. Cracking strategy aims to valorize a maximum from the produced biomass, and not only a short part of it. Lipids can represent up to the half of the dry weight of microalgae [232] and their polarity is similar to those of organo-soluble pigments. The specificity to reach pigments without lipids is low. For applications, which do not need a high purity grade, development of products benefiting from both, some particular lipids (e.g. DHA—Docosahexaenoic acid), and pigments is a good and sustainable strategy. In the framework of a full exploitation of microalgae components, starch and proteins can be interesting beside oil extraction for bioethanol and biodiesel production [374]. Algae are chest of other various valuable metabolites like phenolic compounds, phytoene/terpenoids, phytols, sterols, photoprotective compounds, phytohormones, halogenated compounds, cyanotoxines, phycocolloids, polyhydroxyalkanoates, which may find applications in medicine, in nutraceutics, in cosmeceuticals, in functional food industry, aquaculture, or agriculture [375, 376].

4.6.2 Pigment Analysis

Disruption of a cellular system can lead to rapid degradation of carotenoids and chlorophylls, even at low temperature. Indeed, free pigments may be exposed to organic acids from the extracted cells, which can cause isomerization and rearrangement of 5,6-epoxy- to 5,8-epoxycarotenoids (e.g. in violaxanthin and neoxanthin). Free pigments are more sensitive to oxidations and UV light as well. In order to minimize pigment degradation, it can be advised to add a neutralizer to the sample (calcium carbonate, sodium bicarbonate, or magnesium bicarbonate during the extraction. An antioxidant such as tertbutylhydroquinone, butylhydroxytoluene, pyrogallol, or ascorbyl palmitate can be added to the sample (around 0.1% w/v). Azote can be flushed as well during the process, from extraction to the sample storage to limit oxidations. Then, samples should be protected against light and particularly UV since they can cause transcis-photoisomerization and photodestruction [377]. Microalgae enzymes have to be mentioned as well since some of them can degrade pigments. It is the case with chlorophyllases, which are particularly present in diatoms [378]. They degrade chlorophyll-*a* to chlorophyllide-*a*, which modify the perception of the culture/biomass growth phase (chlorophyll-*a* ratio). A serious issue is that chlorophyllase can be activated by harvesting techniques before the extraction step. Moreover, the latter can amplify the phenomenon except if acetone is used, which inhibits the enzymes. Thus, pigment analyzes should be performed as early as possible or samples should be stored at −80 °C for a short period before analyzes.

4.6.2.1 Spectrophotometric Analysis

Pigment analysis science benefits from oceanography research, which aims to study phytoplankton pigments for the understanding of population dynamic and biogeography. Several techniques have been developed these last decades. With the improvement of the equipment accuracy, high performance liquid chromatography (HPLC) methods hyphenated with UV-VIS detector are the gold standard to detect and quantify known carotenoids and chlorophylls. Amongst all the existing methods, some are kind of reference in oceanography. Eleven international laboratories specialized in the determination of marine pigment concentrations using HPLC (e.g. LOV, CSIRO, DHI, NASA Goddard Space Flight Center) are regularly intercompared using in situ samples within the SeaHARRE studies (SeaWiFS HPLC Analysis Round-Robin Experiment) [379]. These laboratories use the following validated methods:

- Gieskes and Kraay [380]
- Wright et al. [381]
- Egeland et al. [382]
- Pinckney et al. [383]
- Vidussi et al. [384]
- Barlow et al. [385]
- Zapata et al. [386]
- Van Heukelem and Thomas [387]

Each laboratory derived their analysis method from one of the methods mentioned above. The most up to date and used method is those of Van Heukelem and Thomas [387]. This accurate method allows separating most of pigments properly using a liquid chromatography system hyphenated with a diode-array detector. It is nowadays widely used in oceanography for determining the phytoplankton composition and for estimating the biomass of the different algal groups. However, uncertainties may arise because of the partial separation of some derivative pigments (e.g. mono-vinyl and divinyl forms of chlorophyll-*b*). Numerous guidelines regarding sample handling, storage, data reporting, procedures, quality assurance have been settled along the scientific community experience [388, 389]. About detection, in a nutshell, 450 nm is often dedicated to detect specifically carotenoids, 665 nm for chlorophyll related pigments, 405 nm for chlorophyll derivatives (degradation forms), while 436 nm is commonly used because this wavelength provides good detection for most pigments. Regarding quantification, pigments are often reported to the sample dry weight. However, this measurement is strongly dependent of a proper drying which is not so reliable with microalgae, especially for short samples. Oceanographers are used to report pigments as a chlorophyll-a ratio. Even if chlorophyll-*a* allomers can be taken into account for the calculation, a bias can occur according to the growth phase of the species within the sample to study i.e. chlorophyll-*a* and its derivatives can be degraded in various forms like pheophorbides and pheophyins, which escape to the ratio, especially during senescence. Pigment amounts can be reported to carbon unit. An elemental analyzer is required to study the samples. This method is very accurate and reproducible. However, since microalgae have a carbon composition, which is quite different according to the species, interspecies studies should be avoided using a pigment ratio related to carbon.

Regarding quantitation, saponification has to be mentioned since this step can be added just before HPLC analyzes in order to make free carotenoid esters and thus, improve carotenoid content quantitation. Moreover, it is sometimes useful to remove chlorophylls and unwanted lipids which may interfere with the chromatographic carotenoid separation and shorten the column's life [390]. However, some carotenoids (especially xanthophylls) and most chlorophylls are known to be sensitive to oxygen, light, acids, alkaline bases and high temperature. It is crucial to choose a gentle protocol to avoid pigment degradation. For instance, Yuan and Chen developed a method promoting the hydrolysis of astaxanthin esters, minimizing the degradation of astaxanthin, and avoiding loss of lutein, β-carotene and canthaxanthin during saponification [391]. A special care with saponification is important to keep in mind if biological activity assays have to be performed afterwards, since this method can produce non-naturally occurring pigments. Chlorophyllins are semisynthetic derivatives obtained by saponification of chlorophylls and display various potent activities like antigenotoxic, antioxidant, inhibition of cancer initiation, and progression by targeting multiple molecules and pathways involved in the metabolism of carcinogens, cell progression, resistance to apoptosis, metastasis, and angiogenesis [390]. A study with Japanese patients with trimethylaminuria found that a chlorophyllin dietary supplement significantly decreased urinary trim ethylamine concentrations [391].

This is potentially true with some other molecules like lipids and carotenoids even if literature is not hearty yet.

4.6.2.2 Other Techniques for Structural Characterization and Identification

While spectrophotometry is the gold standard to identify major known pigments using standards as reference, pigment diversity/complexity is hidden in the detection baseline and it is not always easy to identify them [334]. If main organosoluble pigments in microalgae are about 40, minor/traces pigments are probably far beyond 150. For instance, Carotenoids Database [392] provides information on 1174 natural carotenoids from 699 source organisms (not only microalgae but also plants, animals, bacteria). Identification based on UV-VIS spectrum can be tricky. In that case, a large panel of physical methods is available. For an acceptable identification, ideally, a complete spectroscopic characterization (UV-VIS, mass spectrometry, ^{1}H and ^{13}C-NMR, circular dichroism for chiral compounds) is required. Total synthesis would confirm the full structure elucidation as well preferably [393]. However, it is not always possible to obtain all of these identification criteria since the purified amount can be low (between few μg to 1 mg at laboratory scale) and the native molecule sensitive to various factors mentioned previously. For subtle details of conformation, some other techniques can be cited like calorimetry, roentgenography, electronography, neutronography, spin electric resonance, molecular polarization, light diffusion or refraction [394]. Most of the time, these techniques are more accessible in academia than in industry. Fluorimetric methods can be considered anecdotal in carotenoid and chlorophyll identification.

Metabolomics approaches are increasingly becoming a powerful strategy in identification of metabolites with the recent advances of high sensitive spectroscopy methods. Amongst the various aspects this field, untargeted metabolomics is challenging. Although identification remains a bottleneck particularly for NMR-based dereplication strategies, MS and MS/MS based metabolomics made interesting progress these last years [395–397]. Some authors developed whole online workflow to analyze metabolomics datasets [398]. Some others developed the Global Natural Products Social Molecular Networking platform so-called GNPS [399]. Based on the use of MS public database and/or customized database, and the creation of molecular networks (ion fragments) using the software Cytoscape, this powerful open-source tool allows making identification hypothesis [400]. However, confidence in identification requires a special care to reach an acceptable level [401]. GNPS authors recommend notably a minimum of 6 MS/MS fragment ions to match in addition to the parent mass to decrease false discovery rates. They advise as well a cross validation of results with additional methods (e.g. retention time analysis, co-migration with standards, or subsequent isolation and NMR analysis) to validate the results. This very promising methodology can be applied to pigments using various MS technologies (MALDI-TOF, Q-TOF) and various ionization sources (APCI, APPI, ±ESI) [104, 402–406]. Coupled for instance with Carotenoids Database mentioned previously, results may be very interesting in a close future.

In conclusion, the panel of extraction techniques available is large. The selection of the most appropriate process depends of the production scale, the targeted pigments, the final application and the characteristics of the biomass to extract (humidity, rigid cell wall, polysaccharides). For instance, water miscible properties of acetone and ethanol helps in the efficient extraction of wet biomass. Some techniques like those using ionic liquids seem very promising but further advancements are needed for full safety applications. When pigments are extracted, analyses can be performed keeping in mind the bias related to identification and quantitation (possible degradation of most sensitive pigments, free and esterified forms, pigment-protein complexes, carbohydrates derivatives like glycosides and glycosyl esters). Cis isomers are also common isolation artefacts although some may be natural.

4.7 Methods for Secondary Metabolites Determination in Microalgae

4.7.1 Introduction

Secondary metabolites, often referred to as natural products, are low molecular weight organic molecules produced by living cells [407] that, unlike primary metabolites, are not directly required for their growth, development and reproduction. They are produced by bacteria, fungi, algae and plants, often for biochemical interactions between organisms [408] and represent a wide variety of chemicals with diverse molecular structures. Many of these chemicals have been found to possess a variety of important pharmacological properties, including anticancer, antiviral, antibiotic, anti-inflammatory and other activities [409, 410]. Frequently, these molecules have contributed to the cure of deadly diseases [409], such that biologically active secondary metabolites are a rich resource as potential new drug candidates.

Around 70,000 microbial metabolites have been described to date. About 33,000 of these exhibit some kind of biological activity; 1290 cyanobacterial bioactive metabolites have been reported, predominantly linked to a group of enzymes referred to as nonribosomal peptide synthases (NRPS), such as cyclic depsipeptides, polyketides derivatives [411] or a hybrid of the two. Other structural classes include alkaloids, shikimate-derivatives, aminoglycosides and more rare, terpenoids [412].

Secondary metabolites have considerable commercial importance. In addition to their use in pharmaceuticals, these compounds find applications in many other areas, including cosmetics, agricultural, animal feed and food products, and chemicals.

4.7.2 Sampling and Preconditioning

In the screening for new metabolites or in their production, cultures from microalgae collections, commercial suppliers or the isolation of indigenous wild type microalgae may be considered. Some of the databases for culture collections worldwide include the CCALA [413] in Trebon, Czech Republic, the SAG [414] in Göttingen, Germany, the ACOI [415] in Coimbra, Portugal, the CCAP [416] in Windermere, U.K., the Chlamydomonas Resource Center [417] in Minneapolis, U.S.A. and the UTEX [418] in Austin TX, U.S.A.

Isolation of microalgae strains into pure cultures include traditional isolation techniques, such as single-cell isolation by a micropipette, isolation of cells on agar plates, dilution method, gravity separation, phototaxis, and automated isolation techniques, such as flow cytometry cell sorting and optical trapping. Single-cell isolation by micropipetting is the most common method for isolation of microalgae strains. Basically, the isolation is performed using a Pasteur pipette or a glass capillary to collect a single cell from a sample and transfer it to a sterile droplet. A detailed description can be found in [419]. Flow cytometry cell sorting is an automated technique that can be used for isolation of small cells, especially those less than about 5 mm, which are much more difficult to isolate by traditional techniques. Single cells can be identified and separated from contaminants and other cells and sorted into multiwell plates for establishing new microalgae cultures [419]. Another automated technique is optical trapping, which is not a common technique, since the reaction centers of photosystems I and II are susceptible to photodamage. However, in a recent report [420], the non-invasive optical manipulation of living cells of the microalgae *Trachydiscus minutus* was performed, using laser wavelengths longer than 935 nm causing no observable photodamage.

For additional information on microalgae isolation the reader is referred to the references [419].

4.7.3 Extraction Techniques

The solvent extraction process is a crucial step in isolation and purification of individual distinct types of metabolites. The method chosen can directly affect the biological activity of the extracted compounds [421] as well as their reproducibility, efficiency and effectiveness. Therefore, when choosing the most appropriate method from amongst the several available options, certain important considerations must be addressed. First, microalgae and Cyanobacteria present a remarkable biodiversity, which is reflected in the chemistry of their unique secondary metabolites, and associated cell wall characteristics and structure. Thick-walled, silicified membranes, multilayered walls, wall-bound exopolysaccharides and armored walls present barriers to permeation by the solvents used in extraction and must be disrupted prior to this to allow the release of compounds of interest [357].

Labile compounds may be subject to degradation upon solvent extraction under elevated temperatures, and light, air and pH are among the factors that should be controlled to prevent the decomposition of these compounds.

4.7.3.1 Cell Disruption

Cell disruption techniques include soaking, pestle and mortar (tissue grinders), maceration, cryogrinding, bead-beating, homogenization, planetary micro milling, sonification [422], mixer milling [357], high-pressure homogenization [423], microwave, autoclaving, and addition of hydrochloric acid, sodium hydroxide or alkaline lysis. Maceration is a cost-effective method and is commonly used in algal cell disruption. Bead beating has been demonstrated to be very effective in disrupting the cell of *Botryococcus braunii* [424] and can be suitable for industrial scale up [422]. Sonification is one of the most commonly used laboratory methods to disrupt the cell wall in microalgae and has been shown to give reliable results in several reports [422, 425, 426]. However, thermolabile compounds may be degraded due to the heat generated by this method, therefore, when low temperatures are required, cells should be placed in an ice bath during the entire sonification process to prevent the formation of artifacts or degradation products and disruption of the cells may be confirmed by optical microscopy.

4.7.3.2 Secondary Metabolite Extraction

The microalgal cell is a complex biological system, with many cellular and sub-cellular structures where several different processes occur and a panoply of chemically distinct metabolites are produced. Because different classes of secondary metabolites have different chemical characteristics, there is currently no standard established extraction procedure. Procedures described in the literature include conventional extraction techniques, supercritical fluid extraction, enzyme extraction, etc.

The presence of thermolabile compounds requires the use of techniques working at low temperature, to prevent thermal degradation, hydrolysis and hydrosolubilization. The cell wall characteristics, the nature of molecules to be extracted from the matrix, extraction times and coextraction of undesirable compounds, as for example, sometimes lipids and pigments, are some of the factors that have to be considered in this.

Conventional Solvent Extraction

Solvent extraction: in conventional solvent extraction, the appropriate solvent or mixture of solvents is added to the microalgal biomass from which the desired metabolite is to be extracted. Once the crude extract is separated from the cell residue and filtered, the solvents, if volatile, can be evaporated.

Soxhlet extraction: the procedure described by Soxhlet [427] has commonly been used for the extraction of primary compounds as lipids from biological samples. Nevertheless, secondary metabolites are also extracted with this methodology from a known quantity of the microalgal material, which is placed in a thimble, by repeatedly washing and leaching (percloration) with the appropriate organic solvent under reflux in the Soxhlet apparatus. The solvent in the flask is heated to boiling. The solvent vapor is condensed in the condenser, flows through the sample and moves back down into the distillation flask. The resulting crude extract should be filtered, to remove any remaining unwanted matter, and can then be concentrated on a rotary evaporator or dried under a nitrogen steam. In the Soxhlet procedure, heat is required to drive the extraction, and this may cause decomposition of thermolabile compounds. In these cases, high temperatures must be avoided by choice of appropriate solvents in order to prevent the development of decomposition products.

The extraction process using organic solvents has come under increasing criticism due to its reliance on solvents that are known to be toxic, carcinogenic or environmental pollutants. Chloroform, dimethyl acetamide, dimethyl formamide, dimethyl sulfoxide and methanol are some of the solvents that are becoming considered unsuitable for such extractions, and there is increasing interest in the development of green solvents or other techniques for this [357].

Regardless of the choice of solvent, the selection of a suitable solvent or solvent system should address some aspects, such as grade and purity, solubility of the compounds of interest in it, safety, and environmental issues.

Non-conventional Extraction Technique

Several novel extraction techniques have been developed in recent years as alternatives to the traditional extraction procedures, to overcome the limitations and inconveniences presented by the conventional methodologies. Methods such as microwave extraction (MAE), ultrasonic extraction (UAE), subcritical water extraction (SWE), supercritical fluid extraction (SFE), microbial-aided extraction (MbAE), and enzyme-aided extraction (EAE), have all been applied for the extraction of secondary metabolites from microalgae [421].

Green extraction methods, such as supercritical fluid extraction (SFE), present advantages over the traditional ones using organic solvents. This eco-friendly process does not use large amounts of organic toxic solvents, is rapid, inexpensive, selective and it is suitable when dealing with thermolabile compounds, since it avoids long extraction times and high temperatures, which may result in degradation of compounds. The most commonly used critical fluid has been supercritical carbon dioxide (SC-CO_2), which is non-flammable, has low critical conditions and is suitable for industrial applications [421].

4.7.4 Chemical Characterization

The separation and detailed identification of secondary metabolites has only really become possible by the development of advanced instrumental methods of analysis, particularly those involving hyphenated techniques (e.g. HPLC-MS). The traditional methods of screening for secondary metabolites lead to the successful identification and development of many drug candidates [428]. For the structural elucidation of metabolites in algae and in other natural products, a repertoire of analytical techniques is currently available including, chromatographic, spectroscopic, hyphenated and genome mining techniques. However, due to the diversity and complexity of microalgal and cyanobacterial metabolites, it is unlikely that one single analytical technique will provide enough information about the metabolites that are present in these organisms and, generally, it is necessary to use various complementary techniques that reveal different facets which enable the chemical structure characterization.

4.7.4.1 Chromatographic Techniques

Chromatographic techniques, such as high performance thin layer (HPTL) and liquid chromatography (HPLC), gas chromatography (GC), liquid chromatography (LC), supercritical fluid chromatography (SFC), have been used to screen various types of secondary metabolites.

Thin Layer Chromatography (TLC)

Although being an old technique, thin layer chromatography provides a simple, rapid and sensitive method for the separation and determination of secondary metabolites in crude extracts. This technique consists of a stationary phase, which involves a layer of adsorbent coated onto a solid support, such as glass, polymeric material or aluminum, and a liquid acting as a mobile phase. The sample is applied in the form of a small spot at one end of the layer and migrates with the solvent along the plate by capillary action. The TLC spots are visualized visible, with ultraviolet light or by treatment with a suitable spray reagent, depending of the characteristics of the compounds and can be compared by simultaneously running the unknown spot with standards to identify compounds present.

TLC has been used for the determination of various bioactive compounds. Metabolites such as nostocionone [332], norharmane [429], microcystins and nodularins have, for example, been detected by this technique.

High Performance Thin Layer Chromatography (HPTLC)

Based on thin layer chromatography separation, high performance thin layer chromatography has emerged as a sophisticated technique with major improvements in resolution and accuracy over standard TLC and has been reported to provide good separation of several biomolecules from biological samples. HPTLC is a simple and fast separation technique, and relatively inexpensive, compared with HPLC. Multiple samples can be analyzed simultaneously; the method is straightforward and does not require complex sample pre-treatment. The main disadvantages are the interferences that can occur, sometimes giving some false positives. HPTLC has been used to determine secondary metabolites such as norharmane in various Cyanobacteria [430].

Gas Chromatography (GC)

Gas chromatography (GC) is the most widely used technique for the separation of volatile mixtures because of the high sensitivity and selectivity it allows. It has been particularly effective when coupled with mass (GC-MS) [431], flame ionization (GC-FID) or electron capture detectors (GC-ECD). The major limitation of GC is that nonvolatile compounds cannot be analyzed directly and require derivatization of the sample. It is limited to samples that are easily volatilized and are thermally stable. Considering that 80% of all known natural compounds are nonvolatile or thermolabile [432], GC is not normally the method of choice in the field of secondary metabolites analysis. Nevertheless, gas chromatography has been used to identify specific secondary metabolites [433] and may provide an effective quality control of samples [434]. One such example is the bioactive exometabolite harmane, extracted from the culture medium of the cyanobacterium *Geitlerinema* sp. The alkaloid harmane can be detected in low volumes of culture media by GC-FID, after a derivatization step [433].

Some microalgae, in particular the Cyanobacteria and the dinoflagellates, are known to produce toxins such as anatoxin-a, saxitoxins and microcystins, which are hazardous to human and animal health, and have been associated with the contamination of blue–green algal food supplements [435]. The microalgal biomass production systems include open ponds, which might be susceptible to contamination by a variety of microalgae species, with some capable of producing a wide range of toxins. A toxicological control guarantees the absence of potential toxins in microalgal food supplements. Toxins that can be derivatized can be successfully detected and quantified by the technique GC-ECD, at pictograms levels, in small amounts of algal samples and in small volumes of water [436].

High Performance Liquid Chromatography (HPLC)

HPLC is a chromatographic technique that can separate complex mixture of compounds and can be used to identify, quantify and purify individual analytes of a mixture [437]. Equipped with ultraviolet (UV) and fluorescence (FLD) detectors, is a widely used method for the analysis of secondary metabolites because of its sensitivity and ease of use. Specific examples of application of this technique include the analysis of secondary metabolic peptides [438] and alkaloids [439].

HPLC with ultraviolet (UV), photodiode array (PDA), and fluorescence (FLD) detectors can be applied to determine different types of cyanotoxins; HPLC-FLD can detect saxitoxins which, although they are neurotoxic, have therapeutic potential in anaesthesia [439]. When confirmation of the molecular structure of a given toxin is required, the hyphenated technique liquid chromatography mass spectrometry (LC-MS) can be used for the identification of the compound [441].

Although UV detectors are limited to molecules having suitable chromophores, HPLC-UV is suitable to the detection and quantification of several secondary metabolites. Other examples include the analysis of the closely related [7.7] paracyclophanes and structural derivatives, extracted from the cyanobacterial strains belonging to the orders Nostocales and Oscillatoriales, which were detected and quantified by HPLC-UV [442]. The same technique was also applied for the determination of fourteen phenolic compounds in the extracts from the diatom *Phaeodactylum tricornutum* [443].

4.7.4.2 Spectroscopic Techniques

This discussion is not exhaustive, and we have not included important spectroscopic techniques such as UV/Vis, FTIR and Raman Spectroscopy. The reader is referred to standard texts for the discussion of these.

Fluorescence Spectroscopy

Fluorescence is the emission of light when electronically excited states relax back to the ground state. Its theoretical basis is closely related to, and involves the same electronic states, as UV-vis absorption spectroscopy. However, the nature of measurement technique means that it may have a sensitivity several orders of magnitude greater than absorption spectroscopy. This is a great advantage in analytical applications, such as the study of microalgal metabolites, which may be present at very low concentrations. Not all molecules have significant fluorescence. Analytically interesting fluorescent organic molecules typically have rigid conjugated structures, such as derivatives of aromatic hydrocarbons. Experimentally, its study requires an exciting light source, the sample, appropriate dispersive elements, such as monohromators, and a detector. It is possible to scan the emission spectrum using a constant excitation wavelength, or to observe the emission at a constant wavelength and study the effect of changing the excitation wavelength. These give the emission and excitation spectra respectively. A number of excellent descriptions of the technique are available [444, 445]. Typical examples of applications include analysis of chlorophylls [446] and alkaloids, such as β-carbolines [447]. It is also possible to add fluorescent probes which are sensitive to particular environments, such as hydrophobic ones, which can provide detailed information on lipid content [448, 449]. The high sensitivity of fluorescence means that it is valuable for combining with chromatographic methods, such as HPLC, in hyphenated techniques. It can also be combined with microscopy in fluorescent imaging and can furnish more detailed information through time-resolved measurements.

Nuclear Magnetic Resonance (NMR)

NMR is a powerful technique that provides both qualitative and quantitative information. It allows the simultaneous detection of diverse groups of secondary metabolites (amino-glycosides, alkaloids, terpenoids and so on). NMR exploits the magnetic properties of certain nuclei (such as ^{1}H, ^{13}C, ^{31}P, which, when placed in a magnetic field, absorb electromagnetic radiation in the radio-frequency region of the spectrum (frequency of resonance). The resonance frequency depends on the chemical environment of the nucleus, and thus, each nucleus in a molecule gives a specific and characteristic signal (given as its chemical shift), allowing structural elucidation. In addition, the technique provides information on the nearest neighbour atoms through "coupling" of the nuclear spins. There are many excellent books on NMR spectroscopy, describing the fundamental theory, applications and the typical values for chemical shifts and coupling constants, and several online databases that provide valuable data resources [450], and useful information regarding organic structures and correspondent spectra are available [450, 451].

Although NMR is a reliable and very robust technique that provides unambiguous information, it has some disadvantages that can limit its application. These include the expensive equipment needed, the time-consuming measurements, the high concentrations of samples required, and the overlapping signals that may be present when applied to complex bioorganic compounds, making the interpretation of the exact structure of the compound difficult.

In general, the NMR technique does not require elaborate sample preparation, and liquid samples, solvent extracts and dried or live microalgal cells can directly be analyzed. It is a straightforward and non-destructive technique, such that samples can be further analyzed by other techniques. NMR spectroscopy has been extensively used in biological studies and significant improvements are continuously being made in the use this high-throughput technique.

Since a large number of NMR experiments are possible in metabolomics, different approaches can be considered when using NMR spectroscopy. One dimensional ^{1}H NMR is one of the most widely used in metabolomics analysis. However, when analyzing complex mixtures, ^{1}H NMR spectra can be very challenging due to overlapping of signals, which can hinder the identification of metabolites.

A relatively new methodology in the field of Metabolomics studies is in vivo High Resolution Magic Angle Spinning Nuclear Magnetic Resonance (^{1}H HR-MAS NMR) that can be applied to live microalgal cells [452].

In ^{13}C NMR spectroscopy, where chemical shifts cover 200 ppm, compared typically with 10 ppm for ^{1}H NMR, the signals are better resolved. However, the major limitation of ^{13}C NMR spectroscopy is the lower sensitivity arising from the low natural abundance of ^{13}C. The use of multidimensional NMR spectrum can help to overcome many of the limitations of one dimensional NMR. Two-dimensional (2D) NMR include correlated spectroscopy (^{1}H-^{1}H COSY), total correlation spectroscopy (^{1}H-^{1}H TOCSY) and heteronuclear single-quantum correlation (^{1}H-^{13}C HSQC).

A customized metabolomics NMR database [453], $^{1}H(^{13}C)$-TOCCATA, contains a complete set of ^{1}H and ^{13}C chemical shift information on individual spin systems and isomeric states of common metabolites, and allows the identification of metabolites in complex mixtures [454].

Because NMR spectroscopy analysis frequently produces a highly complex spectrum which is difficult to interpret, it is useful to employ chemometric analysis, such as Principal Components Analysis (PCA), that allows the reduction of multivariate data into a smaller number of principal components representing the original variables, and Partial Least Squares Discriminant Analysis (PLS-DA), which identify categories, provide a more objective and clear information.

Examples of application of NMR on microalgal secondary metabolites elucidation include the detection of glycosides in powder suspensions of the blue-green algae *Aphanizomenon flos-aquae* [455].

4.7.4.3 Hyphenated Techniques

The combination of individual techniques with NMR and MS has led to the development of hyphenated techniques which have greatly increased the analytical capabilities in metabolomics research. Examples include gas chromatography–mass spectrometry (GC–MS) [456], liquid chromatography–nuclear magnetic resonance spectroscopy LC-NMR [457], ultra-high performance liquid chromatography-two-dimensional mass spectrometry (UHPLC-MS/MS) [458]. GC–MS combines the good resolution of GC and high selectivity and sensitivity of MS. To be detected by GC-MS analysis, the metabolites require a derivatization step in order to create volatile organic compounds, which limits the applicability of this technique [459]. If the unknown sample has derivatizable or volatile components, this technique might be employed to identify and quantitate several secondary metabolites such as different peptides, microcystins and alkaloids [456, 459, 460].

The UHPLC-MS/MS technique provides an ultrasensitive and selective determination of several secondary metabolites, such as cyanotoxins or isoflavones in microalgae and Cyanobacteria. The presence of eight isoflavones such as daidzin, genistin and formononetin was demonstrated by UHPLC-MS/MS in concentrations of ng g^{-1} dry biomass [458]. Another example of the application of a hyphenated technique in the analysis of secondary metabolites is the determination of extracellular diterpenoids by LC-NMR from the terrestrial cyanobacterium *Nostoc commune* [457].

4.7.4.4 Genome Mining

In response to environmental changes such as excessive light, changes in pH, nutrients depletion, etc., many Cyanobacteria activate specific genes that, in turn, synthetize specific proteins; these are involved in the pathways for the production of these metabolites that protect cells from stress or help them to adapt to the new conditions. Genome mining is a powerful technology that has emerged as a new strategy for the discovery of novel secondary metabolites [461]. Based on genetic information, it has become possible to predict and isolate new compounds. The genome sequencing of microorganisms unveils new metabolites with potential therapeutic interest in the treatment of diseases; enhancement of organism growth and increased crop yields and other biotechnology applications are also possible. At present, thousands of microbial genome sequences are available in public repositories containing information on secondary metabolite gene clusters that encode the biosynthetic pathways of secondary metabolites. With the fast development of sequencing methods and bioinformatics, several genome mining methodologies have arised. Currently, computational and web-based analysis platforms such as the Atlas of Biosynthetic Gene Clusters [462] and the antiSMASH [463–467] are used to predict the products of genes whose DNA sequences are known. With the advance of modern technologies such as UHPLC, NMR and various mass spectroscopy methodologies, the products of such genes can be identified and characterized. One such example, for the discovery of new secondary metabolites, involves an interesting combination of techniques; the combination of mass spectrometric metabolic profiling and genomic analysis has resulted in the discovery of a new class of di- and tri-chlorinated acyl amides, the columbamides [468].

4.7.5 Summary

A summary of the main information is given in Table 4.6.

Table 4.6 Information about the biological activity, culture conditions, extraction methods and analytical techniques of the compounds obtained from several species of microalgae and Cyanobacteria

Metabolite	Family/genus/ specie	Biological activity	Culture conditions	Sample treatment	Extraction method/ solvents used	Analytical techniques	References
Aeruginosin 828A	*Planktothrix rubescens*	Anti-Inflammatory	Mineral medium, 20 °C, 6 μmol photon m^{-2} s^{-1}	Freeze/thaw cycle after harvested by centrifugation	50% methanol Water	RP-HPLC	[469]
Anatoxin-a	*Anabaena flos-aquae*	Cytotoxic	BG-11 medium, 18–20 °C, 30–50 μmol photon m^{-2} s^{-1}, 16:8 h	Freeze-dried after harvested by filtration	Dichloromethane, Derivatization with pentafluorobenzylbromide	GC-ECD	[436]
Calothrixins A and B	*Calothrix*	Antimalarial and anticancer	Allen and Arnon medium [470], 25 °C, 10 mM $NaNO_3$ $NiSO_4$, 24:0 h	Lyophilized	Sohxlet/ Dimethylsulfoxide, ethyl acetate, hexane, acetone	EIMS, ^{13}C NMR, ^{1}H NMR, singles-crystal X-ray	[410]
Carriebowmide	*Lyngbya polychroa*	–	Collected sample	Freeze-dried	EtOAc-MeOH, EtOH-H_2O	IR, HPLC, NMR	[438]
Columbamides	*Moorea (Lyngbya)*	Cannabinomimetic	Collected sample, BG-11 medium	Freeze-dried	(2:1, v/v) CH_2Cl_2/MeOH	IR, ^{1}H and ^{13}C NMR, HPLC, HRMS-genomic analysis	[468]
Cyanopeptolin 1020	*Microcystis aeruginosa* UV-006	Anti-Inflammatory	Mineral medium, 26 °C, 40 ± 5 μmol photon m^{-2} s^{-1}	Harvested by centrifugation	60% acetonitrile, Water	RP-HPLC, high-resolution mass spectrometry	[469, 471]
Diterpenoids commostins	*Nostoc commune*	Antibacterial, cytotoxic, molluscicidal	Inorganic culture medium, 24 °C, 29 μmol photon m^{-2} s^{-1}, 24:0 h, 2% CO_2	Harvested by filtration	Methanol	NMR, MS, single-crystal X-ray	[457]
9-Ethyliminomethyl-12-(morpholin—4-ylmethoxy)-5, 8, 13, 16–tetraaza–hexacene–2, 3 dicarboxylic acid	*Nostoc* sp.	Antibacterial	Collected sample, BG-11 medium, 25 °C, 95 μmol photon m^{-2} s^{-1}, 14:10 h	Harvested by centrifugation and lyophilized	Methanol	TLC, HPLC, ESIMS, NMR	[472]

(continued)

Table 4.6 (continued)

Metabolite	Family/genus/ specie	Biological activity	Culture conditions	Sample treatment	Extraction method/ solvents used	Analytical techniques	References
Harmane	*Geitlerinema* sp.	Antimicrobial	Rippka et al. [473], 25 °C, 10 μmol photon m^{-2} s^{-1}, 24:0 h	Harvested by centrifugation	Volk´s procedure, methanol, Derivatization	TLC, LC-MS/MS, GC-FID	[433]
Isoflavones	*Nostoc, Spongiochloris, Scenedesmus*	–	–	Lyofilized	Supercritical fluid extraction	UHPLC-MS/MS	[458]
Lyngbyabellin B	*Lyngbya majuscula*	antifungal	Collected sample	–	(2:1, v/v) CH_2Cl_2/MeOH	1D and 2D NMR, GC-MS	[431]
Mycosporine-like amino acids and glycosides	*Aphanizomenon flos-aquae*	Anticancer	–	Dry powder, sonicated for 1 h	No extraction	NMR, ESI-QTOF-MS, CID-MS/MS	[455]
Norharmane	*Nostoc insulare, Nodularia harveyana*	Algicidal	Pohl et al. medium [474], 27 °C, 25–30 μmol photon m^{-2} s^{-1},	Centrifugation and filtration, freeze-dried	Methanol	HPLC	[475]
Norharmane	*Synechocystis aquatilis*	Algicidal, allelopathic	BG-11 medium, 24 °C, 30 μmol photon m^{-2} s^{-1}, 24:0 h	Harvested by centrifugation	Volk´s procedure	TLC	[429]
Nostocarboline	*Nostoc* 78-12A	Inhibitor of butyrylcholinesterase	Mineral medium, 25°C, continuously illumination, 15 μmol photon m^{-2} s^{-1}, CO_2 enriched air (600 mL min^{-1}, 0.15 vol.% CO_2	Lyophilized	60% methanol $CH_3CN–H_2O$,95:5 V/V Separated by HPLC	2D-NMR	[476, 477]
[7.7] Paracyclophanes	*Nostoc, Cylindrospermum, Anabaena Nodularia, Pseudanabaena*	Antibacterial, cytotoxic	BG-11, BG-11 + 0.5% NaCl, MBL medium, 20–28 °C, 0.5–5% CO_2, 20–80 μmol photon m^{-2} s^{-1}	Freeze-dried after harvested by centrifugation	n-heptane, EtOAc, MeOH, and H2O (5:2:5:2, v/v/v/v)	UV−vis spectrophotometer, HPLC-UV, 2D NMR	[442]

(continued)

Table 4.6 (continued)

Metabolite	Family/genus/ specie	Biological activity	Culture conditions	Sample treatment	Extraction method/ solvents used	Analytical techniques	References
Phenolic compounds	*Phaeodactylum tricornutum*	Antioxidant	f/2 medium [478], 24 °C, 8000 lx, 24:0 h	Freeze-dried after harvested by filtration	solid-phase extraction/ Methanol, acetone:hexane (1:4)	RP-HPLC	[443]
Sulphated polysaccharide	*Nostoc calcicola*	Antiviral	BG-11 medium, 25 °C, 35 µmol photon m^{-2} s^{-1}, 16:8 h	Harvested by centrifugation	[479]	IR, GC	[480]

Acknowledgements This work was supported by the ANR POLYSALGUE project, grant ANR-15-CE21-0013 of the French Agence Nationale de la Recherche.

Benoît Serive was supported by a Marie Skłodowska-Curie International Outgoing Fellowship (grant #622735) within the 7th European Community Framework Program.

References

1. Silkina, A., Flynn, K., Llewellyn, C., Bayliss, C.: Standard Operating Procedures for Analytical Methods and Data Collection in Support of Pilot-Scale Cultivation of microalgae. Public Output report WP1A3.01 of the EnAlgae project, Swansea (2015)
2. https://algaebiomass.org/resource-center/technical-standards/introduction/ (2017)
3. Grobbelaar, J.U.: Quality control and assurance: crucial for the sustainability of the applied phycology industry. J. Appl. Phycol. **15**, 209–215 (2003)
4. Manoylov, K.M.: Taxonomic identification of algae (morphological and molecular): species concepts, methodologies, and their implications for ecological bio-assessment. J. Phycol. **50**, 409–424 (2014)
5. Guiry, M.D., Guiry, G.M.: AlgaeBase. World-wide electronic publication, National University of Ireland, Galway. http://www.algaebase.org (2017). Accessed 20 Feb 2017
6. Tomas, C.R.: Identifying Marine Phytoplankton. Academic Press, San Diego, California (1997)
7. Lang, I., Hodac, L., Friedl, T., Feussner, I.: Fatty acid profiles and their distribution patterns in microalgae; a comprehensive analysis of more than 2000 strains from the SAG culture collection. BMC Plant Biol. **11**(124), 1–16 (2011)
8. Mc Gee, D., Archer, L., Paskuliakova, A., Mc Coy, G.R., Fleming, G.T.A., Gillespie, E., Touzet, N.: Rapid chemotaxonomic profiling for the identification of high-value carotenoids in microalgae. J. Appl. Phycol. **30**, 385–399 (2018)
9. Giordanoa, M., Ratti, S., Domenighini, A., Vogt, F.: Spectroscopic classification of 14 different microalgae species: first steps towards spectroscopicmeasurement of phytoplankton biodiversity. Plant Ecol. Divers. **2**, 155–164 (2009)
10. Driver, T., Bajhaiya, A.K., Allwood, J.W., Goodacre, R., Pittman, J.K., Dean, A.P.: Metabolic responses of eukaryotic microalgae to environmental stress limit the ability of FT-IR spectroscopy for species identification. Algal Res. **11**, 148–155 (2015)
11. Zou, S., Fei, C., Wang, C., Gao, Z., Bao, Y., He, M., Wang, C.: How DNA barcoding can be more effective in microalgae identification: a case of cryptic diversity revelation in *Scenedesmus* (Chlorophyceae). Sci. Rep. (2016). https://doi.org/10.1038/srep36822
12. Ratnasingham, S., Hebert, P.D.N.: Bold: the barcode of life data system. Mol. Ecol. Notes **7** (3), 355–364 (2007)
13. Mann, D.G., Sato, S., Trobajo, R., Vanormelingen, P., Souffreau, C.: DNA barcoding for species identification and discovery in diatoms. Cryptogamie Algol. **31**, 557–577 (2010)
14. Dubrulle, N., Giraud, N.: DNA metabarcoding as a tool to trace plants of interest in ingredients or cosmetics. IFSCC Mag. **2**, 51–54 (2017)
15. Blanken, W., Schaap, S., Theobald, S., Rinzema, A., Janssen, M.: Optimizing carbon dioxide utilization for microalgae biofilm cultivation. Biotech. Bioeng. **114**, 769–776 (2016)
16. Carbone, D.A., Gargano, I., Pinto, G., De Natale, A., Pollio, A.: Evaluating microalgal attachment to surfaces: a first approach towards a laboratory integrated assessment. Chem. Eng. Trans. **57**, 73–78 (2017)
17. Gross, M., Wen, Z.: Yearlong evaluation of performance and durability of a pilot-scale revolving algal biofilm (RAB) cultivation system. Bioresour. Technol. **171**, 50–58 (2014)

18. Zhang, W., McLamore, E.S., Garland, N.T., Chavez Leon, J.V., Banks, M.K.: A simple method for quantifying biomass cell and polymer distribution in biofilms. J. Microbiol. Methods **94**, 367–374 (2013)
19. Laurens, L.M.L., Dempster, T.A., Jones, H.D.T., Wolfrum, E.J., Van Wychen, S., McAllister, J.S.P., Rencenberger, M., Parchert, K.J., Gloe, L.M.: Algal biomass constituent analysis: method uncertainties and investigation of the underlying measuring chemistries. Anal. Chem. **84**, 1879–1887 (2012)
20. Zhu, C.J., Lee, Y.K.: Determination of biomass dry weight of marine microalgae. J. Appl. Phycol. **9**, 189–194 (1997)
21. Li, E., de Ordun, R.M.: A rapid method for the determination of microbial biomass by dry weight using a moisture analyser with an infrared heating source and an analytical balance. Lett. Appl. Microbiol. **50**, 283–288 (2010)
22. Correa-Reyes, J.G., del Pilar Sanchez-Saavedra, M., Viana Maria, T., Flores-Acevedo, N., Vasquez-Pelaez, C.: Effect of eight benthic diatoms as feed on the growth of red abalone (*Haliotis rufescens*) postlarvae. J. Appl. Phycol. **21**, 387–393 (2009)
23. Thyrel, M., Samuelsson, R., Finell, M., Lestander, T.A.: Critical ash elements in biorefinery feedstock determined by X-ray spectroscopy. Appl. Energy **102**, 1288–1294 (2013)
24. Peng, W., Wu, O., Tu, P., Zhao, N.: Pyrolytic characteristics of microalgae as renewable energy source determined by thermogravimetric analysis. Bioresour. Technol. **80**(1), 1–7 (2001)
25. Chen, W., Sommerfeld, M., Hu, Q.: Microwave-assisted Nile red method for in vivo quantification of neutral lipids in microalgae. Bioresour. Technol. **102**, 135–141 (2011)
26. Pane, L., Franceschi, E., De Nuccio, L., Carli, A.: Applications of thermal analysis on the marine phytoplankton, Tetraselmis suecica. J. Therm. Anal. Calor. **66**, 145–154 (2001)
27. Guillard, R.R.L., Sieracki, M.S.: Counting cells in cultures with the light microscope. In: Anderson, R.A. (ed.) Algal Culturing Techniques, pp. 239–252. Elsevier Academic Press, New York (2005)
28. Hyka, P., Lickova, S., Přibyl, P., Melzoch, K., Kovar, K.: Flow cytometry for the development of biotechnological processes with microalgae. Biotechnol. Adv. **31**, 2–16 (2013)
29. Šupraha, L., Gerecht, A.C., Probert, I., Hendericks, J.: Eco-physiological adaptation shapes the response of calcifying algae to nutrient limitation. Sci. Rep. (2015). https://doi.org/10.1038/srep16499
30. Sato, M., Murata, Y., Mizusawa, M., Iwahashi, H., Oka, S.: A simple and rapid dual fluorescence viability assay for microalgae. Microbiol. Cult. Coll. **20**, 53–59 (2004)
31. Chioccioli, M., Hankamer, B., Ross, I.L.: Flow cytometry pulse width data enables rapid and sensitive estimation of biomass dry weight in the microalgae *Chlamydomonas reinhardtii* and *Chlorella vulgaris*. PLoS ONE **9**(5), e97269 (2014). https://doi.org/10.1371/journal.pone.0097269
32. Azevedo, R., Rodriguez, E., Figueiredo, D., Peixoto, F., Santos, C.: Methodologies for the study of filamentous Cyanobacteria by flow cytometry. Fresenius Environ. Bull. **21**(8), 679–684 (2012)
33. Dashkova, V., Malashenkov, D., Poulton, N., Vorobjev, I., Barteneva, N.S.: Imaging flow cytometry for phytoplankton analysis. Methods **112**, 188–200 (2017)
34. D'Haese, E., Nelis, H.J.: Rapid detection of single cell bacteria as a novel approach in food microbiology. J. AOAC Int. **85**, 979–983 (2002)
35. Lisle, J.T., Hamilton, M.A., Willse, A.R., McFeters, G.A.: Comparison of fluorescence microscopy and solid-phase cytometry methods for counting bacteria in water. Appl. Environ. Microbiol. **70**, 5343–5348 (2004)
36. Tobe, K.: Automated detection and enumeration for toxic algae by solid-phase cytometry and the introduction of a new probe for *Prymnesium parvum* (Haptophyta: Prymnesiophyceae). J. Plankton Res. **28**, 643–657 (2006)
37. Cornet, J.F., Dussap, C.G., Cluzel, P., Dubertret, G.: Structured model for simulation of cultures of the cyanobacterium *Spirulina platensis* in photobioreactors: II. Identification of kinetic parameters under light and mineral limitations. Biotech. Bioeng. **40**, 826–834 (1992)

38. Parsons, T.R., Stephens, K., Strickland, J.D.H.: On the chemical composition of eleven species of marine phytoplankters. J. Fish. Res. Bd. Can. **18**, 1001–1016 (1961)
39. Cuellar-Bermudez, S.P., Aguilar-Hernandez, I., Cardenas-Chavez, D.L., Ornelas-Soto, N., Romero-Ogawa, M.A., Parra-Saldivar, R.: Extraction and purification of high-value metabolites from microalgae: essential lipids, astaxanthin and phycobiliproteins. Microb. Biotechnol. **8**, 190–209 (2015)
40. Slocombe, S.P., Ross, M., Thomas, N., McNeill, S., Stanley, M.: A rapid and general method for measurement of protein in microalgal biomass. Bioresour. Technolol. **129**, 51–57 (2013)
41. Stephens, E., Ross, I.L., Mussgnug, J.H., Wagner, L.D., Borowitzka, M.A., Posten, C., Kruse, O., Hankamer, B.: Future prospects of microalgal biofuel production systems. Trends Plant Sci. **15**, 554–564 (2010)
42. Borowitzka, M.A.: High-value products from microalgae—their development and commercialisation. J. Appl. Phycol. **25**, 743–756 (2013)
43. Sánchez, M., Bernal-Castillo, J., Rozo, C., Rodríguez, I.: *Spirulina* (*Arthrospira*): an edible microorganism: a review. Univ. Sci. **8**, 7–24 (2003)
44. Day, J.G., Slocombe, S.P., Stanley, M.S.: Overcoming biological constraints to enable the exploitation of microalgae for biofuels. Bioresour. Technol. **109**, 245–251 (2012)
45. Radakovits, R., Jinkerson, R.E., Fuerstenberg, S.I., Tae, H., Settlage, R.E., Boore, J.L., Posewitz, M.C.: Draft genome sequence and genetic transformation of the oleaginous alga *Nannochlorpsis gaditana*. Nat. Commun. **3**, 686–696 (2012)
46. Gerardo, M.L., Oatley-Radcliffe, D.L., Lovitt, R.W.: Integration of membrane technology in microalgae biorefineries. J. Memb. Sci. **464**, 86–99 (2014)
47. Molina Grima, E., Belarbi, E.H., Acién Fernández, F.G., Robles Medina, A., Chisti, Y.: Recovery of microalgal biomass and metabolites: process options and economics. Biotechnol. Adv. **20**, 491–515 (2003)
48. Gonzales Lopez, C.V.G., Garcia, M.C.C., Fernandez, F.G.A., Bustos, C.S., Chisti, Y., Sevilla, J.M.F.: Protein measurements of microalgal and cyanobacterial biomass. Bioresour. Technol. **101**, 7587–7591 (2010)
49. Barka, A., Blecker, C.: Microalgae as a potential source of single-cell protein. Biotechnol. Agron. Soc. Environ. **2**, 427–436 (2016)
50. Becker, E.W.: Microalgae as a source of protein. Biotechnol. Adv. **25**, 207–210 (2007)
51. Becker, E.W.: Microalgae in human and animal nutrition. In: Richmond, A. (ed.) pp. 312–51. Blackwell Science, Oxford (2004)
52. Richmond, A.: Handbook of Microalgal Culture—Biotechnology and Applied Phycology. Blackwell Science, Oxford (2004)
53. Pulz, O., Gross, W.: Valuable products from biotechnology of microalgae. Appl. Microbiol. Biotechnol. **65**, 635–648 (2004)
54. Bleakley, S., Hayes, M.: Algal proteins: extraction, application, and challenges concerning production. Foods **6**, 33 (2017)
55. Pangestuti, R., Kim, S.K.: Biological activities and health benefit effects of natural pigments derived from marine algae. J. Funct. Foods **3**, 255–266 (2011)
56. Sekar, S., Chandramohan, M.: Phycobiliproteins as a commodity: trends in applied research, patents and commercialization. J. Appl. Phycol. **20**, 113–136 (2007)
57. Batista, A.P., Raymundo, A., Sousa, I.: Rheological characterization of coloured oil-in-water food emulsions with lutein and phycocyanin added to the oil and aqueous phases. Food Hydrocoll. **20**, 44–52 (2006)
58. Spolaore, P., Joannis-Cassan, C., Duran, E., Isambert, A.: Commercial applications of microalgae. J. Biosci. Bioeng. **101**, 87–96 (2006)
59. Chaiklahan, R., Chirasuwan, N., Bunnag, B.: Stability of phycocyanin extracted from *Spirulina* sp.: influence of temperature, pH and preservatives. Process Biochem. **47**, 659–664 (2012)

60. Sarada, R., Pillai, M., Ravishankar, G.: Phycocyanin from *Spirulina* sp.: influence of processing of biomass on phycocyanin yield, analysis of efficacy of extraction methods and stability studies on. Process Biochem. **34**, 795–801 (1999)
61. Romay, C., González, R., Ledón, N., Remirez, D., Rimbau, V.: C-phycocyanin: a biliprotein with antioxidant, anti-inflammatory and neuroprotective effects. Curr. Protein Pept. Sci. **4**, 207–216 (2003)
62. Ejike, C.E., Collins, S.A., Balasuriya, N., Swanson, A.K., Mason, B., Udenigwe, C.C.: Prospects of microalgae proteins in producing peptide-based functional foods for promoting cardiovascular health. Trends Food Sci. Technol. **59**, 30–36 (2017)
63. Udenigwe, C.: Bioinformatics approaches, prospects and challenges of food bioactive peptide research. Trends Food Sci. Technol. **36**, 137–143 (2014)
64. Udenigwe, C., Aluko, R.E.: Food protein-derived bioactive peptides: production, processing, and potential health benefits. J. Food Sci. **77**, 11–24 (2012)
65. Briones, A.M., Touyz, R.M.: Oxidative stress and hypertension: current concepts. Curr. Hypertens. Rep. **12**, 135–142 (2010)
66. Larsen, R., Eilertsen, K.E., Elvevoll, E.O.: Health benefits of marine foods and ingredients. Biotechnol. Adv. **29**, 508–518 (2011)
67. Kovač, D.J., Simeunović, J.B., Babić, O.B., Mišan, A.C., Milovanović, I.L.: Algae in food and feed. Food Feed Res. **40**, 21–32 (2013)
68. Ginzberg, A., Cohen, M., Sod-Moriah, U.A., Shany, S., Rosenshtrauch, A., Arad, S.M.: Chickens fed with biomass of the red microalgae *Porphyridium* sp. Have reduced blood cholesterol level and modified fatty acid composition in egg yolk. J. Appl. Phycol. **12**, 325–330 (2000)
69. Janczyk, P., Halle, B., Souffrant, W.: Microbial community composition of the crop and ceca contents of laying hens fed diets supplemented with *Chlorella vulgaris*. Poult. Sci. **88**, 2324–2332 (2009)
70. Lorenz, R.T., Cysewski, G.R.: Commercial potential for *Haematococcus* microalgae as a natural source of astaxanthin. Trends Biotechnol. **18**, 160–167 (2000)
71. Gatrell, S., Lum, K., Kim, J., Lei, X.: Non-ruminant nutrition symposium: potential of defatted microalgae from the biofuel industry as an ingredient to replace corn and soybean meal in swine and poultry diets. J. Anim. Sci. **92**, 1306–1314 (2014)
72. Lum, K.K., Kim, J., Lei, X.G.: Dual potential of microalgae as a sustainable biofuel feedstock and animal feed. J. Anim. Sci. **4**, 53 (2013)
73. Müller-Fuega, A.: Microalgae for aquaculture: the current global situation and future trends. In: Handbook of Microalgal Culture: Applied Phycology and Biotechnology, 2nd ed. (2013)
74. Kiron, V., Sørensen, M., Huntley, M., Vasanth, G.K., Gong, Y., Dahle, F., Palihawadana, A. M.: Defatted biomass of the microalga, *Desmodesmus* sp., can replace fishmeal in the feeds for Atlantic salmon. Front Mar. Sci. **3**, 67 (2016)
75. Guedes, A.C., Sousa-Pinto, I., Malcata, F.X.: Application of microalgae protein to aquafeed. Handb. Mar. Microalgae Biotechnol. Adv. **14**, 93–125 (2015)
76. Badwy, T.M., Ibrahim, E.M., Zeinhom, M.M.: Partial replacement of fishmeal with dried microalgae (*Chlorella* spp. and *Scenedesmus* spp.) in Nile tilapia (*Oreochromis niloticus*) diets. In: Elghobashy, H., Fitzsimmons, K., Diab, A.S. (eds.) From the Pharaohs to the Future: Proceedings of the 8th International Symposium on Tilapia in Aquaculture, pp 801–810. Egypt Ministry of Agriculture, Cairo (2008)
77. Hempel, F., Lau, J., Klingl, A., Maier, U.G.: Algae as protein factories: expression of a human antibody and the respective antigen in the diatom *Phaeodactylum tricornutum*. PLoS ONE **6**(12), e28424 (2011)
78. Tran, M., Zhou, B., Pettersson, P.L., Gonzalez, M.J., Mayfield, S.P.: Synthesis and assembly of a full-length human monoclonal antibody in algal chloroplasts. Biotechnol. Bioeng. **104**, 663–673 (2009)
79. Yano, A., Maeda, F., Takekoshi, M.: Transgenic tobacco cells producing the human monoclonal antibody to hepatitis B virus surface antigen. J. Med. Virol. **73**, 208–215 (2004)

80. De Muynck, B., Navarre, C., Nizet, Y., Stadlmann, J., Boutry, M.: Different subcellular localization and glycosylation for a functional antibody expressed in *Nicotiana tabacum* plants and suspension cells. Transgenic Res. **18**, 467–482 (2009)
81. Wijffels, R.H., Barbosa, M.J., Eppink, H.M.: Microalgae for the production of bulk chemicals and biofuels. Biofuel Bioprod. Biores. **4**, 287–295 (2010)
82. Sari, Y.W., Mulder, W.J., Sanders, J.P.M., Bruins, M.E.: Towards plant protein refinery: review on protein extraction using alkali and potential enzymatic assistance. Biotechnol. J. **10**, 1138–1157 (2015)
83. Chisti, Y.: Disruption of microbial cells for intracellular products. Enzyme Microb. Tech. **8**, 194–204 (1986)
84. Lee, J.Y., Yoo, C., Jun, S.Y., Ahn, C.Y., Oh, H.M.: Comparison of several methods for effective lipid extraction from microalgae. Bioresour. Technol. **101**, 75–77 (2010)
85. Zheng, H., Yin, J., Gao, Z., Huang, H., Ji, X., Dou, C.: Disruption of *Chlorella vulgaris* cells for the release of biodiesel-producing lipids: a comparison of grinding, ultrasonication, bead milling, enzymatic lysis, and microwaves. Appl. Biochem. Biotechnol. **164**, 1215–1224 (2011)
86. Safi, C.M.: Microalgae Biorefinery : Proposition of a Fractionation Process. Toulouse University (France) Ph.D. thesis (2013)
87. Conde, E., Balboa, E.M., Parada, M., Falque, E.: Algal proteins, peptides and amino acids. In: Functional Ingredients from Algae for Foods and Nutraceuticals, pp. 135–180 (2013)
88. Bajguz, A.: Effect of brassinosteroids on nucleic acids and protein content in cultured cells of *Chlorella vulgaris*. Plant Physiol. Bioch. **38**, 209–215 (2000)
89. Safi, C., Charton, M., Pignolet, O., Silvestre, F., Vaca-Garcia, C., Pontalier, P.Y.: Influence of microalgae cell wall characteristics on protein extractability and determination of nitrogen-to-protein conversion factors. J. Appl. Phycol. **25**, 523–529 (2013)
90. Rausch, T.: The estimation of micro-algal protein content and its meaning to the evaluation of algal biomass I. Comparison of methods for extracting protein. Hydrobiologia **78**, 237–251 (1981)
91. Barbarino, E., Lourenço, S.O.: An evaluation of methods for extraction and quantification of protein from marine macro- and microalgae. J. Appl. Phycol. **17**, 447–460
92. Oliveira, R., Marques, F., Azeredo, J.: Purification of polysaccharides from a biofilm matrix by selective precipitation of proteins. Biotechnol. Tech. **13**, 391–393 (1999)
93. Lourenço, S.O., Barbarino, E., Lanfer Marquez, U.M., Aidar, E.: Distribution of intracellular nitrogen in marine microalgae: basis for the calculation of specific nitrogen-to-protein conversion factors. J. Phycol. **34**, 798–811 (1998)
94. Aschermann, K., Lutter, P., Wattenberg, A.: Current status of protein quantification technologies. Bioprocess Int. **6**, 44–53 (2008)
95. Bradford, M.M.: A rapid and sensitive method for the quantitation of microgram quantities of protein utilizing the principle of protein-dye binding. Anal. Biochem. **72**, 248–254 (1976)
96. Ezeagu, I.E., Petzke, J.K., Metges, C.C., Akinsoyinu, A.O., Ologhobo, A.D.: Seed protein content and nitrogen-to-protein conversion factors from some uncultivated tropical plant seeds. Food Chem. **78**, 105–109 (2002)
97. Lourenço, S.O., Barbarino, E., Lavín, P.L., Lanfer Marquez, U.M., Aidar, E.: Distribution of intracellular nitrogen in marine microalgae: calculation of new nitrogen-to-protein conversion factors. Eur. J. Phycol. **39**, 17–32 (2004)
98. Servaites, J.C., Faet, J.L., Sidhu, S.S.: A dye binding method for measurement of total protein in microalgae. Anal. Chem. **421**, 75–80 (2012)
99. Lowry, O.H., Rosebrough, N.J., Farr, A.L., Randall, R.J.: Protein measurement with the Folin phenol reagent. J. Biol. Chem. **193**(1), 265–275 (1951)
100. Berges, J.A., Fisher, A.E., Harrison, P.J.: A comparison of Lowry, Bradford and Smith protein assays using different protein standards and protein isolated from the marine diatom *Thalassiosira pseudonana*. Mar. Biol. **115**, 187–193 (1993)

101. Everette, J.D., Bryant, Q.M., Green, A.M., Abbey, Y.A., Wangila, G.W., Walker, R.B.: Thorough study of reactivity of various compound classes toward the Folin-Ciocalteu reagent. J. Agric. Food Chem. **58**, 139–8144 (2010)
102. Kruger, N.J.: The Bradford method for protein quantification. In: Walker, J.M. (ed.) The Protein Handbook, pp. 15–21. Humana Press Inc., Totowa, NJ (2002)
103. Lawrence, A.M., Besir, H.U.: Staining of proteins in gels with Coomassie G-250 without organic solvent and acetic acid. J. Vis. Exp. **30**, 1350 (2009)
104. Walker, J.S., Jie, C., Keely, B.J.: Identification of diastereomeric chlorophyll allomers by atmospheric pressure chemical ionisation liquid chromatography/tandem mass spectrometry. Rapid Commun. Mass Spectrom. **17**(11), 1125–1131 (2003)
105. Reichelt, W., Waldschitz, D., Herwig, C., Neutsch, L.: Bioprocess monitoring: minimizing sample matrix effects for total protein quantification with bicinchoninic acid assay. J. Ind. Microbiol. Biotechnol. **43**, 1271–1280 (2016)
106. Connan, S.: Spectrophotometric assays of major compounds extracted from algae. In: Natural Products from Marine Algae: Methods and Protocols, pp. 75–101 (2015)
107. Mæhre, H.K., Dalheim, L., Edvinsen, G.K., Elvevoll, E.O., Jensen, I.J.: Protein determination-method matters. Foods **7**, 5 (2018)
108. Benavides, J., Rito-Palomares, M.: Bioprocess intensification: a potential aqueous two-phase process for the primary recovery of B-phycoerythrin from *Porphyridium cruentum*. J. Chromatogr. B **807**, 33–38 (2004)
109. Jubeau, S., Marchal, L., Pruvost, J., Jaouen, P., Legrand, J., Fleurence, J.: High pressure disruption: a two-step treatment for selective extraction of intracellular components from the microalgae *Porphyridium cruentum*. J. Appl. Phycol. **25**, 983–989 (2012)
110. Bermejo Román, R., Alvárez-Pez, J.M., Acién Fernández, F.G., Molina Grima, E.: Recovery of pure B-phycoerythrin from the microalgae *Porphyridium cruentum*. J. Biotechnol. **93**, 73–85 (2002)
111. Parmar, A., Singh, N.K., Kaushal, A., Madamwar, D.: Characterization of an intact phycoerythrin and its cleaved 14 kDa functional subunit from marine cyanobacterium *Phormidium* sp. A27DM. Process Biochem. **46**, 1793–1799 (2011)
112. Bermejo, R., Ruiz, E., Acien, F.G.: Recovery of B-phycoerythrin using expanded bed adsorption chromatography: scale-up of the process. Enzyme Microb. Technol. **40**, 927–933 (2007)
113. Nui, J., Chen, G., Zhou, B.: Purification of phycoerythrin from *Porphyra yeoensis Ueda* (Bangladesh, Rhodophyta) using expanded bed absorption. J. Appl. Phycol. **22**, 25–31 (2010)
114. Sørensen, L., Hantke, A., Eriksen, N.T.: Purification of the photosynthetic pigment C-phycocyanin from heterotrophic *Galdieria sulphuraria*. J. Sci. Food Agric. **93**, 2933–2938 (2013)
115. Ge, Y., Ning, Z., Wang, Y., Zheng, Y., Zhang, C., Figeys, D.: Quantitative proteomic analysis of *Dunaliella salina* upon acute arsenate exposure. Chemosphere **145**, 112–118 (2016)
116. Guarnieri, M.T., Nag, A., Smolinski, S.L., Darzins, A., Seibert, M., Pienkos, P.T.: Examination of triacylglycerol biosynthetic pathways via de novo transcriptomic and proteomic analyses in an unsequenced microalga. PLoS ONE **6**(10), e25851 (2011)
117. Poirier, I., Pallud, M., Kuhn, L., Hammann, P., Demortière, A., Jamali, A., Bertrand, M.: Toxicological effects of CdSe nanocrystals on the marine diatom *Phaeodactylum tricornutum*: the first mass spectrometry-based proteomic approach. Ecotoxicol. Environ. Saf. **152**, 78–90 (2018)
118. Tran, N.A.T., Padula, M.P., Evenhuis, C.R., Commault, A.S., Ralph, P.J., Tamburic, B.: Proteomic and biophysical analyses reveal a metabolic shift in nitrogen deprived *Nannochloropsis oculata*. Algal Res. **19**, 1–11 (2016)
119. Rodrigues, P.M., Silva, T.S., Dias, J., Jessen, F.: Proteomics in aquaculture: applications and trends. J. Proteom. **75**, 4325–4345 (2002)

120. Li, S., Cao, X., Wang, Y., Zhu, Z., Zhang, H., Xue, S., Tian, J.: A method for microalgae proteomics analysis based on modified filter-aided sample preparation. Appl. Biochem. Biotechnol. **18**, 923–930 (2017)
121. Jia, Y., Xue, L., Li, J., Liu, H.: Isolation and proteomic analysis of the halotolerant alga *Dunaliella salina* flagella using shotgun strategy. Mol. Biol. Rep. **37**, 711–716 (2010)
122. Viswanathan, S., Unlü, M., Minden, J.S.: Two-dimensional difference gel electrophoresis. Nat. Protoc. **1**, 1351–1358 (2006)
123. Samaranayaka, A.G.P., Li-Chan, E.C.Y.: Food-derived peptidic antioxidants: a review of their production, assessment and potential applications. J. Funct. Foods **3**, 229–254 (2011)
124. Sheih, C.I., Fang, T.J., Wu, T.K.: Isolation and characterization of novel angiotensin I-converting enzyme (ACE) inhibitory peptide from the algae protein waste. Food Chem. **115**, 279–284 (2009)
125. Suetsuna, K., Chen, J.R.: Identification of antihypertensive peptides from peptic digest of two microalgae, *Chlorella vulgaris* and *Spirulina platensis*. Mar. Biotechnol. **3**, 305–309 (2001)
126. Hulatt, C.J., Wijffels, R.H., Bolla, S., Kiron, V.: Production of fatty acids and protein by *Nannochloropsis* in flat plate photobioreactors. PLoS ONE **12**(1), e0170440 (2017)
127. Da Silva, T.L., Gouveia, L., Reis, A.: Integrated microbial processes for biofuels and high value-added products: the way to improve the cost effectiveness of biofuel production. Appl. Microbiol. Biotechnol. **98**, 1043–1053 (2014)
128. Ruiz-Ruiz, F., Benavides, J., Rito-Palomares, M.: Scaling-up of a B-phycoerythrin production and purification bioprocess involving aqueous two-phase systems: practical experiences. Process Biochem. **48**, 738–745 (2013)
129. Dumitriu, S.: Polysaccharides: Structural diversity and functional versatility. Polysaccharides: Structural Diversity and Functional Versatility, 2nd edn. Marcel Dekker Inc., New York (2005)
130. Turnbull, J.E., Field, R.A.: Emerging glycomics technologies. Nat. Chem. Biol. **3**, 74–77 (2007)
131. Kamerling, J.P., Gerwig, G.J.: Strategies for the structural analysis of carbohydrates. In: Boons, G.J., Kamerling, J.P., Lee, Y.C., Suzuki, A., Taniguchi, N., Voragen, A.G.J. (eds.) Comprehensive Glycoscience—From Chemistry to Systems Biology, vol. 2, pp. 1–68. Elsevier B.V., Oxford (2007)
132. Delattre, C., Pierre, G., Laroche, C., Michaud, P.: Production, extraction and characterization of microalgal and cyanobacterial exopolysaccharides. Biotechnol. Adv. **34**(7), 1159–1179 (2016)
133. Raposo, M.P.F.J., Morais, R.M.S.C., Morais, A.M.M.B.: Bioactivity and applications of sulphated polysaccharides from marine microalgae. Mar. Drugs **11**, 233–252 (2013)
134. Raposo, M.P.F.J., Morais, R.M.S.C., Morais, A.M.M.B.: Marine polysaccharides from algae with potential biomedical applications. Mar. Drugs **13**, 2967–3028 (2015)
135. Arad, S.M., Levy-Ontman, O.: Red microalgae cell-wall polysaccharides: biotechnological aspects. Curr. Opin. Biotech. **21**, 358–364 (2010)
136. De Philippis, R., Vincenzini, M.: Exocellular polysaccharides from Cyanobacteria. FEMS Microbiol. Let. **22**, 151–175 (1998)
137. De Philippis, R., Sili, C., Paperi, R., Vincenzini, M.: Exopolysaccharide-producing Cyanobacteria and their possible exploitation: a review. J. Appl. Phycol. **13**, 293–299 (2001)
138. Gloaguen, V., Morvan, H., Hoffman, L., Plancke, Y., Wieruszeski, J.M., Lippens, G., Strecker, G.: Capsular polysaccharide produced by the termophilic cyanobacterium *Mastigocladus laminosus*. Eur. J. Biochem. **266**, 1–10 (1999)
139. Pereira, S., Zille, A., Micheletti, E., Moradas-Ferraira, P., De Philippis, R., Tamagnini, P.: Complexity of cyanobacterial exopolysaccharides: composition, structures, inducing factors and putative genes involved in their biosynthesis and assembly. FEMS Microbiol. Rev. **33**, 917–941 (2009)

140. Cheng, Y.S., Labavitch, J.M., Vabdergheynst, J.S.: Elevated CO_2 concentration impacts cell wall polysaccharide composition of green microalgae of the genus *Chlorella*. Lett. Appl. Microbiol. **60**, 1–7 (2015)
141. Magaletti, E., Urbani, R., Sist, P., Ferrari, C.R., Cicero, A.M.: Abundance and chemical characterization of extracellular carbohydrates released by the marine diatom *Cylindrotheca fusiformis* under N- and P-limitation. Eur. J. Phycol. **39**, 133–142 (2004)
142. Underwood, G.J.C., Boulcot, M., Raines, C.A., Waldron, K.: Environmental effects on exopolymer production by marine benthic diatoms—dynamics, changes in composition and pathways of production. J. Phycol. **40**, 293–304 (2004)
143. Azeredo, J., Henriques, M., Sillankorva, S., Oliveira, R.: Extraction of exopolymers from biofilms: the protective effect of glutaraldehyde. Water Sci. Technol. **47**, 175–179 (2003)
144. Bertocchi, C., Navarini, L., Ceshro, A.: Polysaccharides from Cyanobacteria. Carbohydr. Polym. **12**, 127–153 (1990)
145. Liu, H., Fang, H.P.: Extraction of extracellular polymeric substances (EPS) of sludges. J. Biotechnol. **95**, 249–256 (2002)
146. Takahashi, E., Ledauphin, J., Goux, D., Orvain, F.: Optimizing extraction of extracellular polymeric substances (EPS) from benthic diatoms: comparison of the efficiency of six EPS extraction methods. Austr. J. Mar. Freshwat. Res. **60**, 1201–1210 (2009)
147. Li, H., Li, Z., Xiong, S., Zhang, H., Li, N., Zhou, S., Liu, Y., Huang, Z.: Pilot-scale isolation of bioactive extracellular polymeric substances from cell-free media of mass microalgal cultures using tangential-flow ultrafiltration. Process Biochem. **46**(5), 1104–1109 (2011)
148. Zhang, G.J., Ji, S.L., Gao, X., Liu, Z.Z.: Adsorptive fouling of extracellular polymeric substances with polymeric ultrafiltration membranes. J. Membr. Sci. **309**, 28–35 (2008)
149. Liu, B., Sun, Z., Ma, X., Yang, B., Jiang, Y., Wei, D., Chen, F.: Mutation breeding of extracellular polysaccharide-producing microalgae *Crypthecodinium cohnii* by a novel mutagenesis with atmospheric and room temperature plasma. Int. J. Mol. Sci. **16**(4), 8201–8212 (2015)
150. Patel, A.K., Laroche, C., Marcati, A., Ursu, A.V., Jubeau, S., Marchal, L., Petit, E., Djelveh, G., Michaud, P.: Separation and fractionation of exopolysaccharides from *Porphyridium cruentum*. Bioresour. Technol. **145**, 345–350 (2013)
151. Pierre, G., Zhao, J.M., Orvain, F., Dupuy, C., Klein, G.L., Graber, M., Maugard, T.: Seasonal dynamics of extracellular polymeric substances (EPS) in surface sediments of a diatom-dominated intertidal mudflat (Marennes-Oléron, France). J. Sea Res. **92**, 26–35 (2014)
152. Pignolet, O., Jubeau, S., Vacca-Garcia, C., Michaud, P.: Highly valuable microalgae: biochemical and topological aspects. J. Ind. Microbiol. Biotechnol. **40**(8), 781–796 (2013)
153. Charcosset, C.: Membrane processes in biotechnology: an overview. Biotechnol. Adv. **24**, 482–492 (2006)
154. Zeman, L.J., Zydney, A.L.: Microfiltration and ultrafiltration: Principles and Applications, pp. 544–551. Marcel Dekker Inc., New York (1996)
155. Marcati, A., Ursu, A.V., Laroche, C., Soanen, N., Marchal, L., Jubeau, S., Djelveh, G., Michaud, P.: Extraction and fractionation of polysaccharides and B-phycoerythrin from the microalgae *Porphyridium cruentum* by membrane technology. Algal Res. **5**, 258–263 (2014)
156. Zhang, S.J., Santschi, P.H.: Application of crossflow ultrafiltration for isolating exopolymeric substances from a marine diatom (*Amphora* sp.). Limnol. Oceanogr. Methods **7**, 419–429 (2009)
157. Rickman, M., Pellegrino, J., Davis, R.: Fouling phenomena during membrane filtration of microalgae. J. Membr. Sci. **423**, 33–42 (2012)
158. Ye, Y., Le Clech, P., Chen, V., Fane, A.G.: Evolution of fouling during crossflow filtration of model EPS solutions. J. Membr. Sci. **264**, 190–199 (2005)
159. Baruah, G.L., Venkiteshwaran, A., Belfort, G.: Global model for optimizing crossflow microfiltration and ultrafiltration processes: a new predictive and design tool. Biotechnol. Prog. **21**, 1013–1025 (2005)

160. Jhaveri, J.H., Murthy, Z.V.P.: A comprehensive review on anti-fouling nanocomposite membranes for pressure driven membrane separation processes. Desalination **379**, 137–154 (2016)
161. Rossi, N., Jaouen, P., Legentilhomme, P., Petit, I.: Harvesting of cyanobacterium *Arthrospira platensis* using organic filtration membranes. Food Bioprod. Process. **82**, 244–250 (2004)
162. Rossi, N., Petit, I., Jaouen, P., Legentilhomme, P., Derouiniot, M.: Harvesting of cyanobacterium *Arthrospira platensis* using inorganic filtration membranes. Sep. Sci. Technol. **40**, 3033–3050 (2005)
163. Brown, J.M., Lester, N.J.: Comparison of bacterial extracellular polymer extraction methods. Appl. Environ. Microbiol. **40**, 179–185 (1980)
164. Comte, S., Guibaud, G., Baudu, M.: Relations between extraction protocols for activated sludge extracellular polymeric substances (EPS) complexation properties. Part I. Comparison of the efficiency of eight EPS extraction methods. Enz. Microbiol.Technol. **38**, 237–245 (2006)
165. Frølund, B., Palmgren, R., Keiding, K., Nielsen, P.H.: Extraction of extracellular polymers from activated sludge using a cation exchange resin. Water Res. **30**, 1749–1758 (1996)
166. Underwood, G.J.C., Paterson, D.M.: The importance of extracellular carbohydrate production by marine epipelic diatoms. Adv. Bot. Res. **40**, 184–240 (2003)
167. Pierre, G., Graber, M., Orvain, F., Dupuy, C., Maugard, T.: Biochemical characterization of extracellular polymeric substances extracted from an intertidal mudflat using a cation exchange resin. Biochem. Syst. Ecol. **38**, 917–923 (2010)
168. Pierre, G., Graber, M., Rafiliposon, B.A., Dupuy, C., Orvain, F., de Crignis, M., Maugard, T.: Biochemical composition and changes of extracellular polysaccharides (ECPS) produced during microphytobenthic biofilm development (Marennes–Oléron, France). Microb. Ecol. **63**, 157–169 (2012)
169. Underwood, G.J.C., Paterson, D.M., Parkes, R.J.: The measurement of microbial carbohydrate exopolymers from intertidal sediments. Limnol. Oceanogr. **40**, 1243–1253 (1995)
170. Bellinger, B.J., Abdullahi, A.S., Gretz, M.R., Underwood, G.J.C.: Biofilm polymers: relationship between carbohydrate biopolymers from estuarine mudflats and unialgal cultures of benthic diatoms. Aquat. Microb. Ecol. **38**, 169–180 (2005)
171. Chiovitti, A., Molino, P., Crawford, S.A., Ten, R., Spurck, T., Wetherbee, R.: The glucans extracted with warm water from diatoms are mainly derived from intracellular chrysolaminaran and not extracellular polysaccharides. Eur. J. Phycol. **39**, 117–128 (2004)
172. De Brouwer, J.F.C., Stal, L.J.: Does warm-water extraction of benthic diatoms yield extracellular polymeric substances, or does it extract intracellular chrysolaminaran? Eur. J. Phycol. **39**, 129–131 (2004)
173. Delrue, B., Fontaine, T., Routier, F., Decq, A., Wieruszeski, J.-M., Van Den Koornhuyse, N., Maddelein, M.-L., Fournet, B., Ball, S.: Waxy *Chlamydomonas reinhardtii*: Monocellular algal mutants defective in amylose biosynthesis and granule-bound starch synthase activity accumulate a structurally modified amylopectin. J. Bacteriol. **174**, 3612–3620 (1992)
174. Deschamps, P., Haferkamp, I., Dauvillée, D., Haebel, S., Steup, M., Buléon, A., Putaux, J.-L., Colleoni, C., d'Hulst, C., Plancke, C., Gould, S., Maier, U., Neuhaus, E., Ball, S.: Nature of the periplastidial pathway of starch synthesis in the Cryptophyte *Guillardia theta*. Eukaryot. Cell **5**(6), 954–963 (2006)
175. Kobayashi, T., Tanabe, I., Obayashi, A.: On the properties of the starch granules from unicellular green algae. Agric. Biol. Chem. **38**(5), 941–946 (1974)
176. Love, J., Mackie, W., McKinnell, J.W., Percival, E.: Starch-type polysaccharides isolated from the green seaweeds, *Enteromorpha compressa, Ulva lactuca, Cladophora rupestris, Codium fragile*, and *Chaetomorpha capillaris*. J. Chem. Soc. 4177–4182 (1963)
177. Mc Cracken, D.A., Cain, J.R.: Amylose in floridean starch. New Phytol. **88**, 67–71 (1981)

178. Usov, A.I., Zelinsky, N.D.: Chemical structures of algal polysaccharides. In: Dominguez, H., (ed.) pp. 23–86. Woodhead Publishing, Oxford (2013)
179. Rose, R., Rose, C.L., Omi, S.K., Forry, K.R., Durall, D.M., Bigg, W.L.: Starch determination by perchloric-acid vs enzymes—evaluating the accuracy and precision of 6 colorimetric methods. J. Agri. Food Chem. **39**, 2–11 (1991)
180. Yu, S., Blennow, A., Bojko, M., Madsen, F., Olsen, C.E., Engelsen, S.B.: Physico-chemical characterization of floridean starch of red algae. Starch **54**, 66–74 (2002)
181. Eddy, B.P., Fleming, I.D., Manners, D.J.: α-1,4-Glucosans. Part IX. The molecular structure of a starch-type polysaccharide from *Dunaliella bioculata*. J. Chem. Soc. 2827–2830 (1958)
182. Ozaki, H., Maedan, M., Nisizawa, K.: Floridean starch of a calcareous red alga *Joculator maximus*. J. Biochem. **61**(4), 497–503 (1967)
183. Fernandes, B., Dragonne, G., Abreu, A., Geada, P., Teixeira, J., Vicente, A.: Starch determination in *Chlorella vulgaris*—a comparison between acid and enzymatic methods. J. Appl. Phycol. **24**(5), 1203–1208 (2012)
184. Ji, Y., He, Y., Cui, Y., Wang, T., Wang, Y., Li, Y., Huang, W.E., Xu, J.: Raman spectroscopy provides a rapid, non-invasive method for quantitation of starch in live, unicellular microalgae. Biotechnol. J. **9**, 1512–1518 (2014)
185. Takeda, H.: Sugar composition of the cell wall and the taxonomy of *Chlorella* (Chlorophyceae). J. Phycol. **27**, 224–232 (1991)
186. McCleary, B.V.: An integrated procedure for the measurement of total dietary fibre (including resistant starch), non-digestible oligosaccharides and available carbohydrates. Anal. Bioanal. Chem. **389**, 291–308 (2007)
187. Dubois, M., Gilles, K.A., Hamilton, J.K., Rebers, P.A., Smith, F.: Colorimetric method for determination of sugars and related substances. Anal. Chem. **28**, 350–356 (1956)
188. Brückner, J.: Estimation of monosaccharides by the orcinol-sulphuric acid reaction. Biochem. J. **60**, 200–205 (1955)
189. Laurentin, A., Edwards, C.A.: A microtiter modication of the anthrone–sulfuric acid colorimetric assay for glucose-based carbohydrates. Anal. Biochem. **315**, 143–145 (2003)
190. Leyva, A., Quintana, A., Sánchez, M., Rodríguez, E.N., Cremata, J., Sánchez, J.C.: Rapid and sensitive anthrone–sulfuric acid assay in microplate format to quantify carbohydrate in biopharmaceutical products: method development and validation. Biologicals **36**, 134–141 (2008)
191. Albalasmeh, A., Berhe, A.A., Ghezzehei, T.A.: A new method of rapid determination of carbohydrate and total carbon concentrations using UV spectrophotometry. Carbohydr. Polym. **97**, 253–261 (2013)
192. Masuko, T., Minami, A., Iwasaki, N., Majima, T., Nishimura, S.-I., Lee, Y.C.: Carbohydrate analysis by a phenol-sulfuric acid method in microplate format. Anal. Biochem. **339**, 69–72 (2005)
193. Monsigny, M., Petit, C., Roche, A.C.: Colorimetric determination of neutral sugars by a resorcinol sulfuric acid micromethod. Anal. Biochem. **175**, 525–530 (1988)
194. Blumenkrantz, N., Asboe-Hansen, G.: New method for quantitative determination of uronic acids. Anal. Biochem. **54**, 484–489 (1973)
195. Filisetti-Cozzi, T.M., Carpita, N.C.: Measurement of uronic acids without interference from neutral sugars. Anal. Biochem. **197**, 157–162 (1991)
196. Montreuil, J., Spick, G., Chosson, A., Segard, E., Scheppler, N.: Methods of study of the structure of glycoproteins. J. Pharm. Belg. **18**, 529–546 (1963)
197. Craigie, J.S., Wen, Z.C., Van der Meer, J.P.: Interspecific, intraspecific and nutritionally determined variations in the composition of agars from *Gracilaria* sp. Bot. Mar. **27**, 55–61 (1984)
198. Dodgson, K.S., Price, R.C.: A note on the determination of the ester sulfate content of sulfated polysaccharides. Biochem. J. **84**, 106–110 (1962)
199. Jaques, L.B., Ballieux, R.E., Dietrich, C.P., Kavanagh, L.W.: A microelectrophoresis method for heparin. Can. J. Physiol. Pharmacol. **46**, 351–360 (1968)

200. Sloneker, J.H., Orentas, D.G.: Pyruvic acid, a unique component of an exocellular bacterial polysaccharide. Nature **194**, 478–479 (1962)
201. Levigne, S., Thomas, M., Ralet, M.C., Quemener, B., Thibault, J.F.: Determination of the degrees of of methylation and acetylation of pectins using a C18 column and internal standards. Food Hydrocoll. **16**, 547–550 (2002)
202. Voragen, A.G.J., Schols, H.A., Pilnik, W.: Determination of the degree of methylation and acetylation of pectins by HPLC. Food Hydrocol. **1**, 65–70 (1986)
203. Smith, P.K., Krohn, R.I., Hermanson, G.T., Mallia, A.K., Gartner, F.H., Provenzano, M.D., Fujimoto, E.K., Goeke, N.M., Olson, B.J., Klenk, D.C.: Measurement of protein using bicinchoninic acid. Anal. Biochem. **150**, 76–85 (1985)
204. Singlton, V., Orthofer, R., Lamuela-Raventós, R.M.: Analysis of total phenols and other oxidation substrates and antioxidants by means of Folin-Ciocalteu reagent. Methods Enzymol. **299C**, 152–178 (1999)
205. Mishra, A., Jha, B.: Isolation and characterization of extracellular polymeric substances from micro-algae *Dunaliella salina* under salt stress. Bioresour. Technol. **100**, 3382–3386 (2009)
206. Mathlouthi, M., Koenig, J.L.: Vibrational spectra of carbohydrates. Adv. Carbohydr. Chem. Biochem. **44**, 82–89 (1987)
207. Adel, A.M., Abdel-Wahab, Z.H., Ibrahim, A.A., Al-Shemy, M.T.: Characterization of microcrystalline cellulose prepared from lignocellulosic materials, Part I. Acid catalyzed hydrolysis. Bioresour. Technol. **101**, 4446–4455 (2010)
208. Gnanasambandam, R., Proctor, A.: Determination of pectin degree of esterification by diffuse reflectance Fourier transform infrared spectroscopy. Food Chem. **68**, 327–332 (2000)
209. Bouaziz, F., Koubaa, M., Ben Jeddoua, K., Kallel, F., Boisset Helbert, C., Khelfa, A., Ellouz Ghorbel, R., Ellouz Chaabouni, S.: Water-soluble polysaccharides and hemicelluloses from almond gum: functional and prebiotic properties. Int. J. Biol. Macromol. **93**, 359–368 (2016)
210. Maréchal, Y.: Interaction configurations of H_2O molecules in a protein (*Stratum corneum*) by infrared spectrometry. J. Mol. Struct. **416**, 133–143 (1997)
211. Gómez-Ordóñez, E., Rupérez, P.: FTIR-ATR spectoscopy as a tool for polysaccharide identification in edible brown and red seaweeds. Food Hydrocoll. **25**, 1514–1520 (2011)
212. Leal, D., Matsuhiro, B., Rossi, M., Caruso, F.: FT-IR spectra of alginic acid block fractions in three species of brown seaweeds. Carbohydr. Res. **343**, 308–316 (2008)
213. Edwards, H.G.M., Falk, M.J., Sibley, M.G., Alvarez-Benedi, J., Rull, F.: FT-Raman spectroscopy of gums of technological significance. Spectrochim. Acta A. Mol. Biomol. Spectrosc. **54**, 903–920 (1998)
214. Barker, S.A., Bourne, E.J., Stacey, M., Whiffen, D.H.: Infra-red spectra of carbohydrates. Part I. Some derivatives of D-glucopyranose. J. Chem. Soc. **1**, 171–176 (1954)
215. Ponnuswamy, I., Madhavan, S., Shabudeen, S.: Isolation and characterization of green microalgae for carbon sequestration, wastewater treatment and bio-fuel production. Int. J. Bio-Sci. Bio-Technol. **5**, 17–26 (2013)
216. Kamerling, J.P., Gerwig, G.J., Vliegenthart, J.F.G., Clamp, J.R.: Characterization by gas-liquid chromatography-mass spectrometry and proton-magnetic-resonance spectroscopy of pertrimethylsilyl methyl glycosids obtained in the methanolysis of glycoproteins and glycopeptides. Biochem. J. **151**, 491–495 (1975)
217. Bleton, J., Mejanelle, P., Sansoulet, J., Goursaud, S., Tchapla, A.: Characterization of neutral sugars and uronic acids after methanolysis and trimethylsilysation for recognition of plant gums. J. Chromatogr. A **720**, 27–49 (1996)
218. Widmer, W.: Analysis of biomass sugars and galacturonic acid by gradient anion exchange chromatography and pulsed amperometric detection without post-column addition. Biotechnol. Lett. **33**, 365–368 (2011)
219. Fu, D., O'Neill, R.A.: Monosaccharide composition analysis of oligosaccharides and glycoproteins by high-performance liquid chromatography. Anal. Biochem. **227**, 377–384 (1995)
220. Templeton, D.W., Quinn, M., Van Wychen, S., Hyman, D., Laurens, L.M.L.: Separation and quantification of microalgal carbohydrates. J. Chromatogr. A **1270**, 225–234 (2012)

221. Schäffler, K.J., Morel du Boiland, P.G., Walford, S.N.: Some precautious required for the reliable analysis of carbohydrates. Proc. S. Afr. Sug. Technol. Ass. **70**, 241–250 (1996)
222. Hsu, C.L., Cheng, C.Y., Lee, C.T., Ding, W.H.: Derivatization procedures and determination of levoglucosan and related monosaccharide anhydrides in atmospheric aerosols by gas chromatography-mass spectrometry. Talanta **72**, 199–205 (2007)
223. Peña, M.J., Tuomivaara, S.T., Urbanowicz, B.R., O'Neill, M.A., York, W.S.: Methods for structural characterization of the products of cellulose- and xyloglucan-hydrolyzing enzymes. Methods Enzymol. **510**, 121–139 (2012)
224. Klein, G.L., Pierre, G., Bellon-Fontaine, M.-N., Zhao, J.-M., Breret, M., Maugard, T., Graber, M.: Marine diatom *Navicula jeffreyi* from biochemical composition and physico-chemical surface properties to understanding the first step of benthic biofilm formation. J. Adh. Sci. Technol. **28**, 1739–1753 (2014)
225. Depraetere, O., Pierre, G., Deschoenmaeker, F., Badri, H., Foubert, I., Leys, N., Markou, G., Wattiez, R., Michaud, P., Muylaert, K.: Harvesting carbohydrate rich *Arthrospira platensis* by spontaneous settling. Bioresour. Technol. **180**, 16–21 (2015)
226. Hara, S., Okabe, H., Mihashi, K.: Gas-liquid chromatographic separation of aldose enantiomers as trimethylsilyl ethers of methyl 2-(polyhydroxyalkyl)-thiazolidine-4(*R*)-carboxylates. Chem. Pharm. Bull. **35**, 501–506 (1987)
227. Angelis, S., Novak, A.C., Sydney, E.B., Soccol, V.T., Carvalho, J.C., Pandey, A., Noseda, M.D., Tholozan, J.L., Lorguin, J., Soccol, C.R.: Co-culture of microalgae, Cyanobacteria, and macromycetes for exopolysaccharides production: process preliminary optimization and partial characterization. Appl. Biochem. Biotechnol. **167**, 1092–1106 (2012)
228. Gugi, B., Le Costaouec, T., Burel, C., Lerouge, P., Helbert, W., Bardor, M.: Diatom-specific oligosaccharide and polysaccharide structures help to unravel biosynthetic capabilities in Diatoms. Mar. Drugs **13**, 5993–6018 (2015)
229. Duus, J.Ø., Gotfredsen, C.H., Bock, K.: Carbohydrate structural determination by NMR spectroscopy: modern methods and limitations. Chem. Rev. **100**, 4589–4614 (2000)
230. Casu, B.: Structure and conformation of polysaccharides by NMR spectroscopy. Carbohydr. Polym. **2**, 247–253 (1982)
231. Størseth, T.R., Chauton, M.S., Krane, J.: HR MAS NMR spectroscopy of marine microalgae, Part 2: ^{13}C and ^{13}C HR MAS NMR analysis used to study fatty acid composition and polysaccharide structure, In: Webb, G.A. (ed.) pp. 953–957. Springer, The Netherlands (2008)
232. Samuelsen, A.B., Cohen, E.H., Paulsen, B.S., Brüll, L.P., Thomas-Oates, J.E.: Structural studies of a heteroxylan from *Plantago major* L. seeds by partial hydrolysis, HPAEC-PAD, methylation and GC–MS, ESMS and ESMS/MS. Carbohydr. Res. **315**, 312–318 (1999)
233. Feofilova, E.P., Sergeeva, Y.E., Ivasheckhkin, A.A.: Biodiesel-fuel: content, production, producers, contemporary biotechnology (review). Appl. Biochem. Microbiol. **46**(4), 369–378 (2010)
234. Chisti, Y.: Biodiesel from microalgae. Biotechnol. Adv. **25**(3), 294–306 (2007)
235. Kulkarni, M.G., Dalai, A.K.: Waste cooking oil – An economical source for biodiesel: a review. Ind. Eng. Chem. Res. **45**, 2901–2913 (2006)
236. Naghdi, F.G., Gonzalez, L.M.G., Chan, W., Schenk, P.M.: Progress on lipid extraction from wet algal biomass for biodiesel production. Microbiol. Biotechnol. **9**, 718–726 (2016)
237. Wang, C., Chen, L., Rakesh, B., Qin, Y., Renliang, L.V.: Technologies for extracting lipids from oleaginous microorganisms for biodiesel production. Front. Energy **6**(3), 266–274 (2012)
238. Meng, X., Yang, J., Xu, X., Zhang, L., Nie, Q., Xian, M.: Biodiesel production from oleaginous microorganisms. Renew. Energy **34**, 1–5 (2009)
239. Metting Jr., F.B.: Biodiversity and application of microalgae. J. Indust. Microbiol. **17**, 477–489 (1996)
240. Dong, T., Knoshaug, E.P., Pienkos, P.T., Laurens, L.M.L.: Lipid recovery from wet oleaginous microbial biomass for biofuel production: A critical review. Appl. Energy **177**, 879–895 (2016)

241. Lee, S.Y., Cho, J.M., Chang, Y.K., Oh, Y.K.: Cell disruption and lipid extraction for microalgal biorefineries: a review. Bioresour. Technol. **244**, 1317–1328 (2017)
242. Work, V.H., Betley, F.K., Scholz, M.J., D'Adamo, S., Gu, H., Vogler, B.W., Franks, D.T., Stanish, L.F., Jinkerson, R.E., Posewitz, M.C.: Biocommodities from photoshynthetic microorganisms. Environ. Prog. Sustain. Energy **32**(4), 989–1001 (2013)
243. Gerken, H.G., Donohoe, B., Knoshaug, E.P.: Enzymatic cell wall degradation of *Chlorella vulgaris* and other microalgae for biofuels production. Planta **237**, 239–253 (2013)
244. Han, Y., Wen, Q., Chen, Z., Li, P.: Review of methods used for microalgal lipid-content analysis. Energy Procedia **12**, 944–950 (2011)
245. Axelsson, M., Gentili, F.: A single-step method for rapid extraction of total lipids from green microalgae. Plos One **9**(2), e89643 (2014)
246. Ryckebosch, E., Muylaert, K., Foubert, I.: Optimization of an analytical procedure for extraction of lipids from microalgae. J. Am. Oil Chem. Soc. **89**, 189–198 (2012)
247. Nireesha, G.R., Divya, L., Sowmya, C., Venkateshan, N., Babu, M.N., Lavakumar, V.: Lyophilization/freeze drying—a review. Int. J. Novel Trends Pharm. Sci. **3**(4), 87–98 (2013)
248. Yu, X., Dong, T., Zheng, Y., Miao, C., Chen, S.: Investigations on cell disruption of oleaginous microorganisms: hydrochloric acid digestion is an effective method for lipid extraction. Eur. J. Lipid Sci. Technol. **117**, 730–737 (2015)
249. Yang, H.Y., Lu, W.J., Chen, Y.C., Chen, K.T., Teng, J.C., Wan, H.P.: New algal lipid extraction procedure using an amphiphilic amine solvent and ionic liquid. Biomass Bioenergy **100**, 108–115 (2017)
250. Concas, A., Pisu, M., Cao, G.: Microalgal cell disruption through fenton reaction: experiments, modeling and remarks on its effect on the extracted lipids composition. Chem. Eng. Trans. **43**, 367–372 (2015)
251. Zuorro, A., Lavecchia, R., Maffei, G., Marra, F., Migietta, S., Petrangeli, A., Familiari, G., Valente, T.: Enhanced lipid extraction from unbroken microalgal cells using enzymes. Chem. Eng. Trans. **43**, 211–216 (2015)
252. Dvoretsky, D., Dvoretsky, S., Temnov, M., Akulinin, E., Peshkova, E.: Enhanced lipid extraction from microalgae *Chlorella vulgaris* biomass: experiments, modeling, optimization. Chem. Eng. Trans. **49**, 175–180 (2016)
253. Folch, J., Lees, M., Stanley, G.H.S.: A simple method for the isolation and purification of total lipids from animal tissues. J. Biol. Chem. **226**, 497–509 (1957)
254. Bligh, E.G., Dyer, W.J.: A rapid method of total lipid extraction and purification. Can. J. Biochem. Physiol. **37**, 911–917 (1959)
255. Phillipis, K.M., Ruggio, D.M., Amanna, K.R.: Extended validation of a simplified extraction and gravimetric determination of total fat to selected foods. J. Food Lipids **15**, 309–325 (2008)
256. Iverson, S.J., Lang, S.L.C., Cooper, M.H.: Comparison of the Bligh and Dyer and Folch methods for total lipid determination in a broad range of marine tissue. Lipids **36**(11), 1283–1287 (2001)
257. Sheng, J., Vannela, R., Rittmann, B.E.: Evaluation of methods to extract and quantify lipids from Synechocystis PCC 6803. Bioresour. Technol. **102**, 1697–1703 (2011)
258. Ranjan, A., Patil, C., Moholkar, V.S.: Mechanistic assessment of microalgal lipid extraction. Ind. Eng. Chem. Res. **49**, 2979–2985 (2010)
259. Cho, H.S., Oh, Y.K., Park, S.C., Lee, J.W., Park, J.Y.: Effects of enzymatic hydrolysis on lipid extraction from *Chlorella vulgaris*. Renew. Energy **54**, 156–160 (2013)
260. Cavonius, L.R., Carlsson, N.G., Undeland, I.: Quantification of total fatty acids in microalgae: comparison of extraction and transesterification methods. Anal. Bioanal. Chem. **406**, 7313–7322 (2014)
261. Bigelow, N.W., Hardin, W.R., Barker, J.P., Ryken, S.A., MacRae, A.C., Cattolico, R.A.: A comprehensive GC-MS sub-microscale assay for fatty acids and its applications. J. Amer. Oil Chem. Soc. **88**, 1329–1338 (2011)
262. Hara, A., Radin, N.S.: Lipid extraction of tissues with a low-toxicity solvent. Anal. Biochem. **90**, 420–426 (1978)

263. Halim, R., Gladman, B., Danquah, M.K., Webley, P.A.: Oil extraction from microalgae for biodiesel production. Bioresour. Technol. **102**, 178–185 (2011)
264. Kumar, J.S.P., Garlapati, V.K., Dash, A., Scholz, P., Banerjee, R.: Sustainable green solvents and techniques for lipid extraction from microalgae: a review. Algal Res. **21**, 138–147 (2017)
265. Lin, J.H., Liu, L.Y., Yang, M.H., Lee, M.H.: Ethyl acetate/ethyl alcohol mixtures as an alternative to Folch reagent for extracting animal lipids. J. Agric. Food Chem. **52**, 4984–4986 (2004)
266. Guckert, J.B., Cooksey, K.E., Jackson, L.L.: Lipid solvent systems are not equivalent for analysis of lipid classes in the microeukaryotic green alga, *Chlorella*. J. Microbiol. Methods **8**, 139–149 (1988)
267. Christie, W.W.: Lipid Analysis: Isolation, Separation, Identification and Structural Analysis of Lipids, 3rd edn. The Oily Press, Bridgewater (2003)
268. Chen, G.Q., Jiang, Y., Chen, F.: Fatty acid and lipid class composition of the eicosapentaenoic acid-producing microalga, *Nitzschia laevis*. Food Chem. **104**, 1580–1585 (2007)
269. Yao, L., Gerde, J.A., Lee, S.L., Wang, T., Harrata, K.A.: Microalgae lipid characterization. J. Agric. Food Chem. **63**, 1773–1787 (2015)
270. Jones, J., Manning, S., Montoya, M., Keller, K., Poenie, M.: Extraction of algal lipids and their analysis by HPLC and mass spectrometry. J. Amer. Oil Chemists Soc. **89**, 1371–1381 (2012)
271. Graeve, M., Janssen, D.: Improved separation and quantification of neutral and polar lipid classes by HPLC-ELSD using a monolithic silica phase: application to exceptional marine lipids. J. Chrom. B **877**, 1815–1819 (2009)
272. Nordbäck, J., Lundberg, E., Christie, W.W.: Separation of lipid classes from marine particulate material by HPLC on a polyvinyl alcohol-bonded stationary phase using dual-channel evaporative light–scattering detection. Mar. Chem. **60**, 165–175 (1998)
273. Nuzzo, G., Gallo, C., d'Ippolito, G., Cutignano, A., Sardo, A., Fontana, A.: Composition and quantitation of microalgal lipids by ERETIC ^{1}H NMR method. Mar. Drugs **11**, 3742–3753 (2013)
274. Breuer, G., Evers, W.A.C., deVree, J.H., Kleinegris, D.M.M., Martens, D.E., Wijffels, R.H., Lamers, P.P.: Analysis of fatty acid content and composition in microalgae. J. Vis. Exp. **e50628**, 1–9 (2013)
275. Carrapiso, A.I., Garcia, C.: Development in lipid analysis: some new extraction techniques and in situ transesterification. Lipids **35**(11), 1167–1177 (2000)
276. Sattler, W., Puhl, H., Haya, M., Kostner, G.M., Esterbauer, H.: Determination of fatty acids in the main lipoprotein classes by capillary gas chromatography: BF3/methanol transesterification of lyophilized samples instead of Folch extraction gives higher yields. Anal. Biochem. **198**, 184–190 (1991)
277. Ayorinde, F.O., Clifon, Jr., Afolabi, O.A, Shepard, R.L. Rapid transesterification and mass spectrometric approach to seed oil analysis. J. Am. Oil Chem. Soc. **65**(6), 942–947 (1988)
278. Ichihara, K., Shibahara, A., Yamamoto, K., Nakayama, T.: An improved method for rapid analysis of the fatty acids of glycerolipids. Lipids **31**(5), 535–539 (1996)
279. AOAC method.: Official Methods of Analysis of the Association of Official Analytical Chemists, 17th edn. Arlington, Virginia USA (2000)
280. Griffiths, M.J., van Hille, R.P., Harrison, S.T.L.: Selection of direct transesterification as the preferred method for assay of fatty acid content of microalgae. Lipids **45**, 1053–1060 (2010)
281. Yi, T., Li, S.M., Fan, J.Y., Fan, L.L., Zhang, Z.F., Luo, P., Zhang, X.J., Wang, J.G., Zhu, L., Zhao, Z.Z., Chen, H.B.: Comparative analysis of EPA and DHA in fish oil nutritional capsules by GC-MS. Lipids Health Dis. **13**, 190–195 (2014)
282. Zhang, Y., Min, Q., Xu, J., Zhang, K., Chen, S., Wang, H., Li, D.: Effect of malate on docosahexaenoic acid production from *Schizochytrium* sp. B4D1. Electron. J. Biotechnol. **19**, 56–60 (2016)

283. Zhang, K., Li, H., Chen, W., Zhao, M., Cui, H., Min, Q., Wang, H., Chen, S., Li, D.: Regulation of the docosapentaenoic acid/docosahexaenoic acid ratio (DPA/DHA Ratio) in *Schizochytrium limacinum* B4D1. Appl. Biochem. Biotechnol. **182**, 67–81 (2017)
284. Shorupskaite, V., Makareviciene, V., Gumbyte, M.: Opportunities for simultaneous oil extraction and transesterification during biodiesel fuel production from microalgae: a review. Fuel Process. Technol. **150**, 78–87 (2016)
285. Lewis, T., Nichols, P.D., McMeekin, T.A.: Evaluation of extraction methods for recovery of fatty acids from lipid-producing microheterotrophs. J. Microbiol. Methods **43**, 107–116 (2000)
286. Dianursanti, I., Religia, P., Wijanarko, A.: Utilization of n-Hexane as co-solvent to increase biodiesel yields on direct transesterification reaction from marine microalgae. Proc. Environ. Sci. **23**, 412–420 (2015)
287. Vicente, G., Bautista, L.F., Rodriguez, R., Gutierrez, F.J., Sadaba, I., Ruiz-Vazquez, R.M., Martinez, S.T., Garre, V.: Biodiesel production from biomass of an oleaginous fungus. Biochem. Engin. J. **48**, 22–27 (2009)
288. Velasquez-Orta, S.B., Lee, J.G.M., Harvey, A.: Alkaline in situ transesterification of *Chlorella vulgaris*. Fuel **94**, 544–550 (2012)
289. Ehimen, E.A., Sun, Z.F., Carrington, C.G.: Variables affecting the in-situ transesterification of microalgae lipids. Fuel **89**, 677–684 (2010)
290. Tarola, A.M., Girelli, A.M., Lorusso, S.: High performance liquid chromatography determination of fatty acids in drying oils following lipase action. J. Chromat. Sci. **50**, 294–300 (2012)
291. Wu, Z., Zhang, Q., Li, N., Pu, Y., Wang, B., Zhang, T.: Comparison of critical methods developed for fatty acid analysis: a review. J. Sep. Sci. **40**, 288–298 (2017)
292. Hein, M., Isengard, H.D.: Determination of underivated fatty acids by HPLC. Z Lebensm Unters Forsch A. **204**, 420–424 (1997)
293. MacDougall, K.M., McNichol, J., McGinn, P.J., O'Leary, S.J.B., Melanson, J.E.: Triacylglycerol profiling of microalgae strains for biofuel feedstock by liquid chromatography-high-resolution mass spectrometry. Anal. Bioanal. Chem. **401**, 2609–2616 (2011)
294. Samburova, V., Lemos, M.S., Hiibel, S., Hoekman, K., Cushman, J.C., Zielinska, B.: Analysis of triacylglycerols and free fatty acids in algae using ultra-performance liquid chromatography mass spectrometry. JAOCS **90**, 53–64 (2013)
295. Kind, T., Meissen, J.K., Yang, D., Nocito, F., Vaniya, A., Cheng, Y.S., VanderGheynst, J. S., Fiehn, O.: Qualitative analysis of algal secretions with multiple mass spectrometric platforms. J. Chromat. A. **1244**, 139–147 (2012)
296. Tanoi, T., Kawachi, M., Watanabe, M.M.: Effects of carbon source on growth and morphology *Botryococcus braunii*. J. Appl. Phycol. **23**, 25–33 (2011)
297. Baba, M., Kikuta, F., Suzuki, I., Watanabe, M.M., Shiraiwa, Y.: Wavelength specificity of growth, photosynthesis, and hydrocarbon production in the oil-producing green alga *Botryococcus braunii*. Bioresour. Technol. **109**, 266–270 (2012)
298. Yonezawa, N., Matsuura, H., Shiho, M., Kaya, K., Watanabe, M.M.: Effects of soybean wastewater on the growth and hydrocarbon production of *Botryococcus braunii* strain BOT-22. Bioresour. Technol. **109**, 304–307 (2012)
299. Ishimatsu, A., Matsuura, H., Sano, T., Kaya, K., Watanabe, M.M.: Biosynthesis of isoprene units in the C_{34} botryococcene molecule produced by *Botryococcus braunii* strain Bot-22. Proc. Environ. Sci. **15**, 56–65 (2012)
300. Kawachi, K., Tanoi, T., Demura, M., Kaya, K., Watanabe, M.M.: Relationship between hydrocarbons and molecular phylogeny of *Botryococcus braunii*. Algal Res. **2**, 114–119 (2012)
301. Kaya, K.: Chemistry of microalgal oil. Microbiol. Cult. Collect. **26**(1), 1–10 (2010). (written in Japanese)

302. Kaya, K., Nakazawa, A., Matsuura, H., Honda, D., Inouye, I., Watanabe, M.M.: Thraustochytrid *Aurantiochytrium* sp. 18 W-13a accumulates high amounts of squalene. Biosci. Biotechnol. Biochem. **75**(11), 2246–2248 (2011)
303. Fan, K.W., Aki, T., Chen, F., Jiang, Y.: Enhanced production of squalene in the thraustochytrid *Aurantiochytrium mangrovei* by medium optimization and treatment with terbinafine. World J. Microbiol. Biotechnol. **26**(7), 1303–1309 (2010)
304. Jiang, Y., Fan, K.W., Wong, R.T.Y., Chen, F.: Fatty acid composition and squalene content of the marine microalgae *Schizochytrium mangrovei*. J. Agric. Food Chem. **52**(5), 1196–1200 (2004)
305. Li, Q., Chen, G.Q., Fan, K.W., Lu, F.P., Aki, T., Jiang, Y.: Screening and characterization of squalene producing Thraustochytrids form Hong Kong mangroves. J. Agric. Food Chem. **57** (10), 4267–4272 (2009)
306. Suzuki, R., Ito, N., Uno, Y., Nishii, I., Kagiwada, S., Okada, S., Noguchi, T.: Transformation of lipid bodies related to hydrocarbon accumulation in a green alga, *Botryococcus braunii* (Race B). PLoS ONE **8**, e81626 (2013)
307. Nakazawa, A., Kokubun, Y., Matsuura, H., Yonezawa, N., Kose, R., Yoshida, M., Tanabe, Y., Kusuda, E., Thang, D.V., Ueda, M., Honda, D., Mahakhant, A., Kaya, K., Watanabe, M. M.: TLC screening of thraustochytrid strains for squalene production. J. Appl. Phycol. **26**, 29–41 (2014)
308. Guionet, A., Hosseini, B., Teissie, J., Akiyama, H., Hosseini, H.: A new mechanism for efficient hydrocarbon electro-extraction from *Botryococcus braunii*. Biotechnol. Biofuels **10**, 39 (2017)
309. Cavanagh, J.A.E., Juhasz, A.L., Nichols, P.D., Franzmann, P.D., McMeekin, T.A.: Analysis of microbial hydrocarbon degradation using TLC-FID. J. Microbiol. Methods **2**, 119–130 (1995)
310. Okada, S., Murakami, M., Yamaguchi, K.: Hydrocarbon production by the Yayoi, a new strain of the green microalgae *Botryococcus braunii*. Appl. Biochem. Biotechnol. **67**, 79–86 (1997)
311. Wolf, F.R., Nemethy, E.K., Blanding, J.H., Massham, J.A.: Biosynthesis of unusual acyclic isoprenoids in the alga *Botryococcus braunii*. Phytochemistry **24**(4), 733–737 (1985)
312. Dayananda, C., Sarada, R., Srinivas, P., Shamala, T.R., Ravishankar, G.A.: Presence of methyl branched fatty acids and saturated hydrocarbons in botryococcene producing strain of *Botryococcus braunii*. Acta Physiol. Plant. **28**(3), 251–256 (2006)
313. Zhang, Z., Metzger, P., Sachs, J.P.: Biomarker evidence for the co-occurrence of three races (A, B and L) of *Botryococus braunii* in El Junco Lake. Galapagos. Org. Geochem. **38**, 1459–1478 (2007)
314. Dayananda, C., Sarada, R., Rani, M.U., Shamala, T.R., Ravishankar, G.A.: Autotrophic cultivation of *Botryococcus braunii* for the production of hydrocarbons and expolysaccharides in various media. Biomass Bioenergy **31**, 87–93 (2007)
315. De Mesmay, R., Metzger, P., Grossi, S., Derenne, S.: Mono- and dicyclic unsaturated triterpenoid hydrocarbons in sediments from Lake Masoko (Tanzania) widely extend the botryococcene family. Org. Geochem. **39**, 879–893 (2008)
316. Matsuura, H., Watanabe, M.M., Kaya, K.: Squalene quantification using octadecylbenzene as the internal standard. Procedia Environ. Sci. **15**, 43–46 (2012)
317. Gao, M., Simoneit, B.R.T., Gantar, M., Raffe, R.: Occurrence and distribution of novel botryococcene hydrocarbons in freshwater wetlands of the Florida Evarglades. Chemosphere **70**, 224–236 (2007)
318. Metzger, P., Casadevall, E., Pouet, M.J., Pouet, Y.: Structures of some botryococcenes: Branched hydrocarbons from the B-race of the green alga *Botryococcus braunii*. Phytochemistry **24**(12), 2995–3002 (1985)
319. Peyronel, D., Artaud, J., Iatrides, M.C, Rancurel, P., Chevalier, J.L.: Fatty acid and squalene compositions of Mediterranean *Centrophorus* spp. egg and liver oils in relation to age. Lipids **19**(9), 643–648 (1984)

320. Nollet, L.M.L.: Handbook of Food Analysis. Volume 1: Physical Characterization and Nutrient Analysis, 2nd edn. Marcel Dekker Inc., New York (2004)
321. Nara, E., Kubouchi, H., Kobayashi, H., Kotake, M., Suzuki, T., Miyashita, K.: Inhibitory effect of cathodic solution produced by the electrolysis of a dilute NaCl solution on the oxidation of squalene, vitamin A and β-carotene. J. Oleo Sci. **50**(7), 575–581 (2001)
322. Verleyen, T., Verhe, R., Garcia, L., Dewettinck, A., Huyghebaert, A., De Greyt, W.D.: Gas chromatography characterization of vegetable oil deodorization distillate. J. Chromatogr. A. **921**, 277–285 (2001)
323. Wu, W.F., Wang, Q.H., Zhang, T., Mi, S.H., Liu, Y., Wang, L.Y.: Gas chromatography analysis of serum cholesterol synthesis and absorption markers used to predict the efficacy of simvastatin in patients with coronary heart disease. Clin. Biochem. **46**, 993–998 (2013)
324. Popa, O., Babeanu, N.E., Popa, I., Nita, S., Dinu-Parvu, C.E.: Methods for obtaining and determination of squalene from natural sources. BioMed. Res. Int. ID 367202 (2015)
325. Lopes, G., Sousa, C., Valentao, P., Andrade, P.: Chapter 9: sterols in algae and health. In: Bioactive Compounds from Marine Foods: Plant and Animal Sources, 1st edn (2013)
326. Luo, X., Su, P., Zhang, W.: Advances in microalgae-derived phytosterols for functional food and pharmaceutical application. Mar. Drugs **13**, 4231–4254 (2015)
327. Greuzburg, D.M., Merkel, P.: Sterols of freshwater microalgae: potential implications for zooplankton nutrition. J. Plankton Res. **38**(4), 865–877 (2016)
328. Kamenarska, Z.G., Konaklieva, S.D.D., Stefanov, K.L., Popov, S.S.: A comparative study on the sterol composition of some brown algae from the Black Sea. J. Serbian Chem. Soc. **68** (4–5), 269–275 (2003)
329. Fabregas, J., Morales, J.A.E.D., Lamela, T., Otero, A.: Modification of sterol concentration in marine microalgae. Phytochemistry **46**(7), 1189–1191 (1997)
330. Laakso, P.: Analysis of sterols from various food matrices. Eur. J. Lipid Sci. Technol. **107**, 402–410 (2005)
331. Abilasha, R., Geetha, P., Arivazhagan, R.: Evaluation of phytosterol in sesame seed oil and study its effects on fermented dairy product. Eur. J. Food Sci. Technol. **4**(1), 10–17 (2006)
332. Ito, M., Ishimaru, M., Shibata, T., Hatate, H., Tanaka, R.: High-performance liquid chromatography with fluorescence detection for simultaneous analysis of phytosterols (stigmasterol, β-Sitosterol, campesterol, ergosterol, and fucosterol) and cholesterol in plant foods. Food Anal. Methods **10**, 2692–2699 (2017)
333. Mo, S., Dong, L., Hurst, W.J., Breemen, R.B.V.: Quantitative analysis of phytosterols in edible oils using APCI liquid chromatography-tandem mass spectrometry. Lipids **48**(9), 949–956 (2013)
334. Serive, B., Nicolau, E., Bérard, J.-B., Kaas, R., Pasquet, V., Picot, L., Cadoret, J.-P.: Community analysis of pigment patterns from 37 microalgae strains reveals new carotenoids and porphyrins characteristic of distinct strains and taxonomic groups. PLoS ONE **12**(2), e0171872 (2017)
335. Cardenas, D.: Let not thy food be confused with thy medicine: the Hippocratic misquotation. E-SPEN J. **8**(6), e260–e262 (2013)
336. Bishop, W.M., Zubeck, H.M.: Evaluation of microalgae for use as nutraceuticals and nutritional supplements. J. Nutr. Food Sci. **2**, 147 (2012)
337. Serive, B., Bach, S.: Marine Pigment Diversity: Applications and Potential, in Blue Biotechnology: Production and Use of Marine Molecules, pp. 643–682. Hoboken, New Jersey (2018)
338. Sathasivam, R., Ki, J.-S.: A review of the biological activities of microalgal carotenoids and their potential use in healthcare and cosmetic industries. Mar. Drugs **16**(1), 26 (2018). https://doi.org/10.3390/md16010026
339. Galasso, C., Corinaldesi, C., Sansone, C.: Carotenoids from marine organisms: biological functions and industrial applications. Antioxidants **6**(4) (2017)
340. Yaakob, Z., Ali, E., Zainal, A., Mohamad, M., Takriff, M.S.: An overview: biomolecules from microalgae for animal feed and aquaculture. J. Biol. Res.-Thessalon. **21**(1), 6 (2014)
341. www.gminsights.com/pressrelease/carotenoids-market

342. Mata-Gómez, L.C., Montañez, J.C., Méndez-Zavala, A., Aguilar, C.N.: Biotechnological production of carotenoids by yeasts: an overview. Microb. Cell Fact. **13**, 12 (2014)
343. Dufossé, L., Galaup, P., Yaron, A., Arad, S.M., Blanc, P., Chidambara Murthy, K.N., Ravishankar, G.A.: Microorganisms and microalgae as sources of pigments for food use: a scientific oddity or an industrial reality? Trends Food Sci. Technol. **16**(9), 389–406 (2005)
344. Dufossé, L.: Microbial production of food grade pigments. Food Technol. Biotechnol. **44**(3), 313–323 (2006)
345. Guedes, A.C., Amaro, H.M., Malcata, F.X.: Microalgae as sources of carotenoids. Mar. Drugs **9**(12), 625–644 (2011)
346. Richmond, A.: Handbook of Microalgal Culture: Biotechnology and Applied Phycology. Blackwell Publishing, Oxford, Ames, Carlton (2004)
347. Marchal, L., Mojaat-Guemir, M., Foucault, A., Pruvost, J.: Centrifugal partition extraction of β-carotene from *Dunaliella salina* for efficient and biocompatible recovery of metabolites. Bioresour. Technol. **134**, 396–400 (2013)
348. Nobre, B., Marcelo, F., Passos, R., Beirão, L., Palavra, A., Gouveia, L., Mendes, R.: Supercritical carbon dioxide extraction of astaxanthin and other carotenoids from the microalgae *Haematococcus pluvialis*. Eur. Food Res. Technol. **223**(6), 787–790 (2006)
349. Mendes, R.L., Nobre, B.P., Cardoso, M.T., Pereira, A.P., Palavra, A.F.: Supercritical carbon dioxide extraction of compounds with pharmaceutical importance from microalgae. Inorg. Chim. Acta **356**, 328–334 (2003)
350. Kitada, K., Machmudah, S., Sasaki, M., Motonobu, G., Nakashima, Y., Kumamoto, S., Hasegawa, T.: Supercritical CO_2 extraction of pigment components with pharmaceutical importance from *Chlorella vulgaris*. J. Chem. Technol. Biotechnol. **84**(5), 657–661 (2008)
351. Macías-Sánchez, M.D., Mantell Serrano, C., Rodríguez Rodríguez, M., Martínez de la Ossa, E., Lubián, L.M., Montero, O.: Extraction of carotenoids and chlorophyll from microalgae with supercritical carbon dioxide and ethanol as cosolvent. J. Sep. Sci. **31**(8), 1352–1362 (2008)
352. Krichnavaruk, S., Shotipruk, A., Goto, M., Pavasant, P.: Supercritical carbon dioxide extraction of astaxanthin from *Haematococcus pluvialis* with vegetable oils as co-solvent. Bioresour. Technol. **99**(13), 5556–5560 (2008)
353. Hejazi, M.A.: Milking of Microalgae: Production and Selective Extraction of ß-Carotene in Two-Phase Bioreactors. Ph.D. thesis, Wageningen University, The Netherlands (2003)
354. Hejazi, M.A., Wijffels, R.H.: Milking of microalgae. Trends Biotechnol. **22**(4), 189–194 (2004)
355. Castro-Puyana, M., Herrero, M., Urreta, I., Mendiola, J.A., Cifuentes, A., Ibáñez, E., Suárez-Alvarez, S.: Optimization of clean extraction methods to isolate carotenoids from the microalgae *Neochloris oleoabundans* and subsequent chemical characterization using liquid chromatography tandem mass spectrometry. Anal. Bioanal. Chem. **405**(13), 4607–4616 (2013)
356. Pasquet, V., Chérouvrier, J.-R., Farhat, F., Thiéry, V., Piot, J.-M., Bérard, J.-B., Kaas, R., Serive, B., Patrice, T., Cadoret, J.-P., Picot, L.: Study on the microalgal pigments extraction process: performance of microwave assisted extraction. Process Biochem. **46**, 59–67 (2011)
357. Serive, B., Kaas, R., Bérard, J.-B., Pasquet, V., Picot, L., Cadoret, J.-P.: Selection and optimisation of a method for efficient metabolites extraction from microalgae. Bioresour. Technol. **124**, 311–320 (2012)
358. Petrier, C., Jeunet, A., Luche, J.-L., Reverdy, G.: Unexpected frequency effects on the rate of oxidative processes induced by ultrasound. J. Am. Chem. Soc. **114**(8), 3148–3150 (1992)
359. Boussetta, N., Vorobiev, E.: Extraction of valuable biocompounds assisted by high voltage electrical discharges: a review. Comptes Rendus Chim. **17**(3), 197–203 (2014)
360. Dumay, J., Morançais, M., Nguyen, H.P.T., Fleurence, J.: Extraction and purification of R-phycoerythrin from marine red algae. Methods Mol. Biol. Clifton NJ **1308**, 109–117 (2015)

361. Al-Zuhair, S., Ashraf, S., Hisaindee, S., Al Darmaki, N., Battah, S., Svistunenko, D., Reeder, B., Stanway, G., Chaudhary, A.: Enzymatic pre-treatment of microalgae cells for enhanced extraction of proteins. Eng. Life Sci. **17**(2), 175–185 (2017)
362. Lee, A.K., Lewis, D.M., Ashman, P.J.: Disruption of microalgal cells for the extraction of lipids for biofuels: processes and specific energy requirements. Biomass Bioenergy **46**, 89–101 (2012)
363. Carle, R., Schweiggert, R.: Handbook on Natural Pigments in Food and Beverages: Industrial Applications for Improving Food Color. Duxford, UK (2016)
364. Shoaf, W.T., Lium, B.W.: Improved extraction of chlorophyll a and b from algae using dimethyl sulfoxide. Limnol. Oceanogr. **21**, 926–928 (1976)
365. Orosa, M., Franqueira, D., Cid, A., Abalde, J.: Analysis and enhancement of astaxanthin accumulation in *Haematococcus pluvialis*. Bioresour. Technol. **96**(3), 373–378 (2005)
366. Domínguez-Bocanegra, A.R., Guerrero Legarreta, I., Martinez Jeronimo, F., Tomasini Campocosio, A.: Influence of environmental and nutritional factors in the production of astaxanthin from *Haematococcus pluvialis*. Bioresour. Technol. **92**(2), 209–214 (2004)
367. Besombes, C., Berka-Zougali, B., Allaf, K.: Instant controlled pressure drop extraction of lavandin essential oils: fundamentals and experimental studies. J. Chromatogr. A **1217**(44), 6807–6815 (2010)
368. Sarada, R., Vidhyavathi, R., Usha, D., Ravishankar, G.A.: An efficient method for extraction of astaxanthin from green alga *Haematococcus pluvialis*. J. Agric. Food Chem. **54**(20), 7585–7588 (2006)
369. Wang, J., He, Z., Wang, L., Xu, Y., Peng, Y., Liu, X.: Automatic single-step quick, easy, cheap, effective, rugged and safe sample preparation devices for analysis of pesticide residues in foods. J. Chromatogr. A **1521**, 10–18 (2017)
370. Ma, R.Y.N., Chen, F.: Enhanced production of free trans-astaxanthin by oxidative stress in the cultures of the green microalgae *Chlorococcum* sp. Process Biochem. **36**, 1175–1179 (2001)
371. Kang, C.D., Sim, S.J.: Selective extraction of free astaxanthin from *Haematococcus* culture using a tandem organic solvent system. Biotechnol. Prog. **23**(4), 866–871 (2007)
372. Günerken, E., D'Hondt, E., Eppink, M.H.M., Garcia-Gonzalez, L., Elst, K., Wijffels, R.H.: Cell disruption for microalgae biorefineries. Biotechnol. Adv. **33**, 243–260 (2015)
373. Christaki, E., Bonos, E., Florou-Paneri, P.: Innovative microalgae pigments as functional ingredients in nutrition. In: Handbook of Marine Microalgae: Biotechnology Advances, pp. 233–243. London (2015)
374. Odjadjare, E.C., Mutanda, T., Olaniran, A.O.: Potential biotechnological application of microalgae: a critical review. Crit. Rev. Biotechnol. **37**(1), 37–52 (2017)
375. Singh, R., Parihar, P., Singh, M., Bajguz, A., Kumar, J., Singh, S., Singh, V.P., Prasad, S. M.: Uncovering potential applications of cyanobacteria and algal metabolites in biology, agriculture and medicine: current status and future prospects. Front. Microbiol. **8**, 1–37 (2017)
376. Esquivel-Hernández, D.A., Ibarra-Garza, I.P., Rodríguez-Rodríguez, J., Cuéllar-Bermúdez, S.P., Rostro-Alanis, M. de J., Alemán-Nava, G.S., García-Pérez, J.S., Parra-Saldívar, R.: Green extraction technologies for high-value metabolites from algae: a review. Biofuels Bioprod. Biorefin. **11**, 215–231 (2017)
377. Saini, R.K., Keum, Y.-S.: Carotenoid extraction methods: a review of recent developments. Food Chem. **240**, 90–103 (2018)
378. Jeffrey, S., Hallegraeff, G.: Chlorophyllase distribution in ten classes of phytoplankton: a problem for chlorophyll analysis. Mar. Ecol. Prog. Ser. **35**(3), 293–304 (1987)
379. Hooker, S.B., Clementson, L., Thomas, C.S., Schlüter, L., Allerup, M., Ras, J., Claustre, H., Normandeau, C., Cullen, J., Kienast, M., Kozlowski, W.A., Vernet, M., Chakraborty, S., Lohrenz, S., Tuel, M., Redalje, D., Cartaxana, P., Mendes, C.R., Brotas, V., Matondkar, S. G.P., Parab, S.G., Neeley, A., Egeland, E.S.: The fifth SeaWiFS HPLC analysis round-Robin experiment (SeaHARRE-5). In: Hooker, S.B., Firestone, E.R. (eds.) NASA Goddard Space Flight Center, Greenbelt, Maryland (2012)

380. Gieskes, W.W.C., Kraay, G.W.: Estimating the carbon-specific growth rate of the major algal species groups in eastern Indonesian waters by 14C labeling of taxon-specific carotenoids. Deep Sea Res. Part Oceanogr. Res. Pap. **36**(8), 1127–1139 (1989)
381. Wright, S.W., Jeffrey, S.W., Mantoura, R.F.C., Llewellyn, C.A., Bjornland, T., Repeta, D., Welschmeyer, N.: Improved HPLC method for the analysis of chlorophylls and carotenoids from marine phytoplankton. Mar. Ecol. Prog. Ser. **77**, 183–196 (1991)
382. Egeland, E.S., Johnsen, G., Eikrem, W., Throndsen, J., Liaaen-Jensen, S.: Pigments of *Bathycoccus prasinos* (Prasinophyceae): methodological and chemosystematic implications. J. Phycol. **31**, 554–561 (1995)
383. Pinckney, J.L., Millie, D.F., Howe, K.E., Paerl, H.W., Hurley, J.P.: Flow scintillation counting of 14C-labeled microalgal photosynthetic pigments. J. Plankton Res. **18**(10), 1867–1880 (1996)
384. Vidussi, F., Claustre, H., Bustillos-Guzmàn, J., Cailliau, C., Marty, J.-C.: Determination of chlorophylls and carotenoids of marine phytoplankton: separation of chlorophyll a from divinylchlorophyll a and zeaxanthin from lutein. J. Plankton Res. **18**(12), 2377–2382 (1996)
385. Barlow, R.G., Cummings, D.G., Gibb, S.W.: Improved resolution of mono-and divinyl chlorophylls a and b and zeaxanthin and lutein in phytoplankton extracts using reverse phase C-8 HPLC. Mar. Ecol. Prog. Ser. **161**, 303–307 (1997)
386. Zapata, M., Rodriguez, F., Garrido, J.L.: Separation of chlorophylls and carotenoids from marine phytoplankton: a new HPLC method using a reversed phase C8 column and pyridine-containing mobile phases. Mar. Ecol. Prog. Ser. **195**, 29–45 (2000)
387. Van Heukelem, L., Thomas, C.S.: Computer-assisted high-performance liquid chromatography method development with applications to the isolation and analysis of phytoplankton pigments. J. Chromatogr. A **910**(1), 31–49 (2001)
388. Bidigare, R.R., Van Heukelem, L., Trees, C.: HPLC phytoplankton pigments: sampling, laboratory methods, and quality assurance procedures. In: Ocean Optics Protocols for Satellite Ocean Color Sensor Validation Rev. 3 vol. 2, pp. 258–268. NASA Goddard Space Flight Center, Greenbelt, Maryland (2002)
389. Bidigare, R.R., Van Heukelem, L., Trees, C.: Analysis of algal pigments by high-performance liquid chromatography. In: Algal Culturing Techniques, pp. 327–345. Amsterdam, Boston, Heidelberg, London, New York, Oxford, Paris, San Diego, San Francisco, Singapore, Sydney, Tokyo (2005)
390. Rodriguez-Amaya, D.B., Kimura, M.: Handbook for Carotenoid Analysis, 58 pp. HarvestPlus, Washington (2004)
391. Yuan, J.P., Chen, F.: Hydrolysis kinetics of astaxanthin esters and stability of astaxanthin of *Haematococcus pluvialis* during saponification. J. Agric. Food Chem. **47**(1), 31–35 (1999)
392. Carotenoids Database. www.carotenoiddb.jp. Accessed Dec 2018
393. Britton, G., Liaaen-Jensen, S., Pfander, H.: Carotenoids Handbook. Springer, Berlin (2004)
394. Butnariu, M.: Methods of analysis (extraction, separation, identification and quantification) of carotenoids from natural products. J. Ecosyst. Ecography **6**(2) (2016)
395. Lynn, K.S., Cheng, M.L., Chen, Y.R., Hsu, C., Chen, A., Lih, T.M., Chang, H.Y., Huang, C., Shiao, M.S., Pan, W.H., Sung, T.Y., Hsu, W.L.: Metabolite identification for mass spectrometry-based metabolomics using multiple types of correlated ion information. Anal. Chem. **87**(4), 2143–2151 (2015)
396. Allard, P.M., Genta-Jouve, G., Wolfender, J.L.: Deep metabolome annotation in natural products research: towards a virtuous cycle in metabolite identification. Curr. Opin. Chem. Biol. **36**, 40–49 (2017)
397. Bingol, K., Brüschweiler, R.: Knowns and unknowns in metabolomics identified by multidimensional NMR and hybrid MS/NMR methods. Curr. Opin. Biotechnol. **43**, 17–24 (2017)
398. Guitton, Y., Tremblay-Franco, M., Le Corguillé, G., Martin, J.-F., Pétéra, M., Roger-Mele, P., Delabrière, A., Goulitquer, S., Monsoor, M., Duperier, C., Canlet, C., Servien, R., Tardivel, P., Caron, C., Giacomoni, F., Thévenot, E.A.: Create, run, share, publish, and reference your LC–MS, FIA–MS, GC–MS, and NMR data analysis workflows with the

Workflow4Metabolomics 3.0 Galaxy online infrastructure for metabolomics. Int. J. Biochem. Cell Biol. **93**, 89–101 (2017)
399. Global Natural Products Social molecular Networking. http://gnps.ucsd.edu. Accessed June 2017
400. Wang, M., Carver, J.J., Phelan, V.V., Sanchez, L.M., Garg, N., Peng, Y., Nguyen, D.D., Watrous, J., Kapono, C.A., Luzzatto-Knaan, T., Porto, C., Bouslimani, A., Melnik, A.V., Meehan, M.J., Liu, W.-T., Crüsemann, M., Boudreau, P.D., Esquenazi, E., Sandoval-Calderón, M., Kersten, R.D.: Sharing and community curation of mass spectrometry data with global natural products social molecular networking. Nat. Biotechnol. **34**(8), 828–837 (2016)
401. Schymanski, E.L., Jeon, J., Gulde, R., Fenner, K., Ruff, M., Singer, H.P., Hollender, J.: Identifying small molecules via high resolution mass spectrometry: communicating confidence. Environ. Sci. Technol. **48**, 2097–2098 (2014)
402. Grimm, B., Porra, R.J., Rüdiger, W., Scheer, H.: Chlorophylls and Bacteriochlorophylls: Biochemistry, Biophysics, Functions and Applications, Springer, Berlin (2006)
403. van Breemen, R.B., Dong, L., Pajkovic, N.D.: Atmospheric pressure chemical ionization tandem mass spectrometry of carotenoids. Int. J. Mass Spectrom. **312**, 163–172 (2012)
404. Milenković, S.M., Zvezdanović, J.B., Jelković, T.D., Marković, D.Z.: The identification of chlorophyll and its derivatives in the pigment mixtures: HPLC-chromatography, visible and mass spectroscopy studies. Adv. Technol. **1**, 16–24 (2012)
405. Rivera, S.M., Christou, P., Canela-Garayoa, R.: Identification of carotenoids using mass spectrometry. Mass Spectrom. Rev. **33**(5), 353–372 (2014)
406. Wei, J., Li, H., Barrow, M.P., O'Connor, P.B.: Structural characterization of chlorophyll-*a* by high resolution tandem mass spectrometry. J. Am. Soc. Mass Spectrom. **24**(5), 753–760 (2013)
407. Davies, J., Ryan, K.S.: Introducing the parvome: bioactive compounds in the microbial world. ACS Chem. Biol. **7**(2), 252–259 (2012)
408. Zak, A., Kosakowska, A.: Cyanobacterial and microalgal bioactive compounds—the role of secondary metabolites in allelopathic interactions. Oceanol. Hydrobiol. St. **45**(1), 131–143 (2016)
409. Sneader, W.: Drug Discovery: A History, 468 pp. Wiley Ltd., West Sussex, England (2005)
410. Rickards, R.W., Rothschild, J.M., Willis, A.C., de Chazal, N.M., Kirk, J., Kirk, K., Saliba, K.J., Smith, G.D.: Calothrixins A and B, novel pentacyclic metabolites from *Calothrix* Cyanobacteria with potent activity against malaria parasites and human cancer cells. Tetrahedron **55**(47), 13513–13520 (1999)
411. Berdy, J.: Thoughts and facts about antibiotics: where we are now and where we are heading. J. Antibiot. **65**(8), 385–395 (2012)
412. Maschek, J.A., Baker, B.J.: The chemistry of algal secondary metabolism. In: Amsler, C.D. (ed.) Algal Chemical Ecology, p. 313. Springer, Berlin, Heidelberg (2008)
413. Culture Collection of Autotrophic Organisms (CCALA).: http://ccala.butbn.cas.cz/. Accessed 2013
414. Culture Collection of Algae.: http://sagdb.uni-goettingen.de/index.php. Accessed 2013
415. Coimbra Collection of Alage. http://acoi.ci.uc.pt/index.php. Accessed June 2007
416. Culture Collection of algae and protozoa: https://www.ccap.ac.uk/index.htm. Accessed June 2014
417. Chlamydomonas Resource Center: http://www.chlamycollection.org/. Accessed 2017
418. UTEX Culture Collection of Algae: https://utex.org/. Accessed 2010
419. Andersen, R.A. (ed.): Algal Culturing Techniques, p. 578. Elsevier Academic Press, New York (2005)
420. Pilat, Z., Jezek, J., Sery, M., Trtilek, M., Nedbal, L., Zemanek, P.: Optical trapping of microalgae at 735–1064 nm: photodamage assessment. J. Photoch. Photobio. B. **121**, 27–31 (2013)

421. Hammed, A.M., Jaswir, I., Amid, A., Alam, Z., Asiyanbi-H, T.T., Ramli, N.: Enzymatic hydrolysis of plants and algae for extraction of bioactive compounds. Food Rev. Int. **29**(4), 352–370 (2013)
422. Ramachandra, T.V., Madhab, M.D., Shilpi, S., Joshi, N.V.: Algal biofuel from urban wastewater in India: scope and challenges. Renew. Sust. Energy Rev. **21**, 767–777 (2013)
423. Olaizola, M.: Commercial production of astaxanthin from *Haematococcus pluvialis* using 25,000-liter outdoor photobioreactors. J. Appl. Phycol. **12**(3–5), 499–506 (2000)
424. Lee, S.J., Yoon, B.D., Oh, H.M.: Rapid method for the determination of lipid from the green alga *Botryococcus braunii*. Biotechnol. Tech. **12**(7), 553–556 (1998)
425. Prabakaran, P., Ravindran, A.D.: A comparative study on effective cell disruption methods for lipid extraction from microalgae. Lett. Appl. Microbiol. **53**(2), 150–154 (2011)
426. Gouveia, L., Oliveira, A.C.: Microalgae as a raw material for biofuels production. J. Ind. Microbiol. Biot. **36**(2), 269–274 (2009)
427. Soxhlet, F.: Die gewichtsanalytische Bestimmung des Milchfettes. Dinglers Polytechn. J. **232**, 461–465 (1879)
428. Barrios-Llerena, M.E., Burja, A.M., Wright, P.C.: Genetic analysis of polyketide synthase and peptide synthetase genes in Cyanobacteria as a mining tool for secondary metabolites. J. Ind. Microbiol. Biot. **34**(6), 443–456 (2007)
429. Mohamed, Z.A.: Allelopathic activity of the norharmane-producing cyanobacterium *Synechocystis aquatilis* against Cyanobacteria and microalgae. Oceanol. Hydrobiol. St. **42** (1), 1–7 (2013)
430. Volk, R.B.: Screening of microalgae for species excreting norharmane, a manifold biologically active indole alkaloid. Microbiol. Res. **163**, 307–313 (2008)
431. Milligan, K.E., Marquez, B.L., Williamson, R.T., Gerwick, W.H.: Lyngbyabellin B, a toxic and antifungal secondary metabolite from the marine cyanobacterium *Lyngbya majuscula*. J. Nat. Prod. **63**(10), 1440–1443 (2000)
432. Strege, M.A.: High-performance liquid chromatographic electrospray ionization mass spectrometric analyses for the integration of natural products with modern high-throughput screening. J. Chromatogr. B **725**, 67–78 (1999)
433. Caicedo, N.H., Kumirska, J., Neumann, J., Stolte, S., Thoming, J.: Detection of bioactive exometabolites produced by the filamentous marine cyanobacterium *Geitlerinema* sp. Mar. Biotechnol. **14**(4), 436–445 (2012)
434. Rellan, S., Osswald, J., Saker, M., Gago-Martinez, A., Vasconcelos, V.: First detection of anatoxin-a in human and animal dietary supplements containing Cyanobacteria. Food Chem. Toxicol. **47**(9), 2189–2195 (2009)
435. Rawn, D.F.K., Niedzwiadek, B., Lau, B.P.Y., Saker, M.: Anatoxin-a and its metabolites in blue-green algae food supplements from Canada and Portugal. J. Food Protect. **70**(3), 776–779 (2007)
436. Bumke-Vogt, C.: A highly sensitive analytical method for the neurotoxin anatoxin-a, using GC-ECD, and first application to laboratory cultures. Phycologia **35**(6), 51–56 (1996)
437. Garima, A.C., Goyal, P., Kaushik, P.: Antibacterial and anticandidal screening of extracellular and intracellular extracts of *Phormidium*, a cyanobacterium. Int. J. Chem. Life Sci. **2**(3), 1107 (2013)
438. Gunasekera, S.P., Ritson-Williams, R., Paul, V.J.: Carriebowmide, a new cyclodepsipeptide from the marine cyanobacterium *Lyngbya polychroa*. J. Nat. Prod. **71**(12), 2060–2063 (2008)
439. Namikoshi, M., Murakami, T., Watanabe, M.F., Oda, T., Yamada, J., Tsujimura, S., Nagai, H., Oishi, S.: Simultaneous production of homoanatoxin-a, antitoxin-a, and a new non-toxic 4-hydroxyhomoanatoxin-a by the cyanobacterium *Raphidiopsis mediterranea* Skuja. Toxicon **42**(5), 533–538 (2003)
440. Pal, D., Khozin-Goldberg, I., Didi-Cohen, S., Solovchenko, A., Batushansky, A., Kaye, Y., Sikron, N., Samani, T., Fait, A., Boussiba, S.: Growth, lipid production and metabolic adjustments in the euryhaline eustigmatophyte *Nannochloropsis oceanica* CCALA 804 in response to osmotic downshift. Appl. Microbiol. Biot. **97**(18), 8291–8306 (2013)

441. Moreira, C., Ramos, V., Azevedo, J., Vasconcelos, V.: Methods to detect Cyanobacteria and their toxins in the environment. Appl. Microbiol. Biot. **98**(19), 8073–8082 (2014)
442. Preisitsch, M., Niedermeyer, T.H.J., Heiden, S.E., Neidhardt, I., Kumpfmuller, J., Wurster, M., Harmrolfs, K., Wiesner, C., Enke, H., Muller, R., Mundt, S.: Cylindrofridins A-C, linear cylindrocyclophane-related alkylresorcinols from the Cyanobacterium *Cylindrospermum stagnale*. J. Nat. Prod. **79**(1), 106–115 (2016)
443. Rico, M., Lopez, A., Santana-Casiano, J.M., Gonzalez, A.G., Gonzalez-Davila, M.: Variability of the phenolic profile in the diatom *Phaeodactylum tricornutum* growing under copper and iron stress. Limnol. Oceanogr. **58**(1), 144–152 (2013)
444. Valeur B., Berberan-Santos M.N.: Molecular Fluorescence: Principles and Applications, 2nd edn. Wiley-VCH (2012)
445. Lakowicz, J.R.: Principles of Fluorescence Spectroscopy, 3rd edn. Springer, Berlin (2007)
446. Serodio, J., da Silva, J.M., Catarino, F.: Use of in vivo chlorophyll a fluorescence to quantify short-term variations in the productive biomass of intertidal microphytobenthos. Mar. Ecol. Prog. Ser. **218**, 45–61 (2001)
447. Herraiz, T.: Analysis of the bioactive alkaloids tetrahydro-beta-carboline and beta-carboline in food. J. Chromatogr. A **881**(1–2), 483–499 (2000)
448. Chen, W., Zhang, C.W., Song, L.R., Sommerfeld, M., Hu, Q.: A high throughput nile red method for quantitative measurement of neutral lipids in microalgae. J. Microbiol. Meth. **77** (1), 41–47 (2009)
449. Arranja, C.T., Aguiar, A., Encarnacão, T., Fonseca, S.M., Justino, L.L.G., Castro, R.A.E., Benniston, A., Harriman, A., Burrows, H.D., Sobral, A.J.F.N.: Double-tailed long chain BODIPYs—synthesis, characterization and preliminary studies on their use as lipid fluorescence probes. J. Mol. Struct. **1146**, 62–69 (2017)
450. NMRShiftDB: http://nmrshiftdb.nmr.uni-koeln.de/. Accessed Dec 2010
451. Biological Magnetic Resonance Data Bank: http://www.bmrb.wisc.edu/. Accessed Aug 2017
452. Obando, C.Z., Linossier, I., Kervarec, N., Zubia, M., Turquet, J., Fay, F., Rehel, K.: Rapid identification of osmolytes in tropical microalgae and Cyanobacteria by H-1 HR-MAS NMR spectroscopy. Talanta **153**, 372–380 (2016)
453. Campus Chemical Instrument Center (CCIC): http://spin.ccic.ohio-state.edu/index.php/toccata2/index. Accessed 2013
454. Bingol, K., Bruschweiler-Li, L., Li, D.W., Bruschweiler, R.: Customized metabolomics database for the analysis of NMR H-1-H-1 TOCSY and C-13-H-1 HSQC-TOCSY spectra of complex mixtures. Anal. Chem. **86**(11), 5494–5501 (2014)
455. Righi, V., Parenti, F., Schenetti, L., Mucci, A.: Mycosporine-like amino acids and other phytochemicals directly detected by high-resolution NMR on Klamath (*Aphanizomenon flos-aquae*) blue-green algae. J. Agr. Food Chem. **64**(35), 6708–6715 (2016)
456. Méjean, A., Mann, S., Maldiney, T., Vassiliadis, G., Lequin, O., Ploux, O.: Evidence that biosynthesis of the neurotoxic alkaloids anatoxin-a and homoanatoxin-a in the cyanobacterium *Oscillatoria* PCC 6506 occurs on a modular polyketide synthase initiated by L-proline. J. Am. Chem. Soc. **131**(22), 7512–7513 (2009). https://doi.org/10.1021/ja9024353
457. Jaki, B., Orjala, J., Heilmann, J., Linden, A., Vogler, B., Sticher, O.: Novel extracellular diterpenoids with biological activity from the cyanobacterium *Nostoc commune*. J. Nat. Prod. **63**(3), 339–343 (2000)
458. Klejdus, B., Lojkova, L., Plaza, M., Snoblova, M., Sterbova, D.: Hyphenated technique for the extraction and determination of isoflavones in algae: ultrasound-assisted supercritical fluid extraction followed by fast chromatography with tandem mass spectrometry. J. Chromatogr. A **1217**(51), 7956–7965 (2010)
459. Kruger, T., Christian, B., Luckas, B.: Development of an analytical method for the unambiguous structure elucidation of cyclic peptides with special appliance for hepatotoxic desmethylated microcystins. Toxicon **54**(3), 302–312 (2009)

460. Essa, A.M.M., Fathy, S.M.: Sex hormonal disruption by cyanobacterial bioactive compounds. J. Appl. Microbiol. **116**(3), 700–709 (2014)
461. Ziemert, N., Alanjary, M., Weber, T.: The evolution of genome mining in microbes—a review. Nat. Prod. Rep. **33**(8), 988–1005 (2016)
462. Hadjithomas, M., Chen, I.M.A., Chu, K., Ratner, A., Palaniappan, K., Szeto, E., Huang, J. H., Reddy, T.B.K., Cimermancic, P., Fischbach, M.A., Ivanova, N.N., Markowitz, V.M., Kyrpides, N.C., Pati, A.: IMG-ABC: a knowledge base to fuel discovery of biosynthetic gene clusters and novel secondary metabolites. Mbio. **6**(4) (2015)
463. Blin, K., Medema, M.H., Kazempour, D., Fischbach, M.A., Breitling, R., Takano, E., Weber, T.: AntiSMASH 2.0-a versatile platform for genome mining of secondary metabolite producers. Nucleic Acids Res. **41**(W1), W204–W212 (2013)
464. Blin, K., Medema, M.H., Kottmann, R., Lee, S.Y., Weber, T.: The antiSMASH database, a comprehensive database of microbial secondary metabolite biosynthetic gene clusters. Nucleic Acids Res. **45**(D1), D555–D559 (2017)
465. Blin, K., Wolf, T., Chevrette, M.G., Lu, X.W., Schwalen, C.J., Kautsar, S.A., Duran, H.G.S., Santos, E.L.C.D.L., Kim, H.U., Nave, M., Dickschat, J.S., Mitchell, D.A., Shelest, E., Breitling, R., Takano, E., Lee, S.Y., Weber, T., Medema, M.H.: AntiSMASH 4.0-improvements in chemistry prediction and gene cluster boundary identification. Nucleic Acids Res. **45**(W1), W36–W41 (2017)
466. Medema, M.H., Blin, K., Cimermancic, P., de Jager, V., Zakrzewski, P., Fischbach, M.A., Weber, T., Takano, E., Breitling, R.: AntiSMASH: rapid identification, annotation and analysis of secondary metabolite biosynthesis gene clusters in bacterial and fungal genome sequences. Nucleic Acids Res. **39**, W339–W346 (2011)
467. Weber, T., Blin, K., Duddela, S., Krug, D., Kim, H.U., Bruccoleri, R., Lee, S.Y., Fischbach, M.A., Muller, R., Wohlleben, W., Breitling, R., Takano, E., Medema, M.H.: antiSMASH 3.0-a comprehensive resource for the genome mining of biosynthetic gene clusters. Nucleic Acids Res. **43**(W1), W237–W243 (2015)
468. Kleigrewe, K., Almaliti, J., Tian, I.Y., Kinnel, R.B., Korobeynikov, A., Monroe, E.A., Duggan, B.M., Di Marzo, V., Sherman, D.H., Dorrestein, P.C., Gerwick, L., Gerwick, W.H.: Combining mass spectrometric metabolic profiling with genomic analysis: a powerful approach for discovering natural products from Cyanobacteria. J. Nat. Prod. **78**(7), 1671–1682 (2015)
469. Faltermann, S., Hutter, S., Christen, V., Hettich, T., Fent, K.: Anti-inflammatory activity of cyanobacterial serine protease inhibitors aeruginosin 828A and cyanopeptolin 1020 in human hepatoma cell line huh7 and effects in zebrafish (*Danio rerio*). Toxins **8**(7) (2016)
470. Allen, M.B., Arnon, D.I.: Studies on nitrogen-fixing blue-green algae. I. Growth and nitrogen fixation by *Anabaena cylindrica* Lemm. Plant Physiol. **30**(4), 366–372 (1955)
471. Gademann, K., Portmann, C., Blom, J.F., Zeder, M., Jüttner, F.: Multiple toxin production in the cyanobacterium *Microcystis*: isolation of the toxic protease inhibitor cyanopeptolin 1020. J. Nat. Prod. **73**(5), 980–984 (2010)
472. Niveshika, V.E., Mishra, A.K., Singh, A.K., Singh, V.K.: Structural elucidation and molecular docking of a novel antibiotic compound from cyanobacterium *Nostoc sp* MGL001. Front Microbiol. **7** (2016)
473. Rippka, R., Deuelles, J., Waterbury, J.B., Herdman, M., Stainer, R.Y.: Generic assignment strain histories and properties of pure cultures of cynobacteria. J. Gen. Microbiol. **111**, 1–61 (1979)
474. Pohl, P., Kohlhase, M., Krautwurst, S., Baasch, K.H.: An inexpensive inorganic culture medium for the mass cultivation of freshwater microalgae. Phytochemistry **26**, 1657–1659 (1987)
475. Volk, R.B., Furkert, F.H.: Antialgal, antibacterial and antifungal activity of two metabolites produced and excreted by Cyanobacteria during growth. Microbiol. Res. **161**, 180–186 (2006)

476. Becher, P.G., Beuchat, J., Gademann, K., Jüttner, F.: Nostocarboline: Isolation and synthesis of a new cholinesterase inhibitor from Nostoc 78-12A. J. Nat. Prod. **68**(12), 1793–1795 (2005)
477. Bister, B., Keller, S., Baumann, H.I., Nicholson, G., Weist, S., Jung, G., Sussmuth, R.D., Jüttner, F.: Cyanopeptolin 963A, a chymotrypsin inhibitor of *Microcystis* PCC 7806. J. Nat. Prod. **67**(10), 1755–1757 (2004)
478. Guillard, R.R.L., Ryther, J.H.: Studies of marine planktonic diatoms. I. *Cyclotella nana* Hustedt and *Detonula confervacea* Cleve. Can. J. Microbiol. **8**, 229–239 (1962)
479. Filali Mouhim, R., Cornet, J.F., Fontaine, T., Fournet, B., Dubertret, G.: Production, isolation and preliminary characterization of the exopolysaccharide of the cyanobacterium Spirulina platensis. Biotechnol. Lett. **15**, 567–572 (1993)
480. Singh, S., Das, S.: Screening, production, optimization and characterization of cyanobacterial polysaccharide. World J. Microb. Biol. **27**, 1971–1980 (2011)

Chapter 5
Methods to Assess Biological Transformation of Biomass

Hélène Carrère, Georgia Antonopoulou, Céline Druilhe, Eric Latrille, Gerasimos Lyberatos, Julie Jimenez, Ioanna Ntaikou, Konstantina Papadopoulou, Eric Trably and Anne Trémier

Abstract The aim of this chapter is to provide a description of the main methods used to assess biological transformation of biomass. It will address saccharification tests which account for the ability of biomass to release sugars during enzymatic hydrolysis. Biochemical methane potential (BMP) tests, which are widely used to assess anaerobic digestibility of biomass for the production of methane and BMP prediction by near infra-red spectroscopy will be presented. Biohydrogen potential (BHP), used to assess the ability to produce biohydrogen from biomass by dark fermentation will be described. Respirometry tests accounting for assessing aerobic degradability of biomass will be described. Incubation tests used to assess carbon and nitrogen degradation in soils will also be presented.

5.1 Introduction

Biological processes offer various possibilities of biomass valorisation for the production of energy, biofuels and other bio-based products. They may be distinguished into two general groups, anaerobic and aerobic, depending on the availability of oxygen. The former generates a variety of useful end-products as metabolites. Biological processes are generally considered environmentally

H. Carrère (✉) · E. Latrille · J. Jimenez · E. Trably
Laboratoire de Biotechnologie de l'Environnement, INRA, Avenue des Etangs, 11100 Narbonne, France
e-mail: helene.carrere@inra.fr

G. Antonopoulou · G. Lyberatos · I. Ntaikou
Institute of Chemical Engineering Sciences (FORTH/ICE-HT), Stadiou, Platani, 26504 Patras, Greece

G. Lyberatos · K. Papadopoulou
School of Chemical Engineering, National Technical University of Athens, 15780 Athens, Greece

C. Druilhe · A. Trémier
UR OPAALE, Irstea, 17 avenue de Cucillé, CS64427, 35044 Rennes, France

A. Nzihou (ed.), *Handbook on Characterization of Biomass, Biowaste and Related By-products*, https://doi.org/10.1007/978-3-030-35020-8_5

friendly, as they operate under mild conditions of temperature and pressure and a small amount (if any) of chemical additives.

Biological processes imply the use of enzymes and microorganisms; these latter can be used as pure cultures or complex consortia. In the case of biomass transformation, pure culture fermentation generally requires simple sugars as substrate and is consequently combined with a prior saccharification step performed by enzymatic hydrolysis. Second generation bioethanol production, employing enzymatic hydrolysis and fermentation, is the most studied example of biological valorisation of lignocellulosic biomass. Methods to measure the saccharification potential by enzymatic hydrolysis are described in the first part of this chapter.

Complex consortia based biological processes present the advantage of accepting a wide range of feedstocks without the need for (expensive) sterilization. A prominent example is the anaerobic digestion process, where microorganisms convert various organic residues such as the organic fraction of municipal solid wastes, sewage sludge, livestock effluents, food waste, lignocellulosic biomass… into biogas, composed of methane and carbon dioxide, with the parallel production of a digestate which can be used as organic fertiliser. The methane yield of an industrial process depends on the feedstock quality, but also on the process operating parameters, such as the organic loading rate, the retention time and temperature. The so-called "biochemical methane potential" (BMP) test is designed to assess the maximum methane yield that may be given by a particular feedstock (substrate). This test is described in the second part of this chapter.

Hydrogen, which is expected to have an important contribution in the energy transition to renewable sources, is an intermediate of the anaerobic digestion process. Its production is favoured when process conditions (such as the retention time) are set up to avoid methane generation, a process called dark hydrogen fermentation. Similarly to the biochemical methane potential test, a simple protocol allows to estimate the potential of a particular biomass feedstock to generate hydrogen by dark fermentation: the biohydrogen potential (BHP). This test is described in the third part of this chapter.

The other group of biological processes involving complex microorganism consortia is the group of aerobic processes, with composting being the most well-known. They are based on respiration. The aerobic degradation of biomass may be assessed by measuring oxygen consumption using "respirometric" methods, described in the fourth part of this chapter.

Biomass is also degraded by microorganism consortia in soils. This is a key process in the assessment of fertilisers and amendments on cropped soils. Organic fertilizers allow the substitution of mineral fertilizers, nitrogen in particular, and increase the soil carbon stock. Methods for the assessment of organic carbon and nitrogen mineralization are described in the last part of this chapter.

5.2 Enzymatic Hydrolysis (Saccharification)

5.2.1 Fundamentals

Biomass consists largely of biopolymers (macromolecules) such as proteins, lipids and carbohydrates. Carbohydrates comprise the most important fraction of biomass when it comes to production of sugars, liquid fuels such as ethanol and other bio chemicals. Enzymatic hydrolysis (saccharification) may be used to convert the insoluble carbohydrates contained in biomass, such as starch, cellulose and hemi-cellulose to fermentable sugars, with enhanced performance as far as the yield, selectivity and energy requirements are concerned. This process is affected mostly by *substrate-related* and *enzyme-related* factors. The use of specific pretreatments of biomass results to improvement and enhancement of enzymatic hydrolysis, which plays a critical role in the economic production of sugars, liquid fuels and other bio chemicals.

Glucose is produced from various feedstocks. The three main polysaccharides are *starch*, *glycogen*, and *cellulose*, referred to as *homopolymers*, because after complete hydrolysis, only one type of monosaccharides results (glucose). *Heteropolymers* are common in nature (gums, pectins etc.) and contain sugar acids, amino sugars, or non-carbohydrate substances in addition to monosaccharides.

The difference between the main polysaccharides (starch and cellulosic biomass) is their bond structure. Through enzymatic hydrolysis of starch, the structure is easy to breakdown by α-amylase at α-(1–4) acetal linkages. The cellulose is difficult to breakdown on the other hand, because it contains bonds with β-(1–4) acetal linkage and the required degrading enzyme (cellulase) is not so commonly found in degrading species [1, 2].

Consequently, starch enzymatic hydrolysis is easier, and the production process is faster and less complicated compared to cellulose hydrolysis.

In this part of the chapter, the procedures for the assessment of an effective enzymatic hydrolysis in order to produce sugars (saccharification), which may be used for the production of biofuels, from different type of biomass are presented.

The main procedures are:

Compositional analysis of the biomass: determination of moisture, ash, proteins, extractives, carbohydrates (starch, cellulose, hemicellulose, free sugars) and lignin. Based on the determined biomass composition, a theoretical saccharification potential can be estimated.

Need for pretreatment: depending on the biomass composition, it may be necessary to carry out a pretreatment (physical, chemical or biological) before implementing enzyme hydrolysis, (pretreatment methods are not described in this chapter).

Choice of Enzymes: the type of individual enzymes or a mixture of them that needs to be used for enzymatic hydrolysis depends on the composition of the various fractions (cellulose, hemicellulose, lignin and starch) in the biomass substrate. Plenty of enzymes can be found either in bacteria or fungi and many enzyme formulations are available commercially. In order to assess the activity of a particular formulation, the most well-established enzyme assays are (a) total cellulase activity using filter paper (filter paper assay, FPA) and (b) individual cellulase activities: endoglucanases (EGs), exoglucanases, and β-glucosidases.

Finally, once an enzymatic hydrolysis has been actually applied experimentally, the **actual saccharification potential** may be determined and compared with the theoretical one.

Particular *methods*, *assays*, *kinetics* and *models* for assessing the theoretical and actual saccharification potential (given a particular choice of pretreatment and enzyme hydrolysis) will be outlined below and categorized according to the structure of the biomass (starch, cellulose and hemicellulose), after the fundamentals regarding biomass saccharification have been presented.

5.2.2 Lignocellulosic Biomass

Cellulose and ***hemicellulose*** are the most abundant biopolymers on earth. The conversion of biomass to platform sugars is hindered primarily by the complexity of lignocellulosic substrates and the limited hydrolytic enzymes performance. Therefore, efficient, optimized and economically competitive pretreatment methods have been sought in the effort to facilitate their hydrolysis [1].

Pretreatment, before enzymatic hydrolysis is a necessary step, in order to alter some structural characteristics of lignocellulose, increasing glucan and xylan accessibility of the enzymes and thus the enzymatic hydrolysis yield. The choice of pretreatment technology for a particular raw material depends on several factors, such as sugar-release patterns and enzymes, some of them directly related to the chosen enzymatic hydrolysis method. The combination of the composition of the substrate (feedstock), the type of pretreatment, and the dosage and efficiency of the enzymes used for the enzymatic hydrolysis influence the overall biomass utilization [2]. Pretreatment for enhancing enzymatic hydrolysis aims at: (a) solubilization/removal of hemicellulose, (b) removal of lignin, (c) increased surface area for cellulases, (d) reduction of the crystallinity index, and (e) reduction of the degree of polymerization [3–6]. Pretreatment methods are categorized in four main groups: *mechanical*, *physical*, *chemical* and *biological* pretreatment.

Cellulose and hemicellulose can be hydrolyzed by specific enzymes, called cellulases and hemicellulases, respectively. For the enzymatic hydrolysis the major factors are: substrate quality and concentration, applied pretreatment methods, enzyme activity and hydrolysis conditions, i.e. pH (4–5) and temperature (40–50 °C) [7].

The enzymatic hydrolysis reactions involve three groups of enzymes: endo-1,4-β-D-glucanase (EG) (E.C. 3.2.1.4), exo-1,4-β-D-glucanases (or cellobiohydrolase, CBH) (E.C. 3.2.1.91) and β-glucosidase (BG) (E.C. 3.2.1.21) (Fig. 5.1). The EG cuts randomly the cellulose from the inside and CBH hydrolyzes the cellulose from the end and releases mainly cellobiose. Finally, BG hydrolyzes

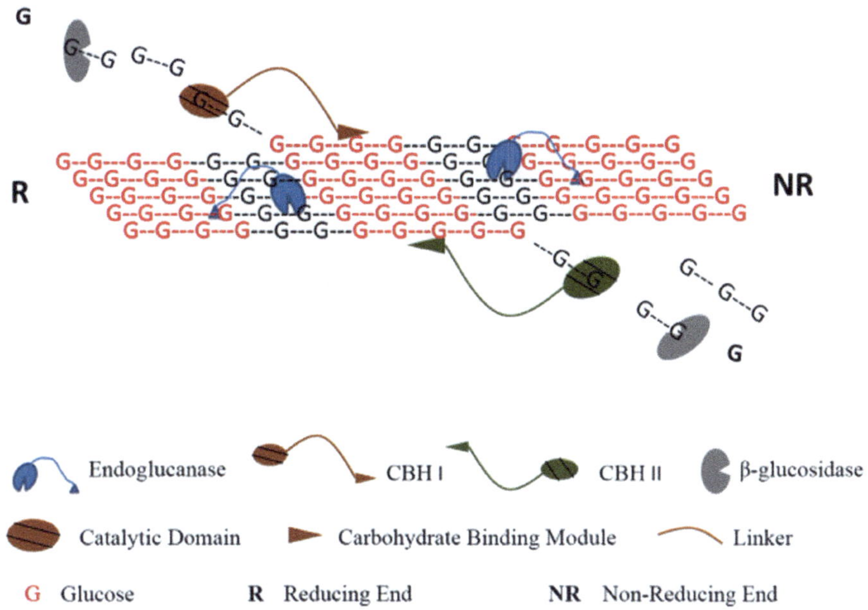

Fig. 5.1 Synergistic model for enzymatic hydrolysis of cellulose. Reprinted with permission from BioMed Central Ltd. [12]

cellobiose into glucose [8]. In *Trichoderma reesei*, the most widely used cellulase filamentous fungus producer, at least two CBHs, five EGs and two BGs have been characterized [9–11].

The hemicellulases, are enzymes that hydrolyze hemicelluloses and consists of endo-β-1,4-xylanase (E.C. 3.2.1.8), exo-β-1,4-xylosidase (E.C. 3.2.1.37), α-L-arabinofuranosidase (E.C. 3.2.1.55), endo-α-1,5-arabinanase (E.C. 3.2.1.99), endo-β-1,4-mannanase (E.C. 3.2.1.78), α-glucuronidase (E.C. 3.2.1.139), α-galactosidase (E.C. 3.2.1.22), exo-β-1,4-mannosidase (E.C. 3.2.1.25), endo-galactanase (E.C. 3.2.1.89), β-glucosidase (E.C. 3.2.1.21), acetyl xylan esterase (E.C. 3.1.1.72), acetyl mannan esterase (E.C. 3.1.1.6), and ferulic and ρ-cumaric acid esterases (E.C. 3.1.1.73) [13].

5.2.3 *Starch-Based Biomass*

Starch is a glucose biopolymer with the general formula $(C_6H_{10}O_5)_n$. It consists of a mixture of two glucose polymers: 25% w/w of the linear polymer amylose, and 75% w/w of a branched polymer, amylopectin. Amylose is a linear polymer made up of about up to 6000 repeating glucose units, connected by α-1–4-glycosidic bonds. Amylopectin is one of the largest naturally occurring biopolymers with about 200,000 glucose units and with a molecular weight more than 108 Da. Unlike amylose, amylopectin is a nonlinear polymer, containing both α-1–4-glycosidic bonds and 4–6% α-1–6 branched side chains.

During the enzymatic hydrolysis of ***starch***, enzymes break the long chains of amylose and amylopectin into shorter molecules-chains. Depending on the enzymes, final products of starch hydrolysis can be maltodextrins, glucose, fructose or maltose maltotriose, and some higher oligomers [14].

There are three main stages in the conversion of starch to sugars with enzymatic hydrolysis: gelatinization, liquefaction and saccharification [15–17].

5.2.3.1 Starch Swelling and Gelatinization

Starch granules are partially insoluble in water. They must typically undergo a multi-stage process called gelatinization in order to solubilize the particles and enhance the hydrolysis rate. Gelatinization, as a term, includes all events occurring as starch is heated in water: absorption of water by the starch granules, increase of the apparent viscosity of the solution, breakage of granules and exposure of amylose and amylopectin, which results to the increase of the rate of the enzymatic reaction by providing an easier enzyme attack. Typically, starch granules suspended in water swell and absorb water up to 30% w/w. When the suspension is heated typically 20–60 °C, hydrogen bonding weakens as a result of the release of hydroxyl groups from the polymer melt, so that the granules start to swell, and the granule structure is lost.

5.2.3.2 Starch Enzymatic Hydrolysis (Liquefaction and Saccharification)

The enzymes used for starch hydrolysis are (Fig. 5.2):

α-amylases

The degradation of starch is mainly initiated by α-amylases (α-1,4-glucanohydrolases), which catalyze the cleavage of α-1,4-glycosidic bonds, causing a decrease in substrate molecular weight and viscosity. These endo-enzymes can be

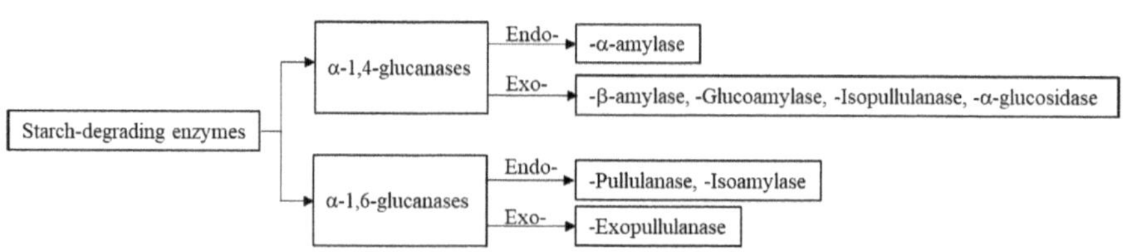

Fig. 5.2 Starch degrading enzymes

divided into liquefying or saccharifying α-amylases, depending on the number of glucose units in the generated oligomers. Prolonged hydrolysis of amylose leads to carbohydrate conversion into maltose, maltotriose and oligosaccharides. The most thermostable α-amylase currently used in biotechnological processes is produced by *Bacillus licheniformis*. It remains active for several hours at temperatures over 90 °C under conditions of industrial starch hydrolysis. For commercial production of amylases *Aspergillus* and *Rhizopus* species are mostly used.

Endo-acting debranching enzymes

There are two main groups of endo-acting debranching enzymes, which can cleave the α-1,6-glycosidic linkages. The first group are pullulanases that attack α-1,6-linkages, liberating linear oligosaccharides of glucose residues linked by α-1,4-bonds. The second group of endo-acting debranching enzymes are neopullulanases and amylopullulanases, which are active on both α-1,6- and α-1,4-linkages. Pullulanases are generally produced by plants, e.g. oat, rice, barley etc., as well as by mesophilic micro-organisms e.g. *Klebsiella*, *Escherichia*, *Streptococcus* etc.

Exo-acting amylases

There are two main types of exo-acting hydrolases used in starch hydrolysis: β-amylases (E.C. 3.2.1.2) and glucoamylases (E.C. 3.2.1.3). Both act on glycosidic linkages at the non-reducing ends of amylose, amylopectin and glycogen molecules, producing low-molecular weight (MW) carbohydrates in the β-anomeric form. The main end-product of hydrolysis catalyzed by β-amylases is maltose, while glucoamylase (amyloglucosidase) produces glucose [18, 19].

Starch Liquefaction (1st Enzymatic Hydrolysis)

Starch solutions with 30–32% dry solids are gelatinized by treatment with α-amylase to break down the starch molecules into shorter-chain maltodextrins, a mixture of poly- and oligosaccharides with a broad molecular weight distribution. This mixture includes mainly maltose, maltooligosaccharides and linear and branched dextrins. Additionally, α-amylase operates at 90 °C (optimum temperature for liquefaction) and pH 6 [14, 15]. Next stage is the saccharification, which is performed at pH 4.2–4.5 (the optimum pH for glucoamylase) (Fig. 5.3).

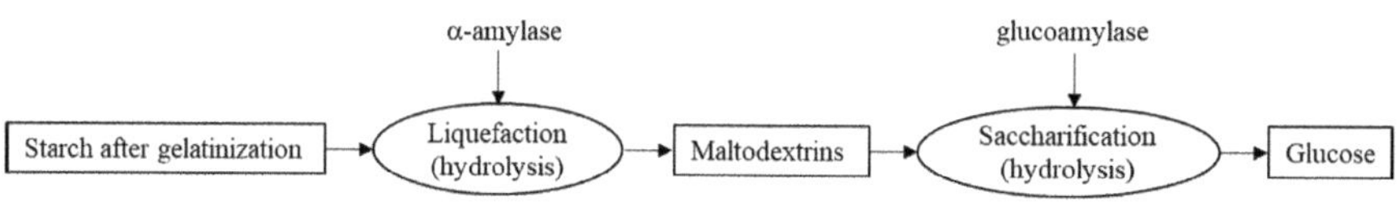

Fig. 5.3 Hydrolysis of starch to glucose

During the first hydrolysis stage (liquefaction), the viscosity of the gelatinized starch mixture is reduced due to partial hydrolysis of the carbohydrate polymers. The first hydrolysis of starch by α-amylase is referred to as liquefaction, because α-amylases lower the viscosity of the gelatinized starch (liquefy) by reducing the length of amylose and amylopectin.

Starch Saccharification (2nd Enzymatic Hydrolysis)

During the second hydrolysis stage (saccharification), these partially hydrolyzed starch chains are broken down into glucose, maltose, maltotriose, and some higher oligomers. Glucoamylase is added by decreasing the temperature to 60 °C and adjusting the pH to 4.5. Glucoamylase removes glucose units from the non-reducing ends until all the oligosaccharides are degraded to glucose units. Starch hydrolysis is expressed in terms of the "dextrose equivalent" (DE), a quantity that indicates the number of dextrose molecules released from the hydrolysis of starch. Dextrose equivalent (DE) values are used to characterize starch hydrolysates. It is a measure of the total reducing power of the sugars in relation to glucose and on a dry mass basis. The higher the value of dextrose equivalent the greater the extent of hydrolysis and the lower the average molecular mass of the oligomeric. Glucose, maltose, maltotriose and starch have DE of 100, 53, 33 and 0 respectively [15]. Maltodextrins are partial hydrolysates of starch with DE values below 20 [6, 16].

5.2.4 *Methods and Assays for Enzymatic Hydrolysis of Biomass*

5.2.4.1 Preparation of Biomass for Compositional Analysis

The NREL for compositional analysis have been optimized using biomass with a specific particle size and moisture content [17]. This procedure is substantially similar to the ASTM Standard Practice E 1757-01, which are similar to TAPPI method number T264 and NFTA method A 1.1.

Three drying methods (to less than 10%) are proposed by NREL:

Method A—*Air-drying*. This method is suitable for the preparation of large quantities (>20 g) of biomass appropriate for compositional analysis. Is also suitable for drying biomass to air-dry to moisture content below 10% as measured using LAP "Determination of Total Solids in Biomass and Total Dissolved Solids in Liquid Process Samples".

Method B—*Convection oven drying*. This drying method is suitable for small samples of biomass (<20 g). An alternative to air-drying, is to dry the biomass in a convection oven at temperatures <45 °C. This method may be used for very wet biomass, with a high risk for microbial growth during drying, wet pretreated biomass, that would not be stable during prolonged exposure to ambient conditions. The sample should have a moisture content below 10%, by using LAP "Determination of Total Solids in Biomass and Total Dissolved Solids in Liquid Process Samples" [20].

Method C—*Lyophilization.* This drying method is suitable for small samples of biomass (<20 g). An acceptable alternative to air-drying or drying in a convection oven, is lyophilization (freeze-drying) of the biomass. This method is suitable for very wet biomass, the optimum moisture content of the biomass should be below 10%, by using LAP "Determination of Total Solids in Biomass and Total Dissolved Solids in Liquid Process Samples" [20].

5.2.4.2 Determination of Total Solids in Biomass

Due to the high variability of moisture content in biomass, all biomass compositional analysis results are reported on dry weight basis. The LAP "Determination of Total Solids in Biomass and Total Dissolved Solids in Liquid Process Samples [20], a %Total Solids determination is required." The LAP describes drying by convection oven at 105 °C and drying by infrared moisture balance and is calculated by the following equation (Eq. 5.1):

$$\%Total\,Solids = \frac{\left(Weight_{dry\,pan\,plus\,dry\,solids} - Weight_{dry\,pan}\right)}{Weight_{sample\,as\,received}} \times 100 \tag{5.1}$$

Calculation of the dry weight of a sample or its Oven Dried Weight (ODW) is described in the following formula.

The oven dry weight (ODW) of the sample is calculated, using the average total solids content as determined by the LAP "Standard Method for the Determination of Total Solids in Biomass" (Eq. 5.2).

$$ODW_{sample} = \frac{\left(Weight_{air\,dried\,sample} \times \%Total\,solids\right)}{100} \tag{5.2}$$

5.2.4.3 Determination of Ash in Biomass

Inorganic materials are present in both whole and structural, or extracted, biomass samples. LAP "Determination of Ash in Biomass" [21], describes methods for the determination of %Ash in biomass. The LAP provides instructions for ash determination in a furnace set to 575 °C. The calculation for determining %Ash is below (Eq. 5.3):

$$\%Ash = \frac{\left(Weight_{crucible\,plus\,ash} - Weight_{crusible}\right)}{ODW_{sample}} \times 100 \quad (5.3)$$

5.2.4.4 Determination of Protein in the Biomass

Biomass used as a feedstock for conversion to fuels and chemicals may contain proteins and other nitrogen containing compounds. In many cases, the nitrogen content of the biomass sample is measured by measuring the nitrogen generated through combustion or, alternately, through the determination of Kjeldahl nitrogen, followed by estimation of the protein content using an appropriate for the particular biomass type Nitrogen Factor (NF). There are published methods that recommend using an NF of 6.25 for all types of biomass, except for wheat grains, where a NF of 5.70 is recommended. Proposed LAP procedure calculates an appropriate NF for most biomass types and covers the determination of nitrogen-to-protein conversion factors that are used to estimate the amount of protein in a biomass sample [22]. Because a portion of the protein is often removed during the extraction process, protein analysis is performed on both whole and extractives-free materials.

5.2.4.5 Compositional Analysis of Biomass

Compositional analysis of lignocellulose is used in almost all second generation (2G) biofuel production studies, except on biogas production that use volatile solids (VS) instead. The product yield, recovery, techno-economical evaluation and feasibility studies are typically based on the carbohydrate content.

Different methods have been developed for the analysis of lignocellulosic biomass. Mainly, the analysis of woody biomass is evaluated by using the Technical Association of Pulp and Paper Institute's (TAPPI) standards. The Association of Official and Analytical Chemists International (AOAC) standards are used for herbaceous and foods analysis [23–26]. An analytical and most accurate procedure for the determination of carbohydrates, acetate, lignin, and ash in biomass, presented by [27] has been provided by the National Renewable Energy Laboratory (NREL), Laboratory Analytical Procedure (LAPs).

5.2.4.6 Determination of Extractives in the Biomass

The following procedure uses a two-step extraction process to remove water soluble and ethanol soluble material [28]. Refer to LAP-010 or ASTM Standard Test Method E 1690 "Determination of Ethanol Extractives in Biomass" for extraction procedures for isolation and characterization of extractives.

An extraction procedure with a toluene/ethanol mixture is applied to lignocellulosic biomass in order to remove the organic-soluble extractives. In this procedure, LC is treated by a mixture of toluene/ethanol (2:1 v/v) in a Soxhlet extractor for 6 h.

The extractive-free solid-residue is then air-dried for 24 h and is dried to constant weight in an oven at 60 °C under vacuum. Subsequently, the solid residue is further extracted with absolute ethanol in a Soxhlet apparatus for 4 h and dried to constant weight.

A further extraction procedure with water under intense magnetic stirring for 5 h, is applied to lignocellulosic biomass in order to remove water-soluble extractives. The sludge is filtered with paper and washed with water. The extractive-free solid residue is dried to constant weight in an oven at 60 °C under vacuum. The yield of the extractives is calculated from the mass loss of the raw starting biomass.

The above procedure covers the determination of soluble non-structural materials in a biomass. The results are calculated on a dry weight basis, as a weight percentage of the biomass. Extractives percentages are measured and used to convert compositions from an extractives-free basis to and as-received basis.

The amount of extractives in the sample is calculated on a percent dry weight basis. If sucrose measurements were performed, the second equation is used to calculate extractives with a correction for the removed sample volume (Eqs. 5.4 and 5.5).

$$\%Extractives = \frac{Weight_{flask\,plus\,extractives} - Weight_{flask}}{ODW_{sample}} \times 100 \tag{5.4}$$

$$\%Extractives = \frac{Weight_{flask\,plus\,extractives} - Weight_{flask}}{ODW_{sample}} \times \frac{Volume_{total}}{Volume_{total} - Volume_{removed\,aliquot}} \times 100 \tag{5.5}$$

5.2.4.7 Determination of Structural Carbohydrates and Lignin in the Biomass

Lignocellulosic feedstocks consist of three main chemical components: cellulose, hemicellulose (xylose, arabinose, mannose etc.) and lignin.

Procedures for solid samples to measure structural carbohydrates (glucose, xylose, galactose, arabinose, and mannose), lignin, extractable materials, protein, and ash can be estimated by ASTM E1758-01 "Standard Method for the Determination of Carbohydrates by HPLC" or NREL-procedure (NREL LAP's) [27].

The NREL procedure is suitable for samples that do not contain extractives. It uses a ***two-step acid hydrolysis to fractionate the biomass***. The lignin is fractionated into acid-insoluble part and acid-soluble part. The acid-insoluble part may include ash and protein, which must be accounted for using gravimetric analysis. The acid-soluble lignin is measured by UV-Vis spectroscopy.

During hydrolysis polymeric carbohydrates are converted into their monomers which may be analyzed using HPLC. Cellulose is mainly hydrolyzed to glucose and converted into fermentation products such as ethanol, lactic acid, propanol, acetone, succinic acid, butanediol and butanol, while hemicellulose can be hydrolyzed to xylose and converted into xylite which is sugar-substitute. Interestingly, hydroxymethylfurfural (HMF) and furfural can be also produced from cellulose and hemicellulose through further conversion process.

Protein may also partition in the liquid fraction. A measure of acetyl content is necessary for biomass containing hemicellulose with a xylan backbone, but not biomass containing a mannan backbone. Acetate is measured by HPLC.

Lignin Content

The conditions of the ***first hydrolysis*** were 72 wt% sulfuric acid, 1:10 of solid-to-liquid ratio, and 30 °C for one hour. The conditions for the ***secondary hydrolysis*** were 4 wt% sulfuric acid and 121 °C for one hour. Sugars in the hydrolyzates are determined by HPLC coupled with a refractive index detector.

For the acid insoluble lignin analysis, the autoclaved hydrolysis solution is vacuum filtered, and the filtered hydrolyzed solid sample is dried and weighed. The dried samples are then combusted in a furnace at 575 ± 25 °C for 3 h to determine the **ash** content.

The difference of the two weights is taken as the acid insoluble lignin. The absorbance of the hydrolysis liquor in the aliquot obtained from the vacuum filter sample at 320 nm on a UV-Visible spectrophotometer measured the acid soluble lignin.

The weight percent acid insoluble residue (AIR) and acid insoluble lignin (AIL) on an extractives free basis are calculated and recorded (Eqs. 5.6 and 5.7):

$$\%AIR = \frac{Weight_{crucible\ plus\ AIR} - Weight_{crucible}}{ODW_{sample}} \times 100 \tag{5.6}$$

$$\%AIL = \frac{(Weight_{crucible\ plus\ AIR} - Weight_{crucible}) - (Weight_{crucible\ plus\ ash} - Weight_{crucible}) - Weight_{protein}}{ODW_{sample}} \times 100 \tag{5.7}$$

where, $Weight_{protein}$ = Amount of proteins present in the acid insoluble residue, as determined in LAP "Determination of Protein Content in Biomass" and ODW_{sample} is presented in Eq. 5.2.

The amount of acid soluble lignin (ASL) on an extractives free basis is calculated (Eqs. 5.8 and 5.9).

$$\%ASL = \frac{UV_{abs} \times Volume_{filtrate} \times Dilution}{\varepsilon \times ODW_{sample} \times Pathlength} \times 100 \tag{5.8}$$

where UV_{abs} = average UV-Vis absorbance for the sample at appropriate wavelength

$$Volume_{hydrolysis\ liquor} = \text{volume of filtrate, } 86.73\ \text{mL}$$

$$Dilution = \frac{Volume_{sample} + Volume_{diluting\ solvent}}{Volume_{sample}} \tag{5.9}$$

where,

ε	Absorptivity of biomass at specific wavelength
ODW_{sample}	Weight of sample in milligrams
Pathlength	Pathlength of UV-Vis cell in cm.

The total amount of lignin on an extractives-free basis is then calculated (Eq. 5.10):

$$\%Lignin_{ext\,free} = \%AIL + \%ASL \tag{5.10}$$

The total lignin value on an "as received" basis is calculated if necessary (Eq. 5.11):

$$\%Lignin_{as\ received} = \left(\%Lignin_{ext\,free}\right) \times \frac{(100 - \%Extractives)}{100} \tag{5.11}$$

where, % Extractives = percent extractives in the prepared biomass sample, as determined in LAP "Determination of Extractives in Biomass".

Hemicellulose and cellulose content

Carbohydrates in the **liquid samples** are determined by secondary acid hydrolysis (conditions: 4% H_2SO_4, 121 °C, 1 h). Dilute acid pretreatment produces a solid and a liquid phase. The liquid phase contains mainly hemicellulose, some lignin, and degradation products from the acid pretreatment. The cellulose typically remains in the solid fraction of the extractives along with the lignin fraction remaining undissolved during the pretreatment. Minor constituents, such as ash and protein, are usually split between the two phases. A series of calibration standards should contain the compounds that are to be quantified by HPLC, for D-cellobiose, D(+) glucose, D(+)xylose, D(+)galactose, L(+)arabinose and D(+)mannose in a suggested concentration range (0.1–4.0 mg/ml).

HPLC conditions: Sample volume: 50 μL Mobile phase: 0.005 M sulfuric acid, 0.2 μm filtered and degassed Flow rate: 0.6 mL/min Column temperature: 55–65 °C Detector temperature: as close to column temperature as possible Detector: refractive index Run time: 50 min.

The concentration of the polymeric sugars from the concentration of the corresponding monomeric sugars is calculated, using an anhydro correction factor which accounts for the water loss resulting from polymerization of the corresponding monomers in each case (Eq. 5.12):

$$C_{anhydro} = C_{HPLC} \times Anhydro \tag{5.12}$$

where, *Anhydro* = 0.88 (or 132/150) for **C-5 sugars** (xylose and arabinose) and 0.90 (or162/180) for **C-6 sugars** (glucose, galactose, and mannose).

The percentage of each sugar on an extractives free basis is calculated by (Eq. 5.13):

$$\%Sugar_{ext\,free} = \frac{C_{anhydro} \times V_{filtrate} \times \frac{1\,\mathrm{g}}{1000\,\mathrm{mg}}}{ODW_{sample}} \times 100 \tag{5.13}$$

where, $V_{filtrate}$ = volume of filtrate, 86.73 mL.

The percentage of each sugar on an "as received" basis, if necessary is calculated by (Eq. 5.14):

$$\%Sugar_{as\,received} = \left(\%Sugar_{ext\,free}\right) \times \frac{(100 - \%Extractives)}{100} \tag{5.14}$$

where, % Extractives = percent extractives in the prepared biomass sample, as determined in LAP "Determination of Extractives in Biomass".

Hemicellulose estimation (Eq. 5.15):

$$\text{xylan} + \text{arabinan} = \text{hemicellulose} \tag{5.15}$$

Cellulose estimation (Eq. 5.16):

$$\text{glucan} = \text{cellulose} \tag{5.16}$$

5.2.5 Theoretical Saccharification Potential

After the above fractionation of the biomass the following theoretical conversion equation for cellulose and hemicellulose can be described from major breakdown products.

Biomass conversion calculations can be formulated to include a simple cellulose conversion equation:

$$Glucose\,yield\,(\%) = \frac{Produced\,glucose\left(\mathrm{gL^{-1}}\right) \times 0.9}{Glucan\,in\,sample\left(\mathrm{gL^{-1}}\right)} \times 100 \tag{5.17}$$

where 0.9 is a correction factor accounting for hydration of glucan to glucose [29].

The maximum theoretical saccharification yield is described by:

$$[glucose]_{max} = F_{glucan} \times [substrate]/0.9 \tag{5.18}$$

where, F_{glucan} is the glucan fraction of the substrate (feedstock).

A more complex cellulose conversion calculation may be found in [30].

5.2.6 *Enzymes Available and Activity Assays*

5.2.6.1 Available Enzymes

The enzymatic hydrolysis of biomass is a result of multiple enzymes. The conditions in which the enzymes start degrading biomass depends on the source of enzyme and the feedstock used. In order to maximize the saccharification yield, a combination of enzymes should be used. The type of individual enzymes or a mixture of them depends on the composition of the various fractions (cellulose, hemicellulose, and lignin) in the biomass substrate. Experiments should be conducted in order to determine and optimize which enzymes and pretreatment methods will attack and degrade a specific feedstock. These enzymes are usually fungal species, such as *Penicillium verruculosum*, *Trichoderma reesei*, *Aspergillus niger* etc.

The most well-known commercial hydrolytic enzymes are: Celluclast, Cellic CTec2, Speczyme CP, Novozyme 188, Cytolase CL, and Accellerase [31].

The components of cellulase are endoglucanases, exoglucanases and β-glucosidases as described in Sect. 5.1. The fraction of the above components in an enzyme mixture dependent on the source of the enzyme. The endoglucanases bind to the cellulose and exposes the reducing and non-reducing ends resulting in the formation of cellooligomers. The exoglucanases bind to the reducing and non-reducing ends of the cellooligomers converting the same to cellobiose. The final component that acts is the β-glucosidase, which converts cellobiose to glucose (Fig. 5.2). A limitation of the hydrolysis reaction could be the possible insufficient quantity of β-glucosidases in the enzyme which leads to the accumulation of cellobiose.

The main parameters affecting enzymatic hydrolysis are the quality and source of enzyme, temperature (some enzymes are sensitive to high temperature), the mixing speed and the pH. Typical maintaining optimal operating temperatures for cellulose hydrolysis range between 40 and 55 °C, while the optimal pH ranges between 4.5 and 5.5 [32].

5.2.6.2 Enzyme Activity Assays

The main parameters, which are considered for assaying enzymes, are temperature, pH, ionic strength and the proper concentrations of the essential components i.e. substrates and type of enzymes.

Different classifications of enzymatic activity assays, especially for cellulase have been used over the years. Cellulase activity is mainly evaluated using a reducing sugar assay to calculate the end products of cellulase hydrolysis activities. The result of such an assay is typically expressed as the hydrolysis capacity of the enzyme.

Total cellulase activity by application of filter paper (filter paper assay, FPA) is the most known cellulase enzyme activity assay, other individual cellulase activities include: EGs, exoglucanases, and β-glucosidases and will be discussed. Moreover, there are also some novel assays which are presented in Fig. 5.4 and are explained analytically in Dashtban et al. [33]: Quartz crystal microbalance (QCM), Miniaturized colorimetric assay, Automated FPA for the measurement of cellulase activity, Fluorescent microfibrils, Amperometric cellobiose dehydrogenase biosensor.

FPA (FPase activity): total cellulase activity

The FPA filter paper activity (FPase) (FPU), is an assay for analysis of total cellulase activity. FPA was first developed by Mandels et al. [34] in 1976 and reviewed by Ghose [35]. The FPA became widely used since 1984, when the

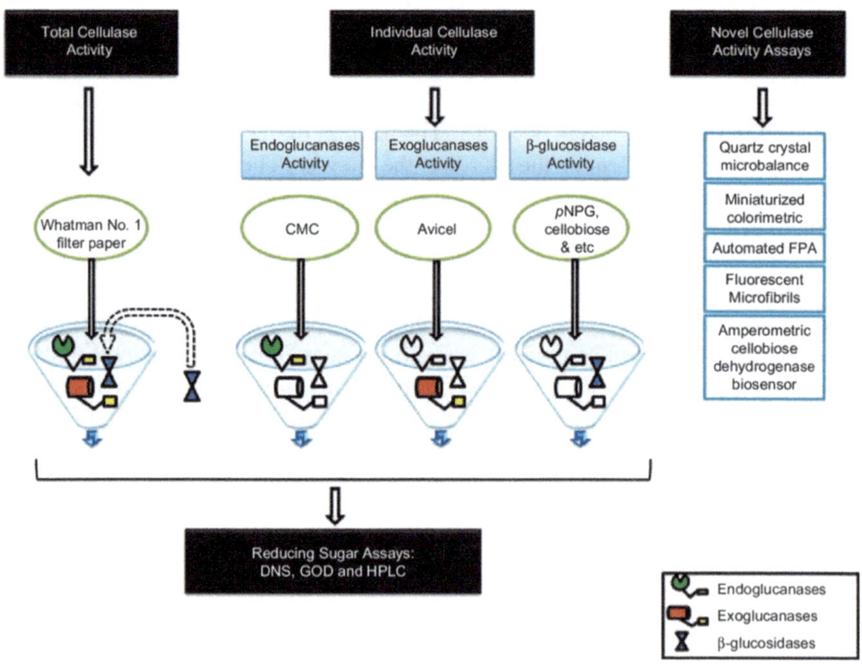

Fig. 5.4 Different cellulase assays that are classified within two groups: (1) total cellulase activity and (2) individual cellulase activity including endoglucanases, exoglucanases, and β-glucosidases and some other novel assays. Reproduced with permission from Taylor & Francis [33]

Commission on Biotechnology of the International Union of Pure and Applied Chemistry (IUPAC) proposed standard procedures for the measurement of cellulase activity. Traditionally, the FPA uses a 1 × 6-cm strip of Whatman no. 1 filter paper as the standard substrate.

The International Unit (IU) of filter paper activity (FPase) (FPU) is evaluated as the micromole of glucose equivalent liberated per minute of culture filtrate under assay conditions (pH and temperature), which are appropriate for the enzymes during the assay and depend largely on the properties of the enzymes. FPA is improved by adding supplemental β-glucosidase. Released reducing sugars can be measured using different reducing sugar assay methods such as 3,5-dinitrosalicylic acid (DNS), glucose oxidase (GOD), and high-performance liquid chromatography (HPLC). Recently, a few novel assays with ease of operation and high reproducibility have been developed [36, 37].

The FPA assay is performed so that 0.5 mL of diluted enzymes release about 2.0 mg of glucose equivalents from 50 mg of filter paper (4% conversion) in 60 min, as determined by the 3,5-dinitrosalicylic acid (DNS) assay [36, 37]. DNS will be added at the last step of the enzyme assay to stop the reaction. The DNS reagent is used in a colorimetric method for the determination of reducing sugars, such as glucose. DNS, contains sodium potassium tartrate, which decreases dissolved oxygen by increasing the ion concentration in the solution. Sodium bisulfite stabilizes the color obtained and reacts with any oxygen in the buffer. Phenol increases the amount of color produced. An alkaline buffer is also required for FPA assay for the redox reaction between DNS and glucose. The final solution is then boiled vigorously, and the absorbance of diluted samples is read at 540 nm.

The FPU is calculated by using the following equation:

$$Fillter\,Paper\,Activity = \frac{0.37}{[enzyme]\;releasing\;2.0\,\text{mg}\;glucosse}\;\text{units/ml} \tag{5.19}$$

where, [enzyme] is the proportion of original enzyme solution present in the directly tested enzyme dilution (that dilution of which 0.5 mL is added to the assay mixture) [38].

Endoglucanase (EGs) activity: carboxymethyl cellulase activity

EGs can randomly hydrolyze internal glycosidic bonds in cellulose chains. One unit (IU) of EG is defined as the amount of enzyme that liberates 1 μmol of glucose per minute under assay conditions. EGs' activities can be measured using soluble cellulose with a high degree of DP such as carboxymethyl cellulose (CMC). Carboxymethyl cellulase (CMCase) is a procedure described by Mandels et al. [34]. In this method, CMCase activity is measured by determining the reducing sugars released after 5 min of enzyme reaction with 0.5% CMC at pH 4.8 and 50 °C. Reducing sugars can be estimated by high-performance liquid chromatography

(HPLC) or a glucose oxidase (GOD)/peroxidase reagent [39] or a colorimetric method i.e. Somogyi-Nelson method that uses alkaline copper as an inorganic oxidant.

Exoglucanases activity: avicellulases

CBH (exoglucanases) are classified as exo-acting based on the assumption that they all cleave β-1,4-glycosidic bonds from chain ends releasing cellobiose and some glucose molecules. Commercial Avicel (also called microcrystalline cellulose or hydrocellulose) is used for measuring exoglucanase activity because it has a low degree of polymerization of cellulose and it is relatively inaccessible to attack by EGs despite some amorphous regions.

β-Glucosidase assay

One unit of β-glucosidase is defined as the amount of enzyme that liberates 1 μmol of p-nitrophenol per minute [40] β-Glucosidase activity can be calculated using various *chromogenic* and *non-chromogenic* substrates [41]. In the *chromogenic* method, p-nitrophenol-β-glucoside (pNPG) is used as the substrate and is measured in order to determine the hydrolysis rate in optimal pH and temperature. Reaction conditions such as temperature and pH for different β-glucosidases vary pNPG, at concentration (usually 1–5 mM), is added to an appropriate buffer with optimal pH, containing the enzyme and incubated at the optimal temperature. After 10-min incubation, the reaction is stopped by adding three volumes of sodium tetraborate saturated solution, and then the absorbance is read at 405 nm. When using *non-chromogenic* substrates, different methods can be used depending on the substrates. If the substrates are oligo- or disaccharides (such as cellobiose), the glucose can be evaluated by the GOD method with a commercial kit. When the substrate is a polysaccharide, reducing sugars are measured by the DNS method. Using polysaccharides, the enzyme unit is determined as the amount of enzyme required for the liberation of 1 μmol of glucose or reducing sugar per minute.

5.2.6.3 Determination of Resistant Starch, Non-resistant Starch and Total Starch

The content of resistant starch, non-resistant starch (solubilized starch) and total starch can be analyzed by using a starch assay kit. The total starch content of the tested samples is calculated as the sum of the resistant starch and solubilized starch fractions.

Amylose and Amylopectin Content

Amylose content is determined after iodine complexation [42] or is calculated by the iodometric method [43, 44].

5.2.7 *Model of Enzymatic Saccharification*

5.2.7.1 Cellulose Model

The mechanism of converting insoluble *cellulose* into soluble glucose by the action of cellulase enzymes has not yet been completely understood due to the complexity of the involved phenomena (such as, adsorption, desorption, enzyme deactivation, accessible area, crystallinity, degree of polymerization, lignin content, enzyme synergism, etc.), which can affect the reaction kinetics. A number of mathematical models for enzymatic hydrolysis have been proposed. Zhang et al. [8] classified the models as (i) non-mechanistic; (ii) semi-mechanistic; (iii) functionally based and (iv) structurally based. Bansal et al. [45] made a review where a collection of diverse mathematical models for biomass enzymatic hydrolysis was presented.

Mainly, the models were classified as: (i) empirical; (ii) Michaelis-Menten based; (iii) adsorption in cellulose hydrolysis and (iv) on soluble cello-oligosaccharides.

Gharpuray et al. [46] derived an empirical model for the relative extent of hydrolysis (REH) of wheat straw, where the specific surface area (SSA—$m^2\ g^{-1}$), crystallinity index (CrI—%) and lignin content (L—%) are presented below:

$$REH = 2.044\,(SSA)^{0.998}(100 - CrI)^{0.257}(L)^{-0.388} \tag{5.20}$$

Semi-mechanistic models are based on adsorption and sometimes include modified Michaelis-Menten kinetics. Kadam et al. [47] proposed a model in relation to substrate and product concentration and their effects on product inhibition. A Langmuir isotherm was used to describe enzyme adsorption on the substrate. Semi-mechanistic models are believed to be more useful for reactor and process design. The mathematical description in Kadam's model [47] is presented below (Eqs. 5.21–5.26):

Enzyme Adsorption

$$\text{Langmuir isotherm} \quad E_{iB} = \frac{E_{imax}K_{iad}E_{iF}S}{1 + K_{iad}E_{iF}} \tag{5.21}$$

Cellulose to Cellobiose Reaction with Competitive Glucose, Cellobiose and Xylose Inhibition

$$r_1 = \frac{k_{1r}E_{1B}R_S S}{1 + \frac{G_2}{K_{1IG2}} + \frac{G}{K_{1IG}}\frac{X}{K_{1IX}}} \tag{5.22}$$

Cellobiose-to-Glucose Reaction with Competitive Glucose and Xylose Inhibition Reaction.

$$r_3 = \frac{k_{3r}E_{2F}G_2}{K_{3M}\left(1 + \frac{G}{K_{3IG}} + \frac{X}{K_{3IX}}\right) + G_2} \quad (5.23)$$

Mass Balances

$$\text{Cellulose:} \quad \frac{dS}{dt} = -r_1 - r_2 \quad (5.24)$$

$$\text{Cellobiose:} \quad \frac{dG2}{dt} = 1.056r_1 - r_3 \quad (5.25)$$

$$\text{Glucose:} \quad \frac{dG}{dt} = 1.111r_2 + 1.053r_3 \quad (5.26)$$

where,

E_{Ti}	Total enzyme concentration (g-protein/L) (i = 1 for CBH + EG; i = 2 for BG)
E_{Bi}	Bound enzyme concentration (i = 1 for cellulase; i = 2 for BG)
E_{Fi}	Concentration of free enzyme in solution (i = 1 for cellulase; i = 2 for BG)
G_i	Glucose (i = 1), cellobiose (i = 2) (g/L)
K_{iad}	Dissociation constant for enzyme adsorption/desorption reaction (L/g-protein) (i = 1 for CBH + EG; i = 2 for BG)
K_{3M}	Substrate (cellobiose) saturation constants (g/L)
K_{iIG2}	Inhibition constant cellobiose (g/L) (i = 1 for r_1; i = 2 for r_2)
K_{iIG}	Inhibition constant glucose (g/L) (i = 1 for r_1; i = 2 for r_2; i = 3 for r_3)
K_{iIX}	Inhibition constant xylose (g/L) (i = 1 for r_1; i = 2 for r_2; i = 3 for r_3)
k_{ir}	Reaction rate constant (i = 1 and 2, L/g h; i = 3, h^{-1})
r_i	Reaction rate (g/L h) (i = 1 for cellulose to cellobiose; i = 2 for cellulose to glucose; i = 3 for cellobiose to glucose)
R_S	Substrate reactivity
S	Substrate concentration (g/L)
X	Xylose concentration (g/L)
α	Dimensionless constant for substrate reactivity.

Although functional and structural models can help to understand the level of substrate structure, multiple enzyme activities and process design can be provided while their applicability is limited to the particular cases for which they were developed [48, 49].

5.2.7.2 Starch Model

Several models of starch hydrolysis have already been presented in the literature, focusing on predicting the concentrations of small carbohydrates (with the degree of polymerization (DP) of up to 10 glucose units) or the dextrose equivalent (DE). Most of the available models describing starch hydrolysis focus only on a fraction of the hydrolysis products [50–54].

An interesting study with a model that simultaneously covers all the separate products and substrates of the starch hydrolysis reaction and predicts the complete product composition and dextrose equivalent (DE) of starch liquefied by *Bacillus licheniformis* α-amylase (BLA) over time, is the stochastic model of Besselink et al. [50]. This model describes starch hydrolysis by predicting both the average dextrose equivalent and the concentrations of small carbohydrates (DP < 8) during hydrolysis of wheat starch, predicts the formation and breakdown of all carbohydrates involved in the amylolytic hydrolysis of amylopectin and amylose in time with a minimum number of experiments.

The subsite maps of Bacillus licheniformis a-amylase (BLA) from Kandra et al. [55] and of BAA from Allen and Thoma [56] were used as input for the model of Besselink et al. [50]. The subsite theory for depolymerizing enzymes [53, 57] was used as the basis for the hydrolysis model. The enzyme is composed of several subsites to which the monomers can bind, each subsite can increase or decrease the free energy of the enzyme–substrate complex. The following reaction scheme is used to describe the hydrolysis of substrate S_n with DP n by the enzyme E into products P_{n-m} and P_m:

$$E + S_n \underset{k_{-1,r,n}}{\overset{k_{+1,r,n}}{\rightleftharpoons}} ES_{r,n} \overset{k_{+2,r,n}}{\rightarrow} E + P_{n-m} + P_m$$

where, r refers to the position of the reducing end of the saccharide in the subsite map. The subsite theory assumes that $K_{+2,r,n} \ll k_{-1,r,n}$, which implies that $K_{r,n}$ (the association constant for the enzyme–substrate complex) is equal $k_{+1,r,n}/k_{-1,r,n}$, $K_{r,n}$ can be calculated from the binding energies of the occupied subsites [46–48] with the following equation:

$$-RT\ln\left(K_{r,n}\right) = \sum_{i=r-n+1}^{r} \Delta G_i + \Delta G_{mix} \tag{5.27}$$

where R is the gas constant, T the absolute temperature in Kelvin, ΔG_i is the binding energy of subsite i and $\Delta G_{mix} = 8.4T\ln(55.5)\,\text{kcal mol}^{-1}$ for bimolecular processes.

The simulations were ended after a pre-defined model time of hydrolysis. The content of the substrate matrix was analyzed to determine the saccharide composition and DE. The DE was calculated with (Eq. 5.28):

$$DE = \frac{180.16 \times n_{re}}{162.14 \times n_{gl} + 18.02 \times n_{re}} \times 100 \tag{5.28}$$

where, n_{re} stands for the number of reducing ends and n_{gl} stands for the number of glucose units in the substrate matrix.

The carbohydrate composition measurements were used to determine the weight fraction of a specific carbohydrate with the following equation (Eq. 5.29):

$$X_w\,(\%) = \frac{C_{DPi}}{C_{tot} + M_{w,w} \cdot \left(\sum_{j=1}^{7} C_{m,DPj}\right)} \cdot 100 \tag{5.29}$$

where C_{DPi} (g l^{-1}) is the mass-based concentration of carbohydrates with degree of polymerization i, (DPi), $C_{m,DPj}$ (mol l^{-1}) is the mole-based concentration of carbohydrates with the degree of polymerization j (DPj), C_{tot} (g l^{-1}) is the total carbohydrate concentration (g l^{-1}), and $M_{w,w}$ is the molar mass of water (18.02 g mol^{-1}).

5.3 Biochemical Methane Potential (BMP)

5.3.1 *Fundamentals: Anaerobic Digestion and BMP*

Anaerobic digestion (AD) is a widely applied and mature technology for biogas production from liquid and solid wastes, which has attracted much interest during the last decades. It was initially employed for the treatment and stabilization of sludge (secondary and primary) in traditional wastewater treatment plants [58], but in the recent years, AD is also applied for the treatment and valorization of different high- strength industrial wastes and wastewaters or for renewable energy production from agricultural wastes or energy crops. The fast growth of the biogas sector can be seen by the considerable number of biogas power plants operating in Europe (17,240 at the end of 2014), corresponding to an installed capacity of 8.3 GW [59]. Biogas upgrading into biomethane is also a sector with a rapid growth (367 biomethane AD plants in Europe at the end of 2014). The increased role of biogas in the EU, is also enhanced by the goals of the European Commission for the reduction of GHG emissions to 40% until 2030 (relative to the levels of 1990) [60] and for 10% replacement of the conventional with renewable energy in the transport sector [61].

AD is a biochemical process, during which complex organic matter is converted into biogas (rich in methane gas) through a series of successive and/or parallel reactions, which are performed, in the absence of oxygen. The biochemical reactions are carried out by a consortium of several groups of microorganisms through four successive steps:

1. hydrolysis by extracellular enzymes of polymers (proteins, lipids, carbohydrates) into their monomers
2. acidogenesis (conversion of monomers to volatile fatty acids),
3. acetogenesis (production of acetic acid from the higher volatile fatty acids) and
4. methanogenesis via acetate consumption and via hydrogen and CO_2 utilization.

As mentioned above, biomasses of different sources such as agricultural wastes, municipal solid waste (MSW), food waste, industrial waste and wastewater or energy crops could be potential feedstocks (substrates) for the AD process. The biomass type influences the potential of the biogas production, and in order to optimize the process, it is of crucial importance to evaluate this potential of a given feedstock [62]. The Biochemical Methane Potential (BMP) test is the most widely used technique to determine the methane yield of different organic substrates (liquid and solids) that can be achieved through AD. It also provides an indication of the biodegradability of the substrate or mixture of substrates used, being also a valuable method for finding proper substrates for AD [63] or for co-digestion configurations [64].

In the last 35 years, a lot of publications on anaerobic biodegradability tests using different substrates have been published, the majority within the last 10 years. Owen et al. [65] were the first to propose a simple and inexpensive batch anaerobic bioassay technique for measuring the biodegradability and the BMP of organic substrates, laying the background for future works. According to this technique, anaerobic serum bottles, containing sample (<2 g/L degradable COD), defined media and seed inoculums, are incubated for 30 d at the desired temperature, and the gas production is continuously monitored volumetrically, using a syringe, and is analyzed through gas chromatography so that the methane content can be assessed. Chynoweth et al. [66] modified this protocol in order to assess the influence of some operational parameters such as the type of inoculum, inoculum-to-feed ratio and particle size of the feedstock on the BMP. Based on the results, it was verified that all parameters are crucial for the BMP assay and it was suggested that an active anaerobic inoculum obtained from a domestic sludge digester with inoculum-to-feed ratio greater than 2, should be used, while regarding the biomass size, a particle size greater than 1 mm is most suitable for BMP determination. The same research group also proposed a modification of Owen's et al. [65] method, which nowadays forms the basis of the BMP measurement of different solid samples (such as solid municipal wastes, or lignocellulosic feedstocks), suggesting an inoculum of 20% v/v and a sample concentration of 2 g VS/L (VS: volatile solids) [67].

Since then, *scopus* shows a large number of published papers with "anaerobic digestion" and "BMP" as keywords (approximately 570 publications at the end of 2017) which present protocols for the assessment of anaerobic biodegradability and methane potential from different kinds of organic wastes and biomasses [68–70]. In addition, there are some technical guidelines such as ISO 1173, 14853, 15985 and VDI 4630, presenting the methodology for biodegradability and ultimate biogas potential determination. Despite the large number of protocols and norms on BMP already published, direct comparison of the biodegradability or BMP of different solid wastes, is very difficult due to the different environmental conditions and methodologies used, different equipment as well as different units expressions of BMP [69]. This was also confirmed by a recent inter-laboratory study [71], in which the BMP of solid organic substrates varied significantly between laboratories, which in turn implies that the BMP protocol should be further standardized. In Fig. 5.5, the theoretical and main experimental methods for BMP determination, are depicted.

5.3.2 *Methods for BMP Measurement*

5.3.2.1 BMP Measurement in Batch Anaerobic Reactors

Biochemical methane potential (BMP) tests are carried in batch reactors or vials. As the BMP value represents the maximal quantity of methane that can be produced per amount of substrate, tests should be carried out in conditions which are favorable to anaerobic digestion, and more particularly to methanogenic archae activity.

Substrate

The amount of substrate introduced in the reactor should be accurately measured. In the case of liquid samples, this amount is generally expressed in terms of chemical oxygen demand (COD). The result is thus expressed as $L_{CH_4}\ kg^{-1}{}_{COD}$ or by default $L_{CH_4}\ L^{-1}{}_{sample}$. In case of solid samples, the amount of substrate is generally expressed in terms of volatile solids (VS, equivalent to organic matter):

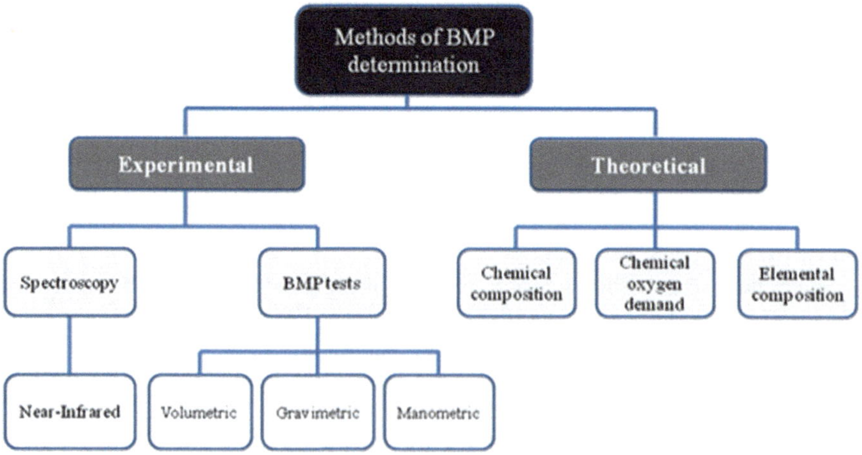

Fig. 5.5 Methods for BMP determination presented in the present handbook

$L_{CH_4}\ kg^{-1}{}_{VS}$ or total solids (TS, equivalent to dry matter): $L_{CH_4}\ kg^{-1}{}_{TS}$ or by default in terms of wet sample: $L_{CH_4}\ kg^{-1}{}_{wetsample}$. Biogas or methane volumes are generally reported in standard conditions (STP: 1 atm, 273.15 K).

Special attention should be paid on sampling heterogeneous solid wastes in order to secure that a representative sample of the waste is taken. Freeze-drying or low temperature drying and milling of a representative amount of waste is a good option to homogenize samples when no volatile compounds are likely to be lost, which is not the case for ensiled biomass for example. It may also be recommended to use as fresh as possible substrates [70]. If needed, samples may be stored for a few days at 4 °C or at −20 °C, but freezing-thawing may significantly impact the BMP as in the case of sewage sludge for example.

Samples are usually ground before being digested in BMP tests, since the particle size has a high impact on anaerobic digestion yield and may influence the BMP value. Particle sizes lower than 10 mm are usually recommended [71].

Inoculum

An inoculum should be collected from an anaerobic digester which is at steady state, well-functioning, and if possible treating the same kind of feedstock as the sample. A good option is to mix various inocula from several anaerobic digesters fed with different feedstocks. If BMP tests are carried out in mesophilic or thermophilic conditions, the inoculum should originate from a mesophilic or thermophilic digester, respectively. The inoculum is generally characterized in terms of volatile solids concentration and should present low endogenous methane production. If it is not the case, a few days incubation at working temperature should allow to degrade the remaining biodegradable material. Inoculum activity ($L_{CH_4}\ kg^{-1}{}_{VSinoculum}\ d^{-1}$) may be assessed by measuring daily methane production from standard substrates such as ethanol, acetate or glucose.

BMP test preparation

Batch reactors consist of bottles or vials closed with septa. According to the available amount of substrate and its heterogeneity, the reactor volume ranges from 100 mL to 2–5 L, 500 mL being usual volume. Small volumes should be used only with homogenous samples. The working volume generally represents 60–80% of the total reactor volume, in order to allow sufficient headspace for biogas accumulation.

One of the most important parameters is the substrate to inoculum ratio (S/I); too high S/I may lead to acidification or inhibition issues, whereas too low S/I ratios increase the part of biogas originating from the inoculum. S/I ratios from 0.25 to 1 ($kg_{VS}\ kg^{-1}{}_{VS}$ or $kg_{COD}\ kg^{-1}{}_{VS}$) are recommended. The COD concentration of the sample in the anaerobic digestion medium should be low in order to facilitate mass transfer, also avoiding inhibition problems. Concentrations ranging from 5 to 10 $g_{COD\ or\ VS_{sample}}$ may be recommended.

The pH should be in the range 6.0–8.5, optimally between 7 and 8, in order to favor methanogen microorganisms activity. Sufficient alkalinity is also required to mitigate risks of acidification; thus a $NaHCO_3$ buffer is usually added (Table 5.1). In order to prevent limitation by any macro and micronutrients, a solution of trace elements is added as shown in Table 5.1.

Before sealing, batch reactors should be free of oxygen. The head space is flushed with a CO_2/N_2 (20/80) mixture or with N_2. Tests should be carried out in triplicate. A negative control prepared as the test reactors but with no substrate addition in order to measure inoculum endogenous methane production which is subtracted from the methane production obtained in reactors fed with substrates. It is also recommended to prepare positive controls fed with well-known standards (acetate, ethanol, glucose or micro-crystalline cellulose) in order to check the inoculum activity.

Incubation

Reactors are incubated under soft agitation at either mesophilic (35–38 °C) or thermophilic (50–55 °C) conditions, until the achievement of a plateau on cumulative methane production. According to Holliger et al. [70], tests should last until the daily methane production during 3 consecutive days, is less than 1% of the accumulated volume of methane.

Biogas production measurement

Biogas volume and composition should be measured every day during the first week of incubation and then measurements can be less frequent (every few days and then once a week) Methane content in biogas (C, expressed as volumic fraction) is usually measured by gas chromatography using thermal conductivity detector (TCD). Two columns may be used, the first one to separate CO_2 and the second one to separate the other gases.

Table 5.1 Concentration in buffer, macro and micronutrients in reactors at the beginning of BMP tests

Compound	Concentration (mg L^{-1})
$NaHCO_3$	2600
NH_4Cl	286
KH_2PO_4	108
$MgCl_2$, $6H_20$	65
$CaCl_2$, $2H_2O$	32
$FeCl_2$, $4H_2O$	20
$CoCl_2$, $6H_2O$	5
$MnCl_2$, $4H_2O$	1
$NiCl_2$, $6H_2O$	1
$ZnCl_2$	0.5
H_3BO_3	0.5
Na_2SeO_3	0.5
$CuCl_2$, $2H_2O$	0.4
Na_2MoO_4, $2H_2O$	0.1

Biogas volume may be measured with different methods, volumetric, manometric or gravimetric.

Volumetric methods usually use a flow meter or gas counter which are on-line connected to the reactors or may be used offline by releasing gas accumulated in the space. A lubricated syringe or a liquid displacement method can also be used for biogas volume measurement. For the latter option, the liquid must be an acidic aqueous solution (pH = 2) to avoid CO_2 solubilization. Biogas can be bubbled into an alkaline solution (NaOH, pH > 12) before going through a flow meter or gas counter. CO_2 is thus dissolved in this solution and methane volume is directly measured as in the case of Automatic Methane Potential Test System (AMPTS) from Bioprocess Control.

If Vi and Ci (%) are the volume and methane contents of biogas measured at sampling i, the methane production between sampling i and i − 1 is expressed, in STP, by:

$$V_{\mathrm{CH_4},i} = [V_i C_i + V_{head}(C_i - C_{i-1})] \frac{P_{atm}}{10^5} \frac{273.15}{T_i} \tag{5.30}$$

where V_{head} is the headspace volume, P_{atm} is the atmospheric pressure and T_i the working temperature (K).

During manometric methods, measurement of methane concentration in biogas (C_i) and reactor pressure before (P_i) and after biogas release is taken place, which is used as initial pressure for the next measurement ($P_{i+1,\,0}$). Normal volume of methane produced between sampling i and i − 1 is expressed by:

$$V_{\mathrm{CH_4},i} = \left[P_i C_i * \frac{V_{head}}{RT_i} - P_{i,0} C_{i-1} * \frac{V_{head}}{RT_{i-1}}\right] \frac{R * 273.15}{10^5} \tag{5.31}$$

where R is the ideal gas constant, R = 8.314 J mol^{-1} K^{-1}.

A gravimetric method was proposed by Hafner et al. [72]. Biogas production is measured by weighing the reactor before and after biogas release and determining the methane content in the biogas (C_i, %). Biogas volume is calculated from mass loss Δm_i

$$V_{biogas,i} = \frac{m_i}{d_{biogas,i} + m_{\mathrm{H_2O},i}} \tag{5.32}$$

where d_{biogas} is density of dry biogas. It is deduced from molar masses of CH_4 and CO_2 and molar volume at STP (22.4 NL)

$$d_{biogas,i} = \frac{16.04\, C_i + 44.01\,(1 - C_i)}{22.4} \tag{5.33}$$

$m_{\mathrm{H_2O}}$ is the mass of water in the biogas, calculated from saturated vapor pressure $P_{\mathrm{H_2O}}$

$$\log_{10} P_{H_2O} = 8.20963 - \frac{2354.731}{T} \tag{5.34}$$

$$m_{H_2O} = \frac{18.02\, P_{H_2O}}{\left(P_{biogas,i} - P_{H_2O}\right)22.4} \tag{5.35}$$

The volume of methane produced is given by:

$$V_{CH_4,i} = C_i\, V_{biogas,i} \tag{5.36}$$

The accumulated volume of methane produced is then calculated:

$$V_{CH_4} = \sum_i V_{CH_4,i} \tag{5.37}$$

Finally the methane potential is given by

$$BMP = \frac{V_{CH_4}}{m_{sample}} \tag{5.38}$$

where m_{sample} is the amount of sample used in the test, expressed in volatile solids or total solids.

Hafner et al. [73] made available a R package and a web-based interface (https://biotransformers.shinyapps.io/oba1/) which provide functions to compute experimental data from volumetric, manometric or gravimetric methods [73].

5.3.2.2 BMP Assessment by NIRS Calibration

Non-destructive measurements using spectroscopic methods such as Near Infrared Reflectance (NIR) spectroscopy are promising. NIR is widely applied by many industries and is generally accepted as an instrument based method giving fast measurement that is used for process control and quality control of raw material [74]. Four pioneering studies regarding the application of NIR to the prediction of BMP and biogas production rate and their correlation to NIR have been reported to date. Two of these are dedicated to specific substrates, like meadow grass [75] or, more recently, urban organic waste [76]. Moreover, two others studies are based on a wide variety of substrates such as municipal solid wastes (MSW) [77] or combined MSW and agro-industrial wastes [78] leading to a global and unique model used for waste characterisation, called Flash BMP® [79].

Near-infrared spectroscopy has the advantage of analysing the organic matter of any type of material globally and quickly in a matter of minutes. Near infrared spectroscopy is an analytical method requiring a training procedure: a calibration set of a few hundreds of samples is defined and used to compute the relationship (i.e. the model) between the spectra and the value of interest (BMP). The model is then tested on a validation set which does not include the calibration samples. The validation is made on a large database which makes the model robust in respect of waste variability. In the global model [79], the sample set includes the following types of material: agro-industrial waste, biowaste, energy crops, agricultural residues, fatty waste, meat waste, co-digestion mix, municipal solid waste, plants and vegetables, agro-industrial sludge, sewage sludge and digestate.

To obtain a low water content, the samples must be freeze-dried. Usually, a simple drying is not enough to guarantee a water content of less than 15%. This is the prerequisite for accurate prediction of BMP value because the presence of water molecules leads to poor NIR prediction accuracy. In addition, grinding between 1 and 2 mm is necessary to improve the homogeneity of the measured sample. Depending on the NIR spectrometer model used, the quantity of organic matter required is between 2 and 30 g. Using a Petri Dish add-on allows the direct analysis of a large and representative quantity of waste. For each sample, a triplicate is recorded on the rotating cup to assess homogeneity and repeatability, which must be less than 1%.

The BMP prediction error on predicting urban organic waste by [76] (i.e. 44 mL CH_4/g VS) was comparable to previously reported studies by Lesteur et al. [77] (i.e. 28 mL CH_4/g VS), Raju et al. [75] (i.e. 37 mL CH_4/g VS), Doublet et al. [78] (i.e. 40 mL CH_4/g VS) and Triolo et al. [80] (i.e. 37 mL CH_4/g VS) for the BMP prediction of MSW, meadow grass, combined MSW and agro-industrial waste and plant biomass, respectively. The prediction error for the global model [79] gives a range of 40–60 mL CH_4/g VS depending on the kind of substrate considered. The model performance is therefore similar to the reproducibility of the reference BMP test.

An advantage of the NIR measurement is the possibility to measure other parameters simultaneously with the same sample and the same instrument: e.g. organic matter characterisation, COD, carbohydrate, protein and lipid contents [81] and kinetic curves of methane production [82]. Near infrared spectroscopy can predict either quantitative (concentrations of products of interest, level of specific properties) or qualitative data (belonging to a class).

Current research is focused on the use of NIR spectrometry to analyse raw substrates that do not require drying and for on-line analysis of anaerobic digestion processes.

5.3.3 Ways to Interpret BMP Measurement (Index of Biodegradability, Kinetics Parameters, Modeling)

Calculation of the BMP of a given substrate can also be performed through theoretical approaches. These methods assume that the substrate is completely degraded anaerobically, resulting in a gas mixture of CH_4 and CO_2. The ratio between CH_4 and CO_2 depends on the oxidation state of the carbon present in the substrate, i.e.

the more reduced organic carbon content results in a biogas production with higher CH_4 content. Thus, carbohydrates result in 50% CH4, proteins in the range of 50–70% and fats between 70 and 75%.

Theoretical BMP calculation

Theoretical BMP (TBMP) can be estimated using the elemental composition, the chemical composition or the chemical oxygen demand (COD) of a given biomass. Theoretical calculations of the BMP of a given substrate is often higher than the respective experimental BMP values. This is due to the fact that a fraction of the substrate is used to synthesize bacterial biomass or to a limited substrate biodegradability, because of the presence of non or hardly biodegradable substances such as lignin. The possibility of microorganisms (methanogens) inhibition due to the presence of toxic or inhibitory compounds in the substrate, is another factor contributing potentially to a discrepancy between the theoretical and experimental BMP.

Buswell and Mauller [83] proposed a formula for the theoretical estimation of the methane yield, based on elemental composition of carbon (C), hydrogen (H) and oxygen (O) (Eq. 5.39)

$$\mathrm{C_cH_hO_o} + \left(c - \frac{h}{2} - \frac{o}{4}\right)\mathrm{H_2O} \rightarrow \left(\frac{c}{2} - \frac{h}{8} + \frac{o}{4}\right)\mathrm{CO_2} + \left(\frac{c}{2} + \frac{h}{8} - \frac{o}{4}\right)\mathrm{CH_4} \quad (5.39)$$

The TBMP, usually expressed as $L_{CH_4}\ g_{VS}^{-1}$ may then be calculated using Eq. 5.40:

$$\mathrm{TBMP}\left(\frac{\mathrm{L}}{g_{VS}}\right) = \frac{22.4\left(\frac{c}{2} + \frac{h}{8} - \frac{o}{4}\right)}{12c + h + 16o} \quad (5.40)$$

An assumption in the calculation, is that there is no ash content in the substrates. This is because ash contains oxygen in the form of metal oxides.

The above chemical reaction can be modified in order to include nitrogen (N) and sulfur (S) in the stoichiometric formula of the organic compound, and as such to obtain the fraction of ammonia (NH_3) and hydrogen sulfide (H_2S) in the produced biogas (Eq. 5.41):

$$C_cH_hO_0N_nS_s + \left(c - \frac{h}{4} - \frac{o}{2} + \frac{3n}{4} + \frac{s}{2}\right)\mathrm{H_2O} \rightarrow \left(\frac{c}{2} - \frac{h}{8} - \frac{o}{4} + \frac{3n}{8} + \frac{s}{4}\right)\mathrm{CO_2} + \left(\frac{c}{2} + \frac{h}{8} - \frac{o}{4} - \frac{3n}{8} - \frac{s}{4}\right)\mathrm{CH_4} + n\mathrm{NH_3} + s\mathrm{H_2S} \quad (5.41)$$

and the TBMP, might then be calculated from Eq. 5.42:

$$\mathrm{TBMP}\left(\frac{\mathrm{L}}{g_{VS}}\right) = \frac{22.4\left(\frac{c}{2}+\frac{h}{8}-\frac{o}{4}-\frac{3n}{8}-\frac{s}{4}\right)}{12\mathrm{c}+\mathrm{h}+16\mathrm{o}+14\mathrm{n}+32\mathrm{s}} \tag{5.42}$$

Again, the generation of microbial biomass is not accounted in this calculation and is a source of error.

The TBMP of typical organic compounds such as carbohydrates, proteins and lipids, based on Eq. 5.42, is presented in Table 5.2.

The TBMP of biomass calculated based on Eq. 5.42 is 474 L CH_4/kg_{vs}. However, in practice the experimental BMP of different lignocellulosic feedstocks is much lower than the theoretical ones, generally not exceeding 60% of TBMPs. This is due to the presence of non-biodegradable compounds (e.g., lignin, peptidoglycan) and the difficulty to hydrolyse some polymers, such as cellulose.

Another approach is to estimate TBMP through the chemical composition. Thus, for complex organic material consisted of carbohydrates, proteins and lipids the TBMP could be estimated using Eq. 5.43:

$$\mathrm{TBMP} = 415 \cdot \%\mathrm{Carbohydrates} + 496 \cdot \%\mathrm{Proteins} + 1014 \cdot \%\mathrm{Lipids} \tag{5.43}$$

The coefficients of this equation correspond to the TBMP of the respective compounds, based on Table 5.2. The different fractions of Eq. 5.43 could be quantified by analytical composition measurements. Apart from the above, a lot of regression models could be used to predict the methane yield of organic matter from chemical composition or structural features [84].

Finally, the TBMP could be calculated by a COD analysis. Theoretically, 0.35 L CH_4 could be obtained at STP from 1 g COD removed (taking into account that 1 mol CH_4 = 22.4 L CH_4 = 16 g CH_4 or 64 g COD CH_4) or 0.395 L at 298.15 °C, which is a typical temperature at mesophilic conditions. However, the quantification of the COD of solid samples is quite difficult and sometimes impossible and thus this theoretical value for BMP is more useful for liquid samples.

Table 5.2 Theoretical characteristics of typical organic compounds

Substrate type	Empirical formula	% CH_4	TBMP (L/g_{vs})
Carbohydrates	$C_6H_{12}O_6$	50	0.415
Proteins	$C_{14}H_{12}O_7N_2$	50–70	0.496
Lipids	$C_{57}H_{104}O_6$	70–75	1.014
Biomass	$C_5H_9O_{2.5}NS_{0.025}$	43	0.474

Anaerobic biodegradability

The TBMP could be used to calculate the anaerobic biodegradability (BD) or digestibility of a substrate, under specific experimental conditions, as a comparison with the respective experimental BMP (Eq. 5.44):

$$BD\,(\%) = \frac{BMP}{TBMP}100 \tag{5.44}$$

and taking into account that 1 g COD produces 0.35 L CH_4, then:

$$BD\,(\%) = \frac{BMP\left(\text{mLCH}_4\text{g VS}^{-1}\right)}{350 * COD_{waste}\left(\text{g CODg VS}^{-1}\right)}100 \tag{5.45}$$

However, direct comparison of both BMP values (theoretical and experimental) would not be accurate, since a 10–15% of the available organic matter is used for microorganism growth, not contributing to methane production. In order to find the real level of BD the experimental BMP should be adjusted (increased), taking into account the yield of biomass ($Y{:}\frac{X}{S}$ fraction of substrate converted to biomass).

BMP mathematical modelling

Kinetic model analysis is an effective and useful tool for predicting methane production as a function of time for a specific substrate. Up to now, numerous kinetic models have been used to describe the methane production from BMP tests [85]. However, the three modeling approaches presented below, are the most widely used for BMP experiments.

First-order model

The first-order kinetic model is the most widely used model due to its simplicity and its ability to reflect the cumulative effect of all the reactions occurring during the real process. It is based on the assumption that hydrolysis is the rate limiting step during the anaerobic digestion process, and as such it could be used for the determination of useful kinetic parameters, such as the first order hydrolysis kinetic constant, k (d^{-1}):

$$\frac{dS}{dt} = -k \tag{5.46}$$

Here, t is the digestion time and S represents the biodegradable substrate concentration. Since the biodegradable substrate concentration, especially for solid wastes, is difficult to measure and taking into account that the biodegradable substrate concentration is related to the methane production, Eq. 5.46 could be written as:

$$B = B_o(1 - \exp(-k\,t)) \tag{5.47}$$

where B and B_o are the cumulative and maximum or ultimate methane yields, ($L_{CH_4}\,kg^{-1}{}_{VS}$), respectively. This model predicts a continually declining rate of biogas production.

The Modified Gompertz equation (Eq. 5.48) is also a modeling approach widely used for batch anaerobic systems such as BMP. This model predicts a lag time and a maximum methane production rate at some point in time. It has been initially proposed for methane production, but it is also used for batch hydrogen production.

$$B = B_o \exp\left(-\exp\left(\frac{R_m e}{B_o}(\lambda - t) + 1\right)\right) \tag{5.48}$$

Here, B and B_o are the simulated cumulative and the maximum simulated cumulative methane yields, respectively ($L_{CH_4}kg^{-1}{}_{VS}$), R_m is the maximum methane production rate ($L_{CH_4}\,kg^{-1}{}_{VS}\,day^{-1}$), λ refers to the lag phase time (days), t is the anaerobic digestion time (days), and e is equal to 2.71828.

In addition, some researchers use the logistic function model, to simulate the methane production during BMP experiments. The model fits to the BMP kinetic, since it corresponds to a sigmoid function (Eq. 5.49) predicting an initial exponential increase and a final stabilization at a maximal production level.

$$B = \frac{B_o}{1 + \exp\left[\frac{4R_m(\lambda - t)}{B_o} + 2\right]} \tag{5.49}$$

Except from the above, some other modeling approaches have already been used by many researchers such as the transference function model [85], Cone, and Fitzhugh models [86] or modified versions of the anaerobic digestion model 1 (ADM1) [87]. Despite the different modeling approaches, the use of Eqs. (5.47) and (5.48), for describing the kinetics of BMP experiments may be considered accurate.

5.4 Biohydrogen Potential (BHP)

5.4.1 Fundamentals

Among all the microbial metabolites that can be found in anaerobic digestion processes, hydrogen gas accumulation in the medium plays a key role by inhibiting the syntrophic acetogenic reactions and also by influencing the metabolic pathways through thermodynamics inhibition [88]. Cazier et al. [89] suggested that high hydrogen partial pressures could even be detrimental to the microbial hydrolytic activity in AD. Therefore, hydrogen-producing fermentation must be avoided when dealing with methane production. In counterpart, hydrogen gas is an interesting energetic vector with a high energy content per mass unit (142 kJ g^{-1}) that is much higher than most of the other usual energy vectors, such as methane. Its use in fuel cells produces only water, making it a serious alternative to fossil fuels, especially in the transportation sector.

The use of organic biomass and residues constitutes the most sustainable way to produce hydrogen [90]. Many studies have dealt with H_2 production from organic materials over the past decade. Most of the substrates that can be used in AD are suitable for H_2 generation, since they can be anaerobically degraded. Similarly to the AD process, a simple protocol to rapidly estimate the potential of a biomass to generate hydrogen by fermentation was recently developed [91]. This test is called BHP (for Biohydrogen Potential) in analogy to the BMP (Biochemical Methane Potential) test.

In BHP test, the conditions are considered as optimal to favour the dark fermentation pathways, but the complexity of metabolic pathways in mixed cultures and subsequent microbial interactions could not preclude the possibility of underestimating the maximum hydrogen potential of a biomass. The BHP test could therefore be considered as a first estimation of the hydrogen yield of a specific biomass. Interestingly, Guo et al. [91] and Monlau et al. [84] showed that only the initial readily accessible soluble carbohydrate content correlated with the maximum hydrogen yield whatever the substrate considered. This observation suggests that the main H_2-producing metabolic pathways are only related to sugar fermentative metabolism even in complex substrates. Hydrogen is a product of a biological process called dark fermentation that is carried out by various anaerobic microorganisms, able to degrade and ferment organic materials, and more specifically carbohydrates. These microorganisms are widely found in natural environments, i.e. sediments, manure, digestive tracts or sewage sludge. In nature, hydrogen is also a source of electrons that can be readily used by various microorganisms such as methanogens and homoacetogens. The presence of hydrogen-consuming microorganisms that might considerably reduce the amount of accumulated hydrogen, has to be carefully considered.

During dark fermentation, the bacterial activity converts organic residues into small organic acids, solvents and gas. Accumulation of reduced biochemical intermediates, e.g. formate, nicotinamide adenine dinucleotide (NADH) and ferredoxin (FDH) inside the cells, leads to the production of hydrogen in order to maintain the intracellular redox potential [92]. Microorganisms with specific hydrogen-producing enzymes, called hydrogenases, have the ability to release their excess electrons through hydrogen gas, by reducing two protons according to the following reaction: $2H^+ + 2e^- \leftrightarrow H_2$. One of the main metabolic pathways that generate enough reduced elements is the acetate pathway. Acetate production generates 2 mol of NADH and 2 mol of FDH. Therefore, as maximum yield for H_2 production, the acetate pathway generates 4 mol of hydrogen per mole of glucose, according to the reaction [88].

$$C_6H_{12}O_6\,(Glucose) + 2H_2O \rightarrow 2CH_3COOH\,(Acetate) + 2CO_2 + 4H_2 \qquad (5.50)$$

The second main metabolic pathway is the butyrate pathway, where 2 mol of FDH are generated per mole of butyrate. Here, a maximum of 2 mol of hydrogen per mole of glucose are produced:

$$C_6H_{12}O_6(Glucose) + 2H_2O \rightarrow CH_3CH_2CH_2COOH\,(Butyrate) + 2CO_2 + 2H_2 \qquad (5.51)$$

In mixed cultures, these pathways are mostly carried out by spore-forming bacteria of the *Clostridium* genus. In a BHP test, the operating conditions are also favouring the selection of clostridium species. However, the combination of different metabolic pathways in mixed cultures usually leads to a maximum of 2.5 mol of H_2 per mole of glucose, according to the overall reaction [93].

$$\begin{aligned} &4C_6H_{12}O_6\,(Glucose) + 2H_2O \rightarrow 3CH_3CH_2CH_2COOH\,(Butyrate) \\ &\quad + 2CH_3COOH\,(Acetate) + 8CO_2 + 10H_2 \end{aligned} \qquad (5.52)$$

The hydrogen produced during the conversion of organic matter represents a maximum of 33% of the COD of the substrate with a maximum of 20% in mixed cultures. The remaining 66–80% of the COD is found in the form of other fermentative metabolites (acetate, butyrate) that should be further valorised as methane (AD) or more H_2 (Microbial Electrolysis Cells). In the case of more complex biomass, such as lignocellulosic material, lower amounts of hydrogen are generally observed [88]. Indeed, other fermentation products can be produced such as ethanol, lactate, propionate as well as acetone and butanol.

Furthermore, H_2 accumulation rapidly results in hydrogen consumption by methanogens and homoacetogenic bacteria. Methanogenesis is carried out by archaeal methanogens present in anaerobic environment that can directly use hydrogen to produce methane according to the following reaction:

$$CO_2 + 4H_2 \leftrightarrow CH_4 + 2H_2O \qquad (5.53)$$

Interestingly, methanogens are not spore-forming microorganisms and heating at 90 °C severely inhibits methanogens, while selecting H_2-evolving spore-forming bacteria [94]. Such an inoculum pretreatment can be used to specifically select H_2-producing microorganisms. Moreover, pH values lower than 6 are detrimental to the growth of methanogens and can also be used to specifically select H_2-forming acidogenic bacteria. Meanwhile, homoacetogenic bacteria constitute the second main group of hydrogen-consuming microorganisms. Homoacetogenic bacteria can consume hydrogen to produce acetate according to the reaction:

$$2CO_2 + 4H_2 \leftrightarrow CH_3COOH\,(Acetate) + 4H_2O \qquad (5.54)$$

In the case of homoacetogens, their presence is difficult to avoid, since several homoacetogens belong to the same group of spore-forming microorganisms, i.e. *Clostridium* sp. Nonetheless, in batch tests, hydrogen-producing bacteria and homoacetogens are sequentially favoured and the maximum H_2 yield can be attained prior to the emergence of the main homoacetogenic activity.

5.4.2 *Methods for BHP Measurement*

Although there are well defined protocols for the estimation of BMP from wastes/ wastewaters, in the case of BHP little has so far been done for the standardization of a specific methodology. This is due to the fact that optimized conditions for hydrogen production from different types of wastes are significantly variable, depending on the type of substrate and the implicated microbial consortium.

5.4.2.1 Bioreactors Set-up

In line with BMP tests, BHP is also evaluated in batch mode reactors. The generation of fermentative hydrogen is achieved mainly by strictly anaerobic as well as facultative anaerobic bacteria under anaerobic conditions. Clearly, ensuring anaerobic conditions in the bioreactor is of extreme importance. The general approach for ensuring such conditions is the sparging of the liquid phase of the culture with CO_2: N_2 (30:70 v/v or 20:80 v/v), whereas the use of pure N_2 [95, 96] or even pure CO_2 has been also reported [97]. However, it must be noted that in the case of CO_2 use, the pH of the culture is affected due to the amount of gas that is inevitably dissolved in the liquid phase, where CO_2 undergoes pH-dependent acid–base reactions [98]. Apart from the removal of dissolved oxygen from the liquid phase, sparging also results to its displacement from the headspace of the vial, which subsequently is sealed using rubber septa. Once the bioreactor is sealed, the generated gases by fermentation i.e. H_2 and CO_2 accumulate in the headspace, leading thus to a built-up of pressure in the vial (bioreactor) and consequently an increase of the partial pressure of H_2 (P_{H_2}). P_{H_2} is among the key factors that affect the potential of biohydrogen and when it builds up in the gaseous phase of the cultures, it leads to end-product inhibition and shifting of metabolic pathways to the production of more reduced substrates [99]. Two are the main methodologies that are reported for the sampling of the gaseous phase and quantification of hydrogen: the intermittent pressure release method (Owen method) and the continuous pressure release (respirometric) method [95].

During the Owen method, the gas phase is expanded periodically into glass (preferably) syringes to equilibrate the internal pressure of the bioreactor with the atmospheric pressure and the total gas production in each sampling is calculated. The content of H_2 in the total gas is estimated via gas chromatography and the cumulative hydrogen produced for each time interval is calculated using the following mass balance equation:

$$V_{H,i} = V_{H,i-1} + C_{H,i}(V_{G,i} - V_{G,i-1}) + V_H(C_{H,i} - C_{H,i-1}) \qquad (5.55)$$

where $V_{H,I}$ and $V_{H,i+1}$ are the cumulative hydrogen gas volumes at the current (i) and previous (i − 1) collections, respectively, $V_{G,I}$ and $V_{G,I+1}$ are the total biogas volumes in the current and next collections, $C_{H,I}$ and $C_{H,i+1}$ are the fraction of hydrogen gas in the headspace of the reactor. The kinetics of the biohydrogen production process are much faster than those of AD and, consequently, the frequency of sampling should be higher, too. Depending on the substrate that is treated, the gaseous samples intervals should be of the order of a few hours and the overall process is completed within 24 h to a few days. Liquid sampling is generally conducted only in the beginning and at the end of the process, in order for the liquid metabolites to be quantified and the bioconversion of the substrate to be assessed.

In the respirometric method, the gas release from the bioreactor is continuous and the estimation of hydrogen production is achieved by measuring gas production in incremental volumes as small as 0.1 ml. The incremental volumes may be measured as frequent evacuations of gas in response to small pressure changes or as the volume of bubbles passed through flow measuring cells. No significant pressure built-up is allowed and the effect of P_{H_2} on the hydrogen potential is negligible [95].

The main disadvantage of the respirometric method is that it requires specialized equipment, and for this reason, the Owen method is most frequently used. A parameter that needs to be considered in this case is the ratio of liquid to gaseous phase volumes of the cultures, which has to be estimated in order to avoid pressure build-up between sampling intervals. A headspace of 2/3 of the total vessel volume is generally considered appropriate [100].

5.4.2.2 Substrate, Nutrients and pH

Substrate loading in the culture can be calculated either in terms of its chemical oxygen demand (COD) or in terms of the solids loading expressed as volatile solids (VS) per volume unit. In the case of solid substrates such as municipal wastes and lignocellulosic biomass, or semi-solid substrates such as olive pulp, g $VS_{subs}.L^{-1}$ is preferably used. By measuring the cumulative H_2 that is generated at the end of the process, the BHP of the substrate can thus be estimated in terms of the initial COD or VS used or in terms of the COD that was bioconverted. The substrate loading can be quite diverse depending on the type of substrate, but the general rule is to avoid high loadings in order to avoid substrate or product inhibition. In general, loadings up to 25 g $COD.L^{-1}$ or up to 5% (w/V) VS are reported as optimal and could be used for assessing the BHP of liquid and solids substrates, respectively.

Constant agitation of the culture is also considered essential for proper BHP estimations for ensuring proper mixing and avoiding sedimentation of solids. Agitation can be performed either mechanically or via magnetic stirrers and the proposed stirring velocity is 150–350 rpms.

The supplementation of the culture with macro- and micronutrients is also of extreme importance so as to prevent nutrient limitation. In the case of substrates with a poor content of essential nutrients, such as N_2 and P, the addition of an organic (yeast extract, tryptone etc.) or inorganic nutrient source has to be made. A trace elements solution of similar composition of those used during AD should also be added to the culture [101].

The activity of hydrogenic microorganisms has proved to be pH dependent. At extreme pH values, ATP is used to ensure cell neutrality rather than to produce H_2, whereas the activity the hydrogenase is inhibited. In general, high H_2 production yields are reported at operating pH values in the range 4.5–6.5. Moreover, the maintenance of acidic conditions has also been shown to favor the hydrolytic activity of the mixed hydrogenic consortia.

5.4.2.3 Inoculum

Origin of Inoculum

Biohydrogen production occurs during the third phase of the AD process i.e. the acidogenesis stage, by the acidogenic microbial communities that constitute the diverse consortium that performs the AD. As such, it is widely accepted that inoculum derived from anaerobic digesters, either mesophilic or thermophilic, can be used for the BHP estimation of various substrates. Mixed cultures of hydrogen producing bacteria and archaea exist however in several other natural environments, such as soil, compost, animal manure, the rumen, and different types of wastewaters [102, 103]. Microbial consortia from those habitats are reported to lead to high hydrogen yields, depending on the substrate tested and are can also be used for BHP estimation. Pure cultures of selective strains have also been widely used for hydrogen generation from different types of biomass and wastes, mainly due to their high selectivity for specific substrates, the easiest manipulation of their metabolism by altering growth conditions, the minimization of undesired by-products generation and the high repeatability of the process. However, by definition, BHP tests do not refer to the optimization of the process but to the estimation of hydrogen production capacity of a certain feedstock compared to others and as such can only be assessed by mixed consortia. This is mainly due to the fact that mixed cultures restrict the sterile requirement that in the case of most types of wastes/wastewater is pointless and also may be able to degrade a broader range of feedstocks [104]. Using mixed consortia as inoculum lurks, though, the possible predominance of hydrogen consuming species which would inevitably lead to false conclusion in terms of the BHP result. Strategies for pretreatment of the maternal inocula are therefore always required to restrict or even terminate hydrogen consuming processes, to ensure that hydrogen will remain as the end product of the metabolic flow and thus to allow for a higher substrate utilization efficiency.

Pretreatment of Inoculum

Many hydrogen-producing bacteria form endospores as a protective mechanism during exposure to adverse environmental conditions, such as high temperature, extreme pH, low humidity irradiation etc. Since, however, methanogenic bacteria and all types of archaea lack such capability, inducing mixed consortia to environmental stress can be an effective pretreatment method for the predominance of hydrogen producers. Various methods for the pretreatment of the inoculum have been reported so far, whereas combinations of different pretreatment methodologies have also been proposed as efficient for enhancing hydrogen production.

Heat treatment is generally the most common practice followed, since its effectiveness has been validated by many researchers and it is not time consuming. By subjecting seed cultures to high temperatures, only the spore-forming microorganisms survive the thermal shock, whereas the number of methanogenic non-spore-forming bacteria, depending on the temperature and time of thermal shock, dramatically decreases or is even eliminated. Reported temperatures are 75 °C, 85 °C, 100 °C or even 121 °C, whereas the duration of the process has been reported to vary between 10 min and 12 h. The exposure to high temperatures could be combined with drying of the seed culture, which however, is performed for lower temperatures only. Microwaving of the inoculum at different powers has also been applied as heat shock pretreatment. In general, the time of exposure to thermal shock increases for lower temperatures, whereas the most common practice is boiling of the seed culture for 15–30 min.

Alternatively, to heat treatment, an acid or base treatment of the inoculum has also been suggested as a possible way for ensuring the predominance of hydrogen producing bacteria. The principle of this method is to maintain the seed microorganisms for a prolonged period at a very acidic or very basic environment, which would result, eventually, to the elimination of non-hydrogen producers that cannot survive in such extreme pH conditions.

Exposure of the inoculum culture to chemical substances that could be highly toxic to methanogens, such as bromoethanesulfonate (BES), and chloroform has also been proposed as an alternate to enhance the predominance of hydrogen producers, whereas subjecting the inoculum to ultrasonic pretreatment and UV irradiation were also employed to achieve the enrichment with H_2-producing bacteria.

Substrate to Inoculum Ratio (S/I), F/M Ratio

For the assessment of the BHP of a substrate. a balanced amount of active inoculum is required, consisting of a diverse microbial community, catalysing the interdependent biochemical reaction of hydrolysis and acidogenesis. The relative quantities of microorganisms added in the cultures and of the substrate that they will treat is commonly expressed as the "substrate to inoculum" ratio (S/I) or the food to microorganisms ratio (F/M) [105]. An excessive availability of substrate could result to imbalanced anabolic and catabolic reactions and consequently to energy spilling, affecting thus the distribution of metabolites during substrate conversion [106]. The S/I can be expressed either as the ratio of the VS of the substrate to the VS of the inoculum (VS_{sub}/VS_{inoc}) or the COD of the substrate to the VS of the inoculum (COD_{sub}/VS_{inoc}) at the beginning of each culture, when solid of liquified/dissolved substrates, respectively, are tested. In the literature, S/I ratios in the range 0.5–10 $g\ VS_{sub}{\cdot}g\text{-}1\ VS_{inoc}$ have been reported and it seems that the optimum ranges between 1 and 8. Pan et al. [107] observed that at mesophilic conditions an S/I of 6 led to higher H_2 yields, when treating food wastes, whereas for S/I > 8 the process failed, and CO_2 was the sole gaseous product of the process. High S/I can indeed lead to substrate inhibition, whereas on the other hand, extremely low S/I could result to prolonged lag phases [108].

5.4.3 Modelling

During dark fermentation by mixed consortia, the generation of hydrogen and soluble metabolites is the result of various interactions between different types of microorganisms. Due the complexity of the metabolic paths, the mathematical description of the processes can be approached in general by empirical models. Different kinetic models have been so far proposed [109], with the modified Gompertz model (Eq. 5.56) being the most widely applied for the description of hydrogen in batch cultures. The modified Gompertz equation describes the batch dark fermentation process using a curve-fitting approach, where three model parameters are adjusted to fit experimentally the measured hydrogen evolution data [110].

$$H = H_{max}\exp\left(-\exp\left[\frac{R_{max^e}}{H_{max}}(\lambda - t) + 1\right]\right) \tag{5.56}$$

where H and $H_{\max}$ are the cumulative and the maximum cumulative hydrogen yields, respectively ($L_{H_2}\,kg^{-1}\,VS$), R_{max} is the maximum hydrogen production rate ($L_{H_2}\,kg^{-1}\,VS\,h^{-1}$) and λ is the lag phase time (h).

The modified logistic model described by Eq. 5.57, has also been used to describe the progress of hydrogen production in the batch tests [111].

$$H = \frac{H_{max}}{1 + \exp\left[\frac{4R_{max}(\lambda - t)}{H_{max}} + 2\right]} \tag{5.57}$$

The classical Monod equation has also been proposed for the description of hydrogen production kinetics, taking into account the effect of substrate concentrations on hydrogen production. Different inhibition functions have also often been added to Monod-based kinetics, in order to describe either the inhibitory effect of crucial operational parameters such as substrate concentration, pH etc., or the inhibition caused to specific microbial types by hydrogen or volatile fatty acids (VFAs) [112].

Different modified versions of the IWA Anaerobic Digestion Model No.1 (ADM1), initially developed for anaerobic digestion, have also been proposed as modeling approaches of the fermentative hydrogen production process [113]. To do this, methanogenic and acetogenic steps were excluded from the original ADM1 model structure and the stoichiometry was changed, in an effort to describe the metabolic products generation from carbohydrates fermentation [114].

5.5 Respirometry

5.5.1 Fundamentals: Respiration and Biodegradation

Aerobic biodegradation, as observed during composting processes, consists in the metabolic use of organic matter by heterotrophic microorganisms in presence of oxygen (E′ = +820 mV [115]). This metabolic use of organic molecules is based on exothermic biochemical reactions, which are part of the biological respiration cycle [116]. This respiration is an inherent process of numerous types of bacteria, fungi and living. A first phase corresponds to the oxidation of the carbon molecules which, through enzymatic catalysis reactions and then use of the obtained pyruvic acid in the Krebs cycle, leads to the production of carbon dioxide, a first product of the respiration, and the release of protons and ATP (adenosine triphosphate). These protons are then transferred to NAD (Nicotinamide Adenine Dinucleotide) molecules, which enter the respiratory chain and participate in the regeneration of ATP via oxidative phosphorylation reactions. At the end of the respiratory chain, protons combine with dissolved oxygen to form water, a second product of respiration. The energy produced along the respiration is used by microorganisms to synthetize cellular constituents, from organic monomers, and develop. The energy required for these processes is provided by the hydrolysis of the stored ATP molecules. Such processes have variable energy yields which generate energy losses in the environment that cause heat generation.

As oxygen is the essential oxidant for the functioning of the biological processes involved in the respiration cycle, the oxygen consumption can be directly correlated to the amount of organic substrate degraded and the amount of microbial cells produced [117]. This observation has suggested the following representation (Fig. 5.6) of the result of the use of organic matter by microorganisms in an aerobic environment.

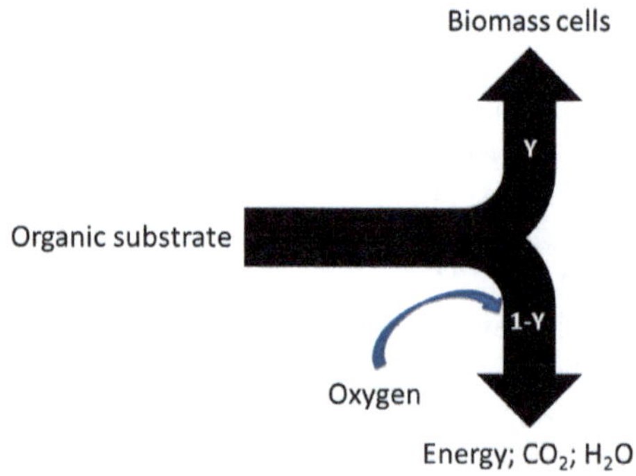

Fig. 5.6 Representation of organic substrate use by aerobic microorganisms where Y is the yield of transformation from organic matter to new biomass cells

Moreover, as biological populations develop preferentially in aqueous media and metabolic reactions take place within microbial cells, it is therefore assumed that biodegraded organic matter is soluble in water and able to cross the cell membrane barrier. Studies conducted on the characterization of organic matter in wastewater have shown that only molecules smaller than or equal to 0.0005 μm are able to diffuse through the cell membrane of bacteria [118]. Therefore, the biological medium must implement size reduction reactions of the solubilized or particulate organic molecules to release substrates directly usable by the microorganisms. This production of directly biodegradable substrate corresponds to enzymatic hydrolysis reactions.

In the case of a solid organic material such as organic waste treated in composting, the above considerations led to conceptually describe the material as solid particles surrounded by an aqueous film. The term "film" here has the meaning of a volume of water closely related to the solid matter and not a biofilm in the common biological understanding. This aqueous film contains microorganisms and soluble organic matter, as well as dissolved oxygen. By colonization of the solid interface and enzymatic attack, the microorganisms solubilize the organic matter that they can then use for their metabolism, i.e. respiration (Fig. 5.7).

Following this description of the aerobic degradation, the measurement of the oxygen consumption of an organic waste medium permits to assess the biodegradation of the organic matter within this medium. The measurement of oxygen consumption via so-called respirometric methods has the advantage of not being destructive for the studied environment, i.e. no sampling is necessary that could disturb the biological reactions. Moreover, on the basis of the schematic representation proposed in Fig. 5.6 [117], a Y fraction of the substrate S is used to produce new microorganisms X, i.e. $rX = Y.rS$, with Y the conversion efficiency of the microorganism substrate, rX the growth kinetics of the microorganisms, rS the kinetics of substrate consumption. The residual fraction of the organic substrate $(1 - Y)$ is oxidized by oxygen for energetic purposes, i.e. $rO_2 = (1 - Y)\ rS$ where rO_2 is the kinetics of oxygen consumption. It is also possible to express the oxygen consumption kinetics rO_2 as a function of the kinetics of growth of the microorganisms rX, i.e. $rO_2 = [(1 - Y)/Y].rX$. The quantity of oxygen consumed and the

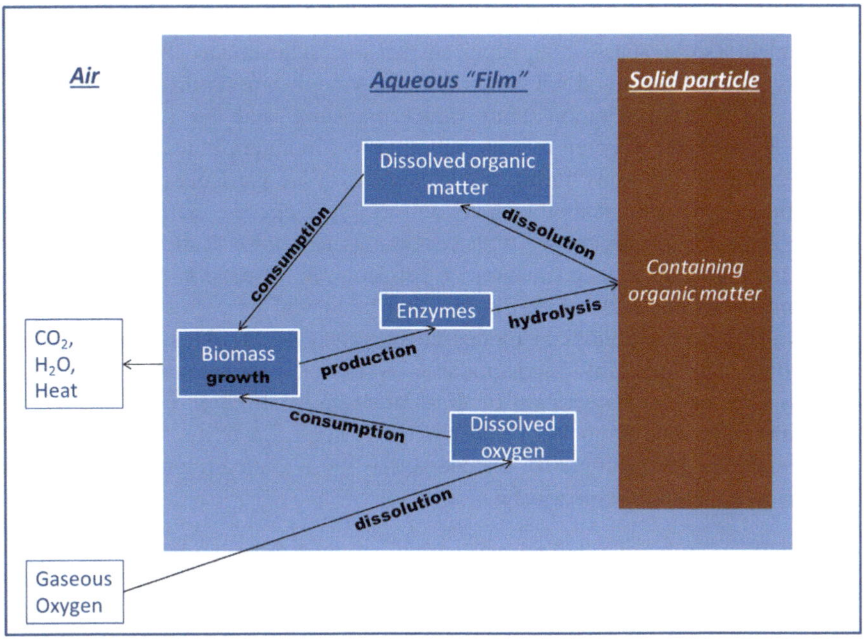

Fig. 5.7 Conceptual representation of aerobic biodegradation for solid or particulate organic matter

oxygen consumption kinetics can thus respectively be related to the quantity of organic matter biodegraded and to the growth kinetics of microorganisms and the kinetics of organic matter consumption: it permits to quantify the biodegradability of an organic substrate, as explained in the following sections.

5.5.2 Respirometric Methods

There is no real standardized respirometric method to measure the biodegradability of solid organic waste or biomass. Nevertheless, there are major currents of thinking among the authors who develop and use these methods that are synthetized in the following sections

5.5.2.1 Classification Criteria of the Existing Respirometric Methods

Numerous respirometric methods are used to study aerobic biodegradability of solid organic biomass. The considered biomasses are mainly solid organic waste, before and/or during their composting treatment, and composts. The respirometric methods differ from one another in many ways, including (1) the sample conditioning and pretreatment (solid state or aqueous suspension, adjustment of moisture, size, nutrients…), (2) the studied volume or mass, (3) the respirometric assay conditions in terms of biomass aeration, temperature, moisture, and test duration, (4) the measured parameter (oxygen consumption, carbon dioxide production).

The aeration conditions during respirometric test are particularly important and the methods have been classified into two different classes: static (without continuous aeration) and dynamic (with continuous aeration). Static methods could induce limitation of oxygen transfer, of diffusion and dispersion through the biomass and then reduce or inhibit the aerobic biological processes and lead to an underestimation of the biomass biodegradability [119–124]. This limit is particularly problematic for highly biodegradable biomass. On the contrary, for dynamic methods, the continuous aeration of the biomass guarantees an optimum oxygen concentration at each moment of time for the biological activity.

Tables 5.3 and 5.4 summarize the characteristics of the main respirometric static and dynamic methods respectively.

5.5.2.2 Static Methods

Two classes of static methods (Table 5.3; Figs. 5.8 and 5.9) exist according to the biomass physical state during the test: methods applied to liquid suspensions of solid biomass and methods applied to solid samples.

Methods using aqueous suspensions [119, 125, 126, 127] do not face the problem of limitation of oxygen transfer as the suspension (mixture of solid sample, water, nutrient solution, pH buffer and nitrification inhibitor) is continuously stirred. These methods also obviate limitation related to the sample structure and moisture, but the experimental conditions for solid waste are far from the real conditions in which the solid waste are biologically treated, and the respiration indices measured in aqueous suspension are affected by the dissolved organic hydrophilic fraction in the studied biomass [122, 125]. Indeed, Adani et al. [125] observed that the oxygen uptake rate peak is more marked with respirometric method using aqueous suspension than with solid state approach, and explained this result by the water medium, responsible for the high availability of the hydrophilic fraction during the initial hours of the test.

Among the methods applied to solid samples, the sample pre-treatment varies from an author to another: screening at 10 or 20 mm, or grinding, and/or moisture adjustment (Table 5.3).

For moisture adjustment, a value or a range of values is fixed by the authors, expressed in moisture content (between 40 and 60% w/w) or in percentage of water holding capacity (60–100%). Water affects biological activity and most authors adjust moisture to an optimum value. According to the biomass composition, this optimum value varies. Adani et al. [142] mentioned that it is better to refer to a percentage of water holding capacity than to a moisture content.

Table 5.3 Static respirometric methods

Sample state and pretreatment	Reaction volume	Temperature	Duration	Measured parameter	Used respiration indices	Studied biomass	References
Aqueous suspension 10 mm screening, mixing of 2 g OM with water, nutrient solution, pH buffer and nitrification inhibitor	1 L + stirring	30 °C	7 days	O_2 consumption (pressure sensor method after CO_2 absorption—Oxitop)	Oxygen uptake rate OUR (mgO_2/gOM/d)	Various finished composts	Aspray et al. [119]
Aqueous suspension 10 mm screening, mixing of 2–5 g (wet weight) with water, nutrient solution, pH buffer and nitrification inhibitor	1 L + stirring	35 °C	Min 20 h	O_2 consumption (dissolved oxygen probe)	Oxygen uptake rate (max) SOUR (mgO_2/gVS/h)	Waste (MSW, OFMSW, biodried MSW) before and during composting treatment	Adani et al. [125]
					Cumulative oxygen consumption (12 and 20 h) OD12 and OD20 (mgO_2/gVS)		
Aqueous suspension Grinding, mixing of 3–8 g (wet weight) with water, nutrient solution, pH buffer and nitrification inhibitor	0.5 L + stirring	30 °C	30–50 h	O_2 consumption (dissolved oxygen probe)	Oxygen uptake rate (max) SOUR (mgO_2/gVS/h)	Sewage sludge composts	Lasaridi and Stentiford [126]
					Cumulative oxygen consumption (20 h) OD20 (mgO_2/gVS)	Waste (dewatered anaerobically digested sewage sludge mixing with bulking agent), before and during composting treatment	Nikaeen et al. [127]

(continued)

Table 5.3 (continued)

Sample state and pretreatment	Reaction volume	Temperature	Duration	Measured parameter	Used respiration indices	Studied biomass	References
Solid 10 mm screening, MC adjustment to 50–60%	0.25 L	30 °C	3 days (after incubation 1 night)	CO_2 production	Carbon dioxide production rate (mean) Static (mgCO_2/gOM/d)	Various finished composts	Aspray et al. [119]
Solid MC adjustment to 40–50%	0.5 L Ianotti et al. [128]	37 °C or Tprocess (in situ temperature at the moment of material sampling)	60 to 100 min (after incubation with aeration during 18 h at 37 °C or 4 h at Tprocess)	O_2 consumption (calculated from le slope of oxygen concentration versus time, on 60 to 100 min)	Oxygen uptake rate (mean) SRI (at 37 °C or Tprocess) (mgO_2/kgOM/h)	Various waste before and during composting treatment	Barrena et al. [129]
Solid MC adjustment to 50–60%	0.5 L Ianotti et al. [128]	37 °C	90 min (after incubation with aeration during 18 h at 37 °C)	O_2 consumption (calculated from le slope of oxygen concentration versus time, on 90 min)	Oxygen uptake rate (mean) SRI (gO2/kgDM/h)	Waste (MSW) before and during MBT (composting)	Barrena et al. [120]
Solid MC adjustment to 40–50%	0.5 L Ianotti et al. [128]	37 °C	90 min (after incubation with aeration during 18 h at 37 °C)	O_2 consumption (calculated from le slope of oxygen concentration versus time, on 90 min)	Oxygen uptake rate (mean) SRI (mgO_2/kgOM/h)	Various waste	Barrena et al. [130]

(continued)

Table 5.3 (continued)

Sample state and pretreatment	Reaction volume	Temperature	Duration	Measured parameter	Used respiration indices	Studied biomass	References
Solid Unspecified pretreatment	30–40 g (unspecified volume)	35 °C	4 days with opening vessel several times	O_2 consumption (pressure sensor method after CO_2 absorption—Oxitop)	Oxygen uptake rate (max) OURmax (gO_2/kgOM/h)	Various manure and agricultural waste before and during composting treatment	Barrena et al. [121]
					Cumulative oxygen consumption (4 days) AT4 (gO_2/kgOM)		
Solid 20 mm screening, MC adjustment to 40–50%	30–40 g (unspecified volume)	20 °C	7 days after lag phase	O_2 consumption (O_2 generation to maintain normal pressure after CO_2 absorption - Sapromat)	Cumulative oxygen consumption (4 and 7 days after lag phase) AT4 and AT7 (mgO_2/gDM)	MBT residual waste after aeration during 2 to 6 weeks in laboratory	Binner and Zack [131]
Solid 20 mm grinding, MC adjustment by "fist test" or "filter method"	0.5 L	20 °C	4 days after lag phase (and after incubation with aeration during 5–7 h at room temperature)	O_2 consumption (O_2 generation to maintain normal pressure after CO_2 absorption - Sapromat)	Cumulative oxygen consumption (4 days after lag phase) AT4 (mgO_2/gDM)	Various waste, composts, digestates…	Binner et al. [132]
Solid 20 mm grinding, MC adjustment by "fist	2.5 L	20 °C	4 days after lag phase (and after incubation with aeration during	O_2 consumption (pressure sensor method after CO_2	Cumulative oxygen consumption	Various waste, composts, digestates…	Binner et al. [132]

(continued)

Table 5.3 (continued)

Sample state and pretreatment	Reaction volume	Temperature	Duration	Measured parameter	Used respiration indices	Studied biomass	References
test" or "filter method"			5–7 h at room temperature)	absorption—Oxitop)	(4 days after lag phase) AT4 (mgO_2/gDM)		
Solid Unspecified pretreatment	1 L	25 °C	3 days (after incubation 1 night)	CO_2 production	Carbon dioxide production rate (mean) Static (mgC-CO_2/gVS/d)	Various manures before and during composting treatment	Changa et al. [133]
Solide 20 mm screening	0.5 L	20 °C	4 days after lag phase	O_2 consumption (O_2 generation to maintain normal pressure after CO_2 absorption - Sapromat)	Cumulative oxygen consumption (4 days after lag phase) RI4 (mgO_2/gDM)	Waste excavated from landfills and waste sampled during MBT	Cossu and Raga [134]
Solid MC adjustment at 100 of WHC	1 L	35 °C	7 days with opening vessel several times	O_2 consumption (pressure sensor method after CO_2 absorption—Oxitop)	Cumulative oxygen consumption and carbon dioxide production (7 days) RI7 (gO_2/kgDM and gC-CO_2/kgDM)	Waste (MSW) before and during MBT (composting and biodrying)	Evangelou et al. [135]
					Oxygen uptake rate (mean over		

(continued)

Table 5.3 (continued)

Sample state and pretreatment	Reaction volume	Temperature	Duration	Measured parameter	Used respiration indices	Studied biomass	References
					the 24 h period of maximum activity) SRI24 (mgO_2/kgDM/h)		
Solid Unspecified pretreatment	0.5 L Ianotti et al. [128]	37 °C or Tprocess (in situ temperature at the moment of material sampling)	90 min (after incubation with aeration during 18 h at 37 °C or 4 h at Tprocess)	O_2 consumption (calculated from le slope of oxygen concentration versus time, on 90 min)	Oxygen uptake rate (mean) SRI (at 37 °C or Tprocess) (gO_2/kgOM/h)	Various waste before and during composting treatment	Gea et al. [136]
Solid MC adjustment at 60 of WHC	1 L	35 °C	7 days	Oxygen consumption and carbon dioxide production	Cumulative oxygen consumption and carbon dioxide production (1, 4 and 7 days) RI1, RI4, RI7 (gO_2/kgDM and gC-CO_2/kgDM)	Various finished composts	Komilis et al. [137]
					Oxygen uptake rate (mean over the 12 h or 24 h period of		

(continued)

Table 5.3 (continued)

Sample state and pretreatment	Reaction volume	Temperature	Duration	Measured parameter	Used respiration indices	Studied biomass	References
					maximum activity) SRI12 and SRI24 (mgO_2/kgDM/h)		
Solid MC adjustment to 50–55%	1 L	20 °C or 35 °C	7 days	Oxygen consumption and carbon dioxide production	Cumulative oxygen consumption and carbon dioxide production (7 days) RI7 (gO_2/kgDM and gC-CO_2/kgDM)	MSW compost	Komilis and Kletsas [123]
					Oxygen uptake rate (mean over the 24 h period of maximum activity) SRI24 (mgO2/kgDM/h)		
Solid MC adjustment to 65 or 70 or 75%	1.9 L with bulk density adjustment system (compaction device)	30 °C	2 days (after incubation 48 h)	O_2 consumption (pressure sensor method after CO_2 absorption—Oxitop)	Oxygen uptake rate (mean) OUR (mgO2/gVS/d)	Waste (apple pomace and woodchips mixtures)	Malinska [138]

(continued)

Table 5.3 (continued)

Sample state and pretreatment	Reaction volume	Temperature	Duration	Measured parameter	Used respiration indices	Studied biomass	References
Solid Unspecified pretreatment	50 g (unspecified volume)	Ambient temperature	6 days	Carbon dioxide production	Carbon dioxide production rate (mean) RI (mgC-CO_2/gCorga/d)	Waste (source separated biowaste) before and during composting treatment	Oviedo-Ocana et al. [139]
Solid Unspecified pretreatment	0.5 L Ianotti et al. [128]	37 °C	90 min (after incubation with aeration during 24 h at 37 °C)	O_2 consumption (calculated from le slope of oxygen concentration versus time, on 90 min)	Oxygen uptake rate (mean) RI (gO_2/kgDM/h)	Waste (MSW and OFMSW) before and during MBT (anaerobic digestion and composting)	Ponsa et al. [140]
Solid 10 mm screening, MC adjustment to 40–60%	0.5 L	37 °C	4 days	O_2 consumption (O_2 generation to maintain normal pressure after CO_2 absorption - Sapromat)	Cumulative oxygen consumption (4 days) AT4 (mgO_2/gVS)	Waste (sewage sludge and bulking agent mixture) before and during composting treatment	Sanchez-Arias et al. [141]
					Oxygen uptake rate (mean over the 24 h period of maximum activity) RI24 (mgO_2/kgVS/h)		Villasenor et al. [124]

DM dry matter; *OM* organic matter; *MC* moisture content; *VS* volatile solids; *WHC* water holding capacity; *OUR* oxygen uptake rate; *SOUR* specific oxygen uptake rate; *OD* oxygen demand; *SRI* static respirometric/respiration index; *AT* respiration activity; *RI* respirometric/respiration index; *MSW* municipal solid waste; *OFMSW* organic fraction of municipal solid waste; *MBT* mechanical-biological treatment

Table 5.4 Dynamic respirometric methods

Sample state and pretreatment	Reaction volume	Temperature	Duration	Aeration conditions	Measured parameter	Used respiration indices	Studied biomass	References
Solid 50 mm grinding, MC adjustment at 75 of WHC	148 L	Adiabatic (insulated reactor)	Between 4 and 14 days	Aeration with adjusted air flow to keep oxygen concentration in the outlet stream higher than 14%	O_2 consumption	Oxygen uptake rate (according to time) DRI (t) (mgO_2/kgVS/h)	Waste (MSW, OFMSW, biodried MSW)	Adani et al. [142]
						Oxygen uptake rate (mean over the 24 h period of maximum activity) DRI24 (mgO_2/kgVS/h)	Waste (MSW, OFMSW, biodried MSW) before and during composting treatment	Adani et al. [125]
Solid 50 mm grinding, MC adjustment at 75 of WHC	148 L	Adiabatic (insulated reactor)	Few hours to 4 days	Aeration with adjusted airflow to keep oxygen concentration in the outlet stream higher than 14%	O_2 consumption	Oxygen uptake rate (max) DRImax (mgO_2/kgVS/h)	Waste (MSW, OFMSW, biodried MSW) before and during composting treatment	Adani et al. [143]
						Oxygen uptake rate (mean over the 24 h period of maximum activity) DRI24 (mgO_2/kgVS/h)	Various finished composts	Adani et al. [144]
						Cumulative oxygen consumption (4 days and 4 days minus the lag phase) DRIcum (mgO_2/gVS) on 4 days DRIcumadj (mgO_2/gVS) on		

(continued)

Table 5.4 (continued)

Sample state and pretreatment	Reaction volume	Temperature	Duration	Aeration conditions	Measured parameter	Used respiration indices	Studied biomass	References
						4 days minus the lag phase		
Solid Unspecified pretreatment	0.5 L	37 °C	5–7 days	Aeration with constant airflow or adjusted airflow according to oxygen concentration in the outlet stream Use of humidified air at 37 °C for aeration	O_2 consumption	Oxygen uptake rate (mean over the 1 h and 24 h period of maximum activity) DRI1 and DRI24 (gO_2/kgDM/h)	Waste (OFMSW) before and during composting treatment	Almeira et al. [145]
						Cumulative oxygen consumption (4 days after lag phase and for the whole experiment) AT4 and ΔO_2 (gO_2/kgDM)		
Solid Unspecified pretreatment	1 kg (unspecified volume)	30 °C	3–5 days	Aeration with adjusted airflow according to oxygen concentration in the outlet stream	O_2 consumption	Oxygen uptake rate (max and mean over the 24 h period of maximum activity) DRImax and DRI24 (mgO_2/kgVS/h)	Waste (broilers poultry manure and turkey poultry manure)	Antognoni et al. [146]
Solid 20 mm screening, mixing with inoculum (well matured compost)	200 gDM mixture (unspecified volume)	30 °C	4 days	Aeration with constant airflow	O_2 consumption and CO_2 production	Oxygen uptake rate and carbon dioxide production rate (mean over the 4 days)	Various finished composts	Aspray et al. [119]

(continued)

Table 5.4 (continued)

Sample state and pretreatment	Reaction volume	Temperature	Duration	Aeration conditions	Measured parameter	Used respiration indices	Studied biomass	References
and nutrients), MC adjustment to 50–57%						DR4 (mgO_2/gOM/d and $gmCO_2$/gOM/d)		
Solid Unspecified pretreatment	20 L	Adiabatic (insulated reactor)	4 days	Aeration with adjusted airflow to keep oxygen concentration in the outlet stream higher than 14%	O_2 consumption	Oxygen uptake rate (max) DRImax (gO_2/kgDM or OM/h)	Waste (MSW) before and during MBT (composting)	Barrena et al. [120]
						Oxygen uptake rate (mean over the 24 h period of maximum activity) DRI24 (gO_2/kgDM or OM/h)	Various manure and agricultural waste before and during composting treatment	Barrena et al. [121]
						Cumulative oxygen consumption (4 days) AT4 (gO_2/kgDM or OM)		
Solid Mixing with bulking agent if necessary	10 L	40 °C	29–54 days	Aeration with constant airflow Use of humidified air at 40 °C for aeration Exhaust air recirculation with constant airflow	O_2 consumption	Oxygen uptake rate (according to time) RI (t) ($mmolO_2$/kgDM/h)	Various waste (sludge and bulking agent mixtures, household waste) before and during composting treatment	Berthe et al. [147]
						Oxygen uptake rate (max) RImax ($mmomO_2$/kgDM/h)		
						Cumulative oxygen consumption (for		

(continued)

Table 5.4 (continued)

Sample state and pretreatment	Reaction volume	Temperature	Duration	Aeration conditions	Measured parameter	Used respiration indices	Studied biomass	References
						the whole experiment) ΔO_2 (mmolO_2/kgDM)		
Solid Unspecified pretreatment	18 L	30 °C	50–120 h	Aeration with constant airflow	O_2 consumption	Oxygen uptake rate (max) RImax (mgO_2/kgVS/h)	Various waste from agricultural activities and food industries	Ciuta et al. [148]
						Oxygen uptake rate (mean over the 24 h period of maximum activity) RI24 (mgO_2/kgVS/h)		
Solid Removal of improper materials if necessary	0.5 L	37 °C	Unspecified	Aeration with constant airflow	O_2 consumption	Oxygen uptake rate (mean over the 24 h period of maximum activity) DRI24 (mgO_2/gDM/h)	Waste (MSW, OFMSW) before and during MBT (anaerobic digestion and composting)	Colon et al. [149]
						Cumulative oxygen consumption (4 days) AT4 (mgO_2/kgDM)		

(continued)

Table 5.4 (continued)

Sample state and pretreatment	Reaction volume	Temperature	Duration	Aeration conditions	Measured parameter	Used respiration indices	Studied biomass	References
Solid 20 mm screening, MC adjustment	1 L	30 °C	7 days	Aeration with constant airflow	CO_2 production	Carbon dioxide production rate (according to time) CO_2 (t) (mgCO_2/gOM/d)	Various composts	Ferrari et al. [150]
Solid MC adjustment to 55% and mixing with bulking agent	1–3.5 L studied material	35 °C	7 days	Aeration with constant airflow (test of different values)	O_2 consumption and CO_2 production	Oxygen uptake rate and carbon dioxide production rate (mean over the 24 h period of maximum activity) DRI24 (mgO_2/kgVS/h) $CO_2$24 (mgC/kgVS/h)	MSW compost	Komilis and Kanellos [151]
						Cumulative oxygen consumption and carbon dioxide production (4 and 7 days) AT4 and AT7 (mgO_2/kgVS and mgC/kgVS)		Komilis and Kletsas [123]
Solid Mixing with bulking agent if necessary	0.5 L	37 °C	5–7 days	Aeration with constant airflow (test of different values) or with adjusted airflow according to oxygen concentration in the outlet stream Use of humidified air at 37 °C for aeration	O_2 consumption	Oxygen uptake rate (max) DRImax (mgO_2/gDM/h)	Various waste (sludge, manure, slurry)	Mejias et al. [152]
						Oxygen uptake rate (mean over the 24 h period of maximum activity) DRI24 (mgO_2/		

Table 5.4 (continued)

Sample state and pretreatment	Reaction volume	Temperature	Duration	Aeration conditions	Measured parameter	Used respiration indices	Studied biomass	References
						Cumulative oxygen consumption (4 days after lag phase) AT4 (mgO_2/kgDM)		
Solid 15 mm grinding and mixing with bulking agent if necessary	0.5 L	37 °C	12 days	Aeration with adjusted airflow according to oxygen concentration in the outlet stream (>10%) Use of humidified air at 37 °C for aeration	O_2 consumption	Oxygen uptake rate (max) DRImax (mgO_2/gDM/h)	Various waste before and during composting treatment	Ponsa et al. [153]
						Oxygen uptake rate (mean over the 24 h period of maximum activity) DRI24 (mgO_2/gDM/h)		
						Cumulative oxygen consumption (4 days, 12 days and over the 24 h period of maximum activity) AT4, AT12 and AT24 h (mgO_2/kgDM) AT4/AT12 (dimensionless)		Ponsa et al. [154]

(continued)

Table 5.4 (continued)

Sample state and pretreatment	Reaction volume	Temperature	Duration	Aeration conditions	Measured parameter	Used respiration indices	Studied biomass	References
Solid MC adjustment at 75 of WHC	30 L	Adiabatic (insulated reactor)	Unspecified	Aeration with adjusted airflow to keep oxygen concentration in the outlet stream higher than 14%	O_2 consumption	Oxygen uptake rate (mean over the 24 h period of maximum activity) DRI24 (mgO_2/kgDM/h)	Waste (MSW) before and during MBT	Scaglia et al. [155] Scaglia et al. [156]
Solid Mixing with bulking agent	10 L	Constant temperature between 19.6 and 56.9 °C	10–20 days	Aeration with constant airflow Use of humidified air for aeration Exhaust air recirculation with constant airflow	O_2 consumption	Oxygen uptake rate (according to time) RI (t) ($mmolO_2$/kgDM/h)	Waste (sewage sludge and bulking agent mixture)	Trémier et al. [157]
Solid 10 mm screening, MC adjustment to 60–65%	0.5 L	37 °C	4 days	Exhaust air recirculation with constant airflow (test of different values) and with air humidification	O_2 consumption (O_2 generation to maintain normal pressure after CO_2 absorption modified Sapromat)	Cumulative oxygen consumption (4 days) AT4 (gO_2/kgDM) Oxygen uptake rate (mean over the 24 h period of maximum activity) RI24 (gO_2/kgVS/h)	Waste (sewage sludge and bulking agent mixture) before and during composting treatment	Villasenor et al. [124]

DM dry matter; *OM* organic matter; *MC* moisture content; *VS* volatile solids; *WHC* water holding capacity; *DRI* dynamic respirometric/respiration index; *AT* respiration activity; *RI* respirometric/respiration index; *MSW* municipal solid waste; *OFMSW* organic fraction of municipal solid waste; *MBT* mechanical-biological treatment

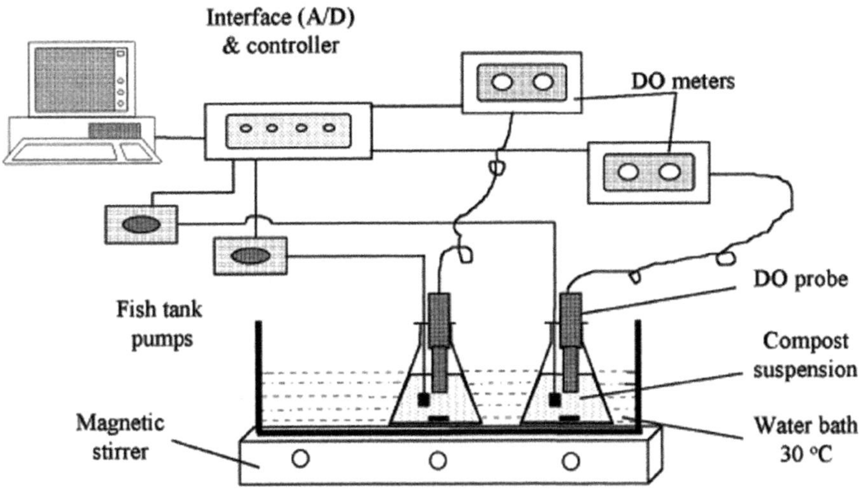

Fig. 5.8 Static respirometer applied on aqueous suspension and used by Lasaridi and Stentiford. Reprinted with permission with Elsevier [126]

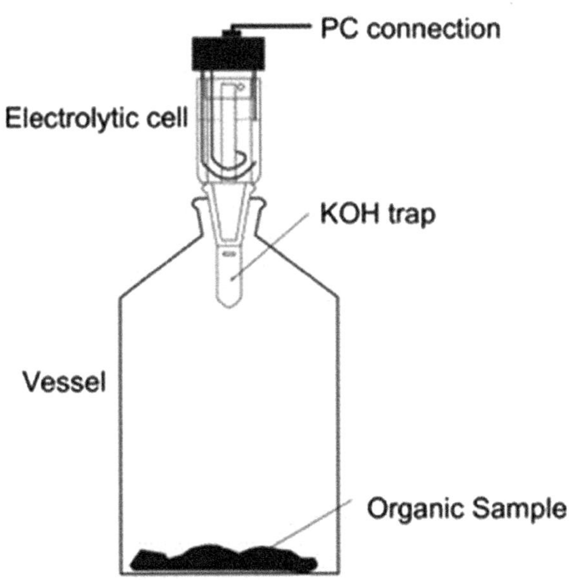

Fig. 5.9 Static respirometer applied on solid sample (with carbon dioxide absorption and electrolytic system for occasional oxygen generation). Reprinted with permission from Elsevier [141]

The temperature of static respirometric assays is generally maintained constant along the experiments but varies highly among the different authors: ambient temperature or equivalent (20 °C or 25 °C), 30 °C, 35 °C or 37 °C (Table 5.3).

In order to study the waste biodegradability along their composting treatment, some authors perform the respirometric test at the temperature that has been reached by the composting process at the moment of the substrate sampling [129, 136]. These authors compared the respirometric results obtained at 37 °C and at the process temperature. They observed that the respiration indices measured at the process temperature were more sensitive to biodegradability variations during composting than those measured at 37 °C. These results suggested that respirometric measurement at process temperature is interesting for the monitoring of the composting process. But the same studies showed that when process temperature falls below mesophilic conditions, the respirometric measurement at process temperature is unable to determine the actual stability of the material. So, respirometric measurement at 37 °C is more suitable for the estimation of material stability at the later stages of a composting treatment. Moreover, Wagland et al. [158] mentioned that the respirometric methods operating at high temperature could present a disadvantage. Indeed, an increase in temperature may limit the oxygen transfer in the studied biomass, which could result in lower indices than expected. Whilst high temperature may accelerate some reactions, the biological activity could be hindered.

Komilis and Kletsas [123] also demonstrated a link between the impact of the applied temperature and the weight of sample considered in the respirometric assay. The reaction volume in static methods varies from 0.25 to 2.5 L, which leads to very different weights of sample. Komilis and Kletsas [123] applied a static respiration test on different solid sample weights (19 to 94 g of dry matter in a 1 L respirometer), at 20 and 35 °C. They observed that, as sample weight increased, the magnitude of respiration indices obtained at 35 °C reduced, but for the respiration indices obtained at 20 °C, all sample weights provided statistically similar results. The respiration activity recorded at 35 °C was higher than at 20 °C for some sample weights, and similar for other weights. The authors explained the effect of sample size observed at 35 °C by the limitation, in these temperature conditions, of oxygen diffusion to the interior of the material with increasing weight, leading to a reduced respiration activity.

The static respirometric tests last for few tens of minutes to 7 days (Table 5.3), sometimes preceded by a few hours incubation stage [119, 120, 129, 130, 132, 133, 136, 138, 140].

In most cases, respiration indices obtained from static respirometric tests are based on an oxygen consumption but few authors based their indices on a carbon dioxide production [119, 123, 133, 137, 139]. In practise, for methods using an aqueous suspension, a dissolved oxygen probe is generally used whereas, for static methods using a solid sample, a pressure sensor technique after carbon dioxide absorption is usually applied, with for some methods an occasional oxygen generation to maintain normal pressure in the respirometric reactor. When a first incubation stage with aeration is used, the drop of oxygen level is then recorded and the slope of oxygen concentration versus time is used for calculation of respiration indices. For the pressure sensor techniques after carbon dioxide absorption and without occasional oxygen generation, the pressure drop is recorded until pressure stabilisation and oxygen depletion. When an occasional oxygen generation is used, this one allows the calculation of an oxygen consumption rate versus time during the respirometric measurement.

5.5.2.3 Dynamic Methods

All dynamic methods (Table 5.4; Figs. 5.10 and 5.11) are applied on solid medium, none on aqueous suspension. But, as for static methods, when the solid sample pretreatment is specified by the authors, this one varies from a study to another: screening at 10 or 20 mm, or grinding, and/or possible addition of bulking agent [123, 147, 151, 152, 153, 154, 157] or inoculum [119], and/or moisture adjustment (Table 5.4). The addition of bulking agent aims at obtaining an adequate physical structure for the studied matrix, allowing an efficient aeration during the dynamic respirometric measurement. Concerning moisture, as previously noticed for static methods, the authors generally adjust the sample moisture to a value or a range of values expressed in moisture content (between 50 and 65% w/w) or in percentage of water holding capacity (75%). Moreover, in order to control the variation of the substrate moisture during the respirometric measurement, some authors applied a humidification of the inlet air (Table 5.4).

The proposed reaction volume for dynamic methods varies more than the one used for static methods: from 0.25 to 148 L. Komilis and Kanellos [151] compared

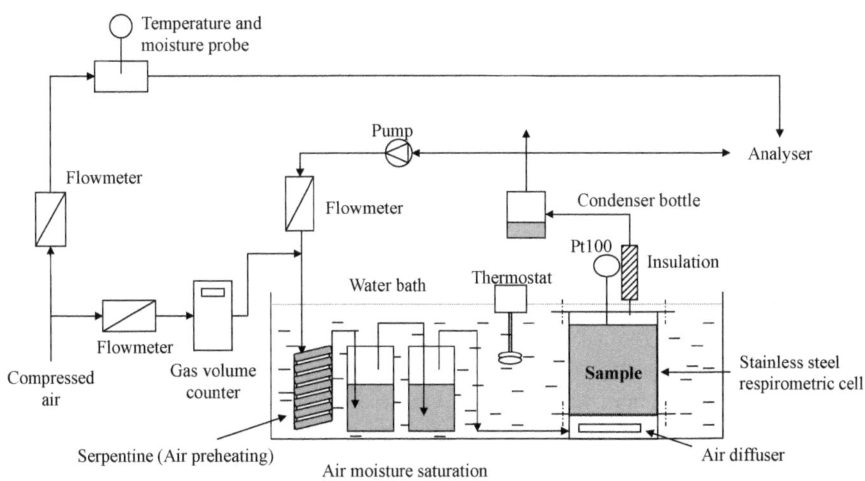

Fig. 5.10 Dynamic respirometer similar to the one used by Berthe et al. Reprinted with permission from Elsevier [147]

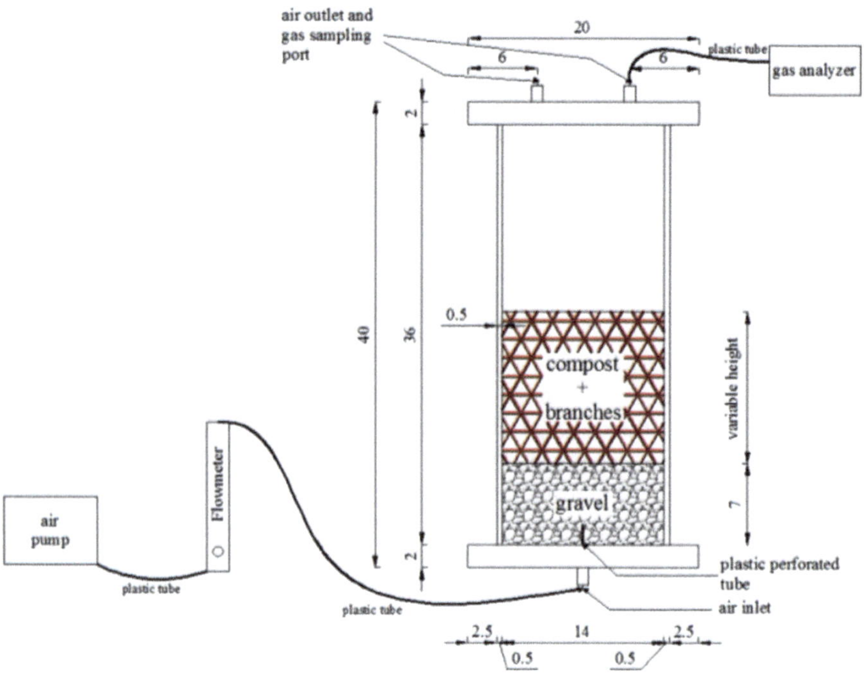

Fig. 5.11 Dynamic respirometer used by Komilis and Kanellos. Reprinted with permission from Elsevier [151]

the impact on respiration indices of the sample size and the applied airflow rate during the respirometric test. They observed that high sample sizes eventually led to reduced oxygen consumption indices. Indeed, in some cases, the oxygen diffusion to the interior of the material could be limited with high sample sizes, but this impact depended on the aeration conditions and the airflow rate.

Concerning the temperature chosen to perform the dynamic respirometric methods, two strategies exist: constant temperature or variable temperature in adiabatic insulated reactor (temperature evolution with biodegradation processes) (Table 5.4). In the case of constant temperature, the chosen value varies from one study to the other: 30, 35, 37 °C or 40 °C. Trémier et al. [157] studied the temperature influence (between 20 and 57 °C) on the biodegradation kinetics. By modelling of the respiration curves, they observed that the kinetic factors (biomass growth and death, hydrolysis of soluble biodegradable organic fraction) first increased with increasing temperature but, beyond approximately 40 °C, kinetics started decreasing. Then, the authors suggested an optimum temperature around 40 °C. Lasaridi et al. [159] also studied respiration activity at different temperatures (between 20 and 66 °C) and observed that the optimal temperature depended on the studied sample and on its biodegradability. Adani et al. [142] thought that the use of a preset temperature during respirometric test could limit biological activity, and then preferred an approach with variable temperature in adiabatic reactor. It could be noticed that the reaction volume depends on the temperature strategy. Indeed, in the case of respirometric study in adiabatic insulated reactor with variable temperature, the reaction volume is comprised between 20 and 148 L, whereas the reaction volume in respirometric study with constant temperature is lower, below to 18 L

To supply aeration during the respirometric test, all authors used fresh ("new") air, except Villasenor et al. [124] that used a system by air recirculation through the organic sample, without fresh air. Berthe et al. [147] and Trémier et al. [157] combined aeration with fresh air and recirculation of exhaust gas during respirometric test. The recirculation airflow rate was adjusted according to a RTD (retention time distribution) measurement method, and was optimum at a value six times higher than the fresh airflow rate [147]. In these conditions, the gas phase in the studied material was considered as perfectly mixed and then the respirometric samples were supplied with homogeneous and sufficient oxygen flow.

Concerning the choice of the aeration airflow rate, two trends appear in literature: use of a constant airflow rate or an adjusted airflow rate according to oxygen concentration in the outlet stream (oxygen-based airflow regime) (Table 5.4). Almeira et al. [145] compared, with different type of biomass, a constant airflow regime and an oxygen-based airflow regime. They observed that, with some very biodegradable biomass, the respiration indices were higher with the oxygen -based regime than those with the constant regime. Nevertheless, no clear difference was obtained with other biomasses, especially with less biodegradable biomass. For their part, Mejias et al. [152] didn't observe statistically significant differences between the respiration indices measured with a constant airflow mode and an oxygen-based airflow mode. Actually, with a constant airflow mode, the respiration indices mainly depend on the applied value of airflow rate. Few authors have studied this dependence. It seems that, when the airflow rate is very low, a limitation of biological activity exists: as a consequence, the respiration indices increase as the airflow rate increase [124, 137, 145, 152]. Some studies have also shown that, beyond a threshold value (depending on the studied biomass), the increase in airflow rate didn't significantly increase the respiration indices [145, 152], or could lead to a decrease in these indices in case of quite stable biomass such as compost [145].

The dynamic respirometric tests last for few hours to 54 days, with only few studies with duration higher than 7 days [125, 142, 147, 153, 154, 157]. Most studies perform respirometric assays with duration from 4 and 7 days.

As for static methods, most of respiration indices obtained from dynamic respirometric methods are based on an oxygen consumption. Few authors use indices based on carbon dioxide production [119, 123, 150, 151]. In practical terms, the oxygen consumption (or carbon dioxide production) is obtained thanks to the oxygen (or carbon dioxide) concentration measurement in the exhaust gas during the dynamic respirometric test.

5.5.3 *Respiration Indices for Biomass Biodegradability Determination*

Different respiration indices, obtained from the previous respirometric techniques, are currently used to determine the level of microbial activity in a sample of solid biomass, and a lot of authors have studied the correlations between respiration indices [120, 121, 126, 132, 135, 144, 145, 146, 148, 149, 152, 153, 155]. Two main classes of indices, mentioned in Tables 5.3 and 5.4, exist:

- the indices based on a kinetics: oxygen uptake rate or carbon dioxide production rate,
- the indices based on a cumulative quantity of consumed oxygen or produced carbon dioxide.

Generally, indices based on an oxygen uptake rate (OUR, OURmax, SOUR, SRI, SRI12, SRI24, DRImax, DRI24…, etc.—Tables 5.3 and 5.4) are calculated as:

- either the mean value reached during the respirometric test,
- or the maximum value reached during the respirometric test,
- or the mean value calculated over the period of maximum activity (1 h or 12 h or 24 h for example).

Among them, the indices considering the respiration peak present the advantage to be obtained rapidly but the disadvantage to traduce mainly the biodegradation of the most rapidly and accessible fraction of the organic matter and not the whole biodegradable organic matter.

For the indices based on cumulative oxygen consumption, the considered duration for the calculation varies widely (Tables 5.3 and 5.4):

- less than 24 h (OD12, OD20…), exclusively in static methods,
- 4 days (AT4, RI4…), which is the most commonly used duration,
- 7 days (AT7, RI7…),
- 12 days (AT12),
- or the whole experiment duration to reach a very low oxygen consumption rate.

The indices considering cumulative oxygen consumption better represent the whole biodegradable organic matter than those based on the oxygen uptake rate at the respiration peak. However, this advantage can be discussed according to the duration taken into account for the calculation and according to the studied biomass. Indeed, for biomass characterized by high content of biodegradable organic matter and/or slow biodegradation kinetics, the respirometric activity should be not completed within few days. In these cases, 4 days cumulative indice might be insufficient to describe the complete respirometric activity. For less biodegradable biomass, this inconvenient of short cumulative indices is less pronounced.

Another factor that could affect the value of the indices based on cumulative oxygen consumption is the possible presence of an initial lag phase without significant respiration activity at the beginning of the assay. It depends on the biomass sample history and nature [143, 145] and on the aeration conditions [145, 152]. Some authors don't specify if they take into account this lag phase in their calculation of cumulative indices, but others mention that the cumulative duration begins effectively after the initial lag phase [131, 132, 134, 143, 144, 145, 152].

It could also be mentioned that the respiration indices unit could also affect the obtained respirometric results. The indices are calculated either on a dry matter basis or on a volatile (or organic) matter basis. Barrena et al. [129] suggested that respiration indices should be expressed on dry matter basis for low biodegradable biomass and on organic matter basis for high biodegradable biomass. For their part, Wagland et al. [158] considered that, taking into account the dry matter content, the respiration indices would include the inorganic material, which is non-biodegradable, meaning that data presented in this form may not be comparable. These authors then advised a volatile matter basis, only taking the organic fraction into account, which would allow, for varying inorganic content, a valid comparison between samples.

Finally, different limits have been established by some authors and few countries for the respiration indices, for their use as a biological stability parameter (for compost) or for a classification of biomasses according to their biodegradability [122]. Nevertheless, there is no consensus on the thresholds of these indices, nor on a common respirometric method to use for their measurement.

5.5.4 Modelling of Respiration Kinetics to Determine Biodegradability

Global respirometric indices as previously described aim at assessing the potential of biodegradable organic matter in a biomass sample but do not permit to distinguish biodegradable compartments of this organic matter and to differentiate their respective biodegradation kinetics. In order to better understand and assess the solid organic matter biodegradability, numerical models describing biodegradation processes have been proposed.

Looking at classical curves of oxygen consumption rate obtained with a dynamic respirometric method performed in non-limiting oxygen conditions with constant temperature and moisture conditions, several phases can be observed in microbial activity (latency, exponential acceleration, stationary, deceleration and decay) and related to the different biodegradable fractions of the material.

Figure 5.12 gives the oxygen consumption kinetics, rO_2, as a function of time, for a mixture of wastewater treatment sludge and pine barks, with moisture of 53%, measured at a temperature of 39.6 °C. The interpretation of the rO_2 profile is based on the fractionation of the substrate organic matter into three fractions:

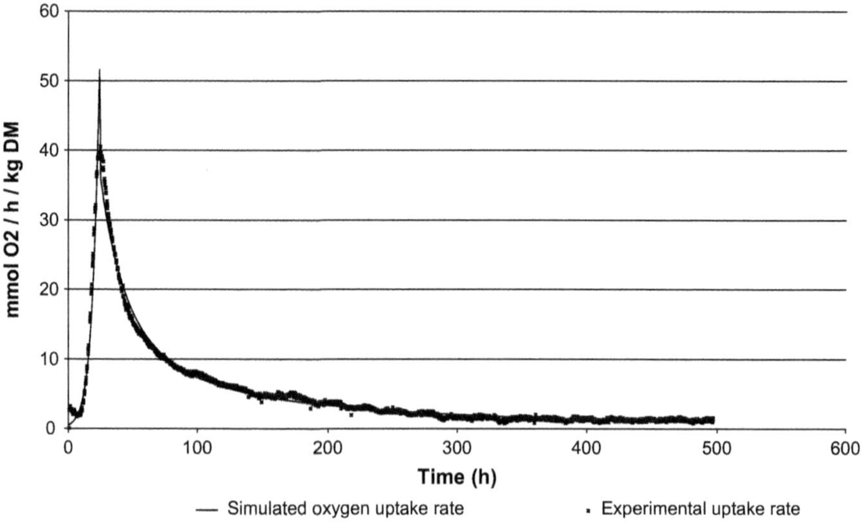

Fig. 5.12 Example of oxygen consumption rate curve obtained with a dynamic respirometric method (wastewater treatment sludge + pine barks; 39.6 °C)

- an easily biodegradable fraction (MB) which is already soluble and available for microorganisms;
- a slowly biodegradable fraction (MH) which is composed of soluble or solid macromolecules that have to be hydrolysed before biological consumption;
- a fraction considered as inert for the biological reaction (MI).

The first noticeable phase in Fig. 5.12 is the exponential growth of oxygen consumption kinetics associated with an exponential development of microorganisms. It corresponds to the consumption by microorganisms of the easily biodegradable fraction (MB). At the end of the growth period of rO_2, all the easily biodegradable MB fraction is supposed to be consumed. The exponential decay of rO_2 then corresponds to the consumption by the microorganisms of an MB fraction resulting from the hydrolysis of the MH fraction. This hydrolysis kinetics is the limiting point for the biodegradation of the substrate. Beyond this decay phase, based on the assumption that MB and MH are depleted, the low kinetics rO_2 is assumed to correspond to the endogenous respiration of the microorganisms. Other authors have chosen the same simple fractionation of the organic matter to model biodegradation behaviour as Ponsa et al. [154], who also proposed to discriminate biodegradable organic matter in a slowly and a rapidly degradable fraction.

Depending on the origin of the substrates studied, the kinetics of biodegradation, expressed via the kinetics of oxygen consumption, can vary considerably. Indeed an initial phase of latency can sometimes be observed. This phase, which can last up to 2 days, is interpreted as an adaptation of the microorganisms to the medium, or

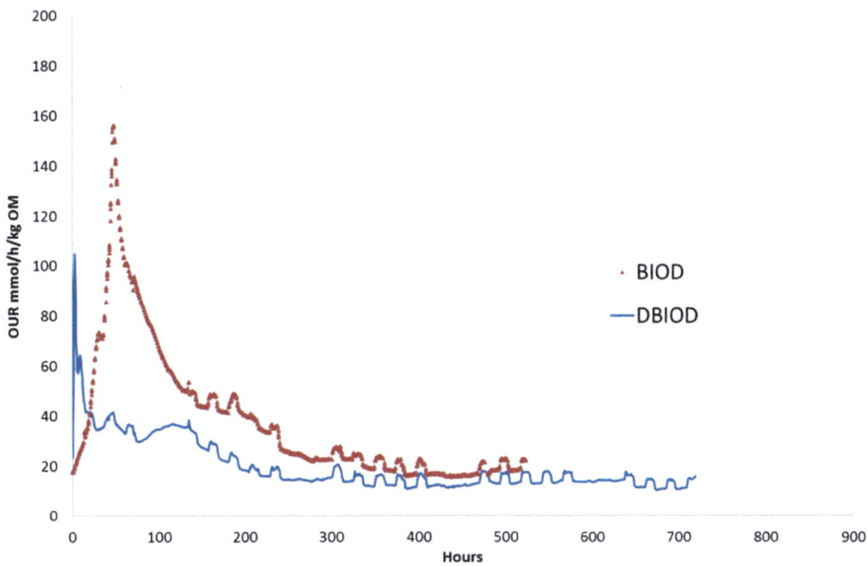

Fig. 5.13 Example of oxygen consumption rate curves obtained with dynamic respirometry for biowaste (BIOD) and for a solid digestate of biowaste (DBIOD)

even to its modification and to the synthesis of enzymes. The growth of the kinetics rO_2 can be more or less rapid. The kinetics of oxygen consumption of certain substrates may also have several consumption peaks or shoulders (Fig. 5.13) showing that the concept of simple fractionation of the organic matter (2 biodegradable fractions and an inert fraction) previously proposed is not always relevant to report respirometric curves obtained on certain wastes. In these latter cases several MB and/or MH fractions as well as several types of active microorganisms have to be considered simultaneously or alternately.

5.5.4.1 Respirometric Modelling Based on a Simple Fractionation of the Biodegradable Organic Matter

Based on the simple fractionation of organic matter previously described, a model, based on ASM models [160], has been developed to simulate the kinetics of biodegradation [157]. This model describes the biodegradation with 4 equations:

1. The kinetics of growth (μ) and mortality (b) of the microbial population X,
2. The kinetics of hydrolysis (Kh) of the MH fraction according to kinetics taking into account the colonization of matter by microorganisms [157],
3. The consumption kinetics of the MB fraction, inversely proportional to the growth yield (Y) of the microorganisms, and its formation via the hydrolysis of MH,

4. Oxygen consumption kinetics required for the oxidation of the easily biodegradable MB fraction (1 − Y) and the partial oxidation (1 − f) of the dead microorganisms (endogenous respiration).

$$\frac{dX(t)}{dt} = \mu_m \times \frac{MB(t)}{K_B + MB(t)} \times X(t) - b \times X(t) \quad (5.58)$$

$$\frac{dMH(t)}{dt} = -K_h \times \frac{MH(t)/X(t)}{MH(t)/X(t) + K_{MH}} \times X(t) \quad (5.59)$$

$$\frac{dMB(t)}{dt} = -\frac{1}{Y} \times \mu_m \times \frac{MB(t)}{K_B + MB(t)} \times X(t) + K_h \times \frac{MH(t)/X(t)}{MH(t)/X(t) + K_{MH}} \times X(t) \quad (5.60)$$

$$rO_2(t) = \frac{(1-Y)}{Y} \times \mu_m \times \frac{MB(t)}{K_B + MB(t)} \times X(t) + b(t) \times (1-f) \times X(t) \quad (5.61)$$

with

X	Microorganisms content (mmol O_2/kg initial dry matter)
MB	Easily-biodegradable organic matter content (mmol O_2/kg initial dry matter)
MH	Slowly biodegradable organic matter content (mmol O_2/kg initial dry matter)
MI	Inert organic matter content (mmol O_2/kg initial dry matter)
μ_m	Growth kinetics of microorganisms (h^{-1})
b	Decay kinetics of microorganisms (h^{-1})
K_h	Hydrolysis kinetics (h^{-1})
Y	Microorganisms' growth yield
K_b	Constant of saturation for MB (mmol O_2/kg initial dry matter)
K_{MH}	Constant of saturation for the hydrolysis
f	Coefficient of transfer of dead microorganisms to the inert organic matter fraction.

The initial parameters of the model are fixed by default (Y, f, K_b, K_{MH} and b) or by identification on the experimental respiratory curve (μ_m, K_h, X, MB, MH). The identification of these parameters is then carried out by minimizing the sum of squared errors between the kinetics of oxygen consumption predicted by this model and the measurement of this kinetics via respirometry. Figure 5.12 gives an illustration of rO_2 simulated with the model developed.

The optimal kinetics described in this model is influenced by environmental conditions such as temperature and humidity. Factors describing the influence of these parameters must therefore be introduced into the equations when experiments are not performed under optimal conditions of temperature and humidity.

In addition to the impossibility of representing several biodegradation peaks with such a model, another limitation lies on the fact that a single experimental data (rO_2 measured in respirometry) is used to determine 5 parameters of the model, including the fractions MB and MH which are not related to an analytical characterization of organic matter. Nevertheless, the quantification of MB and MH contents through the model permits to assess the biodegradability of solid organic substrates. Indeed, the calculation of the ratio of (MH + MB) above the total chemical oxygen demand (COD) of the studied substrate can be interpreted as the percentage of biodegradability of the studied substrate. Moreover, substrates biodegradability can be compared considering the percentage of easily and slowly biodegradable organic matter of each substrate and also kinetics of biodegradation (μ, K_h).

Ponsa et al. [154] modelled CO_2 production rate curves. They proposed first-order kinetics equation for each of the both biodegradable fractions and used carbon content as measuring unit. They also proposed to quantify the biodegradability by calculating the ratio between the biodegradable fractions and the total organic carbon of the studied substrate.

5.5.4.2 Factors Influencing Kinetics

In the case of aerobic biodegradation, three main factors of variation of the biological kinetics can be highlighted: the temperature, the humidity and the oxygen content of the medium.

Thus, the representation of μ_m, b and K_h as a function of temperature shows an acceleration of kinetics with an increase in temperature until reaching a maximum value, then a gradual decrease if the increase in temperature continues. There is also a minimum temperature and a maximum temperature below and beyond which the kinetics are zero. These evolutions have been modelled from the Rosso model [161, 162]. This model has four parameters: three cardinal temperatures (minimum, optimal, maximum) and optimal kinetics.

$$r(T) = \frac{r_{opt} \times (T - T_{max}) \times (T - T_{min})^2}{(T_{opt} - T_{min}) \times \left[(T_{opt} - T_{min}) \times (T - T_{opt}) - (T_{opt} - T_{max}) \times (T_{opt} + T_{min} - 2 \times T)\right]} \tag{5.62}$$

with

r(T)	Considered kinetics (μ_m, b or K_h) as a function of temperature
r_{opt}	Optimal kinetics (μ_{mopt}, b_{opt} or K_{hopt})
$T_{min,\ max,\ opt}$	Cardinal temperatures.

The work developed by Trémier et al. [157] showed that the cardinal temperatures applicable to the aerobic biodegradation kinetics of a mixture of STEP sludge and pine bark were: T_{opt} = 40 °C; T_{max} = 65 °C; T_{min} = 0 °C. These values may vary according to the substrate studied, T_{opt} being able reach up to 50 °C and T_{max} up to 90 °C.

Regarding moisture, it should be noticed that the biological activity tends to stop for moisture of less than 15%. In addition, beyond 70% of moisture, the solid aerobic treatment is no longer possible because the mixture is saturated with water and the oxygen supply is then very limited. To account for these moisture influence levels, several kinetic correction factors have been proposed by Haug [163], Lomander et al. [164], Pommier et al. [165], Richard et al. [166], Stombaugh and Nokes [167]. These factors are defined on the basis of a moisture saturation level and the difference between the studied medium moisture and this saturation threshold.

Finally, the amount of oxygen available to the microorganisms influences their growth kinetics. This influence is most generally expressed with a Monod type factor [168] depending on the concentration of oxygen in the air circulating in the waste matrix:

$$f_{O_2} = \frac{C_{O_2}}{k_{O_2} + C_{O_2}} \tag{5.63}$$

where C_{O_2} is the concentration of oxygen in the air and k_{O_2}, a constant.

5.6 Assessment of the Agronomic Value of Organic Residues: Amendment and Fertilizer Potentials

5.6.1 Fundamentals: Why Assessing Carbon and Nitrogen Fate on Soil?

In a context of environmental biorefinery, organic residues such as sludge, municipal wastes, manure, and crop residues are no more considered as wastes but as potential resources for some targeted services. These services encompass the energy production through the methane produced by their anaerobic digestion, high value molecules production through fermentation, and the use of organic residues as fertilisers and amendments on cropped soils. Moreover, organic fertilizers coming from wastes and organic residues participate to the soil carbon 4 per mille initiative to restore the soil fertility and to trap greenhouse gases [169]. The agronomic knowledge of several organic fertilizers would allow the evaluation of their ability to substitute mineral fertilizer (fertilizing value), increase the soil carbon stock and enhance the soil microbiology.

Concerning the chemical fertilization, mineral elements such as nitrogen, potassium, phosphorous are necessary to the plants growth. These elements can be brought by organic fertilizer though the organic matter mineralization on soil by soil microorganisms. Indeed, concerning soil biological fertility, the organic matter is essential to the ecosystems mechanisms such as soil stability, carbon storage, water quality and biodiversity. Organic matter can be used for enhancing the soil fertility (soil microorganisms) and biodiversity. By using organic matter as substrates, soil microorganisms decompose them and contribute to their mineralisation and several elements such as ammonium, nitrates, and phosphates are freed and can be absorbed by the plants. Another positive impact of organic matter in soil is the soil structure enhancement. Organic matter plays the role of "glue" and aggregate minerals; enhance porosity for a better aeration and enhance water infiltration to prevent runoff and erosion [170].

Soil biological and physico-chemical processes occurring after addition of organic fertilizer are illustrated by Fig. 5.14.

The less stable organic carbon is mineralized into carbon dioxide (under aerobic conditions) or into methane (under anaerobic conditions) such as into volatile organic compounds by soil microorganisms. Humification occurs and the remaining organic carbon is very stable and constitutes humus.

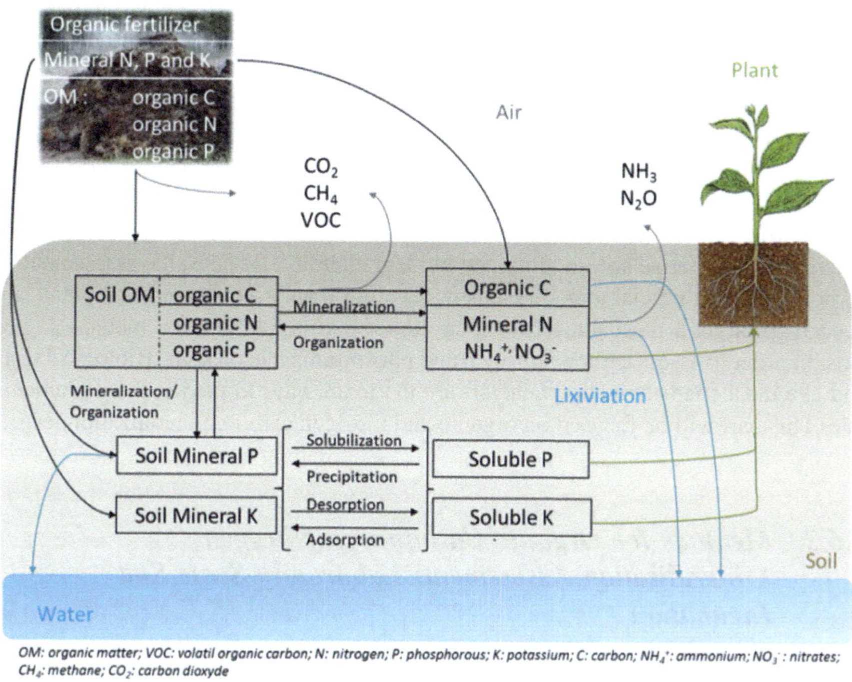

Fig. 5.14 Soil processus after addition of an organic fertilizer

Organic nitrogen is also mineralized into ammonia by ammonification process. Ammonia coming from the organic fertilizer such as the ammonia coming from ammonification is then bioconverted into nitrites and nitrates by autotrophic bacteria through nitrification bioreaction. However, soil microorganisms need nitrogen source to grow up and a part of mineral nitrogen can be immobilized into the soil and will not be available for plant growth. Depending on pH conditions, equilibrium between ammonium and ammonia goes to ammonia and can be transferred into the air by volatilization.

Another bioreaction can occur on anoxic conditions: the denitrification. Heterotrophobic bacteria use organic carbon as substrates such as nitrates as electron acceptor to produce N_2. During this step, nitrous oxide which is a greenhouse effect gas can also be produced and nitrates are not available for plant anymore.

Organic phosphorous is mineralized as carbon and nitrogen. As mineral nitrogen, soil microorganisms need phosphorous source to grow up. Depending on the pH conditions, physico-chemical reactions occur such as solubilisation and precipitation between soil mineral P (i.e. not available phosphorous form) and soil soluble phosphorous (i.e. available phosphorous form). Adsorption and desorption can also occur, as for soil mineral and soluble potassium.

Mineral nitrogen, soluble phosphorous and potassium are available to be absorbed by plants.

Some environmental issues can occur such as lixiviation of soluble compounds (phosphorous, nitrates and dissolved organic carbon) which can pollute underground water and greenhouse gases such as carbon dioxide, methane or nitrous oxide can be released to the air.

Depending on the targeted service, short term or long term use of carbon, nitrogen, phosphorous or potassium is necessary. For example, labile carbon would be used to enhance soil fertility and stable carbon would enhance soil structure and organic matter soil storage. This issue is linked to the accessibility concept of these elements to soil microorganisms and the plants. That means that the mineralization kinetics is crucial to characterize accessibility and biodegradability. Besides, the assessment of carbon and other crucial mineral elements evolution and fate on soils and plants are needed to target a reasoned fertilization strategy. Some laboratory methodologies exist in order to assess carbon, nitrogen and phosphorus fate on soils. Biological tests and chemical characterization analysis are the main ways to predict those elements fate. This part will be focused on organic and nitrogen carbon mineralization tests.

5.6.2 Methods for Organic Carbon and Nitrogen Mineralization Assessment: Laboratory Scale Soil Incubation

Exogenous organic matter decomposition is classically assessed during soil laboratory incubation [171]. As for the biochemical methane potential tests, laboratory soil incubation is carried out in controlled conditions of temperature, aeration and humidity in order to assess the potential of mineralizable organic carbon in soil. Table 5.5 summarizes several incubations conditions found in the literature.

5.6.2.1 Substrate Preparation

Based on the literature, organic fertilizer is dried and grounded before incubation in order to make the sample homogeneous and to reduce the particle size effect in the incubation and plant growth experiment [175]. Table 5.5 presents an overview of the incubation conditions used during this test. Grigatti et al. [175, 176] used air-dried and 2 mm grounded samples. Jimenez et al. [172] and Peltre et al. [174] freeze-dried and ground at 1 mm the samples. Sample preparation norms exist for analytical laboratories. Generally, the fresh product is first grounded and dried at 40 °C until sample mass stabilization as preconized by French norm [178].

5.6.2.2 Soil Preparation

Before its use, soil used has to be sieved at 4 mm [174–176]. According to Grigatti et al. [175, 176], soil is air-dried before 4 mm ground and pre-incubated at 60% of water holding capacity for 4 weeks at 25 °C to remove residual labile organic carbon, soluble nitrogen and phosphorus. Moisture content is restored by weighing, every 2–3 days. Other authors keep the soil under incubation temperature before its use in the test [179, 177].

5.6.2.3 Incubation for C Mineralization

Samples are incubated with an amount of cropped soil (from 25 to 250 g, depending the literature data), and an equivalent dose between 0.8 and 10 g of C kg^{-1} of dry soil) in hermetically closed bottles during, at least 91 days, under moisture and temperature controlled conditions. Generally, the organic fertilizer is added according to nitrogen limitations (140–170 g N ha^{-1}) as on the field [180]. A tube containing 10 mL of deionized water is generally added to prevent soil desiccation [181]. For unlimited microbial growth conditions, mineral nitrogen is added. Inorganic nitrogen was added to the soil as KNO_3 to prevent nitrogen limitation during waste decomposition [182] and to obtain an initial concentration ranging from 30 to 60 µg NO_3^- N g^{-1}dry soil, depending on the type and inorganic nitrogen content of waste [180].

Soil incubation without the addition of organic residues is also performed so as to take into account the mineralization of native soil organic carbon. A blank is also carried out without soil and exogenous matter in order to assess the CO_2 trapped from the ambient air. Soil moisture content is adjusted to a water content equivalent to 75–100% of the soil water holding capacity. Carbon mineralization is monitored

Table 5.5 Soil incubation tests conditions based on literature data

Exogenous organic matter	Temperature	Time of incubation	Dose g C kg^{-1} soil	Soil moisture[a] g kg^{-1} of soil	Sample preparation	References
Wastewater sludge	28	91–168	1.0–2.4	200–240	Air-dried and ground at 1 mm	[171]
Manure (bovine, ovine, poultry, pig)	28	168–219	0.8–5.7	160–250		
Urban composts	28	90–219	0.6–4.6	160–250		
Manure composts	28	168–364	1.1–3.4	160–250		
Plant material	28	158–219	2.0–9.3	160–200		
Digestates from anaerobic digestion	28	91–120	2 –10	250	Freeze-dried and 1 mm grounded	[172]
All	28	175	2	250	40 °C dried and ground at 1 mm	[173]
Municipal solid waste compost	28	21	3	225	Freeze-dried and ground (<1 mm)	[174]
Slurry and compost of slurry and municipal wastes	25	112	0.32–6.78 (200 mgN kg^{-1} soil)	130	Air-dried and ground at 2 mm	[175]
Compost of household waste	25	84	0.07–2.7 (30 mg P kg^{-1} soil)	130	Air-dried and ground at 2 mm	[176]
Non treated and treated animal slurry and manure	15 °C	224	0.35–1.58 (3.5 g DM kg^{-1} soil)	–	40 °C dried and ground at 1 mm for solid wastes Freeze-dried and ground at 1 mm for liquid wastes	[177]

[a]Soil moisture depends on the initial soil moisture

using a 10 mL trap solution of NaOH (0.5 M) for CO_2, which is periodically replaced during the incubations. According to AFNOR [173], sampling times are the following 1, 3, 7, 14, 21, 28, 49, 70, 91, 112, 133, 154, 175 days (13 points) for carbon mineralization assessment. The trapped CO_2 is analysed by titration by continuous flow colorimetry [183] or by titration of the remaining NaOH with 0.25 M HCl after precipitation of the carbonates with a saturated $BaCl_2$ solution [184]. The kinetics of total mineralized carbon are calculated by summing the CO_2–C respired between two sampling occasions.

Another way to analyse trapped CO_2–C uses conductivity [185]. Results are obtained in milisiemens per centimetre, which is negatively proportional to trapped CO_2–C concentration. These results are estimated thanks to a calibration curve obtained with NaOH and Na_2CO_3. CO_2 emission (g CO_2) is calculated according to the Eq. (5.64).

$$CO_2 = \frac{\Delta \times M \times M_{CO_2}}{1000} \tag{5.64}$$

where Δ is the difference between the acid volume during titration for the sample and the blank titration (mL), M is the concentration of the acid (mol L^{-1}), M_{CO_2} is molar mass of CO_2 (44 g mol^{-1}), m is the sample weight (g).

Then, the mineralized carbon percentage is calculated as the Eq. 5.65

$$C\,(\%) = \sum_{t=1}^{tf} \frac{(CO_2 \times t_n \times 0.273)_{sample} - (CO_2 \times t_n \times 0.273)_{control} - (CO_2 \times t_n \times 0.273)_{blank}}{m_{sample} \times C_{sample}} \times 100 \tag{5.65}$$

where 0.273 is the conversion factor from CO_2 to C, m is the sample weight (kg); C_{sample} and m_{sample} are respectively the organic carbon content of the sample (g kg^{-1}) amount of added organic carbon (kg).

5.6.2.4 Incubation for Nitrogen Mineralization

As for carbon mineralization incubation tests, nitrogen mineralization tests can be set-up, following the same methodology of sample and soil preparation as for carbon mineralization, without chemical nitrogen addition. Indeed, agronomy value of organic carbon has to be studied concomitantly with organic nitrogen evolution.

Two experimental apparatus can be used: (i) four replicates of the soil and organic fertilizer mixture (amount of soil between 200 g and 500 g) and (ii) the "tray" apparatus where lower amount of soil (25 g) is mixed with organic fertilizer in three replicates [173]. Each mixture replicate is sacrificed at each sampling time in order to extract mineral nitrogen from all the mixture.

In the first case, dried organic fertilizers are homogeneously mixed with soil samples, (500 g equivalents of dry soil in 2-L glass jars, three replicates per treatment as in Morvan et al. [180]; or 200 g in 500 mL plastic vessels, four replicates as in Grigatti et al. [175]. A chemical reference as ammonium nitrate can also be prepared to compare chemical fertilizer with organic fertilizer behavior [175].

In both cases, nitrogen mineralization is determined periodically by sampling 20 g of soil from the flasks or the entire capsule (tray test case). Soil inorganic nitrogen is extracted by shaking 1 g of dried material with 25 mL KCl (1 M) for 30 min [177]. The $NH_4^+$and NO_3^- contents of the soil extracts are then determined by continuous flow colorimetry using an auto-analyzer and the methods proposed by and Henriksen and Selmer-Olsen [186], respectively as suggested by Morvan et al. [180]. According to AFNOR [173], sampling times are 0, 3, 7, 14, 28, 49, 91, 175 days (8 points) for nitrogen mineralization assessment. As for carbon mineralization, incubation of soil without organic fertilizer is also carried out in order to assess the native part of mineral N.

Equation 5.66 shows the calculation to obtain the nitrogen mineralization percentage.

$$N\,(\%) = \sum_{t=1}^{tf} \frac{(N_{min} \times t_n \times V_{extraction})_{sample} - (N_{min} \times t_n \times V_{extraction})_{control}}{m_{sample} \times N_{sample}} \times 100 \quad (5.66)$$

where N_{min} is the nitrogen content (g L^{-1}) analyzed at t = t_n, $V_{extraction}$ is the KCl volume of extractant used (L), N_{sample} and m_{sample} are respectively the nitrogen content of the sample (g kg^{-1}) and the amount of added sample (kg).

5.6.3 *Modelling of Biological Tests*

5.6.3.1 Carbon Mineralization Curves Models

Carbon mineralization kinetics can be analysed by different types of mathematical models, in order to identify the function best fitting of the carbon mineralization pattern of each specific organic fertilizer.

Table 5.6 presents several models to extrapolate kinetic parameters from mineralization curves, from the simplest kinetics to a more complex model. According to literature, the double exponential model is the model which fit the best with a large panel of organic fertilizer [171, 181, 187]. Grigatti et al. [175] found the same conclusion but some organic fertilizer such as sewage sludge and its compost fitted better with sigmoid models.

Concerning nitrogen mineralization tests, animal wastes and other organic residues contain variable amounts of organic nitrogen, whose behavior in soil depends on the biodegradation of organic pools characterized by different mineralization rates and carbon to nitrogen (C/N) ratios [180]). The C/N ratio is one of the most considered quality criteria but this simple criterion appears sometimes inadequate for predicting the decomposition kinetics [182, 187]. Thus, other quality criteria seems necessary; such criteria can be provided by the sequential analysis of

Table 5.6 Organic carbon mineralization curves models

Model type	Model equation	Definitions	References
Exponential model	$C_m = C_0 \times \left(1 - e^{-k_l \times t}\right)$	C_m represents the percentage of carbon mineralized at time t, C_0 is the maximum mineralizable carbon (% of added C) and k is the mineralization rate (d^{-1})	[171, 175]
Double exponential model	$C_m = C_l \times \left(1 - e^{-k_l \times t}\right) + C_S \times \left(1 - e^{-k_S \times t}\right)$	C_m represents the percentage of carbon mineralized at time t, C_l represents the mineralizable fraction of the labile pool (% of added C), k_l is the mineralization rate constant of the labile pool (d^{-1}), C_S represents the mineralizable fraction of the stable pool (% of added C) and k_S is the mineralization rate constant of the stable pool (d^{-1})	[175, 181, 187]
	$ROC(t) = C_1 \times \left(e^{-k_l \times t}\right) + C_2 \times \left(e^{-k_S \times t}\right) + (1000 - C_1 - C_2)$	ROC is the residual organic carbon at time t, C_1 and C_2 are the two biodegradable pools (gC kg^{-1} organic matter) and k_1 and k_2 their respective mineralization rate constants (d^{-1}) and $(1000 - C_1 - C_2)$ is the fraction of recalcitrant organic carbon (gC kg^{-1} organic matter)	[171]
Sigmoid model	$C_m = \frac{C_0}{\left(1 + e^{-\frac{t - t_0}{k}}\right)}$	C_m represents the percentage of carbon mineralized at time t, C_0 is the maximum mineralizable carbon (% of added C), k is the mineralization rate (d^{-1}) and t0 is the point of inflection (d)	[175]
Logistic model	$C_m = \frac{C_0}{1 + \left(\frac{t}{t_0}\right)^k}$	C_m represents the percentage of carbon mineralized at time t, C_0 is the maximum mineralizable carbon (% of added C), k is the mineralization rate (d^{-1}) and t_0 is the time needed to mineralize half of added carbon (d)	[175]

Van Soest [188] which attempts to divide the organic fertilizer into soluble-, hemicellulose-, cellulose- and lignin-like substances.

According to Recous et al. [182] and Nicolardot et al. [189], incubation time at controlled conditions can be translated into field time through the function described by Eq. 5.67

$$\text{Normalized days} = 1 \times \mathrm{f(T)} \times \mathrm{g}(\theta) \tag{5.67}$$

where f(T) is a temperature function defined by the Eq. 5.68 and $g(\theta)$ is a moisture function defined by Eq. 5.69.

$$\mathrm{f(T)} = \left(\mathrm{a} + \mathrm{b} + \mathrm{e}^{\mathrm{c} \times \frac{\mathrm{T}}{\mathrm{T_{ref}}}}\right) \tag{5.68}$$

where a = −0.566, b = 0.620, c = 0.925, d = 1.026 for Tref of 15 °C and T the incubation temperature (°C)

$$\mathrm{g}(\theta) = (1 - \mathrm{c}) \times \frac{\theta - \theta_{\mathrm{wp}}}{\theta_{\mathrm{fc}} - \theta_{\mathrm{wp}}} + \mathrm{c} \tag{5.69}$$

where c = 0.20, θ is the soil moisture during incubation, θ_{wp} and θ_{fc} are respectively the soil moisture content at wilting point and field capacity.

5.6.3.2 Nitrogen Mineralization Models

Nitrogen mineralization modelling is very difficult, above all in the immobilization case. According to Lashermes et al. [179], six classes of organic fertilizers can be found (Fig. 5.15), depending on the initial nitrogen concentration and depending on

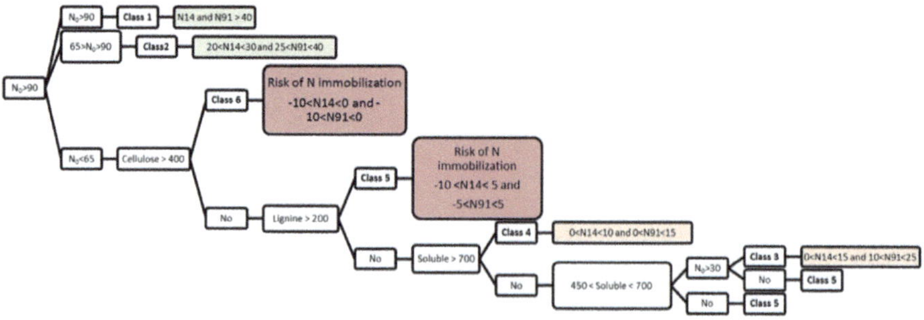

Fig. 5.15 Decision tree based on chemical and biochemical compositions of organic fertilizers adapted from Lashermes et al. [179]: (reproduced with permission from Elsevier): initial nitrogen content N0in g kg^{-1} Dry Matter, biochemical fractions in g kg^{-1} of organic matter, N14 is the mineralized nitrogen after 14 days and N91 the mineralized nitrogen after 91 days in gN kg^{-1} DM. Reproduced with permission from Elsevier [179]

the fibers concentrations (Van Soest basis). Based on this study, 6 classes were defined characterized by a gradient in N availability from class 1 with high potential nitrogen mineralization to the class 6 with a nitrogen immobilization risk.

For example, class 1 concerns organo-fertilizers (43%) and animal by-products (43%), class 2, 3 and 4 contain mainly wastewater sludge (respectively 64, 56 and 89%), class 5 contains composts (46%) and class 6 contains 39% of livestock farming wastes with litter.

Some dynamic models propose to simulate both carbon and nitrogen mineralization by using C/N ratio as Garnier et al. [190] with the dynamic CANTIS model based on Van Soest fractions degradations, and Nicolardot et al. [189].

Indeed, Nicolardot et al. [189] used the C/N ratio to evaluate the capacity of a simple dynamic model to predict the kinetics of mineralization during the decomposition of various crop residues. Despite the fact that carbon mineralization kinetics was overestimated, nitrogen mineralization kinetics was well predicted by this simple model.

5.6.3.3 Fast Analysis and Regression Models for Carbon Mineralization

Indicator of Residual Organic Carbon

A number of authors have applied a chemical characterization of organic fertilizers for soil carbon mineralization prediction. For aerobic biodegradation, an indicator of residual organic carbon (IROC) has been developed by Lashermes et al. [171] and has been standardized as mentioned by Peltre et al. [191]. It is based on fiber fractionation [188] and on the organic carbon fraction mineralized after 3 days of incubation. It has been applied to predict the evolution of soil organic matter after repeated applications of various organic residues in long-term field experiments [174]. Equation 5.70 describes the correlation found.

$$I_{ROC} = 445 + 0.5 \times SOL - 0.2 \times CEL + 0.7 \times LIC - 2.3 \times C_{3d} \quad (5.70)$$

where IROC is the proportion of organic fertilizer contributing to soil organic C storage (expressed as g C $kg^{-1}$of organic matter), SOL, CEL and LIC are the soluble, cellulose-like and lignin-like fractions (g $kg^{-1}$of organic matter) and C_{3d}, the proportion of organic fertilizer mineralized after 3 days of incubation (g C kg^{-1} of organic matter).

However, Van Soest fibers fractionation is not applicable to all types of organic fertilizers such as sludge and liquid manures [172] because of the high part of extracted soluble.

Spectral Techniques

Infra-red spectroscopy

Another common methodology for predicting carbon mineralization on soil is Near InfraRed Spectroscopy (NIRS). NIRS was applied to the IROC prediction [191]. A few outliers were observed and mainly associated to wastewater sludge, liquid manure and organo-mineral fertilisers. However, the NIRS method provides limited qualitative information about chemical composition of organic residues due to overlapping overtones and combination bands [192]. This tool, therefore, does not facilitate a more mechanistic interpretation.

More recently, a Fourier transform mid-infrared photoacoustic spectroscopy technique (FTIR-PAS) was applied to predict mineralizable fraction of carbon in soils from several organic residues [192]. This technique allowed for (i) the influence of sample particle size and absorption of dark and opaque samples to be ignored and (ii) for more information to be produced on the composition of organic matter than with NIRS [192].

Chemical extractions and 3D-fluorescence spectroscopy

Biodegradability of organic residues not only depends on the quantity of organic matter but it also relies on the accessibility and complexity [172]. Given the complex organization of certain organic residues, the term “accessibility” defines the potential access to the molecules by microorganisms responsible for their degradation. It can depend on the physical characteristics of the organic residues (particle size, porosity of the organic residue), on the distribution of the organic fraction within the organic residues matrices in relation with the process duration, the hydrolytic activity or the applied pre-treatment. Molecules with a weight below 1000 Da can pass through the cell wall [172]. Finally, once the organic matter becomes accessible, it can be more or less easily biodegradable by microorganisms, depending on its chemical nature or complexity. The chemical nature of extracted organic fractions have been observed to vary during treatments such as composting, which thus explains the change in biodegradability of given organic fractions [174].

One study successfully predicted both biochemical methane potential and soil mineralizable carbon by considering both accessibility and complexity [172]. The methodology developed was based on the combination of a sequential chemical extraction to assess accessibility and fluorescence spectroscopy to assess the complexity of each extracted organic fraction.

References

1. Himmel, M.E., Ding, S.Y., Johnson, D.K., Adney, W.S., Nimlos, M.R., Brady, J.W., Foust, T.D.: Biomass recalcitrance: engineering plants and enzymes for biofuels production. Science **315**, 804–807 (2007)
2. Chen, H., Liu, J., Chang, X., Chen, D., Xue, Y., Liu, P., Lin, H., Han, S.: A review on the pretreatment of lignocellulose for high-value chemicals. Fuel Process. Technol. **160**, 196–206 (2017)
3. Alexandropoulou, M., Antonopoulou, G., Fragkou, E., Ntaikou, I., Lyberatos, G.: Fungal pretreatment of willow sawdust and its combination with alkaline treatment for enhancing biogas production. J. Environ. Manage. **203**, 704–713 (2017)
4. Chandra, R.P., Bura, R., Mabee, W.E., Berlin, A., Pan, X., Sadler, J.N.: Substrate pretreatment: the key to effective enzymatic hydrolysis of lignocellulosics? In: Olsson, L. (ed.) Biofuels, pp. 67–93. Springer, Berlin, Heidelberg (2007)
5. Fountoulakis, M.S., Dokianakis, S.N., Kornaros, M.E., Aggelis, G.G., Lyberatos, G.: Removal of phenolics in olive mill wastewaters using the white-rot fungus Pleurotus ostreatus. Water Res. **36**, 4735–4744 (2002)
6. Maxwell, K.: Cationic polymer enhanced hydrolysis of cornstarch for the production of biofuels. School of Chemical and Biomolecular Engineering, Georgia Institute of Technology (2012)
7. Taherzadeh, M.J., Karimi, K.: Acid-based hydrolysis processes for ethanol from lignocellulosic materials: a review. BioResources **2**, 707–738 (2007)
8. Zhang, Y.H., Lynd, L.R.: Towards and aggregated understanding of enzymatic hydrolysis of cellulose: non-complexed cellulase systems. Biotechnol. Bioeng. **88**, 797–824 (2004)
9. Foreman, P.K., Brown, D., Dankmeyer, L., Dean, R., Diener, S., Dunn-Coleman, N.S., Goedegebuur, F., Houfek, T.D., England, G.J., Kelley, A.S., Meerman, H.J., Mitchell, T., Mitchinson, C., Olivares, H.A., Teunissen, P.J., Yao, J., Ward, M.: Tran-scriptional regulation of biomass-degrading enzymes in the filamentous fungus *Trichoderma reesei*. J. Biol. Chem. **278**, 31988–31997 (2003)
10. Jørgensen, H., Kristensen, J.B., Felby, C.: Enzymatic conversion of lignocellulose into fermentable sugars: challenges and opportunities. Biofuels Bioprod. Bioref. **1**, 119–134 (2007)
11. Rosgaard, L., Pedersen, S., Langston, J., Akerhielm, D., Cherry, J.R., Meyer, A.S.: Evaluation of minimal *Trichoderma reesei* cellulase mixtures on differently pretreated barley straw substrates. Biotechnol. Prog. **23**, 1270–1276 (2007)
12. Kumar, D., Murthy, G.S.: Stochastic molecular model of enzymatic hydrolysis of cellulose for ethanol production. Biotechnol. Biofuels **2**, 63 (2013)
13. Camassola, M.: Cellulases and hemicellulases, why we need so much of these enzymes?. Ferment. Technol., 2 (2012)
14. Dokic, P., Jakovljevic, J., Dokic-Baucal, L.: Molecular characteristics of maltodextrins and rheological behaviour of diluted and concentrated solutions. Colloids Surf. A: Physicochem. Eng. Aspects **141**, 435–440 (1998)
15. Liu, Z.A., Zhang, F.S.: Effect of various solvents on the liquefaction of biomass to produce fuels and chemical feedstocks. Energ. Convers. Manage. **49**, 3498–3504 (2008)
16. Yunianta, N.H., Fithri, C.N., Mubarok, A.Z., Wulan, S.N.: Variations in dextrose equivalent and dynamic rheology of dextrin obtained by enzymatic hydrolysis of edible canna starch. Int. J. Food Prop. **18**, 2726–2734 (2015)
17. Hames, B., Ruiz, R., Scarlata, C., Sluiter, A., Sluiter, J., Templeton, D.: Preparation of samples for compositional analysis. in: laboratory analytical procedure (LAP). National Renewable Energy Laboratory (2008)
18. Kennedy, J.F., Cabalda, V.M., White, C.A.: Enzymic starch utilization and genetic engineering. Trends Biotechnol. **6**, 184–189 (1988)

19. Synowiecki, J.: The use of starch processing enzymes in the food industry. In: Polaina, J., MacCabe, A.P. (eds.) Industrial Enzymes. Springer, Dordrecht (2007)
20. Sluiter, A., Hames, B., Hyman, D., Payne, C., Ruiz, R., Scarlata, C., Sluiter, J., Templeton, D., Wolfe, J.: Determination of total solids in biomass and total dissolved solids in liquid process samples. In: Laboratory Analytical Procedure (LAP) (Ed.), National Renewable Energy Laboratory (2008)
21. Sluiter, A., Hames, B., Ruiz, R., Scarlata, C., Sluiter, J., Templeton, D.: Determination of ash in biomass. In: Laboratory Analytical Procedure (LAP) (Ed.), National Renewable Energy Laboratory (2005)
22. Hames, B., Scarlata, C., Sluiter, A.: Determination of protein content in biomass. In: Laboratory Analytical Procedure (LAP) (Ed.), National Renewable Energy Laboratory (2008)
23. Agblevor, F.A., Pereira, J.: Progress in the summative analysis of biomass feedstocks for biofuels production. In: Aqueous Pretreatment of Plant Biomass for Biological and Chemical Conversion to Fuels and Chemicals, pp. 335–354. Wiley (2013)
24. Burkhardt, S., Kumar, L., Chandra, R., Saddler, J.: How effective are traditional methods of compositional analysis in providing an accurate material balance for a range of softwood derived residues? Biotechnol. Biofuels **6**, 1754–6834 (2013)
25. Karimi, K., Taherzadeh, M.J.: A critical review of analytical methods in pretreatment of lignocelluloses: composition, imaging, and crystallinity. Biores. Technol. **200**, 1008–1018 (2016)
26. Theander, O., Aman, P., Westerlund, E., Andersson, R., Petersson, D.: Total dietary fiber determined as neutral sugar residues, uronic acid residues, and Klason Lignin (The Uppsala method): collaborative study. J. AOAC Int. **78**, 1030–1044 (1995)
27. Sluiter, R., Hames, B., Ruiz, R., Scarlata, C., Sluiter, J., Templeton, D., Crocker, D.: Determination of structural carbohydrates and lignin in biomass. In: Laboratory Analytical Procedure (LAP), National Renewable Energy Laboratory (2012)
28. Sluiter, R., Ruiz, R., Sluiter, J., Templeton, D.: Determination of extractives in biomass. In: Laboratory Analytical Procedure (LAP) (Ed.), National Renewable Energy Laboratory (2008)
29. Noori, M.S., Karimi, K.: Detailed study of efficient ethanol production from elmwood by alkali pretreatment. Biochem. Eng. J. **105**, 197–204 (2016)
30. Roche, C.M., Dibble, C.J., Stickel, J.J.: Laboratory-scale method for enzymatic saccharification of lignocellulosic biomass at high-solids loadings. Biotechnol. Biofuels **2**, 28 (2009)
31. Fenila, F., Shastri, Y.: Optimal control of enzymatic hydrolysis of lignocellulosic biomass. Resour. Effi. Technol. **2**, S96–S104 (2016)
32. Novozymes Bioenergy.: Novozymes cellulosic ethanol enzyme kit, Enzymes for the hydrolysis of lignocellulosic materials. http://www.bdigital.unal.edu.co/46027/2/2822323.2014%20Anexo.pdf
33. Dashtban, M., Maki, M., Leung, K.T., Mao, C., Qin, W.: Cellulase activities in bio-mass conversion: measurement methods and comparison. Crit. Rev. Biotechnol., 1–8 (2010)
34. Mandels, M., Andreotti, R., Roche, C.: Measurement of saccharifying cellulase. Biotechnol. Bioeng. Symp. **6**, 21–33 (1976)
35. Ghose, T.: Measurement of cellulase activities. Pure Appl. Chem. **59**, 257–268 (1987)
36. Miller, G.L.: Use of dinitrosalicylic acid reagent for determination of reducing sugar. Anal. Chem. **31**, 426–428 (1959)
37. Wood, T.M., Bhat, K.M.: Methods for measuring cellulase activities. Methods Enzymol. **160**, 87–117 (1988)
38. Adney, B., Baker, J.: Measurement of cellulase activities. In: Laboratory Analytical Procedure (LAP) (Ed.), National Renewable Energy Laboratory (1996)
39. Trinder, P.: Determination of blood glucose using 4-amino phenazone as oxygen acceptor. J. Clin. Pathol. **22**, 246 (1969)

40. Chandra, M., Kalra, A., Sharma, P.K., Sangwan, R.S.: Cellulase production by six *Trichoderma* spp. fermented on medicinal plant processing. J. Ind. Microbiol. Biotechnol. **36**, 605–609 (2009)
41. Kubicek, C.P.: Beta-glucosidase excretion by Trichoderma pseudokoningii: correlation with cell wall bound beta-1.3-glucanase activities. Arch. Microbiol. **132**, 349–354 (1982)
42. Morrison, W.R., Laignelet, B.: An improved colorimetric procedure for determining apparent and total amylose in cereal and other starches. J. Cereal Sci. **1**, 9–20 (1983)
43. Kaufman, R.C., Wilson, J.D., Bean, S.R., Herald, T.J., Shi, Y.C.: Development of a 96-well plate iodine binding assay for amylose content determination. Carbohydr. Polym. **115**, 444–447 (2015)
44. McGrance, S.J., Cornell, H.J., Rix, C.J.: A simple and rapid colorimetric method for the determination of amylose in starch products. Starch/Stärke **50**, 158–163 (1998)
45. Bansal, P., Hall, M., Realff, M.J., Lee, J.H., Bommarius, A.S.: Modeling cellulase kinetics on lignocellulosic substrates. Biotechnol. Adv. **27**, 833–848 (2009)
46. Gharpuray, M.M., Lee, Y.H., Fan, L.T.: Structural modification of lignocellulosic by pretreatments to enhance enzymatic hydrolysis. Biotechnol. Bioeng. **25**, 157–172 (1983)
47. Kadam, K.L., Rydholm, E.C., McMil, J.D.: Development and validation of a kinetic model for enzymatic saccharification of lignocellulosic biomass. Biotechnol. Prog. **20**, 698–705 (2004)
48. Tsai, C.T., Meyer, A.S., Johansen, K.S.: Bioethanol from lignocellulose—pretreatment, enzyme immobilization and hydrolysis kinetics. Technical University of Denmark (DTU) (2012)
49. Zheng, Y., Pan, Z., Zhang, R., Jenkins, B.M.: Kinetic modeling for enzymatic hydrolysis of pretreated creeping wild ryegrass. Biotechnol. Bioeng. **102**, 1558–1569 (2009)
50. Besselink, T., Baks, T., Janssen, A., Boom, R.: A stochastic model for predicting dextrose equivalent and sac-charide composition during hydrolysis of starch by alpha-amylase. Biotechnol. Bioeng. **100**, 684–697 (2008)
51. Marchal, L.: Towards a rational design of commercial maltodextrins: a mechanistic approach, p. 197. Wageningen University (1999)
52. Paolucci-Jeanjean, D., Belleville, M.P., Zakhia, N., Rios, G.M.: Kinetics of cassava starch hydrolysis with Termamyl® enzyme. Biotechnol. Bioeng. **68**, 71–77 (2000)
53. Torgerson, E.M., Brewer, L.C., Thoma, J.A.: Subsite mapping of enzymes. Use of subsite map to simulate complete time course of hydrolysis of a polymeric substrate. Arch Biochem Biophys **196**, 13–22 (1979)
54. Wojciechowski, P.M., Koziol, A., Noworyta, A.: Iteration model of starch hydrolysis by amylolytic enzymes. Biotechnol. Bioeng. **75**, 530–539 (2001)
55. Kandra, L., Gyémánt, G., Remenyik, J., Hovánszki, G., Lipták, A.: Action pattern and subsite mapping of Bacillus licheniformis a-amylase (BLA) with modified maltooligosaccharide substrates. FEBS Lett. **518**, 79–82 (2002)
56. Allen, J.D., Thoma, J.A.: Subsite mapping of enzymes: application of the depolymerase computer model to two a-amylases. Biochem. J. **159**, 121–132 (1976)
57. MacGregor, E.A., MacGregor, A.W.: A model for the action of cereal alpha amylases on amylose. Carbohydr. Res. **142**, 223–236 (1985)
58. McCarty, P.L.: One-hundred years of anaerobic treatment. In: Hughes, D.E. (ed.) Anaerobic Digestion, pp. 3–22. Elsevier, Amsterdam (1982)
59. Adib, R.: Renewables 2015 global energy report. In: Ren21 (2015)
60. EU: A Policy Framework for Climate and Energy in the Period from 2020 to 2030 (2014)
61. EU: Directive 2009/28/EC of the European Parliament and the Council of 23 April 2009 on the Promotion of the Use of Energy from Renewable Sources and Amending (2009)
62. Schievano, A., D'Imporzano, G., Malagutti, L., Fragali, E., Ruboni, G., Adani, F.: Evaluating inhibition conditions in high-solids anaerobic digestion of organic fraction of municipal solid waste. Biores. Technol. **101**, 5728–5732 (2010)

63. Kafle, G.K., Chen, L.D.: Comparison on batch anaerobic digestion of five different livestock manures and prediction of biochemical methane potential (BMP) using different statistical models. Waste Manag. **48**, 492–502 (2016)
64. Nielfa, A., Cano, R., Fdz-Polanco, F.: Theoretical methane production generated by the co-digestion of organic fraction municipal solid waste and biological sludge. Bio-technol. Rep. **5**, 14–21 (2015)
65. Owen, W.F., Stuckey, D.C., Healy, J.B., Young, L.Y., McCarty, P.L.: Bioassay for monitoring biochemical methane potential and anaerobic toxicity. Water Res. **13**, 485–492 (1979)
66. Chynoweth, D.P., Turick, C.E., Owens, J.M., Jerger, D.E., Peck, M.W.: Biochemical methane potential of biomass and waste feedstocks. Biomass Bioenerg. **5**, 95–111 (1993)
67. Owens, J.M., Chynoweth, D.P.: Biochemical methane potential of municipal solid-waste (MSW) components. Water Sci. Technol. **27**, 1–14 (1993)
68. Sanders, W., Angelidaki, I.: Assessment of the anaerobic biodegradability of macropollutants. Rev. Environ. Sci. Biotechnol. **3**, 117–129 (2004)
69. Angelidaki, I., Alves, M., Bolzonella, D., Borzacconi, L., Campos, J.L., Guwy, A.J., Kalyuzhnyi, S., Jenicek, P., van Lier, J.B.: Defining the biomethane potential (BMP) of solid organic wastes and energy crops: a proposed protocol for batch assays. Water Sci. Technol. **59**, 927–934 (2009)
70. Holliger, C., Alves, M., Andrade, D., Angelidaki, I., Astals, S., Baier, U., Bougrier, C., Buffiere, P., Carballa, M., et al.: Towards a standardization of biomethane potential tests. Water Sci. Technol. **74**, 2515–2522 (2016)
71. Raposo, F., Fernandez-Cegri, V., De la Rubia, M.A., Borja, R., Beline, F., Cavinato, C., Demirer, G., Fernandez, B., et al.: Biochemical methane potential (BMP) of solid organic substrates: evaluation of anaerobic biodegradability using data from an international interlaboratory study. J. Chem. Technol. Biotechnol. **86**, 1088–1098 (2011)
72. Hafner, S.D., Rennuit, C., Triolo, J.M., Richards, B.K.: Validation of a simple gravimetric method for measuring biogas production in laboratory experiments. Biomass Bioenerg. **83**, 297–301 (2015)
73. Hafner, S.D., Rennuit, C., Triolo, J.M., Azifehkhoran, A.: R package biogas: analyze biogas data and predict biogas production. https://cran.r-project.org/package=biogas (2015)
74. Huang, H.B., Yu, H.Y., Xu, H.R., Ying, Y.B.: Near infrared spectroscopy for on/in-line monitoring of quality in foods and beverages: a review. J. Food Eng. **87**, 303–313 (2008)
75. Raju, C.S., Ward, A.J., Nielsen, L., Moller, H.B.: Comparison of near infra-red spectroscopy, neutral detergent fibre assay and in-vitro organic matter digestibility assay for rapid determination of the biochemical methane potential of meadow grasses. Biores. Technol. **102**, 7835–7839 (2011)
76. Fitamo, T., Triolo, J.M., Boldrin, A., Scheutz, C.: Rapid biochemical methane potential prediction of urban organic waste with near-infrared reflectance spectroscopy. Water Res. **119**, 242–251 (2017)
77. Lesteur, M., Latrille, E., Maurel, V.B., Roger, J.M., Gonzalez, C., Junqua, G., Steyer, J.P.: First step towards a fast analytical method for the determination of bio-chemical methane potential of solid wastes by near infrared spectroscopy. Biores. Technol. **102**, 2280–2288 (2011)
78. Doublet, J., Boulanger, A., Ponthieux, A., Laroche, C., Poitrenaud, M., Rivero, J.A.C.: Predicting the biochemical methane potential of wide range of organic substrates by near infrared spectroscopy. Biores. Technol. **128**, 252–258 (2013)
79. Lesteur, M., Preys, S., Boulanger, A., Latrille, E., Treguer, R.: Fast prediction of biochemical methane potential (BMP) of organic waste by near infrared spectroscopy. Picking up good vibrations. In: NIR2013. La Grande-Motte, France (2013)
80. Triolo, A.J., Ward, L., Pedersen, M.M., Lokke, H., Qu, S.G.: Sommer, near infrared reflectance spectroscopy (NIRS) for rapid determination of biochemical methane potential of plant biomass. Appl. Energy **116**, 52–57 (2014)

81. Charnier, C., Latrille, E., Jimenez, J., Lemoine, M., Boulet, J.C., Miroux, J., Steyer, J.P.: Fast characterization of solid organic waste content with near infrared spectroscopy in anaerobic digestion. Waste Manag. **59**, 140–148 (2017)
82. Charnier, C., Latrille, E., Jimenez, J., Torrijos, M., Sousbie, P., Miroux, J., Steyer, J.P.: Fast ADM1 implementation for the optimization of feeding strategy using near infrared spectroscopy. Water Res. **122**, 27–35 (2017)
83. Buswel, A.M., Sollo, F.W.: The mechanism of the methane fermentation. J. Am. Chem. Soc. **70**(5), 1778–1780 (1948)
84. Monlau, F., Sambusiti, C., Barakat, A., Guo, X.M., Latrille, E., Trably, E., Steyer, J.P., Carrere, H.: Predictive models of biohydrogen and biomethane production based on the compositional and structural features of lignocellulosic materials. Environ. Sci. Technol. **46**, 12217–12225 (2012)
85. Donoso-Bravo, A., Perez-Elvira, S.I., Fdz-Polanco, F.: Application of simplified models for anaerobic biodegradability tests Evaluation of pre-treatment processes. Chem. Eng. J. **160**, 607–614 (2010)
86. Shen, J., Yan, H., Zhang, R.H., Liu, C.Q., Chen, C.: Characterization and methane production of different nut residue wastes in anaerobic digestion. Renew. Energy **116**, 835–841 (2018)
87. Abbassi-Guendouz, A., Brockmann, D., Trably, E., Dumas, C., Delgenes, J.P., Steyer, J.P., Escudie, R.: Total solids content drives high solid anaerobic digestion via mass transfer limitation. Biores. Technol. **111**, 55–61 (2012)
88. Guo, X.M., Trably, E., Latrille, E., Carrere, H., Steyer, J.P.: Hydrogen production from agricultural waste by dark fermentation: a review. Int. J. Hydrogen Energy **35**, 10660–10673 (2010)
89. Cazier, E.A., Trably, E., Steyer, J.P., Escudie, R.: Biomass hydrolysis inhibition at high hydrogen partial pressure in solid-state anaerobic digestion. Biores. Technol. **190**, 106–113 (2015)
90. Dincer, I., Acar, C.: Review and evaluation of hydrogen production methods for better sustainability. Int. J. Hydrogen Energy **40**, 11094–11111 (2015)
91. Guo, X.M., Trably, E., Latrille, E., Carrere, H., Steyer, J.P.: Predictive and explicative models of fermentative hydrogen production from solid organic waste: role of butyrate and lactate pathways. Int. J. Hydrogen Energy **39**, 7476–7485 (2014)
92. Cai, G.Q., Jin, B., Saint, C., Monis, P.: Metabolic flux analysis of hydrogen production network by Clostridium butyricum W5: effect of pH and glucose concentrations. Int. J. Hydrogen Energy **35**, 6681–6690 (2010)
93. Hawkes, F.R., Hussy, I., Kyazze, G., Dinsdale, R., Hawkes, D.L.: Continuous dark fermentative hydrogen production by mesophilic microflora: principles and progress. Int. J. Hydrogen Energy **32**, 172–184 (2007)
94. Mohan, S.V., Babu, V.L., Sarma, P.N.: Effect of various pretreatment methods on anaerobic mixed microflora to enhance biohydrogen production utilizing dairy wastewater as substrate. Biores. Technol. **99**, 59–67 (2008)
95. Logan, B.E., Oh, S.E., Kim, I.S., Van Ginkel, S.: Biological hydrogen production measured in batch anaerobic respirometers. Environ. Sci. Technol. **36**, 2530–2535 (2002)
96. Won, S.G., Lau, A.K.: Effects of key operational parameters on biohydrogen production via anaerobic fermentation in a sequencing batch reactor. Bioresour. Technol. **102**(13), 6876–6883 (2011)
97. Kim, D.H., Shin, H.S., Kim, S.H.: Enhanced H-2 fermentation of organic waste by CO_2 sparging. Int. J. Hydrogen Energy **37**, 15563–15568 (2012)
98. Stumm, W., Morgan, J.: Aquatic Chemistry, Chemical Equilibria and Rates in Natural Waters, 3rd edn. Wiley, New York (1996)
99. Ramirez-Morales, J.E., Tapia-Venegas, E., Toledo-Alarcon, J., Ruiz-Filippi, G.: Simultaneous production and separation of biohydrogen in mixed culture systems by continuous dark fermentation. Water Sci. Technol. **71**, 1271–1285 (2015)

100. Antonopoulou, G., Gavala, H.N., Skiadas, I.V., Lyberatos, G.: Modeling of fermentative hydrogen production from sweet sorghum extract based on modified ADM1. Int. J. Hydrogen Energy **37**, 191–208 (2012)
101. Gavala, H.N., Skiadas, I.V., Ahring, B.K., Lyberatos, G.: Potential for biohydrogen and methane production from olive pulp. Water Sci. Technol. **52**, 209–215 (2005)
102. Ghimire, A., Frunzo, L., Pirozzi, F., Trably, E., Escudie, R., Lens, P.N.L., Esposi-to, G.: A review on dark fermentative biohydrogen production from organic biomass: process parameters and use of by-products. Appl. Energy **144**, 73–95 (2015)
103. Ntaikou, I., Antonopoulou, G., Lyberatos, G.: Biohydrogen production from biomass and wastes via dark fermentation: a review. Waste Biomass Valori. **1**, 21–39 (2010)
104. Boboescu, I.Z., Ilie, M., Gherman, V.D., Mirel, I., Pap, B., Negrea, A., Kondorosi, E., Biro, T., Maroti, G.: Revealing the factors influencing a fermentative biohydrogen production process using industrial wastewater as fermentation substrate. Biotechnol. Biofuels **7**, 139 (2014)
105. Akhlaghi, M., Boni, M.R., De Gioannis, G., Muntoni, A., Polettini, A., Pomi, R., Rossi, A., Spiga, D.: A parametric response surface study of fermentative hydrogen production from cheese whey. Biores. Technol. **244**, 473–483 (2017)
106. Argun, H., Dao, S.: Bio-hydrogen production from waste peach pulp by dark fer-mentation: Effect of inoculum addition. Int. J. Hydrogen Energy **42**, 2569–2574 (2017)
107. Pan, J., Zhang, R., El-Mashad, H.M., Sun, H., Ying, Y.: Effect of food to microorganism ratio on biohydrogen production from food waste via anaerobic fermentation. Int. J. Hydrogen Energy **33**, 6968–6975 (2008)
108. Argun, H., Kargi, F., Kapdan, I.K., Oztekin, R.: Batch dark fermentation of powdered wheat starch to hydrogen gas: effects of the initial substrate and biomass concentrations. Int. J. Hydrogen Energy **33**, 6109–6115 (2008)
109. Wang, J.L., Wan, W.: Kinetic models for fermentative hydrogen production: a review. Int. J. Hydrogen Energy **34**, 3313–3323 (2009)
110. Yuan, X.Z., Shi, X.S., Zhang, P.D., Wei, Y.L., Guo, R.B., Wang, L.S.: Anaerobic biohydrogen production from wheat stalk by mixed microflora: kinetic model and particle size influence. Biores. Technol. **102**, 9007–9012 (2011)
111. Mu, Y., Yu, H.Q., Wang, G.: A kinetic approach to anaerobic hydrogen-producing process. Water Res. **41**, 1152–1160 (2007)
112. Alexandropoulou, M., Antonopoulou, G., Lyberatos, G.: A novel approach of modeling continuous dark hydrogen fermentation. Biores. Technol. **250**, 784–792 (2018)
113. Arudchelvam, Y., Perinpanayagam, M., Nirmalakhandan, N.: Predicting VFA formation by dark fermentation of particulate substrates. Biores. Technol. **101**, 7492–7499 (2010)
114. Gadhamshetty, V., Arudchelvam, Y., Nirmalakhandan, N., Johnson, D.C.: Modeling dark fermentation for biohydrogen production: ADM1-based model vs. Gompertz model. Int. J. Hydrogen Energy **35**, 479–490 (2010)
115. Pelmont, J.: Bactéries et Environnement: Adaptations physiologiques. Presses Universitaires de Grenoble, Grenoble (1993)
116. Bailey, J., Ollis, D.: Biochemical engineering fundamentals, 2nd edn. McGraw-Hill Book Co., Singapore (1986)
117. Spanjers, H., Vanrolleghem, P., Olsson, G., Dold, P.: Respirometry in Control of the Activated Sludge Process: Principles. International Association on Water Quality, London (1998)
118. Janning, K.: Hydrolysis and oxidation of particulate organic matter in biofilters. In: Environmental Science and Engineering. Technical University of Denmark (1998)
119. Aspray, T., Dimambro, M., Wallace, P., Howell, G., Frederickson, J.: Static, dynamic and inoculum augmented respiration based test assessment for determining in-vessel compost stability. Waste Manag. **42**, 3–9 (2015)
120. Barrena, R., d'Imporzano, G., Ponsa, S., Gea, T., Artola, A., Vasquez, F., Sanchez, A., Adani, F.: Search of a reliable technique for the determination of the bio-logical stability of

the organic matter in the mechanical-biological treated waste. J. Hazard. Mater. **162**, 1065–1072 (2009)

121. Barrena, R., Turet, J., Busquets, A., Farrés, M., Font, X., Sanchez, A.: Respirometric screening of several types of manure and mixtures intended for composting. Biores. Technol. **102**, 1367–1377 (2011)
122. Barrena, R., Vasquez, F., Sanchez, A.: The use of respiration indices in the composting process: a review. Waste Manage. Res. **24**, 37–47 (2006)
123. Komilis, D., Kletsas, C.: Static respiration indices to investigate compost stability: effect of sample weight and temperature and comparison with dynamic respiration indices. Biores. Technol. **121**, 467–470 (2012)
124. Villasenor, J., Perez, M.A., Fernandez, F.J., Puchalski, C.M.: Monitoring respiration and biological stability during sludge composting with a modified dynamic respirometer. Biores. Technol. **102**, 6562–6568 (2011)
125. Adani, F., Gigliotti, G., Valentini, F., Laraia, R.: Respiration index determination: a comparative study of different methods. Compost Sci. Util. **11**, 144–151 (2003)
126. Lasaridi, K., Stentiford, E.: A simple respirometric technique for assessing compost stability. Water Res. **32**, 3717–3723 (1998)
127. Nikaeen, M., Nafez, A., Bina, B., Nabavi, B., Hassanzadeh, A.: Respiration and enzymatic activities as indicators of stabilization of sewage sludge composting. Waste Manag. **39**, 104–110 (2015)
128. Ianotti, D., Grebus, M., Toth, B., Madden, L., Hoitink, H.: Oxygen respirometry to assess stability and maturity of composted municipal solid waste. J. Env. Qual. **23**, 1177–1183 (1994)
129. Barrena, R., Vasquez, F., Gordillo, M., Gea, T., Sanchez, A.: Respirometric assays at fixed and process temperatures to monitor composting process. Biores. Technol. **96**, 1153–1159 (2005)
130. Barrena, R., Gea, T., Ponsa, S., Ruggieri, L., Artola, A., Font, X., Sanchez, A.: Categorizing raw organic material biodegradability via respiration activity measurement: a review. Compost Sci. Util. **19**, 105–113 (2011)
131. Binner, E., Zack, A.: Laboratory tests describing the biological reactivity of pre-treated residual wastes. In: W. Bidlingmaier, M. de Bertoldi, L. Diaz, E. Papadimitriou (Eds.) Organic Recovery and Biological Treatment—ORBIT99, pp. 255–261. Weimar (1999)
132. Binner, E., Böhm, K., Lechner, P.: Large scale study on measurement of respiration activity (AT4) by Sapromat and OxiTop. Waste Manag. **32**, 1752–1759 (2012)
133. Changa, C., Wang, P., Watson, M., Hoitink, H., Michel, F.: Assessment of the reliability of a commercial maturity test kit for composted manures. Compost Sci. Util. **11**, 125–143 (2003)
134. Cossu, R., Raga, R.: Test methods for assessing the biological stability of biodegradable waste. Waste Manag. **28**, 381–388 (2008)
135. Evangelou, A., Gerassimidou, S., Mavrakis, N., Komilis, D.: Monitoring the performances of a real scale municipal solid waste composting and a biodrying facility using respiration activity indices. Environ. Monit. Assess. **188**, 302 (2016)
136. Gea, T., Barrena, R., Artola, A., Sanchez, A.: Monitoring the biological activity of the composting process: oxygen uptake rate (OUR), respirometric index (RI), and respiratory quotient (RQ). Biotechnol. Bioeng. **88**, 520–527 (2004)
137. Komilis, D., Kontou, I., Ntougias, S.: A modified static respiration assay and its relationship with an enzymatic test to assess compost stability and maturity. Biores. Technol. **102**, 5863–5872 (2011)
138. Malinska, K.: Application of a modified OxiTop (R) respirometer for laboratory composting studies. Arch. Env. Protect. **42**, 56–62 (2016)
139. Oviedo-Ocana, E., Torres-Lozada, P., Mormolejo-Rebellon, L., Hoyos, L., Gonza-les, S., Barrena, R., Komilis, D., Sanchez, A.: Stability and maturity of biowaste com-posts derived by small municipalities: Correlation among physical, chemical and biological indices. Waste Manag. **44**, 63–71 (2015)

140. Ponsa, S., Gea, T., Alerm, L., Cerezo, J., Sanchez, A.: Comparison of aerobic and anaerobic stability indices through a MSW biological treatment process. Waste Manag. **28**, 2735–2742 (2008)
141. Sanchez-Arias, V., Fernandez, F.J., Rodriguez, L., Villasenor, J.: Respiration indices and stability measurements of compost through electrolytic respirometry. J. Environ. Manage. **95**, S134–S138 (2012)
142. Adani, F., Lozzi, P., Genevini, P.: Determination of biological stability by oxygen uptake on municipal solid waste and derived products. Compost Sci. Util. **9**, 163–178 (2001)
143. Adani, F., Confalonieri, R., Tambone, F.: Dynamic respiration index as a descriptor of the biological stability of organic wastes. J. Environ. Qual. **33**, 1866–1876 (2004)
144. Adani, F., Ubbiali, C., Genevini, P.: The determination of biological stability of composts using the dynamic respiration index: the results of experience after two years. Waste Manag. **26**, 41–48 (2006)
145. Almeira, N., Komilis, D., Barrena, R., Gea, T., Sanchez, A.: The importance of aeration mode and flowrate in the determination of the biological activity and stability of organic wastes by respiration indices. Biores. Technol. **196**, 256–262 (2015)
146. Antognoni, S., Ragazzi, M., Ionescu, G., Passamani, G., Zanoni, S., Rada, E.: Respirometric index as a tool for biogas generation production from poultry manure. Manag. Env. Qual. Int. J. **27**, 269–280 (2016)
147. Berthe, L., Druilhe, C., Massiani, C., Tremier, A., de Guardia, A.: Coupling a respirometer and a pycnometer, to study the biodegradability of solid organic wastes during composting. Biosys. Eng. **97**, 75–88 (2007)
148. Ciuta, S., Antognoni, S., Rada, E., Ragazzi, M., Badea, A., Cioca, L.: Respirometric index and biogas potential of different foods and agricultural discarded biomass. Sustainability **8** (311), 1 (2016)
149. Colon, J., Ponsa, S., Alvarez, C., Vinot, M., Lafuente, F., Gabriel, D., Sanchez, A.: Analysis of MSW full-scale facilities based on anaerobic digestion and/or composting using respiration indices as performance indicators. Biores. Technol. **236**, 87–96 (2017)
150. Ferrari, D., Howell, G., Aspray, T.: Improved precision and efficiency of a modified ORG0020 dynamic respiration test setup for compost stability assessment. Sustainability **9**, 2358 (2017)
151. Komilis, D., Kanellos, D.: A modified dynamic respiration test to assay compost stability: effect of sample size and air flowrate. Biores. Technol. **117**, 300–309 (2012)
152. Mejias, L., Komilis, D., Gea, T., Sanchez, A.: The effect of airflow rates and aeration mode on the respiration activity of four organic wastes: implications on the com-posting process. Waste Manag. **65**, 22–28 (2017)
153. Ponsa, S., Gea, T., Sanchez, A.: Different indices to express biodegradability in organic solid wastes. J. Environ. Qual. **39**, 706–712 (2010)
154. Ponsa, S., Puyuelo, B., Gea, T., Sanchez, A.: Modelling the aerobic degradation of organic wastes based on slowly and rapidly degradable fractions. Waste Manag. **31**, 1472–1479 (2011)
155. Scaglia, B., Confalonieri, R., d'Imporzano, G., Adani, F.: Estimating biogas production of biologically treated municipal solid waste. Biores. Technol. **101**, 945–952 (2010)
156. Scaglia, B., Acutis, M., Adani, F.: Precision determination for the dynamic respirometric index (DRI) method used for biological stability evaluation on municipal solid waste and derived products. Waste Manag. **31**, 2–9 (2011)
157. Trémier, A., de Guardia, A., Massiani, C., Paul, E., Martel, J.: A respirometric method for characterizing the organic composition and biodegradation kinetics and the temperature influence on the biodegradation kinetics, for a mixture of sludge and bulking agent to be composted. Bioresour. Technol. **96**, 169–180 (2005)
158. Wagland, S.T., Tyrrel, S.F., Godley, A., Smith, R.: Test methods to aid in the evaluation of the diversion of biodegradable municipal waste (BMW) from landfill. Waste Manag. **29**, 1218–1226 (2009)

159. Lasaridi, K., Papadimitriou, E., Balis, C.: Development and demonstration of a thermogradient respirometer. Compost Sci. Util. **4**, 53–61 (1996)
160. Henze, M., Gujer, W., Mino, T., van Loosdrecht, M.: Activated Sludge Models ASM1, ASM2, ASM2D and ASM3. IWA Publishing, Londre (2000)
161. Rosso, L., Lobry, J.R., Bajard, S., Flandrois, J.P.: Convenient model to describe the combined effects of temperature and pH on microbial growth. Appl. Environ. Microbiol. **61**, 610–616 (1995)
162. Rosso, L., Lobry, J.R., Flandrois, J.P.: An unexpected correlation between cardinal temperatures of microbial growth highlighted by a new model. J. Theor. Biol. **162**, 447–463 (1993)
163. Haug, R.: The Practical Handbook of Compost Engineering. Lewis Publishers, Boca Raton, FL (1993)
164. Lomander, A., Katterer, T., Andren, O.: Modelling the effects of temperature and moisture on CO_2 evolution from top- and subsoil using a multi-compartment approach. Soil Biol. Biochem. **30**, 2023–2030 (1998)
165. Pommier, S., Chenu, D., Quintard, M., Lefebvre, X.: Modelling of moisture-dependent aerobic degradation of solid waste. Waste Manag. **28**, 1188–1200 (2008)
166. Richard, T.L., Hamelers, H.V.M., Veeken, A., Silva, T.: Moisture relationships in composting processes. Compost Sci. Util. **10**, 286–302 (2002)
167. Stombaugh, D.P., Nokes, S.E.: Development of a biologically based aerobic com-posting simulation model. Trans. ASAE **39**, 239–250 (1996)
168. Mason, I.G.: Mathematical modelling of the composting process: a review. Waste Manag. **26**, 3–21 (2006)
169. Minasny, B., Malone, B.P., McBratney, A.B., Angers, D.A., Arrouays, D., Chambers, A., Chaplot, V., et al.: Soil carbon 4 per mille. Geoderma **292**, 59–86 (2017)
170. Houot, S., Pons, M.N., Pradel, M.: Valorisation des matières fertilisantes d'origine résiduaire sur les sols à usage agricole ou forestier. In: Expertise collective INRA-CNRS-IRSTEA, INRA-CNRS-IRSTEA (2014)
171. Lashermes, G., Nicolardot, B., Parnaudeau, V., Thuries, L., Chaussod, R., Guil-lotin, M.L., Lineres, M., et al.: Indicator of potential residual carbon in soils after exogenous organic matter application. Eur. J. Soil Sci. **60**, 297–310 (2009)
172. Jimenez, J., Lei, H., Steyer, J.P., Houot, S., Patureau, D.: Methane production and fertilizing value of organic waste: organic matter characterization for a better prediction of valorization pathways. Biores. Technol. **241**, 1012–1021 (2017)
173. AFNOR, XPU 44-163: Amendements organiques et supports de culture—Caractérisation de la matière organique par la minéralisation potentielle du carbone et de l'azote (2009)
174. Peltre, C., Dignac, M.F., Derenne, S., Houot, S.: Change of the chemical composition and biodegradability of the Van Soest soluble fraction during composting: a study using a novel extraction method. Waste Manag. **30**, 2448–2460 (2010)
175. Grigatti, M., Di Girolamo, G., Chincarini, R., Ciavatta, C., Barbanti, L.: Potential nitrogen mineralization, plant utilization efficiency and soil CO_2 emissions following the addition of anaerobic digested slurries. Biomass Bioenerg. **35**, 4619–4629 (2011)
176. Grigatti, M., Boanini, E., Cavani, L., Ciavatta, C., Marzadori, C.: Phosphorus in digestate-based compost: chemical speciation and plant-availability. Waste Biomass Valori. **6**, 481–493 (2015)
177. Morvan, T., Nicolardot, B., Pean, L.: Biochemical composition and kinetics of C and N mineralization of animal wastes: a typological approach. Biol. Fertil. Soils **42**, 513–522 (2006)
178. AFNOR, NF: U44-110: Boues-Amendements organiques-Supports de culture-Préparation des échantillons partiellement secs pour essai. Expression des résultats (1982)
179. Lashermes, G., Nicolardot, B., Parnaudeau, V., Thuries, L., Chaussod, R., Guillotin, M.L., Lineres, M., et al.: Typology of exogenous organic matters based on chemical and biochemical composition to predict potential nitrogen mineralization. Biores. Technol. **101**, 157–164 (2010)

180. Morvan, T., Nicolardot, B.: Role of organic fractions on C decomposition and N mineralization of animal wastes in soil. Biol. Fertil. Soils **45**, 477–486 (2009)
181. Thuries, L., Pansu, M., Feller, C., Herrmann, P., Remy, J.C.: Kinetics of added organic matter decomposition in a Mediterranean sandy soil. Soil Biol. Biochem. **33**, 997–1010 (2001)
182. Recous, S., Robin, D., Darwis, D., Mary, B.: Soil inorganic N availability: effect on maize residue decomposition. Soil Biol. Biochem. **27**, 1529–1538 (1995)
183. Alavoine, G., Nicolardot, B.: Comparison of potentiometric titration, IR spectrophotometry and segmented micro-flow analysis to determine inorganic C in alkaline solutions. Anal. Bioanal. Chem. **374**, 354–358 (2002)
184. Zibilske, L.M.: Carbon mineralization. In: Angle, J.S., Bottomley, P.S., Weaver, R.W. (eds.) Methods of Soil Analysis, Part 2, Microbiological and Biochemical Properties, pp. 835–863. ASA-SSSA, Madison, WI (1994)
185. Houot, S.: Laboratory characterization of organic waste produced in the laboratory to predict their behavior in the field. ADEME (2016)
186. Henriksen, A., Selmerol, A.: Automatic methods for determining nitrate and nitrite in water and soil extracts. Analyst **95**, 514–518 (1970)
187. Thuries, L., Pansu, M., Larre-Larrouy, M.C., Feller, C.: Biochemical composition and mineralization kinetics of organic inputs in a sandy soil. Soil Biol. Biochem. **34**, 239–250 (2002)
188. Van Soest, P.: Use of detergents in the analysis of fibrous feeds: II. A rapid method for the determination of fibre and lignin. J. Assoc. Off. Anal. Chem. **46**, 829–835 (1963)
189. Nicolardot, B., Recous, S., Mary, B.: Simulation of C and N mineralisation during crop residue decomposition: a simple dynamic model based on the C:N ratio of the residues. Plant Soil **228**, 83–103 (2001)
190. Garnier, P., Neel, C., Aita, C., Recous, S., Lafolie, F., Mary, B.: Modelling carbon and nitrogen dynamics in a bare soil with and without straw incorporation. Eur. J. Soil Sci. **54**, 555–568 (2003)
191. Peltre, C., Thuries, L., Barthes, B., Brunet, D., Morvan, T., Nicolardot, B., Parnaudeau, V., Houot, S.: Near infrared reflectance spectroscopy: a tool to characterize the composition of different types of exogenous organic matter and their behaviour in soil. Soil Biol. Biochem. **43**, 197–205 (2011)
192. Bekiaris, G., Bruun, S., Peltre, C., Houot, S., Jensen, L.S.: FTIR-PAS: a powerful tool for characterising the chemical composition and predicting the labile C fraction of various organic waste products. Waste Manag. **39**, 45–56 (2015)

Printed by Printforce, the Netherlands